THOMAS KIRBY

PHYSICAL
CHEMISTRY
OF
MACROMOLECULES

CHARLES TANFORD

Professor of Physical Biochemistry
Duke University

JOHN WILEY & SONS, INC., New York · London · Sydney

PHYSICAL
CHEMISTRY
OF
MACROMOLECULES

11 12 13 14 15 16 17 18 19 20

ISBN 0 471 84447 0

LIBRARY OF CONGRESS CATALOG CARD NUMBER: 61–11511

PRINTED IN THE UNITED STATES OF AMERICA

PREFACE

This book represents a comprehensive introduction to the use of physical chemistry in the study of macromolecules. It was written to meet two needs in my laboratory: the need for a textbook in a course for graduate students and the need for a reference book to provide background information for my research program on the physical chemistry of proteins. I hope that it will be useful to others in the same way, both as a textbook and as a reference for self-study.

One of the objectives of the book has been to give consideration to all kinds of macromolecules, both the synthetic polymers and the naturally occurring macromolecules, such as proteins and viruses. I hope that this feature will prove useful to those who may already be familiar with the application of physical chemistry to a particular kind of large molecule but who would like to extend their knowledge to other kinds.

Another objective of the book has been to present the material in such a way that a good undergraduate course in physical chemistry is the only prerequisite. To achieve this end, and at the same time to bring out the theoretical principles and the simplifying assumptions which underlie the various topics treated, has required the presentation of a considerable amount of introductory material, such as the description of the principles of x-ray diffraction in section 3 and the derivation of the Debye-Hückel theory in section 26.

Emphasis throughout the book is on theoretical principles, and on them the organization of the subject matter depends, each chapter containing topics which are based on the same or on related theoretical principles. Chapter 2, for example, includes all the techniques which give direct information about molecular structure; Chapter 4 is concerned primarily with the chemical potential in solution, both from the classical and statistical points of view; Chapter 6 discusses all physical methods which rely ultimately on the equation of motion of the solvent near a dissolved macromolecule; and so forth. One consequence of this scheme of organization is, of course, that different methods for determining the same property of a macromolecule (e.g., molecular weight) appear in different chapters.

A secondary theme develops an intimate knowledge of the structure and behavior of the various kinds of large molecules, and, with this in mind, each chapter is illustrated by experimental data from the literature. Data cited in later chapters are often compared or combined with data cited earlier, so that the picture which the reader acquires of each kind of molecule becomes gradually more detailed as he progresses through the book.

A small part of the subject matter represents new and unpublished material. This is true, for instance, of some of the calculations and inferences made in section 28 and of the general treatment of multiple equilibria in section 29. However, no novelty is claimed for the major part of the book. Many of the topics have been presented in a form which differs from that of the original authors. When this is done, however, it is not intended to imply that greater insight or rigorousness is gained thereby. It merely reflects the fact that this book is intended for readers who may be novices in the field, whereas the original publications were in general directed to experts with knowledge of the previous history of the problem being attacked. If any changes introduced have altered the intent of the original authors cited, it has been done unwittingly, and I shall be grateful if instances of this kind are brought to my attention.

For those who wish to use this book as a textbook, a description of my graduate course at the University of Iowa on the physical chemistry of macromolecules may be helpful: this course has usually been based on the secondary theme, with emphasis on the molecules rather than on the theory. The more difficult derivations have been assigned as optional reading or have been presented in outline form, but the selections of experimental data have been given in full. The areas covered in this manner in one semester have been sections 1, 2, 4, 5, 6a, 7, 8, 9, 10, 11, 12 (in part), 13, 14, 16, 17, 18, 19, 20, 21, 22, 23, 26 (in outline), 27 and 28. Others who wish to teach this kind of course may prefer to substitute all or part of Chapter 9 for some of the foregoing material. I was able to take up much of this material in a course on chemical kinetics, so that it did not need to be repeated in the course on macromolecules.

In writing this book, I have incurred indebtedness to a considerable number of my colleagues who have read portions of the manuscript and offered criticism and advice. Among them have been Drs. R. A. Alberty, N. C. Baenziger, R. L. Baldwin, H. B. Bull, D. K. Chattoraj, J. F. Foster, A. M. Holtzer, W. Kauzmann, J. C. Kendrew, W. R. Krigbaum, Y. Nozaki, L. Peller, W. B. Person, and S. A. Rice. I should point out that I have not always followed the advice I have been given, and any shortcomings in the text are my own responsibility. I am also indebted to the National Science Foundation and to the U. S. Public Health Service for their support of my research program. Much of the stimulus for writing this book was provided by questions which arose as part of this research.

Finally, I should like to express my indebtedness to my wife, Lucia, who not only endured, with minimal complaint, the frequent privations of an author's wife but also shared with me the burden of proofreading and indexing.

CHARLES TANFORD

Durham, North Carolina
May 1961

CONTENTS

APPENDICES

LIST OF IMPORTANT SYMBOLS

a major semi-axis of ellipsoid

a_i thermodynamic activity of component or species i in a solution

B second virial coefficient

b minor semi-axis of ellipsoid

C concentration, moles/liter or moles/cc.

c concentration, g/cc.

$\tilde{c}$ velocity of light

D diffusion coefficient

D dielectric constant

E internal energy

$\mathscr{E}$ electric field strength

F Gibbs' free energy

f frictional coefficient

G general symbol for any measurable property

g_i grams of component i in a mixture or solution

H enthalpy

h_{av} average end-to-end distance of a polymer chain

h_{ij} distance from segment i to segment j of a polymer chain

I ionic strength

J_i flux of component i across unit cross section per second

K equilibrium constant

K constant relating observed displacement in Schlieren optical system to dc/dx

$\mathscr{K}$ optical constant in light scattering

k rate constant

k equilibrium constant for association reactions

k Boltzmann's constant

L phenomenological coefficient in thermodynamic treatment of flow processes

l_{av} average bond length in a polymer chain

M molecular weight

m concentration, moles/kg. of solvent

N concentration, molecules/cc.

$\mathscr{N}$ Avogadro's number

n_i number of moles of component or species i in a solution

$\tilde{n}$ refractive index

P, p	pressure
q	electrostatic charge in esu
R	radius
R_G	radius of gyration
R_θ	experimental quantity in light scattering
$\mathscr{R}$	gas constant
r	symbol for radial distance
S	entropy
s	sedimentation coefficient
T	temperature
t	time
$\mathbf{u}, u$	flow velocity in a fluid
V	volume
$\overline{V}_i$	partial molal volume of component or species i in a solution
v_i	partial specific volume of component or species i in a solution
W	statistical probability
W_{el}	electrostatic free energy
X_i	mole fraction of component or species i in a solution

x	degree of polymerization of a polymer chain
y_i	activity coefficient (molarity scale) of component or species i in a solution
Z, z	electrostatic charge in units of protonic charges
α	polarizability
α	empirical expansion factor for polymer dimensions
β	effective bond length in polymer chains
γ_i	activity coefficient (molal scale) of component or species i in a solution
ϵ	proton charge in esu
η	viscosity
$[\eta]$	intrinsic viscosity
λ	wavelength
u_i	chemical potential of component or species i in a solution
ρ	density
σ	number of elements in a polymer chain

1

INTRODUCTION

1. A BRIEF HISTORY

Macromolecular physical chemistry was born in 1861 when Thomas Graham[10] discovered that a number of substances exist which differ markedly in their diffusion properties from such ordinary materials as inorganic salts and sugars. These substances, Graham found, diffuse much more slowly in a column of water than do ordinary materials and do not diffuse at all through parchment paper and other membranes, through which ordinary materials pass with ease. Graham correctly deduced that these unusual properties ("They appear like different worlds of matter") were the result of a high molecular weight. Gelatin was a typical example of this group of substances, and, accordingly, Graham called these substances *colloids*, after the Greek word for glue.

Unfortunately, many of the best-known colloids of the day were what Kraemer, Williams, and Alberty[1] have termed *accidental colloids*. They are substances, such as the hydrated oxides of iron and aluminum and colloidal gold, which are well known to exist ordinarily as molecules of small size. That these substances are sometimes found in the form of colloidal particles, i.e., particles of high molecular weight, is due to aggregation by secondary bonds, stabilization of microcrystals with a surface layer of ions, etc. It is not due to an intrinsic difference between these substances and other ordinary chemical compounds. Physical chemists began therefore to talk of the "colloidal state of matter" rather than of "colloidal substances." The belief grew that *all* substances could under the proper conditions be transformed to a colloidal state and, conversely, that *all* colloidal substances were physical aggregates of simpler molecules.

This belief persisted well into the twentieth century. In 1922 cellulose was still regarded as a cyclic tetrasaccharide, its natural occurrence as a high-molecular-weight substance being due to physical association. Emil Fischer held a similar view regarding proteins, the intrinsic molecules of which were supposedly polypeptides with about twenty amino acid units.

1

A well-known textbook of colloid chemistry, published as late as 1934, states categorically "... nor are there any particular elements or compounds which have the specific nature of being 'colloids'."

This statement, of course, is not true, and it is precisely with such substances that this book is concerned. The molecules with which we shall deal are large because they contain a large number of atoms joined together by primary chemical bonds. They cannot be reduced to "small" molecules without irreversible destruction of their chemical identity. They are *molecular colloids*, or *macromolecules*, first so called by Staudinger.[11]

The recognition that these are substances which must exist intrinsically as giant molecules and cannot have a simpler form came during the 1920's. Staudinger[11] pointed out that polymers such as polystyrene and natural rubber maintain their high molecular weight in all solvents, in contrast to colloids which are the result of physical association, which generally exist as such only in selected solvents. Svedberg[12] produced his spectacular sedimentation diagrams of proteins, which showed that even the relatively poorly purified proteins available to him consisted of molecules which were essentially all identical, something not expected of loose association compounds.

Also of importance from the historical point of view is the first crystallization of pure protein enzymes, that of urease by Sumner[13] and that of pepsin by Northrop.[14] For the previous non-crystallizability of colloids as such had been an important argument in favor of their existence as loose aggregates. Most colloids which could be induced to form crystals did so only with loss of their colloidal nature and reversion to their intrinsic low-molecular-weight form. A few proteins (notably hemoglobin) were known to crystallize without loss of their colloidal nature, but they were believed to be fortuitous aggregates of polypeptide chains of low molecular weight. (The reader is referred to Edsall[25] for an elegant review of this subject.) The crystallization of urease and pepsin, and, subsequently, of numerous other proteins, and the fact that the unit cell weights of such crystals are always simple multiples of the molecular weight determined in solution, and not small fractions thereof, are two of the important facets of the proof of their intrinsically macromolecular nature.

It is recognized today that a majority of colloids are macromolecules, and a brief enumeration of the various types is given in the following section.

2. VARIETIES OF MACROMOLECULES

2a. Synthetic linear organic polymers.[2,3] The simplest of all macromolecules are the synthetic linear organic polymers, which consist of long unbranched chains of small identical subunits or, sometimes, of two or three different kinds of subunits. They are formed by polymerization of simple substances known as *monomers*. Examples are *polymethylene*

$$\cdots -CH_2-CH_2-CH_2-CH_2-\cdots$$

prepared from diazomethane, *polystyrene*

$$\cdots -CH_2-CH-CH_2-CH-CH_2-CH-\cdots$$

prepared from styrene, *polyesters*

$$\cdots -O-(CH_2)_x-O-\underset{O}{\overset{\|}{C}}-(CH_2)_y-\underset{O}{\overset{\|}{C}}-O-(CH_2)_x-O-\underset{O}{\overset{\|}{C}}-(CH_2)_y-\underset{O}{\overset{\|}{C}}-\cdots$$

prepared from the condensation of $HO-(CH_2)_x-OH$ and $HOOC-(CH_2)_y-COOH$, *polyamides*

$$\cdots -HN-(CH_2)_x-NH-\underset{O}{\overset{\|}{C}}-(CH_2)_y-\underset{O}{\overset{\|}{C}}-HN-(CH_2)_x-NH-\underset{O}{\overset{\|}{C}}-(CH_2)_y-\underset{O}{\overset{\|}{C}}-\cdots$$

prepared from the condensation of $H_2N-(CH_2)_x-NH_2$ with $HOOC-(CH_2)_y-COOH$, and *polypeptides* .

$$\cdots -NH-\underset{R}{CH}-CO-NH-\underset{R}{CH}-CO-NH-\underset{R}{CH}-CO-\cdots$$

prepared from the corresponding N-carboxy anhydrides.

Polystyrene is an example of a large group of polymers known as *vinyl polymers* or *addition polymers*, all of which are obtained by successive addition of monomers of the type $CH_2=CHR$. Polyamides and polyesters are examples of another large group known as *condensation polymers*, preparation of which involves the elimination of water or some other simple substance in the condensation of functional groups such as $-COOH$ and $-OH$.

Long chain molecules of this kind may be prepared so as to contain almost any number of subunits, from just ten or twenty to nearly a million. Moreover, a given preparation may contain the entire spectrum of possible molecular weights; molecules containing all possible numbers of subunits, from one to many thousand, may all be present together. Fractionation procedures may be carried out to separate the larger from the smaller molecules in such a preparation, but this can serve only to reduce the range of molecular size in a given sample. A fundamental difference between such synthetic polymers and ordinary "classical" molecules always remains: in small molecules all the individual molecules of a given preparation have the same mass; molecules of synthetic polymers do not.

Similar variation between individual molecules is found if we prepare polymer molecules containing more than one kind of subunit, as, for example, a polymer of vinyl acetate and vinyl chloride

$$\cdots-CH_2-CH-CH_2-CH-CH_2-CH-CH_2-CH-\cdots$$

with substituents below: O, Cl, O, Cl; where the O groups continue to $C=O$ then CH_3.

Such polymers, called *copolymers*, may be represented by a formula such as ABABAAABBA ..., where A and B represent two different subunits. The arrangement along the chain of the units A and B will depend on the relative quantities of the monomers originally present, and on the relative rate constants for formation of A—A, A—B, and B—B bonds. However, again, individual molecules will differ from one another in the precise number of subunits of each kind and in their order.

These simple considerations of the nature of synthetic polymers indicate at once that their physical chemistry will be concerned with a vital factor which does not arise to nearly the same extent in the physical chemistry of small molecules, nor in the physical chemistry of many of the naturally occurring macromolecules. This is the factor of *statistics*. Statistical considerations occur in classical physical chemistry primarily when we consider the distribution of energy among an assembly of molecules, but, in most situations, all the individual molecules are alike in size and in structure. When synthetic polymers are considered, on the other hand, statistics pervade every facet of our investigations. Even fundamental properties such as the molecular weight are necessarily governed by statistical distributions, and the measurement of such properties will always determine an average quantity. Usually several different averages may be evaluated.

2b. Non-linear or branched polymers. Carothers,[15] the originator of nylon and one of the pioneers of polymer chemistry, first introduced the concept of *functionality* to characterize the points on a monomer molecule at which polymerization may occur. The monomers of linear polymers are all *bifunctional*; they contain two and only two such points, e.g., the hydroxyl groups of $HO—(CH_2)_x—OH$ or the two unsatisfied valences potentially present in a vinyl monomer, $RHC{=}CH_2 \rightarrow R\overset{|}{H}C—\overset{|}{C}H_2$.

When a monomer molecule is *polyfunctional*, i.e., when it contains more than two locations susceptible to attack leading to polymerization, the resulting polymer will not be linear. Thus the polymer resulting from the condensation of glycerol with a dicarboxylic acid will necessarily be a solid network, extending inflexibly and greatly entangled into three dimensions.

Less highly branched polymers may be obtained by introducing a few cross-links into an essentially linear structure, for instance by introducing just a small amount of glycerol into a polymerizing mixture of a glycol and a dicarboxylic acid.

Branched polymer molecules are less easy to characterize statistically than linear ones, and they have been much less extensively studied. We shall refer to them in this book only infrequently.

2c. Synthetic polyelectrolytes.[4] If the subunit of a linear polymer contains an ionic group, the polymer is known as a *polyelectrolyte*. Examples are *polyacrylate*

$$\cdots —CH_2—\underset{\underset{COO^-}{|}}{CH}—CH_2—\underset{\underset{COO^-}{|}}{CH}—CH_2—\underset{\underset{COO^-}{|}}{CH}—\cdots$$

and *poly-N-butyl-4-vinyl-pyridinium salts*

The first is the anion of a weak acid, and it can be converted to an uncharged polymer by titration with acid. The second, however, is a strong electrolyte and charged at all pH values.

Polymers of this kind differ from non-electrolyte polymers in the same way that low-molecular-weight electrolytes differ from non-electrolytes. They are soluble in polar solvents, conduct electricity, and are profoundly affected by Coulomb forces between the charges they possess.

An interesting class of synthetic polyelectrolytes of which just a few examples have been prepared to date are *polyampholytes*, containing both positive and negative charges. Examples are the copolymer of vinylpyridine and methacrylic acid

$$\cdots -CH_2-CH-CH_2-\underset{\underset{COO^-}{|}}{\overset{\overset{CH_3}{|}}{C}H}-CH_2-CH-\cdots$$

and the polypeptide *copoly*-L-*lysine*-L-*glutamic acid*

$$\cdots -NH-\underset{\underset{NH_3^+}{|}}{\underset{(CH_2)_4}{|}}{CH}-CO-NH-\underset{\underset{COO^-}{|}}{\underset{(CH_2)_2}{|}}{CH}-CO-NH-\underset{\underset{NH_3^+}{|}}{\underset{(CH_2)_4}{|}}{CH}-CO-\cdots$$

2d. Simple natural polymers. Among naturally occurring macromolecules there are a few which resemble the synthetic organic polymers in simplicity and in the heterogeneity of chain lengths. Among them are *natural rubber*[3]

$$\cdots -CH=\underset{\underset{CH_3}{|}}{C}-CH_2-CH_2-CH=\underset{\underset{CH_3}{|}}{C}-CH_2-CH_2-CH=\underset{\underset{CH_3}{|}}{C}-CH_2-CH_2-\cdots$$

and the bacterial polypeptide, *poly-γ-D-glutamate*,[21]

$$\cdots -NH-\underset{\underset{COO^-}{|}}{CH}-CH_2-CH_2-\underset{\underset{O}{\|}}{C}-NH-\underset{\underset{COO^-}{|}}{CH}-CH_2-CH_2-\underset{\underset{O}{\|}}{C}-\cdots$$

2e. Proteins.[5,6] Proteins are naturally occurring macromolecules, each consisting of one or a few (in some of the larger proteins perhaps of many) *polypeptide chains*. These chains are built up from about twenty different amino acids, with the structural formula

$$\cdots -NH-\underset{\underset{R_1}{|}}{CH}-CO-NH-\underset{\underset{R_2}{|}}{CH}-CO-NH-\underset{\underset{R_3}{|}}{CH}-CO-\cdots$$

where R_1, R_2, R_3, etc., represent the side chains of the constituent amino acids. These side chains may be non-polar (e.g., —H, —CH_3), polar but uncharged (e.g., —CH_2OH), negatively charged (e.g., —CH_2COO^-), or positively charged (e.g., —CH_2—CH_2—CH_2—$CHNH_3^+$). Most proteins contain nearly all of the possible side chains in different proportions.

In terms of constitution, proteins resemble synthetic polyampholytes. There is, however, a vast difference, for, in their synthesis in living tissues, nature has accomplished what man has failed to do: the construction of macromolecules all of which are identical in mass, in the number of side chain groups, and in their order along the chain. As a result, the physical chemistry of proteins is in some ways more akin to that of small molecules than is the physical chemistry of synthetic polymers, especially in that no distributions and averages have to be considered when we talk of molecular weight and structure.

The elucidation of the structural formulae of common proteins is a problem currently occupying many laboratories. At the present time the formula of one protein, *insulin*, is known in its entirety. This protein has a molecular weight of 5733 and consists of two short chains, linked to one another by disulfide bonds, as shown in Fig. 2–1. Of the two polypeptide chains one is composed of twenty-one, the other of thirty amino acid residues, the term *residue* being applied to the structural unit —NH—CHR—CO—. The order of the residues has been established in a series of pioneering researches by F. Sanger.[16]

The structure of a somewhat larger protein, *ribonuclease*, of molecular weight 13,683, has also been completely elucidated.[19] This protein consists of a single polypeptide chain of 124 residues, and it is cross-linked by four disulfide bonds. The location of these cross-links is shown in Fig. 2–2.

Some proteins are not composed entirely of polypeptide chains of the form —NH—CHR_1—CO—NH—CHR_2—CO— but are associated with small amounts (rarely, with large amounts) of other types of structures. *Hemoglobin* is a well-known example: it contains an iron-porphyrin complex. Other proteins, for instance *ovalbumin*, are associated with a few molecules of phosphoric acid. Proteins also exist which are combined with carbohydrate (*glycoproteins*), steroids (*lipoproteins*), or nucleic acids (*nucleoproteins*). The nucleoproteins are of particular importance in biology, and they are described briefly in section 2g.

Over 100 different proteins have been isolated in pure form, most of them from animal sources. Many of them are enzymes, i.e., potent specific catalysts of biological reactions.

The chemistry of animals of the same order is by and large the same; for instance, all mammals possess hemoglobin. However, there are small differences in the amino acid residue content from one species to another (e.g., for insulin, see Harfenist[20]). There may furthermore be even smaller differences (e.g., a single amino acid residue) between individuals of a given species, caused by differences in the genetic make-up of each individual. Finally, a given individual may possess two or more of these slightly different protein molecules. The differences do not affect most physico-chemical properties to any appreciable extent, so that we shall usually not find it necessary to distinguish between the various known forms of a given protein. It may be assumed that properties ascribed to "hemoglobin" will, within experimental error, apply to hemoglobin from all species and to all known genetic variants. Only where this assumption would be erroneous will a more detailed name be given to the protein concerned. This was done, for instance, in Fig. 2–1. The formula there given applies only to beef insulin (type A).

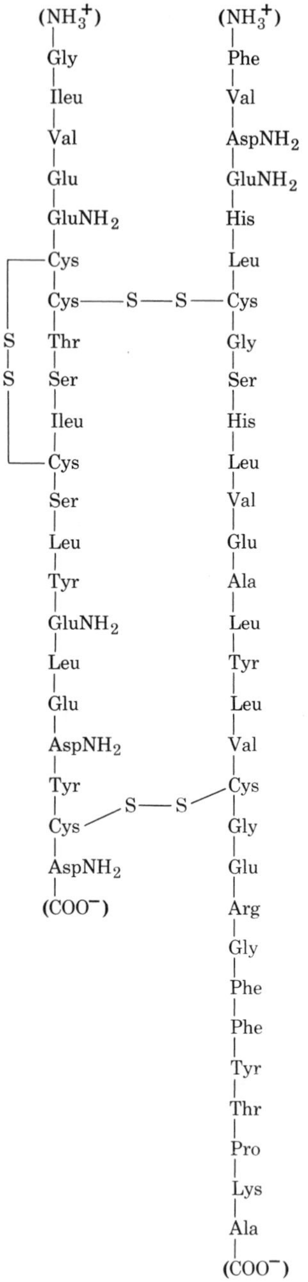

```
(NH₃⁺)              (NH₃⁺)
  |                   |
 Gly                 Phe
  |                   |
 Ileu                Val
  |                   |
 Val                AspNH₂
  |                   |
 Glu                GluNH₂
  |                   |
 GluNH₂             His
  |                   |
 Cys                Leu
  |                   |
 Cys——S——S——Cys
  |                   |
S Thr               Gly
|                     |
S Ser               Ser
  |                   |
 Ileu               His
  |                   |
 Cys                Leu
  |                   |
 Ser                Val
  |                   |
 Leu                Glu
  |                   |
 Tyr                Ala
  |                   |
 GluNH₂             Leu
  |                   |
 Leu                Tyr
  |                   |
 Glu                Leu
  |                   |
 AspNH₂             Val
  |                   |
 Tyr                Cys
  |    S——S     /    |
 Cys              Gly
  |                   |
 AspNH₂             Glu
  |                   |
(COO⁻)              Arg
                      |
                     Gly
                      |
                     Phe
                      |
                     Phe
                      |
                     Tyr
                      |
                     Thr
                      |
                     Pro
                      |
                     Lys
                      |
                     Ala
                      |
                    (COO⁻)
```

Code

Ala — alanine
Arg — arginine
AspNH₂ — asparagine
Cys — ½ cystine
Glu — glutamic acid
GluNH₂ — glutamine
Gly — glycine
His — histidine
Ileu — isoleucine
Leu — leucine
Lys — lysine
Phe — phenylalanine
Pro — proline
Ser — serine
Thr — threonine
Tyr — tyrosine
Val — valine

Each symbol in the formula corresponds to a residue

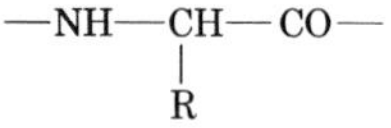

$$-NH-CH-CO-$$
$$\quad\quad\ \ |$$
$$\quad\quad\ \ R$$

derived from the parent amino acid listed in the code. The terminal residues have the structures

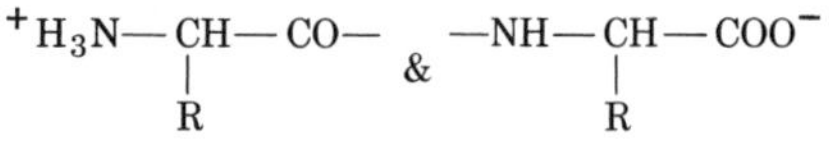

$$^{+}H_3N-CH-CO- \quad\&\quad -NH-CH-COO^{-}$$
$$\qquad\quad | \qquad\qquad\qquad\quad |$$
$$\qquad\quad R \qquad\qquad\qquad\quad R$$

Fig. 2–1. The structural formula of beef insulin A.[16]

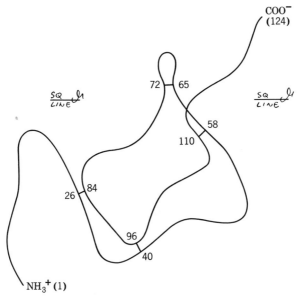

Fig. 2–2. Schematic representation of the structure of ribonuclease. The molecule is a single chain of 124 monomer units. Eight of them are "half residues," derived from the amino acid cystine, NH_2—$CH(COOH)$—CH_2—S—S—CH_2—$CH(COOH)$—NH_2, and they lead to the formation of four cross-links shown in the figure. The numbers represent the locations of the half residues along the chain, the terminal monomer unit with NH_3^+ being counted number 1.

2f. Nucleic acids.[7] Nucleic acids, like proteins, occur in all forms of living matter. They are *polynucleotides*, a *nucleotide* being a compound of phosphate, a pentose sugar, and a purine or pyrimidine, as shown in Fig. 2–3. The sugar may be ribose or deoxyribose, leading to a division of nucleic acids into ribonucleic acids and deoxyribonucleic acids, the former containing ribose, the latter deoxyribose. The names of these acids are generally abbreviated to RNA and DNA. DNA molecules make use of five different bases in their structure, these being adenine, guanine, cytosine, thymine, and 5-methylcytosine. RNA molecules contain only four bases in appreciable amounts, the particular bases in this case being adenine, guanine, cytosine, and uracil. The organic structure of nucleic acids (specifically of DNA) is shown in Fig. 2–3. It is to be noted that the natural form is the anion, in which each phosphate group bears a negative charge.

One of the most important sources of DNA is the cell nucleus of living organisms, and it has been established that DNA molecules are in fact the crucial constituents of the nucleus, carrying in their structure (presumably

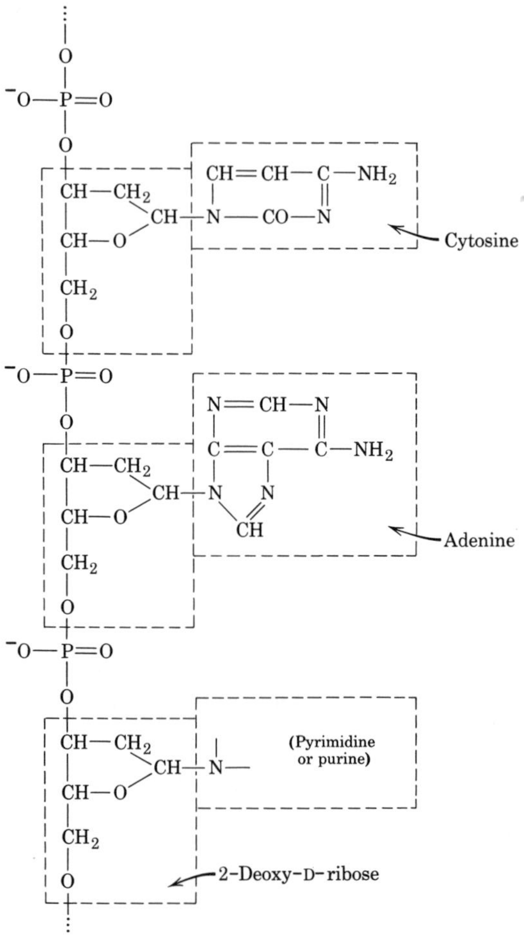

Fig. 2–3. Structural formula of a portion of a nucleic acid molecule (DNA).

in the sequence of bases) the key to inherited characteristics of each organism. This finding implies that each cell must contain a variety of DNA molecules and, furthermore, that each individual among higher organisms must possess a unique combination of DNA molecules. There is at present no experimental confirmation of this supposition. All DNA samples which have been investigated, originating from a variety of sources, are mixtures of different molecules, with molecular weight averages generally in the range of 4 to 7 million.

RNA is just beginning to be investigated on an intensive scale. In contrast to DNA, physically distinguishable preparations of RNA molecules have been obtained, and it may well be that a large variety of RNA's will be characterized in the near future. Present indications are that the molecular weights of many RNA's will be smaller than the molecular weight of DNA, of order 100,000.

2g. Nucleoproteins.[8] Nucleic acids ordinarily occur in nature in combination with proteins, and it is most probable that they nearly always exist as definite compounds called nucleoproteins. There are both DNA and RNA nucleoproteins.

By far the best characterized nucleoproteins, because they are self-replicating and can thus be "grown" in quantity, are the *viruses*. A number of viruses have been isolated in pure form, and some have been crystallized. One of the best known is tobacco mosaic virus. It consists of 6% RNA and 94% protein, and its molecular weight is about 40,000,000. Molecular weights of other viruses range from 2 million to 1 billion.

The nucleic acid and protein portions of nucleoproteins are easily separated and recombined. This suggests that the linkage between them is weak and that it may not consist of covalent bonds. Nevertheless, at least in the case of virus, they appear to combine in definite proportions, so that nucleoproteins are properly regarded as true compounds.

2h. Polysaccharides.[9,3] Polysaccharides are among the most familiar of the naturally occurring macromolecules. They are long chain polymers of simple sugars, the best known being starch and cellulose, both of which are 1,4-polyglucoses. The glucose units are joined by an α-glucosidic linkage in starch and by a β-glucosidic linkage in cellulose, as shown in Fig. 2–4.

Cellulose is the major component of vegetable fibers (cotton is almost pure cellulose). The molecular weight obtained for cellulose from most sources is in excess of one million. It is, however, subject to degradation, and there is no really convincing evidence to indicate what the molecular weight of a truly undegraded cellulose molecule might be. Many chemical derivatives, such as cellulose acetate (rayon) and cellulose nitrate (gunpowder), are, of course, familiar substances. Degradation occurs during the preparation of these derivatives. Their molecular weights are lower than that of native cellulose and vary from sample to sample.

Starch is the chemical form in which carbohydrate is stored in potatoes, wheat, corn, etc. Two forms of starch appear to be present in the natural sources: amylose, a linear polymer with molecular weight ranging from

10,000 to 400,000, and amylopectin, a branched polymer with a much higher molecular weight. The method of chain branching in amylopectin is by a 1,6-linkage as shown in Fig. 2–4. It is likely that both of these forms are degraded during preparation, and it is possible that all starch in an individual starch granule is one giant molecule.

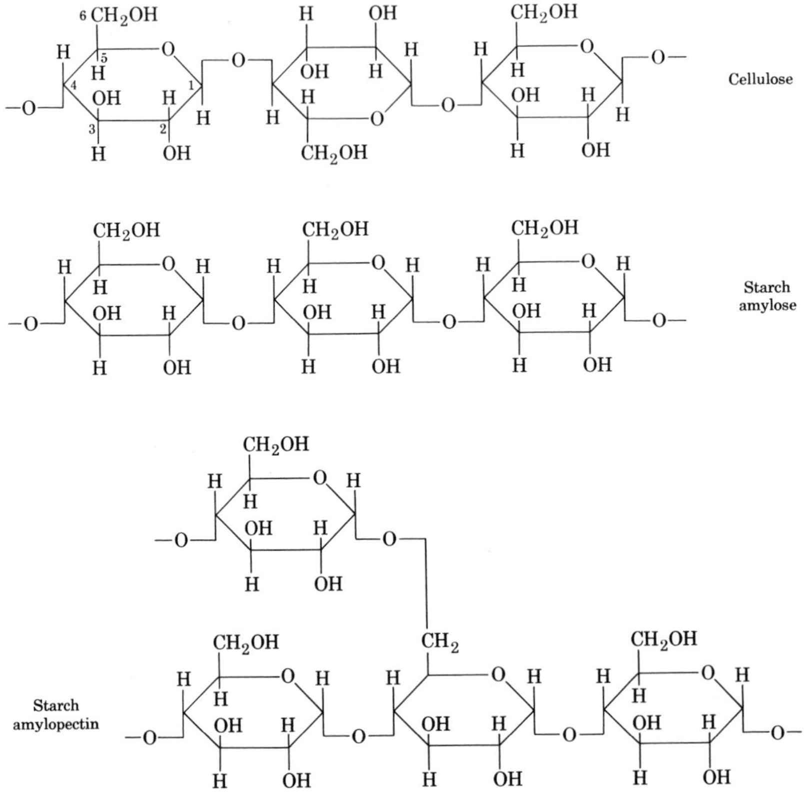

Fig. 2–4. Structural formulae for cellulose and starch. The first ring in the cellulose chain shows the system used for numbering the carbon atoms of the D-glucose ring.

Among other important polysaccharides are glycogen, an animal product resembling starch; bacterial polysaccharides, e.g., dextran, a 1,6-linked polyglucose; chitin, from the skeletons of crustaceans, which is a polymer of acetylated glucosamine; gums and mucilages, which are copolymers of a variety of sugars; and hyaluronic acid, a polyelectrolyte which is a copolymer of glucuronic acid and N-acetyl glucuronic acid. Most of these substances have so far received relatively little attention from physical chemists.

2i. Inorganic polymers. All the macromolecules mentioned so far have been from the realm of organic chemistry. Very few inorganic substances are known which fall logically into the macromolecular class. Any crystal may, of course, be said to be a single very large molecule, but the term macromolecule is generally used only if the large size is maintained under a variety of conditions: in solution, as well as in the solid state. A few inorganic substances do seem to be truly macromolecular, however, among them being the polyphosphates[22]

$$\cdots -\underset{\underset{O^-}{|}}{\overset{\overset{O}{\|}}{P}}-O-\underset{\underset{O^-}{|}}{\overset{\overset{O}{\|}}{P}}-O-\underset{\underset{O^-}{|}}{\overset{\overset{O}{\|}}{P}}-O-\cdots$$

and the polysilicates

$$\cdots -\underset{\underset{O^-}{|}}{\overset{\overset{OH}{|}}{Si}}-O-\underset{\underset{O^-}{|}}{\overset{\overset{OH}{|}}{Si}}-O-\underset{\underset{O^-}{|}}{\overset{\overset{OH}{|}}{Si}}-O-\cdots$$

which exist as polymeric ions in solution. An inorganic polymer $(PNCl_2)_x$ is formed as a result of reaction between NH_4Cl and PCl_5.[23] Perhaps the best-known inorganic polymers are the silicones,[24] which have the structure

$$\cdots -O-\underset{\underset{R}{|}}{\overset{\overset{R}{|}}{Si}}-O-\underset{\underset{R}{|}}{\overset{\overset{R}{|}}{Si}}-O-\underset{\underset{R}{|}}{\overset{\overset{R}{|}}{Si}}-\cdots$$

where R represents an organic group, such as a methyl group.

2j. Micellar and accidental colloids. The particles of colloidal size which are formed by intrinsically small molecules of course obey the same kind of physico-chemical laws as true macromolecules. Among the substances which form such particles are the hydrated oxides of many metals; finely divided gold, silver, sulfur, etc.; and soaps and other detergents. Their physical chemistry has been quite intensively studied, with major emphasis on the conditions required to maintain them in the colloidal state.[17,18] The present work, being limited to true macromolecules, will make no specific reference to micellar and accidental colloids of this kind.

General References

1. E. O. Kraemer, J. W. Williams, and R. A. Alberty, "The Colloidal State and Surface Chemistry" in H. S. Taylor and S. Glasstone (ed.), *A Treatise on Physical Chemistry*, 3rd ed., vol. 2, D. Van Nostrand Co., Princeton, N.J., 1951.

2. P. J. Flory, *Principles of Polymer Chemistry*, Cornell University Press, Ithaca, 1953·
3. *High Polymers*, a series of monographs, Interscience Publishers, New York, 1940 et seq.
4. H. Eisenberg and R. M. Fuoss, "Physical Chemistry of Synthetic Polyelectrolytes" in J. O'M. Bockris and B. E. Conway (ed.), *Modern Aspects of Electrochemistry*, Academic Press, New York, 1954.
5. H. Neurath and K. Bailey (ed.), *The Proteins*, Academic Press, New York, 1953–1954.
6. S. W. Fox and J. F. Foster, *Introduction to Protein Chemistry*, John Wiley and Sons, New York, 1957.
7. E. Chargaff and J. N. Davidson (ed)., *The Nucleic Acids*, Academic Press, New York, 1958.
8. R. Markham and J. D. Smith, "Nucleoproteins and Viruses," Ref. 5: vol. IIA, p. 1.
9. R. L. Whistler and C. L. Smart, *Polysaccharide Chemistry*, Academic Press, New York, 1953.

Specific References

10. T. Graham, *Phil. Trans. Roy. Soc. (London), Ser. A*, **151**, 183 (1861).
11. H. Staudinger, *Ber.*, **53**, 1073 (1920); **57**, 1203 (1924); H. Staudinger and J. Fritschi, *Helv. Chim. Acta.*, **5**, 785 (1922).
12. T. Svedberg and K. O. Pedersen, *The Ultracentrifuge*, Oxford University Press, 1940, p. 406.
13. J. B. Sumner, *J. Biol. Chem.*, **69**, 435 (1926).
14. J. H. Northrop, *J. Gen. Physiol.*, **13**, 739 (1930).
15. W. H. Carothers, *Chem. Revs.*, **8**, 353 (1931).
16. F. Sanger and H. Tuppy, *Biochem. J.*, **49**, 481 (1951); F. Sanger and E. O. P. Thompson, *ibid.*, **53**, 353, 366 (1953); F. Sanger, E. O. P. Thompson, and R. Kitai, *ibid.*, **59**, 509 (1955).
17. E. J. W. Verwey and J. Th. G. Overbeek, *Theory of the Stability of Lyophobic Colloids*, Elsevier Publishing Co., Amsterdam, Netherlands, 1948.
18. H. R. Kruyt (ed.), *Colloid Science*, Elsevier Publishing Co., Amsterdam, Netherlands, 1952.
19. C. H. W. Hirs, W. H. Stein, and S. Moore, *J. Biol. Chem.*, **221**, 151 (1956); **235**, 633 (1960); D. H. Spackman, W. H. Stein, and S. Moore, *ibid.*, **235**, 648 (1960).
20. E. J. Harfenist, *J. Am. Chem. Soc.*, **75**, 5528 (1953).
21. H. Edelhoch and J. B. Bateman, *J. Am. Chem. Soc.*, **79**, 6093 (1957).
22. C. F. Callis, J. R. van Wazer, and P. G. Arvan, *Chem. Revs.*, **54**, 777 (1954).
23. F. Patat and F. Kollinsky, *Makromol. Chem.*, **6**, 292 (1951); F. Patat and K. Frombling, *Monatsh.*, **86**, 718 (1955).
24. E. Rochow, *An Introduction to the Chemistry of the Silicones*, 2nd ed., John Wiley and Sons, New York, 1951.
25. J. T. Edsall, *Arch. Biochem. Biophys.*, Supplement 1, 12 (1962).

2

MOLECULAR STRUCTURE

In classical organic chemistry a molecule is completely characterized if its structural formula is known, i.e., if the arrangement of the atoms and the primary bonds joining them are known. This is not true for macromolecules. Two macromolecules identical in structural formula may be conceived which may yet be entirely different in their physical and chemical properties, for one may be stretched into a long chain, for example, and the other coiled into a tight ball. It is with this new dimension to our concept of "structure" that this chapter will be concerned. We shall find large differences between different macromolecules in this respect. At one extreme will be found molecules with strong secondary intramolecular bonds which lead to a single preferred configuration, adopted by all the molecules. At the other extreme are substances in which specific secondary attractions are almost completely absent, so that all possible configurations have intrinsically the same probability.

Of course, the physical chemist is not ordinarily content with the organic chemist's structural formula. He wants to know more precise details of the structure: the distance between atomic nuclei, the angles between bonds, etc. This aspect of molecular structure, however, is a relatively unspecific one. The distance between adjacent carbon atoms joined by a single bond is about 1.54 Å regardless of the molecule examined; the four bonds issuing from an aliphatic carbon atom are always closely directed towards the corners of a tetrahedron, with bond angles of about 109.5°. We do not expect, nor do we find, appreciable differences between small and large molecules in this regard. By and large, therefore, interatomic distances and angles are, from the point of view of this chapter, parameters whose values are already known. Our attention will be focused chiefly on the relation between parts of a macromolecule which, according to the structural formula, are separated by several intervening bonds.

In the study of small molecules the method of x-ray diffraction has proved to be the most useful tool in the elucidation of structural details.

15

It has not only provided distances and bond angles between the atoms of individual molecules but also given precise information concerning the arrangement of molecules with respect to one another in the solid state. It is therefore a logical method by which to investigate the arrangement of different portions of a macromolecule with respect to each other. Accordingly, x-ray diffraction will be the major subject discussed in this chapter. Section 3 will take up the basic theory which connects x-ray diffraction with the structural features of molecular aggregates. Section 4 will give examples of the application of this method to macromolecular solids.

Among other methods which give information regarding molecular structure, spectroscopy is the most important. Its application to macromolecules is discussed in section 5. Other methods of more limited utility will be summarized in section 6, and this chapter will conclude, in section 7, with general conclusions based on all available data.

3. THE PRINCIPLES OF X-RAY DIFFRACTION

3a. X-ray scattering by a single atom. If an electromagnetic wave impinges upon any particle of matter, a part of the radiation will be scattered. The reason is that the incident radiation sets up a vibration between the positively charged and negatively charged elements of the particle, creating an oscillating dipole. Such a dipole is itself a source of electromagnetic radiation. The theory of this effect is discussed in detail in Chapter 5. It is sufficient to note here that this secondary radiation has the same frequency as the incident radiation and that it is exactly 180° out of phase with it. It is scattered in all directions with an intensity which depends uniquely on the scattering angle. The scattered intensity also depends strongly on wavelength and on the nature of the scattering particle. We are interested here in scattering by an individual atom. The intensity of scattering in this case increases with the number of electrons which the atom possesses, approximately as the square of this number.

3b. X-ray scattering by a pair of identical atoms. Consider, as in Fig. 3–1a, two atoms at P_1 and P_2 in the path of an electromagnetic ray. Let s_0 represent a unit vector in the direction of propagation. Let an observer or a detecting device be placed at a direction specified by the unit vector s_1 from the two points. (The distance of the observer from the two atoms is to be very much greater than the distance P_1P_2, which is to be of the order of a few Ångstrom units, so that the direction of the observer from P_1 is for all practical purposes the same as his direction from P_2.) The points P_1 and A represent equal positions in the direction of propagation of the incident

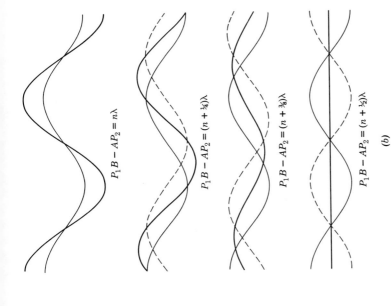

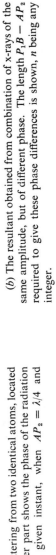

(b) The resultant obtained from combination of x-rays of the same amplitude, but of different phase. The length $P_1B - AP_2$ required to give these phase differences is shown, n being any integer.

$P_1B - AP_2 = n\lambda$

$P_1B - AP_2 = (n + \frac{1}{4})\lambda$

$P_1B - AP_2 = (n + \frac{3}{8})\lambda$

$P_1B - AP_2 = (n + \frac{1}{2})\lambda$

(b)

(a)

Fig. 3-1. (a) X-ray scattering from two identical atoms, located at P_1 and P_2. The lower part shows the phase of the radiation at each position at a given instant, when $AP_2 = \lambda/4$ and $P_1B = \lambda/2$.

beam and, therefore, positions of equal phase. At P_2, the incident radiation has progressed a distance AP_2 from the line of equal phase, AP_1. Its phase angle has therefore changed by $2\pi(AP_2/\lambda)$, where λ is the wavelength of the x-ray. Because the atoms at P_1 and P_2 are identical and because the angle of observation is the same at both points, the ray scattered from P_2 has the same intensity as that scattered from P_1. Each ray, however, is 180° out of phase with the incident ray at the point of scattering, so that the phase angle of the ray scattered from P_2 (measured at P_2) differs by $2\pi(AP_2/\lambda)$ from the phase angle of the ray scattered from P_1 (measured at P_1).

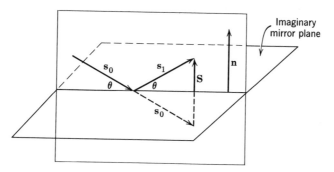

Fig. 3–2. Representation of x-ray scattering from a pair of identical atoms in terms of an imaginary mirror plane.

To investigate the way this scattered radiation appears to the observer we consider the line P_2B, at which both scattered rays are at equal positions in the direction of the observer. The scattered ray from P_1 has a phase angle (relative to the position P_1) of $2\pi(P_1B/\lambda)$, which will in general be different from that at P_2, the difference in phase angle being $(2\pi/\lambda)(P_1B - AP_2)$. As Fig. 3–1*b* shows, the resultant of the two scattered rays will depend on this difference in phase angle. The resultant intensity will be a *maximum* if the phase angles differ by an integral multiple of 2π; the resultant wave disappears entirely if they differ by an odd integral multiple of π. The direction s_1 will clearly represent a maximum in the scattered intensity if $P_1B - AP_2$ is an integral multiple of the wavelength λ.

If $\mathbf{r}$ is the vector from P_1 to P_2, the distance P_1B is $\mathbf{r} \cdot \mathbf{s}_1$ and the distance AP_2 is $\mathbf{r} \cdot \mathbf{s}_0$. The condition that s_1 represents a direction for maximum intensity is therefore

$$\mathbf{r} \cdot (\mathbf{s}_1 - \mathbf{s}_0) = \mathbf{r} \cdot \mathbf{S} = n\lambda \qquad (n = 0, 1, 2, 3, \cdots) \qquad (3\text{–}1)$$

where $\mathbf{S}$ is the vector $\mathbf{s}_1 - \mathbf{s}_0$. Equation 3–1 is known as von Laue's condition.

It will prove convenient to consider von Laue's condition in a different way, in terms of the angle between s_1 and s_0, which we shall call 2θ. Let us imagine a *mirror* plane, so placed that a ray which strikes it in the direction s_0 will be reflected in the direction s_1. Since reflection from a mirror (specular reflection) occurs such that the incident and reflecting angles are equal, the desired plane will clearly be at an angle θ with respect to both s_0 and s_1, as shown in Fig. 3–2. Moreover, the vector S must be perpendicular to the mirror plane and have a length of $2 \sin \theta$. Von Laue's condition may therefore be restated as specifying that a diffraction maximum may be formally represented as a reflection, from an imaginary plane which makes an angle θ with the incident ray, such that

$$2 \sin \theta \, (\mathbf{r} \cdot \mathbf{n}) = n\lambda \quad (n = 0, 1, 2, 3, \cdots) \tag{3–2}$$

where $\mathbf{n}$ is a unit vector normal to the plane in question. Each value of n, with its corresponding value of θ, defines a different possible reflecting plane.

3c. Scattering from a linear array of identical atoms. It is of interest to apply equation 3–1 to scattering from a linear array of equidistant, identical atoms. Let $\mathbf{d}$ be the vector joining an atom to its immediate neighbor. Then the vector to the next following atom will be $2\mathbf{d}$, that to the next one $3\mathbf{d}$, etc. A diffraction maximum will occur if equation 3–1 is satisfied for *all* possible pairs of atoms, i.e., for $\mathbf{r} = \mathbf{d}$, $\mathbf{r} = 2\mathbf{d}$, etc. This clearly means that it need be satisfied explicitly for nearest neighbors only, for, if $\mathbf{d} \cdot \mathbf{S}$ is an integral number of wavelengths, then $2\mathbf{d} \cdot \mathbf{S}$, $3\mathbf{d} \cdot \mathbf{S}$, etc., are automatically also integral numbers of wavelengths. Thus the diffraction maxima for a linear array of atoms, spaced at a distance d, appear at the same direction, regardless of the number of atoms in the array.

The intensity of scattering at directions which are intermediate between maxima does, however, depend on the number of atoms. If there are only two atoms, the intensity decreases gradually as the phase difference between the two individual scattered rays increases (cf. Fig. 3–1). If there are many atoms, however, the intensity decreases very sharply. For, if adjacent atoms produce scattered rays only slightly out of phase, atoms separated by a distance $2d$ produce rays differing twice as much in their phase, etc. Even if the difference in phase between rays from adjacent atoms is only, say, $5°$, the difference in phase between atoms separated by a distance of $36d$ will be $180°$. If there are several thousand atoms in a linear array, all but a very small number can be paired off such that the rays completely cancel one another. As a result such a linear array produces diffraction essentially *only* at the maxima given by equation 3–1 or 3–2.

One of the consequences of this fact is that only x-rays (or electromagnetic waves of even shorter wavelengths) will produce any diffraction at all from arrays of atoms separated by distances of the order of a few

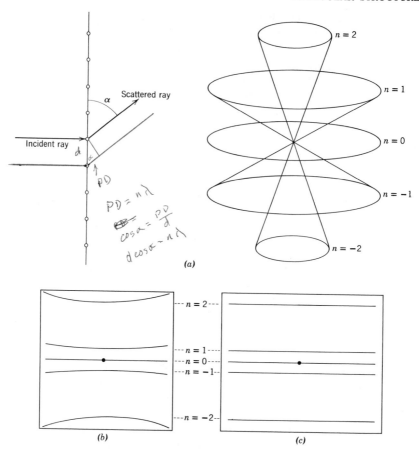

Fig. 3–3. Location of maxima in scattered intensity when an x-ray beam strikes a linear array of equidistant atoms at right angles. The letter n is the integer of equation 3–1. The photograph obtained on a film mounted vertically behind the linear array is shown in diagram b; that obtained on a film placed cylindrically about the sample is shown in diagram c.

Ångstrom units. As equation 3–2 shows (note that $\mathbf{n}$ is a *unit* vector) the condition for a diffraction maximum cannot be met at all unless $\lambda < 2d$, for $\sin \theta$ cannot be greater than unity.

As an example of the experimental manifestation of the von Laue condition, we consider an x-ray beam perpendicular to a row of equally spaced atoms (Fig. 3–3). Since $\mathbf{d}$ and $\mathbf{s}_0$ are perpendicular, $\mathbf{d} \cdot \mathbf{s}_0 = 0$, and $\mathbf{d} \cdot \mathbf{s}_1 = d \cos \alpha$, where α is the angle between the vectors $\mathbf{d}$ and $\mathbf{s}_1$, as shown in Fig. 3–3a. Equation 3–1 then becomes

$$\cos \alpha = n\lambda/d \quad (n = 0, 1, 2, 3, \cdots) \tag{3–3}$$

Diffraction thus occurs along cones, as shown in Fig. 3–3a. If a photographic plate were mounted behind the scattering sample, perpendicular to the incident beam, diffraction lines would occur along the intersections of these cones with the photographic plate. These intersections are known as *layer* lines, and they are shown in Fig. 3–3b. If we replace the photographic plate by a strip of film wrapped cylindrically about the scattering sample, the intersections between the film and the cones are circles, and, on unfolding of the film, the layer lines appear as the straight lines shown in Fig. 3–3c. The distance between layer lines, in either case, enables us to calculate the distance d. If the layer lines are sharp, it means that there are many atoms in the linear array; if they are broad and diffuse, it means that there are just a few. It should be noted that closely spaced layer lines correspond to a large interatomic distance, d, and vice versa. This kind of reciprocal relation between an actual structure and the corresponding diffraction pattern is a feature common to all x-ray diffraction patterns.

The layer lines would not become infinitely sharp for an array of real atoms, even if the number of atoms were to become very large, because real atoms undergo vibrations which would create small fluctuations in interatomic distances and, hence, in the precise direction of the diffraction cones of Fig. 3–3a. For a quantitative discussion of all factors which influence diffraction intensity, the reader is referred to standard textbooks and especially to that of James.[3]

3d. Diffraction of x-rays by crystals with a single atom per unit cell. Crystals are solid structures in which molecules or atoms are arranged in a regular, ordered three-dimensional pattern. The entire structure can be built up by multiple repetition, in three dimensions, of a small three-dimensional building block known as the *unit cell*. For a given structure the unit cell is the same, regardless of the size of the whole crystal. The unit cell is always a parallelepiped and may thus be described by three vectors **a**, **b**, and **c**. Unit cells are classified on the basis of symmetry rather than shape, but there is usually a correspondence between symmetry and shape. The vectors **a**, **b**, and **c** are equal in length and mutually perpendicular in a *cubic* cell. They are unequal in length, but still perpendicular, in an *orthorhombic* cell. The vectors **a** and **b** are perpendicular, but **c** is usually inclined at an angle other than 90° to the *ab* plane in a *monoclinic* cell. None of the axes are normally perpendicular in a *triclinic* cell. (See any standard textbook for other descriptive terms for crystal structures.)

If a vector $\mathbf{r} = u\mathbf{a} + v\mathbf{b} + w\mathbf{c}$, with u, v, and w integers, is drawn from a corner of a unit cell, its terminus will be a corresponding corner of another unit cell. Moreover, since all unit cells are identical, any such vector drawn from any position within a unit cell will terminate at a corresponding position in another unit cell.

The positions of atoms within the unit cell may be specified by co-ordinates x, y, and z. These coordinates are generally defined in terms of the unit cell axes, such that $x = 1$ means a distance a in the **a** direction, $y = 1$ a distance b in the **b** direction, etc. The origin of the unit cell, $x = y = z = 0$, is arbitrary. In a crystal of NaCl, for example, it may be placed on a Na^+ ion, or a Cl^- ion, or anywhere in between. Often the crystal structure may be described in terms of several different repeating cells; the choice between them is largely one of convenience. Once the unit cell and its origin are decided upon, the positions of all atoms within it are fixed.

We consider first a crystal containing only one kind of atom, with only a single atom per unit cell. A simple monatomic cubic structure would be an example of such a crystal. The origin of the unit cell may be placed on the atom, in which case its coordinates would be 0,0,0; alternatively it might be placed so that the atom lies at its exact center with coordinates $\frac{1}{2},\frac{1}{2},\frac{1}{2}$. Regardless of this choice, a vector $\mathbf{r} = u\mathbf{a} + v\mathbf{b} + w\mathbf{c}$ (with u, v, w integers) will describe the relative positions of any pair of atoms. Diffraction maxima will occur whenever every possible $\mathbf{r}$ satisfies von Laue's condition, equation 3–1. For this to occur it is clearly necessary that each of the following relations be simultaneously satisfied,

$$\mathbf{a} \cdot \mathbf{S} = h\lambda$$
$$\mathbf{b} \cdot \mathbf{S} = k\lambda \qquad (3\text{–}4)$$
$$\mathbf{c} \cdot \mathbf{S} = l\lambda$$

with h, k, and l any integers.

As in going from equation 3–1 to equation 3–2, equation 3–4 may be restated in terms of reflection from imaginary planes through the crystal. A diffraction maximum may be thought of as a reflection from a plane which makes an angle θ with the incident x-ray, such that equation 3–2 is satisfied simultaneously for each of the three dimensions, giving

$$2 \sin \theta \, (\mathbf{a} \cdot \mathbf{n}) = h\lambda$$
$$2 \sin \theta \, (\mathbf{b} \cdot \mathbf{n}) = k\lambda \qquad (3\text{–}5)$$
$$2 \sin \theta \, (\mathbf{c} \cdot \mathbf{n}) = l\lambda$$

where $\mathbf{n}$ is a unit vector normal to the reflecting plane. Any combination of integral values of h, k, and l defines a different plane. The possible reflecting planes are generally identified in terms of the values of h, k, and l as (hkl) planes. The (100) plane, for example, is that imaginary reflecting plane which gives the diffraction maximum with $h = 1$, $k = 0$, and $l = 0$. The values of h, k, and l are known as the *Miller indices* of the plane.

Equation 3–5 shows that a property of the imaginary reflection planes is that $(\mathbf{a}/h) \cdot \mathbf{n} = (\mathbf{b}/k) \cdot \mathbf{n} = (\mathbf{c}/l) \cdot \mathbf{n} = \lambda/2 \sin \theta$. This property is satisfied by real planes which intersect the a axis of the unit cell at intervals of a/h, the b axis at intervals of b/k, and the c axis at intervals of c/l. For such planes

$$(\mathbf{a}/h) \cdot \mathbf{n} = (\mathbf{b}/k) \cdot \mathbf{n} = (\mathbf{c}/l) \cdot \mathbf{n} = d_{hkl} \qquad (3\text{--}6)$$

where d_{hkl} is the interplanar spacing along the normal $\mathbf{n}$ to the plane. Substituting equation 3–6 into equation 3–5 gives the Bragg relation

$$\sin \theta = \lambda/2d_{hkl} \qquad (3\text{--}7)$$

Equation 3–7 is a simpler representation of diffraction maxima than equations 3–4 or 3–5. Accordingly diffraction maxima are generally thought of as reflections arising from (hkl) planes. The angles θ at which maxima are observed lead to values for the interplanar spacings d_{hkl}. It should be noted that the (100), (010), and (001) spacings give values for the lengths a, b, and c of the unit cell axes.

For rectangular unit cells, $d_{100} = a$, $d_{010} = b$, $d_{001} = c$. If the axes are inclined to one another at angles other than 90°, the perpendicular interplanar distances involve both the unit cell axis lengths and these angles. For example, if γ is the angle between $\mathbf{a}$ and $\mathbf{b}$ in a monoclinic unit cell, $d_{100} = a \sin \gamma$, $d_{010} = b \sin \gamma$.

Bragg's condition is often given as $2d_{hkl} \sin \theta = n\lambda$, where n is any integer. If h, k, and l are allowed to take on any integral values, this is a redundant expression. Consider the planes $(h'k'l')$ where $h' = nh$, $k' = nk$, $l' = nl$. These planes are parallel to the (hkl) planes but have $d_{h'k'l'} = d_{hkl}/n$. The value of θ corresponding to these planes is given by equation 3–7 as $\sin \theta = \lambda/2d_{h'k'l'} = n\lambda/2d_{hkl}$, which is identical to the value predicted for the parallel (hkl) planes by the relation $2d_{hkl} \sin \theta = n\lambda$. Thus the latter relation does not lead to new diffraction maxima not already included in equation 3–7.

At first sight, equation 3–7 appears to predict an infinite number of diffraction maxima. However, as h, k, and l become larger, the corresponding interplanar distances become smaller, and $\sin \theta = \lambda/2d_{hkl}$ becomes larger. Since $\sin \theta$ must be less than unity, diffraction maxima are thus limited, at a given wavelength, to relatively small values of h, k, and l. As the unit cell increases in size, the interplanar distances for given Miller indices also become larger, with the result that the permissible range of h, k, and l increases. (The reciprocal relation between diffraction and the actual structure is observed again. A large unit cell leads to many diffraction maxima at closely spaced values of θ; a small unit cell produces just a few maxima at widely separated values of θ.)

Equations 3–4 to 3–7 describe the conditions for the directions which correspond to maxima in the intensity of scattered radiation. However, as in the case of scattering by a row of atoms, these directions are essentially the only directions in which scattering can be observed if the crystalline order persists, in each of the three dimensions, over a distance of several hundred unit cells.

3e. Diffraction of x-rays by crystals with several identical atoms per unit cell.
The fact that the vector $\mathbf{r} = u\mathbf{a} + v\mathbf{b} + w\mathbf{c}$ joins any point of a unit cell to an identical point in another unit cell does not depend on the number of atoms in the unit cell. In all crystals, therefore, each such vector, for all integral values of u, v, and w, represents a vector from any atom to an identical atom. Clearly, therefore, Bragg's condition (equation 3–7) continues to apply regardless of the number of atoms per unit cell. The existence of more than one atom in a unit cell, however, imposes additional requirements which limit the possible values of h, k, and l to which Bragg's condition applies.

Consider, for instance, a *body-centered* unit cell, containing an atom at $x + \frac{1}{2}, y + \frac{1}{2}, z + \frac{1}{2}$ in addition to that at x, y, z. (Such a cell is called body centered because the origin of the unit cell may be placed at one of the atoms. The other will then have the coordinates $\frac{1}{2}, \frac{1}{2}, \frac{1}{2}$, i.e., it will be in the center of the unit cell.) If the two atoms are identical, the vector $\mathbf{r} = \frac{1}{2}\mathbf{a} + \frac{1}{2}\mathbf{b} + \frac{1}{2}\mathbf{c}$ is a vector which must satisfy von Laue's condition; i.e., to obtain a diffraction maximum, we must have

$$\tfrac{1}{2}(\mathbf{a} + \mathbf{b} + \mathbf{c}) \cdot \mathbf{S} = n\lambda \tag{3–8}$$

where n is any integer. At the same time equations 3–4 must hold true, as before. Clearly, equations 3–4 are compatible with equation 3–8 if, and only if, $h + k + l$ is an even number. It can be concluded, therefore, that Bragg's relation applies to body-centered unit cells with the additional restriction that reflections will be obtained from only those (hkl) planes which have $h + k + l$ an even number.

Similarly, for a face-centered unit cell (atoms at x, y, z; $x + \frac{1}{2}, y + \frac{1}{2}, z$; $x + \frac{1}{2}, y, z + \frac{1}{2}$; $x, y + \frac{1}{2}, z + \frac{1}{2}$) the corresponding condition is that h, k, and l must either be all odd or all even.

The reader will note that a body-centered cubic structure with two identical atoms per cell can also be described in terms of a small monoclinic unit cell, containing only a single atom per cell. The body-centered structure has a higher symmetry and is thus the more convenient choice. A similar alternative description exists in the case of face-centered cubic structures.

The general extension of this treatment is usually made in terms of the theory of *space groups*. A space group describes the symmetry of a solid structure in terms of a set of operations which will transform the structure into a form identical to that originally present. Among the operations which can produce this result are those of rotation, inversion, reflection, a combination of rotation and translation, etc. Various ways of combining such operations lead to a total of 230 possible space groups. Their significance from the present point of view is that each space group has associated with it a characteristic set of conditions on possible (hkl) reflections.

The space group $P2_12_12$, for instance, describes an orthorhombic figure with the following symmetry properties:

(1) A *two-fold rotation axis* in the z direction, which means that rotation by 180° about this axis is an identity operation.

(2) A *two-fold screw axis* in the x direction, which means that rotation by 180° about this axis, coupled with translation by half the unit cell repeat along the axis, is an identity operation.

(3) Another *two-fold screw axis* in the y direction.

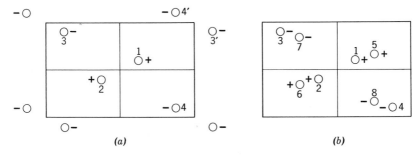

Fig. 3–4. Location of atoms in space group $P2_12_12$. The figures show the x,y projection. A + sign represents an atom located at $+z$, a − sign an atom at $-z$. In diagram *a* the asymmetric unit is a single atom; in diagram *b* it consists of two atoms.

Reflection, inversion, rotation alone about the x or y axis, etc., are specifically not symmetry operations. (Other space groups would describe a structure with these additional symmetry features.)

A unit cell with four atoms at positions x, y, z; $-x$, $-y$, z; $\frac{1}{2} + x$, $\frac{1}{2} - y$, $-z$; $\frac{1}{2} - x$, $\frac{1}{2} + y$, $-z$, as shown in Fig. 3–4a, would belong to this space group. The symmetry axes could occupy any one of a number of possible positions. The rotation axis in the z direction could be at the origin, for instance; the two screw axes could be at $x = \frac{1}{4}$, $z = 0$ and $y = \frac{1}{4}$, $z = 0$. If the axes are so placed, symmetry operation (1) sends atom 1 (Fig. 3–4a) to position 2, and vice versa, and atom 3 to position 4, and vice versa. Symmetry operation (2) sends atom 1 to position 3', atom 2 to position 4', atom 3 to position 1, and atom 4 to the position that atom 2 would have in the unit cell which would lie immediately above atom 3' of Fig. 3–4a. And similar relations exist for operation (3). *Any* of the symmetry operations, performed on *any* atom of the unit cell, leads to another atom of the unit cell.

Corresponding to the space group $P2_12_12$ there are the following specific restrictions on (hkl) reflections: of the $(h00)$ reflections only those for which h is even will be observed; of the $(0k0)$ reflections only those for which k is even will be observed. There are in general no other conditions.

All 230 space groups, with the corresponding restrictions on observable reflections are listed in volume I of the *International Tables for X-Ray Crystallography*.[13]

It is important to note that a unit cell may contain atoms which, even though they may be identical, are not connected by the same set of symmetry operations. As a result, many different unit cells may be constructed which belong to the same space group. Figure 3–4b, for example, shows a unit cell with atoms in the same positions as in Fig. 3–4a but containing also an additional set of four atoms (nos. 5 to 8). The same symmetry operations apply; e.g., the operation which sends atom 1 to position 2 also sends atom 5 to position 6. But none of these operations nor any other symmetry operation sends any of the four original atoms to any of the positions 5 to 8. The figure could be rotated about an axis in the z direction, placed between atoms 1 and 5, which would send atom 1 to position 5 and vice versa. This operation would not, however, be an identity operation, for it does not send other atoms to equivalent positions. Thus Fig. 3–4b, like Fig. 3–4a, belongs to space group $P2_12_12$. The additional atoms therefore produce no new systematic disappearance of (hkl) reflections. (They will, however influence the intensities of all reflections, as discussed in section 3h.)

Another way of looking at Fig. 3–4 is this: That the space group $P2_12_12$ requires the presence of four identical structural units (each alone without symmetry with respect to the unit cell coordinates) in the unit cell. In Fig. 3–4a the structural unit consists of a single atom; in Fig. 3–4b it consists of a pair of atoms.

3f. Diffraction by crystals with more than one kind of atom in a unit cell. The presence of more than one kind of atom in a unit cell in no way influences the requirement that the vectors $\mathbf{r} = u\mathbf{a} + v\mathbf{b} + w\mathbf{c}$, with u, v, and w integral, represent displacements between identical points in different unit cells; i.e., Bragg's law is still the basic law specifying the reflections that will be obtained. The presence of atoms of more than one kind has qualitatively the same effect as the presence of several atoms of the same kind. However, such atoms can never produce any *missing* reflections, because they can never lead to new symmetry operations. An operation which sends an atom of kind A to the position of an atom of kind B cannot be a symmetry operation. An operation which sends all atoms of kind B to other positions occupied by B atoms, without doing the same for atoms of kind A, is also not a symmetry operation.

As an example, consider a body-centered type of unit cell with, however, a lighter atom at $\frac{1}{2}, \frac{1}{2}, \frac{1}{2}$ than that at 0, 0, 0. Then, when $h + k + l$ is an odd number, the ray scattered from the central atom is exactly 180° out of phase with that of a corner atom. If the two atoms were identical, complete elimination of the corresponding (hkl) reflections would result (see p. 24). If the atoms are different, however, the intensity of the ray scattered from the lighter atom is less than that scattered from the heavier one. The two rays, therefore, even though completely out of phase, will differ in amplitude. The (hkl) reflections with $h + k + l$ odd will be reduced in intensity but will not disappear entirely.

It was stated in section 3a that the intensity of scattering of an atom is roughly proportional to the square of its atomic number. If, therefore, the atoms of a particular crystal differ greatly in atomic number, the observed intensity of diffracted x-rays comes almost entirely from the atoms of higher atomic number. In PbO, for instance, the diffraction pattern would arise almost entirely from the arrangement of the Pb^{++} ions. This is advantageous from one point of view, since it reduces the complexity of the diffraction pattern. On the other hand it also means that the diffraction pattern can yield information on the location of oxygen atoms only if intensity measurements are made with greatest accuracy, and then only in terms of small differences between the observed intensity distribution and that calculated on the basis of the arrangement of Pb^{++} ions alone. (One consequence of this situation is that, when x-ray diffraction studies of organic crystals are made, it is rarely feasible to try to determine the positions of hydrogen atoms.)

3g. The experimental determination of crystal structure by x-ray diffraction. Numerous experimental procedures exist by which the particular reflection angles corresponding to Bragg's law may be determined. One of them is the *powder method.* A powdered sample of a crystalline substance is placed in the path of an x-ray beam. The sample is rotated or agitated so that in the course of an experiment all planes of the crystals pass through all possible orientations with respect to the incident beam, including the particular orientation (incident angle θ as given by the Bragg condition) required for a diffraction maximum. Planes making an appropriate angle θ with the incident beam reflect x-rays at the angle 2θ. The corresponding diffracted beam may thus lie anywhere along a cone, with axis parallel to the beam and apex at the sample, as shown in Fig. 3–5. If a photographic plate is placed behind the sample, perpendicular to the incident beam, the diffraction maxima will appear as concentric circles. A cylindrical strip of film placed about the powdered sample will show arcs of maximum intensity, as shown in Fig. 4–3a. The powder method will give all observable maxima with the restriction that $2\theta < 90°$ if a photographic plate is used.

The sharpness of powder lines is subject to the same factors discussed earlier in connection with diffraction of x-rays from a linear array of identical scattering units. A pure crystalline solid, even when it is in powdered form, ordinarily consists of many billions of unit cells, and sharp diffraction patterns are thus always obtained.

Another common technique is to use a relatively large (~1 mm) single crystal and to mount it on the sample stage so that a well-defined axis is

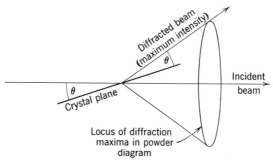

Fig. 3–5. Location of a diffraction maximum in the scattering of x-rays from crystals. The maximum is obtained only when the crystal plane is correctly oriented with respect to the incident beam, the values of θ required being given by Bragg's law.

perpendicular to the incident beam. The crystal is then rotated about this axis. During the rotation different crystal planes in turn arrive at the correct position for a maximum in scattering intensity. However, the fact that the rotation keeps one axis of the crystal fixed means that the orientation of the planes is also fixed. As a result scattering maxima appear on the film as points, rather than circles or arcs.

A consequence of the fact that a particular axis of the crystal is always perpendicular to the incident radiation is that the von Laue condition (equation 3–3) for scattering from a regular perpendicular array of scattering points must be obeyed. All scattered radiation must therefore lie along layer lines (Fig. 3–3). That this is indeed so may be seen, for instance, from Figs. 4–3b and c. (The spots of a single crystal diffraction pattern may be thought of as the points of intersection of appropriate powder diagram cones with appropriate layer lines.) Furthermore, the spots in a given layer line always have one Miller index in common. For instance all planes parallel to the axis of rotation are always perpendicular to the plane of the incident x-ray beam. As a result the scattered beam (obeying the laws for mirror reflection) is also perpendicular to the axis of rotation. Thus all spots in the plane of the incident x-ray are spots derived from crystal planes parallel to the axis of rotation. If the axis of rotation is, say, the c axis, this means that reflections in the plane of the incident x-ray are all reflections with $l = 0$, i.e., from $(hk0)$ planes. Similarly, reflections in the first layer line above (or below) the plane of the incident beam come from $(hk1)$ planes, those in the second layer line from $(hk2)$ planes, etc.

The most sophisticated methods of all are those in which the camera, the film, the crystal, and a slotted screen all move in concert during exposure, in such a way that not only one but all three of the Miller indices of a particular reflection are automatically determined. (See Buerger[2] for a description of such methods.)

In actual practice the two-dimensional diffraction pattern is used to generate a three-dimensional pattern closely related to what is known as the *reciprocal lattice*. This is an array of points, each of which represents one of the (hkl) planes. The distance from the origin of the lattice is proportional to sin θ, i.e., by equation 3–7, it is a measure of the reciprocal of d_{hkl}. The advantage of this procedure is that a unique relationship exists between the real lattice and the reciprocal lattice, so that the unit cell axes of the real lattice, the angles between them, and the symmetry can be immediately deduced. For the purpose of the discussion in this chapter, it is sufficient to know that each observable reflection can be indexed; i.e., its (hkl) values can be assigned.

When indexing has been accomplished, the following information can at once be obtained.

(1) *The unit cell dimensions.* If rotation diagrams have been obtained with rotation about each unit cell axis, the separations a, b, and c can be evaluated directly, by equation 3–3, from layer line spacings. For rectangular cells they can also be evaluated from the relations $d_{h00} = a/h$, etc. Most generally they are obtained by mathematical transformation of the axes of the reciprocal lattice.

(2) *The space group.* As was mentioned in the preceding section there is a correlation between the space group and missing reflections. On the basis of missing reflections, therefore, the structure can be assigned to one of the 230 space groups, or at least, in unfavorable situations, to a choice among a small number of them. In addition, certain aspects of the unit cell symmetry will be reflected in the symmetry of the diffraction pattern and can be obtained on immediate inspection.

(3) *Number of molecules per unit cell.* Ordinarily the atoms of an individual molecule cannot be related to one another by symmetry operations, even if they are identical. Furthermore, several molecules can be in

relative positions without special symmetry, analogous to the situation described for atoms by Fig. 3–4b. The total number of atoms or molecules in a unit cell is thus unknown. The space group, however, will limit the possibilities which are allowed. If the unit cell belongs, for example, to the space group $P2_12_12$ (Fig. 3–4), any atom must be identically duplicated four times in each unit cell. Ordinarily this means that the number of molecules in the unit cell must be an integral multiple of four. An exception arises if the individual molecule consists of several identical parts. Then each part can act as an individual structural unit. (In Table 4–1, for instance, a dimer of serum albumin is listed as belonging to space group $P2_12_12$, yet it contains only two dimer molecules per unit cell. The reason is that the individual halves of each molecule, being identical, can, and in this case do, form the basis for the symmetry.)

If the molecular weight of the substance under consideration is known, the number of molecules per unit cell can be evaluated from a measurement of the density of the crystal. If we let j be the number of molecules per unit cell and V the unit cell volume, as determined from the cell dimensions, we can write the following expression for the crystal density ρ

$$\rho = \frac{jM}{\mathcal{N}V} \qquad\qquad (3\text{–}9)$$

where M is the molecular weight and $\mathcal{N}$ is Avogadro's number. If M is known, a measurement of the density clearly provides us at once with a value for j.

It will be seen later that most true macromolecular crystals contain a considerable amount of the solvent from which the macromolecular substance is crystallized. The molecular weight M which appears in equation 3–9 must then include an appropriate share of this solvent. The amount of solvent present is readily determined by analysis, so that its presence causes no great difficulty.

(4) *Molecular weight.* If the molecular weight of a substance is unknown, and this is often the case for macromolecules, then a combination of the unit cell dimensions with a measured density leads to a limited number of possible values for the molecular weight. By use of equation 3–9 and after correction for the presence of incorporated solvent, a value for jM is obtained. The symmetry of the structure ordinarily imposes limitations on the possible values of j, so that, if an approximate value for M is known from other considerations, the x-ray data will provide a precise value. For the protein lysozyme, for example (cf. Table 4–1), jM is equal to 111,200. The space group requires that j be a multiple of four. Possible values for M (estimated error less than 1 %) are therefore 27,800, 13,900, 9270, 6950, etc. Approximate values for M obtained by other methods lie between 14,000

and 17,000. The true molecular weight can at once be assigned the value 13,900.

(5) *Positions of atoms within the unit cell.* In particularly simple crystals, e.g., NaCl, the combination of symmetry, unit cell dimensions, and the number of molecules per unit cell may be compatible with a single unique arrangement of the atoms within the unit cell. In general, however, this information can be obtained only from determination of the intensities of the various reflections, as discussed in the following section.

3h. The intensity of diffraction patterns. The intensity of a given diffraction spot or line depends on two types of factors: those which arise from variation in the intensity of scattering from individual atoms, and those which depend on the interrelation between rays scattered from different atoms within the unit cell. (The interrelation between rays scattered from different unit cells is not involved; it is manifested in the fact that only the reflections predicted by Bragg's law are observed.)

The factors influencing the intensity which do not depend on the arrangement of atoms within the unit cell are the size of the sample, its distance from the photographic film, the wavelength of the x-ray and the scattering angle, 2θ. The influence of the last two factors on any scattered radiation is discussed in Chapter 5. As has been mentioned, the amplitude of the radiation scattered from any atom also depends on an atomic structure factor. All of these factors are readily calculated. The atomic structure factor, f_j, is defined as the ratio of the amplitude of radiation scattered from a particular atom to the amplitude of radiation scattered from a single electron at the same position. As we have mentioned, f_j is roughly proportional to the atomic number of the jth atom; the intensity, being proportional to the square of the amplitude, is thus roughly proportional to the square of the atomic number. (For exact computation of f_j see James.[3])

The most interesting factor, however, is that which depends on the relative positions of the atoms within the unit cell. Let x_j, y_j, z_j be the location of the jth atom relative to an arbitrary origin. Let the phase angle of a ray scattered from an imaginary atom at the origin be zero. The path difference between an x-ray scattered in a given direction from the jth atom and that which would have been scattered from the origin is given in section 3b as $\mathbf{r} \cdot \mathbf{S}$ where $\mathbf{r}$ is the vector $x_j\mathbf{a} + y_j\mathbf{b} + z_j\mathbf{c}$ which leads from the origin to the jth atom and $\mathbf{S}$ is the difference between the unit vector $\mathbf{s_1}$, in the direction of scattering, and the unit vector $\mathbf{s_0}$, in the incident direction. We are concerned here, however, only in the particular directions specified by the Bragg relation. Thus equation 3–4 applies, and for any (hkl) reflection we have a path difference

$$\mathbf{r} \cdot \mathbf{S} = (x_j\mathbf{a} + y_j\mathbf{b} + z_j\mathbf{c}) \cdot \mathbf{S} = (x_jh + y_jk + z_jl)\lambda \qquad (3\text{–}10)$$

The phase angle ϕ_j of the reflection from the jth atom is $2\pi(\mathbf{r} \cdot \mathbf{S})/\lambda$, i.e.,

$$\phi_j = 2\pi(x_j h + y_j k + z_j l) \tag{3–11}$$

A convenient representation of an electromagnetic wave in space (at constant time) is

$$\mathscr{E} = \mathscr{E}_0 e^{i\phi} \tag{3–12}$$

where $\mathscr{E}$ is the magnitude of the electric field vector (which is always perpendicular to the direction of propagation) at any position along the direction of propagation, $\mathscr{E}_0$ is the maximum value of this magnitude, known as the *amplitude*, ϕ is the phase angle, and i is $(-1)^{1/2}$. Since

$$e^{i\phi} = \cos \phi + i \sin \phi \tag{3–13}$$

equation 3–12 is seen to consist of a real part and an imaginary part, each of which alone can be used to represent the sinusoidal variation of $\mathscr{E}$ with ϕ. An advantage of representing the wave by equation 3–12 is that it is easier to combine exponentials than trigonometric functions. Another advantage lies in the fact that the intensity, I, of the wave (equal to $\mathscr{E}_0^2$; cf. Chapter 5) is easily obtained as

$$I = \mathscr{E}\mathscr{E}^* \tag{3–14}$$

where $\mathscr{E}^*$ is the complex conjugate of $\mathscr{E}$, i.e., the same expression with i replaced by $-i$.

The superposition of the waves scattered by all the atoms of a unit cell, in the direction of any particular (hkl) reflection, can be evaluated by means of equations 3–11 and 3–12 as

$$\mathscr{E} = KF_{hkl} \tag{3–15}$$

where K includes all the factors which influence the amplitude (or intensity) other than the phase angles due to the individual atomic positions and the atomic scattering factors (f_j) of the atoms in those positions. These are included in the term F_{hkl} (called the *structure factor*), i.e.,

$$F_{hkl} = \sum_j f_j e^{i\phi_j} = \sum_j f_j e^{2\pi i(hx_j + ky_j + lz_j)} \tag{3–16}$$

The summation in equation 3–16 extends over all atoms in a unit cell. [The factor K of equation 3–15 is, of course, not a constant. It has a different value for each (hkl) reflection.]

The intensity of any (hkl) reflection, by equations 3–14 and 3–15, is

$$I_{hkl} = K^2 F_{hkl} F_{hkl}^* \tag{3–17}$$

If the positions of all atoms are known, then the intensity can be computed at once with the aid of equation 3–16. Alternatively, if the F values could be experimentally determined, then the entire crystal structure could be obtained. (See the next paragraph.) Unfortunately we can observe only intensities, and, after calculation of K^2, they give us values of FF^*. Suppose now that F were replaced by $Fe^{i\alpha}$. Then the complex conjugate of this new structure factor would be $F^*e^{-i\alpha}$, and $(Fe^{i\alpha})(Fe^{i\alpha})^*$ would reduce to FF^*. Thus F is in fact determinable (from experiment) only to within the factor $e^{i\alpha}$. There is an indeterminable phase angle α associated with each reflection.

It should be noted that F_{hkl} will always be real if the unit cell has a center of symmetry. For in that case, for every atom at x_j, y_j, z_j, there will be an identical atom at $-x_j$, $-y_j$, $-z_j$. All terms of equation 3–16 can then be paired off into pairs of the form $f_j(e^{-i\phi_j} + e^{i\phi_j})$, which reduces to $2f_j \cos \phi_j$. Thus F_{hkl} will be entirely real, and α must be zero. Even in this special case, however, equation 3–17 gives $I_{hkl} = K^2 F_{hkl}^2$, and F_{hkl} can be determined only within a factor of ± 1; i.e., its magnitude is known, but not its sign.

It should also be noted that some of the F_{hkl}, but not all of them, will generally be real even if there is no center of symmetry. The unit cell represented by space group $P2_12_12$, for example (Fig. 3–4a), has a two-fold rotation axis in the z direction, so that, for any z, there are always identical pairs of atoms at x, y and $-x$, $-y$. Thus all structure factors of the type F_{hk0}, since they do not involve z, will be entirely real.

We wish to show now that the entire structure could be determined if the F_{hkl} factors were known. To do this we first write a new expression for F_{hkl} in terms of a continuous distribution of scattering electrons. In equation 3–16 the distribution of scattering electrons is described in terms of discretely located atoms, each with f_j times the scattering power of a single electron. We now replace this with a continuous function $\rho(x, y, z)$ giving the apparent density of electrons per cubic centimeter at all points in the unit cell; i.e., ρ is defined so that its value integrated over the space occupied by the jth atom is f_j. (This function of course has intense maxima at the positions of the atoms, but this in no way influences the mathematical development.) In place of equation 3–16 we thus get

$$F_{hkl} = \int_0^1 \int_0^1 \int_0^1 \rho(x, y, z)e^{2\pi i(hx + ky + lz)} \, dx \, dy \, dz \qquad (3\text{–}18)$$

the integration extending from $x = 0$ to $x = 1$, etc., i.e., over the volume of one unit cell.

(For the sake of simplicity we are confining the mathematical treatment of this section to unit cells in which the three unit cell axes are mutually perpendicular, so that $dx \, dy \, dz$ is the element of volume obtained by arbitrary displacements, dx, dy, and dz, and $\int_0^1 \int_0^1 \int_0^1 dx \, dy \, dz$ is the volume of the unit cell. For monoclinic or triclinic unit cells the corresponding terms in equation 3–18 and all subsequent equations would in addition include the angles of inclination of the unit cell axes. Since they are constants they would not affect the expressions for relative intensity.)

Now, over the entire crystal, ρ is a periodic function of x, y, and z. It can equally well be expressed as a periodic function of $2\pi x$, $2\pi y$, and $2\pi z$. In terms of these variables it would repeat itself at intervals of 2π (since it repeats itself at unit intervals in terms of x, y, and z). But a function which repeats itself at intervals of 2π can be expressed as a Fourier series, i.e., as an infinite series in terms of the functions $\cos\phi$, $\cos 2\phi$, $\cos 3\phi$, $\cdots$, etc.; $\sin\phi$, $\sin 2\phi$, $\sin 3\phi$, $\cdots$ etc. (For an elementary discussion of Fourier series see Sokolnikoff and Sokolnikoff[33] or any suitable advanced calculus textbook.) In three dimensions a convenient representation is in terms of the exponentials, $\exp[-2\pi i(hx + ky + lz)]$; i.e., we can write

$$\rho(x, y, z) = \sum_{h,k,l=-\infty}^{h,k,l=\infty} C_{hkl} e^{-2\pi i(hx+ky+lz)} \tag{3-19}$$

That this is an expansion in terms of cosines and sines follows from equation 3–13.

The coefficients C_{hkl} of equation 3–19 remain to be determined. A simple transformation[33] shows that they are given by the relation

$$C_{hkl} = \frac{\int_0^1 \int_0^1 \int_0^1 \rho(x, y, z) e^{2\pi i(hx+ky+lz)} \, dx \, dy \, dz}{\int_0^1 \int_0^1 \int_0^1 dx \, dy \, dz} \tag{3-20}$$

The denominator of equation 3–20 is the volume V of a unit cell. (This is equal to unity in the present units, but it will be retained in the equations so as to facilitate comparison with some of the references cited at the end of this chapter, in which somewhat different units are employed for x, y, and z.) The numerator of equation 3–20 is just F_{hkl}, as given by equation 3–18. Thus equation 3–19 becomes

$$\rho(x, y, z) = \frac{1}{V} \sum_{h,k,l=-\infty}^{h,k,l=\infty} F_{hkl} e^{-2\pi i(hx+ky+lz)} \tag{3-21}$$

If all the F_{hkl} were known, we could at once obtain the value of ρ as a function of x, y, and z. The maxima of this function represent the locations of atoms; the more intense the maximum, the larger the f_j value of that atom.

Much easier to determine than $\rho(x, y, z)$ itself is the projection of ρ upon a plane, i.e., the sum of all the ρ values along a given dimension as a function of the other two dimensions. For example, the projection $\rho(x, y)$ upon the x, y plane is, for a rectilinear cell,

$$\rho(x, y) = \int_{z=0}^{1} \rho(x, y, z) \, dz \tag{3-22}$$

Replacing $e^{-2\pi i l z}$ by the corresponding trigonometric expression (equation 3–13), we see at once that $\int_0^1 e^{-2\pi i l z}\, dz$ vanishes unless $l = 0$ and is equal to unity when $l = 0$. Thus when equation 3–21 is introduced into equation 3–22, only the terms with $l = 0$ appear and

$$\rho(x, y) = \frac{1}{A} \sum_{h,k=-\infty}^{h,k=\infty} F_{hk0} e^{-2\pi i(hx+ky)} \qquad (3\text{–}23)$$

Fewer reflections are thus needed to evaluate the projection. Furthermore, the F values are more likely to be real, i.e., known except for sign (cf. p. 33). For example, it follows from the discussion on p. 33 that all the structure factors which occur in $\rho(x, y)$ for the space group $P2_12_12$ would be real. To compensate for its greater simplicity, the projection of ρ also gives less information. For a simple structure containing just a few atoms per unit cell, each atom may appear separately in the projection, and knowledge of all the x and y coordinates may make possible a good guess of the total structure. These conditions can never be satisfied for macromolecular crystals (cf. Fig. 4–11).

As we have stated, the F_{hkl} values are in fact not known. The structure problem must therefore be solved by a trial-and-error procedure. An intuitive guess is made as to the positions of the atoms. F_{hkl} values are calculated by equation 3–18. The observed intensities, equal to $K^2 F_{hkl} F_{hkl}^*$, are computed. If they agree reasonably well with observed intensities, equation 3–21 may be used to calculate the electron densities. The densities will show maxima at somewhat different positions from the initial guesses. These improved atomic positions are used to compute a new set of F_{hkl}, etc. If a unit cell contains as many as thirty or forty atoms, it may take several years to evaluate the entire structure. For macromolecular crystals the procedure is essentially impossible with present calculating devices. That is, there are macromolecular crystals with thousands of recorded and indexed (hkl) reflections, but the corresponding structures are as yet unsolved.

One other auxiliary procedure should be mentioned. This is the evaluation of the Patterson function, $P(x, y, z)$. This function describes the electron density relative to the *centers of atoms*. It is the superposition of all electron density distribution functions which could be obtained if each atom in turn were made the origin of the coordinate system, with the further qualification that each individual distribution is weighted in proportion to the scattering power of the atom at the origin. It always has an intense maximum at the position $(0, 0, 0)$ since the electron density is high at the center of *every* atom. It has lesser maxima at numerous other values of the coordinates (x, y, z). These mean that there are in each unit cell one or more atoms *at unknown positions* (x', y', z') and that there are then other atoms at the positions $(x' + x, y' + y, z' + z)$.

It is readily shown that the Patterson function is given by

$$P(x, y, z) = \frac{1}{V} \sum_{h,k,l=-\infty}^{h,k,l=\infty} F_{hkl}^* F_{hkl} e^{-2\pi i(hx + ky + lz)} \qquad (3\text{-}24)$$

and it is seen to differ from equation 3–21 only in having $F_{hkl}^* F_{hkl}$ in place of F_{hkl}. This, however, makes it immediately determinable, since the product of the structure factor and its complex conjugate is the observed intensity.

The information which the Patterson function provides is obviously not sufficient to determine the crystal structure. For simple structures it may greatly limit the number of allowable positions of the atoms and may thus eliminate many bad guesses in the trial-and-error procedure used in evaluating the electron density distribution. For more complex structures such as macromolecular crystals it does not sufficiently limit the number of possibilities to be very useful. However, one method for investigating such crystals, which will be cited in the next section, is by comparing the diffraction pattern of the crystal with that of an isomorphous crystal containing a small number of heavy metal atoms per unit cell. The major difference between the two patterns will be due to the diffraction maxima arising from x-rays scattered by the heavy atoms: since the number of these is small, the Patterson function is of real value in determining their positions.

Two examples of Patterson analyses are given in the next section (Figs. 4–9 and 4–10). Both of them are projections upon a plane; i.e., the function given is $P(x, y)$, and it differs from $P(x, y, z)$ in the same way that equation 3–22 differs from equation 3–21.

3i. The diffraction of x-rays by liquids and amorphous solids.[4]

Glasses, carbon blacks, and other amorphous solids are not structureless materials. In fact, the atoms or molecules of these substances tend to arrange themselves just as regularly as the atoms or molecules in crystals. The principal difference is (*a*) that ordered structures persist over relatively minute regions of space and (*b*) that the ordered regions are not regularly oriented with respect to one another. Even a liquid is not structureless. Over a small volume and short period of time regularly ordered structures are maintained.

The statements just made follow from the fact that amorphous solids and liquids exhibit the phenomenon of x-ray diffraction. The patterns obtained (e.g., Fig. 4–2*b*) when an x-ray beam is directed at a stationary sample of an amorphous solid or a liquid resemble the powder diagrams which we obtain from crystals, except that the lines are broad and diffuse. This is just what we would expect from randomly oriented crystals of very small size. Usually only one or a very small number of diffraction maxima can be measured, and they usually correspond to spacings of the order of the expected intermolecular distances; i.e., they suggest that the structures observed are usually close-packed arrays of single molecules.

Many solids are intermediate between true crystals and thoroughly disordered structures. They give x-ray diffraction patterns with several maxima, with lines sharper than those for liquids. Considerable information about the structure of such solids has been obtained, much of which is described in a review by Randall.[4] As will be shown in the next section, many macromolecular solids are of this type.

4. DIFFRACTION OF X-RAYS BY MACROMOLECULAR SOLIDS

4a. Synthetic linear polymers.[12] It is ordinarily not possible to prepare synthetic polymers in the crystalline state. When liquid polymers are cooled, they invariably form amorphous solids. As has been pointed out, however, even amorphous solids have crystalline regions, and amorphous polymers are typical in this respect. It will be seen that x-ray diffraction can provide useful information about the structure of synthetic polymers despite the fact that precise single crystal studies cannot be performed.

The growth of very minute single crystals of polyethylene has been reported recently.[142] The crystals have not been sufficiently large for x-ray analysis.

It is frequently observed that the physical properties of solid polymers depend on the rate of cooling from the liquid state, the precise conditions of the original synthesis, and similar factors. Parallel variations are observed in the sharpness of the x-ray diffraction patterns. From these observations it has been concluded that at least some polymers, in their solid state, may be prepared so as to possess more extensive crystalline regions than most non-polymeric amorphous materials. Such polymers are often referred to as "crystalline," this term being used to indicate a relatively high content of relatively extensive crystalline regions.

An especially useful technique is to stretch polymer samples into long fibers. This process tends to align extended, unentangled portions of polymer chains parallel to the direction of stretching. As a result, the formation of crystalline regions by bundles of adjacent polymer chains is favored. Moreover, the crystalline regions so formed will be uniquely oriented; the unit cell direction which is parallel to the extended polymer chains will also be parallel to the fiber axis. As a result the x-ray diffraction patterns obtained for such fibers resemble single crystal rotation diagrams rather than the powder diagrams ordinarily obtained for amorphous materials. Each fiber contains numerous crystalline regions, each with one axis uniquely oriented and the other two randomly oriented. The diffraction pattern is thus similar to that obtained when a single crystal is mounted with one axis fixed and is then rotated so as to bring the other axes successively through all possible orientations. Figures 4–2c and d provide a striking example of the change in x-ray diffraction pattern which follows fiber formation.

The x-ray diffraction patterns obtained in this way are known as *fiber diagrams*, and they are, of course, not quite as sharp or detailed as rotation diagrams of true crystals. Four aspects of this difference may be noted. (1) The reflections are more diffuse than those obtained from true crystals, which means that reflection occurs over a wider range about the Bragg angles (equation 3–7). This is due to the fact that the crystalline order persists over a shorter distance than in true crystals. (2) The reflections are short arcs, rather than spots; i.e., each (*hkl*) reflection is broadened in the direction of the circle which represents the corresponding powder diagram reflection. This must clearly be a result of imperfect alignment of the crystalline regions parallel to the fiber axis. (3) Fewer reflections are observed than for a true crystal. This is a consequence of the smaller size of the crystalline regions, which prevents the appearance of reflections corresponding to the larger interplanar distances. (4) Some uniform background scattering is always observed, indicative of the presence of non-crystalline regions in the fiber.

The interpretation of fiber diagrams resembles in principle the interpretation of single crystal rotation or oscillation diagrams, as discussed in section 3. However, because fewer spots are available than in single crystal patterns, the determination of the unit cell symmetry and dimensions becomes less certain. The unit cell dimension parallel to the fiber axis can always be determined from layer line spacings by equation 3–3. As Fig. 4–2 shows, however, the number of spots in each layer line may be small, so that there is not necessarily a unique way of indexing them. The simple unit cells usually proposed are therefore not necessarily correct in every case.

The use of intensity measurements to determine the positions of atoms within the unit cells, as described in section 3h, is completely impossible except for unusually clear fiber diagrams. This does not mean that detailed structures cannot be proposed, for even the sparsest x-ray data may be combined with reasonable assumptions to lead to reasonable hypothetical structures. In particular, use may be made of the fact that interatomic distances and bond angles vary little from one molecule to the next. These parameters may therefore be assumed to have in polymeric molecules values which are closely similar to those which have been rigorously established for appropriate small molecules. Combining these values with the unit cell dimensions (one of which, at least, is unequivocal) often leaves little choice among possible structures.

In interpreting fiber diagrams, such as those of Fig. 4–2, it should be borne in mind that *equatorial* reflections (those in the layer line which lies in the plane of the incident beam) represent reflections from planes parallel to the fiber axis; i.e., if the fiber axis is the c axis, they are ($hk0$) reflections; those in the first layer line above the equator are ($hk1$) reflections, etc. Spots close to the origin represent low values of θ, which, by equation 3–7, correspond to large interplanar distances. Those near the periphery represent short spacings.

Fiber diagrams that have been obtained for a variety of synthetic linear polymers (but not for all; cf. sections 4d and e) share a number of general features, and from these certain general conclusions may be drawn.

(1) The most important feature is that the unit cells observed are of the same order of magnitude as unit cells in crystals of low-molecular-weight substances. The organization in the crystalline regions of these polymers is thus one that involves not the polymer molecules as a whole, but only their individual chain segments.

(2) From the width of layer lines we can estimate roughly the number of successive unit cells along the fiber direction of the crystalline regions. From the width of the reflections in the equatorial plane we can make a similar estimate for the directions normal to the fiber axis. Such calculations clearly show that not only the unit cells but also the extent of entire crystalline regions are generally smaller than the length of a single polymer molecule. Thus it appears that individual polymer molecules will have short portions aligned with portions of other polymer molecules so as to form small crystalline regions. In between such portions there is no regularity in structure, perhaps because the molecules are so entangled as to prevent formation of an orderly array. In polymer molecules that are not completely linear the occurrence of branching points will also always lead to interruption of the ordered structure.

It may be noted that the small single crystals of polyethylene, referred to above, were obtained from polyethylene samples entirely free from branching and from dilute solutions in which entanglement of individual molecules is virtually absent.

(3) The presence of general background scattering, indicative of the presence of amorphous regions of the fiber, was noted above. By comparison of background intensity with the intensity of the ordered diffraction rough estimates of the fraction of polymer material in the crystalline

regions can be obtained. For polymers of relatively high crystallinity this figure tends to lie between 50 and 85%.

(4) The variation among a considerable number of different kinds of synthetic polymers in their ability to form crystalline regions and in the structures they adopt (see following section) appears in most cases to be determined primarily by the *spatial* relation between any segment and its immediate neighbors. There is ordinarily no evidence for special forces between segments, other than the covalent bonds which tie each segment to that which follows it in the polymer chain.

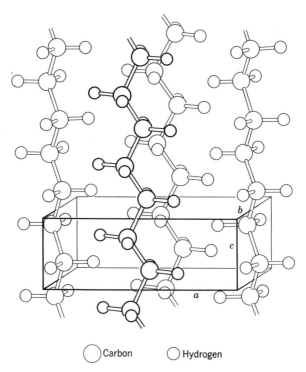

◯ Carbon   ◯ Hydrogen

Fig. 4–1. The structure of crystalline regions of polyethylene. The solid figure enclosed by the lines drawn is the orthorhombic unit cell. (Bunn.[14])

4b. Structures of some selected polymers. One of the simplest of all polymers is polyethylene, $—(CH_2—CH_2—)_x$. It is also highly crystalline. The unit cell is orthorhombic,[14] with the dimensions $a = 7.40$ Å, $b = 4.93$ Å, $c = 2.53$ Å. The c axis is in the direction of the extended chain. The distance of 2.53 Å in this direction is very close to the repeat distance expected for a fully extended chain (cf. Fig. 4–1), with repetition of the pattern at every second carbon atom. For a normal C—C distance of

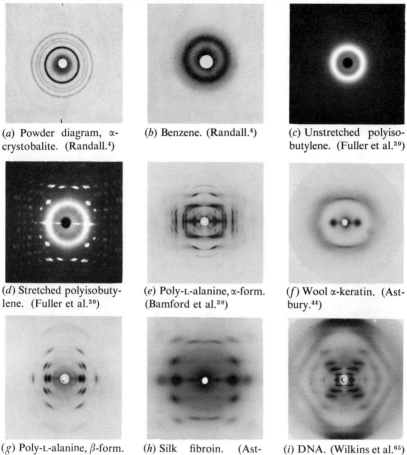

(a) Powder diagram, α-crystobalite. (Randall.[4])

(b) Benzene. (Randall.[4])

(c) Unstretched polyisobutylene. (Fuller et al.[20])

(d) Stretched polyisobutylene. (Fuller et al.[20])

(e) Poly-L-alanine, α-form. (Bamford et al.[28])

(f) Wool α-keratin. (Astbury.[44])

(g) Poly-L-alanine, β-form. (Bamford et al.[28])

(h) Silk fibroin. (Astbury.[44])

(i) DNA. (Wilkins et al.[65])

Fig. 4–2. Some x-ray diffraction patterns, all taken on photographic plates mounted perpendicular to the incident beam, behind the sample.

1.54 Å and an included (tetrahedral) angle of 109° 28′ the distance between alternate carbon atoms would be 2.51 Å, but crystals of hydrocarbons such as n-$C_{29}H_{60}$ appear to have a C—C distance of 1.52 Å and an included angle of 114°, leading to a slightly larger repeat distance. Presumably the structure of polyethylene is similar. The over-all unit cell is shown in Fig. 4–1. The x-ray diffraction pattern of course reveals only the position of the carbon atoms, since the total intensity scattered from hydrogen atoms is small.

When mono-substituted vinyl polymers, of the type —(CH$_2$—CHR—)$_x$, are considered, a new complication arises. For in a polymer of this type

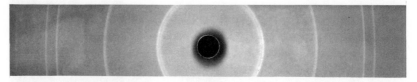

(*a*) Powder diagram, NdN. (Courtesy of Dr. K. Vorres.)

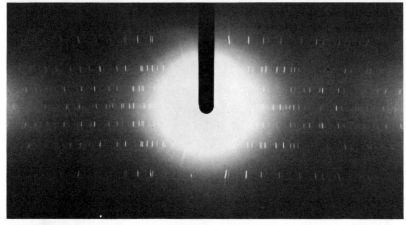

(*b*) Rotation diagram, cycloheptadiene–PdCl₂. (Courtesy of Dr. C. L. Carpenter.)

half of the carbon atoms are asymmetric, i.e., capable of existing in *d* and *l* configurations. Ordinarily the configuration of each such asymmetric carbon atom is determined during synthesis by a random process, and the polymers are optically inactive. It then becomes impossible to form a regular repeating pattern of the kind found in polyethylene, for, if the polymer chains are fully extended, the R side chains will occur without regularity on opposite sides of the polymer chain. Alternatively a regular arrangement of the R groups can be achieved, if at all, only by highly irregular gyrations of the polymer chain. If the side chain groups are large, the polymer chains cannot be aligned in a parallel arrangement even if the R groups are allowed to dispose themselves at random, for a parallel arrangement would require a large spacing between the parallel chains so as to allow room on either side for the R groups. Thus large holes would be left in the structure where hydrogen atoms appear in place of R groups. The existence of such holes is incompatible with the principles underlying the structure of solids in general.

As a result polymers such as polystyrene, $-(CH_2-CHC_6H_5-)_x$, and polyvinyl acetate, $-(CH_2-CHOOCCH_3-)_x$, have extremely low crystallinity. When stretched they do *not* give fiber diagrams such as Fig. 4–2*d*.

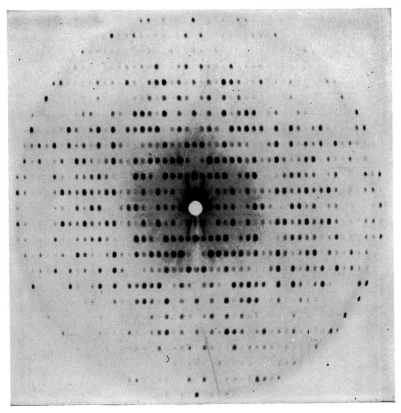

(c) Precession diagram, myoglobin. (Kendrew et al.[5]; courtesy of Dr. J. C. Kendrew.)

Fig. 4–3. X-ray diffraction patterns taken on photographic film forming a cylinder about the sample (a and b), or by means of a precession camera (c).

On the other hand, polyvinyl alcohol and copolymers of vinyl alcohol and ethylene do produce typical fiber diagrams,[15,16] with the same axial repeat distance (2.53 Å) as polyethylene. This must mean that hydroxyl groups and hydrogen atoms are sufficiently close in size so that one can replace the other without leaving a hole large enough to destroy the stability of the ordered structure. As might be expected, the unit cell of polyvinyl alcohol, in the directions perpendicular to the fiber axis, is somewhat larger than the polyethylene unit cell.

It should be noted that the oxygen atoms of polyvinyl alcohol have intrinsically a somewhat greater scattering power than the carbon atoms. Their irregular positions, however, lead to scattering in all directions; i.e., they contribute to the background scattering. The fiber diagram spots are due solely to the regularly arranged carbon atoms.

Apart from the possibility of d and l configurations at half the carbon atoms, there is another difference between mono-substituted vinyl polymers and polyethylene. This is the fact that they could polymerize in a "head to tail" arrangement, $\cdots$ —CH_2—CHR—CH_2—CHR— $\cdots$, or in an alternating arrangement, $\cdots$ —CH_2—CHR—CHR—CH_2—CH_2—CHR— $\cdots$,

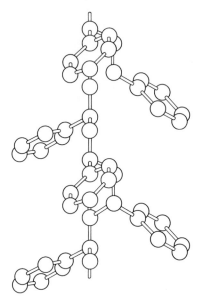

Fig. 4–4. Configuration of the polystyrene chain in crystalline regions of isotactic polystyrene. (Bunn and Howells.[19])

or in a mixture of both arrangements. It appears certain, largely on the basis of chemical evidence,[17] that the "head to tail" arrangement is always favored.

It was stated above that the asymmetric carbon atoms of mono-substituted vinyl polymers ordinarily take on d and l configurations at random. This is due to the fact that such polymers are ordinarily synthesized by successive addition of monomer molecules to free radicals of the type —$(CH_2$—CHR—$)_x$—CH_2—$CHR\cdot$ (cf. Chapter 9). It is also possible, however, to synthesize these polymers catalytically, under conditions where the addition is stereospecific.[18] The resulting polymers (called *isotactic*) thus have a *unique* configuration at each asymmetric carbon atom. As a result they are optically active. For such polymers the irregularity which ordinarily prevents crystallization is absent. Thus isotactic polystyrene, in contrast to the stericly random form, has a high degree of crystallinity.[18] The bulky styrene side chains favor a crystal structure differing from that of polyethylene. The unit cell is monoclinic, $a = b = 12.64$ Å, c (along the fiber axis) $= 6.65$ Å, with an angle of inclination between the ab plane and the c axis of 120°. The structure of a single polystyrene chain in this configuration is helical, as shown in Fig. 4–4.[18,19]

Symmetrically di-substituted vinyl polymers, such as polyisobutylene, $—(CH_2—C(CH_3)_2—)_x$, contain no asymmetric carbon atoms and should thus be easily crystallized. This proves to be the case.[20] Fibers produced by stretching polyisobutylene show a high degree of crystallinity. The repeat distance along the fiber axis is relatively large, 18.6 Å. Presumably the methyl side chains are unfavorable to an alignment of fully extended polymer chains. Instead, the polymer chain takes on a twisted or helical configuration (with perhaps eight isobutylene monomers per turn of the helix), with better accommodation for the methyl groups. The over-all details of the structure have not yet been elucidated, even though the fiber diagram, shown in Fig. 4–2d, is unusually detailed.

Many other synthetic polymers,[12] as well as natural rubber and some related natural polymers,[21] show x-ray diffraction patterns similar to those that have been discussed. For each such polymer there are differences in structural details, but there is a structural principle common to all, namely, that the structure is determined largely by steric considerations. This principle does not, however, apply to certain synthetic polymers such as the polyamides and polypeptides (sections 4d and e). Nor does it apply to most of the naturally occurring macromolecules apart from rubber.

4c. Cellulose and other polysaccharides. Cellulose, as it occurs in nature, is already fibrous and highly crystalline. The x-ray diffraction pattern appears to be essentially independent of the source from which the cellulose is obtained. The generally accepted unit cell is that of Meyer and Misch,[22] and it is shown in Fig. 4–5. The crystalline regions appear to be several hundred Ångstrom units in length and somewhat less than 100 Å in width. The degree of crystallinity has been estimated to range from 40 to 70%.

An interesting feature of the structure shown in Fig. 4–5 is the closeness of adjacent glucose rings in the a direction. In this direction the centers of nearest oxygen atoms are only 2.5 Å apart. The van der Waals radius of an oxygen atom, however, is about 1.5 Å; i.e., organic molecules containing oxygen are normally packed so that these atoms are no closer than 3.0 Å to one another. (The shortest O—O distance in cellulose, in the c direction, is 3.1 Å, in agreement with this figure.) Relatively strong lateral forces must be invoked to account for the shortened distance in the a direction, and it is reasonable[23] that these forces are due to hydrogen bonds, —O · · · H—O—, between hydroxyl groups of adjacent glucose rings.

A new principle is thus involved in the cellulose structure, which does not arise in any of the polymers discussed in the preceding section. The crystal structure of cellulose is not solely the result of efficient three-dimensional packing of the polymer chains but is determined in part by

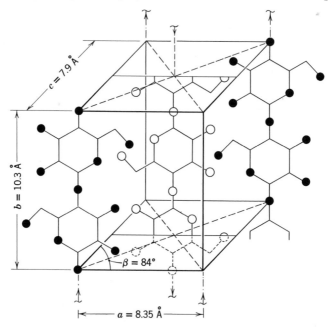

Fig. 4–5. Unit cell structure in crystalline regions of cellulose. (Meyer and Misch.[22])

attractive forces (other than primary bonds) between segments not adjacent in the chain sequence. We shall see that forces of this kind take on even greater importance for synthetic polypeptides, proteins, and nucleic acids.

The Meyer-Misch structure has been criticized by a number of authors, although there is not at present any agreement as to how it should be modified. In a recent paper on this subject Jones[143] suggests a more completely hydrogen-bonded structure than that given in Fig. 4–5.

Another characteristic of cellulose fibers is their ability to absorb water and other substances with a resultant swelling of the structure. In many instances part or all of this swelling occurs in the non-crystalline regions of the structure and is thus not pertinent to the present discussion. Strong acids and bases, however, as well as amines, enter into the crystalline regions and lead to enlargement of the unit cell. The spacing along the fiber axis usually remains unchanged when this occurs; i.e., the parallel extended configuration of cellulose molecules is not altered. The lateral spacings, however, may be increased by a factor of two or three.

A detailed account of this swelling phenomenon, as well as x-ray diffraction patterns for alternative crystalline modifications of cellulose and its chemical derivatives have been presented in a review by Howsmon and Sisson.[24] Polysaccharides other than cellulose are discussed by Meyer.[25] One of them, *chitin*, a polymer of the amino-sugar glucosamine, gives particularly sharp fiber diagrams. Its structure, like that of cellulose, involves lateral hydrogen bonding.[26]

4d. Fiber diagrams for helical structures.[6] The fiber diffraction patterns discussed so far have resembled rotation photographs for single crystals; i.e., the fibers examined have apparently contained numerous small three-dimensional crystalline regions, uniquely oriented with respect to the fiber axis. Some of the fiber diagrams to be considered in the following sections, on the other hand, have a much poorer appearance (e.g., Figs. 4–2f, 4–2i). The spots (or smears) occur in layer lines, but there is a very marked variation in intensity and lateral spacing from one layer line to the next, as compared, for instance, with the more normal type of picture shown for isobutylene in Fig. 4–2c. Some layer lines may be completely absent. Usually a distant layer line (e.g., the tenth layer line in Fig. 4–2i) shows a relatively intense meridional (i.e., along the vertical axis) smear. Because of the lack of sharp definition, these diagrams would seem to make unpromising material for detailed analysis, but a mathematical analysis by Cochran, Crick, and Vand[38] has shown that the broad features of these diffraction patterns are just those to be expected of an array of parallel helices with little or no regular lateral arrangement and, most important of all, that two dimensions of the helix can be obtained simply from the layer line spacing and from observing which of the layer lines have high intensity.

The results of Cochran et al. may be stated in terms of Fig. 4–6a. If identical atoms are arranged on a helix of pitch P (the pitch is the spacing between successive turns of the helix) and if the spacing between such atoms in the direction of the helix axis is p, layer lines will occur only at heights $n/P + m/p$ above or below the equator, n and m being integers. Furthermore, the intensities are expressible as Bessel functions, and only low order Bessel functions make an appreciable contribution, so that many layer lines allowed by this general formula will not occur at all or only faintly. In any event, if a structure can be assumed to be of this type, P and p can be determined even from poorly resolved fiber patterns.

It should be noted that the true crystallographic repeat (the distance c in

Fig. 4–6a) has virtually no effect on the observed pattern. Figure 4–6a is drawn so as to contain 3.33 atoms per helix turn, and it repeats exactly after three turns (ten atoms). A helix of the same pitch P that would make a complete turn for each 3.35 atoms, for which the tenth atom comes after

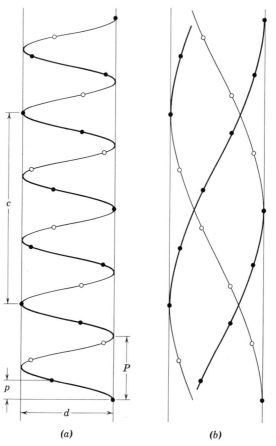

(a)   (b)

Fig. 4–6. Helical arrangement of repeating units of a fibrous molecule. The arrangement of repeating units is identical in (a) and (b), but they are shown in (a) as they would be connected if part of a single chain and in (b) if part of three intertwined chains. (Patterned after drawings of Crick and Kendrew.[5])

2.985 turns instead of 3.00 turns, and that would have a true repeat only after twenty turns (sixty-seven atoms) would nevertheless show an almost identical pattern because p would have almost the same value.

One additional feature of diffraction patterns that are to be interpreted in this way must be pointed out: All that the diffraction pattern reveals is that like atoms are regularly spaced in the manner shown (for example) by Fig. 4–6a. The diffraction pattern does not necessarily indicate that these atoms belong to the same molecular chain. In other words, it is possible that the basic unit for helix formation is not a single molecular chain but consists of two or more intertwined chains. Figure 4–6b, for example, shows how atoms placed exactly as in Fig. 4–6a could be part of a three-stranded helix composed of three separate intertwined chains. To decide among the various possibilities thus left open we make use of the fact already noted earlier in this chapter, that interatomic distances in macromolecules must be essentially the same as those which occur in corresponding small molecules. The distances p and P, characteristic of the helical structure, must be compatible with interatomic distances and angles which are not allowed to vary appreciably from accepted values. Usually only a single choice of the number of molecular chains in a helix makes this possible. (In addition the cylinder diameter d of Fig. 4–6a can sometimes be determined, at least approximately. It should be noted, however, that different atoms of a macromolecule may all lie on helices of the same p and P but at different distances from the axis, so that no unique value of d may exist.)

It is reasonable at this point to ask whether the basic assumption of the foregoing theory, i.e., helical structure in the direction of the fiber and little or no lateral order, is likely to be a realistic representation of any real macromolecular structure. The answer is that it is very likely. If a long chain molecule is to be ordered in the fiber direction, without benefit of interaction with its lateral neighbors, then a helical structure is almost inevitable. As a matter of fact we have already seen such a structure adopted by isotactic polystyrene and by polyisobutylene even when three-dimensional crystalline order has existed.

In a three-dimensional crystalline structure screw axes must be two-fold, three-fold, four-fold, or six-fold, i.e., each turn of a helix must consist of two, three, four, or six identically placed repeating units. No other kind of symmetry will fall into one of the 230 space groups which allow all space to be filled in three directions by repetition of a unique unit cell. This restriction does not, of course, apply here. The helix need not even be integral and, in fact, generally is not integral, e.g., that shown in Fig. 4–6a has 3.33 atoms per turn. The reason for this, of course, is that the symmetry of the helix applies only to the individual helix and not to the structure as a whole. (In general, as we have said, there need be no real structure in the fiber at right angles to the fiber axis.)

4e. Synthetic polypeptides. Each monomer unit (or "residue") of a polypeptide, $—(NH—CHR—CO—)_x—$, contains an asymmetric carbon atom.

These asymmetric carbon atoms are present before polymerization, and their configuration is not affected by the polymerization process. Thus, if we start with an optically pure amino acid, the resulting polymer will automatically have the same configuration at each asymmetric atom, and a regular structure will become possible. The polypeptides which have been examined by x-ray methods have been of this type. They have a high degree of crystallinity and exhibit well-defined fiber diagrams.[27,28] The interpretation of these data leads to the conclusion that specific forces between non-adjacent segments of polymer chains play an all-important role in determining the structure of polypeptides in the solid state. It appears that a considerable free energy decrease can be achieved, at least in the solid state, by formation of hydrogen bonds between monomer units of the type

$$
\begin{array}{cc}
\mid & \mid \\
NH \cdots O{=}C \\
\mid & \mid \\
RCH & HCR \\
\mid & \mid \\
C{=}O \cdots HN \\
\mid & \mid
\end{array}
\qquad
\begin{array}{cc}
\mid & \mid \\
NH \cdots O{=}C \\
\mid & \mid \\
RCH & NH \cdots \\
\mid & \mid \\
\cdots O{=}C & RCH \\
& \mid \\
NH \cdots O{=}C \\
& \mid
\end{array}
\qquad , \text{ etc.,}
$$

and that the solid polypeptides choose among different sterically feasible structures those which can best accommodate the maximum possible number of such hydrogen bonds, either between peptide carbonyl and imino groups of the same polypeptide chain or between such groups on different polypeptide chains.

The importance of hydrogen bonds between peptide links is not confined to large molecules. The crystal structure of small molecules which contain peptide bonds (e.g., diketopiperazine, $\overline{CH_2{-}CO{-}NH{-}CH_2{-}CO{-}NH}$; and glycylglycine, $^+H_3N{-}CH_2{-}CO{-}NH{-}CH_2{-}COO^-$) involves similar (intermolecular) hydrogen bonds. It is found that the six atoms of each peptide group, in such crystals, always lie in a plane, because of resonance between the structures

$$
\begin{array}{ccc}
& H & \\
\mid & \mid & \mid \\
{-}C{-}C{-}N{-}C{-} & \text{and} & {-}C{-}C{=}N{-}C{-} \\
\mid & \| & \mid \\
& O &
\end{array}
$$

One experimental manifestation of this fact is that the C—N distance between the CO and NH groups is always found to be 1.32 Å, which is much shorter than the normal single-bond C—N distance of 1.47 Å. (A complete summary of data for appropriate small molecules is given by Vaughan and Donohue.[29]) This planarity of the peptide group must also be expected to occur in polypeptides. Its occurrence will not, of course make the polypeptide chains inflexible, since each planar unit can still rotate with respect to its neighbor about the outer C—N bond.

If the peptide hydrogen bonds in a polypeptide structure are to be formed *intramolecularly*, i.e., between monomer units of the same molecule, and if the maximum number of such bonds is to be formed, the most likely structure is a helical one, as first suggested by Taylor[30] and by Huggins.[31] Numerous possible helical configurations have been suggested. The most rigorous study of this problem is that of Pauling, Corey, and Branson,[32] who were able to show that one particular helical structure (involving only one polypeptide chain) can be constructed which will best satisfy certain requirements which they considered essential. This particular structure has become known as the 3.6-residue helix, or the α-helix. It is shown in Fig. 4–7.

The requirements which Pauling and coworkers[32] considered essential, apart from the requirement for unique optical configuration, were the following: (1) That every CO and NH group form a hydrogen bond. (2) That the peptide group be planar, as found in the crystals of appropriate small molecules,[29] and that the bond lengths be the same as in these crystals. (3) That the hydrogen-bonding H atom lie not more than 30° from the vector joining the nitrogen and oxygen atoms involved. (4) That the progression from one residue to the next be the same for every residue. (5) That the orientation about C—N and C—C single bonds be near the potential energy minimum for rotation about these bonds. Only the α-helix will satisfy these requirements. Other sterically possible helices can be formed only by deviation in bond angles from their best values, with a resulting increase in energy.[34]

If —C≡O ⋯ H—N— hydrogen bonds are to be *intermolecular*, i.e., between segments of adjoining polypeptide chains, sheetlike structures will be formed. Two of them, which satisfy the criteria of Pauling and coworkers,[35] are shown in Fig. 4–8.

Bamford and coworkers[27,28] have examined polypeptides such as poly-L-alanine, —(NH—$CHCH_3$—CO—)$_x$, and poly-γ-methyl-L-glutamate, —(NH—$CHCH_2CH_2COOCH_3$—CO—)$_x$, by x-ray diffraction. They have found that two kinds of crystalline regions may be obtained, as shown in Figs. 4–2e and g. One of them, the α-form, is characterized by layer lines of just the type predicted in the preceding section for a helical structure. When analyzed by the theory of Cochran et al.,[38] the layer line structure is found to correspond to a helix with 3.6 residues per turn and with the spacings p and P (Fig. 4–6a) equal to 1.5 and 5.4 Å respectively. These parameters agree exactly with those predicted by the structure of Pauling and coworkers shown in Fig. 4–7. Other x-ray evidence for this structure, using fiber diagrams with a fiber tilted relative to the vertical direction, i.e.,

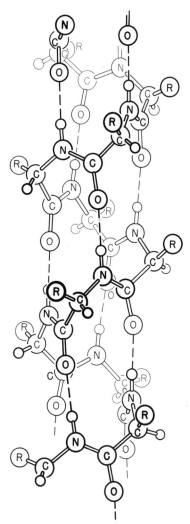

Fig. 4–7. The α-helical structure for polypeptide chains. (Pauling and co-workers.[32] This representation provided by courtesy of Professors Pauling and Corey.)

not at 90° with respect to the incident rays, has been obtained by Perutz.[37]

The β-form of the synthetic polypeptides (Fig. 4–2g) does not have the helix-type pattern at all. Analysis of the pattern by conventional means shows that the structure in this case must be very close to one of the pleated sheet structures proposed by Pauling and Corey, and shown in Fig. 4–8.

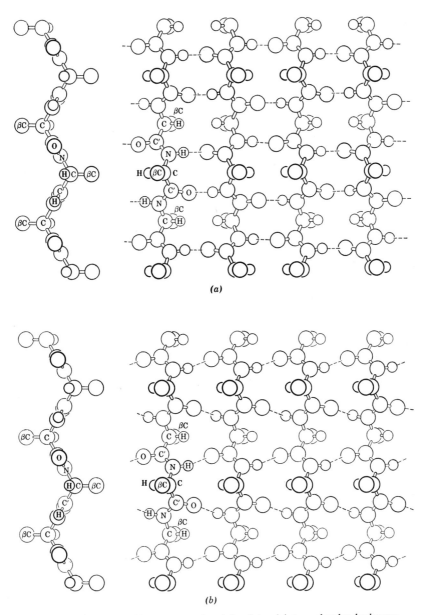

(a)

(b)

Fig. 4–8. Two possible structures involving lateral intermolecular hydrogen bonds. They have been called "pleated sheet" structures. (Pauling and Corey[35]. This representation provided by courtesy of the authors.)

It is important to emphasize the role which assigned molecular dimensions play in this type of structure determination. We have mentioned that models are used to determine which of several possible structures can, with minimum distortion, best accommodate the x-ray observations. If the model structures used are based on false molecular dimensions, a "best" structure will still be found, but it will bear little relation to the real structure. In the present situation we can have considerable confidence in the structure of Pauling and coworkers for two reasons. First, this structure and the "best" structures based on different intermolecular distances and bond angles were all suggested prior to the experimental discovery of the 1.5 Å spacing. This feature of the structure is incompatible with any of the other suggested structures. Second, the α-helical structure for these polypeptides is supported by numerous other methods of investigation, notably the infrared absorption (p. 87), and light scattering and hydrodynamic properties of the same molecules dissolved in appropriate solvents.

4f. Synthetic polyamides.[39]   The crystalline portions of solid synthetic polyamides, such as nylon, also assume structures based on intermolecular hydrogen bonds between CO and NH groups. The structures differ, of course, from those of polypeptides because the amide groups are further apart in the chains and because they alternate in direction, the general structural formula being

$$\cdots -NH-CO- \cdots -CO-NH- \cdots -NH-CO- \cdots$$

Some proposed structures are described by Bunn and coworkers.[39]

4g. Fibrous proteins.[40]   Among the proteins found in nature are those which participate in the formation of the structural elements of living tissue. They are the fibrous proteins. As their name implies they are fibrous in nature and give typical x-ray fiber diagrams, two of which are shown in Fig. 4–2f and h. Belonging to this class of proteins are the *keratins* of wool, hair, horn, and feathers; several muscle proteins, such as *myosin*; the *fibroin* of silk; the *collagens* of cartilage; and *fibrin* formed in clotted blood.

Since these proteins are polypeptides and fibrous, we should expect their structure to resemble that of synthetic polypeptide fibers, and this proves to be the case. A number of the fibrous proteins do in fact have x-ray diffraction patterns quite similar to those of poly-L-alanine or the esters of poly-L-glutamic acid. For instance, the various keratins[41,44] (Fig. 4–2f) have similar diffraction patterns which resemble the pattern (Fig. 4–2e) found for the α-helical form of poly-L-alanine. The corresponding structure is known as the α-keratin structure, and it presumably is closely related to the α-helical structure. On the other hand, silk fibroin[45] (Fig. 4–2h) has an x-ray pattern which clearly resembles that of the β-form of poly-L-alanine (Fig. 4–2g). (Many of the keratins, when stretched, adopt a similar diffraction pattern.[41]) The corresponding structure is unquestionably very close to one of the pleated sheet structures shown in Fig. 4–8.

The fibrous protein collagen,[46,47] which occurs universally in connective tissue, has a structure which differs from the two hydrogen-bonded structures of the polypeptide chain which have been discussed so far. Its x-ray diffraction pattern is clearly of the helix type, but the two characteristic spacings are $p = 2.86$ Å and $P = 9.5$ Å. These spacings cannot be accounted for on the basis of a single helical chain, if the bond distances are to be normal. The simplest structure, and the one generally accepted as true, involves three intertwined helices of the type shown in Fig. 4–6b.

It is important to note in this connection that collagen is a very unusual protein in terms of composition. Over 20% of its residues are proline or hydroxyproline, derived from the corresponding *cyclic* amino acids. The peptide nitrogen atom of these residues does not contain a hydrogen atom, the residue formula being

$$\cdots -N\text{————}CH-CO- \cdots$$

with side groups:

$$\begin{array}{ccc} & | & | \\ & CH_2 & CH_2 \\ & \searrow & \swarrow \\ & CH_2 & \end{array}$$

(or CHOH)

The peptide group of this residue can clearly form only one hydrogen bond of the type shown on p. 50, instead of the normal two hydrogen bonds. The failure of collagen to adopt the maximally hydrogen-bonded structures of Pauling et al. is thought to be due to this fact. (It is of interest that the synthetic polypeptide, poly-L-proline, also does not adopt either the α or β configuration typical of other polypeptides. Its structure contains no hydrogen bonds at all but is nevertheless helical.[42] The structure is not the same as that of collagen, however.)

A reservation which applies to some of the x-ray studies of fibrous proteins, especially to the studies on keratin and silk fibroin, is that they were made on fibrous structures which are obtained from silk, hair, etc., simply by dissolving away outer coverings, soluble proteins, etc. The residual fibrous structures do not necessarily represent a particular chemical compound. In fact, the opposite is true, so that the conclusion to be drawn from the data which have been cited is that there are polypeptide chains in these fibers which take on the configurations given but that they can form only part of the fiber content. The precise nature of the molecules to which they belong has not been established.

This reservation has particular strength in the case of the keratins. For the proteins of keratins contain an abnormally large amount of the amino acid cystine. This amino acid forms residues containing disulfide bonds, which form cross-links between distant residues of the same or different polypeptide chains. (See, for instance, the disulfide cross-links in insulin, shown in Fig. 2–2.) If many of these cross-links are between portions of the same chain, the formation of continuous α-helices can clearly not be expected, yet it is the α-helical structure which x-ray diagrams of keratin normally suggest. The explanation undoubtedly lies in the recent discovery[43] that wool keratin consists of several chemically separable proteins and that some of them (comprising 30 to 40% of the total amount of protein) have a very low sulfur content, in contrast to the high sulfur content of keratin as a whole. It is undoubtedly these proteins which are responsible for the observation of α-helical structure in wool keratin. A similar situation presumably prevails in the other keratins.

4h. Globular proteins. The common proteins of the body fluids of living systems, the globular proteins, consist, like the fibrous proteins, of poly-peptide chains, with varied side chains in definite order, often including some disulfide cross-links. However, these proteins differ sharply in their solid states, not only from the fibrous proteins but from any of the macro-molecular substances thus far discussed in this chapter. The outstanding difference is that many of these proteins can exist in a true crystalline state. The crystals can be grown to large size and can be examined by single crystal x-ray methods. The resulting diffraction patterns, as illustrated by Fig. 4–3c, consist of sharp spots, like those for crystals of low-molecular-weight substances. From these diffraction patterns unit cell dimensions and symmetries can be *unequivocally* determined (section 3g). Invariably, the unit cells are found to be large, larger than individual molecules. This leads to a second difference between crystals of globular proteins and the crystalline regions of most other macromolecular substances; the basic unit of the crystal structure is the whole molecule and not the individual monomer unit.

The fact that regular crystals of these substances can be obtained, with individual molecules as building blocks, means that the constituent mole-cules must be substantially identical in size, shape, and molecular con-figuration. Whereas laboratory synthesis of large molecules invariably results in molecules which differ from one another, at least in size, the synthesis of globular proteins by living systems must lead to molecules which are all essentially alike.

Small variations between individual molecules are by no means excluded, provided that they do not change the over-all size and shape and provided that they do not interfere with any bonding which may be essential to maintenance of an ordered array. Crystal-line proteins have in fact often been shown, by sensitive methods of chemical separation, to contain molecules which differ in some way from one another. The nature of the difference, in every case investigated, has proved slight, e.g., the substitution of one —COOH side chain for a —CONH$_2$ side chain. Since a protein molecule typically contains several thousand atoms (exclusive of hydrogen atoms) and since the x-ray pattern is a measure of the distribution of electrons, a small deviation from regularity of this kind will clearly have no observable effect.

The protein crystals with which we are here concerned invariably contain water, derived from the solution from which they were crystallized. The percentage of water is typically near 50%. The crystals may be partially dried in air, but they tend to lose their regularity if too much of the water is removed. The water content thus appears to be a necessary part of the crystalline structure, but there is clearly a wide range of water content which is compatible with a given crystal structure. The removal of water, within these limits, causes a shrinking of the unit cell but usually no change in symmetry. A few examples of unit cell dimensions are shown in Table 4–1. For three of the proteins dimensions of both wet and relatively dry crystals are given.

Table 4–1 also lists the space groups to which the crystals belong. In most cases the space group is one requiring the presence of several equivalent positions, and, therefore, the presence of several identical structural units, in the unit cell. Each such structural unit could be an individual molecule, or it could consist of several molecules arranged unsymmetrically with respect to one another.

As was shown in section 3g, the x-ray data can be combined with measured densities to evaluate the molecular weight of the structural unit. This will be j times the molecular weight of a single molecule (equation 3–9) where j is an integer which must be a multiple of the number of equivalent positions in the unit cell. The calculations of the last three columns of Table 4–1 show that it is always possible to choose an allowed value of j so as to obtain molecular weights for the individual protein molecules which coincide within the precision of the determinations with molecular weights obtained by chemical analysis (insulin and ribonuclease) or by measurements in solution. All the molecular weights of Table 4–1 refer to *dry* molecules; i.e., the measured water content has been subtracted from the experimental value determined directly by the x-ray method.

Of special interest are the molecular weights evaluated for two of the proteins of Table 4–1, insulin and hemoglobin. The structural unit of the insulin crystal was found to have a molecular weight of about 12,000. The molecular weight of the same protein in solution, however, is about 36,000. At the time the x-ray study was performed this discrepancy was not understood. It is now known, however, that the chemical molecular weight of insulin is less than 6000. The structural unit in the crystal is thus actually a dimer. The molecule in solution is an aggregate of three such dimers. Its dissociation into smaller units under suitable conditions has been observed in recent years by numerous investigators. A somewhat similar situation exists for hemoglobin. Its molecular weight in solution at neutral pH is about 67,000, but it dissociates readily into molecules of half

TABLE 4–1. X-Ray Diffraction Data for

Protein[a]	H_2O Content, %	Crystal Density grams/cc	Space Group[b]
Zinc insulin[48]	5.4	1.306	$R3$
Ribonuclease[49]	12.5	1.341	$P2_12_12_1$
Lysozyme[50]	9.0[e]	1.305	$P4_12_1$
β-lactoglobulin[51]	46.2	1.144	$P2_12_12_1$
	9.8	1.259	$P2_12_12_1$
Serum albumin[52]	52.6	1.145	$F2$
(human)	(19.0)	1.26	$F2$
Serum albumin[52]	55.4	1.135	$P2_12_12$
(human, Hg dimer)	(11.1)	1.29	$P2_12_12$
Hemoglobin[53]	52.4	1.16	$C2$
(horse I)			
Hemoglobin[53]	—	—	$P2_12_12_1$
(horse II)			
Hemoglobin[54]	—	—	$P3_121$
(Horse, reduced)			
Hemoglobin[54]	—	—	$P4_12_1$
(human I)			
Hemoglobin[54]	—	—	$P2_1$
(human, reduced)			
Tobacco necrosis virus[55]	21.4	1.317	$P1$
Tomato bushy stunt virus[56]	55.0	1.286	I

  [a] The symbols I, II refer to different crystalline forms. Most of the proteins listed crystallize in a variety of forms. Usually we have given data for only one form.
  [b] For description see Ref. 13.
  [c] Most of the space groups of this table are monoclinic or orthorhombic. The angle β represents the angle of inclination of the c axis. The unit cell $P1$ of tobacco necrosis virus is triclinic with angles $\alpha = 100°$, $\beta = 110°$, $\gamma = 120°$. The unit cells $R3$ and $P3_121$ are hexagonal with an angle of 60° between the a and b axes and an angle of 90° between the c axis and the ab plane.

that size, in agreement with the maximum molecular weight compatible with the x-ray data.

One of the general conclusions which can be drawn from the x-ray data on these proteins is that their molecules must be relatively compact and symmetrical. The individual molecules must be able to fit into the unit cell with symmetry appropriate to the crystallographic space group. This leaves

Some Crystalline Proteins and Viruses

Unit Cell Dimensions,[e] $a \times b \times c$ (Å); β (degrees)	Molecules/unit cell		Mol. Wt.[d]	
	Possible $(n = 1, 2, 3, \cdots)$	Assumed	X-Ray	Other
74.7 × 74.7 × 30.6; —[c]	$9n$	18	6,200	5,733
36.6 × 40.5 × 52.3; 90°	$4n$	4	13,700	13,683
71.2 × 71.2 × 31.4; 90°	$4n$	8	13,900	14,000
69.3 × 70.4 × 156.5; 90°	$4n$	16	17,700	17,500
60.7 × 61.0 × 112.4; 90°	$4n$	16	17,800	17,500
178 × 54 × 166; 91°	$4n$	8	65,200	65,000
168 × 38 × 134; 95°	$4n$	—	—	—
165 × 83 × 63; 90°	$4n$	4	65,600	65,000
148 × 50 × 51.5; 90°	$4n$	—	—	—
109 × 63.2 × 54.4; 111°	$4n$	4	33,350	33,500
122 × 82.4 × 63.7; 90°	$4n$	—	—	—
56.1 × 56.1 × 354; —[c]	$6n$	—	—	—
53.7 × 53.7 × 193.5; 90°	$4n$	—	—	—
62.5 × 83.2 × 52.8; 98°	$2n$	—	—	—
157 × 154 × 147; —[c]	n	1	1,600,000	1,850,000
386 × 386 × 386; 90°	$2n$	2	10,800,000	10,600,000

[d] The molecular weight is that of the smallest asymmetric unit. Thus the serum albumin dimer has an actual molecular weight of 130,000, but each molecule consists of two identical halves. All figures have been corrected for the water content; i.e., the molecular weights are those which would obtain in the complete absence of water.

[e] The crystals used in this study consisted of lysozyme hydrochloride, with an HCl content of 2.45%. The molecular weight has been corrected so as to exclude the HCl present.

little choice in the dimensions, and, moreover, the required dimensions are invariably those of a relatively compact figure. In no case is it possible to fit long fully extended polypeptide chains or helical chains, such as those which occur in the synthetic polypeptides and the fibrous proteins, into the unit cell. In hemoglobin, for instance, each polypeptide chain (there are two in each molecule of molecular weight 33,500) contains about 150 amino acid residues. There are, moreover, no disulfide cross-links which might prevent these chains from taking on fully extended or helical configurations. However, the length of each such chain, in an α-helical configuration would be about 230 Å; if fully extended it would be 500 Å. These distances are clearly incompatible with the unit cell dimensions given in Table 4–1. (Insulin is perhaps the only exception to this generalization. Its individual polypeptide chains are short enough to enable them to fit into the unit cell as extended helices.[58]) The data here given do not, of course, exclude the possibility that portions of these protein molecules are helical. In fact, we shall see below that the polypeptide chains of hemoglobin and of the closely related protein myoglobin do exist as mixtures of helical and non-helical portions, the helix content amounting to about 60% of the total.

Another general conclusion which we can draw is that the structures of these protein molecules are relatively rigid, i.e., that they cannot easily alter their configuration from one form to another. That this is so is indicated, for instance, by the data of Bragg and Perutz,[54] who showed that in several chemical modifications of hemoglobin, differing in the oxidation state of the iron atom and in the biological source (human and horse varieties) and crystallizing in structures of quite different symmetry and unit cell dimensions (some of these are listed in Table 4–1), the individual molecules must nevertheless have close to identical molecular dimensions. An oblate ellipsoid of revolution, with axes 53 × 53 × 71 Å, will fit any of the unit cells. The possible variation in these dimensions is just a few per cent. Such an observation strongly indicates that there is a definite fairly rigid structure, common to all hemoglobin molecules, which is not affected by minor chemical differences, including those which differentiate hemoglobins from different species, nor by the manner of crystallization.

The dimensions given are for a hemoglobin molecule of molecular weight 67,000. It appears that the rigid structure[57] consists of a dimer of the molecule which forms the basis of the crystal structure. (Hemoglobin in neutral solutions also exists as a dimer.) The volume of the molecule includes about 30% by weight of water. (Proteins such as hemoglobin also behave in solutions as if they contain about this amount of tightly bound water.)

Especially convincing evidence for this same conclusion comes from the Patterson projections for zinc insulin[59] shown in Fig. 4–9. The two diagrams show the projections upon the *ab* plane of the unit cell for wet and for partially dried crystals. As was discussed in section 3h, such a projection includes the projected length and direction of *all* interatomic vectors of the unit cell. The high density peaks represent those interatomic vectors which occur frequently in the unit cell. Of course, some of the vectors are between atoms on the same molecule, whereas others are between atoms on different molecules within the unit cell. There is no way of distinguishing between them. It is reasonable to suppose, however, that short interatomic distances mostly represent distances between atoms on the same molecule, while the longer ones represent those between atoms on different molecules. It is therefore significant that the drying of the crystal and the accompanying shrinkage of the unit cell do not affect the short interatomic distances shown in Fig. 4–9 (apart from slight rotation about the c axis). Only the longer interatomic distances (beyond about 20 Å) are affected.

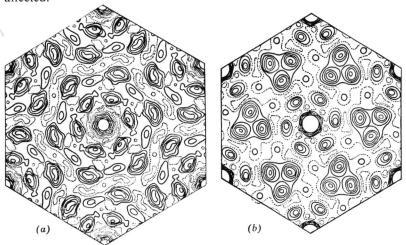

(a)   (b)

Fig. 4–9. Patterson projection along the c axis upon the ab plane for the unit cell of insulin. The diagram is a contour map of electron density, each peak representing a maximum. In the case of a Patterson projection each atom is in turn placed at the origin, and the resulting contour maps are superimposed (but with unequal weight, see text), so that the maxima represent prominent projected interatomic distances and not absolute positions. In this figure (a) represents a normal wet crystal and (b) an air-dried crystal. The central peaks, representing short interatomic distances, are essentially the same in both diagrams, but the outer peaks are appreciably different. (Hodgkin.[59])

Another conclusion to be drawn from Fig. 4–9 and similar data for other globular proteins is that the molecules themselves must be compactly folded so as to contain relatively little of the water content of the unit cell. Removal of water would otherwise alter both short and long interatomic distances. A similar conclusion was reached in another way by Bragg and Perutz.[60] From the effect of substituting salt solutions for water on the diffraction patterns of wet hemoglobin crystals, they concluded that volumes essentially identical with the volume of the molecule with molecular weight 67,000 were inaccessible to the added salt ions.

Experiments such as those discussed above suggest that globular proteins are held fairly rigidly in particular spatial configurations by relatively strong intramolecular bonds. The possible nature of such bonds will be discussed in section 7.

4i. Globular proteins. Detailed structure.[5] The x-ray diffraction patterns of single crystals of globular proteins are sufficiently detailed to permit, in principle, the complete determination of their structure. This fact has spurred on a number of investigators in spite of the seeming insuperability of the problem.

A recent innovation which has made more realistic the hope that the complete structure of some protein molecules may eventually be solved by x-ray diffraction is the application to proteins of the method of isomorphous replacement.[64] This method rests on the discovery that *heavy atoms* may be introduced at specific positions of a protein molecule without changing the size, symmetry, etc., of the unit cell of the corresponding crystals and without appreciable alteration in the coordinates of any of the atoms.[57] The heavy metal atom is usually in the form of an ion and presumably replaces H^+ or OH^- ions and perhaps one or more interstitial water molecules from the original structure. Preferably only a single heavy atom is introduced per protein molecule. The introduction of such a heavy metal atom is of tremendous assistance in the elucidation of a crystal structure because of its great scattering power. Thus even a handful of such atoms in a unit cell which may contain several thousand C, N, O, and S atoms will exert an observable effect.

To understand why the heavy metal substitution is so greatly helpful we must return to the procedure described in section 3h for the elucidation of detailed structure from x-ray diffraction patterns. The primary objective, it will be recalled, is the evaluation of the electron density everywhere in the cell by use of equation 3–21, or, as a preliminary step, the projection of the electron density upon one of the unit cell faces by use of equation 3–22. The difficulty is that the structure factors F_{hkl} which occur in this equation are known only as to absolute magnitudes. In favorable situations the F_{hkl} may be known to be real quantities, so that it is simply the sign (+ or −) which is in question. More often F_{hkl} is unknown to within a factor $e^{-i\alpha}$ where α may take on any value.

Confining ourselves to F_{hkl} values which are real, we shall explain the role of the heavy atoms by use of the simple line of argument given by Crick and Kendrew.[5] The basis for the argument is equation 3–16 which shows that each structure factor is a simple sum of contributions from all the atoms of the unit cell. If F_P is the value of a particular F_{hkl} in the absence of the heavy atom and F_{P+H} is its value in the presence of the heavy atom, then

$$F_{P+H} = F_P + F_H \tag{4-1}$$

where F_H is the contribution of the heavy atoms of the unit cell to equation 3–16. Clearly F_H is also the total value of F_{hkl} which would be obtained for a unit cell containing only the heavy atoms and none of the protein atoms.

Each heavy atom, of course, contributes about ten times as much to the structure factor as does a C, N, or O atom. If we were dealing with a *small* molecule containing a heavy atom, the magnitude of F_H would be greater than that of F_P. When, however, there are hundreds C, O, and N atoms for each heavy atom, then $|F_H|$, although sufficiently great to cause F_{P+H} to be different from F_P, will still in general be smaller than $|F_P|$, even allowing for the fact that the contributions of the C, O, and N atoms will in part be out of phase and thus cancel one another. (Of course, this assumption will be wrong once in a while. But the values of the electron density calculated by equation 3–21 or 3–22 will not be too seriously affected if a few F_{hkl} values get assigned the wrong sign.) Assuming then that $|F_H| < |F_P|$ we can at once evaluate the magnitude of F_H, but not its sign. For, we know $|F_{P+H}|$ and $|F_P|$ directly from the intensities (for real structure factors, they are just the square roots of I_{hkl}/K^2; cf. equation 3–17), and $|F_H|$ could in general be either the sum or the difference of them. In the present situation, requiring $|F_H| < |F_P|$, it must be the difference. (For example, if the experimental intensities give in arbitrary units $F_{P+H}^2 = 64$, $F_P{}^2 = 100$, then F_H could be ± 18 or ± 2. Only the latter value is allowed.)

We have said that F_H is the structure factor that would have been obtained from a unit cell containing only the heavy atoms and none of the protein atoms. Thus $F_H{}^2$ for every (hkl) reflection (temporarily for every real reflection) is known, and the Patterson functions (equation 3–24) for this hypothetical unit cell can be evaluated. A typical plot, taken from Kendrew and coworkers' data for myoglobin,[62] is shown in Fig. 4–10. (Sperm whale myoglobin: a single chain of molecular weight 17,000, two molecules per unit cell, space group $P2_1$, which means that a two-fold screw axis in the b direction is the only symmetry axis. From the discussion on p. 33 it is clear that all F_{h0l} will be real, so that projections upon the ac plane are feasible within the framework of the present treatment.) There being only two heavy atoms per unit cell, only a single vector between them is represented. The symmetry of the cell allows just a single pair of positions joined by such a vector, and *the x and z coordinates of the two atoms are thus immediately established.* Equation 3–16 then permits us at once to evaluate the contribution F_H to all F_{h0l} (the sum is taken over just the two heavy atoms), so that we now know the *absolute value* of F_H for these reflections.

Returning to the example given above, $F_{P+H}^2 = 64$, $F_P{}^2 = 100$, $F_H = \pm 2$, suppose that $F_H = +2$ turns out to be correct. Clearly F_P must be

-10 and not $+10$ since $F_H + F_P$, by equation 4–1, must be equal to ± 8. Thus the *absolute value* of F_P is established [for each $(h0l)$ reflection], and the Fourier projection upon the ac plane can be evaluated by equation 3–22. The result for myoglobin[62] is shown in Fig. 4–11. As can be seen, it is not easy to interpret. The diagram looks down the b axis of the unit cell, through a distance of 31 Å, and projects all the electron densities which it

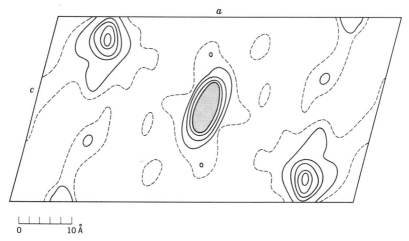

0   10 Å

Fig. 4–10. Patterson projection obtained by equation 3–26 from the difference between myoglobin and its isomorphous HgI_4^{--} derivative. There are two molecules per unit cell and one heavy ion per molecule. What is plotted is a contour map of electron density, relative to the centers of atoms, projected along the b axis upon the ac plane. The origin is in the middle of the figure. The vector between heavy atoms produces the two peaks in the upper left and lower right corner, both of which are 23 Å from the origin. This means that the heavy atom positions in the unit cell, projected upon the ac plane, are 23 Å apart, in the direction given by a line from the origin to either of the two intensity peaks. Since the unit cell has a two-fold screw axis, which, in the projection becomes a two-fold rotation axis, each HgI_4^{--} ion must be 11.5 Å from the center of the unit cell, in the direction given. (Bodo, Dintzis, and Kendrew[62]; taken from Crick and Kendrew.[5])

sees upon a single plane. Clearly the result is not capable of yielding any definite information about the structure. It does show, however, that there are no regularities such as parallel helices in the b direction, for, had any been present, intense maxima in electron density, with cylindrical symmetry, would have appeared.

This analysis shows that a treatment based solely on (hkl) reflections with real F_{hkl} will not suffice. Kendrew and coworkers have thus proceeded to apply to myoglobin the full three-dimensional treatment. It is much more complicated, because the intensities of so many more reflections have to be measured (at least 20,000 of them in the case of myoglobin if atomic positions are to be revealed) and because not only the sign but also the phase angle α has to be determined for each F_{hkl} factor. (To do this requires several independent isomorphous replacements.) The task is

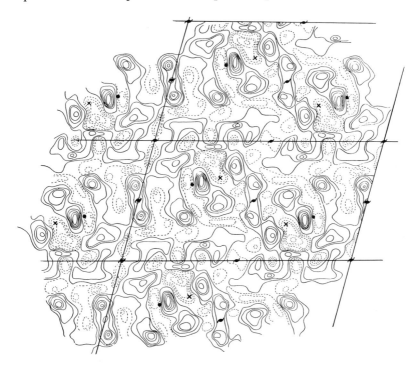

Fig. 4–11. Fourier projection of myoglobin, type A. This shows the projected electron density of the monoclinic unit cell (which contains two molecules), looking down the b axis, i.e., through 31 Å of protein and mother liquor. Resolution 6 Å. The positions of two heavy groups are shown: $\cdot HgI_4^{--}$; $\times$ Hg in p-chloro-mercuribenzene sulfonate. The two molecules overlap, and it is not easily possible to tell where one begins and the other ends. (Bodo, Dintzis, and Kendrew.[62] Figure taken from Crick and Kendrew.[5])

clearly a forbidding one, but excellent progress has been made, and the complete elucidation of the structure may be expected in the very near future.

Progress to date[148] is summarized in Fig. 4–12. The left diagram represents the result of using the intensities of about 400 reflections, those which correspond to spacings of 6 Å or greater. Use of these reflections allows the evaluation of the electron density with a resolution of 6 Å, i.e., it tells where, in a general way, the regions of high electron density are located, but it cannot resolve the positions of individual atoms. The diagram shows the high electron density regions of one myoglobin molecule (half a unit cell), and these regions, aside from the disk which corresponds to the heme group, are taken to represent the location of —CO—NH— groups of the polypeptide chain, since most of the side

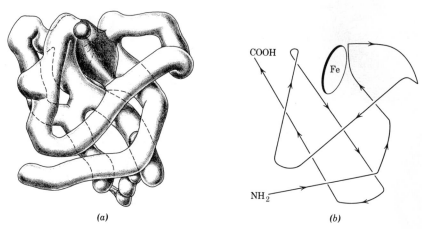

(a)   (b)

Fig. 4-12. The structure of the myoglobin molecule, as deduced from x-ray diffraction. (a) shows the regions of high electron density obtained from an analysis of 4000 reflections and represents a resolution of 6 Å. (b) is obtained from an analysis of 9600 reflections, giving a resolution of 2 Å. The line shows the course of the polypeptide backbone, straight portions representing α-helices. The disk-shaped object in both diagrams represents the heme group. (Kendrew and coworkers.[148])

chains have a relatively low electron density. The diagram shows several apparent bridges between parts of the chain. Since myoglobin contains no disulfide bonds, these bridges presumably represent places of entanglement or locations of bulky side chains (e.g., those containing COO^- groups). The diagram also shows many "holes" in the structure. These do not represent solvent-filled spaces; they show the location of side chains of relatively low electron density.

The right half of Fig. 4-12 is the result of a refined analysis, employing the intensities of 9600 reflections, each of which had to be measured for each of several isomorphous crystals containing different heavy atom substituents. These reflections bring the resolution down to 2 Å, i.e., almost to the point where individual atom positions can be seen. The figure shows the course of the polypeptide backbone, and of particular importance are the straight-line portions which have been definitely established to represent α-helices. These helical portions include more than 60% of the total polypeptide chain.

Perutz and coworkers[149] have carried out an x-ray diffraction study of hemoglobin which has progressed to about the same stage as Kendrew's study of myoglobin. Each of the four polypeptide chains of hemoglobin is quite similar in size and arrangement of amino acid residues to a single myoglobin molecule, and it turns out that the polypeptide chain folding in the crystal is also quite similar.

Aside from the presence of relatively long helical regions, the outstanding feature of the myoglobin and hemoglobin structures is the over-all lack of regularity. This lack of regularity is presumably the result of the lack of regularity in the sequence of side chains. As each side chain has a marked preference for being adjacent to certain other side chains (see section 7), only an irregular structure can hope to satisfy these preferences.

X-ray diffraction studies of other crystalline globular proteins have not yet progressed to the stage reached in the studies with hemoglobin and myoglobin. Preliminary results[63,68,69] indicate that the lack of regularity which has been observed with hemoglobin and myoglobin will prove typical of globular proteins in general. It is likely, moreover, that the helix content of other proteins may be considerably lower. (Perutz.[156])

4j. Nucleic acids. Deoxyribose nucleic acid (usually abbreviated to DNA) from various sources has been obtained in fibrous form and examined by x-ray diffraction. As was explained on p. 9, nucleic acids exist as anions under ordinary conditions, and, in the fibers that have been examined, the DNA has always been present as the sodium salt. Two types of diffraction pattern have been observed:[65,66,70] one, obtained at low relative humidity, is a sharp pattern indicative of long-range order; the other, obtained at high relative humidity, is a poorly defined pattern (shown in Fig. 4–2i) suggesting the presence of crystalline regions of exceedingly small dimensions. The structures which correspond to these patterns may be reversibly converted into one another, simply by changing the relative humidity.

It is the less well-defined pattern which has led to the currently accepted structure for DNA. The pattern is of the helix type and suggests a structure involving a slowly twisting helix of pitch 34 Å, with ten residues per turn. To achieve such a structure requires the presence of two parallel polynucleotide chains coiled about the same axis. The separation between individual nucleotides of the same chain, projected on the fiber axis, is 3.4 Å. The over-all diameter is about 20 Å. A diagrammatic representation of the structure, as visualized by Watson and Crick,[67] is shown in Fig. 4–13. The structure has a partially hollow core, which presumably contains water. The purine and pyrimidine groups are on the inside perpendicular to the helix axis, the phosphate ions on the outside.

To account for the parallel alignment of the two polynucleotide chains of this structure, Watson and Crick suggested the existence of hydrogen bonds between the purine and pyrimidine groups facing each other in the

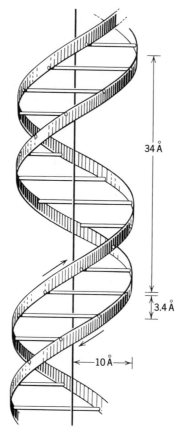

Fig. 4–13. The structure for DNA proposed by Watson and Crick.[67]

center of the helix. Because the purine groups are considerably larger than the pyrimidine groups, this arrangement can occur only if a purine group from one of the molecular chains is always paired with a pyrimidine group from the other. When models were constructed, a further restriction was discovered, in that hydrogen bonds were able to be formed in the correct positions only between an adenine side chain and a thymine side chain and between a guanine side chain and a cytosine side chain. Thus, if one of the coiling chains of DNA has a particular sequence of basic side chains, the sequence of the second chain is completely determined. There must always be an adenine on the second chain opposite a thymine on the first, and vice versa, and there must always be a guanine on the second chain opposite a cytosine on the first, and vice versa.

If the structure proposed by Watson and Crick is correct and if it applies to the entire DNA molecule, the number of guanine groups per molecule must be equal to the number of cytosine groups, and the number of adenine groups must be equal to the number of thymine groups. Analytical data indicate that these conditions are actually closely, though not perfectly, satisfied by most DNA samples which have been examined.

Watson and Crick have suggested that their complementary structure provides a possible mechanism for the self-duplication of genetic structures in living systems. Such structures consist largely of DNA and, when a living cell splits, are themselves split into two duplicate copies of the original. If the process of duplication is thought of as beginning with a splitting of the DNA double helix into its separate strands, followed by synthesis of new companion strands, then the new strands, if they are to form a new double helix, must inevitably be identical with the original companion strands.

The Watson-Crick proposal is a long, long way from the information provided by Fig. 4–2i. It illustrates, however, the important contribution which exact knowledge of molecular structure can make to an understanding of biological processes.

4k. Viruses.[5] A few viruses have been obtained in truly crystalline form, and single crystals of them have been examined by x-ray diffraction. Unit cell dimensions and molecular weights can be calculated for these viruses, and data for two of them are included with similar data for globular proteins in Table 4–1. The most striking feature is the enormous size of the unit cells and the correspondingly large molecular weight. Another conclusion which may be drawn from the data is that the virus particles (of the two viruses listed) must, like globular proteins, be compact and close to spherical. Tomato bushy stunt virus, crystallizing in a body-centered cubic lattice, must, in fact, be taken to have a packing unit with perfectly spherical symmetry, although this packing unit of course contains solvent in addition to the virus particle.

Only fragmentary information is available at present on the more detailed structure of these viruses, but such information has been obtained for tobacco mosaic virus (TMV), a virus that can form true crystals, which, however, have never been obtained in sufficiently large size for x-ray diffraction studies. Tobacco mosaic virus easily forms paracrystalline gels,

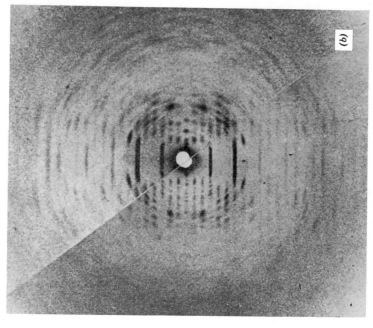

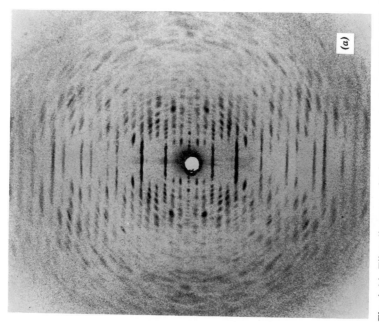

Fig. 4-14. Fiber diagrams of tobacco mosaic virus (TMV). (*a*) is the pattern obtained for oriented gels of TMV. (*b*) shows a comparison between TMV and the nucleic acid-free protein obtained from it, the upper half being TMV and the lower half being the protein. (Franklin.[119,121])

however, which produce well-defined fiber diagrams, one of which is shown in Fig. 4–14a. They suggest that this virus is rod shaped, rather than spherical, the gels consisting of parallel rods whose ends are not lined up. That tobacco mosaic virus is, in fact, rod shaped is actually well known from electron micrographs, as will be discussed in section 6a. The probable length of the rods is 3000 Å; their diameter about 150 Å; and the molecular weight about 40 million.

Figure 4–14a suggests, as do other fiber diagrams, that TMV molecules have a helical structure,[118],[119] each turn of the helix corresponding to a progression of 23 Å along the longitudinal axis. There are now believed to be $16\frac{1}{3}$ identical (more or less identical) repeating units per turn.[123] Chemical analysis shows that TMV contains 94% protein and 6% RNA. If the protein constituent is helically arranged, with $16\frac{1}{3}$ identical units for each 23 Å of molecular length, there must be 2100 such units per rod-shaped molecule of length 3000 Å. Each unit would have a molecular weight of about 18,000.

This conclusion agrees remarkably well with chemical evidence. It is possible to disaggregate TMV by means of dilute alkali and to separate the resulting molecules into protein and nucleic acid fractions.[120] The protein fraction is found to consist of molecules of molecular weight about 100,000, but each molecule consists of several polypeptide chains; the molecular weight of each of which is 18,000, in agreement with the molecular weight of the repeating unit of the x-ray structure. Moreover, the limited chemical evidence available strongly suggests that all the polypeptide chains are identical. (The existence of protein molecules of molecular weight 18,000 in solution has been demonstrated by very recent work of Lauffer.[144])

An interesting feature of TMV protein is that it may be repolymerized even in the absence of nucleic acid. The resulting particles have an x-ray diffraction pattern which is very similar to that of TMV itself,[121] as shown by Fig. 4–14b. There are minor differences in the over-all pattern, of the kind that might be expected between two identical structures, one of which (the nucleic acid-free protein) contains holes, presumably solvent filled, in place of material which is rigid in the other. For instance, gradual drying leads to some shrinkage of the helix dimensions of TMV protein, whereas in the case of intact virus only solvent between particles appears to be removed and the helix dimensions remain constant.

The symmetry of TMV about its longitudinal axis is such that the equatorial structure factors of the inner part are real. Furthermore, isomorphous substitution with heavy atoms has been found possible (Caspar,[122] using lead, and Franklin,[123] using mercury) so that a projection of the electron density upon a cross-section of the cylindrical rod is readily obtained, using the method described on p. 63. In sharp contrast to what was found in the case of myoglobin (Fig. 4–11) such a projection proves to provide substantial information, despite the fact that the projection includes the x and y coordinates of the electron density of all the matter through a height of 69 Å (taking three helix turns as a true repeat distance). The reason it does so, of course, is that the entire structure, in projection,

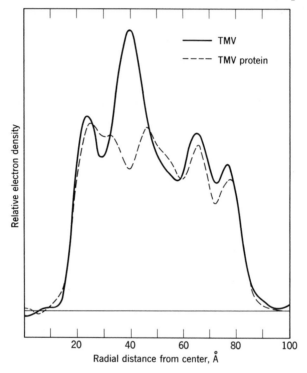

Fig. 4–15. The radial distribution of electron density in tobacco mosaic virus (TMV) and in nucleic acid-free protein obtained from it. The ordinate shows the difference between the electron density in the particle and that of water. (Franklin, Klug, and Holmes.[123])

has cylindrical symmetry, so that like atoms always fall at the same radial distance from the center.

The material examined is, as we have said, an oriented gel and not a true crystal. The disorientation, however, consists of vertical displacement of individual rods. The *packing* of rods about one another is regular, with hexagonal symmetry, so that the structure behaves in projection much like a regular crystal.

The isomorphous substitution method has been applied not only to intact TMV but also to TMV protein freed from nucleic acid. The results, shown in Fig. 4–15, are strikingly similar, with one noteworthy exception, and clearly indicate the following features of the structure of the individual virus particle.[122,123]

(1) There is a large hole down the center of each rod, the projected electron density of which is equal to that of the solvent (water) which permeates the gel. The diameter of the hole is 40 Å. It must be assumed to reflect the presence of a solvent-filled hole in the structure of the virus, as well as in that of TMV protein particles.

(2) There is a pronounced peak in the electron density of the virus at a distance of 40 Å from the center. At the same position, in the nucleic acid-free protein, there is a minimum in the electron density. This is the only significant difference between TMV protein and intact TMV, and it must therefore indicate the position at which the nucleic acid is located in the latter.

(3) The electron density does not decrease sharply at a distance of 75 Å from the center but decreases gradually from 60 Å to about 85 Å. This is taken to indicate that the rod-shaped particles cannot have smooth walls but that they are grooved, the structure extending at some places to only 60 Å from the center, whereas the maximum extension is about 85 Å.

It should be noted in this connection that the diameter of 150 Å mentioned earlier is obtained from the distance between centers of adjacent virus particles when packed close together. The diameter of maximum extension (about 170 Å) is not incompatible with the 150 Å value if the grooved surfaces can interlock when adjacent particles are brought together.

5. SPECTROSCOPY

The internal energy of a molecule cannot vary continuously. Instead each molecule possesses a series of discrete quantized energy states, which differ from one another in the energy of electrons in the electrostatic field of the atomic nuclei, in the energy of vibration of these nuclei relative to one another, and in the energy of rotation of the molecule as a whole. Additional energy levels may be created if a molecule is placed into an external electric or magnetic field, owing to the interaction of such a field with the rotational motion of the molecule as a whole, or with the spin of electrons and nuclei.

In typical small molecules the separation between electronic energy levels is of the order of 10^{-11} erg or larger; the separation between vibrational levels is of the order of 10^{-13} to 10^{-12} erg; the separation between rotational levels is of the order of 10^{-16} erg; and the separation between energy states induced by electric or magnetic fields is even smaller.

If the correct amount of energy required for transition from one energy state to another is supplied to a molecule, it *may* absorb this energy and undergo the corresponding transition. Whether it actually does so depends on the *transition probability*, a quantity which, in simple cases, can be calculated by quantum mechanics. In most cases numerous transitions cannot occur at all; they are said to be *forbidden* transitions. Simple formulas are often available to determine which transitions are allowed and which are forbidden under a given set of experimental conditions. They are known as *selection rules*.

One way of supplying the energy necessary for a transition is by means of electromagnetic radiation. Electromagnetic radiation behaves, from the standpoint of energy transfer, as if it were composed of discrete "particles" of energy, called *photons*. The energy E of each photon is related to the wave characteristics of the radiation by Planck's constant, $h = 6.624 \times 10^{-27}$ erg-sec,

$$E = h\nu = h\tilde{c}\omega = h\tilde{c}/\lambda \qquad (5\text{--}1)$$

where ν is the frequency in $\sec^{-1}$, λ is the wavelength in cm, $\omega = 1/\lambda$ is the wave number in cm^{-1}; and $\tilde{c} = \nu\lambda = 2.9979 \times 10^{10}$ cm/sec is the velocity of propagation of all electromagnetic waves in vacuum. (The wavelength is also that measured in vacuum.)

If equation 5–1 is combined with the typical energy level separations given above, it is seen that radiation in the ultraviolet range of the spectrum ($\lambda \sim 2000$ Å) will provide photons of the right energy to induce typical electronic transitions. In some molecules these levels are sufficiently closely spaced so that visible light ($\lambda \sim 5000$ Å) will suffice; in others they are so far apart that radiation in the inaccessible range between ultraviolet and x-radiation would be necessary. Infrared radiation [$\lambda = 1$ to $20\ \mu$ (microns)] will provide photons which can induce typical vibrational transitions. Radiation of radio frequency ($\lambda = 1$ mm to 1000 meters) will provide photons which can induce rotational transitions and the transitions between energy levels in electric and magnetic fields. Molecules will therefore absorb characteristic frequencies in these ranges of the spectrum, corresponding to the exact separations between their energy levels.

It is possible to induce transitions between vibrational levels in another way, by using visible light. Photons of visible light are scattered by all molecules, generally without loss of energy; i.e., the scattered radiation has the same frequency as the incident radiation. (This subject was mentioned in section 3 and is discussed in detail in Chapter 5.) A small fraction of the scattered photons, however, may interact with the molecules so as to give up enough of their energy to induce a vibrational transition. The scattered radiation then has a frequency less than that of the incident radiation. The frequency difference corresponds, by equation 5–1, to the separation of vibrational energy levels. This effect is known as the *Raman* effect (after its discoverer, Sir C. V. Raman), and, in the study of small molecules, it plays an important role, especially since selection rules for Raman absorption differ from those for infrared absorption, so that vibrational transitions observed by one method supplement those observed by the other. However, the method has so far not been applied to macromolecules because the ordinary (Rayleigh) scattering by these molecules is relatively more intense, and it will therefore not be discussed here. (Preliminary experiments with a synthetic polypeptide and a protein have been reported by Garfinkel and Edsall[124].)

Exact theoretical treatment of the dynamics of molecules by quantum mechanics is possible for the very simplest molecules only. Thus the theoretical equations which relate the separation between energy levels to structural parameters (e.g., Herzberg,[71] Duncan,[72] Matsen[73]) are largely of academic interest in macromolecular chemistry, and even in the chemistry of relatively complex small molecules. The application of spectroscopy to the study of such molecules is largely empirical. It is based on the principle that energy levels of many kinds (e.g., electronic) depend on the "structure" in the immediate vicinity of only one or a few atomic nuclei, so that structural features of localized regions (a few atoms in extent) of large molecules may be inferred from similarity of their absorption spectra to spectra of appropriate small molecules containing similar groupings of atoms.

5a. Infrared spectra of organic molecules.[7,8] Absorption bands which lie in the infrared region of the spectrum measure the separation between vibrational energy levels. If molecular vibrations were truly harmonic (i.e., if the restoring force were truly proportional to the square of the displacement from an equilibrium position), the vibrational energy levels would have energy values given by the expression $E = h\nu_0(n + \frac{1}{2})$ where ν_0 is the fundamental vibration frequency and n may be 0, 1, 2, 3, etc. (This fact is demonstrated in any quantum mechanics textbook.) The selection rules allow transitions only between adjacent levels, and, as a result, absorption frequencies ν corresponding to vibrational transitions would correspond exactly to the fundamental frequencies ν_0. Actual molecular vibrations, however, are not truly harmonic, and this correspondence is therefore only approximate. Nevertheless, one absorption band should be observed for each fundamental mode of vibration, and ν should be close to ν_0. An additional selection rule for infrared absorption is that only those vibrations which alter the dipole moment of the molecule can be observed.

The selection rule $\Delta n = \pm 1$, i.e., the rule which states that only a single quantum of vibrational energy can be absorbed at a time, is not strictly true for complex molecules. Weak absorption will also be observed which corresponds to a change of 2 or more in the vibrational quantum number n or which corresponds to simultaneous change of energy level in two or more fundamental vibrations. The corresponding absorption bands are known as *overtones* and *combination bands*, respectively. These bands are generally of much lower intensity than the fundamental bands, i.e., those which correspond to absorption of a single quantum of vibrational energy.

In a complex molecule the number of different fundamental vibrations is enormous. (The dynamics of a molecule with n atoms predicts the existence of $3n - 6$ fundamental modes of vibration.) The utility of infrared spectroscopy in the study of complex molecules rests on the fact that some of these vibrations, those of peripheral atoms with respect to their neighbors, should have energies and corresponding frequencies which closely match corresponding vibrations in small molecules. Furthermore, the skeletal vibrations, those which involve displacement of larger groups of atoms relative to one another, would normally be expected to have low frequencies and low intensities, so that they do not interfere seriously with the observation of the absorption bands which are easily assignable to known types of vibrations.

Two recent publications give detailed summaries of the characteristic absorption frequencies which correspond to group vibrations in organic molecules. One summary is given in a chapter by R. N. Jones and C. Sandorfy in the treatise, *Chemical Applications of Spectroscopy*[7]; the other is in a book by Bellamy.[8] A few frequencies of interest are listed in Table 5–1. These and the frequencies discussed in the text are all taken from the two compilations, which should be consulted for the original references.

The vibrations shown in Table 5–1 are classified as to type: *stretching* vibrations are those involving lengthening of the interatomic bond distance; *deformation* vibrations are those in which bond angles are changed. Deformation vibrations are also called *bending* vibrations.

There is often more than one kind of stretching or deformation vibration characteristic of a given atomic group. For the —CH_3 group, for example, one stretching vibration will be observed in which all three C—H bonds are lengthened and shortened in phase with one another, whereas another vibration will have two C—H bonds lengthened while the third is shortened and vice versa. (The corresponding frequencies in saturated hydrocarbons are 2962 cm^{-1} and 2872 cm^{-1}.) As another example, there are four different deformation vibrations for a —CH_2— group. In one of them the two C—H bonds bend towards one another in the HCH plane (*scissoring* vibration, near 1470 cm^{-1}), in another each C—H bond bends in the same direction in the HCH plane; i.e., the net effect is one of torsional oscillation about a C—C bond (*rocking* vibration, near 720 cm^{-1}). In the other two deformation vibrations the C—H bonds bend out of the HCH plane, either in the same direction (*wagging* vibration) or in opposing direction (*twisting* vibration). The last two are less intense than the other two, and both appear near 1300 cm^{-1}.

TABLE 5–1. Approximate Location of Some Infrared Absorption Bands in Organic Molecules[a]

Type of Vibration		Wave Number, cm^{-1}	Wavelength μ
—O—H	Stretching	3600	2.8
—N—H	Stretching	3400	2.9
—C—H	Stretching	2900	3.4
—C=O	Stretching	1700	5.9
—C=N—	Stretching	1670	6.0
—C=C—	Stretching	1650	6.1
—N—H	Deformation	1650	6.1
—CH$_2$	Deformation (scissor)	1470	6.8
—CH$_3$	Deformation (bending)	1380	7.2
—CH$_2$	Deformation (wagging)	1305	7.7
—CH$_3$	Deformation (rocking)	1135	8.8
—C—O—	Stretching	1100	9.0
—C—C—	Stretching	1000	10.0
H—C=C—H	Deformation (out of plane)	900	11.0
—CH$_2$	Deformation (rocking)	720	13.9

[a] As is discussed in the text, some of these frequencies are subject to considerable variation. Variations of 100 or 200 cm^{-1} are common, so that this table should not be regarded as providing an immediate identification of an atomic group from a given vibration frequency.

The frequencies given in Table 5–1 are intended to give only the approximate locations of the absorption bands. Wide variations from the figures given are observed, depending on a number of factors, as follows:

(1) *Covalent structure.* The precise frequency depends strongly on the kind of atom to which the vibrating group is attached and on the immediately adjacent covalent bonds. The —C=O stretching frequency, for example, occurs near 1710 cm^{-1} in saturated ketones, near 1690 cm^{-1} in completely aromatic ketones, near 1730 cm^{-1} in aliphatic aldehydes, near 1750 cm^{-1} in aliphatic carboxylic acids (but near 1600 cm^{-1} in the corresponding carboxylate anion), and near 1800 cm^{-1} in carbonyl halides.

(2) *Physical state and nature of solvent.* For a given compound the vibration frequency depends on whether it is in the pure state (gaseous, liquid, or solid) or in solution. If in solution it depends on the nature of the solvent. For example, the —C=O stretching vibration in dilute solutions of amides occurs near 1690 cm^{-1}, in the solid state the corresponding absorption occurs near 1650 cm^{-1}. Similarly the —C=O stretching frequency in methyl acetate vapor is at 1774 cm^{-1}. For the same compound dissolved in cyclohexane it occurs at 1756 cm^{-1}; in dioxane it occurs at 1747 cm^{-1}, and in bromoform at 1738 cm^{-1}. These differences, of course, are due to intermolecular interactions, between absorbing molecules in the solid state and between such molecules and the solvent in the dissolved state. (If the vibration can be treated as an oscillating dipole and if the interaction with the solvent is a simple dipole-dipole interaction, a simple theoretical relation exists between frequency and the dielectric constant of the solvent. This relation is applicable to some, but by no means to all solvent effects.[74])

(3) *Hydrogen bonding.* Specific intermolecular (or intramolecular) interactions may be expected to be reflected by especially prominent changes in infrared spectra, and this proves to be the case. The best-known example of such bonding is hydrogen bonding, and its effects have been observed on numerous occasions. For instance, a 0.0007 molar solution of pyrrole in CCl_4 shows a fairly sharp absorption band at 3496 cm^{-1}, assignable to the —N—H stretching vibration. A 1.26M solution, however, shows two bands, one at 3494 cm^{-1} and one at 3411 cm^{-1}, the latter undoubtedly due to the same stretching vibration in an —N—H · · · N— hydrogen-bonded configuration. In a 3.8M solution the second band becomes the most prominent. The free vibration is seen only as a shoulder on the band at the lower frequency. Even greater reductions are observed in *dilute* solutions when the solvent is capable of forming hydrogen bonds. Examples of such solvents for pyrrole are acetone (3395 cm^{-1}) and pyridine (3219 cm^{-1}).

The strongest of all hydrogen bonds are those which occur in the (crystalline) solid state. They may be accompanied by frequency shifts for —O—H or —N—H stretching vibration of as much as 1900 cm^{-1}. That these shifts are not the result of faulty assignment is indicated by x-ray determinations of interatomic distances in the same crystalline solids. There is excellent correlation between the frequency shifts and the corresponding shortening of O—H · · · O, N—H · · · O, etc., bond distances.[75]

5b. Infrared spectra of macromolecules. Absorption spectra of two simple polymers are shown in Fig. 5–1, together with the spectrum of ethyl

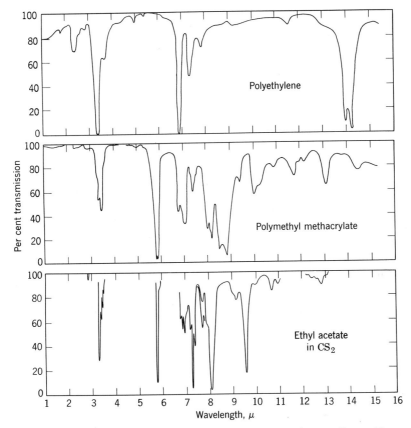

Fig. 5–1. Infrared absorption spectra of two polymers (from Billmeyer[10]). The spectrum of ethyl acetate in CS_2 solution (from Weissberger[7]) is included for comparison.

acetate. These spectra illustrate the fact that the complexity of the pattern of peaks depends on the variety of atomic groups rather than on the size of the molecule. There are more major peaks in the spectrum of ethyl acetate than in that of polyethylene. On the other hand the spectrum of polymethyl methacrylate

$$-\left(CH_2-\underset{\underset{COOCH_3}{|}}{\overset{\overset{CH_3}{|}}{C}}\text{------} \right)_x$$

is somewhat more complex than that of ethyl acetate. Both for low-molecular-weight organic compounds and for simple polymers, many of the absorption bands are easily assigned to particular vibrations. At the same time, in both cases, there invariably remain absorption bands for which an assignment has not yet been made.

To illustrate the method used to arrive at an assignment we may consider the infrared absorption in the region of 1500 to 1800 cm^{-1} for the

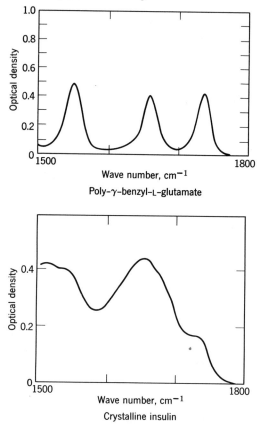

Fig. 5–2. Absorption spectra of poly-γ-benzyl-L-glutamate (α-form) and of insulin in the region of 1500 to 1800 cm^{-1}. (Ambrose and Elliott.[76,79])

simple synthetic polypeptide, poly-γ-benzyl-L-glutamate.[76] The pattern obtained for solid films (in the α-form) of this polymer is shown in Fig. 5–2a. It is seen to consist of three well-defined peaks, at 1737, 1659, and 1530 cm^{-1}. Table 5–1 shows only two kinds of pertinent atomic groups which are strong absorbers in this frequency range, these being the C=O group (stretching vibration) and the N—H group (deformation vibration). Since the polymer molecule under consideration has an N—H group and two different C=O groups, one in the peptide bond and one in the ester side chain, the three bands are tentatively assigned to these three groups.

Comparison with model compounds serves to confirm this assignment and to indicate which of the three bands belongs to which group.

Thus the N—H deformation absorption occurs in general in the neighborhood of 1600 cm^{-1}. An appropriate model compound would be a secondary amide, R—CO—NHR', in the solid state. In such amides[77] the N—H deformation frequency is found at 1560 cm^{-1}; i.e., the polymer absorption band at 1530 cm^{-1} can be assigned to the N—H group. X-ray diffraction studies described earlier have indicated that all the N—H groups of the α-form of synthetic polypeptides are involved in intramolecular C=O · · · H—N hydrogen bonds, so that the corresponding N—H deformation frequency is not expected to be that of a free N—H group. However, N—H groups of amides, in the solid state, are also hydrogen bonded, so that the use of solid amides as model compounds is an appropriate one.

Similarly, the C=O stretching frequency, which has been shown above to be subject to considerable variation from one compound to another occurs in solid amides close to 1640 cm^{-1} (Richards and Thompson[77]) and in saturated carboxylic esters at 1735 cm^{-1}. Thus the polymer absorption band at 1659 cm^{-1} is assigned to the peptide C=O group and that at 1737 cm^{-1} to the ester C=O group.

Another absorption band of poly-γ-benzyl-L-glutamate for which an assignment is easily made is the strong band near 3300 cm^{-1} (shown, for example, in Fig. 5–5a, which will be discussed later). Only O—H and N—H stretching frequencies occur in this region, and, since the polymer possesses no hydroxyl groups, the band must be assigned to its N—H group.

It is worth noting that even the apparently straightforward assignment problem just discussed is not free from controversy. It has been suggested, for instance, that the absorption occurring at 1560 cm^{-1} in solid amides and at 1530 cm^{-1} in synthetic polypeptides is not an N—H deformation absorption at all but is ascribable instead to a C=N stretching vibration in the resonance form —C(OH)=N— of the amide group. (A summary of the arguments has been presented by Bellamy.[78]) The same resonance form introduces an OH group which could absorb in the 3400 cm^{-1} region, thereby making less certain the assignment of *all* the absorption in this region (in simple secondary amides and in poly-γ-benzyl-L-glutamate) to N—H stretching vibrations.

The absorption bands of macromolecules which are not polymers of a single kind of monomer are much more difficult or even impossible to assign to particular vibrations. Figure 5–2b, for example, shows the absorption spectrum, in the region of 1500 to 1800 cm^{-1}, of crystalline insulin.[79] The spectrum is seen to differ markedly from that for poly-γ-benzyl-L-glutamate. Instead of well-defined peaks with essentially zero absorption between them there is continuous absorption through most of the region. The continuous pattern shows peaks, but they are broad and irregular. It is not hard to find the reason for this. Insulin contains peptide N—H and C=O groups, as does poly-γ-benzyl-L-glutamate, and it is possible that the majority of them may be hydrogen bonded. However, it also contains side chains of a considerable variety (cf. section 2) many of which contain C=O or N—H groups or both. The molecule contains amino groups (lysine side chains), imidazole imino groups (histidine side chains), guanidine amino groups (arginine side chains), side chain amide groups which are not necessarily hydrogen bonded, etc. In addition insulin possesses aromatic side chains (e.g., tyrosine) which also absorb in the 1500 to 1800 cm^{-1} region, and, finally, its crystals invariably contain water, which has a bending vibration near 1650 cm^{-1} in solid hydrates. The spectrum of Fig. 5–2b is the resultant of all these separate absorptions, and any assignment made for the peaks which exist must be largely speculative.

Spectra of proteins in the region of 1500 to 1800 cm^{-1} can be partially simplified if they are studied in D_2O solution or as solid films prepared from D_2O solutions. Under these conditions all hydrogen atoms attached to nitrogen are replaced by deuterium (see, however, section 36d) and the corresponding deformation frequencies are shifted to lower frequencies. The D—O—D deformation absorption is also moved to lower frequencies. The 1500 to 1800 cm^{-1} region is thus freed of all bands except those due to C=O stretching vibrations. There is still a variety of them, but a tentative assignment of them can be made with the aid of studies at different D$^+$ ion activities which allow identification of the bands due to COOD (prominent at low pD) and to COO$^-$ (prominent at high pD).[80,81]

5c. Structural information from infrared spectra. The most obvious use of infrared spectra is as a tool in the qualitative identification of atomic groups present in macromolecules and in the quantitative measure of their prevalence. For example, when butadiene is polymerized by addition polymerization, two kinds of product might be obtained

$$\cdots -CH_2-CH-CH_2-CH-\cdots \qquad \cdots -CH_2-CH=CH-CH_2-\cdots$$
$$\qquad\quad | \qquad\quad\;\; |$$
$$\qquad\quad CH \qquad\;\; CH$$
$$\qquad\quad \| \qquad\quad\;\; \|$$
$$\qquad\quad CH_2 \qquad\;\; CH_2$$
$$\qquad\qquad\quad \text{I} \qquad\qquad\qquad\qquad\qquad\qquad\quad \text{II}$$

form I by 1,2 addition to the double bond and form II by 1,4 addition. These products are easily distinguished by their infrared spectra, the best mode of vibration for this purpose being the deformation, perpendicular

to the 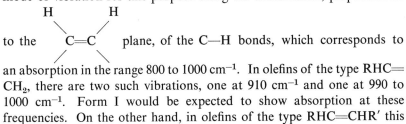 plane, of the C—H bonds, which corresponds to

an absorption in the range 800 to 1000 cm^{-1}. In olefins of the type RHC=CH$_2$, there are two such vibrations, one at 910 cm^{-1} and one at 990 to 1000 cm^{-1}. Form I would be expected to show absorption at these frequencies. On the other hand, in olefins of the type RHC=CHR' this

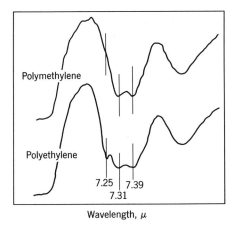

Wavelength, μ

Fig. 5–3. Infrared spectra of polymethylene and polyethylene in the 7 μ region. The peak at 7.25 μ is due to the presence of methyl groups. (Bryant and Voter.[83])

absorption occurs at 965 cm^{-1}. (This is for the *trans-* form of these olefins; *cis-* forms show relatively weak absorption at several frequencies) Form II would be expected to absorb at this frequency. Actual samples of poly-butadiene show absorption at all three frequencies (990, 962, and 910 cm^{-1}) so that the polymer must contain segments of both forms.[82]

Another example is provided by the studies of Bryant and Voter[83] (and earlier studies, e.g., Fox and Martin[84]) on polyethylene. They found that polymethylene (polymerization product of CH$_2$N$_2$) has a pair of peaks at 1368 and 1354 cm^{-1} (7.31 and 7.39 μ in Fig. 5–3), ascribable to the "wagging" vibration of —CH$_2$— groups. (The same peaks occur in catalytically synthesized polyethylene). Polyethylene prepared by high pressure polymerization of ethylene, however, contains an additional peak, at 1379 cm^{-1} (7.25 μ), shown in Fig. 5–3. This frequency is characteristic of the symmetrical bending vibration of CH$_3$ groups. It is found at exactly 1379 cm^{-1} in all liquid saturated hydrocarbons, and its identity is established by its diminished intensity (relative to —CH_2— bands) as the chain length of normal hydrocarbons, i.e., the ratio of —CH_2— groups to —CH_3 groups, is increased. Ordinary polyethylene is therefore not a pure polymer of the form —$(CH_2$—CH_2—$)_x$ but must also contain methyl groups. From the observed intensity of the CH_3 band it was concluded that some samples of polyethylene contain as many as 4.6 CH_3 groups per 100 carbon atoms. Similar examination of another part of the spectrum showed that the same polymers contain —CH=CH_2 double bonds (from the presence of the 910 cm^{-1} deformation discussed earlier; absorption at this frequency disappeared when the polymer was hydrogenated).

Infrared spectra may also be used to determine the extent of crystallinity in polyethylene, by use of the 720 cm^{-1} absorption associated with the CH_2 rocking vibration.[85] In crystalline regions this band is split into two parts, presumably reflecting interaction between CH_2 groups on adjacent polymer chains. One absorption peak corresponds to the situation where such neighboring CH_2 groups are vibrating in phase, and one where they are vibrating out of phase. The amount of crystallinity can be determined from the extent of splitting. Another band sensitive to the physical state of the polyethylene is a band at 1303 cm^{-1} (7.67 μ; this band can be seen on the right-hand side of Fig. 5–3). It is present only in amorphous regions of the polymer, or in the liquid state. The ratio of the intensity of the second peak at 720 cm^{-1} to the intensity of the 1303 cm^{-1} peak provides an accurate measure of crystallinity.

The most important information that can in principle be obtained from infrared spectra is information on local interaction between atomic groups. Especially important is the fact that infrared spectra can be obtained for very dilute solutions of macromolecules, in which these molecules are isolated from one another. Infrared spectra can thus provide a measure of interactions within single molecules, which is exactly the kind of knowledge most desirable. Unfortunately there are many difficulties which interfere. Of them the difficulty of isolating easily assignable absorption bands in relatively complicated macromolecules is a major one. Another problem is to find a suitable solvent. Proteins, and nucleic acids, for instance, dissolve most readily in water, but this solvent, as has already been mentioned, is an intense absorber of infrared radiation over many of the interesting frequency regions. Moreover, water dissolves sodium chloride and other materials which are the best (i.e., most transparent) materials for construction of absorption cells or windows for some frequency ranges.

In any event, the infrared spectroscopy of dilute solutions of macromolecular substances is still in its infancy, and this technique has so far made only isolated contributions to our knowledge of molecular structure.

One interesting example is provided by the studies of Ambrose and Elliott[76] on poly-γ-benzyl-L-glutamate, to which reference has already been made. This polymer was shown by x-ray diffraction to be able to exist in a helical form (the α-form) in which each NH group provides an intramolecular hydrogen bond to a CO group located 3.6 monomer units further along the polypeptide chain. The infrared spectra of solid films containing this structure show, as expected, that the NH absorption band has a frequency characteristic of hydrogen-bonded NH groups. However, this observation does not provide any real support for the α-helical configuration because all amides appear to have intermolecular hydrogen bonds in the solid state, and even in concentrated solutions. Proof that the hydrogen bonds of poly-γ-benzyl-L-glutamate are intramolecular (and that they persist in solution) is, however, provided by Ambrose and Elliott's spectra of dilute solutions of this polymer in chloroform and in carbon tetrachloride. They found that the NH stretching absorption continued to appear at 3300 cm^{-1}. In simple amides, on the other hand, where hydrogen bonding is definitely intermolecular, absorption at or below 3300 cm^{-1} disappears on dilution and a single band at 3440 cm^{-1}, characteristic of free NH groups, appears in its place.

5d. Infrared dichroism. In the study of solid materials a considerable increase in the usefulness of infrared spectra can be achieved by the use of polarized incident radiation. In such radiation the electric field vector, instead of being oriented at random in the plane perpendicular to the direction of propagation, is directed along a particular line in this plane. The energy which polarized radiation provides is therefore also directed along this line.

In an oriented solid, the direction of the dipole moment change accompanying a particular vibration may be uniquely fixed. If this direction coincides with the direction of polarization of incident radiation of the correct frequency for absorption, transfer of energy from the radiation to the solid will occur. If the direction of the dipole moment change is at an angle to the direction of polarization, the probability for energy transfers, i.e., for absorption, will be decreased as the component of the electric vector of the radiation in the direction of the dipole moment change is decreased. If the two directions are perpendicular, no absorption will occur at all. It is thus clear that oriented solids exposed to polarized infrared radiation may exhibit *dichroism*, which is a term signifying a dependence of the absorption characteristics on the direction of orientation of the solid.

This property may be utilized in two ways. If the structure of a macromolecular solid has been established and if it predicts that some vibrations will be uniquely oriented, dichroism will facilitate assignment of particular absorption frequencies to these vibrations. If, on the other hand, assignment has been possible without the use of dichroism, dichroism will give important structural information; i.e., it will tell whether a particular vibration is uniquely oriented and, if so, in what direction.

As an example of the many instances of the use of dichroism in the assignment of absorption frequency (for others see Liang and coworkers[86]) we may consider an as yet unsolved problem in the spectrum of solid films

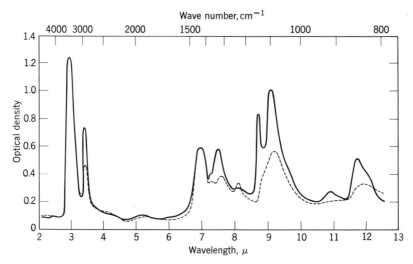

Fig. 5–4. Infrared dichroism of polyvinyl alcohol. The electric vector of the radiation is perpendicular to the direction of molecular chains in the solid curve and parallel in the broken curve. (Tadokoro et al.[88])

of polyvinyl alcohol.[87,88] As shown in Fig. 5–4 this polymer has two strong absorption bands in the region of 1000 to 1200 cm^{-1}, one occurring at 1096 cm^{-1}, the other at 1144 cm^{-1}. In this frequency range only two kinds of vibration (compatible with the atomic groups of polyvinyl alcohol) absorb infrared radiation; they are the C—C and C—O stretching vibrations. The former are normally quite weak in long chain polymers, and the latter are normally strong, so that the preferred assignment is for the C—O vibration. In that case, however, why are there two bands for just one kind of C—OH side chain?

Investigation of this region of the spectrum by polarized radiation confirms the assignment of both bands to the C—O vibration, for both bands are diminished in intensity when the radiation is parallel to the direction of extended polymer chains in the crystalline regions of the polymer, and they are enhanced when the radiation is perpendicular to this direction. This is, of course, the expected effect for C—O bonds extending laterally from the polymer chains. It is noted further that the dichroism is much more complete for the 1144 cm^{-1} band than for that at 1096 cm^{-1}. This suggests that the 1144 cm^{-1} band is associated only with C—O bonds in the crystalline regions of the polymer. (The polymers used were about 50% crystalline.) In the non-crystalline regions there would be no preferred direction for the polymer chain even in an oriented or stretched film, and therefore no dichroism would be expected. This still does not explain, of course, why two kinds of C—O bonds should be present in the crystalline regions and only one in the non-crystalline regions. Krimm, Liang, and Sutherland[87] have suggested that some dehydration occurs in formation of the crystalline regions, with the appearance of —C—O—C— cross-links, and that these cross-links are responsible for the 1144 cm^{-1} absorption. In support of this suggestion they show that the 1144 cm^{-1} band is unaffected by deuteration, which converts all OH groups into OD groups and which shifts the 1096 cm^{-1} band to 1052 cm^{-1}. However, alternative suggestions have been made by Tadokoro and coworkers.[88]

For one of the best examples of the use of dichroism to investigate structure we turn again to the work of Ambrose and Elliott[76] on poly-γ-benzyl-L-glutamate. Figure 5–5a shows the infrared spectra for the α-form of this polymer. It shows that the 1530 cm^{-1} band, previously assigned to an N—H deformation vibration, is greatly enhanced if the electric vector of the radiation is *perpendicular* to the direction of extended polypeptide chains. Since the change in dipole moment which accompanies a deformation vibration is at right angles to the direction of the bond, the N—H bond itself must be *parallel* to the direction of the polypeptide chains. The 3300 cm^{-1} band, assigned to the N—H stretching vibration, is greatly enhanced if the polarization is *parallel* to the molecular orientation, again showing that N—H bonds are parallel to this direction. The 1659 cm^{-1} band, earlier assigned to the C=O stretching vibration of the peptide bond, is also enhanced by parallel polarization, showing that most of the peptide C=O bonds are also parallel to the molecular orientation. Finally, the 1737 cm^{-1} band, previously shown to belong to the ester C=O group in the side chain, shows no appreciable dichroism at all. These conclusions are clearly strong supporting evidence for the α-helical structure shown in Fig. 4–7, in which all CO and NH groups participate in —C=O $\cdots$ H—N— $\cdots$ hydrogen bonds parallel to the helix axis. The side chains in this structure would be expected to pack between the helices without preferred orientation.

In sharp contrast are the spectra obtained from silk, which are shown in Fig. 5–5b. The peaks are not as clearly defined, but it may be taken for

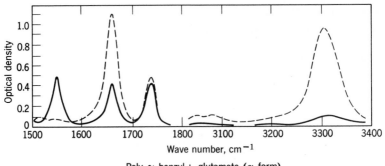

Poly-γ-benzyl-L-glutamate (α-form)

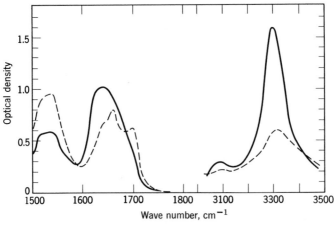

Silk fibroin (degummed silk)

Fig. 5–5. Infrared dichroism of the α-form of a synthetic polypeptide and of degummed silk (consisting principally of the protein silk fibroin). The solid curves show the absorption spectra when the electric field vector of the incident radiation is perpendicular to the direction of the molecular chains. The dashed curves show the corresponding spectra with the electric field vector parallel to the direction of the molecular chain. (Data of Ambrose and Elliott.[76,89])

granted that much of the absorption near 3300 cm⁻¹ is again assignable to the stretching vibration of peptide N—H groups, that near 1550 cm⁻¹ to the deformation of the same groups, and that near 1650 cm⁻¹ to the peptide C=O stretching vibration. The dichroism is exactly the opposite of that found in Fig. 5–5a, so that it can be concluded that most of the C=O and N—H bonds in silk fibroin are perpendicular to the extended peptide chains. This confirms the assignment of the β-structure (Fig. 4–8) to silk fibroin, in agreement with x-ray diffraction data.

It should be noted that silk fibroin is unique among proteins in that 75% of its constituent amino acid residues are glycine and alanine, so that 75% of the side chains are simply H atoms or CH_3 groups, which do not absorb in the 1500 to 1800 cm^{-1} or 3300 cm^{-1} regions. It is for this reason that its spectrum is simpler, and easier to interpret, than the spectra of most proteins.

Ambrose and Elliott have also studied keratins[89] and some globular proteins.[79] As expected, the results are not as conclusive as those shown in Fig. 5–5.

5e. Ultraviolet and visible spectra of organic molecules.[7] Ultraviolet and visible spectra are a measure of the energy involved in electronic transitions. In the gaseous state such spectra often consist of a series of sharp peaks, extending over a wavelength range of as much as 1000 Å. The fact that a series of peaks is obtained may be due partly to the existence of several closely related excited states. Part of the reason also lies in the fact that electronic transitions may occur with or without simultaneous changes in vibrational energy. The resulting "fine structure" can be observed as a series of sharp peaks because, in the gaseous state, individual molecules absorb light without interference by neighboring molecules.

Electronic spectra are extremely sensitive to intermolecular interaction. One consequence of this is that ultraviolet and visible spectra in the liquid state invariably consist of broad bands   The fine structure is washed out by the variability in the energy of interaction with neighboring molecules. In addition, intermolecular interaction may also lead to sizable differences in the locations of absorption bands; e.g., a particular band may lie at a different frequency in solution in different solvents.

A certain amount of sharpening of electronic spectra in solution can be obtained under special conditions. One of them involves variation in the length of exposure of dilute solutions to the incident light[92]; the more obvious technique is the use of a very low temperature.[93]

There have been very few studies of electronic spectra of solids. Infrared spectra of solids are easily obtained, by suspending powdered crystals in a suitable medium (e.g., in a pellet of KCl, a substance which is transparent to infrared radiation over a considerable range of frequency). Such suspensions, however, consisting of very large particles, much larger than even the largest macromolecules, would be powerful *scatterers* of radiation. As will be shown in Chapter 5, the intensity of scattering varies with wavelength as $1/\lambda^4$ and is thus enormously greater for visible and ultraviolet light than for infrared. In fact scattering of light is sufficiently strong so as to obliterate any absorption which might occur in such suspensions of powdered solids. Electronic spectra of low-molecular-weight solids can thus be measured on single crystals only, and this fact has greatly limited experimental investigations in this field.

Electronic spectra are as sensitive to intramolecular changes as to intermolecular interactions. The spectrum of toluene, for example, is very different from that of benzene.[90] There are distinctly observable differences[90] even between molecules as similar as

$$
\begin{array}{ccc}
& & \text{CH}_3 \\
& & / \\
\bigcirc\!\!-\text{CH}_2\!-\!\text{CH} & \quad\text{and}\quad & \bigcirc\!\!-\text{CH}_2\!-\!\text{CH}_2\!-\!\text{CH}_2\!-\!\text{CH}_3 \\
& & \backslash \\
& & \text{CH}_3
\end{array}
$$

although the electronic transitions are the same in both cases and involve only the electrons of the benzene ring. No quantitative theoretical interpretation of such changes is at present possible.

Among small organic molecules, those which have the most intense observable electronic spectra are molecules with conjugated double-bond systems. They ordinarily absorb ultraviolet light below 3000 Å in wavelength. They are usually studied in liquid solution and in that state possess absorption bands so broad as to extend to the shortest wavelengths experimentally attainable, which is about 2000 Å for most spectrophotometers (most substances, including air and quartz begin to absorb near 2000 Å). Differences in the shapes of absorption bands are not easily related to structure. Small changes in structure often lead to as pronounced differences as exist between compounds of quite different structure, as is shown, for example, by the spectra of cytosine and tryptophan, below.

Many small molecules of interest to the present discussion are acids or bases, and their spectra are greatly affected by pH. The pyrimidine cytosine, for example, has two hydrogen ions which dissociate between pH 2 and pH 13,

$$
\begin{array}{ccccc}
\text{NH}_2 & & \text{NH}_2 & & \text{NH}_2 \\
| & & | & & | \\
\text{C} & & \text{C} & & \text{C} \\
^{+}\text{HN}\diagup\quad\diagdown\text{CH} & \rightleftharpoons & \text{N}\diagup\quad\diagdown\text{CH} & \rightleftharpoons & \text{N}\diagup\quad\diagdown\text{CH} \\
| \qquad \| & & | \qquad \| & & | \qquad \| \\
\text{O}\!=\!\text{C}\quad\text{CH} & & \text{O}\!=\!\text{C}\quad\text{CH} & & ^{-}\text{O}\!-\!\text{C}\quad\text{CH} \\
\diagdown\diagup & & \diagdown\diagup & & \diagdown\diagup \\
\text{NH} & & \text{NH} & & \text{N} \\
\text{I} & & \text{II} & & \text{III}
\end{array}
$$

The three possible forms I, II, and III all have different spectra, as Fig. 5–6 shows. This figure also shows the close similarity between the

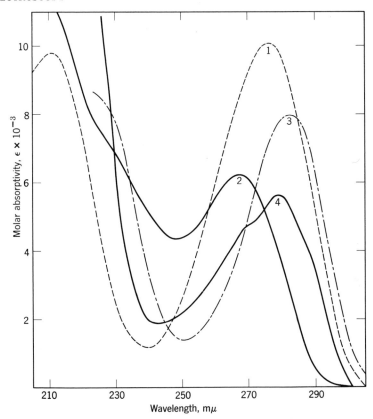

Fig. 5–6. Absorption spectrum of cytosine in three different ionization states. Curve 1 is form I (pH 1); curve 2 is form II (pH 7.2); curve 3 is form III (pH 14). Curve 4 is the absorption spectrum of the entirely unrelated compound tryptophan. (Cytosine spectra from Shugar and Fox.[91])

spectrum of form II and that of the completely unrelated compound, tryptophan

Even weak electrostatic forces may affect the ultraviolet absorption spectrum. For example, tryptophan[92] has two acidic groups, but they are far removed from the conjugated double-bond system. Nevertheless, when they are titrated, small but easily observable changes in spectrum occur. This is presumably due to the influence of the electrostatic charges of the $-NH_3^+$ and $-COO^-$ groups on the energy levels of the indole ring system.

Among organic molecules which absorb visible light are the porphyrins and many complexes of transition metals. Of particular interest, because they may occur in nature incorporated into protein molecules, are the *hematin* compounds. They are iron complexes of porphyrins, and their spectra in the visible and ultraviolet region are intense and complicated, being derived in part from the characteristic absorption of the porphyrin conjugated double-bond system and in part from the absorption of the iron atom.

A very complete account of the properties of these compounds, including their spectra, has been given by Lemberg and Legge.[102] The spectra vary in an erratic manner, as do most electronic spectra, but Theorell[103] has suggested that the compounds may be divided into four classes, the division being based on the valence of the iron atom and on the kind of bonding in the two coordination positions of the iron atom which are not used by the iron-porphyrin bonds. Certain similarities characterize each class. The four classes are (I) Fe^{+++} with ionic bonding, (II) Fe^{+++} with covalent bonding, (III) Fe^{++} with ionic bonding, and (IV) Fe^{++} with covalent bonding.

All porphyrin compounds also have a strong absorption band near 4000 Å. This band is generally called the *Soret band*, after J. L. Soret who discovered its existence in hemoglobin.

5f. Ultraviolet and visible spectra of macromolecules. It is clear from the examples of the preceding paragraph that electronic spectra cannot be used, as infrared spectra are, to give direct information concerning the structure of macromolecules. This is unfortunate because ultraviolet and visible spectra can be determined with high experimental precision, both as to wavelength and intensity. Moreover, macromolecules ordinarily contain just a few kinds of atomic groups which give intense absorption in the visible or ultraviolet range, so that the assignment of absorption bands to particular atomic groups is ordinarily easy even for complicated molecules. It is therefore unfortunate that we cannot interpret, in terms of structure, the differences which are observed between macromolecular spectra and spectra of small molecules containing similar atomic groups.

The inability to interpret electronic spectra of macromolecules theoretically has not prevented the accummulation of a large amount of experimental data, for these spectra prove useful in a number of ways unrelated or only indirectly related to structure. They may, for instance, be used as indicators of purity or as a measure of concentration. Rates and equilibria of many reactions have been studied by using intensities of appropriate absorption bands as a measure of concentration. Furthermore, changes in absorption spectrum may often be one of the first indications of structural change, even though they are unable to give a clue as to the nature of the change. A few examples are presented below to illustrate the type of information which has been obtained in this way.

Nucleic acids are polymers of nucleotides, these being molecules obtained by combination of a purine or pyrimidine base with a sugar (ribose or deoxyribose) and a phosphate group (cf. Fig. 2–3). The spectra of the nucleotides are very close to those of the parent purines or pyrimidines, e.g., the spectrum of the nucleotide cytidylic acid is very close to that of cytosine. However, the nucleic acid spectrum is not the same as that of the sum of its constituent nucleotides. In typical DNA preparations the intensity of absorption may be 40% less than is observed in a mixture of the corresponding nucleotides.[94,95] This effect is known as *hypochromism*, and a theoretical explanation for it has been given by Tinoco.[157] This explanation is based on the interaction between the absorbing purine and pyrimidine bases when these bases are in a regularly ordered array, such as the helical structure of Watson and Crick (Fig. 4–13).

The ultraviolet absorption spectra of proteins, between 2500 and 3000 Å, rest on an even simpler basis.[92] Absorption in this region is due almost entirely to the indole side chains of tryptophan and to the phenolic side chains of tyrosine. The phenyl side chains of phenylalanine also absorb in this region, but their molar absorbance is much lower. The spectra of proteins are much closer to those computed from the sums of the spectra of the contributing side chains than is true in the case of nucleic acids. The protein spectrum is generally shifted slightly (ca. 30 Å) towards longer wavelengths, but individual differences are so small that the absorbance at suitable wavelengths may actually be used as the basis for analysis for the number of tryptophan and tyrosine side chains.[92] The relatively small difference between the spectra of proteins and those of their constituent side chains probably means that these side chains are not in a regularly ordered array, in agreement with the conclusion reached earlier on the basis of x-ray diffraction. The small spectral differences which are observed may simply reflect the difference in the environment of the chromophoric side chains, many of which are likely to be in the interior of the native protein molecule, removed from intimate contact with the solvent.[158,159]

One of the protein side chains which absorb in the ultraviolet, the phenolic side chain, has a dissociable proton. Ultraviolet absorption spectra of proteins therefore generally show a sizable transition with pH which occurs somewhere in the alkaline pH range (in water typically between pH 10 and 13) and corresponds to the reaction

$$-CH_2-\!\!\bigcirc\!\!-OH \;\rightleftharpoons\; -CH_2-\!\!\bigcirc\!\!-O^- + H^+$$

The spectral change, illustrated by Fig. 5–7, is very similar to that which occurs when small molecules containing phenolic groups lose a proton.

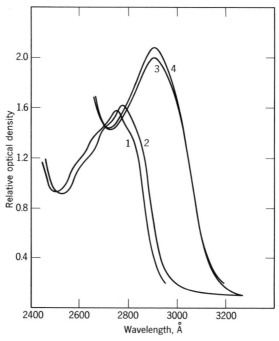

Fig. 5–7. Ultraviolet absorption spectrum of a typical protein molecule (serum albumin, curves 2 and 4) and of the amino acids obtained after degradation (curves 1 and 3). Curves 1 and 2 were obtained in acid solution, where the phenolic groups are un-ionized; curves 3 and 4 in alkaline solution where the phenolic groups exist as phenolate ions. (Beaven and Holiday.[92])

The ionization of phenolic groups and the accompanying spectral change provide important indirect information concerning structure differences between different proteins. In ovalbumin, for example,[96] this spectral change is not observed at all under the conditions typically used. It occurs only if the protein is exposed for several minutes to a pH of 13. Under these conditions drastic alterations in the structure of the protein are known to occur. It must be concluded that the phenolic side chains of native ovalbumin are different from those found in other proteins in that they are buried in the interior of the molecule or are in some other way made inaccessible to the water molecules or hydroxyl ions which are needed to pull off the dissociable proton. In another protein, ribonuclease,[97] there are six phenolic groups, three of which behave normally,

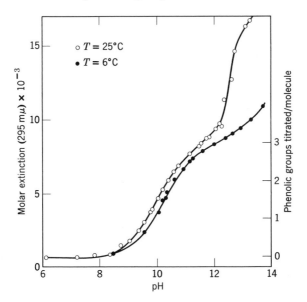

Fig. 5–8. Ionization of the phenolic groups of ribonuclease, as measured by ultraviolet light absorption at a wavelength (295 mμ) at which the ionized form has an absorption maximum, as shown in Fig. 5–7. The ionization curves show that three of the six phenolic groups ionize far less readily than the other three. (Tanford et al.[97])

whereas the other three behave like those of ovalbumin. At 6° C the latter three are not ionized even at pH 14. The data for ribonuclease are shown in Fig. 5–8.

The ultraviolet absorption spectra of proteins sometimes undergo small alterations at pH values other than those characteristic of the ionization of phenolic groups.[98] Such alterations might reflect structural changes, for, as has been mentioned, such changes are not likely to alter the spectrum greatly. On the other hand an alteration in the spectrum may merely indicate that a nearby ionic side chain is being titrated, in the same way as the titration of the amino and carboxyl groups of tryptophan (p. 91) affects the spectrum of the indole ring.

Mention should be made, finally, of the heme proteins,[102] all of which absorb light in the visible range of the spectrum, owing to the fact that metal-porphyrin complexes form part of their structure. Best known among them is hemoglobin. The spectra of this protein and of four of its derivatives are shown in Fig. 5–9. The portions of these spectra which lie between 5000 and 7000 Å fall more or less into the following classes of

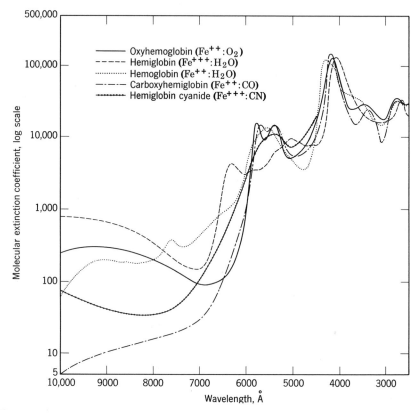

Fig. 5–9. Spectra of hemoglobin and some of its derivatives. (Lemberg and Legge.[102])

the Theorell classification (p. 92). Hemoglobin itself is in class III; its oxygen and CO complexes are both in class IV; the oxidized form, ferrihemoglobin, falls into class I; and the CN^- complex of this form is in class II. This classification is in accord with the known conditions of the iron atoms in these compounds. However, the differences between these spectra and those of hematin compounds which fall into the same classes are considerable. As a result the absorption spectra are not regarded as a reliable guide to the structure of heme proteins in general.

Three features of Fig. 5–9 should be noted which are as yet unexplained in terms of structure. They are the pronounced differences in the near-infrared region (above 7000 Å), the band near 3400 Å in the CO and O_2 complexes, and the small differences in the Soret band near 4000 Å. If a structural interpretation of these features were possible, it would add considerably to our knowledge of hemoglobin structure and would provide a powerful tool for the investigation of the many other heme proteins which are known.

5g. Visible and ultraviolet dichroism. Ordered macromolecular solids would normally be expected to show dichroism, in their visible and ultraviolet absorption, and this proves to be the case (e.g., single crystals of hemoglobin show marked dichroism[99]). However, there have been very few studies of this effect, not only because of experimental difficulties in the determination of electronic spectra of solids but also because of the difficulty in interpreting the results. If, as is usually the case, the dipole moment change which accompanies an electronic transition has an unknown direction with respect to molecular structure, a measure of dichroism cannot be related to molecular structure.

For very asymmetric molecules dichroism may be studied in solution. Such molecules may be oriented if the solutions containing them are placed between two closely spaced cylinders, rotating with respect to one another (the apparatus is the same as for flow birefringence; see section 25). The velocity gradient so set up will orient highly asymmetric molecules so that their long axes are parallel to the lines of flow. We can then measure the absorption of light, polarized so that the electric vector is also parallel to the lines of flow and, hence, parallel to the long axes of the molecules. The observed absorption can then be compared with that obtained when the direction of polarization is perpendicular to the long axes.

One such study has been made, for example, on the blue complex formed between I_2 and the amylose fraction of starch (p. 12).[100] Enhancement of absorption was observed when the electric vector was parallel to the flow lines. The fact that dichroism was observed at all shows that the molecules in the solution must be rodlike, and Rundle and Baldwin[100] suggested that they are long helical rods. Furthermore, studies of the dichroism of I_2 crystals indicated that the dipole moment change in the electronic transition of this molecule which causes the absorption is in the direction of the I—I bond (i.e., the excited electronic state has an I—I bond with partial ionic character). Thus the fact that absorption by the amylose-I_2 complex is enhanced when the electric vector of the light is parallel to the helix axis means that I_2 molecules must also be parallel to the helix axis. Rundle and coworkers suggested the over-all structure shown in Fig. 5–10. The starch helix contains six glucose units per turn and has a hole down its center into which I_2 molecules will fit if they are placed lengthwise.

The amylose-I_2 complex has also been examined by x-ray methods.[101] The unit cell is hexagonal and its dimensions support the structure suggested in Fig. 5–10.

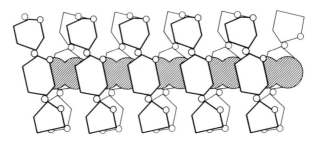

Fig. 5–10. Proposed helical structure for the amylose–iodine complex. The dark spheres represent iodine atoms. (Rundle and Baldwin.[100])

5h. Nuclear magnetic resonance. Three kinds of energy transitions may be observed by the absorption of low frequency radiation in the range of $\lambda = 1$ cm and above. They are (1) the pure rotational transitions between rotation levels of molecules in the gas phase, (2) the transitions between the energy levels created by the interaction of an external magnetic field with the spin of unpaired electrons (*electronic magnetic resonance* or *paramagnetic resonance*), and (3) the corresponding transitions resulting from the interaction of an external magnetic field with the spin of nuclei (*nuclear magnetic resonance*).

Three kinds of spectra can thus be obtained with microwave or radio-frequency radiation. The utility of pure rotational spectra, however, is confined to gaseous substances, and that of electron magnetic resonance spectra to free radicals. Neither of them is applicable to ordinary organic molecules or to macromolecules, although electronic magnetic resonance spectra can be obtained if free radicals are produced by irradiation with x-rays. Nuclear magnetic resonance spectra, however, are of general usefulness, and they are becoming increasingly important in the elucidation of the structure of all kinds of molecules.

Nuclear magnetic resonance absorption can occur only for those nuclei which have non-zero spin. Among such nuclei are H^1 ($I = \frac{1}{2}$), H^2 ($I = 1$), C^{13} ($I = \frac{1}{2}$), N^{14} ($I = 1$), O^{17} ($I = \frac{5}{2}$), and P^{31} ($I = \frac{1}{2}$). (In this listing the superscript represents the isotopic mass number, and I is the nuclear spin.) Nuclei which have zero spin include C^{12}, C^{14}, O^{16}, and O^{18}. Because the

C^{13} and O^{17} content of natural carbon and oxygen is extremely low, nuclear magnetic resonance can give information about H, N, and P atoms but not about C or O atoms.

The energy of interaction between a nucleus of spin I and a magnetic field H may take on $2I + 1$ values, given by equation 5–2,

$$E = M\mu H/I \qquad (5\text{-}2)$$

where μ is the nuclear magnetic moment and M is a quantum number which can have the values $I,\ I - 1, \cdots,\ 0,\ \cdots,\ -(I - 1),\ -I$. The selection rule for absorption is that $\Delta M = 1$, so that the absorption frequency, at a given magnetic field strength, is, by equation 5–1, $\nu = \mu H/hI$. In the experimental detection of absorption the frequency ν of the incident radiofrequency is usually kept constant, and H is varied instead. Absorption occurs when

$$H = \nu hI/\mu \qquad (5\text{-}3)$$

Nuclear magnetic resonance is of interest to the chemist because of three easily observable effects:

(1) The magnetic field to which the nucleus is subject is not the same as the applied field because of shielding by the surrounding electrons. The shielding depends, however, on the chemical bonding of the nucleus under consideration. As a result, for example, the protons of $-NH_2$, $-CH_2-$, $-CH_3$, etc., all absorb at slightly different applied fields. The resulting *chemical shifts* in the critical value of H for absorption may be used as a measure of the type of bonding of the absorbing atom to its neighbors.

(2) If nearby nuclei both possess spin and if they are joined by covalent bonds, interaction between the spins occurs. This splits individual absorption bands into several peaks, the number of peaks for any one nucleus being equal to the number of possible spin orientations of interacting nuclei. This effect can provide a sensitive measure of structure near absorbing nuclei.

(3) Nuclear magnetic resonance absorption peaks in liquids are very sharp, in contrast to what is observed for infrared and electronic spectra. The reason for this difference is that the lifetime of a nuclear spin state is very long in comparison with electronic or vibrational energy states. The effect of interaction with nearby molecules, which, in liquids, perturbs the energy in a given state in a manner which varies from instant to instant, is therefore averaged out to a constant value during the lifetime of a nuclear spin state, whereas this is not true for an electronic or vibrational energy state. In amorphous solids the absorption peaks will be as sharp

as in liquids. In crystalline solids which have immobile molecules, however, there is no corresponding averaging out of interactions, and a broadening of absorption bands occurs. This effect is clearly of great importance in the investigation of the solid state of macromolecules and suggests, for instance, the use of nuclear magnetic resonance to determine the extent of crystallinity in solid polymers.

The last effect is complicated by the fact that sharp peaks may still be observed in solids which are crystalline by all other criteria. The molecules in these crystals apparently can rotate or invert without disturbing the crystal structure. In crystals of this type there are usually transitions at low temperature to a completely immobile state, which are accompanied by sudden broadening of the absorption peaks.

Nuclear magnetic resonance was not discovered until 1946, and, to date, only isolated studies of macromolecules have been reported.[104,105,106] None of them has led to conclusions of general interest. It seems certain, however, that this method will in the future make important contributions to our knowledge of macromolecular structure.

6. OTHER METHODS FOR STRUCTURE DETERMINATION

6a. Electron microscopy.[9] The electron microscope is a device which utilizes the wave properties of matter. A beam of electrons is accelerated in an electric field until the electrons have a velocity of about 10^9 cm/sec or higher, and a corresponding momentum of about 10^{-18} gram cm/sec (the mass of an electron being 9×10^{-28} gram). By the well-known de Broglie relation this beam has associated with it wave properties with a wavelength of

$$\lambda = \frac{h}{mv} = 6.6 \times 10^{-9} \text{ cm (or smaller)}$$

where h is Planck's constant, 6.6×10^{-27} erg-sec. These waves can be focused and can be used to form images of objects with a resolution of about the same order of magnitude as the wavelength λ. Theoretically, then, an electron microscope can "see" objects of atomic dimensions. This theoretical resolving power cannot be achieved in practice, for the electron beam is non-selective. It provides what is essentially a picture of electron density. Since objects to be examined must be placed on some kind of supporting medium, the real factor which limits resolution is the ability to distinguish the object from its supporting surface.

To observe objects of molecular dimensions at all it is necessary either that they possess higher electron density than the supporting surface, which for particles containing chiefly carbon and hydrogen atoms is impossible, or that replicas of the objects be prepared which possess high electron density. The latter goal can be attained by the method of shadow-casting. In this procedure a beam of metal atoms evaporated from a heated filament is directed obliquely at the specimen (Fig. 6–1). A uniform film of metal is formed on the surface, except where the macromolecules (or other objects of interest) rise above the surface. Here there will be a heavy accummulation of metal atoms behind the object and a shadow in front of it. The object thus stands out in the electron micrograph, and its dimensions can be computed from the known direction of the incident beam of metal atoms. If this method is to be successful, it is, of course, essential that the surface on which the particles of interest are mounted be perfectly smooth. The best data on individual molecules have been obtained by using a surface of freshly cleaved mica.

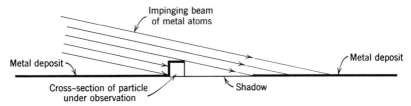

Fig. 6–1. Shadow casting in electron microscopy. The length of the shadow in the direction of the beam of metal atoms is equal to $h \cot \theta$, where h is the height of the particle and θ the angle at which the metal atoms strike it.

Electron microscopy initially found its greatest use in the examination of structures that are larger than single molecules, such as fibers of muscle and connective tissue. The electron micrographs of such structures form the basis for hypotheses concerning the arrangement of muscle protein or collagen molecules in these living tissues. However, as improvements in resolution have occurred, some of the larger individual molecules have fallen within the scope of the technique. Some representative electron micrographs are shown in Fig. 6–2.

Figure 6–2a, for example, shows a micrograph of tobacco mosaic virus. The rod-shaped virus molecules appear to have a uniform diameter of about 150 Å and a length of about 3000 Å. Tomato bushy stunt virus (Fig. 6–2b) and tobacco necrosis virus (Fig. 6–2d), by contrast, appear to be spherical, and tobacco ringspot virus (Fig. 6–2c) is clearly polyhedral. It will be noted that x-ray diffraction (Table 4–1) also indicated that tomato bushy stunt and tobacco necrosis virus are spherical, or close to it. However, this conclusion was reached only from the unit cell dimensions and symmetry. A polyhedral shape could not have been distinguished from a spherical shape in this way.

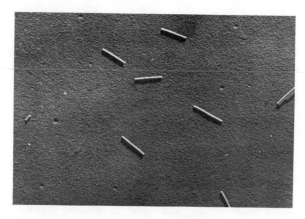

(*a*) Tobacco mosaic virus. (R. C. Williams.[147])

(*b*) Tomato bushy stunt virus. (R. C. Williams and R. C. Backus.[107])

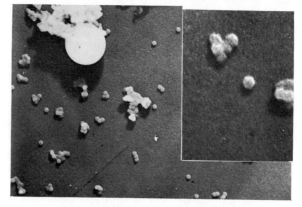

(*c*) Tobacco ringspot virus. (R. L. Steere.[109])

(*d*) Tobacco necrosis virus. (R. W. G. Wyckoff.[110])

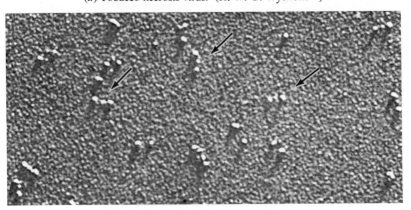

(*e*) Fibrinogen. (C. E. Hall and H. S. Slayter.[160])

Fig. 6-2. Representative electron micrographs. The large particles in parts *b* and *c* are polystyrene latex particles of diameter 2500 Å. The wedge-shaped sector of part *b* and the insert in part *c* are magnified portions of the micrographs. All the figures are reproduced from photographs provided through the courtesy of the authors.

Figure 6–2d is of special interest because it is an electron micrograph of a single crystal. It provides visual proof for the type of crystal structure which ordinarily can be envisaged only indirectly from x-ray diffraction patterns. In the case of this virus, x-ray diffraction patterns do in fact lead to a structure very similar to that shown in Fig. 6–2d.

It is only quite recently that the technique of electron microscopy has been refined to the point of being able to pick out individual molecules smaller than the virus particles just discussed. Hall however has recently been able to photograph considerably smaller particles. For example, electron micrographs of collagen[108] show that collagen molecules are rods, with an average length of about 2500 Å and a diameter of only 15 Å. These dimensions are compatible with the molecular structure proposed in section 4 on the basis of x-ray diffraction. Another example is provided by the electron micrographs of fibrinogen[160] shown in Fig. 6–2e, which lead to the molecular dimensions given in Fig. 6–3.

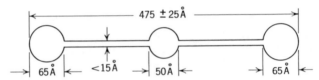

Fig. 6–3. The dimensions of the fibrinogen molecule, as deduced from the electron micrograph of Fig. 6–2e. (Hall and Slayter.[160])

Electron microscopy can be used to determine molecular weights, simply by counting the number of particles obtained from a known volume of solution containing a known weight concentration of the substance under consideration. The major source of error in this technique is the possibility of fragmentation or aggregation of molecules during transfer to the supporting medium. (This transfer involves spraying a dilute solution onto the supporting surface, followed by freezing and drying.) If fragmentation or aggregation occurs, it is, however, easily discovered by direct observation of anomalous particles in the electron microgram. Williams and Backus[107] used the counting technique on the data of Fig. 6–2b and computed a molecular weight of 9.4 ± 0.7 million for bushy stunt virus. The x-ray molecular weight for the same virus (Table 4–1) is 10,800,000. Solution measurements (Table 21–3) give 10,700,000.

Another valuable use of electron microscopy is direct observation of the extent of homogeneity of a preparation of macromolecules. Figure 6–4, for example, shows the distribution of lengths found in a sample of tobacco mosaic virus, 201 particles of which were measured.[125] Seventy-five per cent of them had a length between 2900 and 3100 Å, and 25% had lengths outside this range, more or less randomly distributed. This is not the kind of distribution which is expected from a molecule which is *naturally* heterogeneous (see Chapter 3). It may be assumed therefore that the active virus is represented by the 75% of the particles with lengths near 3000 Å and that the remaining particles are inactivated ones, damaged in the living plant from which they were taken or during the process of isolation and electron microscopy. Whether or not the variation between 2900 and 3100 Å, among the major (75%) fraction of the molecules, is a

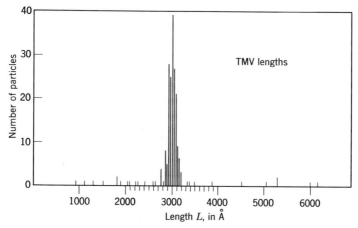

Fig. 6–4. The distribution of particle lengths in a sample of 201 molecules of tobacco mosaic virus. (Hall.[125])

real allowed variation in active virus particles cannot at present be decided. It, too, could represent minor artefacts produced during laboratory manipulation.

6b. Dielectric constants and dipole moments.[11] The dielectric constant of a medium is a measure of its *polarizability*, i.e., of the polarization produced in the medium by the action of an electric field. As is customary, we interpret it at the molecular level, attributing the polarizability to an average contribution α from each molecule, α being called the *molecular polarizability*. We shall be interested particularly in the molecular polarizability of macromolecules in solution.

Molecular polarizability may be ascribed to the following phenomena:

(1) Distortion of the charge distribution within the molecule, i.e., perturbation of the location of electrons relative to nuclei. This factor makes a relatively small contribution to the dielectric constant. It is the

sole cause of polarizability for molecules that possess no permanent dipole moment, and material consisting of such molecules thus always has a low dielectric constant.

(2) For molecules that possess a permanent dipole moment the major contribution to α arises from the orientation of the dipoles by the electric field. In the absence of the field the dipoles are randomly oriented, producing, on the average, no polarization. In the presence of the electric field, orientation occurs, with resulting polarization.

(3) In ionic media an additional contribution is due to the displacement of ions relative to one another, and this factor must also make a contribution to the *molecular polarizability* of molecules that contain actual charges, such as aspartic acid,

$$^+NH_3 - CH_2 - COO^-$$
$$|$$
$$CH_2 - COOH$$

in which a proton can move back and forth between the two carboxyl groups. It is generally believed that this factor makes a much smaller contribution to the dielectric constant of polar molecules (the only kind of molecule in which it can arise) than does the orientation factor.

We shall be concerned here with the use of dielectric measurements to determine the dipole moments of polar macromolecules; i.e., we shall be interested only in the second of the three phenomena enumerated above. The contribution which this factor makes to the molecular polarizability can be computed, using a method first employed by Debye.[127] Let the dipole be represented by a rod of length $2a$, with a charge $+q$ at one end and $-q$ at the other, so that the dipole moment μ is

$$\mu = 2aq \tag{6-1}$$

Let ϕ represent the angle between the axis of a particular dipole and the direction of the electric field $\mathscr{E}$, as shown in Fig. 6–5. The field will exert a force $\mathscr{E}q$ on each charge, but in opposite directions. Each force may be divided into a component $\mathscr{E}q \cos \phi$ along the dipole axis, and a tangential component $\mathscr{E}q \sin \phi$. Figure 6–5 shows that the axial components cancel, but the tangential components act in concert, producing a torque $\mathscr{T}$ in the direction of decreasing ϕ, given by

$$\mathscr{T} = -2\mathscr{E}qa \sin \phi = -\mathscr{E}\mu \sin \phi \tag{6-2}$$

The potential energy of a dipole at any orientation ϕ, with the orientation $\phi = 0°$ designated arbitrarily as the position of zero energy, can be computed as the work done in rotation of the dipole from $\phi = 0$ to $\phi = \phi$, i.e.,

$$W(\phi) = \int_{\phi=0}^{\phi} \mathscr{T} \, d\phi = \mathscr{E}\mu(\cos \phi - 1) \tag{6-3}$$

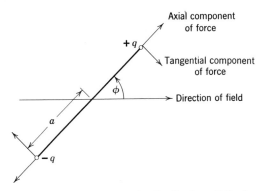

Fig. 6–5. Model used for computing the distribution of dipole orientations in an electric field.

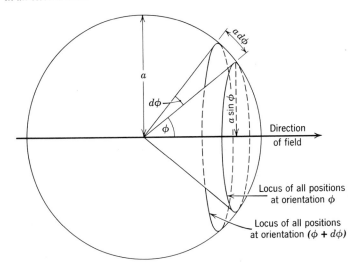

Fig. 6–6. Calculation of the purely spatial factor determining the relative probability of a given value of ϕ. The probability of a value of ϕ between ϕ and $\phi + d\phi$ is proportional to the surface area (of the sphere of radius a) which lies between the loci corresponding to orientations ϕ and $\phi + d\phi$.

The relative probability of the occurrence of any particular value of ϕ between ϕ and $\phi + d\phi$ can now be calculated, if we assume the system to be *at equilibrium*. This probability will depend on two factors. (*a*) The number of positions in three-dimensional space which are available at any value of ϕ, this being proportional to the area of surface of a sphere of radius a which is subtended between the angle ϕ and $\phi + d\phi$. This factor is clearly proportional to $2\pi a^2 \sin \phi \, d\phi$, as shown by Fig. 6–6. (*b*) The

Boltzmann factor $\exp[-W(\phi)/kT]$. The probability that the orientation lie between ϕ and $\phi + d\phi$ is thus

$$f(\phi)\,d\phi = (\text{constant})e^{-\mathscr{E}\mu\cos\phi/kT}\sin\phi\,d\phi \qquad (6\text{-}4)$$

all factors independent of ϕ, including the term $\exp(+\mathscr{E}\mu/kT)$ which arises from equation 6–3, having been incorporated in a single constant.

We wish to use equation 6–4 to calculate the average molecular polarization in the direction of the field, i.e., the average moment m in that direction. Clearly $m = \mu\cos\phi$ at any value of ϕ, and the average value of m is given at once as

$$\bar{m} = \frac{\displaystyle\int_0^\pi \mu\cos\phi\, f(\phi)\,d\phi}{\displaystyle\int_0^\pi f(\phi)\,d\phi} \qquad (6\text{-}5)$$

Using equation 6–4 for $f(\phi)$ and substituting $\cos\phi = x$, $-\sin\phi\,d\phi = dx$, we get

$$\bar{m} = \frac{\displaystyle\mu\int_1^{-1} xe^{-\mathscr{E}\mu x/kT}\,dx}{\displaystyle\int_1^{-1} e^{-\mathscr{E}\mu x/kT}\,dx} \qquad (6\text{-}6)$$

which, by Appendix A, gives $\bar{m} = \mu\mathscr{L}(\mathscr{E}\mu/kT)$, where $\mathscr{L}$ represents Langevin's function. In practice, $\mathscr{E}\mu/kT$ is always much less than unity. Under these conditions $\mathscr{L}(\mathscr{E}\mu/kT) = \mathscr{E}\mu/3kT$, and

$$\bar{m} = \mathscr{E}\mu^2/3kT \qquad (6\text{-}7)$$

The corresponding contribution to the polarizability is just $\bar{m}/\mathscr{E}$, so that the total molecular polarizability may be written as

$$\alpha = \alpha_d + \mu^2/3kT \qquad (6\text{-}8)$$

where α_d is the contribution made by the distortion factors (1) and (3) given above.

It is essential to note that the derivation of equation 6–8 depends on the assumption of *equilibrium* orientation of the dipoles. It requires that the frequency of the alternating electric field which is used to measure the dielectric constant, and, hence, α be relatively low. If too high a frequency is used, the dipoles will have insufficient time to acquire the equilibrium distribution of orientations at each stage of the alternating cycle. At sufficiently high frequencies no orientation will take place at all, and the molecular polarizability, α_∞, will become

$$\alpha_\infty = \alpha_d \qquad (6\text{-}9)$$

It remains to relate the molecular polarizability to the observable macroscopic dielectric constant, D. The basic relation between these quantities, applicable when each molecule is an isolated entity in an unpolarizable medium, is

$$D - 1 = 4\pi N\alpha \tag{6-10}$$

where N is the number of molecules per cubic centimeter. In a real gas or liquid, however, an applied field will polarize the vicinity of the molecule of interest, and the electric field which results from this polarization will be superimposed on the applied field. If α is to represent the polarizability of an individual molecule, this modification of the actual local field must be taken into account, and this can be accomplished for gases or liquids of relatively low dielectric constant by the substitution of the equation of Debye,[127]

$$\frac{D - 1}{D + 2} = \tfrac{4}{3}\pi N\alpha$$

for equation 6-10. In polar liquids, where the interaction between the field and the medium is stronger, an equation of Onsager[128] and Kirkwood[129] must be employed, it being

$$\frac{(D - 1)(2D + 1)}{9D} = \tfrac{4}{3}\pi N\alpha \tag{6-11}$$

It will be noted that both of these relations approach equation 6-10 as D approaches unity.

Any of these relations may be used to measure the molecular dipole moments in pure substances. By plotting the left-hand side versus $1/T$, we effectively plot α versus $1/T$ and, by equation 6-8, obtain μ^2 from the slope.

Another method applicable to pure substances is to use Maxwell's relation between dielectric constant and refractive index, $\tilde{n}$,

$$D = \tilde{n}^2 \tag{6-12}$$

This equation applies when D and $\tilde{n}$ are measured at the same frequency. The refractive index for visible light will thus give the value of D at very high frequency, and, hence, the value of α_∞ or α_d. This may be subtracted from α to give μ^2 by equation 6-8.

By using equation 6-11 in place of 6-10 we take into account the effect which a polar environment has on the electric field acting on an individual molecule. The polar environment must also act on the molecule itself, affecting the relation between orientation and potential energy, as given by equation 6-3. This effect has been considered by Kirkwood,[129] who has shown that it leads to a modification of equation 6-8, the term μ^2 in that expression being replaced by $\mu\tilde{\mu}$, where $\tilde{\mu}$ includes the effect of interaction with the environment. The difference between μ and $\tilde{\mu}$ is sufficiently small so that it can be ignored in the present context.

The dielectric constant measurements to be discussed below were made on dilute solutions of macromolecules in polar solvents. Equation 6–11 must be used with the right-hand side replaced by $\frac{4}{3}\pi(N_1\alpha_1 + N_2\alpha_2)$ where subscript 1 refers to solvent and subscript 2 to solute molecules. The term $N_1\alpha_1$ in this expression will be a function of N_2: first, because interaction between solute and solvent molecules may alter the average value of α_1, and, second, because the presence of solute molecules decreases the concentration N_1. In a dilute solution each solute molecule may be expected to exert an independent effect, so that $N_1\alpha_1$ varies linearly with N_2; i.e., we may write $N_1\alpha_1 = N_1{}^0\alpha_1{}^0 - bN_2$, where the superscript zero refers to the pure solvent and b is a constant. Thus we get, in place of equation 6–11,

$$\frac{(D-1)(2D+1)}{9D} = \frac{4\pi}{3}(N_1{}^0\alpha_1{}^0 - bN_2 + \alpha_2 N_2) \qquad (6\text{–}13)$$

Writing $N_2 = \mathcal{N}c/M$, where c is the macromolecular concentration in grams per cubic centimeter, M the molecular weight, and $\mathcal{N}$ Avogadro's number, and then differentiating equation 6–13, we get

$$\frac{1}{4.5}\left(1 + \frac{1}{2D^2}\right)\frac{\partial D}{\partial c} = \frac{4\pi\mathcal{N}}{M}(-b + \alpha_2) \qquad (6\text{–}14)$$

all variables other than c being kept constant during the differentiation. The term $\partial D/\partial c$ is known as the *dielectric increment*, and it is the quantity of interest. For most polar liquids $1/2D^2 \ll 1$ and may be neglected so that, with introduction of equation 6–8, we get for low frequency fields,

$$\frac{\partial D}{\partial c} = 4.5\frac{4\pi\mathcal{N}}{3M}\left(-b + \alpha_d + \frac{\mu^2}{3kT}\right) \qquad (6\text{–}15)$$

where α_d is the molecular distortion polarization and μ the dipole moment, of the macromolecules in the solution. At high frequencies,

$$\left(\frac{\partial D}{\partial c}\right)_{\text{high }\nu} = 4.5\frac{4\pi\mathcal{N}}{3M}(-b + \alpha_d) \qquad (6\text{–}16)$$

which may be positive or negative, depending on the relative magnitudes of b and α_d, and, by subtraction,

$$\frac{\partial D}{\partial c} - \left(\frac{\partial D}{\partial c}\right)_{\text{high }\nu} = 4.5\mu^2\frac{4\pi\mathcal{N}}{9MkT} \qquad (6\text{–}17)$$

This is the equation used to determine the dipole moments of dissolved macromolecules. The dielectric increments are independent of concentration when c is sufficiently small, so that the dielectric constant of a solution may be written as $D = D^0 + (\partial D/\partial c)c$, where D^0 is the dielectric constant of the solvent. Thus $\partial D/\partial c = (D - D^0)/c$, which is often written as $\Delta D/c$.

It should be noted that the temperature dependence of $\partial D/\partial c$ (equation 6–15) cannot be used to evaluate μ^2 because the term b contains interaction parameters which depend on temperature, and also because μ itself may be temperature dependent in the case of a complex macromolecule. Nor can the refractive increment $\partial \tilde{n}/\partial c$ be used to determine $\partial D/\partial c$ at high ν because the term b contains the molecular polarizability α_1 of the solvent, which will itself be frequency dependent, depending on the ability of solvent molecules to orient themselves in response to the alternating electric field. For reasons which will appear below, solvent molecules maintain

TABLE 6–1. Dipole Moments of Protein Molecules[a]

	Mol. Wt.	$\partial D/\partial c\,(\times 10^{-3})$ Low ν	High ν	$\mu \times 10^{18}$ (esu)
β-Lactoglobulin	35,500	1.51	−0.07	700
Ovalbumin	45,000	0.10	−0.07	250
Hemoglobin	67,000	0.33	−0.09	480
Serum albumin	69,000	0.17	−0.07	380

[a] The measured quantity is the square root of the average value of μ^2.

an equilibrium distribution of orientations to a much higher frequency than do macromolecular solute molecules, so that a wide frequency range exists in which the latter are unable to be oriented at all whereas equilibrium orientation of solvent molecules is maintained (cf. Fig. 6–7). It is a frequency within this range which must be used to determine $(\partial D/\partial c)_{\text{high }\nu}$. The frequencies of light waves would be too high and would lead to random orientation of both solvent and solute molecules.

It should be remarked that the theoretical treatment presented here has been quite sketchy, so that several difficulties have been glossed over. It is by no means certain that Kirkwood's treatment represents the final answer to the problem, and some workers have in fact preferred to use an empirical treatment of Wyman's,[130] which differs from equation 6–17 in having a constant 8.5 in place of Kirkwood's constant of 4.5. This changes the values of μ determined from the experimental data by about 35%, which figure might then be taken as a measure of the accuracy with which μ is determinable.

The most interesting experimental studies are those of Oncley and coworkers[131] on proteins in aqueous solution. A schematic diagram illustrating the type of data obtained by them is shown in Fig. 6–7, and representative dipole moments obtained by equation 6–17 or equivalent procedures are listed in Table 6–1. The observed dipole moments are, of course, very large when compared with the moments of smaller molecules, but they are not nearly as large as might be predicted. The molecules listed in Table 6–1 all contain from 50 to 100 charges of each sign (e.g., —NH$_3^+$, —COO$^-$) and, moreover, must have dimensions extending 50 Å or more in each direction. (Hemoglobin, for instance, was represented on p. 60 as an ellipsoid with axes 53 × 53 × 71 Å.) A single pair of

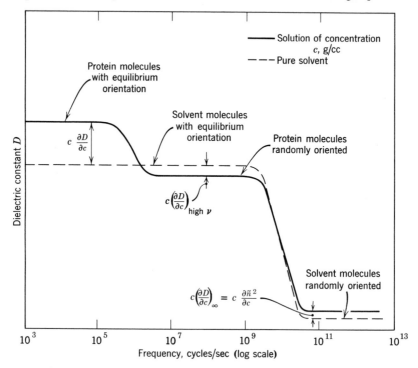

Fig. 6–7. Schematic diagram illustrating the data obtained in measurement of the dielectric properties of protein solutions (concentration c in grams per cubic centimeter). (Patterned after Oncley.[131])

charges separated by a distance of 50 Å will lead to a value of μ of $(4.18 \times 10^{-10}$ esu$)(50 \times 10^{-8}$ cm$) = 209 \times 10^{-18}$ esu, a figure already of the order of magnitude of the observed moments. It is clear that the charges of the molecules listed in Table 6–1 must be very evenly distributed to yield the observed result.

Kirkwood and Shumaker[132] have in fact suggested that the distribution is on the average almost perfectly symmetrical and that the observed moments arise wholly or largely from fluctuations in the charge distribution. The charges on protein molecules are due to acidic and basic groups such as

$$-COOH \rightleftharpoons -COO^- + H^+$$
$$-NH_3^+ \rightleftharpoons -NH_2 + H^+$$

and, at any pH, there are likely to be present both —COOH and —COO⁻ groups, or both —NH_3^+ and —NH_2 groups, etc. All of these acidic groups will be in dynamic equilibrium with one another and with the solvent, and H^+ ions will migrate freely from one site to another and from one molecule to another. It has already been pointed out that the mobility of H^+ ions will make a contribution to the polarizability, but what is important here is that it will produce, at any given instant, small differences between individual solute molecules. Differences in the total molecular charge Z will arise, for example. Thus we may be studying a solution in which the average number of positive charges per molecule is equal to the average number of negative charges, so that the average net charge, $\overline{Z}$, is zero. Such a solution will, however, contain molecules with total charges $+1, +2, \cdots, -1, -2, \cdots$, etc., as well as molecules with $Z = 0$. One result of this is that the average of the *square* of the charge is never zero. (The difference between $\overline{Z^2}$ and $(\overline{Z})^2$ is, in fact, readily calculated, as will be shown in section 30, where the equilibrium between H^+ ions and protein molecules will be discussed in detail.)

What is pertinent to the present discussion is that the same considerations apply to the dipole moment. There will be present at any instant molecules which will differ from one another in the distribution of charges. Thus all the solute molecules in a protein solution will not possess the same dipole moment, and any measurement must represent an average value. Dielectric constant measurements, by equation 6–17, will clearly measure the average of the *square* of the dipole moment, i.e., $\overline{\mu^2}$, and not the square of the average moment $\bar{\mu}$. The values of μ in Table 6–1 thus represent values of $(\overline{\mu^2})^{1/2}$. Kirkwood and Shumaker have calculated values of this parameter with the assumption that the charge distribution is such that $\bar{\mu} = 0$. They find that the root-mean-square moment due to fluctuations in charge distribution is of order 500×10^{-18} esu and will account for the entire observed moments of the molecules in Table 6–1, except for that of β-lactoglobulin.

Takashima and Lumry[133] have found that the combination of O_2 or CO with the iron atoms of hemoglobin changes the dielectric properties of the molecule, as illustrated by the data of Fig. 6–8. This indicates that structural changes accompany this reaction, a conclusion which is also reached from roughly parallel changes in the absorption spectrum. The exact nature of the structural change has not yet been elucidated. Measurement of dielectric properties would seem to be a powerful tool for investigating structural changes in solutions of proteins, viruses, etc., but very little work of this kind has been reported.

There has also been little study of the dipole moments of weakly polar polymer molecules in non-polar solvents. One such study is that of Marchal and Benoit[134] on polyoxyethylene glycol, a polymer with the repeating unit

$$-CH_2-CH_2-O-$$

The polymer was studied in benzene solution, and it was found that for chain lengths of eighty and above $\overline{\mu^2}$ varies linearly with chain length. The significance of this result will be discussed on p. 161.

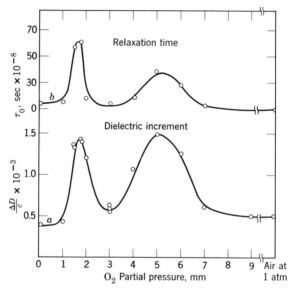

Fig. 6–8. The dielectric increment and mean relaxation time of hemoglobin in aqueous solution, as a function of the partial pressure of oxygen. The dielectric dispersion in this case requires a wide distribution of relaxation times, τ_0 being the mean of the distribution. (Takashima and Lumry.[133])

6c. Dielectric dispersion and relaxation time. The variation of dielectric constant with the frequency of the electric field is known as *dielectric dispersion*. It follows from the earlier discussion that the critical frequencies at which D undergoes a change are related to the rate with which the dipoles can undergo rotation. It is customary to express this rate in terms of a *relaxation time*, τ. This term is defined as follows. Suppose that we have a static electric field with a particular equilibrium distribution of orientations, i.e., that given by equation 6–4. There will be a corresponding average value of cos ϕ, which we may call $\overline{(\cos \phi)}_0$. Suppose that an instantaneous change in the field occurs. There will be a new equilibrium average value, but it will take some time to achieve it; we shall call it $\overline{(\cos \phi)}_\infty$, indicative of the fact that it will be the average value of cos ϕ

at infinite time. (It should be noted, however, that 10^{-6} sec may be equivalent to infinite time in this kind of process.) The transition from one equilibrium to the other is an exponential function of time, such that at any time t during the approach to equilibrium

$$\frac{\overline{\cos \phi} - (\overline{\cos \phi})_\infty}{(\overline{\cos \phi})_0 - (\overline{\cos \phi})_\infty} = e^{-(\text{const})t} \tag{6-18}$$

The constant occurring in this relation is the reciprocal of the relaxation time; i.e., the right-hand side of equation 6–18 may be written as $\exp(-t/\tau)$. The relaxation time is thus the time during which $1/e$ of the change in $\overline{\cos \phi}$ from one equilibrium position to another has occurred.

In the simplest case there is only a single mode of rotation and a single relaxation time. In that case the change in dielectric constant with frequency may be described by the relation (due to Debye[127])

$$\frac{D - D_\infty}{D_0 - D_\infty} = 1 + \frac{\nu^2}{\nu_c{}^2} = 1 + 4\pi^2 \tau^2 \nu^2 \tag{6-19}$$

where D_0 and D_∞ represent the dielectric constants at frequencies $\nu = 0$ and $\nu = \infty$, respectively, and $\nu_c = 1/(2\pi\tau)$ is called the *critical frequency*.

Usually equation 6–19 provides only an approximation to the actual variation of D with frequency in one of the transition regions because molecules are generally capable of rotation about different axes, each characterized by a different relaxation time. Moreover, if the molecules are not rigid, portions of them may rotate independently. Finally, in liquids, there will be variation from one molecule to another in the local environment. In all these situations several relaxation times or a continuous distribution of relaxation times is required to describe the dielectric dispersion. We shall not describe the resulting modifications of equation 6–19; they are discussed in some detail by Smyth.[11]

It was remarked earlier that the critical frequency (or average critical frequency) for water molecules in protein solutions is much larger than that for the dissolved protein molecules (see Fig. 6–7). This means, according to equation 6–19, that τ is much smaller for the water molecules, and, according to the foregoing physical picture, this means that water molecules rotate at a faster rate. This is the expected result, since the smaller water molecules encounter much less frictional resistance to rotation than do the bulky protein molecules.

This analysis can be extended to a comparison between different protein molecules and other macromolecules. Their relative relaxation times in similar solvents can be evaluated and used to give an indication of relative "bulkiness." We shall see in Chapter 6 that a variety of properties may in fact be used for this purpose, and we shall consider dielectric relaxation times again in this connection in section 25.

It may be noted that the data of Takashima and Lumry (Fig. 6–8) indicate parallel changes in dielectric increment (i.e., dipole moment) and in relaxation time. This observation serves to exclude one of the possible explanations for the changes in dielectric increment, which is that these changes are due to alterations in charge distribution without over-all structural change, for the frictional properties depend largely on the over-all shape of the molecule and only trivially, if at all, on the charge distribution.

Several studies of the dielectric dispersion and relaxation times of pure synthetic polymer preparations have been reported (e.g., Strella and Zand,[135] who studied polyalkyl methacrylates). These studies give information about rotational freedom in the solid polymer.

6d. Refractive index. It is evident from section 6b that the refractive index, or the corresponding quantity for solutions, the *refractive increment*, $\partial \tilde{n} / \partial c$, is a measure of *local* polarizability due to deformation of the electron configuration about nuclei. It can thus clearly give little or no information about the long-range structure of macromolecules with which this chapter is concerned, for, as was pointed out in the introduction on p. 15, the nature of individual chemical bonds is essentially invariant in this aspect of molecular structure.

This insensitivity of the refractive index to detailed molecular structure actually makes it a useful analytical tool, for it means that the refractive index of a solution of macromolecules of a given type (e.g., polyethylene or any protein) is a measure of concentration regardless of the detailed molecular structure. The chief use of refractive index measurements for this purpose is described in Appendix C.

6e. Birefringence. All but the simplest molecules are optically anisotropic, so that their optical properties will depend on their orientation relative to the plane of polarization of the light used to examine them. This anisotropy is not necessarily observed when an actual sample containing many molecules is examined. If the individual molecules are randomly oriented, as in a liquid, the sample as a whole will be isotropic. However, anisotropy will usually be present when the sample as a whole has an ordered structure, and an example of this has already been seen in section 5, where the dichroism of ordered films of polymers was considered.

We shall briefly take up here the anisotropy which results from differences in polarizability along different directions, which will clearly lead to corresponding differences in refractive index for light that is polarized so that electric vibrations occur along these directions. The phenomenon is known as *birefringence* or *double refraction*.

The simplest case corresponds to a *uniaxial* crystal, which has a single optic axis in a given direction, all directions at right angles to the optic axis being equivalent. The refractive index, $\tilde{n}_1$, of a polarized beam of light with the electric vibrations perpendicular to the optic axis will differ from the refractive index, $\tilde{n}_2$, observed when the plane of polarization is parallel to the axis. (If $\tilde{n}_2 > \tilde{n}_1$, the birefringence is said to be positive, and vice versa.) Any light with plane of polarization neither parallel nor perpendicular to the optic axis will split into two components, with polarizations perpendicular and parallel, respectively, to the optic axis. A difference in refractive index means a difference in the velocity of light, so that the two components with refractive indices $\tilde{n}_1$ and $\tilde{n}_2$ will travel with different speeds in the birefringent medium and become out of phase with each other. This creates a number of interesting optical effects which need not be discussed here. (One of them is the process of producing polarized light from unpolarized light by means of a Nicol prism.) It is sufficient to note that birefringence is easily detected and the difference $\tilde{n}_2 - \tilde{n}_1$ readily measured.

Of course, uniaxial anisotropy is a special, though common, case. The general description (biaxial anisotropy) would require two optic axes and three different refractive indices. It will not be considered here.

An important aspect of birefringence is that the refractive index is a property of the sample as a whole and does not, like dichroism, reflect the behavior of selected molecules or chemical groupings. If, therefore, we examine ordered structures consisting of an isotropic medium in which an ordered array of "particles" is embedded, the sample as a whole may be anisotropic and exhibit birefringence even though the "particles" as well as the surrounding medium are themselves isotropic. For this to occur, the refractive index of the "particles" must be different from that of the medium, for the light waves would not otherwise be able to distinguish the "particles" from the surrounding medium.

Wiener[136] has evaluated the birefringence to be expected in such a situation. Where the embedded "particles" are thin rods in parallel array, the major axis of the "particles" will become the optic axis and

$$\tilde{n}_2^2 - \tilde{n}_1^2 = \frac{\phi_p \phi_m (\tilde{n}_p^2 - \tilde{n}_m^2)^2}{(1 + \phi_p)\tilde{n}_m^2 + \phi_m \tilde{n}_p^2} \qquad (6\text{-}20)$$

where $\tilde{n}_p$ and $\tilde{n}_m$ are, respectively, the refractive indices of the "particles" and the medium, and ϕ_p and ϕ_m are the corresponding volume fractions. It will be noted that $\tilde{n}_2^2 - \tilde{n}_1^2$ is positive regardless of the sign of $n_p^2 - n_m^2$, so that the birefringence is in this case necessarily positive. The opposite situation exists for an array of thin discs, for which Wiener obtained

$$\tilde{n}_2^2 - \tilde{n}_1^2 = -\frac{\phi_p \phi_m (\tilde{n}_p^2 - \tilde{n}_m^2)^2}{\phi_p \tilde{n}_p^2 + \phi_m \tilde{n}_m^2} \qquad (6\text{-}21)$$

In this case the birefringence is necessarily negative. In both situations the birefringence disappears, as it must, when $\tilde{n}_p = \tilde{n}_m$.

The birefringence produced in this way by anisotropy of the sample as a whole is called *form birefringence*. If the ordered embedded "particles" are themselves anisotropic, an *intrinsic birefringence* must be added to the form birefringence. It will be independent of the refractive index of the surrounding medium.

Structures containing macromolecules which will exhibit birefringence include:

(*a*) Single crystals. The only macromolecular crystals (section 4) are wet crystals, consisting of ordered macromolecules embedded in solvent. The form birefringence will vanish when the refractive index of the solvent is equal to the refractive index of the macromolecules, so that the intrinsic birefringence may be determined. Perutz[137] has made measurements on hemoglobin crystals which suggest that hemoglobin molecules have negative intrinsic birefringence. (However, the crystals were not stable except in solvents for which $\tilde{n}_m < \tilde{n}_p$. A long extrapolation was needed to reach a state of $\tilde{n}_m = \tilde{n}_p$.)

(*b*) Ordered gels, fibers, and films. Polyethylene,[138] for example, is ordinarily isotropic but becomes birefringent when stretched, as might be expected from the x-ray diffraction studies cited earlier. The "particles" in this example are crystalline regions of the polymer embedded in a medium of randomly oriented chains. The amount of birefringence can be used as a measure of crystallinity.

(*c*) Solutions of asymmetric macromolecules which have been ordered in some way, most commonly by having the liquid flow in a narrow channel. The resulting birefringence is called *flow birefringence* (in analogy with flow dichroism, discussed on p. 97). Lauffer[139] has examined tobacco mosaic virus by flow birefringence, observing strong positive birefringence when the solvent refractive index was 1.3. The birefringence approached zero asymptotically as $\tilde{n}_m$ approached 1.6; i.e., the intrinsic birefringence of the virus particles was found to be zero. This is an unexpected result as x-ray data shown in Fig. 4–15 indicate that the virus particles have a definite structure with relation to the long axis which presumably orients itself parallel to the lines of flow. No clear-cut structural information has yet emerged from this and similar measurements of intrinsic birefringence.

It should be noted, however, that the extent of orientation which occurs in flow birefringence is an important tool for evaluation of the rotatory frictional properties of macromolecules, and it will be considered in that connection in section 25.

A special case of form birefringence in a solution arises when the orienting force is an electric field. (The production of birefringence in this way is known as the *Kerr effect*.) The birefringence under these conditions is clearly a direct measure of the distribution of dipoles and, hence, provides a means for measurement of dipole moments which is a useful alternative to that considered earlier. The theory is discussed by Benoit[140] and Tinoco.[141]

6f. Optical rotation.[111,150] When electromagnetic radiation interacts with matter, it may produce several observable effects, these being absorption, scattering, refraction, and, when the incident radiation is polarized, rotation of the plane of polarization. All of these effects are intimately related, and their theoretical basis has been discussed in a number of places. A particularly instructive discussion of the theory, from both the classical and quantum-mechanical point of view, may be found in Kauzmann's textbook of quantum chemistry.[112]

This section concerns only the last of these effects, the rotation of the plane of polarization of polarized radiation. The phenomenon is commonly called *optical rotation*, and it is a quite general result of the interaction between radiation and matter. However, the rotation produced by a given molecule is equal in magnitude, but opposite in sign, to that produced by its mirror image, so that an assembly of molecules with freedom of rotation, such as that which occurs in a solution, will produce an observable net effect only if the individual molecules are *optically asymmetric*, i.e., if they are molecules which cannot be converted by rotation into their mirror image.

The direction and magnitude of optical rotation are both very sensitive to the detailed structure in the vicinity of the asymmetric center of a molecule. Such structural effects are understood in principle, but they are ordinarily not amenable to quantitative computation. Quantitative aspects of optical rotatory power are therefore at present largely empirical.

Macromolecules will usually contain a large number of asymmetric centers per molecule, or none at all. Often all the asymmetric centers will be chemically identical. In that case each asymmetric center will make essentially the same contribution to the total rotation, as dictated by its own structure and that in its vicinity. Only those asymmetric centers which are near the chain ends will have a different environment, leading to a different contribution to the rotation. For sufficiently long molecules the rotation produced by the chain ends will be so small a fraction of the whole as to be negligible. The rotation produced by a molecule will thus be proportional to its length, and the *specific rotation*, i.e., the rotation per unit mass, will be independent of molecular weight.

An interesting application of optical rotation to the determination of molecular structure has been made in the case of the polypeptide chains of proteins and of synthetic polypeptides prepared from optically active amino acids. Considering the synthetic polypeptides first, it was shown in sections 4e and 5d that these substances may be prepared in the solid state in highly crystalline forms in which the individual polypeptide chains are in a helical configuration as illustrated by Fig. 4–7. Spectroscopic evidence was presented in section 5c which indicated that the helical configuration may be retained when the molecules are dispersed in a solution. We shall make frequent reference to these synthetic polypeptides in the subsequent portions of this book and will be able to prove conclusively that the α-helical configuration is indeed the stable configuration of these molecules in many solvents. We shall find, however, that this is not true in all solvents. In solvents which have a strong tendency to form hydrogen bonds the carbonyl and imino groups of the polypeptide chain will tend to form hydrogen bonds to the solvent rather than to one another, and the α-helix will not be formed. Instead, the polypeptide chains become randomly coiled, without preferred structure.

A helical configuration of unique sense, i.e., either right handed or left handed, is necessarily optically asymmetric. The right-handed and left-handed forms are mirror images, and they cannot be interconverted by rotation of the helix in space. Polypeptides in the α-helical configuration preferentially form right-handed helices if the constituent amino acids have the L-configuration.[150] Such polypeptides are therefore optically active in the helical form.

Polypeptides prepared from L-amino acids are, of course, also optically active in the randomly coiled form. The basis for the activity is quite different, however, each peptide bond, $\cdots$ NH—$\overset{*}{\text{C}}$HR—CO— $\cdots$, making an independent contribution. (Asterisk indicates asymmetric carbon atom.)

Clearly the two possible configurations of synthetic polypeptides in solution will both have optical rotatory power, but, in view of the entirely different structures of the asymmetric centers, the optical rotatory properties should be quite different. This supposition is confirmed by experimental studies on polypeptides known from other measurements to be definitely in one or the other of the two possible configurations, as the discussion which follows will demonstrate.

There has been a considerable amount of theoretical work on the expected optical rotatory properties of polypeptide chains in the randomly coiled and helical configurations, and this subject has been summarized and reviewed by Moffitt, Kirkwood, and Fitts.[114] Since some of the parameters which enter into the theory are not determinable with any precision and since actual applications are based on empirical considerations, we shall present here only the final relations for practical use (Moffitt and Yang[126]).

The parameter determined from rotation measurements is the specific rotation, ordinarily defined in terms of the concentration c of solute, in grams per cubic centimeter, and the length d of the light path in *decimeters*. If α is the observed angle of rotation at wavelength λ, the specific rotation $[\alpha]$ at that wavelength is

$$[\alpha] = \frac{\alpha}{dc} \qquad (6\text{--}22)$$

Since we are interested in the rotation per peptide link rather than the specific rotation, we shall substitute the molar concentration of monomer units (residues) of the polypeptide chains for the weight concentration. If C is defined as the concentration of residues, in moles per 100 cc, and M_0 is the mean residue mass, then $C = 100c/M_0$, and the mean residue rotation $[m]$ becomes

$$[m] = \frac{\alpha}{dC} = \frac{M_0[\alpha]}{100}$$ (6-23)

In defining $[m]$ in this way the concentration has purposely been expressed in moles per 100 cc so that $[m]$ will have the same order of magnitude as $[\alpha]$. One other factor that must be taken into account is the variation in $[m]$ or $[\alpha]$ with refractive index, which occurs even when all configurational features remain unchanged. The rotation turns out to be proportional to $(\tilde{n}^2 + 2)$, where $\tilde{n}$ is the refractive index. It is convenient to reduce all data to the values they would have in a medium of unit refractive index; i.e., we define an effective residue rotation, $[m']$, as

$$[m'] = \frac{3}{\tilde{n}^2 + 2} [m] = \frac{3M_0}{100(\tilde{n}^2 + 2)} [\alpha]$$ (6-24)

In considering the optical rotation of polypeptides not only $[\alpha]$ or $[m']$ but also its dispersion (or wavelength dependence) will be of interest. The simplest equation for the dispersion of optical rotation in general is a simple *Drude equation*, of the form

$$[m'] = \frac{a_0\lambda_0^2}{\lambda^2 - \lambda_0^2}$$ (6-25)

where λ_0 and a_0 are constants. If this equation is obeyed, the dispersion is said to be *simple*. If the wavelength dependence requires a more complex equation, the dispersion is said to be *anomalous*. Several types of equations (empirical, semi-empirical, and theoretical) have been suggested to describe anomalous dispersion. The one favored by workers in this field has the form

$$[m'] = \frac{a_0\lambda_0^2}{\lambda^2 - \lambda_0^2} + \frac{b_0\lambda_0^4}{(\lambda^2 - \lambda_0^2)^2}$$ (6-26)

which contains an additional constant b_0.

If the dispersion is simple, then, by equations 6-24 and 6-25, $\lambda^2[m']$ is a linear function of $[m']$, and $\lambda^2[\alpha]$ is a linear function of $[\alpha]$. In both cases the slope is λ_0^2. If the dispersion is not simple, a value of λ_0^2 is sought which will produce a linear plot of $[m'](\lambda^2 - \lambda_0^2)$ or $[\alpha](\lambda^2 - \lambda_0^2)$ versus $1/(\lambda^2 - \lambda_0^2)$, the parameter b_0 being obtained from the slope. In both cases the small variation of $\tilde{n}$ with λ is neglected.

Figure 6-9 shows experimental data of Yang and Doty[113] on aqueous solutions of poly-L-glutamic acid at two pH values, pH 5 and pH 8. At the lower pH the polymer molecules are known from other measurements (e.g., infrared spectra) to be helical, whereas at pH 8 they are randomly coiled. (The effect of pH can be ascribed to the fact that the —COOH side chains of the polymer molecules are largely converted to —COO⁻ at the higher pH. Mutual repulsion between nearby groups of like charge then favors the randomly coiled form, in which there is greater average separation between near neighboring side chains). It is seen that the randomly coiled form has

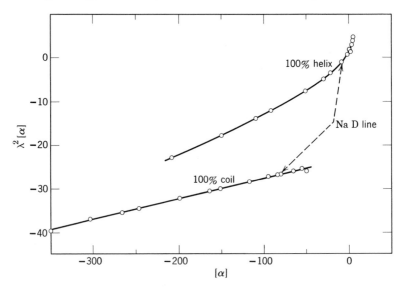

Fig. 6–9. Optical rotatory data for poly-L-glutamic acid in the helical and randomly coiled forms. (Yang and Doty.[113])

simple dispersion, with $\lambda_0 = 2120$ Å, but that a linear plot is not obtained with the helical form. The dispersion of the helical form must thus be represented by equation 6–26. It was found that a linear plot of $[\alpha](\lambda^2 - \lambda_0^2)$ versus $1/(\lambda^2 - \lambda_0^2)$ could be obtained by using the same value of λ_0 (2120 Å) as applies to the coiled form. With this value of λ_0 we find that $b_0 = -630°$. Another way of stating this result is that both the randomly coiled form and the helical form obey equation 6–26, with $\lambda_0 = 2120$ Å, b_0 being zero for the former and $-630°$ for the latter. (The value of a_0 is $-600°$ in the randomly coiled form and close to zero in the helical form. The value of $[m']$ for the D line of the Na lamp (5890 Å) may be used as an alternative parameter to distinguish the two forms. We obtain $[m']_D = -90°$ and $+10°$, respectively, for the randomly coiled and helical forms.)

The results here reported apply to a particular synthetic polypeptide. In view of the fact that the optical rotatory properties arise entirely from the $\cdots$—NH—CHR—CO—$\cdots$ backbone, however, there being no reason to expect any major influence from the R side chain, the results should be applicable to *any* molecule containing polypeptide chains and should make it possible to determine the extent to which such molecules are helical in configuration. (Some reservations to this statement will appear on p. 124.) This contention is supported by the theoretical work, previously cited, of Moffitt, Fitts, and Kirkwood[114], whose calculations

TABLE 6–2. Apparent Helix Content (% of Maximum Possible) for Various Proteins in Aqueous Solution

	b_0	Per Cent Helix
Tropomyosin[152]	−650°	100
Serum albumin[115]	−290°	46
Insulin[115]	−240°	38
Ovalbumin[115]	−195°	31
Lysozyme[115]	−180°	29
Ribonuclease[115,153]	−100°	16
Chymotrypsin[154]	−95°	15
β-Lactoglobulin[151]	−70°	11

Water-containing crystals of hemoglobin[149] and myoglobin[148] contain about 65% helix. These proteins are colored, and the b_0 value characteristic of the peptide chromophore in solution cannot be determined.

assume no involvement of the R groups at all. The calculated changes in optical properties, for the transformation of a random coil into an α-helix, agree well with the experimental results of Fig. 6–9.

We may assume then that any polypeptide chain (with no optically active R groups) will have $\lambda_0 = 2120$ Å. If entirely helical, $b_0 = -630°$; if randomly coiled, $b_0 = 0°$. It is further assumed that a molecule which has portions of its polypeptide chains helical, while other portions have no particular ordered structure, will have a b_0 value which is directly proportional to the helix content; i.e., 50% helix content corresponds to $b_0 = -315°$, etc.

It appears that a_0 and $[m']_D$ cannot be used in the same way. Present indications are that these parameters are strongly influenced by solvent interactions.

The results obtained for synthetic polypeptides have been applied by several workers to the study of protein solutions. Data obtained in aqueous solution are shown in Table 6–2, and they indicate that the helix content of these molecules is subject to considerable variation. For most of the globular proteins it appears to be quite small. We may therefore conclude that the compact structure of these molecules in aqueous solution is not primarily the result of intramolecular hydrogen bonds. (A discussion of possible forces responsible for the compact structure of globular proteins is given in section 7.)

It is important to note, however, that data obtained for globular proteins in several organic solvents lead to a different result. As Table 6–3 shows, the helix content is usually high. In $8M$ urea, on the other hand, the optical behavior of most proteins corresponds closely to that of randomly coiled polypeptides.

The figures for percentage helix content contained in Tables 6–2 and 6–3 should be regarded with considerable reservation. There are at least three

TABLE 6–3. Helix Content of Globular Proteins in Various Solvents

	Solvent	b_0	Per Cent Helix
Serum albumin[115]	2-chloroethanol	$-470°$	75
Insulin[115]	2-chloroethanol	$-285°$	45
Ovalbumin[115]	2-chloroethanol	$-535°$	85
Ribonuclease[115,153]	2-chloroethanol	$-390°$	62
β-Lactoglobulin[151]	75% dioxane, 25% H_2O	$-390°$	62
	70% ethanol, 30% H_2O	$-380°$	60
	80% DMF, 20% H_2O	$-330°$	52
Most proteins[111,115]	$8M$ aq. urea	$\sim0°$	0

reasons for believing that the optical rotatory properties of protein molecules may not correspond exactly to those of the synthetic polypeptide to which the data of Fig. 6–9 apply. (1) If only a fraction of each polypeptide chain is in helical form, the continuous lengths of helix must be quite short. End effects may then become appreciable, and the rotation per residue may no longer be independent of the length of the helix. (2) Although a left-handed helix is intrinsically less stable than a right-handed helix, other structural features may, in the protein molecules, favor the existence of some portions of left-handed helical chain. (The polymer of β-benzyl-L-aspartate has been shown to form left-handed instead of right-handed helices.[155]) (3) In a similar way other structural features of the protein molecules may lead to helical structures somewhat different from the Pauling α-helix, or to limited regions where the β-structure (Fig. 4–8) is formed by adjacent lengths of the same peptide chains, etc. Any of these phenomena may lead to serious error in the interpretation put upon the observed data.

6g. Density and other physical properties of solids. All physical properties of matter are, of course, a reflection of molecular structure. The particular properties which have been discussed in this chapter in some detail were chosen because their relation to molecular structure is a direct one, easily understood in terms of simple theory. Other properties which are easily measured, such as density, are less direct consequences of molecular structure and less easily adapted to the elucidation of quantitative structural information.

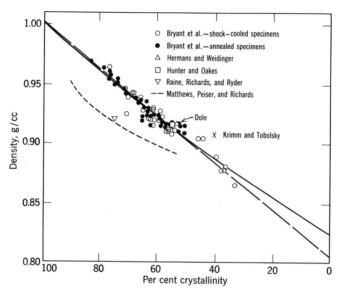

Fig. 6–10. The density of polyethylene as a function of the per cent crystallinity. (Nichols.[85])

These properties may, however, be used as a measure of structural parameters if they have been previously calibrated by use of a more direct method. Figure 6–10 provides an example. It shows the density of polyethylene as a function of the per cent crystallinity, as determined directly by one or more of the methods outlined in sections 4 and 5. Once it has been established that such a relation is reproducible, the density may, of course, be used as a measure of crystallinity.

Secondary physical properties of this kind, especially properties such as tensile strength and flexibility, are, of course, of the utmost importance in the adaptation of macromolecular solids to commercial use. The relation between these properties and molecular structure is vital to the technology of synthetic polymers. A very readable account of this work has been given by Billmeyer.[10]

7. GENERAL CONCLUSIONS

7a. Summary of structural information. On the basis of information derived from x-ray diffraction and spectral data it is reasonable to divide macromolecules very broadly into three classes, depending upon the extent to which they tend to acquire a definite spatial configuration in the solid state.

Class I, consisting of the majority of synthetic polymers, comprises molecules with little or no tendency to take on a special configuration. These molecules were found to form only amorphous solids or partially crystalline solids in which the principal factor influencing crystal structure was to provide the most efficient way of filling space.

Class II comprises molecules of definite structure, the structural pattern of which has been at least tentatively identified. Included in this group are synthetic polypeptides, some of the fibrous proteins, starch amylose (iodine complex), and deoxypentose nucleic acid. In each case the crystal structure appears to be determined by a strong tendency to form hydrogen bonds, which may be intramolecular, leading to helical structures, or intermolecular, leading to many-stranded helices or sheetlike structures.

Class III consists of molecules which clearly have a definite structure, the basis for which, however, has not been established. Into this group fall the globular proteins and the viruses. Molecules of these substances take on a unique shape, but they do not fall into the known patterns of the molecules of Class II. Speculation on the structure of these molecules is summarized in section 7d.

This classification is, of course, an oversimplification, and some molecules must be considered as occupying intermediate positions. Cellulose, for instance, has a fairly high degree of crystallinity, and the crystal structure is *partially* determined by hydrogen bonding. These bonds, however, involve only side chains of the cellulose molecule and are quite weak. Cellulose probably belongs more nearly to Class I than to Class II.

7b. Molecular structure in solution. In the remainder of this book we shall be primarily concerned with macromolecular solutions, and it is therefore of interest to make some guesses, based on the *exact* information derived from studies on solids, of what structures these molecules might take on in the dissolved state.

Molecules which in the solid state fall into Class I would be expected to have no preferred structure in solution. The long molecular chains would be expected to progress at random through the solution. Such molecules will be said to be *randomly coiled*. (Cf. Fig. 7–1.)

Molecules which in the solid state fall into Class II must be divided into two subclasses. Those which form hydrogen-bonded helical structures (single or multiple strand) may well be expected to maintain such a structure in solvents that have little tendency to form hydrogen bonds between solvent and solute molecules. On the other hand, these molecules may be expected to become randomly coiled in solvents that possess molecules

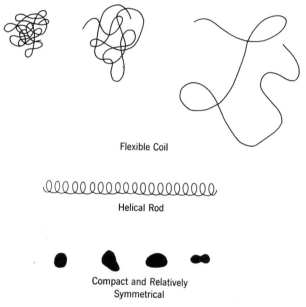

Flexible Coil

Helical Rod

Compact and Relatively
Symmetrical

Fig. 7-1. Schematic diagrams of the possible simple configurations of macromolecules in solution. The relative sizes of the illustrations give an indication of the relative space occupied in the solution for molecules of equal chain length.

with a strong tendency for hydrogen bond formation. For such solvent molecules would compete with the hydrogen-bond-forming groups of the macromolecules; formamide, $H—CO—NH_2$, for instance, might be expected to break $—C=O \cdots H—N—$ hydrogen bonds.

Data to support this guess have been given earlier. Infrared spectra of polypeptides in carbon tetrachloride or chloroform, for example (cf. p. 85), indicate that the α-helical structure is present in these solvents. In the section on optical rotation (section 6f) evidence was presented that in other solvents such polypeptides may be randomly coiled.

The second subclass of Class II consists of molecules which form sheet-like structures in the solid state. Such structures can clearly not be maintained in solution. All molecules in this subclass are fibrous proteins (e.g., silk fibroin) or synthetic polypeptides; many of them are molecules which may form either helical or sheetlike structures in the solid state, and all of them contain the same polypeptide backbone chain. They may therefore well be transformed to a helical form in some solvents. In others they may be randomly coiled.

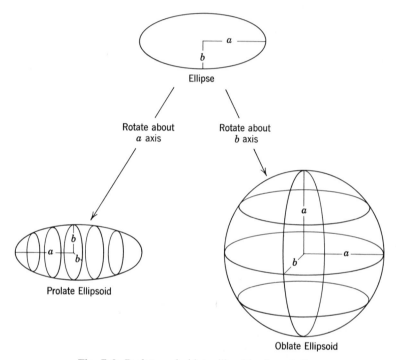

Fig. 7–2. Prolate and oblate ellipsoids of revolution.

The globular proteins and viruses which fall into Class III have a unique over-all molecular shape, often compact and symmetrical, in the crystalline state. It has been stated that the forces leading to this configuration have not been established, and therefore any estimate of what might happen to these molecules in solution is pure conjecture. It is significant, however, that these molecules have all been derived from the aqueous media of living systems, that they are usually crystallized from aqueous solutions, and that the crystals invariably contain water, in amounts which may vary considerably without interference with the crystal structure. These facts suggest that the structure which these molecules possess in the crystalline state may well remain unchanged in *water* solution, where the immediate environment of the molecules is similar to that in the crystalline state. Non-aqueous solvents, of course, would change the molecular environment and may thus be expected to lead to the adoption of new (unpredictable) structures.

7c. Mathematical models. Theoretical consideration of macromolecular solutions will ordinarily require that the dissolved molecules be represented as some simple particle, the boundaries of which can be described by a simple mathematical equation. Two possible representations of this kind are a sphere and a long thin rod. The latter might be used to represent helical rod-shaped molecules, the former to represent globular proteins and spherical viruses.

Since it is easy to write equations for ellipsoidal boundaries, dissolved molecules are often represented as ellipsoids of revolution. Two such ellipsoids can be constructed from an ellipse with major semi-axis a and minor semi-axis b, as shown by Fig. 7–2. One, obtained by revolution about the major axis is called a *prolate* ellipsoid, the other, obtained by revolution about the minor axis, is called an *oblate* ellipsoid. When a/b becomes large, a prolate ellipsoid becomes a good representation for a long thin rod, whereas an oblate ellipsoid approaches the shape of a flat circular disk.

A randomly coiled structure cannot be represented by any of these solid figures. Instead it must be considered as a singly connected chain of mass points, the configuration of which is determined purely on the basis of statistical reasoning.

7d. Speculations regarding the structure of globular proteins.[116,145,146] It has been stated that the polypeptide chains of globular protein molecules take on unique configurations in the solid state and that these configurations may be maintained in aqueous solution. The structural basis for such configurations is not known, but considerable speculation on this subject has been made, and the purpose of this section is to summarize the conclusions which have been reached.

Peptide hydrogen bonds. Since the polypeptide chains of fibrous proteins and of synthetic polypeptides show a strong tendency for formation of —C=O $\cdots$ H—N— hydrogen bonds, globular proteins may also be expected to show such a tendency. However, the optical rotation data of Table 6–2 clearly show that in an aqueous medium only a fraction of the peptide bonds can be part of a helical structure. A molecule that is partly helical and has no other elements of internal structure would, however, have long lengths of randomly coiled chain and would not be expected to have the unique compact form characteristic of the globular protein molecules. Thus some other strong force must be sought to explain the observed configuration.

It is very likely that the principal reason for the failure of globular protein molecules to take complete advantage of the hydrogen-bonding ability of its C=O and NH groups lies in the hydrogen-bonding ability of the water molecules with which these protein molecules are ordinarily in contact. Water molecules would be expected to interfere with —C=O $\cdots$ H—N— bonding by forming C=O $\cdots$ H—O—H and $H_2O \cdots$ H—N— hydrogen bonds. This explanation is in accord with the optical rotation data shown in Table 6–3. These data indicate that there is an increased tendency for formation of —C=O $\cdots$ H—N— bonds when globular protein molecules are placed into a solvent with a very weak hydrogen-bonding tendency (chloroethanol) whereas in a concentrated urea solution, which contains molecules (i.e., urea) most likely to be equally as strong as peptide bonds in their tendency to form hydrogen bonds, all —C=O $\cdots$ H—N— bonds appear to be destroyed.

Hydrophobic bonds.[117,145] The hydrogen-bonding ability of H_2O molecules, which interferes with hydrogen bonding within protein molecules, may act in another way so as to be perhaps the major force causing globular protein molecules to be collapsed into a compact, fairly symmetrical shape. The best example of the type of force envisaged occurs in solutions of soaps and detergents. These molecules consist of relatively long chain hydrocarbons terminated by an ionic group. When dissolved in water, which is a liquid with a high degree of structure impressed upon it by —O—H $\cdots$ O— hydrogen bonds, the ionic end of the detergent molecule, like all ions, is well accommodated into this structure (in fact, polar bonds leading to the hydration of ions are stronger than the —OH $\cdots$ O— hydrogen bonds, otherwise ions in general, and detergent molecules especially, would not be soluble.) The hydrocarbon portion of the detergent molecule, on the other hand, is an unwelcome guest in the aqueous medium,* for all water molecules which lie adjacent to it are deprived of their usual share of hydrogen bonds. In other words, the hydrocarbon portion makes a "hole" in the water structure. To avoid this, the detergent molecules simply associate, forming *micelles*, in which the hydrocarbon portions are in mutual contact in the center, with the ionic groups on the surface. In this way the interference with the water structure is, if not eliminated entirely, at least minimized. (It has been estimated[145] that there is a gain in free energy of about 1000 cal for each CH_2 group transferred from an aqueous medium to the interior of a non-polar medium.)

The same kind of bonding is a vital factor in the structure of polymeric soaps, i.e., polyelectrolyte molecules which contain charged sites and long non-polar side chains, as will be discussed in section 28.

* The word "hydrophobic" is really a misnomer. It implies that the dissolved substance dislikes water, whereas, in fact, it is the water that dislikes the dissolved substance.

Hydrophobic bonding must also be of importance in at least some of the soluble proteins, and perhaps in all. Although polar and non-polar groups of protein molecules are not nearly as conveniently arranged for hydrophobic bond formation as are those of detergents, some configurations of protein molecules must expose a relatively large number of non-polar side chains to contact with water, whereas other ways of folding must leave a reduced number of non-polar groups at the surface. The second type will be greatly favored.

In this connection it is of interest to recall the spectroscopic evidence on p. 95 which indicates that some globular proteins, in aqueous solution, may have part or all of their phenolic ($-CH_2-C_6H_5OH$) side chains buried in the interior of the molecule so as to be inaccessible to the approach of water molecules and unable to ionize at the pH where ionization would normally be expected to occur. This observation may well be a reflection of the existence of hydrophobic regions inside these molecules. For phenol is about equally soluble in water and in non-polar solvents. If hydrophobic regions exist, the phenolic side chains would be just as likely to enter these regions as to project from the surface of the molecule into the solvent. If so placed, they would produce the phenomenon actually observed.

Side chain hydrogen bonds. It has often been suggested that hydrogen bonds between side chain groups of protein molecules, e.g., bonds such as

provide one of the forces leading to a definite compact structure. The argument in favor of such bonds, in aqueous media, would appear to be weak, because of the already mentioned effect of H_2O on intramolecular hydrogen bonds in general. If, however, side chains containing phenolic, carboxyl, or other groups capable of hydrogen bond formation are trapped in the hydrophobic regions postulated above, then, freed from the competition of water molecules, they would presumably tend to form hydrogen bonds if it is sterically possible to do so. However, the ability to do so would be a secondary factor in the formation of a compact structure.

Intramolecular ion pairs. In a medium of dielectric constant as high as that of water univalent positive and negative ions have only a weak tendency to associate to form ion pairs. In water even this weak tendency toward association is entirely overcome by the formation of stabilizing bonds between water and free ions (hydration), so that electrolytes such as KCl are essentially 100% dissociated in aqueous solutions. Even guanidinium and acetate ions, association of which would be aided by resonance stabilization and hydrogen bonding,

$$CH_3-C \begin{matrix} \overset{\ominus}{O} \cdots H_2N \overset{\oplus}{} \\ \\ O \cdots H_2N \end{matrix} C = NH_2$$

associate to only a negligibly small extent. We would therefore not expect any attraction between positively and negatively charged side chain groups of a protein molecule, such as carboxylate and guanidinium or amino groups, if there is opportunity for these charged groups to lie at the surface of the molecule, where the dielectric constant would be that of the solvent and where the charged groups would be able to take advantage of the stabilizing effect of hydration.

Bonds between oppositely charged side chain groups may, however, occur if these groups are partially surrounded by non-polar groups, so that they do not have their full measure of hydration, and, especially, if the formation of the bond results in association of many non-polar groups, i.e., in simultaneous formation of a hydrophobic bond. As in the case of hydrogen bonds between side chain groups, the actual abundance of this type of bond is at present still unknown.

Long-range electrostatic forces.[146] Coulombic forces between charged side chain groups will exist even if the groups are relatively far apart. In the case of a protein molecule with about equal numbers of positive and negative charges the attractive forces between unlike charges will predominate over the repulsive forces between like charges because (*a*) the sum of $+-$ pairs formed by n positive and n negative charges is n^2, whereas the sum of $++$ and $--$ pairs is only $n^2 - n$, and (*b*) the dipole moment measurements cited in section 6 indicate essentially uniform distribution of charges, so that positive charges are, on the average, likely to be a little closer to negative charges than to other positive charges. This subject will be considered in some detail in section 28.

It should be noted, finally, that many of the globular proteins possess disulfide bonds, which may influence the structure in two ways, by preventing formation of a fully extended random coil and by interfering with the ability to form a helical structure or to form the optimum arrangement of side chains for hydrophobic bonding.

General References

1. C. W. Bunn, *Chemical Crystallography*, Oxford University Press, 1945.
2. M. J. Buerger, *X-Ray Crystallography*, John Wiley and Sons, New York, 1942.
3. R. W. James, *The Optical Principles of the Diffraction of X-Rays*, G. Bell and Sons, London, 1948.
4. J. T. Randall, *The Diffraction of X-Rays and Electrons by Amorphous Solids, Liquids, and Gases*, John Wiley and Sons, New York, 1934.
5. F. H. C. Crick and J. C. Kendrew, "X-Ray Analysis and Protein Structure," *Advances in Protein Chem.*, **12**, 133 (1957).
6. A. R. Stokes, "The Theory of X-Ray Fibre Diagrams," *Progress in Biophysics and Biophysical Chemistry*, **5**, 140 (1955).
7. A. Weissberger (ed.), *Technique of Organic Chemistry*, Vol. IX, "Chemical Applications of Spectroscopy," Interscience Publishers, New York, 1956.
8. L. J. Bellamy, *The Infra-red Spectra of Complex Molecules*, Methuen and Co., London, 1954.
9. C. E. Hall, *Introduction to Electron Microscopy*, McGraw-Hill Book Co., New York, 1953; R. W. G. Wyckoff, *Electron Microscopy*, Interscience Publishers, New York, 1949.
10. F. W. Billmeyer, Jr., *Textbook of Polymer Chemistry*, Interscience Publishers, New York, 1957.
11. C. P. Smyth, *Dielectric Behavior and Structure*, McGraw-Hill Book Co., New York, 1955.

Specific References

12. C. S. Fuller, *Chem. Revs.*, **26**, 143 (1940).
13. *International Tables for X-Ray Crystallography*, Vol. I, The Kynock Press, Birmingham, England, 1952.
14. C. W. Bunn, *Trans. Faraday Soc.*, **35**, 482 (1939).
15. C. W. Bunn, *Nature*, **161**, 929 (1948).
16. R. C. L. Mooney, *J. Am. Chem. Soc.*, **63**, 2828 (1941).
17. P. J. Flory, *Principles of Polymer Chemistry*, Cornell University Press, Ithaca, 1953, p. 231.
18. G. Natta, *J. Polymer Sci.*, **16**, 143 (1955); G. Natta and P. Corradini, *Makromol. Chem.*, **16**, 77 (1955).
19. C. W. Bunn and E. R. Howells, *J. Polymer Sci.*, **18**, 307 (1955).
20. C. S. Fuller, C. J. Frosch, and N. R. Pape, *J. Am. Chem. Soc.*, **62**, 1905 (1940).
21. C. W. Bunn, *Advances in Colloid Sci.*, **2**, 95 (1946).
22. K. H. Meyer and L. Misch, *Ber.*, **70B**, 266 (1937); *Helv. Chim. Acta.* **20**, 232 (1937).
23. H. Mark, *Chem. Revs.*, **26**, 169 (1940).
24. J. A. Howsmon and W. A. Sisson in E. Ott, H. M. Spurlin, and M. W. Grafflin (ed.), *Cellulose and Cellulose Derivatives*, 2nd ed., Part I, Interscience Publishers, New York, 1954, p. 231 et seq.
25. K. H. Meyer, *Natural and Synthetic High Polymers*, 2nd ed., Interscience Publishers, New York, 1950.
26. K. H. Meyer and G. W. Pankow, *Helv. Chim. Acta.*, **18**, 589 (1935).
27. C. H. Bamford, W. E. Hanby, and F. Happey, *Proc. Roy. Soc.*, **A 205**, 30 (1951).

28. C. H. Bamford, L. Brown, A. Elliott, W. E. Hanby, and I. F. Trotter, *Nature*, **173**, 27 (1954).
29. P. Vaughan and J. Donohue, *Acta Cryst.*, **5**, 530 (1952).
30. H. S. Taylor, *Proc. Am. Phil. Soc.* **85**, 1 (1941).
31. M. L. Huggins, *Chem. Revs.* **32**, 195 (1943).
32. L. Pauling, R. B. Corey and H. R. Branson, *Proc. Natl. Acad. Sci., U.S.*, **37**, 205 (1951).
33. I. S. Sokolnikoff and E. S. Sokolnikoff, *Higher Mathematics for Engineers and Physicists*, 2nd. ed., McGraw-Hill Book Co., New York, 1941, Ch. 2.
34. J. Donohue, *Proc. Natl. Acad. Sci., U.S.*, **39**, 470 (1953).
35. L. Pauling and R. B. Corey, *Proc. Natl. Acad. Sci., U.S.*, **37**, 251, 729 (1951).
36. L. Pauling and R. B. Corey, *Proc. Natl. Acad. Sci., U.S.*, **37**, 241 (1951).
37. M. F. Perutz, *Nature*, **167**, 1053 (1951).
38. W. Cochran, F. H. C. Crick, and V. Vand, *Acta Cryst.*, **5**, 581 (1952).
39. C. W. Bunn and E. V. Garner, *Proc. Roy. Soc.*, A **189**, 39 (1947); D. R. Holmes, C. W. Bunn and D. J. Smith, *J. Polymer Sci.*, **17**, 159 (1955).
40. J. C. Kendrew in H. Neurath and K. Bailey, *The Proteins*, vol. IIB, Academic Press, New York, 1954, Ch. 23; K. Bailey, *ibid.*, Ch. 24.
41. W. T. Astbury and H. J. Woods, *Nature*, **126**, 913 (1930); *Phil. Trans. Roy. Soc.*, A **232**, 333 (1933); W. T. Astbury and A. Street, *ibid*, A **230**, 75 (1931); W. T. Astbury and W. A. Sisson, *Proc. Roy. Soc.*, A **150**, 533 (1935).
42. P. M. Cowan and S. McGavin, *Nature*, **176**, 501 (1955); *Comm. to 3rd Intern. Congr. Biochem.*, **2**, 64 (1955).
43. J. M. Gillespie, *Biochim. et Biophys. Acta*, **27**, 225 (1958).
44. W. T. Astbury, *Fundamentals of Fibre Structure*, Oxford University Press, 1933.
45. R. Brill, *Ann.*, **434**, 204 (1923); C. H. Bamford, L. Brown, A. E. Elliott, W. E. Hanby, and I. F. Trotter, *Nature*, **171**, 1149 (1953).
46. A. Rich and F. H. C. Crick, *Nature*, **176**, 915 (1955).
47. L. Pauling and R. B. Corey, *Proc. Natl. Acad. Sci., U.S.*, **37**, 272 (1951).
48. D. Crowfoot, *Nature*, **135**, 591 (1935); D. Crowfoot and D. Riley, *ibid.*, **144**, 1011 (1939).
49. I. Fankuchen, *Advances in Protein Chem.*, **2**, 387 (1945).
50. K. J. Palmer, M. Ballantyne, and J. A. Galvin, *J. Am. Chem. Soc.*, **70**, 906 (1948).
51. F. R. Senti and R. C. Warner, *J. Am. Chem. Soc.*, **70**, 3318 (1948).
52. B. W. Low, *J. Am. Chem. Soc.*, **74**, 4830 (1952).
53. J. Boyes-Watson, E. Davidson, and M. F. Perutz, *Proc. Roy. Soc.*, A **191**, 83 (1947).
54. W. L. Bragg and M. F. Perutz, *Acta Cryst.*, **5**, 323 (1952); W. L. Bragg, E. R. Howells, and M. F. Perutz, *Proc. Roy. Soc.*, A **222**, 33 (1954).
55. P. Cowan and D. C. Hodgkin, *Acta Cryst.*, **4**, 160 (1951).
56. J. D. Bernal and I. Fankuchen, *J. Gen. Physiol.*, **25**, 111, 147 (1941).
57. D. W. Green, V. M. Ingram, and M. F. Perutz, *Proc. Roy. Soc.*, A **225**, 287 (1954).
58. B. W. Low, *Nature*, **169**, 955 (1952).
59. D. C. Hodgkin, *Cold Spring Harbor Symposia Quant. Biol.*, **14**, 69 (1949).
60. W. L. Bragg and M. F. Perutz, *Acta Cryst.*, **5**, 277 (1952).
61. J. C. Kendrew, G. Bodo, H. M. Dintzis, R. G. Parrish, H. Wyckoff, and D. C. Phillips, *Nature*, **181**, 662 (1958).
62. G. Bodo, H. M. Dintzis, and J. C. Kendrew, unpublished data, quoted by Crick and Kendrew, Ref. 5.

63. R. B. Corey, J. Donohue,, K. N. Trueblood, and K. J. Palmer, *Acta Cryst.*, **5**, 701 (1952).
64. F. H. C. Crick and B. S. Magdoff, *Acta Cryst.*, **9**, 901 (1956).
65. M. H. F. Wilkins, A. R. Stokes, and H. R. Wilson, *Nature*, **171**, 738 (1953).
66. R. E. Franklin and R. G. Gosling, *Nature*, **171**, 740 (1953).
67. J. D. Watson and F. H. C. Crick, *Nature*, **171**, 737, 964 (1953); F. H. C. Crick and J. D. Watson, *Proc. Roy. Soc.*, A **223**, 80 (1954).
68. B. S. Magdoff, F. H. C. Crick, and V. Luzzati, *Acta Cryst.*, **9**, 156 (1956).
69. F. H. C. Crick, *Acta Cryst.*, **6**, 600 (1953).
70. M. Feughelman et al., *Nature*, **175**, 834 (1955).
71. G. Herzberg, *Molecular Spectra and Molecular Structure I. Spectra of Diatomic Molecules.* D. Van Nostrand Co., Princeton, 1950; *Infrared and Raman Spectra of Polyatomic Molecules*, D. Van Nostrand Co., Princeton, 1945.
72. A. B. F. Duncan, Ref. 7, p. 187, 581.
73. F. A. Matsen, Ref. 7, p. 629.
74. R. N. Jones and C. Sandorfy, Ref. 7, p. 306.
75. K. Nakamoto, M. Margoshis, and R. E. Rundle, *J. Am. Chem. Soc.* **77**, 6480 (1955); R. C. Lord and R. E. Merrifield, *J. Chem. Phys.* **21**, 166 (1953).
76. E. J. Ambrose and A. Elliott, *Proc. Roy. Soc.*, A **205**, 47 (1951).
77. R. E. Richards and H. W. Thompson, *J. Chem. Soc.*, **1947**, 1248.
78. Ref. 8, p. 186.
79. E. J. Ambrose and A. Elliott, *Proc. Roy. Soc.*, A **208**, 75 (1951).
80. E. R. Blout and H. Lenormant, *J. Opt. Soc. Am.*, **43**, 1093 (1953).
81. G. Ehrlich and G. B. B. M. Sutherland, *J. Am. Chem. Soc.*, **76**, 5268 (1954).
82. E. J. Hart and A. W. Meyer, *J. Am. Chem. Soc.*, **71**, 1980 (1949).
83. W. M. D. Bryant and R. C. Voter, *J. Am. Chem. Soc.*, **75**, 6113 (1953).
84. J. J. Fox and A. E. Martin, *Proc. Roy. Soc.*, A **175**, 208 (1940).
85. J. B. Nichols, *J. Appl. Phys.*, **25**, 840 (1954).
86. S. Krimm, C. Y. Liang, and G. B. B. M. Sutherland, *J. Chem. Phys.*, **25**, 549 (1956); C. Y. Liang and S. Krimm, *ibid*, **25**, 563 (1956).
87. S. Krimm, C. Y. Liang, and G. B. B. M. Sutherland, *J. Polymer Sci.*, **22**, 227 (1956).
88. H. Tadokoro, S. Seki, and I. Nitta, *J. Polymer Sci.*, **22**, 563 (1956).
89. E. J. Ambrose and A. Elliott, *Proc. Roy. Soc.*, A **206**, 206 (1951).
90. American Petroleum Institute Research Project 44, Carnegie Institute of Technology, *Catalog of Ultraviolet Spectral Data*, Pittsburgh, 1950.
91. D. Shugar and J. J. Fox, *Biochim. et Biophys. Acta*, **9**, 199 (1952).
92. G. H. Beaven and E. R. Holiday, *Advances in Protein Chem.*, **7**, 319 (1952).
93. R. L. Sinsheimer, J. F. Scott, and J. R. Loofbourow, *J. Biol. Chem.*, **187**, 299, 313 (1950).
94. G. H. Beaven, E. R. Holiday, and E. A. Johnson, in E. Chargaff and J. N. Davidson (ed.), *The Nucleic Acids*, Vol. I. Academic Press, New York, 1955.
95. B. Magasanik and E. Chargaff, *Biochim. et Biophys. Acta*, **7**, 396 (1951).
96. J. L. Crammer and A. Neuberger, *Biochem. J.*, **37**, 302 (1943).
97. C. Tanford, J. D. Hauenstein, and D. G. Rands, *J. Am. Chem. Soc.*, **77**, 6409 (1955).
98. M. Laskowski, Jr., and H. A. Scheraga, *J. Am. Chem. Soc.*, **76**, 6305 (1954).
99. M. F. Perutz, *Acta. Cryst.*, **6**, 859 (1953).
100. R. E. Rundle and R. R. Baldwin, *J. Am. Chem. Soc.*, **65**, 554 (1943).
101. R. E. Rundle and D. French, *J. Am. Chem. Soc.*, **65**, 1707 (1943).

102. R. Lemberg and J. W. Legge, *Hematin Compounds and Bile Pigments*, Interscience Publishers, New York, 1949.
103. H. Theorell, *Arkiv. Kemi, Mineral. Geol.*, **16A**, no. 3 (1943). See also ref. 102, p. 162.
104. N. Fuschillo, E. Rhian, and J. A. Sauer, *J. Polymer Sci.*, **25**, 381 (1957).
105. J. G. Powles, *J. Polymer Sci.*, **22**, 79 (1956).
106. M. Saunders, A. Wishnia, and J. G. Kirkwood, *J. Am. Chem. Soc.*, **79**, 3289 (1957).
107. R. C. Williams and R. C. Backus, *J. Am. Chem. Soc.*, **71**, 4052 (1949); R. C. Williams, *Exptl. Cell Research*, **4**, 188 (1953).
108. C. E. Hall and P. Doty, *J. Am. Chem. Soc.*, **80**, 1269 (1958).
109. R. L. Steere, *Phytopathology*, **46**, 60 (1956).
110. R. W. G. Wyckoff, *Acta Cryst.*, **1**, 292 (1948).
111. C. Schellman and J. A. Schellman, *Compt. rend. trav. lab. Carlsberg, Sér. chim.*, **30**, 463 (1958).
112. W. Kauzmann, *Quantum Chemistry*, Academic Press, New York, 1957, Ch. 15 and 16.
113. J. T. Yang and P. Doty, *J. Am. Chem. Soc.*, **79**, 761 (1957).
114. W. Moffitt, *J. Chem. Phys.*, **25**, 467 (1956); *Proc. Natl. Acad. Sci., U.S.*, **42**, 736 (1956); D. Fitts and J. G. Kirkwood, *ibid*, **43**, 33 (1957); W. Moffitt, D. Fitts, and J. G. Kirkwood, *ibid*, **43**, 723 (1957).
115. K. Imahori, E. Klemperer, and P. Doty, Abstracts of Papers presented at 131st Meeting of the American Chemical Society, Miami, Florida, April 1957.
116. W. Kauzmann in W. D. McElroy and B. Glass (ed.), *The Mechanism of Enzyme Action*, John Hopkins Press, Baltimore, 1954.
117. P. Debye, *Ann. N.Y. Acad. Sci.*, **51**, 575 (1949).
118. J. D. Watson, *Biochim. et Biophys. Acta*, **13**, 10 (1954).
119. R. E. Franklin, *Nature*, **175**, 379 (1955); *Biochim. et Biophys. Acta*, **19**, 203 (1956).
120. G. Schramm, *Z. Naturforsch*, **2b**, 112, 249 (1947).
121. R. E. Franklin, *Biochim. et Biophys. Acta*, **18**, 313 (1955).
122. D. L. D. Caspar, *Nature*, **177**, 928 (1956).
123. R. E. Franklin, *Nature*, **177**, 928 (1956); R. E. Franklin, A. Klug, and K. C. Holmes in G. E. W. Wolstenholme and E. C. P. Millar (ed.), *The Nature of Viruses*, Little, Brown and Co., Boston, 1957.
124. D. Garfinkel and J. T. Edsall, *J. Am. Chem. Soc.*, **80**, 3818 (1958).
125. C. E. Hall, *J. Am. Chem. Soc.*, **80**, 2556 (1958).
126. W. Moffitt and J. T. Yang, *Proc. Natl. Acad. Sci., U.S.*, **42**, 596 (1956).
127. P. Debye, *Polar Molecules, Chemical Catalog Co.*, New York, 1929.
128. L. Onsager, *J. Am. Chem. Soc.*, **58**, 1486 (1936).
129. J. G. Kirkwood, *J. Chem. Phys.*, **7**, 911 (1939).
130. J. Wyman, Jr., *J. Am. Chem. Soc.*, **58**, 1482 (1936); *Chem. Revs.*, **19**, 213 (1936).
131. J. L. Oncley in E. J. Cohn and J. T. Edsall, *Proteins, Amino Acids and Peptides*, Reinhold Publishing Corp., New York, 1943.
132. J. G. Kirkwood and J. B. Shumaker, *Proc. Natl. Acad. Sci., U.S.*, **38**, 855 (1952).
133. S. Takashima and R. Lumry, *J. Am. Chem. Soc.*, **80**, 4238, 4244 (1958).
134. J. Marchal and H. Benoit, *J. chim. phys.*, **52**, 818 (1955).
135. S. Strella and R. Zand, *J. Polymer Sci.*, **25**, 97, 105 (1957).
136. O. Wiener, *Abhandl. math. Kl. sächs. Akad. Wiss.*, **32**, 509 (1912). See also W. L. Bragg and A. B. Pippard, *Acta Cryst.*, **6**, 865 (1953).
137. M. F. Perutz, Acta Cryst., **6**, 859 (1953).
138. S. M. Crawford and H. Kolsky, *Proc. Phys. Soc., London B*, **64**, 119 (1951).

139. M. A. Lauffer, *J. Phys. Chem.*, **42,** 935 (1938).
140. H. Benoit, *Ann. Phys.*, **6,** 561 (1951).
141. I. Tinoco, *J. Am. Chem. Soc.*, **77,** 4486 (1955).
142. P. H. Till, Jr., *J. Polymer Sci.*, **24,** 301, (1957).
143. D. W. Jones, *J. Polymer Sci.*, **32,** 371 (1958).
144. M. Lauffer, *Nature*, **183,** 1601 (1959).
145. W. Kauzmann, *Advances in Protein Chem.*, **14,** 1 (1959).
146. C. Tanford in A. Neuberger, (ed.), *Symposium on Protein Structure*, Methuen and Co., London, 1958.
147. R. C. Williams in T. M. Rivers and F. L. Horsfall, (ed.), *Viral and Rickettsial Infections of Man*, 3rd ed., J. B. Lippincott Co., Philadelphia, 1959.
148. J. C. Kendrew, R. E. Dickerson, B. E. Strandberg, R. G. Hart, D. R. Davies, D. C. Phillips, and V. C. Shore, *Nature*, **185,** 422 (1960).
149. M. F. Perutz, M. G. Rossmann, A. F. Cullis, H. Muirhead, G. Will, and A. C. T. North, *Nature*, **185,** 416 (1960).
150. E. R. Blout in C. Djerassi, *Optical Rotatory Dispersion*, McGraw-Hill Book Co., New York, 1960, Ch. 17.
151. C. Tanford, P. K. De, and V. G. Taggart, *J. Am. Chem. Soc.*, **82,** 6028 (1960).
152. C. M. Kay and K. Bailey, *Biochim. et Biophys. Acta*, **31,** 20 (1959).
153. R. E. Weber and C. Tanford, *J. Am. Chem. Soc.*, **81,** 3255 (1959).
154. B. Jirgensons, *Arch. Biochem. Biophys.*, **74,** 57, 70 (1958); **78,** 235 (1958); **85,** 89, 532 (1959).
155. R. H. Karlson, K. S. Norland, G. D. Fasman, and E. R. Blout, *J. Am. Chem. Soc.*, **82,** 2268 (1960).
156. M. F. Perutz, discussion of paper by H. A. Scheraga, Brookhaven Symposia in Biology, No. 13, 86 (1960).
157. I. Tinoco, Jr., *J. Am. Chem. Soc.*, **82,** 4785 (1960).
158. E. J. Williams and J. F. Foster, *J. Am. Chem. Soc.*, **82,** 242 (1960).
159. S. Yanari and F. A. Bovey, *J. Biol. Chem.*, **235,** 2818 (1960).
160. C. E. Hall and H. S. Slayter, *J. Biophys. Biochem. Cytol.*, **5,** 11 (1959).

3

STATISTICS OF
LINEAR POLYMERS

8. MOLECULAR WEIGHT AVERAGES AND DISTRIBUTIONS

Simple polymer molecules may have two characteristics which require statistical analysis. (1) A sample may contain a mixture of molecules with different numbers of monomer units per molecule, so that we are led to ask such questions as how many molecules there are of each kind or what the average number of monomer units is. (2) Intramolecular forces may be so weak that all possible configurations of a single molecule possess essentially equal energy and are thus equally probable. We then wish to know how likely any given configuration is, what the average distance from one end of the molecule to the other will be, etc. A discussion of these questions is the objective of this chapter. Section 8 deals with molecular weights, sections 9 and 10 with configuration.

8a. Degree of polymerization and molecular weight. The *degree of polymerization* of a given polymer molecule is the number of structural units which it contains. It will be designated by the symbol x. A molecule of degree of polymerization x is called an x-mer. This definition may be used for branched as well as linear polymers. Its value will depend only on the definition of the structural unit, e.g., the straight chain hydrocarbon, $C_{100}H_{202}$, may be regarded as polymethylene of degree of polymerization 100, or as polyethylene of degree of polymerization 50, the structural units being $—CH_2—$ and $—CH_2—CH_2—$, respectively. It should be noted that the terminal structural units always differ from those in the interior of the chain (e.g., $CH_3—$ instead of $—CH_2—$); where branching occurs, the structural unit at the point of branching differs from the rest (e.g., $—CH—$ instead of $—CH_2—$). Thus the x structural units of an x-mer are never completely identical, though in the simple linear polymers all but the two terminal units will be identical.

The molecular weight M_x of an x-mer is the sum of the contributions from the x structural units which it contains. When x becomes large (greater than 100, say), we can ordinarily ignore the small difference in the two terminal structural units and assume that each structural unit makes the same contribution M_0 to the molecular weight. We can then relate M_x to the degree of polymerization,

$$M_x = xM_0 \qquad (8\text{--}1)$$

with M_0 a constant for a given kind of polymer, independent of the value of x. This permits us to use the terms *degree of polymerization* and *molecular weight* essentially interchangeably, since, for a given kind of polymer, they are related by a known numerical constant.

There arises here the question: when is a polymer a macromolecule? A trimer or tetramer is legitimately called a polymer but is obviously not a macromolecule. The present discussion suggests that a convenient definition of a macromolecular polymer might be one with a majority of its molecules having at least 100 structural units. The question will arise again in several places, especially in this chapter, where numerous statistical calculations will be made which require the number of structural units to be "large." Again a minimum value of $\bar{x}$ of 100 (and, preferably, higher) is suggested. Thus, although we wish to draw no hard-and-fast boundaries, we shall usually, when we discuss polymers, have this definition in mind.

8b. Molecular weight distributions. Typical synthetic polymer preparations contain molecules with values of x varying over a wide range. *A molecular weight distribution* is a description of the frequency of occurrence of a given degree of polymerization. This might be expressed in terms of mole fractions, e.g., $X_1 = 0.001$, $X_2 = 0.0015, \ldots$, where X_x represents the mole fraction of x-mer. More usually there is systematic variation of mole fraction of x-mer as a function of x, which can be expressed in terms of an equation, of the form $X_x = f(x)$. Such an equation is known as a *distribution function*.

We shall derive here (after Flory[5]) the distribution function for a simple case, condensation polymerization in which there is no change in reactivity of functional groups with chain length. To serve as a model we might choose a substance HO—R—COOH, which would polymerize to molecules of the form

$$\text{HO—R—CO(—O—R—CO)}_{x-2}\text{—O—R—COOH}$$

where x is the degree of polymerization. The nature of the functional groups, however, will have no bearing on the final result.

Let there be initially N_0 monomer molecules in our sample, and therefore also N_0 functional groups (i.e., N_0 hydroxyl groups and N_0 carboxyl groups). Let polymerization proceed until a fraction p of all functional groups has reacted. The remaining number of functional groups of either

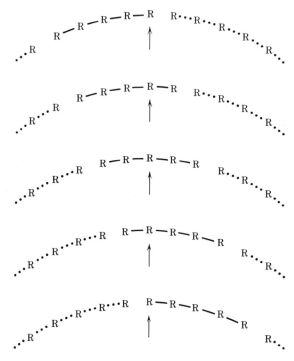

Fig. 8-1. Derivation of the molecular weight distribution for condensation polymerization. All monomer units, R, are arranged in a giant circle, a portion of which is shown in the figure. The figure shows five different ways in which a monomer unit chosen at random, and indicated by the arrow, may be part of a molecule with $x = 5$. A solid line represents a —CO—O— bond; a space represents unbonded functional groups, —COOH HO—; a dotted line represents a position which can be either bonded or unbonded.

kind is then $N_0(1 - p)$. This same number will also represent the final number of independent molecules since each molecule must have an unreacted functional group at each end and no other uncondensed hydroxyl or carboxyl groups.

In terms of the extent of reaction, p, we may at once obtain a value for the *number-average degree of polymerization*, $\bar{x}_n$, which is defined as the average number of monomer units in the polymer molecules present. This is clearly equal to the initial number of monomer molecules, divided by the final number of independent molecules, i.e.,

$$\bar{x}_n = N_0/N_0(1 - p) = 1/(1 - p) \tag{8-2}$$

It is clear that p must be greater than 0.99 if $\bar{x}_n$ is to be greater than 100; i.e., we shall in general be interested in values of p between 0.99 and 1.00.

Let us now consider all the molecules in the final mixture placed end-to-end along a closed path, such as a giant circle. We may consider this circle as composed of N_0 structural units. At pN_0 places along the circle these structural units will be joined to one another by —CO—O— bonds; at the remaining $(1 - p)N_0$ places they will not be joined. They will be the places where one molecule ends and another begins. If the reactivity of the functional groups is independent of chain length, the location of these places will be completely random.

Let us focus our attention on a single structural unit, chosen at random. What is the probability that this structural unit forms part of an x-mer? To calculate this probability we first compute the probability that the chosen structural unit is part of an x-mer beginning at a given position along the circle (cf. Fig. 8-1). This probability is clearly equal to the probability that $x + 1$ particular positions along the circle shall be successively: first an unbonded position, then $x - 1$ —CO—O— bonds, and finally another unbonded position. The remaining positions along the circle may be either bonds or spaces. Since what occurs at each position is an independent event, the probability that we have an x-mer beginning at a given location is the product of the probabilities of the desired events at each position. This probability is p for having a bond and $1 - p$ for not having a bond, for the fraction of all functional groups which are bonded is clearly also a measure of the probability that there is a bond at a given position. Thus the desired probability is the product $(1 - p)p^{x-1}(1 - p)$, i.e., $(1 - p)^2 p^{x-1}$. (For all positions other than the $x + 1$ positions of immediate interest, the probability is unity; it makes no difference whether they are bonded or not.)

If the randomly chosen structural unit is to be part of *any* x-mer, there are x positions along the chain where we may begin, as illustrated for $x = 5$ in Fig. 8-1. Thus the over-all probability that a given structural unit is part of an x-mer is x times the expression of the last paragraph; i.e., it is $x(1 - p)^2 p^{x-1}$.

The probability of finding a given structural unit incorporated in an x-mer is equivalent to the fraction of all structural units present in x-mers. Since each structural unit, neglecting the small difference in terminal units, has the same mass, this will also represent the weight fraction of the material present as x-mer. This fraction is represented by W_x and is thus given by

$$W_x = xp^{x-1}(1 - p)^2 \qquad (8\text{–}3)$$

(The total weight of material present is, of course, less than the original amount of monomer added because of the water split out in the polymerization.)

The total number of structural units is equal to the original number of monomer molecules, previously designated as N_0; the total mass present is thus $N_0 M_0 / \mathcal{N}$, where M_0 is the molecular weight of the structural unit —O—R—CO— and $\mathcal{N}$ is Avogadro's number, and again the slight difference in the terminal units is neglected. The mass present as x-mer is thus $W_x N_0 M_0 / \mathcal{N}$. The mass of each x-mer molecule is $x M_0 / \mathcal{N}$, so that the number of x-mer molecules is $W_x N_0 / x$. The total number of all molecules was previously shown to be $N_0(1 - p)$, so that the mole fraction of x-mer is

$$X_x = W_x / x(1 - p) = p^{x-1}(1 - p) \qquad (8\text{–}4)$$

The desired distribution function has thus been derived, in two forms. In equation 8–3 we have what is known as the *weight distribution*;* i.e., the fraction of the total weight present as x-mer. In equation 8–4 we have the *number distribution*; i.e., the fraction of the number of molecules present as x-mer, or, in other words, the mole fraction. For many purposes the weight distribution is the more important. A fairly large number of molecules of low degree of polymerization may represent only a negligibly small fraction of all the material present. Combining equations 8–3 and 8–4 with equation 8–2 we see that $W_x > X_x$ when $x > \bar{x}_n$ and that $W_x < X_x$ when $x < \bar{x}_n$.

The sum of all the weight fractions and the sum of all the mole fractions must, of course, be equal to unity. It is readily seen that this condition is satisfied for both equations 8–3 and 8–4: $\sum\limits_{x=1}^{\infty} X_x = (1 - p)(1 + p + p^2 + \cdots) = 1$, $\sum\limits_{x=1}^{\infty} W_x = (1 - p)^2(1 + 2p + 3p^2 + \cdots) = 1$. To evaluate these summations use has been made of the binomial expansion as shown in Appendix A.

Some typical plots according to equations 8–3 and 8–4 are shown in Figs. 8–2 and 8–3. Figure 8–2 is especially noteworthy since it shows that in the particular distribution which we have here calculated $X_1 > X_2 > \cdots > X_{x-1} > X_x > X_{x+1} > \cdots$, a relation which follows at once from the requirement that p in equation 8–4 must be less than unity. The figure

---

* The term *mass distribution* would be preferable, but *weight distribution* is customarily used.

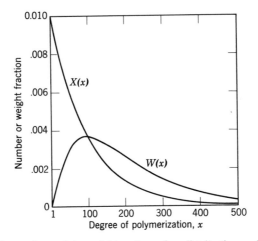

Fig. 8–2. Comparison of the weight and number distributions calculated by equations 8–3 and 8–4 with $p = 0.99$. (The corresponding value of $\bar{x}_n = 100$.)

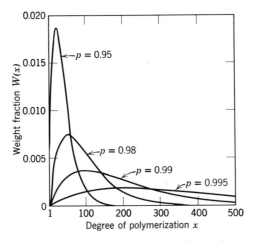

Fig. 8–3. Progress of the weight distribution with increasing p, as calculated by equation 8–3.

also shows, however, that this relation does not prevent the peak of the distribution on a weight basis from appearing at high values of x. Experimental verification of the distribution given by these equations has been obtained (by fractionation) for at least one condensation polymer ("nylon 66").[6]

Equations 8–3 and 8–4 have meaning only for integral values of x. In other words, the curves of Figs. 8–2 and 8–3 are actually not continuous.

Where the distribution reaches large values of x, however, we may substitute for these discontinuous functions a continuous distribution function. Thus we may designate by dW_x the infinitesimal fraction of the weight of polymer having a degree of polymerization between x and $x + dx$ and by dX_x the infinitesimal mole fraction over the same range and write, in place of equations 8–3 and 8–4,

$$dW_x = xp^{x-1}(1 - p)^2 \, dx \qquad (8\text{–}5)$$

$$dX_x = p^{x-1}(1 - p) \, dx \qquad (8\text{–}6)$$

The advantage of distribution functions of this type is that they may be integrated between finite limits. We may use them, for example, to calculate the fraction of polymer with x lying between given limits x_1 and x_2; the summation of equations 8–3 and 8–4 over such limits would be a more difficult mathematical problem. The error introduced by substitution of the integral for the summation is given by the Euler-Maclaurin formula, a discussion of which is given by Mayer and Mayer.[7] The error becomes negligible if the sum or integral is taken over a sufficiently wide range so that the total amount of polymer is very much greater than that occurring as molecules with any given integral value of x.

The molecular weight distribution given by equations 8–3 to 8–6 is, of course, only a particular example of the type of molecular weight distribution that might be obtained in the synthesis of a polymeric substance. Numerous other distributions are possible, the type of distribution obtained depending on the kinetics of synthesis of the polymeric substance, as will become evident in Chapter 9 where we shall consider the kinetics of polymerization reactions. Throughout this book we shall be concerned with general principles of the physical chemistry of macromolecules, including synthetic polymers. For most purposes it will be necessary to know only that a molecular weight distribution of some kind exists and that it can be described by appropriate equations.

8c. Fractionation of polymers.[3] A typical heterogeneous sample of polymeric material may be separated into fractions of considerably less heterogeneity. To accomplish this we may make use of the change in solubility of chemically identical molecules differing in chain length (see section 15), or we may use techniques such as ultrafiltration through graded membranes. The various methods have been reviewed by Cragg and Hammerschlag.[3] A polymer sample resulting from such fractionation is still heterogeneous, but the molecular weights are distributed over a much smaller range.

When fractionation has been especially thorough it becomes a fairly good approximation to consider the resulting sample homogeneous with respect to molecular weight. Such fractionated samples are especially useful when accurate physico-chemical data are desired for testing of particular theories, for they eliminate the necessity for faith in distribution functions derived on the basis of idealized assumptions.

8d. Molecular weight averages. Molecular weight distributions such as that described in section 8b may be treated statistically to yield several possible means or medians which could serve as a single over-all description of the average degree of polymerization or average molecular weight. Unfortunately, it is not possible to choose by convention a certain type of average and to label a polymer mixture consistently by this average, for it will be found that the laws governing the dependence of a given physical property on molecular weight also dictate the kind of averaging procedure which must be applied. Thus in considering different physical properties we are inevitably compelled to use different molecular weight averages. Especially important are the number average and weight average. Other averages encountered are the viscosity average and the "z-average."

It should be noted that in the approximation discussed in section 8a, applicable to polymer samples of relatively high degree of polymerization, there is a constant relation between average molecular weight and average degree of polymerization. When $\bar{M}$ and $\bar{x}$ represent, respectively, any of the averages (the particular type will be designated by subscripts), equation 8–1 gives

$$\bar{M} = M_0 \bar{x} \tag{8–7}$$

where M_0, as previously indicated, is considered a constant independent of $\bar{x}$ or $\bar{M}$ for any given kind of polymer.

The *number-average molecular weight*, $\bar{M}_n$, is obtained by adding the number of molecules, each multiplied by its molecular weight, and dividing by the total number of molecules; i.e., for *any* mixture of molecules,

$$\bar{M}_n = \frac{\sum_i N_i M_i}{\sum_i N_i} \tag{8–8}$$

where N_i is the number of molecules of kind i present in the mixture, and M_i is their molecular weight. The summation extends over all the kinds of molecules in the mixture.

In place of the number of molecules, N_i, we could just as well use the number of moles, n_i, or the number of molecules or moles per unit volume, or the mole fraction. The mole fraction is especially useful since $\sum_i X_i = 1$ so that

$$\bar{M}_n = \sum_i X_i M_i \tag{8–9}$$

When the mixture with which we are dealing is a pure polymer sample, the molecules differing only in degree of polymerization, the various kinds of molecules are the molecules with x ranging from one to infinity, and their molecular weights, within our approximation, are $M_x = xM_0$ so that

$$\bar{M}_n = \sum_{x=1}^{\infty} X_x M_x = M_0 \sum_{x=1}^{\infty} xX_x \qquad (8\text{--}10)$$

The *number-average degree of polymerization* is defined in a similar way,

$$\bar{x}_n = \sum_{x=1}^{\infty} xN_x \bigg/ \sum_{x=1}^{\infty} N_x = \sum_{x=1}^{\infty} xX_x \qquad (8\text{--}11)$$

where N_x is the number of x-mer molecules, or their number per unit volume.

The *weight-average molecular weight*, $\bar{M}_w$, is obtained by adding the number of grams, g_i, of material with molecular weight M_i, each multiplied by this molecular weight, and dividing by the total number of grams. Again, we can equally well use grams per milliliter or weight fraction, W_i, in place of the number of grams so that

$$\bar{M}_w = \sum_i g_i M_i \bigg/ \sum_i g_i = \sum_i W_i M_i \qquad (8\text{--}12)$$

Once again, when we are dealing with a pure polymer,

$$\bar{M}_w = \sum_{x=1}^{\infty} W_x M_x = M_0 \sum_{x=1}^{\infty} xW_x \qquad (8\text{--}13)$$

In a similar way we can define the *weight-average degree of polymerization*,

$$\bar{x}_w = \sum_{x=1}^{\infty} xg_x \bigg/ \sum_{x=1}^{\infty} g_x = \sum_{x=1}^{\infty} xW_x \qquad (8\text{--}14)$$

where g_x is the number of grams of x-mer.

The general definition of $\bar{M}_w$ may be written another way. Since the weight of the ith kind of molecule present in a mixture, g_i, is clearly $N_i M_i / \mathcal{N}$, where $\mathcal{N}$ is Avogadro's number,

$$\bar{M}_w = \sum_i N_i M_i^2 \bigg/ \sum_i N_i M_i = \sum_i X_i M_i^2 \bigg/ \sum_i X_i M_i \qquad (8\text{--}15)$$

The number-average and weight-average molecular weights of a given mixture can be quite different. If we have equal numbers of two kinds of molecules, with molecular weights 1000 and 100,000, respectively, for example, then $\bar{M}_n = 50,500$, but $\bar{M}_w = 99,020$. If we had equal weights of the same molecules, $\bar{M}_w$ would be 50,500, but $\bar{M}_n$ would be only 1980.

The distribution described in section 8b also shows a large difference between $\bar{M}_n$ and $\bar{M}_w$. Combining equations 8–3 and 8–4 with equations 8–10 and 8–13 and using the equations of Appendix A, we get

$$\bar{M}_n = M_0(1 - p) \sum_{x=1}^{\infty} xp^{x-1} = M_0/(1 - p) \qquad (8\text{--}16)$$

and

$$\bar{M}_w = M_0(1 - p)^2 \sum_{x=1}^{\infty} x^2 p^{x-1} = M_0(1 + p)/(1 - p) \qquad (8\text{--}17)$$

so that $\bar{M}_w/\bar{M}_n = 1 + p$. Since p is very close to unity if appreciable polymerization has occurred, the weight-average molecular weight in this case must be twice as great as the number average.

The ratio $\bar{M}_w/\bar{M}_n$ is a useful measure of the spread of a polymer distribution. For a completely homogeneous macromolecular substance, all the molecules of which have the same mass, $\bar{M}_w = \bar{M}_n$. For synthetic polymers of the type chosen here as an example, we have seen that $\bar{M}_w/\bar{M}_n = 2$. Larger values indicate a very wide spread in molecular weights, with substantial amounts of material at both extremes.

It is suggested at this point that reference be made to Table 17–1 in which the two kinds of molecular weight averages are compared for a variety of large molecules. It is seen that both highly homogeneous and highly heterogeneous samples are encountered in practice. Both kinds of molecules which form true crystals, the globular proteins and the viruses, are seen to be essentially homogeneous, as is to be expected from the fact that true crystals can be formed.

Weight-average and number-average molecular weights are the most frequently encountered. Higher averages are of interest only occasionally; they have been given the names z-average,[8] (z + 1)-average, etc., the definitions of them, with equations 8–8 and 8–15 included for comparison, being

$$\bar{M}_n = \sum_i N_i M_i \Big/ \sum_i N_i$$
$$\bar{M}_w = \sum_i N_i M_i^2 \Big/ \sum_i N_i M_i$$
$$\bar{M}_z = \sum_i N_i M_i^3 \Big/ \sum_i N_i M_i^2 \qquad (8\text{--}18)$$
$$\bar{M}_{z+1} = \sum_i N_i M_i^4 \Big/ \sum_i N_i M_i^3$$

the summation in each case extending over all possible values of i. If a sample is completely homogeneous, then $\bar{M}_n = \bar{M}_w = \bar{M}_z = \bar{M}_{z+1}$. If it is not homogeneous, then invariably

$$\bar{M}_n < \bar{M}_w < \bar{M}_z < \bar{M}_{z+1}$$

8e. Experimental determination of molecular weight averages. We have defined above several kinds of molecular weight averages, in relation to the distribution of molecules among the various possible molecular weights. Experimentally, of course, the distribution functions are much less easy to evaluate than the molecular weight averages. Many physical properties of macromolecular solutions are always or sometimes dependent on molecular weight. This is true, for example, of colligative properties (especially osmotic pressure), light scattering, sedimentation, and viscosity. Each of these experimental methods, when applied to a pure homogeneous macromolecular substance, may yield a value for its molecular weight; when applied to a heterogeneous mixture, such as a synthetic polymer preparation, it will yield a molecular weight average. The underlying theory will be discussed in due course in subsequent chapters. In the present chapter we shall merely show how different methods can lead to different kinds of averages.*

The simplest method of determining molecular weight is by end group analysis. It is also a hazardous one, for it depends on the certain knowledge of the number of end groups possessed by each molecule. Usually there is some doubt concerning this factor, for, for example, we can never completely eliminate the possibility that chain branching might have occurred during the preparation of a linear polymer. By assuming that no such undesired reaction has occurred, a measurement of the number of end groups is a count of the number of molecules, and the total mass present then gives us an average molecular weight. This is clearly a number average. In the case of the polyester discussed in section 8b, for example, the factor p was defined as the fraction of end groups which had reacted. It is easily measurable by titration. Equation 8–2 showed that this factor immediately gives us a measure of the number-average degree of polymerization and, hence, the number-average molecular weight, without knowledge of the distribution. From equation 8–2 we obtain at once, $\bar{M}_n = M_0/(1 - p)$. This answer, of course, is identical with equation 8–16, which was evaluated from the definition of $\bar{M}_n$ *after* we had decided what the distribution in this particular case (based on the same assumptions) should be.

Physical methods are more unequivocal tools for molecular weight determination than end group analysis because they require no assumptions about what has occurred during synthesis, just as they give no direct information about the distribution. For example, the colligative properties of dilute solutions are proportional to the total concentration of all molecules, here expressed in terms of moles per liter. Let C_i be the molar

---

* A method for the experimental study of molecular weight distributions will be described in section 16.

concentration of a given kind of molecule, and let G be any measured colligative property. Then,

$$G = k \sum_i C_i \tag{8-19}$$

where k is an appropriate constant, and the summation is extended over all the different kinds of molecules present. We do not, of course, know the values of the various C_i; we know usually only the weight concentration of dissolved material. If c_i is the concentration of each different species in grams per milliliter so that $c_i = M_i C_i / 1000$, the only quantity known is $c = \sum_i c_i = \sum_i M_i C_i / 1000$. Dividing both sides of equation 8–19 by c, we obtain

$$\frac{G}{c} = 1000k \frac{\sum_i C_i}{\sum_i C_i M_i} = 1000k \frac{\sum_i N_i}{\sum_i N_i M_i} \tag{8-20}$$

By equation 8–8, however, the quotient of summations in equation 8–20 is the reciprocal of $\bar{M}_n$; so that

$$G/c = 1000k/\bar{M}_n \tag{8-21}$$

i.e., G/c is a measure of *number-average* molecular weight.

Alternatively, we may consider the intensity of scattered light, which is proportional to the product of weight concentration and molecular weight. (See Chapter 5.) Again using G to represent the measured quantity and k as the constant relating it to concentration, we thus have in this case

$$G = k \sum_i c_i M_i \tag{8-22}$$

Again dividing both sides by c, we have

$$\frac{G}{c} = k \frac{\sum_i c_i M_i}{\sum_i c_i} = k \frac{\sum_i g_i M_i}{\sum_i g_i} = k\bar{M}_w \tag{8-23}$$

Clearly, in this case the measured property gives the *weight-average* molecular weight.

Certain experimental procedures may yield more cumbersome averages than those defined by equation 8–18. It will be shown in Chapter 6, for instance, that the difference between the viscosity of a polymer solution and that of the corresponding solvent may often be proportional to the product of weight concentration and M^a, where a is a constant lying between 0.5 and 2.0. Thus it is $(G/c)^{1/a}$ which would yield the molecular

weight of a homogeneous preparation. The value of $(G/c)^{1/a}$ for a heterogeneous sample in this case yields the viscosity-average molecular weight,

$$\bar{M}_v = \left(\sum_i W_i M_i^a\right)^{1/a} = \left(\sum_i N_i M_i^{1+a} \middle/ \sum_i N_i M_i\right)^{1/a} \qquad (8\text{--}24)$$

$\bar{M}_v$ is equal to $\bar{M}_w$ when $a = 1.0$, but, for other values of a, $\bar{M}_v$ is equal to
none of the averages defined by equations 8–18. For $a < 1.0$, $\bar{M}_v$ is
intermediate in magnitude between $\bar{M}_n$ and $\bar{M}_w$.

9. AVERAGE DIMENSIONS

It has been pointed out that most synthetic linear polymer molecules,
and some natural ones, may take on numerous configurations of essentially
identical energy. Some of them may be densely coiled, and others may be
greatly extended in space. We shall call such molecules *flexible polymers*,
and it is the purpose of this section to describe suitable average dimensions
by which such molecules can be characterized. The average dimensions
will be evaluated for certain simple models.

There is, of course, no necessary connection between molecular weight
heterogeneity and configurational flexibility; i.e., molecules which are
homogeneous with respect to molecular weight may possess a flexible
structure and vice versa. The commonest molecules exhibiting flexible
structures are, however, the synthetic organic polymers, and they, as we
have seen, also show molecular weight heterogeneity. The discussion in
this section will be based on the possible configurations of a *single molecule*,
so that the results apply as they stand only when all molecules in a sample
have the same molecular weight. Extension to typical organic synthetic
polymers thus requires an additional averaging process. It does not
introduce any great difficulty, as the brief discussion in section 9k will show.

9a. Average end-to-end distance. Two average dimensions are commonly
used to describe the spatial extension of a flexible polymer molecule, one of
them being the average end-to-end distance, the other the average radius of
gyration.

The average end-to-end distance (h_{av}) is simply the root-mean-square
average of the separation between the two ends of the polymer chain; i.e.,
where h is the distance between the ends in a given configuration,

$$h_{av} = (\overline{h^2})^{1/2}$$

the averaging process being carried out over all possible configurations.
To evaluate this quantity theoretically for simple models, it is best expressed

in terms of vectors. If **h** is the vector drawn from one end of the polymer chain to the other, in a particular configuration, then $\mathbf{h} \cdot \mathbf{h} = h^2$, and

$$h_{av} = (\overline{h^2})^{1/2} = (\overline{\mathbf{h} \cdot \mathbf{h}})^{1/2} \qquad (9\text{-}1)$$

9b. Average radius of gyration. To define the radius of gyration we consider a polymer chain (or, for that matter, any other kind of molecule) as an assembly of mass elements of mass m_i, each located a distance r_i from the *center of mass*. The radius of gyration, R, for a given configuration, is then defined as the square root of the weight average of r_i^2 for all the mass elements; i.e.,

$$R^2 = \sum_i m_i r_i^2 \Big/ \sum_i m_i \qquad (9\text{-}2)$$

For flexible chains R depends on the configuration. We shall be interested again in the root-mean-square average over all configurations. Calling this average R_G, we have

$$R_G = (\overline{R^2})^{1/2} = \left(\overline{\sum_i m_i r_i^2 \Big/ \sum_i m_i} \right)^{1/2}$$

where $\sum_i m_i$ is, of course, independent of configuration, so that only the numerator of the right-hand side of equation 9-2 need be averaged.

It will be noted that $\overline{\sum_i m_i r_i^2}$ is the sum of $m_i r_i^2$ over all values of i and over all configurations, divided by the total number of configurations. The order in which these mathematical operations are carried out does not affect the result so that $\overline{\sum_i m_i r_i^2}$ can be replaced by $\sum_i \overline{m_i r_i^2}$, giving

$$R_G = (\overline{R^2})^{1/2} = \left(\sum_i \overline{m_i r_i^2} \Big/ \sum_i m_i \right)^{1/2} \qquad (9\text{-}3)$$

A further simplification is possible for most flexible polymer chains (and sometimes for rigid molecules, too, but we are not concerned with these here), which is that they can generally be considered as assemblies of mass elements of *identical* mass, so that all m_i of equation 9-3 are the same, and

$$R_G = (\overline{R^2})^{1/2} = \left(\sum_i \overline{r_i^2} / \sigma \right)^{1/2} \qquad (9\text{-}4)$$

where σ is the total number of mass elements per chain.

The symbol σ will be used throughout the discussion of the dimensions of polymer molecules. It will be used interchangeably to mean the number of mass elements per chain or the number of bonds between them. The latter quantity is less by 1 than the former, but this difference is within the limits of error for all values of σ with which we shall be concerned. In simple situations $\sigma = x$, the degree of polymerization, but this is not always true, for the logical structural unit for dimensional considerations is often different from the logical monomer unit for designating the degree of polymerization. For polystyrene, for example, as will be seen below, the logical unit for computing dimensions is a C—C single bond; i.e., there are two such units for each styrene monomer.

The reader may question the statement made above, that flexible chains can generally be considered as having mass elements of equal mass, since individual atoms or groups of atoms clearly differ in mass. It is easy to see, however, that this will not affect the result provided that the actual elements of different mass occur regularly or randomly throughout the polymer chain. For in that case every different variety of atom will on the average have the same root-mean-square displacement from the center of mass. Thus, in a polystyrene chain,

$$\cdots -CH_2-CH-CH_2-CH- \cdots$$
$$\hspace{2.2cm} | \hspace{1.6cm} |$$
$$\hspace{2.3cm} C_6H_5 \hspace{1.1cm} C_6H_5$$

the value of R_G *based on the backbone C atoms alone* is the same as that based on the phenyl ring C atoms, or on the H atoms, and we can, without error, consider all the mass of each monomer unit concentrated at the position of any one of these atoms or at any other desired point (or several points) within the space actually occupied by the monomer unit. Only if the different mass units of a polymer chain are arranged in a regular non-repeating pattern, e.g., if all units of one kind are in the center of the polymer chain and those of another kind near the ends, will the assumption of equal mass units be invalid.

For subsequent calculation it will be desirable, as in the case of h_{av}, to express the radius of gyration in vector notation. Thus, where r_i is the vector from the center of mass to the ith mass element,

$$R_G = (\overline{R^2})^{\frac{1}{2}} = \left[\sum_i \overline{(r_i \cdot r_i)}/\sigma \right]^{\frac{1}{2}} \tag{9-5}$$

It will be shown in section 9j that a unique relation exists between h_{av} and R_G for all flexible linear (unbranched) chains. For such chains it will therefore be necessary to discuss in detail only one of these quantities. We shall choose the end-to-end distance, and the following sections will be devoted to this subject.

9c. The completely unrestricted polymer chain. It will be instructive to consider first a hypothetical problem, that of a completely unrestricted linear polymer chain. Let there be $\sigma + 1$ equivalent elements, joined by σ bonds. Let these bonds be all of fixed but different lengths, and let all bond angles be equally probable. Such a chain is illustrated by Fig. 9–1.

Let the length and direction of each bond be described by a vector $\mathbf{l}_i$, so that the end-to-end vector in a given configuration becomes $\mathbf{h} = \sum\limits_{i=1}^{\sigma} \mathbf{l}_i$. The square of the average end-to-end distance, by equation 9–1 is then

$$\overline{h^2} = \overline{\mathbf{h} \cdot \mathbf{h}} = \overline{\left(\sum_{i=1}^{\sigma} \mathbf{l}_i\right) \cdot \left(\sum_{j=1}^{\sigma} \mathbf{l}_j\right)}$$

The subscript j in the second sum has the same meaning as the subscript i. A different symbol is used merely to indicate that each term in the first

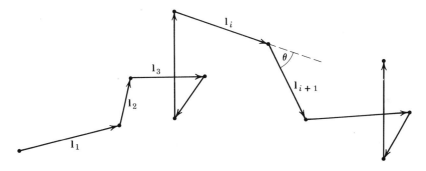

Fig. 9–1. The completely unrestricted polymer chain. The bond lengths l_i are arbitrary, and the angles θ may take on all possible values with equal probability.

sum is to be multiplied by each term in the second sum. Carrying out this multiplication, we see that

$$\overline{h^2} = \sum_{i=1}^{\sigma} \sum_{j=1}^{\sigma} \overline{(\mathbf{l}_i \cdot \mathbf{l}_j)} \qquad (9\text{--}6)$$

Let θ designate the angle between the positive directions of any two successive bonds. The scalar product $\mathbf{l}_i \cdot \mathbf{l}_{i+1}$ is then equal to $l_i \, l_{i+1} \cos \theta$, where l_i and l_{i+1} are the absolute values of the lengths of the two successive bonds (i.e., $l_i^2 = \mathbf{l}_i \cdot \mathbf{l}_i$). For a completely unrestricted polymer chain, however, the angle θ may take on, in different configurations, all possible values with equal probability. Averaging over all configurations, therefore, we have

$$\overline{\mathbf{l}_i \cdot \mathbf{l}_{i+1}} = l_i l_{i+1} \overline{\cos \theta} = 0 \qquad (9\text{--}7)$$

for positive and negative values of θ are equally likely.

Similarly, all other terms of the sum of equation 9–6, with i different from j, must vanish. The only terms remaining are therefore the terms

$\mathbf{l}_i \cdot \mathbf{l}_i$, each of which, of course, is merely the square of the length of the ith bond, $l_i{}^2$. Thus equation 9–6 becomes

$$\overline{h^2} = \sum_{i=1}^{\sigma} l_i{}^2 = \sigma \overline{l_i{}^2} = \sigma l_{av}{}^2 \tag{9–8}$$

where $(\overline{l_i{}^2})^{\frac{1}{2}} = l_{av}$ is the average (root-mean-square) bond length in the molecule.

If the bond lengths in an otherwise unrestricted chain are all identical, and equal to l, equation 9–8 would be altered only to the extent that l^2 would appear in place of $l_{av}{}^2$.

9d. The polymethylene chain, assuming free rotation about the bonds. The long chain polymethylene hydrocarbons, as obtained by polymerization of ethylene or diazomethane consist of chains of carbon atoms, with identical bond lengths, l, and a fixed angle, θ, between the positive directions of successive bonds (Fig. 9–1). The values of these parameters are $l = 1.54$ Å and $\theta = 180° - 109° 28' = 70° 32'$, $109° 28'$ being the tetrahedral angle included between successive bonds.

A hydrocarbon chain of $\sigma + 1$ carbon atoms may again be represented by σ vectors, $\mathbf{l}_i$, as in Fig. 9–1; and $\overline{h^2}$ may again be evaluated by means of equation 9–6, with the restriction of constant bond length and angle. The restriction of constant bond angle clearly makes equations 9–7 and 9–8 invalid. Figure 9–2 shows three successive bonds in such a chain, the first two being in the plane of the paper. With the aid of this diagram, the individual terms of the summation of equation 9–6 may be evaluated. There will be, first of all, σ terms with $i = j$, i.e., terms of the form $\mathbf{l}_i \cdot \mathbf{l}_i$. Each of these is equal to l^2, where l is the constant bond length. There will then be $2(\sigma - 1)$ terms of the form $\overline{\mathbf{l}_i \cdot \mathbf{l}_{i+1}}$, $\sigma - 1$ of them arising from values of i from 1 to $\sigma - 1$ with j from 2 to σ, the other $\sigma - 1$ arising from values of j from 1 to $\sigma - 1$ with i from 2 to σ. Each of these terms is clearly equal to $l^2 \cos \theta$.

Next come $2(\sigma - 2)$ terms of the form $\overline{\mathbf{l}_i \cdot \mathbf{l}_{i+2}}$, arising from $i = 1$ to $\sigma - 2$ and $j = 3$ to σ and from $j = 1$ to $\sigma - 2$ and $i = 3$ to σ. To evaluate these terms we split $\mathbf{l}_{i+2}$ into two components, $\mathbf{l}_a$ parallel to $\mathbf{l}_{i+1}$ and $\mathbf{l}_b$ perpendicular to $\mathbf{l}_{i+1}$, so that

$$\overline{\mathbf{l}_i \cdot \mathbf{l}_{i+2}} = \overline{\mathbf{l}_i \cdot \mathbf{l}_a} + \overline{\mathbf{l}_i \cdot \mathbf{l}_b} \tag{9–9}$$

Referring now to Fig. 9–2, we see that the lengths of $\mathbf{l}_a$ and $\mathbf{l}_b$ are fixed, $l_a = l \cos \theta$, $l_b = l \sin \theta$. However, the terminal point of the vector $\mathbf{l}_b$ may lie anywhere along the circumference of the circle C in that figure. The assumption of free rotation which we are here making is equivalent to the assumption that all values of the angle ϕ in Fig. 9–2 are equally probable. Thus $\overline{\mathbf{l}_i \cdot \mathbf{l}_b}$ is clearly zero, for if we imagine the vector $\mathbf{l}_i$ drawn so as to terminate at the center of the circle C, the value of $\mathbf{l}_i \cdot \mathbf{l}_b$ for a given configuration will be equal to $l_i l_b \cos \alpha$, where α is the angle between $\mathbf{l}_i$ and $\mathbf{l}_b$; i.e., it will be equal to $l^2 \sin \theta \cos \alpha$, where only α varies from configuration to configuration. For any configuration with a given value of

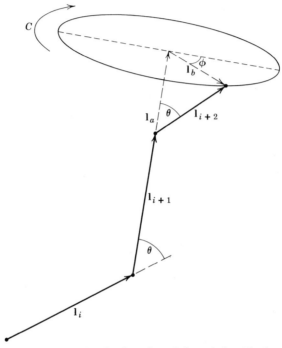

Fig. 9–2. Three successive bonds of a polymethylene chain. The first two are in the plane of the figure. The terminus of $\mathbf{l}_{i+2}$ may lie anywhere on the circle C.

α, however, there is an equally probable configuration for which this angle has the value $180° - \alpha$, i.e., for which $\mathbf{l}_i \cdot \mathbf{l}_b$ is equal to $-l^2 \sin \theta \cos \alpha$. The average value of $\mathbf{l}_i \cdot \mathbf{l}_b$ must therefore be zero.

Since $\mathbf{l}_i \cdot \mathbf{l}_a$ is equal to $l_i l_a \cos \theta$ and l_a itself is equal to $l \cos \theta$, equation 9–9 becomes

$$\overline{\mathbf{l}_i \cdot \mathbf{l}_{i+2}} = l^2 \cos^2 \theta \qquad (9\text{–}10)$$

We may consider in the same way the $2(\sigma - 3)$ products of the form $\overline{\mathbf{l}_i \cdot \mathbf{l}_{i+3}}$. Again $\mathbf{l}_{i+3}$ may be split into components of length $l \cos \theta$ parallel to $\mathbf{l}_{i+2}$ and of length $l \sin \theta$ perpendicular to $\mathbf{l}_{i+2}$. The average value of the product of the latter with $\mathbf{l}_i$ is again zero. The component parallel to $\mathbf{l}_{i+2}$ may then be split again into a component of length $l \cos^2 \theta$ parallel to $\mathbf{l}_{i+1}$ and a component of length $l \cos \theta \sin \theta$ perpendicular to $\mathbf{l}_{i+1}$. The average product of the perpendicular component with $\mathbf{l}_i$ again vanishes, whereas the product of the parallel component with $\mathbf{l}_i$ becomes $l^2 \cos^3 \theta$, so that

$$\overline{\mathbf{l}_i \cdot \mathbf{l}_{i+3}} = l^2 \cos^3 \theta$$

In general, there will be $2(\sigma - k)$ terms of the type $\overline{\mathbf{l}_i \cdot \mathbf{l}_{i+k}}$, the value of each term being $l^2 \cos^k \theta$. Thus equation 9–6 becomes

$$\overline{h^2} = l^2[\sigma + 2(\sigma - 1) \cos \theta + 2(\sigma - 2) \cos^2 \theta + \cdots$$
$$+ 2(\sigma - k) \cos^k \theta + \cdots + 2 \cos^{\sigma - 1} \theta] \quad (9\text{–}11)$$

Equation 9–11 was first derived by Eyring.[9]

For a polymethylene chain, as was mentioned above, $\theta = 70° 32'$ and therefore $\cos \theta = 0.333$. Thus all but the first few terms of equation 9–11 must be negligibly small. We shall furthermore be dealing usually with chains where σ is quite large, say several hundred or greater. Under these conditions we may replace $\sigma - k$ for all the terms which are not negligible (i.e., for terms with small k) by σ. In addition, since all but the first few terms are negligible, the finite series of equation 9–11 may be replaced by the corresponding infinite series, so that we obtain

$$\overline{h^2} = l^2\sigma(1 + 2 \cos \theta + 2 \cos^2 \theta + 2 \cos^3 \theta + \cdots) \quad (9\text{–}12)$$

which, since $\cos \theta$ must be smaller than unity, is equivalent (Appendix A) to

$$\overline{h^2} = \sigma l^2 \frac{1 + \cos \theta}{1 - \cos \theta} \quad (9\text{–}13)$$

For a polymethylene chain, substituting $\cos \theta = 0.333$, we thus get $\overline{h^2} = 2.00\sigma l^2$. Comparing this result with equation 9–8 we see that the effect of fixing the bond angle, instead of allowing it to take on any value with equal probability, has been the introduction of a numerical constant, in this case equal to 2.00.

The result here given for a polymethylene chain will of course apply to any chain of singly bonded carbon atoms, regardless of lateral substituents, for only the bond lengths and angles of the backbone of the polymer chain have entered into the calculation.

Equation 9–13, with appropriate values of l and θ, would also be applicable to any other long flexible chain of identical elements, joined by bonds of fixed length and fixed bond angle. If the bond angle is $90°$, the equation reduces to equation 9–8. If the angle included between successive bonds (i.e., $180° - \theta$) is *obtuse*, as in the case of the polymethylene chain, h_{av} always becomes larger than equation 9–8 would predict; if the angle is *acute*, h_{av} becomes smaller than equation 9–8 would predict.

Equation 9–13 will be incorrect if $\cos \theta$ is close to unity, i.e., if θ is close to zero ($\theta \sim 180°$ being physically impossible), for then the derivation of equation 9–12 from 9–11 becomes invalid. However, a chain with θ close to zero will not be representative of a flexible molecule with numerous configurations of equivalent energy but will correspond more closely to a rigid rodlike molecule with little flexibility. Needless to say, equation 9–13 is also inapplicable to short chains, as, for example, saturated normal hydrocarbons with fewer than 100 carbon atoms. In this case equation 9–11 must be used.

9e. The effect of restricted rotation. The equations of the preceding section were based on the hypothesis of freedom of rotation about the bonds of the polymer chain. In actual fact, it is well known that rotation about single bonds is generally restricted, at least to the extent that certain values of the angle ϕ (Fig. 9–2) become more probable than others. It has been shown by Benoit and Sadron[10] and by others that this leads to the introduction of an additional multiplying factor into equation 9–13. For example, where the restriction of rotation is not severe, the equation for $\overline{h^2}$ becomes

$$\overline{h^2} = \sigma l^2 \frac{1 + \cos \theta}{1 - \cos \theta} \frac{1 + \overline{\cos \phi}}{1 - \overline{\cos \phi}} \tag{9–14}$$

where $\overline{\cos \phi}$ is the average value of $\cos \phi$. For completely free rotation, of course, $\overline{\cos \phi} = 0$, and equation 9–14 reduces to equation 9–13.

Equation 9–14 is not of general utility, however, because the theoretical calculation of $\overline{\cos \phi}$ is ordinarily not possible. In general, therefore, an empirical treatment is used to take the effects of restricted rotation into account. This treatment will be discussed in section 9g.

9f. Other simple flexible polymer chains. A treatment similar to that of section 9d may be applied to other types of flexible polymer chains. The repeating unit of a cellulose chain, for example, may be thought of as successive bonds of length a, b, and a, as shown in Fig. 9–3, with fixed angles α and θ as indicated there. Allowing free rotation about the C—O bonds, Benoit[11] has shown that

$$\overline{h^2} = \sigma \left[(b \sin \alpha)^2 + (2a + b \cos \alpha)^2 \frac{1 + \cos \theta}{1 - \cos \theta} \right] \tag{9–15}$$

Other expressions of this type have been derived for silicone and other chains.[1]

9g. General ideal expression for the dependence of h_{av} on chain length.
Examination of the equations of sections 9c to 9f shows that an important
generalization can be made. Each of the equations for $\overline{h^2}$ is seen to consist
of two factors, one containing terms depending only on the *nature of the
polymer*, such as the bond length *l*, the bond angle *θ*, or the degree of
restriction of rotation. The other factor is independent of the nature of
the polymer and depends only on chain length. In all the cases examined,
for long flexible chains, this factor is merely the number of chain elements,
σ. Thus we may write, *for all long flexible chains*,

$$\overline{h^2} = \beta^2 \sigma \tag{9–16}$$

where β^2 is a constant characteristic of the nature of the polymer. This

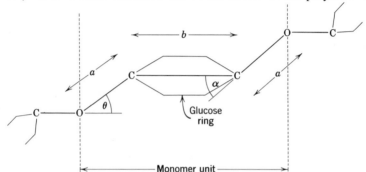

Fig. 9–3. The cellulose chain, showing the parameters used in equation 9–15.

constant should be independent of the solvent in which the polymer may
be dissolved and, since it reflects the relative configuration of immediately
adjacent polymer segments, should not depend on molecular weight. It
will, however, depend on temperature, because freedom of rotation in
molecules is temperature dependent.

The reservation should be made here that interaction between *distant* polymer
segments and between polymer segments and solvent molecules, both of which will be
considered in section 9h, will require further modification of equation 9–16. Polymer
solutions in which such interactions produce no net effect are called *ideal*, so that the
generalization inherent in equation 9–16 applies, without the introduction of additional
terms, to ideal solutions only.

The essence of equation 9–16 is that flexible chains restricted by rotation
may be treated mathematically as if they were completely unrestricted
chains with a bond length *β* which will be larger than the true average
bond length l_{av}. We have seen that $\beta = 2^{\frac{1}{2}} l_{av}$ for a carbon–carbon chain
with free rotation and that its value increases as further restrictions are

imposed. We shall see in Chapters 5 and 6 that experimental values of β can be determined from appropriate measurements, and it is worth noting for subsequent discussion that typical synthetic organic polymers are found to have $\beta \simeq 3l_{av}$. Quite generally, β/l_{av} will be regarded as a measure of the lack of rotational freedom, or of the stiffness of a polymer chain.

That equation 9–16 should be generally applicable to all flexible polymers, regardless of restriction upon rotation, has been demonstrated by W. Kuhn.[12] The proof requires that we inquire a little more closely into what we mean by a "flexible" chain; for our purposes it may be defined as a chain, any part of which is randomly situated with respect to another part. For the completely unrestricted polymer chain (section 9c) "any part" meant adjacent mass units. For polymethylene chains (section 9d), however, adjacent mass units are a definite distance apart and their bonds at a fixed angle to one another. As Fig. 9–2 shows, randomness in position occurs only when the relation between a mass unit and the third following unit is considered. (The position of the tenth following unit, say, would be essentially completely random with respect to the starting point.) Generalizing this picture, we can define a "flexible" chain as a chain in which a vector drawn from any one structural unit to another structural unit m positions removed from it is capable of having any length or orientation if m is chosen sufficiently large. Chain molecules may thus be flexible despite considerable restriction in the relation between a structrual unit and its more immediate neighbors.

The m structural units taken together are called a statistical segment. A polymer molecule, if it is flexible by our definition, can then be divided into such statistical segments, as shown by Fig. 9–4. The number of vectors required to represent a configuration will be $\sigma' = \sigma/m$, where $\sigma + 1$ is the number of fundamental structural units, as before. The length of each vector, l_i', will, in general, be different, and the angles between them will be completely arbitrary, as is illustrated by Fig. 9–4, which shows as example two alternative positions of a vector l_2' following a vector l_1'. Provided only that the chain is long and m reasonably small, i.e., that $\sigma' = \sigma/m$ is still sufficiently large for statistical analysis, the polymer chain described in terms of the vectors l_i' becomes a completely unrestricted chain. Equation 9–8 is therefore applicable, and we obtain

$$\overline{h^2} = \sigma'\overline{l_i'^2} = \sigma'l_e^2 = \sigma l_e^2/m \qquad (9\text{–}17)$$

where l_e is the root mean square of the length of one of the statistical segments, often called the *Kuhn statistical segment length*.

Equation 9–17 is clearly identical to equation 9–16, with $\beta^2 = l_e^2/m$, and thus constitutes a proof of that equation.

It has been noted that equation 9–16 defines a hypothetical effective unrestricted chain which has the same number of segments, σ, as the real chain but with a greater segment length. An alternative hypothetical unrestricted chain is sometimes used, which has the same *contour length* as the real chain rather than the same number of segments. The contour length is defined as the length of all bonds laid end-to-end, so that, if the hypothetical chain has σ_K links of length l_K,

$$\sigma l_{av} = \sigma_K l_K \qquad (9\text{–}18)$$

This hypothetical chain differs, of course, from the statistical chain

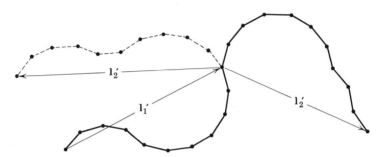

Fig. 9–4. Schematic two-dimensional diagram illustrating the representation of relatively stiff chains in terms of statistical lengths. All bonds have the same bond length and bond angle, and the stiffness is represented schematically in terms of a small angle between successive bonds, so that the direction of each bond is necessarily close to that of its predecessor. Despite this limitation, the vectors joining a mass element to the tenth following element are seen to be essentially randomly disposed. Two alternative allowed positions are shown for the vector $\mathbf{l}_2'$.

pictured in Fig. 9–4, since for the latter $\sigma' l_e$ is clearly considerably smaller than σl_{av}, and it also differs from the hypothetical chain of σ segments of length β, for which $\sigma\beta$ is much larger than $\sigma l_{av}'$. The relation between l_K and β is easily determined. By definition, $\overline{h^2} = \sigma\beta^2 = \sigma_K l_K{}^2$, which, together with equation 9–18, gives

$$l_K/l_{av} = (\beta/l_{av})^2$$

so that l_K/l_{av} gives exactly the same information as the ratio β/l_{av} which we shall use as a measure of rotational restriction.

One further aspect of the derivation of equation 9–16, i.e., equation 9–17, must be mentioned, and this is that the derivation has required the presence of a sufficiently large number of *statistical* segments to permit application of equation 9–11. Since each statistical segment replaces several ordinary monomer units, it follows that a more restricted chain must be longer than a more flexible one before equation 9–16 can be used. With greater restriction on rotation, it becomes increasingly likely that molecules available in practice will not be sufficiently large for applicability of equation 9–16. Molecules of this type are called stiff chains. For stiff chains $\overline{h^2}$ increases with σ more rapidly than σ, as will be discussed briefly in section 9i.

It has not been possible to design an experiment which will provide a direct measurement of h_{av}, so that the confirmation of the validity of equation 9–16 (modified, where necessary, by the effects of intramolecular interactions, as discussed in the following section) rests on measurement of the radius of gyration which will be described in later portions of this book. It should be noted, however, that the total average *dipole moment* of the polymer of a polar monomer is a property which should behave just like h_{av}. If the moment of each monomer unit is μ_0 and if there is sufficient randomness in the succession of monomers along the chains, $\overline{\mu^2}$ should be proportional to $\sigma\mu_0{}^2$. Thus the result of Marchal and Benoit, cited on p. 114, which showed that $\overline{\mu^2}$ for polyoxyethylene glycol is proportional, for long chains, to the chain length, is in effect a confirmation of equation 9–16.

To conclude this section, it may be noted that we shall often be interested in the relation between $\overline{h^2}$ and molecular weight rather than that between $\overline{h^2}$ and σ. Since $M = \sigma M_0$, for all σ large enough to be of interest, we can write equation 9–16 as

$$\overline{h^2} = \beta^2 M/M_0 \qquad\qquad (9\text{–}19)$$

The value of M_0 in this equation depends on the choice of the structural unit used to define σ (see p. 151), and it is not necessarily equal to the molecular weight of the ordinary monomer unit. For polystyrene, for example, it is logical to choose each C—C bond as the basic unit of known length and bond angle, so that each C atom along the chain backbone represents a structural unit. Thus σ is twice the number of styrene monomer units, and M_0 is half the molecular weight of such a unit, i.e., $M_0 = 52$.

9h. Interaction between polymer segments and solvent molecules, and its effect on the end-to-end distance. The equations of the preceding section are analogous to the equation of state for ideal gases in that they neglect secondary interaction between different polymer molecules and between segments of a single polymer chain, just as the ideal gas equation neglects secondary interaction between gas molecules. It will be recalled that in the case of gas molecules interaction leading to non-ideality is of two kinds: attraction due to dispersion forces, and repulsion (primarily at high pressure) due to the inherent inability of two molecules to occupy the same point in space. A parallel classification may be made for polymers with the difference that they are usually encountered in solution, so that forces between solvent and polymer molecules must be considered along with those between polymer molecules (or segments) alone.

There is one major difference between non-ideal gases and non-ideal polymer solutions. Secondary interaction in gases can be made insignificant by dilution to low pressures. In polymer solutions interaction between different polymer molecules can be reduced in the same way, by reducing concentration. However, interaction between different segments of the same polymer molecule can never be eliminated, for there is no way of "diluting" these segments. Intramolecular interaction of this type will obviously have an important effect on the dimensions of polymer molecules, even in the ideal limit of infinite dilution.

Considering first the inability of different segments of the same molecule to occupy the same space, it is easy to see that this will have two consequences.

(1) Our statistical treatment has represented a polymer molecule as dimensionless mass points joined by vectors. All possible configurations have been considered equally probable in evaluating $\overline{h^2}$. Among the configurations considered there must be some in which one or more real segments of finite volume would occupy space already occupied by other segments. Such configurations are impossible. Configurations of this kind will clearly occur more frequently among compact configurations with small end-to-end distance than among the more extended ones, so that the net effect is to increase $\overline{h^2}$. If the ideal value calculated by the preceding methods is called $\overline{h_0^2}$, the true mean square end-to-end distance, $\overline{h^2}$, will clearly be larger than $\overline{h_0^2}$. Flory and Fox[13] have introduced an empirical function, α, to take this into account. In terms of this function we may write, with the aid of equations 9–16 and 9–19,

$$\overline{h^2} = \alpha^2 \overline{h_0^2} = \alpha^2 \beta^2 \sigma = \alpha^2 \beta^2 M / M_0 \qquad (9\text{–}20)$$

(2) It is further obvious that the probability that any segment occupies space already filled by another segment increases with the number of segments in the chain; i.e., the empirical function α, unlike β, depends on the molecular weight.

Turning now to attractive forces, we note that they will have no effect in a pure liquid or amorphous polymer, for here each segment of a polymer molecule is completely surrounded by like segments, either from the same or other polymer molecules. There is therefore no preferred orientation to upset the statistical analysis of the preceding sections, and only the effect incorporated in equation 9–20 will occur. In polymer solutions, however (which we shall assume sufficiently diluted to make *inter*molecular forces unimportant), a preference for certain configurations will occur. We may distinguish between *good solvents*, in which attraction between polymer segments and solvent molecules is greater than that between polymer segments themselves, *poor solvents* in which the reverse is true, and *indifferent solvents*. For indifferent solvents we again need no further addition to equation 9–20. For good solvents and poor solvents, however, further modification is necessary.

This modification may be introduced without change in equation 9–20. In that equation α represented the effect of a physical volume from which one segment physically excludes all others. In a good solvent the presence of segments adjacent to one another, though not forbidden, is made less likely. Since α is just an empirical function, this effect can be taken care of by redefining α in terms of an effective exclusion volume which is larger than the physical volume; i.e., in a good solvent α is larger than otherwise. In the same way, in a poor solvent, the presence of segments adjacent to one another is preferred, and, since adjacency occurs more often in compact configurations, such configurations are preferred, partially canceling the effect of the physical excluded volume. Clearly this effect may be incorporated into equation 9–20 by making α smaller than otherwise. We thus conclude that equation 9–20 can describe all possible long-range effects in flexible polymers.

It is possible that the effect of attractive forces may exactly cancel the effect of the physical excluded volume. In this case $\alpha = 1$ and $\overline{h^2} = \overline{h_0^2}$. A solvent in which this occurs is called an *ideal solvent*. It is also possible that the attractive forces more than compensate for the physical excluded volume, so that $\overline{h^2} < \overline{h_0^2}$ and $\alpha < 1$. This requires a very poor solvent, however, so that the solution formed will become unstable, i.e., solvent-polymer solutions for which $\alpha < 1$ are in general unstable and separate into two phases, as is discussed in section 15.

Thus $\alpha = 1$ is essentially the minimum value of α. For both indifferent and good solvents α is an increasing function of molecular weight. A theoretical treatment (p. 201) will indicate that the maximum possible variation with M is described by the relation

$$\alpha = (\text{constant})M^{0.10} = (\text{constant})\sigma^{0.10} \qquad (9\text{--}21)$$

A typical polymer solution in a good solvent, which will be analyzed in detail in section 23, has

$$\alpha = 0.492M^{0.086} = 0.734\sigma^{0.086} \qquad (9\text{--}22)$$

In the solution described by equation 9–22, h_{av} exceeds the ideal value by 9% at molecular weight 10,000, by 32% at molecular weight 100,000, and by 61% at molecular weight 1 million.

A special case of intramolecular interaction exists for polyelectrolytes. Here electrostatic forces play a predominant role, excluding in extreme cases all but highly extended configurations. These polymeric substances will be discussed in detail in Chapter 7, where ways of calculating the electrostatic free energy will be considered.

Because of the omission of intermolecular effects, the discussion here is limited to very dilute solutions. Intermolecular interaction in such solutions, which will be seen to be closely related to the function α describing intramolecular interaction, will be discussed in section 12.

9i. Completely rigid chain polymers. Stiff chains. The preceding sections have introduced successively greater restrictions to the flexibility of polymer chains. Until interaction between distant portions of a chain was taken into account, however, the proportionality between the average end-to-end distance and the square root of the molecular weight or degree of polymerization was maintained. Consideration of such interaction brought about an increase in molecular weight dependence, such that

$$h_{av} = (\text{constant})M^{0.50-0.60} \qquad (9\text{–}23)$$

It is logical to inquire at this stage what the relation between $\overline{h^2}$ and M would be when all flexibility is lost, when a polymer chain becomes completely rigid. In this case each successive segment would clearly make an equal contribution to the end-to-end distance, as, for example, in a fully extended hydrocarbon chain. Thus the end-to-end distance, for which there would now be a unique value, h, since only a single configuration is allowed, would now be directly proportional to the molecular weight or degree of polymerization, so that

$$h_{av} = (\text{constant})M \qquad (9\text{–}24)$$

This relation applies both to fully extended rigid polymer chains and to perfectly rigid helical chains (e.g., Figs. 4–7 and Fig. 4–13).

Wada[22] has observed that the dipole moment of poly-γ-benzyl-L-glutamate is directly proportional to the molecular weight, which is the expected result if this polymer has the helical configuration illustrated by Fig. 4–7.

To complete the analysis of the dependence of h_{av} on M it is convenient to return to the concept of "stiff" chains introduced on p. 161. Such chains may be considered as intermediate between completely rigid and flexible chains. Although a treatment of such chains is beyond the scope of this book, it may reasonably be concluded that for chains of this type h_{av} is also intermediate in its behavior; i.e., we may expect that

$$h_{av} = (\text{constant})M^{0.60-0.99} \qquad (9\text{-}25)$$

In all cases of this kind, however, the exponent of M may be expected to decrease slowly with increasing molecular weight, since, as explained on p. 161, any flexibility at all will lead to an equation of the form of 9–16 (or 9–23 when long-range interactions are taken into account) if the molecular chains are sufficiently long.

9j. The radius of gyration. It was pointed out in section 9a that a unique relation exists between the average end-to-end distance of linear polymers

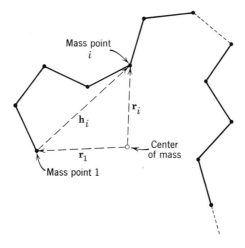

Fig. 9–5. Representation of the vectors used in relating the radius of gyration to the end-to-end distance.

and their average radius of gyration, as defined by equation 9–5. In this section this relation will be derived, using the method developed by Zimm and Stockmayer.[14] A similar relation was previously computed in another way by Debye.[15]

In Fig. 9–5 let the points represent the centers of mass of the individual identical elements of a flexible polymer chain. They may be the successive monomer units, fractions thereof, or statistical units of the kind discussed in section 9g. In other words, as long as it remains valid to concentrate the mass of such a statistical unit at a single point, the derivation here given will be quite general.

Let $\mathbf{h}_i$ represent a vector from the first mass point of the chain to the ith mass point. Let O in Fig. 9–5 represent the center of mass of the polymer molecule, and let $\mathbf{r}_i$ be the vector from O to the ith mass point. Then, for any value of i,

$$\mathbf{r}_i = \mathbf{r}_1 + \mathbf{h}_i \tag{9-26}$$

The center of mass is defined in such a way that $\sum_i m_i \mathbf{r}_i = 0$, where m_i is the mass of the ith mass point. Since all mass points are to be considered identical, this definition reduces to

$$\sum_i \mathbf{r}_i = \sigma \mathbf{r}_1 + \sum_i \mathbf{h}_i = 0 \tag{9-27}$$

where σ is the number of elements. From equation 9–27 we get

$$\mathbf{r}_1 = -\frac{1}{\sigma} \sum_i \mathbf{h}_i \tag{9-28}$$

The square of the radius of gyration, $R^2 = \sum_i r_i^2/\sigma$, for a given configuration, can now be evaluated. Using equation 9–26,

$$R^2 = \frac{1}{\sigma} \sum_{i=1}^{\sigma} (\mathbf{r}_1 + \mathbf{h}_i) \cdot (\mathbf{r}_1 + \mathbf{h}_i)$$

$$= r_1^2 + \frac{1}{\sigma} \sum_{i=1}^{\sigma} h_i^2 + \frac{2}{\sigma} \mathbf{r}_1 \cdot \left(\sum_{i=1}^{\sigma} \mathbf{h}_i \right) \tag{9-29}$$

By equation 9–28, however, $r_1^2 = \left(\sum_{i=1}^{\sigma} \sum_{j=1}^{\sigma} \mathbf{h}_i \cdot \mathbf{h}_j \right) \Big/ \sigma^2$ and $(2/\sigma)\left(\mathbf{r}_1 \cdot \sum_{i=1}^{\sigma} \mathbf{h}_i \right)$

$= -(2/\sigma^2) \left(\sum_{i=1}^{\sigma} \sum_{j=1}^{\sigma} \mathbf{h}_i \cdot \mathbf{h}_i \right)$, so that equation 9–29 becomes

$$R^2 = \frac{1}{\sigma} \sum_{i=1}^{\sigma} h_i^2 - \frac{1}{\sigma^2} \sum_{i=1}^{\sigma} \sum_{j=1}^{\sigma} \mathbf{h}_i \cdot \mathbf{h}_j \tag{9-30}$$

To evaluate the last term of this equation we apply the cosine rule to the triangle formed by mass points 1, i, and j. Where h_{ij} is the distance between the ith and jth mass point, and recalling that $\mathbf{h}_i \cdot \mathbf{h}_j$ is by definition the product of the lengths of the vectors h_i and h_j multiplied by the cosine of the angle between them, we get

$$h_{ij}^2 = h_i^2 + h_j^2 - 2\mathbf{h}_i \cdot \mathbf{h}_j \tag{9-31}$$

Solving for $\mathbf{h}_i \cdot \mathbf{h}_j$ and substituting in equation 9–30 thus gives

$$R^2 = \frac{1}{\sigma} \sum_{i=1}^{\sigma} h_i^2 - \frac{1}{2\sigma^2} \sum_{i=1}^{\sigma} \sum_{j=1}^{\sigma} (h_i^2 + h_j^2 - h_{ij}^2) \tag{9-32}$$

Since $\displaystyle\sum_{i=1}^{\sigma} \sum_{j=1}^{\sigma} h_i{}^2 = \sum_{i=1}^{\sigma} \sum_{j=1}^{\sigma} h_j{}^2 = \sigma \sum_{i=1}^{\sigma} h_i{}^2$, equation 9–32 reduces to

$$R^2 = \frac{1}{2\sigma^2} \sum_{i=1}^{\sigma} \sum_{j=1}^{\sigma} h_{ij}{}^2 \tag{9–33}$$

Averaging over all possible configurations, we obtain for the square of the *average radius of gyration*

$$R_G{}^2 = \overline{R^2} = \frac{1}{2\sigma^2} \sum_{i=1}^{\sigma} \sum_{j=1}^{\sigma} \overline{h_{ij}{}^2} \tag{9–34}$$

The average value of $h_{ij}{}^2$, however, is just the average end-to-end distance of a chain containing $|j - i|$ elements, where the vertical bars signify the absolute numerical value regardless of sign. For the ideal case, where physical interference is neglected, this average is given by equation 9–16,

$$\overline{h_{ij}{}^2} = \beta^2 |j - i| \tag{9–35}$$

Since the average end-to-end distance of the entire chain of σ elements is $\beta^2 \sigma$,

$$\overline{h_{ij}{}^2} = |j - i|\, \overline{h^2}/\sigma \tag{9–36}$$

It remains to evaluate $\displaystyle\sum_{i=1}^{\sigma} \sum_{j=1}^{\sigma} |j - i|$. Consider first the sum over j. This may be split into two parts: the terms with $j < i$ and those with $j > i$, $|j - i|$ being equal to $i - j$ for the former and to $j - i$ for the latter. Thus

$$\sum_{j=1}^{\sigma} |j - i| = \sum_{j=1}^{i} (i - j) + \sum_{j=i+1}^{\sigma} (j - i)$$
$$= i^2 - \tfrac{1}{2}i(i + 1) + \tfrac{1}{2}(\sigma - i)(\sigma + i + 1) - i(\sigma - i)$$
$$= i^2 - i\sigma + \tfrac{1}{2}\sigma^2 + \tfrac{1}{2}\sigma - i$$

Summing now over all values of i, using the fact that $1^2 + 2^2 + \cdots + \sigma^2 = \sigma(\sigma + 1)(2\sigma + 1)/6$, we get

$$\sum_{i=1}^{\sigma} \sum_{j=1}^{\sigma} |j - i| = (\sigma^3 - \sigma)/3 \tag{9–37}$$

We are interested only in large values of σ and may neglect σ in comparison with σ^3. Combining with equations 9–34 and 9–36 then leads to

$$R_G{}^2 = \overline{R^2} = (\overline{h^2}/2\sigma^3) \sum_{i=1}^{\sigma} \sum_{j=1}^{\sigma} |j - i| = \overline{h^2}/6 \tag{9–38}$$

The desired relation between radius of gyration and end-to-end distance, for the ideal situation where there is no physical interference, has thus been established.

If we wish to take into account intramolecular interaction of the kind discussed in section 9h, the value of $\overline{h_{ij}{}^2}$ given by equation 9–35 will be too small. Introducing an empirical factor α' we can write instead $\overline{h_{ij}{}^2} = \alpha'^2\beta^2|j - i|$. In place of the relation $\overline{h^2} = \beta^2\sigma$, we use, according to equation 9–20, $\overline{h^2} = \alpha^2\beta^2\sigma$, so that equation 9–36 becomes

$$\overline{h_{ij}{}^2} = (\alpha'^2/\alpha^2)|j - i|\overline{h^2}/\sigma \qquad (9\text{--}39)$$

Since both α'^2 and α^2 are empirical parameters, we have no way of deciding how their ratio might behave. It seems reasonable, however, to suppose that it is unity, i.e., that $\alpha'^2 = \alpha^2$. Since α^2 varies with chain length, this is equivalent to saying that the effect of intramolecular interaction on $\overline{h_{ij}{}^2}$ does not depend on the separation between the ith and jth mass points, but only on the over-all length of the polymer chain. With this assumption equation 9–36 holds for all flexible chains, so that the result given by equation 9–38 holds for all flexible polymer chains.

Combining equation 9–38 with the general relations for end-to-end distance, equation 9–20, we get

$$R_G{}^2 = \alpha^2\beta^2\sigma/6 = \alpha^2\beta^2M/6M_0 \qquad (9\text{--}40)$$

We shall make frequent use of this relation in later chapters, and it may thus be convenient to repeat here the meaning of the empirical parameters α and β. Thus β is the effective length per segment, typically equal to $3l_{av}$. It is a constant for a given kind of polymer, independent of the solvent and molecular weight, but dependent on temperature. The factor α, on the other hand, is strongly solvent dependent, being slightly less than or about equal to unity for a poor solvent, and greater than unity for indifferent or good solvents. It is weakly dependent on molecular weight.

Wall and Erpenback[23] have tested the preceding considerations in an ingenious manner by actually summing the effects of a succession of random steps on an electronic computer. Interaction between segments was included by giving appropriate instructions to the computer. The results in general agree with those given here, except that the relation between R_G and h_{av} given by equation 9–38 was found not to be exactly valid. In one series of trials which simulated a hydrocarbon chain with interaction, it was found that $R_G{}^2 = h_{av}{}^2/6.37$.

9k. Application to heterogeneous polymer mixtures. To apply the equations of the preceding pages to typical polymer mixtures with molecules possessing a wide range of molecular weights clearly requires an additional averaging process. Number averages, weight averages, etc., may be used as in the case of molecular weight; i.e., where $\langle \ \rangle$ represents the averaging process over molecules of different molecular weights (N_i molecules with molecular weight M_i), we get, by analogy with equations 8–18,

$$\langle \overline{h^2} \rangle_n = \sum_i N_i \overline{h^2} / \sum_i N_i$$

$$\langle \overline{h^2} \rangle_w = \sum_i N_i M_i \overline{h^2} / \sum_i N_i M_i \qquad (9\text{–}41)$$

$$\langle \overline{h^2} \rangle_z = \sum_i N_i M_i^2 \overline{h^2} / \sum_i N_i M_i^2$$

etc.

where the subscripts n, w, and z represent number, weight, and z averages, as before. In the ideal case (equation 9–19), where $\overline{h^2}$ is proportional to M_i,

$$\langle \overline{h^2} \rangle_n = \beta^2 \sum_i N_i M_i / \sum_i N_i = \beta^2 \overline{M}_n$$

$$\langle \overline{h^2} \rangle_w = \beta^2 \sum_i N_i M_i^2 / \sum_i N_i M_i = \beta^2 \overline{M}_w$$

etc.

but such simple relations are not obtained for the general case; e.g., if $\overline{h^2}$ is proportional to $M_i^{1.20}$,

$$\langle \overline{h^2} \rangle_n = \beta^2 \sum_i N_i M_i^{1.20} / \sum_i N_i$$

The failure in this case to be able to relate $\langle \overline{h^2} \rangle$ to a corresponding molecular weight average does not, of course, create any real difficulties but simply means that, if $\langle \overline{h^2} \rangle$ is to be calculated, the molecular weight distribution must be known sufficiently well to permit the evaluation of an abnormal average, such as $\sum_i N_i M_i^{1.20} / \sum_i N_i$.

The particular kind of average of h_{av} or R_G or $\overline{h^2}$ or $R_G{}^2$ which is obtained experimentally depends, of course, on the procedure used. Many of the best experiments pertaining to molecular dimensions of polymers have been performed on exceptionally well-fractionated samples. For such experiments the approximation has usually been made that all the polymer molecules are identical in degree of polymerization, i.e., the existence of single characteristic values of h_{av} and R_G has been assumed.

91. Non-linear polymers. The concept of end-to-end distance is clearly ambiguous and inapplicable to branched polymers. These macromolecules must therefore be described by means of the average radius of gyration. Furthermore, the average radius of gyration will be smaller than for corresponding linear molecules. For an ideal unrestricted chain, for example, combination of equations 9–8 and 9–38 would lead for a *linear* chain to $R_G{}^2 = \sigma l_{av}{}^2/6$. Zimm and Stockmayer[14] have shown by calculation that one point of random branching in the chain would reduce $R_G{}^2$ to about $0.9\sigma l_{av}{}^2/6$, two points of branching to $0.83\sigma l_{av}{}^2/6$, etc. Zimm and Stockmayer have also calculated the effect of the formation of cyclic chains on the value of R_G.

10. DISTRIBUTION FUNCTIONS FOR POLYMER CONFIGURATION

It is often desirable to know not only the average dimensions of flexible polymer molecules but also a detailed description of their distribution among all possible configurations. It is the object of sections 10a and 10b to arrive at such a description, i.e., to evaluate a distribution function, $W(h)\,dh$, which will give the probability that the end-to-end distance of a polymer molecule lies between h and $h + dh$, regardless of direction.

We shall consider only the ideal situation where the secondary interaction between polymer segments, or between polymer segments and solvent molecules, can be ignored. The effect of such interaction on average dimensions was discussed in section 9h by means of the empirical function α, but this simple treatment cannot be extended to the present problem. The effect of non-ideality on the distribution function can be treated only in terms of the detailed thermodynamic theory of polymer solutions which will be presented in section 12.

The calculation of the ideal distribution, however is a problem of considerable interest, for it is common to a variety of physical phenomena. It is known as the problem of "random flight," first treated by Lord Rayleigh.[16] Its first application to the present problem was made by Kuhn[17] and by Guth and Mark.[18] We shall make use of the solution not only in connection with the distribution of end-to-end distances but also in the treatment of diffusion (section 21).

In section 10c we shall briefly discuss another kind of distribution, that of polymer segments relative to the center of mass of a polymer molecule.

10a. The problem of random flight. Consider a large number, σ, of steps of equal length L, in three-dimensional space, such that the direction of one step with respect to the preceding one is completely random. What is the probability for the occurrence of a given distance separating the beginning of the first step from the end of the σth step? This is the problem with which we are concerned.

We shall first treat this problem as reduced to one dimension, by considering the projection of each step along an arbitrary axis, which we shall designate as the z axis. The projection of each step may be characterized by a positive or negative sign, plus its length, L_z. If the random flight is performed numerous times, we shall only very rarely find the majority of the projections with the same sign. The problem here is the same as that involved in tossing a coin; in a large number of tosses the difference between the frequency of "heads" and "tails" is nearly always very much smaller than the total number of tosses. Accordingly, if we designate as σ_+ the number of steps with positive projections along the z axis, and as σ_- the number with negative projections, such that

$$\sigma_+ + \sigma_- = \sigma \qquad (10\text{--}1)$$

both σ_+ and σ_- will be quite close to $\sigma/2$ in the vast majority of trials. If we designate by $W(\sigma_+)$ the probability of finding a given value of σ_+ (and, hence, of σ_-) in a particular flight of σ steps, we may express this conclusion another way by saying that $\Sigma W(\sigma_+)$ for all values of σ_+ close to $\sigma/2$ will be very close to unity, at least if σ is large. Accordingly we shall confine ourselves to values of σ_+ and σ_- close to $\sigma/2$. We place

$$\begin{aligned} \sigma_+ &= \sigma/2 + \xi \\ \sigma_- &= \sigma/2 - \xi \end{aligned} \qquad (10\text{--}2)$$

where $\xi \ll \sigma/2$. Clearly, this will produce a result correctly expressing the distribution over the more probable configurations. We shall not be able to use the result, however, to compute the frequency of improbable configurations, such as a fully extended chain. (An alternative solution not involving this assumption is given below.)

The value of $W(\sigma_+)$ is now computed simply as the number of ways of achieving that value of σ_+, which is $\sigma!/\sigma_+!\,\sigma_-!$, divided by the total number of possible arrangements, which is 2^σ. Thus

$$W(\sigma_+) = \sigma!/2^\sigma\,\sigma_+!\,\sigma_-! \qquad (10\text{--}3)$$

By the assumption inherent in equation 10–2 both σ_+ and σ_- must be large, so that Stirling's approximation applies. This relation, valid for large values of any variable x, is

$$\ln x! = \tfrac{1}{2}\ln 2\pi + (x + \tfrac{1}{2})\ln x - x \qquad (10\text{--}4)$$

Taking logarithms of equation 10–3 we thus obtain

$$\begin{aligned} \ln W(\sigma_+) = (\sigma + \tfrac{1}{2})\ln \sigma - (\sigma_+ + \tfrac{1}{2})\ln \sigma_+ \\ - (\sigma_- + \tfrac{1}{2})\ln \sigma_- - \sigma \ln 2 - \tfrac{1}{2}\ln 2\pi \quad (10\text{--}5) \end{aligned}$$

Introducing equations 10–2, and taking cognizance of the fact that there is a unique relation between σ_+ and ξ, so that $W(\sigma_+) = W(\xi)$, we get

$$\begin{aligned} \ln W(\xi) = (\sigma + \tfrac{1}{2})\ln \sigma - (\sigma/2 + \tfrac{1}{2} + \xi)\ln(\sigma/2)(1 + 2\xi/\sigma) \\ - (\sigma/2 + \tfrac{1}{2} - \xi)\ln(\sigma/2)(1 - 2\xi/\sigma) - \sigma \ln 2 - \tfrac{1}{2}\ln 2\pi \quad (10\text{--}6) \end{aligned}$$

Equation 10–6 may be separated into terms containing ln σ, ln 2, ln π, ln $(1 + 2\xi/\sigma)$ and ln $(1 - 2\xi/\sigma)$. Since $2\xi/\sigma$ is $\ll 1$, we may further expand these terms and retain only the first two terms of the expansion; i.e., we may write

$$\ln (1 + 2\xi/\sigma) = 2\xi/\sigma - 2\xi^2/\sigma^2$$
$$\ln (1 - 2\xi/\sigma) = -2\xi/\sigma - 2\xi^2/\sigma^2 \qquad (10\text{--}7)$$

Equation 10–6 then becomes

$$\ln W(\xi) = \ln (2/\pi\sigma)^{1/2} - 2\xi^2/\sigma \qquad (10\text{--}8)$$

ignoring terms in ξ^2/σ^2, which are small compared to those retained. Thus

$$W(\xi) = (2/\pi\sigma)^{1/2} e^{-2\xi^2/\sigma} \qquad (10\text{--}9)$$

Equation 10–9 is, of course, the familiar Gaussian error function. For this reason the distribution computed in this section is commonly known as the Gaussian distribution.

Equation 10–9 applies only to integral values of σ_+ and σ_-, i.e., only to integral values of ξ if σ is even, and to half-integral values of ξ if σ is odd (cf. equation 10–2). For long chains, however, we may substitute, as discussed on p. 144, a continuous distribution function, $W(\xi)$, such that $W(\xi) \, d\xi$ gives the probability of finding a value of ξ between ξ and $\xi + d\xi$. The equation will clearly have the same form as equation 10–9,

$$W(\xi) \, d\xi = (2/\pi\sigma)^{1/2} e^{-2\xi^2/\sigma} \, d\xi \qquad (10\text{--}10)$$

It is now a simple matter to introduce the value of h_z, the z projection of the end-to-end distance, corresponding to any value of ξ. For, with both σ_+ and σ_- very large, the average length of the z projection of all steps with positive sign will be the same as the average length of the z projection of all steps with negative sign, both being equal to the average projection of a line of length L at an angle θ to the z axis, where θ may take on all values with equal probability. This length is $(\overline{L^2 \cos^2 \theta})^{1/2} = L/3^{1/2}$, since $\overline{\cos^2 \theta} = \frac{1}{3}$. Thus the value of h_z corresponding to a given value of ξ, using equations 10–2, is

$$h_z = (\sigma_+ - \sigma_-)L/3^{1/2} = 2\xi L/3^{1/2} \qquad (10\text{--}11)$$

By differentiation,

$$dh_z = 2L \, d\xi/3^{1/2} \qquad (10\text{--}12)$$

If now $W(h_z) \, dh_z$ represents the probability of a value of the z projection of the end-to-end distance between h_z and $h_z + dh_z$, then, by equation 10–12, the probability of finding a value of ξ between ξ and $\xi + d\xi$ is

$(2L/3^{1/2})W(h_z)\,d\xi$. This expression, however, must be identical to $W(\xi)\,d\xi$, as given by equation 10–10, so that

$$W(h_z) = (3^{1/2}/2L)\,W(\xi)$$
$$= (3/2\pi\sigma L^2)^{1/2} e^{-2\xi^2/\sigma} \qquad (10\text{–}13)$$

Rearranging, with the aid of equation 10–11, we get

$$W(h_z)\,dh_z = (b/\pi^{1/2})e^{-b^2 h_z^2}\,dh_z \qquad (10\text{–}14)$$

where

$$b^2 = 3/2\sigma L^2 \qquad (10\text{–}15)$$

No given direction is unique in this problem; i.e., the probability of finding a projection along an x axis (perpendicular to the z axis) between h_x and $h_x + dh_x$ must be given by a relation similar to equation 10–14. Similarly, the same must be true for the third perpendicular axis of a rectangular coordinate system; i.e.,

$$W(h_x)\,dx = (b/\pi^{1/2})e^{-b^2 h_x^2}\,dh_x$$
$$W(h_y)\,dy = (b/\pi^{1/2})e^{-b^2 h_y^2}\,dh_y \qquad (10\text{–}16)$$

We may now calculate the probability of an over-all distance of h between the beginning and end of a random flight by the common procedure of transformation to spherical coordinates.

The probability of simultaneously finding a value of h_x between h_x and $h_x + dh_x$, h_y between h_y and $h_y + dh_y$, and h_z between h_z and $h_z + dh_z$ is obtained by multiplying together the probabilities for each independent projection; i.e., it is given by

$$W(h_x)\,W(h_y)\,W(h_z)\,dh_x\,dh_y\,dh_z = (b/\pi^{1/2})^3 e^{-b^2 h^2}\,dh_x\,dh_y\,dh_z \qquad (10\text{–}17)$$

where $h^2 = h_x^2 + h_y^2 + h_z^2$, i.e., the square of the end-to-end distance.

An infinite set of different combinations of h_x, h_y, and h_z all lead to the same value of h^2. We are interested here in calculating the probability of a given value of h^2, regardless of the actual values of h_x, h_y, and h_z. This probability can be calculated by transposing equation 10–17 into spherical coordinates. The volume element $dh_x\,dh_y\,dh_z$, in these coordinates, becomes $h^2 \sin\theta\,dh\,d\theta\,d\phi$, so that the probability of an end-to-end distance of h, regardless of direction, becomes

$$W(h)\,dh = \int_{\theta=0}^{\pi} \int_{\phi=0}^{2\pi} \left(\frac{b}{\pi^{1/2}}\right)^3 e^{-b^2 h^2} h^2 \sin\theta\,d\theta\,d\phi\,dh$$
$$= 4\pi \left(\frac{b}{\pi^{1/2}}\right)^3 e^{-b^2 h^2} h^2\,dh$$

With the value of b^2 given by equation 10–15 this becomes

$$W(h)\,dh = 4\pi \left(\frac{3}{2\pi\sigma L^2}\right)^{3/2} e^{-3h^2/2\sigma L^2} h^2\,dh \qquad (10\text{–}18)$$

An alternative expression for $W(h)$ is in terms of the root-mean-square path length of the random flight,

$$\overline{h^2} = \int_0^\infty h^2\, W(h)\, dh \bigg/ \int_0^\infty W(h)\, dh$$

which, with equation 10–18 and the integrals of Appendix A, reduces to $\overline{h^2} = \sigma L^2$. This is, of course, the same result obtained by direct evaluation of $\overline{h^2}$ for the same model, as given by equation 9–8. The result may be substituted into equation 10–18, giving

$$W(h)\, dh = 4\pi \left(\frac{3}{2\pi \overline{h^2}}\right)^{3/2} e^{-3h^2/2\overline{h^2}} h^2\, dh \qquad (10\text{–}19)$$

As stated earlier, the derivation here used is applicable to only relatively small values of h, where $W(h)$ is relatively large. A more precise answer has been obtained by Kuhn and Grün[19] and James and Guth,[20] which, in the present terminology, is

$$W(h)\, dh = (\text{constant}) e^{-\frac{1}{L}\int_0^h \mathscr{L}^*(h/\sigma L)\, dh} h^2\, dh \qquad (10\text{–}20)$$

where $\mathscr{L}^*$ represents the inverse of the Langevin function, $\mathscr{L}(x) = \coth x - 1/x$. Equation 10–20 must be used in place of equation 10–18 or 10–19 if it is desired to calculate the probability of an inherently unlikely value of h.

10b. Distribution function for end-to-end distances. A flexible polymer molecule with σ segments of average length l_{av} may be represented as a succession of σ steps of length l_{av}, but the steps will not be in random directions, because of the requirement of particular bond angles and because of restricted rotation. Thus the solution to the random flight problem cannot be applied directly to give the distribution of end-to-end distances in polymer molecules.

It was shown in section 9g, however, that the behavior of a polymer molecule (in ideal solution) can always be described in terms of an equivalent chain with randomly disposed segments. It was shown rigorously (Fig. 9–4) that this can always be done if the model chain is chosen to have fewer segments (of number σ') than the actual polymer molecule, the average length of each segment (l_e) being much larger than l_{av}. Equation 9–16 suggested that an alternative model chain is one containing the true number of segments (σ), with a segment length β which would always be larger than l_{av}, but, of course, not as large as l_e.

The random flight treatment should be applicable to either of these model chains; i.e., equation 10–18 should correctly represent the ideal solution behavior of real polymer chains if we place $L = \beta$ or if we place

$\sigma = \sigma'$ and $L = l_e$. Equation 10–19 is even more convenient, for it should clearly be applicable as it stands, regardless of the choice of model chains, since $\overline{h^2} = \sigma\beta^2 = \sigma'l_e^2$.

A typical calculation, based on equation 10–19, is shown in Fig. 10–1. It represents the distribution function for an organic polymer molecule, the backbone of which consists of a chain of 2000 single-bonded C atoms; i.e., the chain is a succession of 2000 C—C bonds. As was mentioned on p. 159, β for such a chain is typically of the order of $3l_{av}$; i.e., $\beta \simeq 4.6$ Å, which, by equation 9–16, leads to $h_{av} = 206$ Å.

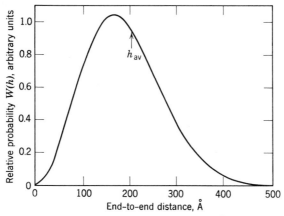

Fig. 10–1. The distribution function for end-to-end distances, as given by equation 10–18 or 10–19. The calculation is for a molecular chain of 2000 C—C bonds, with $\beta = 4.6$ Å (or $h_{av} = 206$ Å), in an ideal solvent.

10c. Distribution of segments relative to the center of mass. Equation 10–18 provides, in effect, a measure of the probability that any given segment (in particular, the terminal segment) lie within a distance h of the first segment of a polymer chain. By placing σ equal to $j - i$ it could be used to compute the probability that the jth segment lies a distance h_{ij} from the ith segment. Finally if we choose for the ith segment one which is located very close to the center of mass, equation 10–18 would give the probability that any given segment lie at a given distance from the center of mass.

We are interested in the present section in a slightly different problem, to calculate the average distribution of all segments relative to the center of mass. We may guess that this, too, can be represented by an expression of the form of equation 10–18, i.e., that the relative number of segments lying in a volume $4\pi r^2\, dr$ from the center of mass is proportional to $e^{-B^2r^2}r^2\, dr$ or, in other terms, that the segment density, ρ, i.e., the number of segments per cubic centimeter, at any distance r from the center of mass, can be represented by

$$\rho = Ae^{-B^2 r^2} \tag{10-21}$$

where A and B are arbitrary constants to be determined.

Debye and Bueche[21] have outlined an exact treatment of this problem and have shown that equation 10–21, though not exact, is indeed a good approximation for the distribution of the segments of the polymer chain about the center of mass in the case of ideal solutions. If it is a good approximation for ideal solutions, it will be a good approximation for all flexible polymer solutions, for, as we have discussed on p. 168, the non-ideality parameter α is such that all distances within a polymer coil are increased proportionately as α increases. The functional relationship between ρ and r will then not change, only the constants A and B will be affected.

The arbitrary parameters A and B of equation 10–21 can be evaluated in terms of any two constants of the polymer chain, the values of which are known. Since we shall want to apply the equation to both ideal and non-ideal solutions, we must choose constants which are known for both. Two such constants are the total number of segments (σ) per molecule and the radius of gyration, R_G, which is given for all solutions by equation 9–40. Evaluating σ first, we note that any spherical layer of volume $4\pi r^2 \, dr$ will contain $4\pi r^2 \rho \, dr$ segments, so that

$$\sigma = \int_{r=0}^{\infty} 4\pi r^2 \rho \, dr = 4\pi A \int_{r=0}^{\infty} r^2 e^{-B^2 r^2} \, dr = \pi^{3/2} A/B^3 \tag{10-22}$$

Similarly R_G^2 is the average value of r^2 for all segments. Since there are $4\pi r^2 \rho \, dr$ segments for which r^2 has a given value, we get

$$R_G^2 = \int_{r=0}^{\infty} 4\pi r^4 \rho \, dr \Big/ \int_{r=0}^{\infty} 4\pi r^2 \rho \, dr = 3\pi^{3/2} A/2\sigma B^5 \tag{10-23}$$

Combining equations 10–22 and 10–23, we can solve for A and B, obtaining

$$B^2 = \tfrac{3}{2} R_G^2 \tag{10-24}$$

$$A = \sigma(3/2\pi R_G^2)^{3/2} \tag{10-25}$$

so that

$$\rho = \sigma(3/2\pi R_G^2)^{3/2} e^{-3r^2/2R_G^2} \tag{10-26}$$

Equation 10–26 indicates that the segment density decreases with increasing chain length. At the center of mass ($r = 0$), for instance, with equation 9–40 for R_G, we have ρ inversely proportional to $\alpha^3 \sigma^{1/2}$, which, for the molecular weight dependence of α given in section 9h, means that ρ varies as $1/\sigma^{0.5}$ to $1/\sigma^{0.8}$. The equation also shows, as expected, that ρ must be larger in a poor solvent ($\alpha \simeq 1$) than in a good solvent ($\alpha > 1$).

To get an idea of actual values of ρ we shall make a calculation for a typical polymer chain. The dimensions used will apply approximately to polymers such as polyisobutylene or polystyrene,

$$\left(\begin{array}{c} CH_3 \\ | \\ -CH_2-C- \\ | \\ CH_3 \end{array}\right)_x \quad \text{or} \quad \left(\begin{array}{c} C_6H_5 \\ | \\ -CH_2-C- \\ | \\ H \end{array}\right)_x$$

For a chain of 1000 monomer units (i.e., $\sigma = 2000$ in terms of C—C bonds), and with $\beta = 4.6$ Å as before, the radius of gyration given by

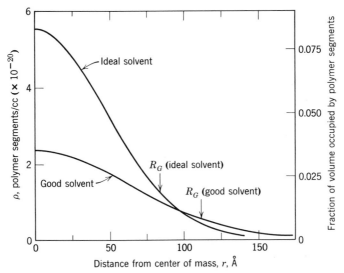

Fig. 10–2. Distribution of segments about the center of mass, as calculated by equation 10–26. The data apply to a molecular chain of 2000 C—C bonds, with $\beta = 4.6$ Å.

equation 9–40 is 84 Å in an ideal solvent ($\alpha = 1$) and 112 Å in a typical good solvent, equation 9–22 having been used for α. The resulting density distributions, computed from equation 10–26, are shown in Fig. 10–2. They show that the density of segments decreases as one moves outward from the center of mass, and they illustrate strikingly the effect of a good solvent, which is to decrease greatly the maximum density near the center of the coil and to extend the general domain of the molecule over a greater volume.

It should be noted that σ was taken to mean the number of C—C bonds in the foregoing calculations. Equation 10–26 thus gives the number of such bonds per unit volume. To obtain the true density of monomer segments, which is what is plotted in Fig. 10–2, this number has simply been divided by 2.

By multiplying the segment density by the volume, τ, per segment, we can obtain the fraction of space, $f = \rho\tau$, which the segments occupy. For the most common organic polymers τ will lie in the range of 100 to 200 Å^3. Using a figure of 150 Å^3 we then obtain the values of f which are shown as the right-hand ordinate of Fig. 10–2. They reveal an important property of flexible polymer coils, namely that the polymer molecule itself occupies only a very small fraction of the volume which it pervades. In the example chosen this fraction never exceeds 6% in poor solvents or $2\frac{1}{2}$% in good solvents. For longer chains, as indicated above, it would be even smaller, so that, for molecular weights above 1 million, the maximum value of f may become less than 1% even in a poor solvent. It is clear then that a flexible polymer molecule may be coiled over a region of space which is several hundred times greater than the volume actually occupied by the molecule's segments. In a dilute solution, this space would be filled by solvent. In a sample of pure liquid polymer it would be filled by segments of other molecules.

The figures here given are based on an approximate equation, and the actual calculations apply to just a particular kind of chain. Nevertheless, the figures may be taken as typical of flexible polymers in general. They will prove very useful in guiding us to a realistic treatment of the thermodynamic and hydrodynamic properties of such molecules in subsequent chapters.

General References

1. P. J. Flory, *Principles of Polymer Chemistry*, Cornell University Press, Ithaca, 1953. Chapters VIII, IX, X, and XIV.
2. E. M. Frith and R. F. Tuckett, *Linear Polymers*, Longmans, Green and Co., London, 1951.
3. L. H. Cragg and H. Hammerschlag, "The Fractionation of High-Polymeric Substances," *Chem. Revs.*, **39**, 79 (1946).
4. H. V. Craig, *Vector and Tensor Analysis*, McGraw-Hill Book Co., New York, 1943; or any other textbook on this subject. Also I. S. and E. S. Sokolnikoff, *Higher Mathematics for Engineers and Physicists*, 2nd ed., Chapter IX ("Vector Analysis"), McGraw-Hill Book Co., New York, 1941.

Specific References

5. P. J. Flory, *J. Am. Chem. Soc.*, **58**, 1877 (1936).
6. G. B. Taylor, *J. Am. Chem. Soc.*, **69**, 638 (1947).
7. J. E. Mayer and M. G. Mayer, *Statistical Mechanics*, John Wiley and Sons, New York, 1940. Appendix A III.

8. E. O. Kraemer in T. Svedberg and K. O. Pedersen, *The Ultracentrifuge*, Oxford University Press, 1940, p. 325 et seq.; E. O. Kraemer and W. D. Lansing, *J. Am. Chem. Soc.*, **55**, 4319 (1933).

9. H. Eyring, *Phys. Revs.*, **39**, 746 (1932).

10. H. Benoit and C. Sadron, *J. Polymer Sci.*, **4**, 473 (1949).

11. H. Benoit, *J. Polymer Sci.*, **3**, 376 (1948).

12. W. Kuhn, *Kolloid-Z.*, **76**, 258 (1936); **87**, 3 (1939).

13. P. J. Flory, *J. Chem. Phys.*, **17**, 303 (1949); P. J. Flory and T. G. Fox, Jr., *J. Am. Chem. Soc.*, **73**, 1904 (1951).

14. B. H. Zimm and W. H. Stockmayer, *J. Chem. Phys.*, **17**, 1301 (1949).

15. P. Debye, *J. Chem. Phys.*, **14**, 636 (1946).

16. Lord Rayleigh, *Phil. Mag.*, [6], **37**, 321 (1919).

17. W. Kuhn, *Kolloid-Z.*, **68**, 2 (1934).

18. E. Guth and H. Mark, *Monatsh.*, **65**, 93 (1934).

19. W. Kuhn and F. Grün, *Kolloid-Z.*, **101**, 248 (1942).

20. H. M. James and E. Guth, *J. Chem. Phys.*, **11**, 470 (1943).

21. P. Debye and F. Bueche, *J. Chem. Phys.*, **20**, 1337 (1952).

22. A. Wada, *J. Chem. Phys.*, **29**, 674 (1958); **30**, 328, 329 (1959).

23. F. T. Wall and J. J. Erpenback, *J. Chem. Phys.*, **30**, 634, 637 (1959).

4

THERMODYNAMICS

11. FUNDAMENTAL PRINCIPLES OF THE THERMODYNAMICS OF SOLUTIONS

Systems containing macromolecules are, of course, subject to exactly the same thermodynamic laws as systems containing only small molecules. Their behavior may be described in the same general way in terms of the phases of the system, the components of each phase, the first and second laws, etc.

It was once fashionable to treat macromolecules in solution as a separate "dispersed" phase, perhaps a natural procedure at a time when macromolecules were believed to be aggregation colloids. Today, however, this procedure is rarely followed, and macromolecular solutions are regarded as homogeneous (single-phase) systems with no particular regard being paid to the fact that the solute molecules are larger than those of the solvent. The thermodynamics of such solutions rests then on precisely the same principles as those which apply to solutions containing only small molecules, and we shall briefly review these principles here. For greater detail the reader is referred to any thermodynamics textbook[1] or to the treatment of thermodynamics in books on physical chemistry in general.

11a. Partial molal and partial specific quantities. There are two kinds of thermodynamic quantities: those, like temperature, density, and pressure, which have the same value throughout a homogeneous phase and are called *intensive*; and those, like volume, the various kinds of energy, and entropy, which are proportional to the amount of material in the phase and are called *extensive*. An extensive quantity may be converted to an intensive one by dividing by the amount of material present; the *molal volume* and the *specific volume*, i.e., the volume per mole or per gram, are examples of intensive quantities obtained in this way.

In terms of these intensive properties the total volume of n_1 moles of a particular substance in the pure state is $n_1 V_1$ where V_1 is the molal volume of the substance, or that of g_1 grams is $g_1 v_1$ where v_1 is the specific volume. In a solution containing several substances a corresponding equation could be written, $V = n_1 V_1 + n_2 V_2 + \cdots = g_1 v_1 + g_2 v_2 + \cdots$, where the V_i are the molal volumes of the various components in their pure form and the v_i are the corresponding specific volumes. It is elementary knowledge, however, that such a relation does not exist and that if we wish to express the total volume of a solution in terms of intensive properties we must define new quantities, the partial molal volume, $\bar{V}_i$, and the partial specific volume, $\bar{v}_i$,

$$\bar{V}_i = (\partial V / \partial n_i)_{T,P,n_j} \quad (j \neq i) \tag{11-1}$$

$$\bar{v}_i = (\partial V / \partial g_i)_{T,P,g_j} \quad (j \neq i) \tag{11-2}$$

The use of the symbols $\bar{V}_i$, $\bar{v}_i$, $\bar{F}_i$, etc., to indicate partial molal or partial specific quantities may lead to some confusion because a bar over a letter is also used to denote average quantities. This duplicate meaning of a bar over a letter is, however, so universal that greater confusion would result if we were to adopt a new set of symbols. It is hoped that confusion between the two meanings will be minimized by the fact that the kinds of quantities for which we frequently seek average values are usually quite different from the thermodynamic properties which we differentiate to obtain partial molal or partial specific properties.

That the total volume of a solution can be expressed in terms of partial molal or partial specific volumes is easily proved. For, a solution, the *final* composition and total amount of which are given in terms of the mole numbers $n_1, n_2, \cdots$, can be considered as being prepared by addition of successive increments $n_1 \, d\lambda, n_2 \, d\lambda, \cdots$, such that the relative composition remains always the same. The volume change on each successive addition is $dV = \bar{V}_1 n_1 \, d\lambda + \bar{V}_2 n_2 \, d\lambda + \cdots$, the $\bar{V}_i$ being defined by equation 11-1. Since partial molal and partial specific quantities are intensive quantities with values independent of the total amount of solution, they remain constant throughout the addition. We may therefore integrate from $\lambda = 0$ to $\lambda = 1$ to obtain the final volume as

$$V = n_1 \bar{V}_1 + n_2 \bar{V}_2 + \cdots \tag{11-3}$$

or, in a similar fashion, in terms of partial specific volumes

$$V = g_1 \bar{v}_1 + g_2 \bar{v}_2 + \cdots \tag{11-4}$$

The equations here derived for volume would apply equally well for any other extensive property. Particularly important is the corresponding relation for free energy

$$F = \sum_i n_i \bar{F}_i = \sum_i n_i \mu_i \tag{11-5}$$

where the summation extends over all the components of the solution. The partial molal free energy, $\bar{F}_i$, is so important a quantity that it is given a special symbol, μ_i, and a special name, *chemical potential*;

$$\mu_i = \bar{F}_i = (\partial F/\partial n_i)_{T,P,n_j} \quad (j \neq i)$$

An important general relation applicable to all partial molal quantities is the Gibbs-Duhem equation. It follows from the fact that changes dn_1, dn_2, etc., in the mole numbers at constant temperature and pressure must result in a change in the total property (designated by G) given by $dG = \sum_i \bar{G}_i \, dn_i$. The complete differential of G at constant temperature and pressure, by equation 11–3 or 11–5, is $dG = \sum_i \bar{G}_i \, dn_i + \sum_i n_i \, d\bar{G}_i$. Since these expressions for dG must be identical, it follows that

$$\sum_i n_i \, d\bar{G}_i = 0 \qquad (11\text{–}6)$$

A similar relation may be written for partial specific quantities. Taking partial specific volume as an example, we have

$$\sum_i g_i \, d\bar{v}_i = 0 \qquad (11\text{–}7)$$

Finally, it should be pointed out that it is readily shown[1] that partial molal and partial specific quantities are related to one another in the same manner as the parent extensive properties, e.g.,

$$\mu = \bar{H} - T\bar{S} \qquad (11\text{–}8)$$

$$\left[\frac{\partial(\mu/T)}{\partial T}\right]_P = -\frac{\bar{H}}{T^2} \qquad (11\text{–}9)$$

11b. Components of a solution. The various kinds of molecules or ions present in a solution are most generally given the name *species*. For example, a water solution of NaCl will contain the species H_2O, Na^+, and Cl^-. It is sometimes convenient to express any extensive property of a solution in terms of the partial molal or partial specific properties of these species. For aqueous NaCl, for example, we could write $F = n_{H_2O}\mu_{H_2O} + n_{Na^+}\mu_{Na^+} + n_{Cl^-}\mu_{Cl^-}$. When this procedure is followed, however, it should be done with the realization that it has no real meaning in terms of any conceivable experiment. Na^+ and Cl^- ions, for example, cannot be independently added to a solution. Thus μ_{Na^+} and μ_{Cl^-} are quantities which are not representative of any possible experimental operation. In other words, the implication in the equation, $F = n_{H_2O}\mu_{H_2O} + n_{Na^+}\mu_{Na^+} + n_{Cl^-}\mu_{Cl^-}$, that n_{Na^+} and n_{Cl^-} are independent variables is clearly false. Thus, although it may be convenient at times to use this equation, it should never be used without realization of the restrictions accompanying it, and, in general, it is preferable not to use it at all but to express extensive quantities in terms of the partial molal or partial specific properties of the *components* of a solution rather than the *species*. The *components* are defined as those substances which can be added independently of one another; e.g., the components of an aqueous NaCl solution are H_2O and NaCl, and we may use the corresponding equation for free energy, $F = n_{H_2O}\mu_{H_2O} + n_{NaCl}\mu_{NaCl}$, without restriction.

It should be pointed out that the enumeration of the components of a solution cannot necessarily be achieved in a unique way. For example, a solution containing Na^+, K^+, Cl^-, and NO_3^- consists of three components in addition to water. Any three of the neutral substances NaCl, KCl, $NaNO_3$, KNO_3 can be chosen as the three components. This arbitrariness in the choice of components is of great importance when we consider macro-ions (section 14a).

In a solution in general all components are equivalent. We are largely confining ourselves, however, to *dilute* solutions, which implies that we have one component known as the *solvent* which is always present in large excess and thus occupies a special position. We shall always use the subscript 1 to indicate this component.

11c. Equilibrium between phases.[2] If two phases are brought in contact with one another there will be transfer of heat and of matter from one to the other until equilibrium is reached. The conditions for equilibrium were worked out in the classic work of Willard Gibbs in the last century.[2] We shall here summarize his results for a fluid at rest, in the absence of applied forces, and neglecting surface effects. In so doing we should recall that a phase is defined as *any part* of a system which is homogeneous in all respects; i.e., a phase is not necessarily a large region visually distinguishable from its neighbor.

The state of equilibrium between two phases of any chemical system will be unaffected by the presence of other portions. Without loss of generality we may therefore suppose that all phases considered are part of a larger isolated system, of fixed volume, unable to exchange energy or matter with its surroundings. The first law of thermodynamics requires that the energy of such an isolated system remain constant. The second law defines the state of equilibrium as that for which the entropy has reached a maximum value. If, therefore, an infinitesimal exchange of heat or matter between phases occurs, then, if the system is at equilibrium, no change in total energy or entropy may occur. If $E^{(p)}$ and $S^{(p)}$ represent the energy and entropy of a phase, then

$$\sum_{(p)} dE^{(p)} = \sum_{(p)} dS^{(p)} = 0 \qquad (11\text{–}10)$$

the summation extending over all phases.

It follows at once that each phase must have the same temperature. A small amount of heat dq may be transferred from one phase ($p = 1$) to another ($p = 2$), all other phases of the over-all system remaining unaltered. At equilibrium $dS^{(p)} = dq^{(p)}/T^{(p)}$, so that

$$dS^{(1)} + dS^{(2)} = -dq/T^{(1)} + dq/T^{(2)} \qquad (11\text{–}11)$$

Since dq is a completely arbitrary quantity of heat, equation 11–10 is satisfied only if $T^{(1)} = T^{(2)}$.

It is also clear that each phase must have the same pressure. Consider a cylindrical volume element of cross-section dA at the interface between two phases and normal to the interface. The force on one face of the element is $P^{(1)} dA$; that on the other face is in the opposite direction and equal to $P^{(2)} dA$. At equilibrium these forces must be equal, or else the volume element would move, so that $P^{(1)} = P^{(2)}$.

It remains to consider the transfer of matter from one phase to another. To do so, we must write the complete expression for the change in energy of a phase. In the absence of external applied forces (other than the pressure) and if no loss or gain of matter occurs, we have $dE^{(p)} = dq^{(p)} - dw^{(p)}$, which, at equilibrium, becomes $dE^{(p)} = T\,dS^{(p)} - P\,dV^{(p)}$. If, in addition, the numbers of moles $n_1^{(p)}$, $n_2^{(p)}$, $\cdots$, of components 1, 2, $\cdots$, become variables, then

$$dE^{(p)} = T\,dS^{(p)} - P\,dV^{(p)} + \sum_i \mu_i^{(p)}\,dn_i^{(p)} \qquad (11\text{–}12)$$

where $\mu_i^{(p)}$ is the chemical potential of the ith component in phase (p), and the summation extends over all components.

Equation 11–12 is the one which originally defines chemical potential. The definition is clearly

$$\mu_i = (\partial E/\partial n_i)_{S,V,n_j} \quad (j \neq i)$$

If the definitions of the Helmholtz and Gibbs free energies, $A = E - TS$ and $F = E + PV - TS$, are combined with equation 11–12, we at once obtain the somewhat more common alternative definitions of μ_i, namely,

$$\mu_i = (\partial A/\partial n_i)_{V,T,n_j} = (\partial F/\partial n_i)_{T,P,n_j} \quad (j \neq i)$$

If dn_i moles of any component are now transferred from any phase 1 to another phase 2, without other change, we get, for the total energy of the isolated system,

$$dE^{(1)} + dE^{(2)} = -\mu_i^{(1)}\,dn_i + \mu_i^{(2)}\,dn_i \qquad (11\text{–}13)$$

Since dn_i is arbitrary, equation 11–10 is satisfied only if $\mu_i^{(1)} = \mu_i^{(2)}$. This gives the most important property of the state of equilibrium between phases: that *each chemical potential μ_i must have the same value in every phase of the system*. It is important to note that this condition can apply only to *components* and not to species which cannot be transferred independently from one phase to another. The condition cannot apply, for example, to single ion species since we cannot transfer them from one phase to another without having them be accompanied by ions of opposite charge to maintain neutrality.

The reader may be more familiar with other methods of deriving the condition $\mu_i^{(1)} = \mu_i^{(2)}$, which do not employ the technique of having the two phases of interest form part of a larger isolated system. If the two phases are considered alone, with the understanding that they can exchange energy, but not matter, with their surroundings, then a transfer of dn_i moles of component i from one phase to another will not necessarily occur at constant energy; nor will a maximum value of S be a criterion for equilibrium. Instead the criterion for equilibrium is that the transfer of matter obeys the equation $dE = T\,dS - P\,dV$. If the total volume is fixed, it reduces to $dA^{(1)} + dA^{(2)} = 0$; if the pressure is fixed, it reduces to $dF^{(1)} + dF^{(2)} = 0$. With $T^{(1)} = T^{(2)}$, the condition $\mu_i^{(1)} = \mu_i^{(2)}$ follows at once. That all of these methods lead to the same result is a necessary consequence of the fact that the state of equilibrium is a uniquely defined state, regardless of the perturbation we choose to impose upon the system in order to establish that equilibrium exists.

It is possible now to investigate numerous additional conditions which may be imposed upon an assembly of phases and to deduce their effect upon the equilibrium conditions. (Most familiar is the condition that some of the components may be able to react to form other components, which leads to the laws of chemical equilibrium. We shall not take up this particular condition.) One such condition which will be of interest in this chapter is the presence of a rigid semi-permeable membrane between two phases. Unless such a membrane is a thermal insulator it will not impede the transfer of heat from one phase to another, and the condition of equal temperature between the phases follows at once from equation 11–11. However, the pressures $P^{(1)}$ and $P^{(2)}$ may now be different, for a volume element between the phases will now be located in the rigid membrane and will remain in position regardless of the pressures on the two sides. Similarly, the relation $\mu_i^{(1)} = \mu_i^{(2)}$ continues to hold for all components which can pass freely through the membrane. In applying this condition we must, however, take into account the fact that $P^{(1)}$ may differ from $P^{(2)}$ and, where necessary, compute the effect this will have on the chemical potentials. For a component to which the membrane is impermeable no transfer can occur. For such a component the condition $\mu_i^{(1)} = \mu_i^{(2)}$ is clearly inapplicable, even if the component is present on both sides.

One further modification of the conditions for equilibrium will be considered in section 16, namely, that which is applicable to equilibrium under the influence of an external force.

11d. The chemical potential in ideal solutions. We shall be interested primarily in the behavior of liquid solutions. The thermodynamics of such solutions is determined by the manner in which the total free energy, or the chemical potentials of the components, depend on composition.

It is customary to define an *ideal solution* as one in which for each component

$$d\mu_i = \mathcal{R}T\, d \ln X_i$$

where X_i is the mole fraction of the ith component. Upon integration this gives

$$\mu_i = \mu_i^0 + \mathcal{R}T \ln X_i \tag{11-14}$$

where μ_i^0 is a constant, called the *standard chemical potential.* If the ith component is miscible with the other components of a solution at all possible concentrations, including $X_i = 1$, and if equation 11–14 is valid for all values of X_i, then $\mu_i^0 = F_i^0$, the molar free energy of the ith component in its pure state.

It is, of course, extremely rare to find liquid solutions in which equation 11–14 is even approximately obeyed for any appreciable range of X_i. However, all dilute solutions tend to approach ideality as they approach infinite dilution; i.e., equation 11–14 becomes valid as the solvent mole fraction, X_1, approaches unity, while all other X_i approach zero. In this situation $\mu_1^0 = F_1^0$, but all other μ_i^0 are arbitrary constants which bear no relation to the molar free energies of the pure substances. The standard chemical potential will depend, in general, on the solvent being used; e.g., μ_i^0 has a different value for I_2 dissolved in water and for I_2 dissolved in CCl_4.

In dilute solutions only the solvent concentration is customarily expressed in terms of its mole fraction. Solute concentrations are more often expressed in terms of *molarity*, C_i (moles per liter of solution), or *molality*, m_i (moles per kilogram of solvent). In a solution which approaches infinite dilution (i.e., one to which the ideal equations for μ_i will be applicable) the total volume approaches $n_1 V_1^0$ where n_1 is the number of moles of solvent and V_1^0 its molar volume. At the same time the total number of moles in the solution becomes insignificantly different from n_1. Thus X_i becomes n_i/n_1, where n_i is the number of moles of the ith solute, and C_i becomes $1000\, n_i/n_1 V_1^0$ if V_1^0 is expressed in milliliters. Thus

$$X_i = C_i V_1^0/1000 \tag{11-15}$$

and equation 11–14 can be written as

$$\mu_i = \mu_i^0 + \mathcal{R}T \ln C_i \tag{11-16}$$

where the standard chemical potential differs from that defined by equation 11–14 by the constant term $\mathscr{R}T \ln V_1^0/1000$. In a similar way equation 11–14 can be converted to the form

$$\mu_i = \mu_i^0 + \mathscr{R}T \ln m_i \tag{11–17}$$

where μ_i^0 differs from that of equation 11–14 by the term $\mathscr{R}T \ln M_1/1000$. (Some authors employ different symbols to designate the different standard chemical potentials. There is no particular advantage to this here, however, since in most applications it is necessary only to know that μ_i^0 is a constant.)

For dilute solutions it is also profitable to rephrase equation 11–14 for the chemical potential of the solvent, since X_1 in such solutions is very close to unity. For a two-component system with $X_1 \gg X_2$, $\ln X_1 = \ln (1 - X_2) = -X_2 - \frac{1}{2}X_2^2 \cdots$, with subsequent terms falling off rapidly. Thus (with F_1^0 replacing μ_1^0)

$$\mu_1 - F_1^0 = -\mathscr{R}T(X_2 + \tfrac{1}{2}X_2^2 + \cdots) \tag{11–18}$$

Alternatively, we could use the molal or molar concentrations of solute in place of X_2. An especially useful procedure is to express μ_1 in terms of the weight concentration of solute, expressed as c_2 grams per milliliter. Confining ourselves again to a two-component system $c_2 = C_2 M_2/1000$, so that, by combination with equations 11–15 and 11–18, we get

$$\mu_1 - F_1^0 = -\mathscr{R}T V_1^0[(1/M_2)c_2 + (V_1^0/2M_2^2)c_2^2 + \cdots] \tag{11–19}$$

For dilute solutions containing several solutes we would substitute $\sum_i X_i$ or $\sum_i c_i$ ($i = 2, 3, \cdots$) for X_2 or c_2 respectively.

11e. The chemical potential in real solutions. Real solutions are rarely ideal except at the limit of infinite dilution. For general application, therefore, the expressions for μ_i of the preceding section are invalid. To obtain valid expressions for μ_i it is necessary to have an accurate knowledge of the structure of a solution and to use statistical mechanics to derive thermodynamic properties from the structure, as is done, for example, in section 12.

Often, however, it is desired to have completely general expressions in which the non-ideal behavior of the solution is expressed in terms of empirical parameters. One way in which this may be done is to invent a quantity, the *activity*, so defined that the form of the ideal equations is maintained; i.e., we write

$$\mu_i = \mu_i^0 + \mathscr{R}T \ln a_i \tag{11–20}$$

This equation contains *two* variable parameters, μ_i^0 and a_i, to represent a *single* experimental quantity, μ_i. To obtain values for these parameters we need an additional condition, and it is provided by the requirement that all solutions become ideal at the limit of infinite dilution, i.e., as X_1 approaches unity. Under these conditions, equation 11–20 must asymptotically approach equation 11–14, 11–16, or 11–17, according to our choice of concentration units, and it must do so without change in the constant $\mu_i{}^0$. Since this constant has a different value in each of the three ideal equations, the $\mu_i{}^0$ of equation 11–20 and, hence, also the activities do not have unique values. Their values must depend on the concentration units employed for the reference state.

If numerical values for a_i are desired, it is customary to replace a_i by the product of the corresponding concentration unit and an activity coefficient. The symbol used for the latter depends on the choice of concentration unit; i.e., we write $a_i = f_i X_i$ or $a_i = \gamma_i m_i$ or $a_i = y_i C_i$, with the condition that f_i, γ_i, or y_i, whichever is used, approaches unity as X_1 approaches unity.

In applications of the activity concept, we shall use the symbols $\mu_i{}^0$ and a_i without specifying the concentration units of the reference state, since, ordinarily, any concentration unit may be used. In equations in which the symbols refer to specific numerical values, a_i will be replaced by the product of concentration and activity coefficient, and the concentration units are then specifically indicated.

An alternative procedure, most useful for the solvent in dilute solutions, is to express the chemical potential as a power series in the concentration. This procedure is analogous to that used in the theory on non-ideal gases. The PV product per mole of gas is expressed in terms of the *virial equation*,

$$PV = \mathscr{R}T(1 + BP + CP^2 + \cdots)$$

The coefficients B, C, etc., are known as the second, third, etc., *virial coefficients*. The equation reduces at zero pressure to the ideal gas equation. The corresponding equation for the solvent in a dilute solution containing a single solute can, for instance, be written, after the form of equation 11–19, as

$$\mu_1 - F_1{}^0 = -\mathscr{R}TV_1{}^0 c_2(1/M_2 + Bc_2 + Cc_2{}^2 + \cdots) \qquad (11\text{–}21)$$

the first term being the same as that of equation 11–19 so that ideal behavior is attained as c_2 becomes very small.

Equation 11–21 is often written with $\bar{V}_1$ in place of $V_1{}^0$ and is, in fact, more versatile in that form. (It would still reduce to equation 11–19 when $c_2 = 0$ since $\bar{V}_1$ becomes equal to $V_1{}^0$ at infinite dilution.) In practice, however, values of $\bar{V}_1$ are rarely determined. Moreover, we shall nearly always use equation 11–21 only to consider initial deviations from ideality at low c_2, and under these conditions $\bar{V}_1$ and $V_1{}^0$ are in any event always taken to be identical.

McMillan and Mayer[10] have proved that an equation of the form of equation 11–21 is completely general for non-electrolyte solutions, with the coefficients B, C, etc., functions of temperature and pressure and of the specific nature of solvent and solute. For electrolyte solutions equation 11–21 is not sufficient, a term in $c_2^{1/2}$ being required to describe the limiting behavior at low concentration. (If a third component is present, which provides a high concentration of ions, then equation 11–21 again becomes valid even if component 2 is an electrolyte.)

The coefficients B, C, etc., are not always defined exactly as was done in equation 11–21. For example it may be preferred to take $1/M_2$ outside the parentheses and write

$$\mu_1 - F_1{}^0 = -(\mathscr{R}TV_1{}^0/M_2)c_2(1 + B'c_2 + C'c_2{}^2 + \cdots)$$

11f. Colligative properties and molecular weight. A number of experimental measurements provide essentially a measure of the chemical potential of the solvent in a solution, relative to that of pure solvent, i.e., a measure of $\mu_1 - F_1{}^0$. It is clear from equation 11–21 that these properties always permit the determination of the molecular weight, M_2, of the solute. Even if the solution being studied is quite non-ideal, an extrapolation of $(\mu_1 - F_1{}^0)/\mathscr{R}TV_1{}^0c_2$ to infinite dilution will always yield the value of M_2. The experimental quantities which can be used in this way are known as the *colligative* properties of a solution. Falling into this class are the vapor pressure, the freezing point depression, the boiling point elevation, and the osmotic pressure. Of these only the osmotic pressure is commonly used for large solute molecules. It is discussed in detail in section 13.

11g. The total free energy of a solution. It is often desirable to have an expression for the total free energy of a solution, $F = \sum_i n_i \mu_i$. The previous section has shown, however, that the chemical potentials μ_i are known only relative to their standard chemical potentials, and F can therefore in general be expressed only relative to the various $\mu_i{}^0$. It is therefore customary to define a quantity known as the *free energy of mixing*,

$$\Delta F_{\text{mix}} = \sum_i n_i(\mu_i - \mu_i{}^0) = F - \sum_i n_i \mu_i{}^0 \qquad (11\text{–}22)$$

the summation extending over all components of the solution. This quantity cannot in general be thought of as the free energy change which accompanies the mixing of a solution from its components in their pure state. For the free energy change of that process would be $\sum_i n_i(\mu_i - F_i{}^0)$, where $F_i{}^0$ is the molar free energy of the ith component in the pure form. As was discussed earlier, the $\mu_i{}^0$ which occur in equation 11–22 are equal to the corresponding $F_i{}^0$ only in special circumstances. (Some authors prefer, for this reason, to use a symbol such as ΔF^* for the quantity defined by equation 11–22.)

It is often convenient, however, to consider that equation 11–22 represents the free energy change in a hypothetical mixing process in which the μ_i^0 are regarded as the molar free energies of pure components in some hypothetical state. Suppose, for instance, that an expression is desired for ΔF_{mix} in a dilute solution of two components and that expressions for μ_1 and μ_2 are available in which, formally, μ_1^0 and μ_2^0 represent the values of μ_1 and μ_2 when X_1 and X_2, respectively, become unity. The expressions for μ_1 and μ_2 will be ones valid for dilute solutions and will encompass the composition $X_1 = 1$. Thus μ_1^0 is in fact equal to F_1^0. However, the expression for μ_2 will be valid only when X_2 is small (component 2 may even have limited solubility in the solvent) so that μ_2^0 is clearly not equal to F_2^0. We may suppose, however, that there is a hypothetical form of component 2 which obeys the expression for μ_2 in dilute solution and continues to obey it at all values of X_2. Thus μ_2^0 would be the molar free energy of component 2 in this hypothetical form. As long as the final result which we seek is to be applicable only to the dilute solution range in which the expressions for μ_1 and μ_2 are applicable, it makes no difference whether we think of the solution as having been derived from the components in their real pure state or in suitable hypothetical states.

Corresponding to equation 11–22 other thermodynamic properties of the mixing process can be defined; e.g.,

$$\Delta H_{mix} = \sum_i n_i(\bar{H}_i - \bar{H}_i^0) = H - \sum_i n_i \bar{H}_i^0$$
$$\Delta V_{mix} = \sum_i n_i(\bar{V}_i - \bar{V}_i^0) = V - \sum_i n_i \bar{V}_i^0 \qquad (11\text{–}23)$$

where H and V are the total heat content and volume, respectively, of the solution. The quantities $\bar{H}_i^0$, $\bar{V}_i^0$, etc., are equivalent to the corresponding molar quantities H_i^0, V_i^0, etc., for the pure components under the same conditions as those which make $\mu_i^0 = F_i^0$. Otherwise they refer, like μ_i^0, to a hypothetical state of the pure component.

Quite generally, for any extensive thermodynamic property G, we can write $\Delta G_{mix} = G - \sum_i n_i \bar{G}_i^0$, from which it follows that

$$[\partial(\Delta G_{mix})/\partial n_i]_{T,P,n_j(j \neq i)} = \partial G/\partial n_i - \bar{G}_i^0 = \bar{G}_i - \bar{G}_i^0 \qquad (11\text{–}24)$$

It is worth noting that the thermodynamic quantities ΔG_{mix} take on particularly simple forms for ideal solutions. Combining equations 11–14 and 11–22 we get

$$\Delta F_{mix} = \mathscr{R}T \sum_i n_i \ln X_i \qquad (11\text{–}25)$$

Differentiating equation 11–25 we obtain, since $(\partial F/\partial P)_{T,n_i} = V$ and $[\partial(F/T)/\partial T]_{P,n_i} = -H/T^2$,

$$\Delta V_{\text{mix}} = [\partial(\Delta F_{\text{mix}})/\partial P]_{T,n_i} = 0$$
$$\Delta H_{\text{mix}} = -T^2[\partial(\Delta F_{\text{mix}}/T)/\partial T]_{P,n_i} = 0$$

and, since $\Delta F = \Delta H - T\,\Delta S$,

$$\Delta S_{\text{mix}} = -\mathscr{R} \sum_i n_i \ln X_i \qquad (11\text{–}26)$$

Equation 11–26 is the same as the entropy of mixing for ideal gases.

The fact that ideality requires zero values for ΔH_{mix} and ΔV_{mix} explains why ideality is more likely to be approached in dilute solutions. In dilute solutions solute particles are surrounded on all sides by solvent so that each solute particle forms essentially a single region of the solution, separated from other such regions by pure solvent. Adding additional solvent to such a solution will only increase the extent of the regions of pure solvent but will not alter the environment of the solute molecules which are already completely surrounded by solvent. These are conditions under which ΔV_{mix} and ΔH_{mix} would be expected to vanish.

11h. Thermodynamics and statistical mechanics. One of the features of the thermodynamic relations which have been summarized above is that they are independent of any physical picture which we might have of the chemical system to which they are to be applied. In this book, however, it is the physical picture in which we are primarily interested; i.e., we shall want to know what structural features are responsible for observed values of thermodynamic parameters.

The bridge between these two aspects of physical chemical behavior is provided by statistical mechanics and, in particular, by the relation between entropy and probability,

$$S = k \ln W \qquad (11\text{–}27)$$

where k is Boltzmann's constant and W is the "probability" of a particular state of an assembly of molecules. This probability is quite generally measured as the number of equivalent "microscopic" states of the assembly. It is large, for instance, in a gas, where each molecule might with equal likelihood be anywhere in a large vessel, and it is small in a crystal where each molecule is confined to an exceedingly small space about a particular lattice site. To take another example, there is a configurational contribution to the entropy which is large for a flexible polymer molecule of the type discussed in the preceding chapter but which will be vanishingly small for a rigid molecule (e.g., for a helical rod).

The following section will take up in some detail the application of statistical mechanics to assemblies of macromolecules.

12. STATISTICAL TREATMENT OF THE CHEMICAL POTENTIAL IN NON-ELECTROLYTE SOLUTIONS (ESPECIALLY FLEXIBLE POLYMERS)[3,4,5,6]

The ideal equation for the chemical potential of the solvent in dilute solutions, equation 11–19, suggests that the second virial coefficient, B, should decrease with increasing molecular weight and therefore become very small for macromolecular solutions. This prediction is contrary to experience; macromolecular solutions, more often than not, show a large concentration dependence of μ_1. They also tend to show other features indicative of non-ideal behavior, such as separation into immiscible phases. An inquiry into the reasons for such non-ideal behavior is therefore called for. This section will present the present status of work in this field as it applies to non-electrolytes, with the major emphasis on flexible coiled polymers. For electrolytes the presence of charged particles introduces entirely separate considerations, which will be taken up in section 14.

12a. Excluded volume treatment for dilute solutions.[15,16] The statistical evaluation of the chemical potential in liquids is a formidable problem regardless of whether the constituent molecules are small or large. It is possible to obtain a simple solution only under simple limiting conditions. One such simple situation exists in a very dilute solution of macro-molecules, so dilute that the solute molecules are far apart and essentially independent of one another. Under such conditions, as was pointed out on p. 191, the partial molal heat content and volume ($\bar{H}_i$ and $\bar{V}_i$) are independent of concentration, so that ΔH_{mix} and ΔV_{mix} are zero and only ΔS_{mix} contributes to the relative chemical potential, $\mu_i - \mu_i^0$. We shall be interested only in the chemical potential of the solvent (designated by subscript 1); i.e., we shall seek to evaluate

$$\mu_1 - F_1^0 = -T(\bar{S}_1 - S_1^0) \tag{12-1}$$

where S_1^0 is the molar entropy of pure solvent.

It is important that the assumption $\Delta H_{\text{mix}} = 0$ be clearly understood. This assumption does *not* imply the absence of interaction between solute and solvent molecules. In fact, the present treatment will be applied to situations in which such interaction is strong, so that the partial molal enthalpy $\bar{H}_i$ of the macromolecular solute in the solution will in general be quite different from the molar enthalpy H_i^0 of pure solute. What the statement $\Delta H_{\text{mix}} = 0$ does imply is that solute molecules are at such low concentration that $\bar{H}_i$ is independent of concentration and thus equal to the value $\bar{H}_i^0$ which it would have at infinite dilution. It will be noted that equation 11–23 (for dilute solutions) defines ΔH_{mix} as $\sum_i n_i (\bar{H}_i - \bar{H}_i^0)$ and not as $\sum_i n_i(\bar{H}_i - H_i^0)$.

When the various terms which might contribute to ΔS_{mix} or to $\bar{S}_i - \bar{S}_i^0$ are considered, the majority can at once be set equal to zero. As long as individual macromolecules are far apart and independent of one another all vibrational, rotational, and configurational freedom per molecule will be the same regardless of concentration; i.e., these factors make no contribution to $\bar{S}_i - \bar{S}_i^0$. Only the freedom of translational motion remains as a variable factor. This particular type of freedom is also the simplest type to consider statistically, for the number of available microscopic states is simply the number of available positions; i.e., it is simply proportional to the free volume. We shall call the proportionality factor A.

Consider now the number of ways, ν_1, of introducing the first of N_2 macromolecules into a solution of volume V. More specifically, we wish to know the number of ways of placing the center of mass of the molecule, the positions of other parts of the molecule relative to the center of mass being a measure of the rotational, vibrational, and configurational properties. Clearly the entire volume V is free for occupation so that

$$\nu_1 = AV$$

Considering next the number of ways, ν_2, of introducing the center of mass of the second molecule, the available volume is clearly less than V because the first molecule already present will exclude the center of mass of the second one from a certain volume about it. We shall call this volume the *excluded volume, u*. The volume available for the second molecule is thus $V - u$ and

$$\nu_2 = A(V - u)$$

As for the third molecule, it will be excluded from a volume $2u$; i.e., $\nu_3 = A(V - 2u)$, and, in general, for the ith molecule

$$\nu_i = A[V - (i - 1)u]$$

It is important to note that this manner of computing the excluded volume can be valid only for dilute solutions. As a solution becomes more concentrated, with the introduction of a greater number of molecules, we shall begin to have an overlap of the volume excluded by individual molecules. The total excluded volume for the ith molecule will then become less than $(i - 1)u$, as is illustrated by Fig. 12–1. For very concentrated solutions the excluded volume per molecule will approach the volume actually occupied.

The total number of ways, W, of introducing all N_2 solute molecules is now simply

$$W = \frac{1}{N_2!} \prod_{i=1}^{N_2} \nu_i = \frac{A^{N_2}}{N_2!} \prod_{i=1}^{N_2} [V - (i - 1)u] \qquad (12\text{–}2)$$

The factor $1/N_2!$ in equation 12–2 arises from the fact that the N_2 molecules which have been labeled as molecules 1, 2, 3, etc., are, in fact, indistinguishable, so that Πv_i alone counts each distinguishable build-up of the solution $N_2!$ times.

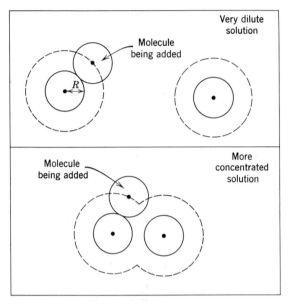

Fig. 12–1. The excluded volume for solid spheres. In very dilute solutions each molecule contributes a spherical volume of radius $2R$. In more concentrated solutions the sum of the contribution of several molecules becomes less than this.

To simplify equation 12–2 we note first that

$$\prod_{i=1}^{N_2} [V - (i-1)u] \equiv \prod_{i=0}^{N_2-1} (V - iu) \equiv \prod_{i=0}^{N_2-1} V(1 - iu/V)$$

Furthermore, as long as we deal with very dilute solutions, $iu/V \ll 1$ even when $i = N_2 - 1$, so that $\ln(1 - iu/V) = -iu/V$. Thus equation 12–2 may be written in logarithmic form as

$$\ln W = N_2 \ln A - \ln N_2! + N_2 \ln V - (u/V) \sum_{i=0}^{N_2-1} i \qquad (12\text{–}3)$$

The summation in equation 12–3 is equal to $N_2(N_2 - 1)/2$, which, since N_2 is the final number of molecules in the solution and therefore a very large number even in a dilute solution, may be written as $N_2^2/2$. A further substitution is to write $N_2 = n_2\mathcal{N}$ where n_2 is the number of moles of solute and $\mathcal{N}$ is Avogadro's number. We may also write $V = n_1\bar{V}_1 + n_2\bar{V}_2$, which, for the very dilute solutions here under consideration becomes $V = n_1 V_1^0 + n_2 \bar{V}_2^0$, where V_1^0 is the molar volume of pure solvent and $\bar{V}_2^0$ the partial molal volume of solute at infinite dilution, or, if we prefer, in the hypothetical pure state (p. 190) on the basis of which ΔS_{mix} will be computed. Thus

$$\ln W = \mathcal{N} n_2 \ln A - \ln (\mathcal{N} n_2)! + \mathcal{N} n_2 \ln (n_1 V_1^0 + n_2 \bar{V}_2^0)$$
$$- (\mathcal{N} n_2)^2 u / 2(n_1 V_1^0 + n_2 \bar{V}_2^0) \quad (12\text{--}4)$$

Equation 12–4, as it stands, represents the value of $\ln W$ for the final solution. If we use the same equation with the same value of n_2 but with $n_1 = 0$, we shall get the value of $\ln W$ for the pure solute in the hypothetical state in which its internal energy, molar volume, etc., are identical to what they are at infinite dilution in solution. If we use the same equation with the same value of n_1 but with $n_2 = 0$, we shall get $\ln W$ for the pure solvent before it is mixed with solute. (The latter $\ln W = 0$.) We can then use equation 11–27 to calculate ΔS_{mix},

$$\Delta S_{mix} = k \ln W \text{ (final soln.)} - k \ln W \text{ (pure solute)}$$
$$- k \ln W \text{ (pure solvent)} \quad (12\text{--}5)$$

Where $\mathcal{R}$, the gas constant, is equal to $\mathcal{N} k$, we get

$$\Delta S_{mix} = \mathcal{R} n_2 \ln \frac{n_1 V_1^0 + n_2 \bar{V}_2^0}{n_2 \bar{V}_2^0} - \frac{\mathcal{R} n_2^2 \mathcal{N} u}{2(n_1 V_1^0 + n_2 \bar{V}_2^0)} + \frac{\mathcal{R} n_2 \mathcal{N} u}{2 \bar{V}_2^0} \quad (12\text{--}6)$$

It is now a simple matter to evaluate $\bar{S}_1 - S_1^0$ by equation 11–24. We get

$$\bar{S}_1 - S_1^0 = \frac{\mathcal{R} n_2 V_1^0}{n_1 V_1^0 + n_2 \bar{V}_2^0} + \frac{\mathcal{R} n_2^2 \mathcal{N} u V_1^0}{2(n_1 V_1^0 + n_2 \bar{V}_2^0)^2} \quad (12\text{--}7)$$

Placing $n_2/(n_1 V_1^0 + n_2 \bar{V}_2^0) = c_2/M_2$, where c_2 is the concentration of solute in grams per cubic centimeter and M_2 is the solute molecular weight, and converting to chemical potential by equation 12–1, we get

$$\mu_1 - F_1^0 = -\mathcal{R} T V_1^0 c_2 \left(\frac{1}{M_2} + \frac{\mathcal{N} u}{2 M_2^2} c_2 \right) \quad (12\text{--}8)$$

Equation 12–8 is seen to be identical with equation 11–21 with

$$B = \mathcal{N} u / 2 M_2^2 \quad (12\text{--}9)$$

It is to noted that, if u is of the same order of magnitude as a solvent molecule, then $B \sim \frac{1}{2} V_1^0 / M_2^2$, which is precisely the value of B for an ideal solution (equation 11–19). Macromolecular solutions are clearly intrinsically non-ideal on account of their size ($\mathcal{N} u \gg V_1^0$), although we shall see that attractive forces between them may produce an effect equivalent to a reduction in volume; i.e., the excluded volume is not in general to be equated with the physical volume.

12b. The excluded volume for solid spheres and rods.

In a dilute solution the center of mass of a solid spherical solute molecule of radius R may clearly approach to within a distance $2R$ of another solute molecule, as shown by Fig. 12–1. The excluded volume thus becomes

$$u = (32/3)\pi R^3 = 8M_2 v_2/\mathcal{N} \qquad (12\text{–}10)$$

where v_2 is the specific volume of the solute particle. The corresponding equation for the second virial coefficient is

$$B = 16\pi \mathcal{N} R^3/3M_2{}^2 = 4v_2/M_2 \qquad (12\text{–}11)$$

The evaluation of the excluded volume for rigid rods is a considerably more involved problem, which has been solved independently by Zimm,[15] Onsager,[16] and Schulz.[17] The procedure is as follows. The rods are taken as cylinders of length L and diameter d. Their orientation in space may be specified by suitable angles; for an isolated rod the entire spherical solid angle of 4π is available. Two rods whose centers are separated by a distance L or more are quite independent of one another. The available directions for orientation may be specified as the product $4\pi \times 4\pi = 16\pi^2$. When the rods are separated by less than a distance L, however, certain simultaneous orientations are not allowed because no part of space may be simultaneously occupied by portions of both rods. At any distance a between centers the product of available orientations, Ω, is thus less than $16\pi^2$. The intrinsic probability of locating a second rod at a distance a from another rod is thus $\Omega/16\pi^2$. This is equivalent to saying that, of the volume $4\pi a^2\, da$ which exists at a distance a from the center of molecule 1, a fraction $\Omega/16\pi^2$ is available, whereas a fraction $1 - \Omega/16\pi^2$ is unavailable. The total excluded volume is then $\int(1 - \Omega/16\pi^2)4\pi a^2\, da$. The limits of integration are from $a = 0$ to $a = \infty$, but the region from $a = L$ to $a = \infty$, of course, makes no contribution. The result obtained is that

$$u = \tfrac{1}{2}\pi dL^2 = \frac{2LM_2 v_2}{d\mathcal{N}} \qquad (12\text{–}12)$$

where v_2 is again the specific volume of the solute. The corresponding expression for B is

$$B = Lv_2/dM_2 \qquad (12\text{–}13)$$

Many of the best known macromolecules which may form essentially solid particles in solution include within such particles, either by chemical combination (e.g., hydration) or by "trapping," a certain amount of the solvent. This solvent must be included in the excluded volume. Thus, if M_2 is the molecular weight of dry solute, v_2 must be a composite factor including not only the volume actually occupied per gram by the dry molecule but also the volume occupied by the solvent associated with each gram of the dry molecule. A full discussion of this topic will be given in section 20, when hydrodynamic properties are considered. (However, the effective thermodynamic specific volume, v_2, is not necessarily exactly identical with the effective hydrodynamic specific volume.)

12c. The excluded volume for flexible polymers. Simplified theory. In section 10 (Fig. 10–2) we computed the density of polymeric material within the region of space occupied by a flexible polymer molecule. It was found to be very small, of the order of 1%. Thus it would seem that one polymer molecule interferes virtually not at all with the location of the center of mass of a second molecule. In mathematical terms, let ρ, as in section 10c, represent the density of segments per cubic centimeter, and let τ be the volume of a single segment. Then $f = \rho\tau$ is the fraction of the volume at any place filled by segments of the first molecule. The probability of locating the center of mass of a second molecule within the domain of the first, as compared to the probability of locating it far from the first molecule, which is clearly proportional to the fraction of free volume, becomes $1 - f = 1 - \rho\tau$. Since f is very small, this probability is essentially unity.

Further thought, however, drastically revises this opinion, for, if the center of mass of the second molecule is placed within the domain of the first molecule, then all the segments of the second molecule must likewise be within the domain of the first molecule. Thus, once the center of mass has been placed in position, the probability of finding a suitable location for each subsequent segment is also $1 - f$ relative to the probability of finding such a location far from the first molecule. The total relative probability thus becomes of order of magnitude $(1 - f)^\sigma$, and this is very small even if $1 - f$ is of order 0.99 or more.

Using the general concept just introduced, we shall compute an order of magnitude value for the *physical* excluded volume of a flexible polymer molecule. Suppose we have one such molecule fixed in place. Let it be represented by its time-average distribution of polymer segments, as given by equation 10–26 (or Fig. 10–2), with R_G given by equation 9–40.

We wish to place a second polymer molecule near the first. The probability of being able to do so depends both on the configuration of the second molecule and on where it is placed. Suppose that we decide on a given location and configuration for the second molecule. The probability that we shall be able to place the first segment at the chosen position will be given by $1 - f_1$, where $f_1 = \rho_1 \tau$ is the volume fraction filled by segments of the first molecule at this position, ρ_1 being the value of ρ appropriate to the position. Similarly for the second segment we obtain a probability $1 - f_2$, etc. The probability of being able to introduce the entire second molecule in its desired location and configuration is thus $\prod_{i=1}^{\sigma} (1 - f_i) = \prod_{i=1}^{\sigma} (1 - \rho_i \tau)$. Each of the ρ_i will, in general, be different because each segment will be located at a different distance from the first molecule's center of mass. The product will have a different value for each configuration. We now assume that the *average* value of the product for all configurations may be replaced by $(1 - \rho \tau)^{\sigma}$, where $\rho \tau$ is evaluated at the *average* position of segments of the second molecule, i.e., at its center of mass. Thus, if $W(r)$ represents the probability of locating the center of mass of the second molecule at a distance r from the center of mass of the first molecule and $W(\infty)$ represents the probability of locating it an infinite distance away,

$$W(r)/W(\infty) = (1 - \rho \tau)^{\sigma} \tag{12–14}$$

Taking logarithms we get $\ln [W(r)/W(\infty)] = \sigma \ln (1 - \rho \tau) = -\sigma \rho \tau$ since $\rho \tau$ (cf. Fig. 10–2) is of the order 0.01, and higher terms in the expansion of the logarithm may thus be ignored. Thus, introducing equation 10–26,

$$\ln \frac{W(r)}{W(\infty)} = -\sigma \rho \tau = -\sigma^2 \tau \left(\frac{3}{2\pi R_G^2} \right)^{3/2} e^{-3r^2/2R_G^2} \tag{12–15}$$

A graphical representation of the function $W(r)/W(\infty)$, for the same polymer as was used in computing the data of Fig. 10–2, is shown in Fig. 12–2. It shows that the probability of locating the center of mass of a second polymer molecule within the volume encompassed by the radius of gyration of the first polymer molecule is essentially zero. As the distance from the center of mass increases, the freedom to locate the second polymer molecule increases steeply, and, at $r = 2R_G$, $W(r)$ is close to $W(\infty)$. The value of r/R_G where $W(r)/W(\infty)$ rises steeply from zero to unity is about 1.7.

The important idea which we wish to draw here from Fig. 12–2 is that the behavior of a flexible polymer with respect to the physical exclusion of other polymer molecules approaches that of a solid sphere. Such a sphere of radius R (Fig. 12–1) excludes the center of mass of a second sphere completely to a distance $2R$ from its center and has no effect at all outside this distance; i.e., $W(r)/W(\infty) = 0$ for $r < 2R$ and $W(r)/W(\infty) = 1$ for $r > 2R$, as shown by the dashed line of Fig. 12–2.

We are thus led to the important idea that a polymer molecule may be represented by an *equivalent solid sphere*, of radius R_e, where R_e in the example chosen lies somewhere near 0.85 of the radius of gyration. It suggests that we may quite generally be able to write

$$R_e = \gamma R_G \qquad (12\text{--}16)$$

the parameter γ being for the present undetermined. The excluded volume is then calculated by equation 12–10 as $(32/3)\pi R_e^3$; i.e.,

$$u = (32/3)\pi\gamma^3 R_G^3 \qquad (12\text{--}17)$$

We have so far taken into account only the volume from which a given flexible polymer molecule excludes all others because of the physical

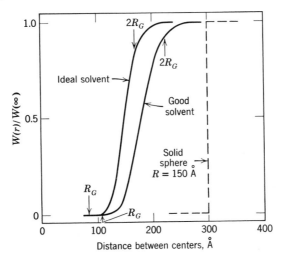

Fig. 12–2. The probability of being able to locate a flexible polymer molecule near another one. The corresponding behavior of a solid sphere is shown for comparison, and it is seen that the polymer molecules behave much like spheres of radius $0.8R_G$. (Attractive forces are not taken into account. See text for their effect.)

volume of its segments. A little consideration shows, however, that the excluded volume must have two other essential properties:

(1) The volume from which the segments of a polymer molecule exclude the segments of other polymer molecules must be intimately related to the volume from which these segments exclude other segments *on the same molecule*. In other words there must be an intimate relation between the excluded volume u and the empirical factor α introduced in section 9h to allow for the effect of intersegment interaction on $\overline{h^2}$ or R_G^2.

(2) When this factor α was considered, it was pointed out that the physical volume of exclusion is not the sole factor in the interaction

between segments. In a *good solvent* there will be a preference for having solvent molecules adjacent to polymer segments, which will in effect increase the excluded volume due to each segment. In such a solvent α is larger than otherwise, and there must be a corresponding increase in the excluded volume u. Similarly, in a *poor solvent*, there is a preference for segment–segment contacts, which has the same effect as if an increased free volume existed near each segment. This effect decreases α and must produce a corresponding decrease in the excluded volume u. In particular, as was noted in section 9h, a solvent may be sufficiently poor so that $\alpha = 1$; i.e., the polymer molecule behaves as if its segments occupy no volume at all. Under these circumstances we should expect the polymer segments to behave in the same manner towards a second polymer molecule; i.e., *we should expect the excluded volume to become zero.* Such a solvent was referred to in section 9h as an *ideal solvent*, a name which is equally appropriate here, since, if $u = 0$, then, by equation 12–9, $B = 0$, and, this, we have seen, is an apt criterion for ideality in macromolecular solutions.

The relation between u and α proposed in the preceding paragraphs may be incorporated into equation 12–16 and 12–17 if γ is taken to be an appropriate increasing function of α which becomes zero when $\alpha = 1$; i.e., we can expect that there will be a relationship of the type $\gamma^3 = f(\alpha^2 - 1)$ between these two empirical parameters.

The discussion of the excluded volume of flexible polymer molecules which has been presented here has been entirely qualitative. Two objectives must be met by an exact quantitative treatment of the problem: (*a*) the correct relation between the excluded volume and α must be established, and (*b*) both of these parameters must be related to suitable thermodynamic functions characterizing the interaction between polymer segments and solvent molecules. To meet these objectives turns out to be a difficult task, and the problem is at present not completely solved. A number of theories have been proposed, all involving some assumption of uncertain validity, and none of them is completely successful in predicting exactly the experimental behavior of flexible polymers, though most of them come quite close to doing so.

We shall make use of only one of these theories, that of Flory and Krigbaum.[18] In their theory the interaction between segments and solvent molecules can be characterized by heat and entropy parameters κ and ψ (a positive value of κ representing a poor solvent with segment–segment contacts energetically not favored, whereas a negative value indicates a good solvent). These parameters are defined so that the corresponding free energy becomes $T(\kappa - \psi)$, and, for practical purposes, it proves illuminating to write $\kappa = \psi\Theta/T$, where Θ is a function which has the dimensions of temperature. In terms of this function the free energy becomes $T\psi(1 - \Theta/T)$ and the expression for the excluded volume is

$$u = \frac{2\bar{v}_2{}^2 M_2{}^2}{V_1{}^0 \mathcal{N}} \, \psi\left(1 - \frac{\Theta}{T}\right)\left(1 - \frac{X}{2! \, 2^{3/2}} + \frac{X^2}{3! \, 3^{3/2}} - \cdots\right) \quad (12\text{–}18)$$

where

$$X = \frac{2\bar{v}_2{}^2 M_2{}^2}{V_1{}^0 \mathcal{N}} \, \psi\left(1 - \frac{\Theta}{T}\right)\left(\frac{3}{4\pi R_G{}^2}\right)^{3/2} \quad (12\text{–}19)$$

In these equations M_2 is the molecular weight of the polymer and $\bar{v}_2$ its partial specific volume; $V_1{}^0$ is the molar volume of solvent and T the temperature.

The intramolecular expansion factor α may be expressed in terms of the same parameters, the result being

$$\alpha^5 - \alpha^3 = \frac{27\bar{v}_2{}^2 M_0{}^{3/2}}{(2\pi)^{3/2}\beta^3 V_1{}^0 \mathcal{N}} \, \psi(1 - \Theta/T)M_2{}^{1/2} \quad (12\text{–}20)$$

where β is the effective bond length which is defined by equation 9–16.

Equations 12–18 and 12–20 may be combined so as to yield an expression for u in terms of α,

$$u = 2(4\pi/3)^{3/2}(\alpha^2 - 1)R_G{}^3(1 - X/2! \, 2^{3/2} + X^2/3! \, 3^{3/2} \cdots) \quad (12\text{–}21)$$

$$X = 2(\alpha^2 - 1) \quad (12\text{–}22)$$

Equation 12–21 is seen to have the form predicted by the qualitative treatment given in this section, being of the form of equation 12–17 with γ^3 a function of $\alpha^2 - 1$.

The expressions for B which correspond to these relations for the excluded volume need not be given explicitly. They are obtained at once by combining equations 12–18 or 12–21 with equation 12–9.

It is to be noted that all the parameters of equation 12–20, except M_2, are independent of molecular weight, so that α is seen to be an increasing function of M_2, as predicted qualitatively in section 9h. If α is large so that $\alpha^5 \gg \alpha^3$, α becomes proportional to $M_2^{0.10}$. This is the upper limit in the dependence of α on M_2 which the theory permits.

(As mentioned earlier, the Flory-Krigbaum theory represents only one of several alternative approaches to the problem stated. Further developments and alternative procedures are given, for instance, by Zimm, Stockmayer, and Fixman[20] and by Orofino and Flory.[21])

12d. The effect of temperature on the chemical potential in solutions of flexible polymers. An increase in temperature decreases the effect of intermolecular and intramolecular attractive forces. Temperature should therefore have a pronounced effect on the chemical potential of polymer solutions in poor solvents, where such attractive forces are important. Specifically, poor solvents should become better solvents, with both α and B increasing, as the temperature is raised.

The effect is given quantitatively by the equations of Flory and Krigbaum, particularly in the form in which they are given in equations 12–18 and 12–20. These equations show that, in a good solvent, where Θ is always negative (Θ being proportional to κ), u is always positive and α always >1. In a poor solvent, however, where Θ is positive, there will always be a temperature, $T = \Theta$, at which $u = 0$ (i.e., $B = 0$) and $\alpha = 1$; i.e., the solution will behave ideally at this temperature. When $T > \Theta$, on the other hand, $u > 0$ and $\alpha > 1$, and the solution will progressively approach good solvent behaviour as T increases.

The reason for the substitution $\kappa = \psi\Theta/T$ made above is now clear, for the parameter so introduced is an "ideal" temperature, characteristic of any poor solvent-polymer pair, at which $B = 0$ and $\alpha = 1$. It is worth noting the parallel between this ideal temperature and the Boyle point in non-ideal gases, this being a point at which attractive forces between gas molecules exactly compensate for the repulsion due to the finite volume of the molecules so that $PV/\mathscr{R}T = 1$.

Experimental studies on the effect of temperature on the chemical potential in polymer solutions will be reported in section 13, and they will bear out the conclusions reached here.

12e. Attractive forces in solutions of solid spheres. We have seen in the preceding section that attractive forces between flexible polymer molecules may effectively decrease the excluded volume to the point where it vanishes. Can the same situation arise in solutions of spheres (section 12b)? The answer is that it can do so only in special circumstances. The weak interactions which lead to the difference between poor and good solvents for flexible polymers can have no appreciable effect because the segments of solid spheres cannot interpenetrate one another as the segments of flexible chains can. However, electrostatic forces, which are long-range forces, can produce the same effect and can in fact lead to negative values of the second virial coefficient B. They will be discussed in section 14 and again in section 28. Another way in which solid particles can display attractive forces is if they possess a tendency to form association products: if, for example, there is a tendency for formation of dimers. We shall briefly consider this latter possibility here.

Suppose that solid spherical macromolecules, A, of molecular weight M and radius R can coalesce to form spherical dimers, A_2, of molecular weight $2M$ and radius $2^{1/3}R$ (i.e., the specific volume is the same for A

and A_2). Let K be the equilibrium constant in the usual units for the reaction, $2A \rightleftharpoons A_2$; i.e.,

$$K = C_{A_2}/C_A{}^2 = Mc_{A_2}/2000c_A{}^2 \qquad (12\text{--}23)$$

where C_i represents concentration in moles per liter and c_i concentration in grams per cubic centimeter. If the total concentration in grams per cubic centimeter is c_2, i.e., $c_2 = c_A + c_{A_2}$, then

$$c_A = \frac{-1 + (1 + 8000Kc_2/M)^{\frac{1}{2}}}{4000K/M}$$

$$c_{A_2} = \frac{1 + 4000Kc_2/M - (1 + 8000Kc_2/M)^{\frac{1}{2}}}{4000K/M} \qquad (12\text{--}24)$$

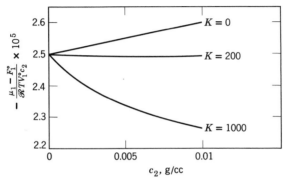

Fig. 12–3. The effect of dimerization on the chemical potential of solid spherical molecules. K represents the equilibrium constant in (moles/liter)$^{-1}$ for the reaction $2A \rightleftharpoons A_2$. The molecular weight of A is 40,000.

By equation 11–21, adding the contributions of the two solutes gives

$$\mu_1 - F_1{}^0 = -\mathscr{R}TV_1{}^0(c_A/M + c_{A_2}/2M + B_e c_2{}^2 + \cdots) \qquad (12\text{--}25)$$

where $B_e c_2$ represents the physical excluded volume term of section 12b. By equation 12–11 this term will fall to half its value as we go from complete dissociation ($c_2 = c_A$) to complete association ($c_2 = c_{A_2}$). In Fig. 12–3 we have made a plot of $(c_A/M + c_{A_2}/2M + B_e c_2{}^2)/c_2$ versus c_2 for a molecule of molecular weight 40,000 for three values of K: $K = 0$, 200, and 1000. $K = 200$ corresponds to 8% dimerization at the highest concentration ($c_2 = 0.01$) used; $K = 1000$ corresponds to 27% dimerization at that concentration. In view of the fact that the majority of all solute molecules are still in the monomer state we have used the same value of B_e throughout, as given by equation 12–11, with $v_2 = 1$ and $M = 40,000$.

By equations 11–21 and 12–25 we see that the quantity plotted would normally be interpreted as $1/M_2 + Bc_2 + \cdots$, and it is seen that association does indeed have the expected effect on the apparent value of B, permitting B to go to zero and even to become negative.

12f. The Flory-Huggins lattice theory.[6] Before concluding this section, it is desirable to mention the theory of Flory[12] and Huggins[13] for calculation of the chemical potential in relatively concentrated polymer solutions. It is the most widely used treatment of such solutions but is interesting also for another reason; it is the basic treatment from which several of the more advanced theories of polymer solutions (including the Flory-Krigbaum theory discussed earlier) have been derived.

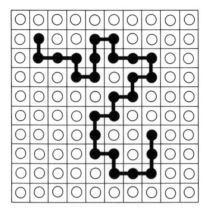

Fig. 12–4. Schematic diagram of the lattice used to represent a polymer solution in the Flory-Huggins theory. (Taken from Flory.[6])

In the Flory-Huggins theory each polymer molecule is represented as a chain of σ segments, each exactly equal in volume to a solvent molecule. (This is not a serious restriction.) A polymer solution may now be represented by a lattice, each point of which may be occupied either by a solvent molecule or by a segment of the polymer molecule. A two-dimensional representation of such a lattice is shown in Fig. 12–4. We assume that $\Delta V_{\text{mix}} = 0$. The total volume of the solution is then $(n_1 + \sigma n_2)V_1^0$, where n_1 and n_2 represent the numbers of moles of solvent and solute, respectively, in the final solution. The total number of sites in the lattice is $N_1 + \sigma N_2$, where $N_1 = \mathcal{N}n_1$ and $N_2 = \mathcal{N}n_2$. We also define a *coordination number*, z, as the number of nearest neighboring lattice sites at any one site; e.g., $z = 6$ for a regular cubic lattice, or 12 for a hexagonal lattice.

Let us consider the process of filling these lattice sites. Let the solute molecules be introduced first. At any stage of the addition let N_i be the number of solute molecules already introduced; i.e., during the addition N_i will vary from zero to N_2. Let f_i designate the fraction of all lattice sites occupied at a particular stage of the addition; i.e.,

$$f_i = \sigma N_i / (N_1 + \sigma N_i) \qquad (12\text{–}26)$$

Our first object will be to calculate the number of possible ways of adding the $(N_i + 1)$st solute molecule.

The first segment may be set down in any of the free sites, i.e., in $(1 - f_i)(N_1 + \sigma N_2)$ ways. The second segment may be placed in any of the z adjacent lattice sites which are unoccupied. Since, on the average, $1 - f_i$ is the fraction of sites unoccupied, then, on the average, $(1 - f_i)z$ will be the number of sites available for the second segment. The number of ways of placing the third segment will depend on the flexibility of the solute molecule. If the molecule is completely flexible, any of the z neighboring sites may be used, except that already occupied by segment number 1. The fraction of the accessible sites which is free is on the average $1 - f_i$; i.e., the number of ways of placing the third segment becomes $(1 - f_i)(z - 1)$. In general we may write for the number of ways of introducing the third segment $(1 - f_i)y$, where y has a maximum value of $z - 1$ but is less for molecules with more restricted flexibility and would become equal to unity for rigid solute molecules, for which, however, the present model would no longer be suitable.

The number of ways of placing the fourth and all subsequent segments is taken as equal to the number of ways of placing the third segment.

This method of calculating the number of ways of placing the successive segments of each polymer molecule clearly involves the crucial assumption that the segment density f_i has a unique value independent of position. This assumption can be valid only for concentrated solutions, in which the domains of individual molecules overlap so that the segment density is more or less uniform. The assumption must be invalid for dilute solutions. For such solutions f_i will be large in the vicinity of solute molecules already present, but it will be essentially zero in the spaces between solute molecules. The value calculated by equation 12–26 will then correctly evaluate the number of positions available to the *first* segment of the $(N_i + 1)$st molecule. The second segment, however, will necessarily have to be adjacent to the first segment: if the first segment is in one of the "vacant" regions, f_i will be essentially zero; if the first segment is near an existing molecule, f_i is likely to be larger than the average value given by equation 12–26. Moreover, for some of the later segments of a given molecule, f_i must always be larger than the average value as the preceding segments have necessarily created a region of relatively high segment density.

Another assumption, inherent in the entire method, is that all accessible lattice positions have an equal probability of being occupied by a segment being introduced. This will not be true if there are significant attractive forces between segments (cf. section 9h). This assumption will be discussed again on p. 209.

The number of ways of introducing the $(N_i + 1)$st solute molecule, ν_{i+1}, is the product of the number of ways of placing each of the σ successive segments, divided by 2; i.e.,

$$\nu_{i+1} = \tfrac{1}{2}(1 - f_i)^\sigma (N_1 + \sigma N_2) z y^{\sigma - 2} \qquad (12\text{–}27)$$

The reason for dividing by 2 is that the two terminal ends of the polymer molecule are indistinguishable. Thus an arrangement having the first segment at site 1, the second at site 2, and so on, with the last segment at site σ, is identical with an arrangement using exactly the same sites but having the first segment at site σ and the last at site 1. In our counting process, however, these identical arrangements have been counted separately.

The total number of ways of introducing all N_2 solute molecules may now be computed as the product of N_2 functions of the form of equation 12–27 with N_i ranging from zero to $N_2 - 1$.

The N_1 solvent molecules are introduced next to complete the construction of the solution. All the solvent molecules are, however, indistinguishable, so that there is only one distinguishable way of arranging them once the positions of the solute molecules are fixed. Thus the total number of distinguishable ways of arranging the entire mixture is

$$W = \frac{1}{N_2!} \prod_{i=0}^{N_2 - 1} \nu_{i+1} \qquad (12\text{–}28)$$

The factor $1/N_2!$ takes into account the indistinguishability of the N_2 solute molecules.

Taking logarithms and substituting equation 12–27 we get

$$\ln W = -\ln N_2! + N_2 \ln \tfrac{1}{2}(N_1 + \sigma N_2) z y^{\sigma - 2} + \sigma \sum_{i=0}^{N_2 - 1} \ln (1 - f_i) \quad (12\text{–}29)$$

Since any real solution will contain a large number of molecules we may replace the summation in equation 12–29 by an integral. Using equation 12–26 for f_i we get

$$\sum_{i=0}^{N_2 - 1} \ln (1 - f_i) = \int_{N=0}^{N_2} \ln \left(1 - \frac{\sigma N}{N_1 + \sigma N_2}\right) dN$$

$$= -\frac{N_1}{\sigma} \ln \frac{N_1}{N_1 + \sigma N_2} - N_2 \qquad (12\text{–}30)$$

[To obtain the value of the integral let $1 - \sigma N/(N_1 + \sigma N_2) = x$ so that the integral becomes a simple multiple of $\int \ln x \cdot dx$.]

Applying Stirling's formula (equation 10–4) to $\ln N_2!$ we may ignore the terms $\frac{1}{2} \ln 2\pi$ and $\frac{1}{2}$ in comparison with N_2 since for a real solution N_2 must approach to within a few magnitudes of Avogadro's number. Thus $\ln N_2! = N_2 \ln N_2 - N_2$. With this substitution and introduction of equation 12–30, equation 12–29 becomes

$$\ln W = N_2 \ln \frac{zy^{\sigma-2}}{2N_2} - N_2(\sigma - 1)$$
$$+ (N_1 + N_2) \ln (N_1 + \sigma N_2) - N_1 \ln N_1 \quad (12\text{--}31)$$

The entropy of mixing can be calculated from equation 12–31 by use of equation 12–5, as in the derivation of equation 12–6. We get

$$\Delta S_{\text{mix}} = \mathcal{R} n_2 \ln \frac{n_1 + \sigma n_2}{\sigma n_2} + \mathcal{R} n_1 \ln \frac{n_1 + \sigma n_2}{n_1} \quad (12\text{--}32)$$

Again applying equation 11–24 we get

$$\bar{S}_1 - S_1{}^0 = -\mathcal{R} \ln \phi_1 - \mathcal{R}(1 - 1/\sigma)\phi_2 \quad (12\text{--}33)$$

where ϕ_1 and ϕ_2 are the volume fractions of solvent and solute,

$$\phi_1 = \frac{n_1}{n_1 + \sigma n_2} \qquad \phi_2 = \frac{\sigma n_2}{n_1 + \sigma n_2} \quad (12\text{--}34)$$

Equation 12–33 may also be derived by other methods, as, for instance, by Hildebrand.[11]

It is interesting to note that the flexibility of the polymer chain, as expressed by the parameter y in equation 12–27, does not enter into the final result. The reason for this is that the flexibility is not altered by the process of solution; i.e., that part of the configurational entropy which is contributed by the multiple configurations of a single molecule is not really a part of the entropy of mixing.

The energy or heat of mixing may be computed on the basis of the same model (i.e., for the more concentrated solutions that we are here considering ΔH_{mix} cannot be taken as zero). We assume that there are appreciable attractive forces only between nearest neighbors, i.e., that our model need consider for each solvent molecule or solute segment only the interaction with the occupants of the z immediately adjacent lattice sites. Let the attractive energy between two solvent molecules be $-\epsilon_{11}$ (i.e., ϵ_{11} is the energy required to break two adjacent solvent molecules apart). Similarly let the attractive energy between two segments of solute molecules be $-\epsilon_{22}$, and that between a solute segment and a solvent molecule $-\epsilon_{12}$. The total energy content (due to weak intermolecular forces) of N_1 solvent molecules and σN_2 solute segments is then obtained as follows. Surrounding each solvent molecule there will be on the average $zN_1/(N_1 + \sigma N_2)$ other solvent molecules and $z\sigma N_2/(N_1 + \sigma N_2)$ solute segments. These give rise to an energy content for all N_1 solvent molecules of $-\frac{1}{2}zN_1^2\epsilon_{11}/(N_1 + \sigma N_2) - z\sigma N_1 N_2\epsilon_{12}/(N_1 + \sigma N_2)$. The factor of $\frac{1}{2}$ in the first term arises because our method of counting solvent–solvent bonds has included each pair of neighbors twice. Considering next the solute segments; each of these participates again, on the average, in $zN_1/(N_1 + \sigma N_2)$ weak interactions with solvent molecules and $z\sigma N_2/(N_1 + \sigma N_2)$ with other polymer segments. The interaction of the segments with solvent molecules has already been taken care of; that of solvent segments with one another will lead to an energy contribution of $-\frac{1}{2}z\sigma^2 N_2^2\epsilon_{22}/(N_1 + \sigma N_2)$, so that the complete expression for the energy becomes

$$- \frac{z}{N_1 + \sigma N_2} (\tfrac{1}{2}N_1^2\epsilon_{11} + \sigma N_1 N_2\epsilon_{12} + \tfrac{1}{2}\sigma^2 N_2^2\epsilon_{22}) \qquad (12\text{--}35)$$

We are ignoring here the fact that two nearest neighbor positions in the coordination sphere of a polymer segment are of necessity other segments. This increases the probability of polymer–polymer neighbors especially in dilute solutions. There is also the likelihood that the weak interaction ϵ_{22} for two segments joined by primary bonds will not be the same as ϵ_{22} for two segments not so joined.

Using equation 12–35, the heat mixing n_1 moles of solvent with n_2 moles of solute becomes

$$\Delta H_{\text{mix}} = - \frac{z\mathscr{N}}{n_1 + \sigma n_2} (\tfrac{1}{2}n_1^2\epsilon_{11} + \sigma n_1 n_2\epsilon_{12} + \tfrac{1}{2}\sigma^2 n_2^2\epsilon_{22})$$

$$+ \frac{z\mathscr{N}}{2} (n_1\epsilon_{11} + \sigma n_2\epsilon_{22}) \quad (12\text{--}36)$$

which, with equations 12–34, may be rewritten as

$$\Delta H_{\text{mix}} = n_1\phi_2\,\Delta\epsilon \qquad (12\text{--}37)$$

where $\Delta\epsilon$ is z times the energy of formation of $\mathscr{N}$ segment-solvent bonds,

$$\Delta\epsilon = \mathscr{N}z(\tfrac{1}{2}\epsilon_{11} + \tfrac{1}{2}\epsilon_{22} - \epsilon_{12}) \qquad (12\text{--}38)$$

The parameter $\Delta\epsilon$ is a constant, depending only on the nature of the solute-solvent system and independent of the size of the solute molecules. It will clearly be negative for a good solvent and positive for a poor solvent.

By use of equations 11–24 and 12–34 we get, from equation 12–36,

$$\bar{H}_1 - H_1^0 = \Delta\epsilon\phi_2^2 \qquad (12\text{--}39)$$

and, by combination with equation 12–33 and the fact that $\phi_1 + \phi_2 = 1$,

$$\mu_1 - F_1^0 = \mathscr{R}T[\ln(1 - \phi_2) + (1 - 1/\sigma)\phi_2 + (\Delta\epsilon/\mathscr{R}T)\phi_2^2] \quad (12\text{--}40)$$

(It should be noted that the method here used to evaluate the heat content is essentially the same as that used in the statistical mechanics of solutions of low-molecular-weight compounds. See, for instance, van Laar,[14] Guggenheim,[19] and Hildebrand and Scott.[4])

As has been pointed out previously there are two major assumptions involved in the derivation of the equations of this section. One of these is the assumption of a uniform segment distribution which need not cause concern if we avoid the use of equation 12–40 for very dilute solutions. The other assumption, however, is important at all concentrations. It is the assumption that the placement of segments in the solution is governed solely by statistical considerations. This cannot be true if $\Delta\epsilon$ is appreciably different from zero, for, if $\Delta\epsilon$ is positive, segments will tend to avoid positions adjacent to other segments; if $\Delta\epsilon$ is negative, they will seek them out. As a consequence the $\Delta\epsilon$ which appears in equation 12–40 cannot have exactly the meaning assigned to it by equation 12–38. Moreover, when $\Delta\epsilon$ is not zero, the first two term in brasckets become incorrect. There must be an additional term to ΔS_{mix} which reflects the existence of preferred sites among the total number of unoccupied sites. It can be shown, however, that the net effect may be taken into account approximately by considering $\Delta\epsilon$ as an empirical parameter which is *not* simply the heat of mixing (e.g., as specified by equation 12–38) but the sum of an energy and an entropy term. Flory and Krigbaum[18] have expressed $\Delta\epsilon$ in terms of the parameters ψ, κ, and Θ used earlier (p. 200) to describe segment-solvent interaction, obtaining

$$\tfrac{1}{2} - \Delta\epsilon/\mathscr{R}T = \psi - \kappa = \psi(1 - \Theta/T) \qquad (12\text{–}41)$$

12g. Comparison of second virial coefficients and concluding remarks. To complete this section some values of the second virial coefficient have been computed for each of the equations applicable to dilute solutions, and they are shown in Table 12–1. It is seen that rod-shaped and randomly coiled molecules can be expected to have B values which are 10 to 100 times those expected of solid spheres. To visualize the magnitude of B, Table 12–1 also includes a figure for the per cent difference in $(\mu_1 - F_1^0)/c_2$, between $c_2 = 0$ and $c_2 = 0.01$ gram/cc, which each value of B represents. In the case of rod-shaped and randomly coiled molecules this difference is seen to be of the order of 100%, whereas for spheres it is only 4%. (A calculated plot of $(\mu_1 - F_1^0)/c_2$ for spheres is shown in Fig. 12–3 as the line labeled $K = 0$.)

Table 12–1 also shows that B values which would be calculated by the ideal equation, equation 11–19, are insignificantly small, justifying the definition adopted on p. 200 for an ideal solution as one for which $B = 0$.

It will be shown in sections 13 and 17 that experimental values of B determined from osmotic pressure or light scattering experiments have the magnitudes which Table 12–1 predicts.

It should be mentioned, in conclusion, that, although this section has discussed in some detail the thermodynamics of macromolecular solutions, there are a number of important topics which have been omitted. They include estimates of the third virial coefficient C (equation 11–21), which can become appreciably large in flexible polymers of high molecular

TABLE 12–1. Calculated Values of the Second Virial
Coefficient for a Molecular Weight of 100,000[a]

	B(mole-cc/gram^2)[b]	Per Cent Change in $(\mu_1 - F_1^0)/c_2$ between $c_2 = 0$ and $c_2 = 0.01$
Ideal solutions (Eq. 11–19)	5×10^{-9}	0.0005
Solid spheres (Eq. 12–11)	4×10^{-5}	4.0
Long rods ($L/d = 100$) (Eq. 12–13)	1.0×10^{-3}	100
Flexible polymers, good solvent (Eq. 12–21)	5×10^{-4}	50
Flexible polymers, poor solvent, $T = \Theta$	0	0

[a] The values of α and R_G used in equation 12–21 are the typical values used in computing the data of Fig. 10–2 and 12–2. In equation 11–19 V_1^0 has been taken as 100 cc/mole, in equations 12–11 and 12–13 v_2 has been taken as 1.0 cc/gram.

[b] The units of B (equation 11–21) are (grams/cc)^2/(cc/mole).

weight (Stockmayer and Casassa[25]); discussion of the expected values of B for polymers with molecular weight heterogeneity (Flory and Krigbaum[18]); and review of experimental determinations of heats of mixing and heats of dilution (Newing,[22] Gee and coworkers[23,24]). Another topic omitted is that of electrostatic effects which arise in solutions of macromolecular ions. It will be treated in sections 14, 27, and 28.

13. OSMOTIC PRESSURE

13a. The colligative properties of macromolecular solutions.[7] It was noted in section 11f that the colligative properties of solutions are a measure of the chemical potential of the solvent and that they can therefore be used as a measure of the molecular weight of the solute (in the ideal limit of infinite dilution) and as a test for various models of macromolecular solutions from the behavior at finite concentrations.

Equation 11–21, shows, however, that the effect of solute on the chemical potential decreases as M_2 increases. The experimental method used to measure $\mu_1 - F_1{}^0$ must therefore increase in sensitivity as higher molecular weight solutes are studied. It turns out that osmotic pressure is the most sensitive method available. In water, for example, a solution which would show a freezing point depression of 0.01° C would have a vapor pressure differing from that of pure water by less than 0.1 mm Hg, but it would have an osmotic pressure of 10 mm Hg. Osmotic pressure is thus clearly the preferred property.

However, even osmotic pressure fails for really high molecular weights. Keeping in mind the fact that we shall want to extrapolate experimental data to zero concentration, in order to obtain M_2 and B from the data (cf. equation 11–21), we will have to make measurements at concentrations below $c_2 = 0.01$ gram/cc. Equation 13–8, which will be derived below, shows that the osmotic pressure at this concentration is approximately $(200,000/M_2)$ mm Hg. With a manometric fluid such as toluene, an osmotic pressure of 1 mm Hg corresponds to a liquid level of about 15 mm, so that measurements with $M_2 = 200,000$ are easily made. However, accurate measurements will become difficult as M_2 rises much above this value.

Another advantage of osmotic pressure measurements is that, whereas other colligative properties are a measure of the sum of the concentration of all solutes, osmotic pressure is a measure of the concentration of only those solutes which cannot pass through a suitable semi-permeable membrane. Such membranes, prepared from cellophane or collodion, usually contain pores large enough to pass molecules with molecular weights about 10,000 or less but not those with molecular weights greater than 10,000. Low-molecular-weight impurities are therefore automatically excluded and do not contribute to the osmotic pressure.

What we are discussing here should, strictly speaking, be called the *colloid osmotic pressure*. The *total osmotic pressure* of a solution is defined as the osmotic pressure in which the membrane separating solution from solvent is impermeable to all solutes, regardless of size (i.e., it might be a vapor barrier if only the solvent is volatile). This type of osmotic pressure, however, is rarely measured. The term *osmotic pressure* here will always be used to signify only colloid osmotic pressure.

13b. Van't Hoff's limiting law. Figure 13–1 illustrates schematically the apparatus used for osmotic pressure measurement. (For detailed description of apparatus, see the review of R. H. Wagner.[26] Osmometers used in recent work on macromolecular solutions are described in many of the papers referred to in Table 13–1.) The membrane is permeable to solvent and to low-molecular-weight solutes. The latter are present in equal concentration on both sides or rapidly attain this condition. (We assume no chemical interaction between solutes and, for the present, exclude macro-ions, which give rise to the special effect described in section 14c.)

It is evident that the system illustrated by Fig. 13–1, with both sides open to the atmosphere, is not at equilibrium. However, the usual path to equilibrium is blocked. If the membrane were not present the macromolecules would diffuse from side II to side I (Fig. 13–1), and solvent would diffuse from side I to side II, until the concentrations are everywhere equal. The membrane permits only the second of these processes

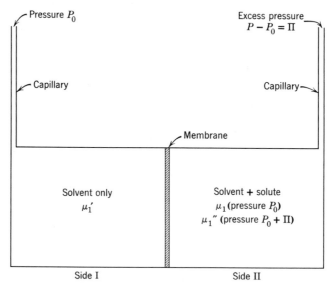

Fig. 13–1. Schematic diagram of the apparatus for osmotic pressure determination. The symbols used for the chemical potential of the solvent on the two sides of the membrane are shown.

to occur. Uniform concentrations throughout cannot exist until all the solvent on side I of the membrane has diffused to side II.

Thermodynamic equilibrium, however, is reached, even in the absence of any applied pressure, before this happens. Passage of solvent from side I to side II causes the liquid level of side II to rise in the capillary tube, setting up a hydrostatic pressure which *opposes* the flow of liquid through the membrane. When this pressure becomes sufficiently high further net flow is prevented.

Experimentally it is often preferable to apply an external hydrostatic pressure to bring about equilibrium more rapidly, i.e., to oppose the liquid flow through the membrane at once. The correct pressure required to attain equilibrium may be obtained by observing the rate of rise or fall of the liquid level in the capillary on side II at various pressures and determining graphically the pressure required to make this rate zero.

(The capillary may equally well be on side I, and the pressure there *reduced* below that on side II.) The pressure difference, $P - P_0$, at equilibrium, is defined as the *osmotic* pressure Π.

In order to calculate the osmotic pressure in terms of other thermo-dynamic parameters, we consider first of all a two-component system, consisting of solvent and a single macromolecular non-electrolytic solute. The condition for equilibrium then is (section 11c) that the chemical potential of the solvent must be the same on both sides of the membrane.

We shall designate with a single prime (e.g., μ_1') the thermodynamic parameters on side I of the membrane. We shall designate by unprimed symbols (e.g., μ_1) the thermodynamic parameters of the solution on side II at the prevailing atmospheric pressure, P_0, and we shall use doubly primed symbols (μ_1'') for the solution on side II when its pressure has been raised to bring it to the equilibrium value, $P = P_0 + \Pi$. The condition for equilibrium thus becomes

$$\mu_1' = \mu_1'' \tag{13-1}$$

In the present treatment the solution on side I is pure solvent at atmos-pheric pressure, so that $\mu_1' = F_1{}^0$. The chemical potential μ_1'' is related to the chemical potential μ_1 of the macromolecular solution at atmospheric pressure by the relation

$$\mu_1'' = \mu_1 + \int_{P_0}^{P_0+\Pi} \left(\frac{\partial \mu_1}{\partial P}\right)_T dP \tag{13-2}$$

The partial derivative $(\partial \mu_1/\partial P)_T$ is equal to the partial molal volume of the solvent, $\bar{V}_1$. This quantity is virtually independent of the pressure, certainly for the small pressure changes encountered in osmotic pressure work, and may therefore be taken outside the integral sign. Thus equation 13-2 becomes

$$\mu_1'' = \mu_1 + \bar{V}_1 \Pi \tag{13-3}$$

and the condition for equilibrium, equation 13-1, gives

$$\Pi = -\frac{\mu_1 - F_1{}^0}{\bar{V}_1} \tag{13-4}$$

We see therefore that osmotic pressure is a direct measure of the chemical potential of the solvent in the macromolecular solution at the original atmospheric pressure, P_0. It is thus given immediately by the equations discussed in sections 11 and 12, in particular by the general relation, equation 11-21.

At the limit of zero macromolecular concentration (where $\bar{V}_1 = V_1{}^0$) this leads at once to van't Hoff's limiting law for osmotic pressure,

$$\mathscr{L}_{c_2=0} \frac{\Pi}{c_2} = \frac{\mathscr{R}T}{M_2} \tag{13-5}$$

i.e., osmotic pressure, like other colligative properties, is a measure of solute molecular weight.

For a mixture of macromolecules with molecular weights M_2, M_3, $\cdots$, present at concentrations c_2, c_3, $\cdots$, the right-hand side of equation 13–4 becomes $\mathscr{R}T \sum_i c_i/M_i$, or, where $c = \sum_i c_i =$ total macromolecular concentration in grams per cubic centimeter,

$$\mathop{\mathscr{L}}_{c=0} \frac{\Pi}{c} = \mathscr{R}T \frac{\sum_i c_i/M_i}{\sum_i c_i} \tag{13–6}$$

Since c_i is proportional to $n_i M_i$ where n_i is the number of moles of the ith solute, equation 13–6 becomes

$$\mathop{\mathscr{L}}_{c=0} \frac{\Pi}{c} = \mathscr{R}T \frac{\sum_i n_i}{\sum_i n_i M_i} = \frac{\mathscr{R}T}{\bar{M}_n} \tag{13–7}$$

i.e., osmotic pressure is a measure of the *number-average molecular weight* $\bar{M}_n$ as defined by equation 8–8.

As will be shown in section 14 the limiting equations 13–5 and 13–7 hold true for macro-ions as well as for neutral macromolecules. They are also valid if, in addition to the macromolecular solute, non-reacting low-molecular-weight solutes are present in the solution. (To treat solutions of the latter kind it is only necessary to redefine the word "solvent" so as to include the diffusible solutes; i.e., "one mole" of "solvent" is defined so as to include, in addition to one mole of *pure* solvent, appropriate fractional numbers of moles of these solutes.)

It is important to note that the limiting form of the osmotic pressure equation can give no information regarding possible solvation of the solute whose molecular weight is being measured. Like all colligative properties, in the limit of infinite dilution, we measure in effect the number of solute particles (per cubic centimeter) formed by c_2 grams (per cubic centimeter) of solute. This number is unchanged if solvation occurs. (Looking at it another way, we measure c_2/M_2. If the solute component is weighed out as dry substance, the molecular weight will also be that of dry substance. If we *assume* a certain amount of solvation and express c_2 in terms of solvated solute, M_2 will be the molecular weight of the solute solvated to the extent originally assumed. In neither case do we get information about actual solvation.)

By contrast, association between solute molecules, or their dissociation, will affect the number of solute particles and is therefore measurable. (If association is weak and reversible, it will decrease in extent as the concentration is lowered, and its effect will disappear entirely when $c_2 = 0$. This situation was discussed on p. 202.)

Osmotic pressure measurement is one of the standard methods for molecular weight determination for all but the largest macromolecules, and some molecular weights obtained by its use are given in Table 13–1. It will be noted that typical synthetic organic polymers can be prepared so as to possess almost any desired molecular weight. Furthermore,

TABLE 13–1. Some Molecular Weights Determined
by Osmotic Pressure[a,b]

Polystyrene, various fractions[29]	31,000	to	612,000
Polyisobutylene, various fractions[29]	38,000	to	722,000
Natural rubber, various preparations[27]	165,000	to	350,000
Nitrocellulose, various fractions[41]	62,000	to	2,500,000
Potato starch amylose, various preparations[42]	32,000	to	150,000
Bacterial dextrans, various preparations[42]	44,000	to	100,000
Pectins, various preparations[42]	100,000	to	140,000
Insulin (in acid solution)[40]			12,000
β-Lactoglobulin[43]			35,000
Ovalbumin[36]			45,000
Hemoglobin[37]			67,000
Serum albumin[28]			69,000

[a] The osmotic pressure becomes too small for accurate work when the molecular weight exceeds 1,000,000. Osmotic pressure determinations of very high molecular-weight substances are therefore not included in the table.

[b] Without auxiliary measurements (see especially Table 17–1) we cannot decide whether these numbers represent unique molecular weights of homogeneous preparations or whether they are number averages, $\bar{M}_n$, of preparations which contain molecules distributed over a range of molecular weights.

preparations can ordinarily be separated into fractions differing widely in molecular weight. These observations confirm that these polymers are heterogeneous, possessing broad molecular weight distributions of the type discussed in section 8. Molecular weights listed for these polymers in Table 13–1 are thus number averages, $\bar{M}_n$.

Among the naturally occurring macromolecules listed in Table 13–1 are many which also possess molecular weights which vary from one preparation to another. Where fractionation has been carried out (e.g., nitrated cellulose), molecular weight heterogeneity has usually been found in these preparations.

The globular proteins, on the other hand, five of which are included in Table 13–1, are *expected* to be molecularly homogeneous because of their ability to form true crystals. The osmotic pressure molecular weights by themselves provide no conclusive evidence which bears on this question. Such evidence is provided, however, by Table 17–1.

One word of caution should be added concerning the use of osmotic pressure to determine $\bar{M}_n$ for heterogeneous mixtures such as unfractionated organic polymers. Samples of such polymers may be expected to

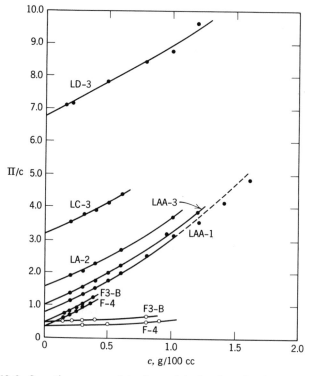

Fig. 13–2. Osmotic pressure data for polyisobutylene fractions in cyclohexane (filled circles) and in benzene (open circles). The osmotic pressure was measured in grams per square centimeter, and c in grams per 100 cubic centimeters. (Data of Krigbaum and Flory[29].)

contain molecules which contain just a few monomer units. These molecules, however, can pass through the osmometer membrane and do not contribute to the osmotic pressure. The molecular weight observed by osmotic pressure is thus $\bar{M}_n$ for only those molecules which do *not* pass through the membrane, and, consequently, its value can be expected to depend on the nature of the membrane which is used. This expectation is confirmed in a study of a single unfractionated sample of polystyrene by several different investigators.[53] One investigator, using a polyvinyl alcohol membrane, capable of retaining molecules of molecular weight 2000 or even less, observed $\bar{M}_n = 270,000$. The average of the results of seven other laboratories was $\bar{M}_n = 480,000$.

13c. Concentration dependence of osmotic pressure. Because of the relation, equation 13–4, between osmotic pressure and chemical potential, the equations already derived for the concentration dependence of the latter in section 12 apply directly to osmotic pressure. In general, it is customary

TABLE 13–2. Osmotic Pressure Data of Krigbaum and Flory[29] for Polyisobutylene in Cyclohexane or Benzene at 30° C

Fraction	$\bar{M}_n$	Virial Coefficient, $B \times 10^4$	
		In Cyclohexane	In Benzene
LD-3	37,900	8.17	—
LC-3	81,400	7.22	—
LA-2	169,000	6.62	—
LAA-3	254,000	6.31	—
LAA-1	339,000	5.97	—
F-3B	555,000	5.95	0.83
F-4	720,000	5.32	0.71

The values of B have been recalculated to cgs units, i.e., refer to Π measured in dynes per square centimeter and c_2 in grams per cubic centimeter.

to plot Π/c_2 versus c_2. By equation 11–21 the most general function describing such a plot in dilute solutions would be

$$\Pi/c_2 = \mathscr{R}T(1/M_2 + Bc_2 + Cc_2{}^2 + \cdots) \qquad (13\text{–}8)$$

The limiting slope of such a plot would give a value for the second virial coefficient, B.

We are particularly interested here in the value of B for linear polymers and, accordingly, show in Fig. 13–2 some appropriate plots for well-fractionated samples of polyisobutylene. (The effect of heterogeneity is sufficiently small so that these samples may be considered completely homogeneous.) The molecular weights obtained from the intercepts of these plots and the values of B obtained from the limiting slopes are shown in Table 13–2. Data are given for a good solvent, cyclohexane, and a poor solvent, benzene. (Polyisobutylene has aliphatic side chains (methyl groups), so that an aliphatic solvent is a good solvent and an aromatic one is a poor solvent.)

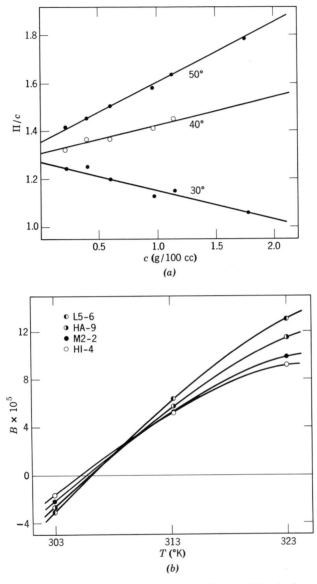

Fig. 13–3. The effect of temperature on thermodynamic behavior in a poor solvent. Part *a* shows experimental data for a polystyrene fraction in cyclohexane at 30, 40, and 50°C. Part *b* shows plots of the second virial coefficient versus T for a series of polystyrene fractions in the same solvent. (Taken from Krigbaum.[30])

These data show at once that the magnitude of B in the good solvent is of the order predicted by Table 12–1. (The calculations there were based on values of the various constants reasonably close to those that would apply to a polyisobutylene-cyclohexane solution.) In the poor solvent, benzene, the values of B are much smaller than in a good solvent, as predicted.

To show that the difference observed between cyclohexane and benzene is not just some property peculiar to these solvents, these data may be compared to those obtained by the same authors for polystyrene.[29],[30]

TABLE 13–3. Osmotic Pressure Data for Polystyrene in Cyclohexane, Benzene or Toluene at 30° C

Virial Coefficient, $B \times 10^4$

$\bar{M}_n$	In Cyclohexane	In Benzene	In Toluene
30,900			6.30
41,700			6.03
50,500	−0.28		
61,500			5.02
120,000			4.82
125,000	−0.24		
328,000			3.70
359,000	−0.20		
566,000	−0.18		
612,000			3.31
220,700[a]		3.7	7.3

[a] Hookway and Townend, *J. Chem. Soc.*, **1952**, 3190; unfractionated sample of polystyrene. All other data from Krigbaum and Flory.[29],[30]

For this polymer, with aromatic side chains, benzene is a good solvent (toluene is even better), and cyclohexane is a poor solvent. The experimental results given in Table 13–3 then show, as expected, a reversal in the relative roles of the aromatic and aliphatic solvents. In toluene and benzene values of B of the order of 5×10^{-4} are obtained, while in cyclohexane at 30° C they are actually negative.

It was pointed out in section 12 that temperature should have a pronounced effect on the value of B in poor solvents. This, too, has been confirmed for both polyisobutylene and polystyrene. In both cases negative values of B are observed in the poor solvent at sufficiently low temperature, and they rise to small positive values with an increase in temperature. In each case there is an ideal temperature, at which $B = 0$. As explained on p. 202, this is the temperature at which $T = \Theta$. The data for polystyrene are shown in Fig. 13–3.

Next we shall examine the effect of molecular weight on the coefficient B. The plots of Fig. 13–4 show that log B in good solvents decreases linearly with molecular weight, such that, for polyisobutylene in cyclohexane, B varies approximately as $1/M^{0.14}$ and, for polystyrene in toluene, as $1/M^{0.22}$. This variation is of the order of magnitude expected for good solvents. Combining equations 9–40, 12–9, and 12–17, we get

$$B = \left(\frac{2}{3}\right)^{5/2} \frac{\pi \mathcal{N}}{M_0^{3/2}} \frac{\alpha^3 \beta^3 \gamma^3}{M^{1/2}}$$

where M is the molecular weight of the polymer and γ the empirical parameter defined by equation 12–16. Of the parameters which occur in

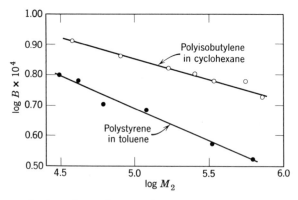

Fig. 13–4. The dependence of B on molecular weight in good solvents. Data from Tables 13–2 and 13–3.

this equation, β and M_0 are independent of M, but α and γ both increase with M in good solvents, α typically as $M^{0.05-0.10}$, γ somewhat less so. Thus B is predicted to vary roughly as $1/M^{0.05-0.25}$.

If the empirical equation 12–17 is replaced by the quantitative relation for the excluded volume given in equation 12–21, and the result combined with values of α determined experimentally from viscosity measurements (p. 404), the agreement between experiment and prediction is not as good. For polyisobutylene in cyclohexane Krigbaum and Flory compute that B should vary as $1/M^{0.08}$, and for polystyrene in toluene they compute that it should vary as $1/M^{0.05}$.

There is also a minor discrepancy between calculated and observed behavior in poor solvents. From data such as those of Fig. 13–3 we can calculate the entropy and heat parameters ψ and κ. For polyisobutylene in benzene, for instance, $\psi = 0.34$ and $\kappa = 101$ (corresponding to $\Theta = 297.5°$ K). The parameter ψ is close but, as predicted, not equal, to the value of $\frac{1}{2}$ for ideal solvents; the value of κ is positive as expected for a poor solvent. But, again, when the same parameters are calculated in other ways (e.g., from precipitation experiments of the type discussed in section 15), somewhat different values are obtained.

These discrepancies are probably the result of the use of a continuous (cloudlike) distribution of segments in deriving the equations of section 12f. As was pointed out there, the use of such a distribution is an approximation.

Regardless of these minor inconsistencies, the data here presented provide proof that linear polymers actually have the configuration of randomly coiled chains. We might, for example, compare the variation of B with molecular weight, as given in Tables 13–2 and 13–3 or in Fig. 13–4, with the expected behavior of solid spheres or of rigid rods with constant diameter. The values of B predicted for such models are given by equations 12–11 and 12–13. We note that for spheres B would vary as $1/M$, whereas for rods, with L proportional to M, B would be independent of M. For solid spheres, moreover, the magnitude of B would be much smaller than is observed.

It should be noted, finally, that the curvature of the plots of Fig. 13–2 indicates that even in the low concentration region studied the third coefficient $Cc_2{}^2$ of equation 13–8 is not entirely negligible. This result agrees with the prediction of Stockmayer and Casassa,[25] previously mentioned.

14. MACRO-IONS

We shall discuss in this section the special influence which the electrostatic charge of macromolecular ions exerts upon their thermodynamic behavior in solution. Two types of effect may be distinguished:

(1) The classical effect produced by the requirement that solutions of macro-ions must be electrically neutral. This means that solutions containing only a single kind of macro-ion must always contain a sufficient number of small ions to neutralize the macro-ion charge, and it leads to important consequences which will be discussed in considerable detail.

(2) True non-ideality, resulting from the electrostatic interaction between the charges on macro-ions and those of other macro-ions and small ions in its vicinity. This effect will be discussed only at an elementary level because a detailed discussion entails a knowledge of the potential field about a macro-ion. Methods for calculation of this field will be described in Chapter 7, but they will be found exceedingly complex.

14a. Enumeration of components. A solution containing macro-ions of charge Z necessarily contains small ions also, since the solution as a whole must be neutral. Furthermore, it is desirable to allow for variation in the concentration of the small ions, for the total ion concentration (or ionic strength) is known to be one of the most important variables in any ionic solution. Thus we shall find it necessary throughout this section to consider systems of at least three components. To keep the treatment simple we shall suppose that the small ions present are all univalent ions, M^+ and X^-. Extension to small ions of higher charge is usually not difficult. Much of the time it is not necessary; since our interest lies in the macro-ions there is no particular reason why we should ordinarily use other than univalent ions to build up the ionic strength.

We shall express our concentrations in terms of *molalities*, but moles per liter would do just as well. In fact, in dilute aqueous solutions, which serve as the commonest vehicle for macro-ions, the two are almost equal.

Suppose a solution contains a concentration m_p of macro-ions of charge Z (expressed in protonic charges) and also concentrations m_+ and m_- of the ions M^+ and X^-. To fulfil the requirement of neutrality it is necessary that

$$Zm_p = m_- - m_+ \qquad (14\text{--}1)$$

The value of Z may, of course, be positive or negative. (We shall use the symbol $|Z|$ to designate the absolute numerical value of Z; i.e., $Z = \pm |Z|$.)

In aqueous solutions hydrogen and hydroxyl ions are inevitably present. If M^+ and X^- are to refer to the ions of an added salt, equation 14–1 is limited to solutions in which the concentrations of hydrogen and hydroxyl ions are negligible compared to those of the added salt. For solutions which are pronouncedly acidic or basic (say pH < 4 or >10) and in which the concentration of added salt is low (say <0.01 molal), equation 14–1 must be rewritten as $Zm_p = m_- + m_{OH^-} - m_+ - m_{H^+}$, and the subsequent treatment revised accordingly.

As was pointed out in section 11b, we wish to express the composition of a solution not in terms of the ion species just enumerated but in terms of *neutral components*, since the chemical potentials of ion species have no real meaning. As was also pointed out in section 11b, a choice is often available in the definition of the neutral components. In the present situation, for example, we may define these components as follows:

> Comp. 1—Solvent (usually H_2O)
> Comp. 2—Macro-ion $+ Z$ ions X^-
> Comp. 3—Neutral salt, MX

If m_2 and m_3 represent the molalities of these components, the concentrations of ion species are

$$m_p = m_2$$
$$m_+ = m_3 \qquad\qquad (14\text{--}2)$$
$$m_- = m_3 + Zm_2$$

and equation 14–1 is, of course, automatically satisfied, as it must be if neutral components are chosen.

The definition of components here given is the one that would "naturally" occur to us if Z were positive; i.e., component 2 would then be the neutral salt PX_Z, where P represents the macro-ion. However, the definition would apply equally well if Z were negative, in which case the addition of one mole of the second component would involve the addition of one mole of macro-ions and the *removal* of $|Z|$ moles of X^-, or, in terms of actual experiment, the addition of one mole of the neutral salt $M_{|Z|}P$ (recall that Z is negative) and the removal of $|Z|$ moles of MX.

The definition of components here given [and the corresponding one which would "naturally" occur to us if Z were negative: component 2 containing the macro-ion and $-Z$ (i.e., $+|Z|$) ions M^+] is a perfectly valid way of defining the components of a solution containing macro-ions, but it has one important disadvantage. For, by this definition, the macromolecular component contains $|Z|$ times as many small ions as macro-ions. Thus μ_2, for example, would represent the chemical potential of the macro-ion together with that of $|Z|$ ions of opposite charge. In order that the macromolecular component may more nearly represent the macro-ion alone, Scatchard[31] has suggested that the following definition of components be adopted:

> Comp. 1—Solvent
> Comp. 2—Macro-ion + $Z/2$ ions X^- − $Z/2$ ions M^+
> Comp. 3—Neutral salt, MX

In terms of these components

$$m_p = m_2$$
$$m_+ = m_3 - Zm_2/2 \qquad\qquad (14\text{--}3)$$
$$m_- = m_3 + Zm_2/2$$

Equation 14–1 is again automatically satisfied.

The operational definition of component 2, with Z positive, is the addition of one mole of PX_Z and the removal of $Z/2$ moles of MX. With Z negative, it is the addition of one mole of $M_{|Z|}P$ and the removal of $|Z|/2$ moles of MX. The chemical potential of component 2, μ_2, is that of one mole of macro-ions plus $Z/2$ times the *difference* between μ_+ and μ_- for the small ions, a difference which for ions such as K^+ and Cl^- is usually thought to be very small.*

The Scatchard definition will be used throughout the present discussion.

14b. The conditions for equilibrium across a semi-permeable membrane. We are interested in the equilibrium between a solution containing macro-ions and another solution not containing such ions: especially in the equilibrium across a semi-permeable membrane, permeable to small ions but not to the macro-ions. As before (section 13) we shall use primed symbols to indicate the side of the membrane free from macro-ions; unprimed symbols will designate the solution containing macro-ions, at the pressure P_0; doubly primed symbols the same solution after the equilibrium pressure has been attained. The conditions for equilibrium are then simply that

$$\mu_1' = \mu_1''$$
$$\mu_3' = \mu_3'' \tag{14-4}$$

(It is *not* true, of course, that $\mu_+' = \mu_+''$ or $\mu_-' = \mu_-''$; cf. section 11c.)

The first condition, as before, is the one from which the osmotic pressure is calculated. In place of equation 13–4, since μ_1' is not now equal to μ_1^0,

$$\Pi = -\frac{\mu_1 - \mu_1'}{\bar{V}_1} \tag{14-5}$$

This equation is examined further in section 14d.

The second condition, since we are dealing with a simple 1–1 electrolyte, is most conveniently transformed into a relation between activities. Using equation 11–20 we get

$$\mu_3' = \mu_3^0 + \mathscr{R}T \ln a_3'$$
$$\mu_3 = \mu_3^0 + \mathscr{R}T \ln a_3 \tag{14-6}$$

and, as in equation 13–2,

$$\mu_3'' = \mu_3 + \int_0^{\Pi} \bar{V}_3 \, dP \tag{14-7}$$

so that equation 14–4 becomes

$$\mathscr{R}T \ln a_3' = \mathscr{R}T \ln a_3 + \int_0^{\Pi} \bar{V}_3 \, dP \tag{14-8}$$

---

* Thermodynamically, of course, we can say nothing about the chemical potential of single ions. The Debye-Hückel theory (Chapter 7), however, would predict that the activities and hence the variable parts of the chemical potentials of ions such as K^+ and Cl^- are essentially the same, just as their transport numbers in conductivity are each essentially equal to 0.5.

The integral in equation 14–8 can be neglected for any ordinary electrolyte. Osmotic pressures are usually desired at as low concentrations of the macromolecule as possible, so that values of Π are nearly always kept below 0.1 atm; since values of $\bar{V}_3$ are generally below 0.1 liter for ordinary electrolytes, the integral has a value of less than 0.01 liter-atm; dividing by $\mathscr{R}T$ gives a value of <0.0004 to be added to $\ln a_3$. This represents a 0.05% change in a_3, which is well within the experimental error in any possible determination of this quantity. Thus we can neglect the effect of the osmotic pressure on μ_3, and consider μ_3 and $\mu_3{''}$ the same quantity. Equation 14–8 therefore becomes

$$a_3 = a_3{'} \tag{14–9}$$

14c. The Donnan effect. The most significant consequence of equation 14–9, as was first pointed out in 1911 by F. G. Donnan,[32] is that the small ions must be unequally distributed on the two sides. This is easily shown, as follows.

The activity of an electrolyte is the product of the constituent ion activities, i.e., $a_3 = a_+a_-$, so that equation 14–9 becomes

$$a_+a_- = a_+{'}a_-{'} \tag{14–10}$$

Replacing a_+ by $m_+\gamma_+$ and a_- by $m_-\gamma_-$ and writing $\gamma_+\gamma_- = \gamma_\pm{}^2$, where $\gamma_\pm$ is the usual mean ion activity coefficient, we get

$$m_+m_-\gamma_\pm{}^2 = m_+{'}m_-{'}\gamma_\pm{'}^2 \tag{14–11}$$

(For a review of these standard expressions for the activities of electrolytes see any thermodynamics or physical chemistry textbook.[1]) The ion concentrations appearing in equation 14–11 are not independent but are related to one another by the neutrality condition, equation 14–1, which for the solution not containing macro-ions, is simply $m_+{'} = m_-{'}$. Combination of the neutrality condition with equation 14–11 at once leads to

$$
\begin{aligned}
(m_+{'})^2 &= (\gamma_\pm/\gamma_\pm{'})^2 m_+(m_+ + Zm_p) \\
(m_-{'})^2 &= (\gamma_\pm/\gamma_\pm{'})^2 m_-(m_- - Zm_p)
\end{aligned}
\tag{14–12}
$$

It is at once apparent that, if Z is positive, $m_+{'} > m_+$ and $m_-{'} < m_-$, and vice versa if Z is negative. In other words, the charge of the macro-ion on one side of the membrane is compensated for in part by a reduced concentration on that side of the small ion of like charge and in part by an increased concentration of the small ion of opposite charge.

Some calculated values of the molalities on each side are given in Table 14–1. In making these calculations, the activity coefficients have been assumed the same on both sides, and values of 0.002 and 0.02 have been chosen for Zm_2. The lower value corresponds to a charge of $+2$ for a macromolecular ion of molecular weight 10,000, at a concentration of 1 gram per 100 grams of solvent; or to a charge of $+20$ for a macromolecular ion of molecular weight 100,000, at the same weight concentration. Such values of the charge are encountered routinely in solutions of proteins near their isoelectric points. In more strongly acid or basic solutions charges ten times as great may be obtained, and these charges would lead then to the more uneven distribution of the ions, as calculated for $Zm_2 = 0.02$.

TABLE 14–1. Donnan Equilibrium

Zm_2	$m_+' = m_-'$	m_+	m_-	$\dfrac{m_+'}{m_+} = \dfrac{m_-}{m_-'}$
0.002	0.0010	0.00041	0.00241	2.41
	0.0100	0.00905	0.01105	1.10
	0.100	0.0990	0.1010	1.01
0.02	0.0010	0.00005	0.02005	20.05
	0.0100	0.00414	0.02414	2.41
	0.100	0.0905	0.1105	1.10

For negative values of Zm_2 the same results apply but with positive and negative signs interchanged.

It is seen from Table 14–1 that the relative concentration differences are decreased if the total electrolyte concentration is sufficiently high. Equation 14–12 shows that this is a general result; if Zm_2 is much smaller than m_+ and m_-, the ion activities on the two sides become equal. This is a very important conclusion, for it is often desirable to suppress the Donnan effect, and we can see that this goal can be achieved in the presence of a sufficiently high concentration of neutral salt.

If many different diffusible ions of different charge type are present, the Donnan equilibria might be expected to become very complicated. Fortunately, this is not so. Equation 14–9 must hold true for any diffusible electrolyte which can be formed from the ions present. Let us suppose, for example, that, in addition to the ions M^+ and X^-, we have present other small ions B^{Z_B} and A^{Z_A}. Let B be a cation, so that Z_B is positive, and A an anion, so that Z_A is negative. Equation 14–9 must then be applicable to the salts BX_{Z_B} and $M_{|Z_A|}A$, i.e.,

$$a_{BX_{Z_B}} = a_B{}^{Z_B}(a_-)^{Z_B} = a_B'{}^{Z_B}(a_-')^{Z_B} = a_{BX_{Z_B}}'$$

$$a_{M_{|Z_A|}} = (a_+)^{|Z_A|}a_A{}^{Z_A} = (a_+')^{|Z_A|}a_A'{}^{Z_A} = a_{M_{|Z_A|}A}'$$

(14–13)

and, combining with equation 14–10, we get

$$\frac{a_+'}{a_+} = \frac{a_-}{a_-'} = \left(\frac{a_A{}^{Z_A}}{a_A'{}^{Z_A}}\right)^{1/|Z_A|} = \left(\frac{a_B'{}^{Z_B}}{a_B{}^{Z_B}}\right)^{1/Z_B}$$

(14–14)

In general, if Z_i denotes the *algebraic* charge on *any diffusible ion,*

$$\left(\frac{a_i}{a_i'}\right)^{1/Z_i} = \text{constant} \tag{14-15}$$

equation 14-14 being the special form of equation 14-15 for ions with Z_i equal to $+1$, -1, $+Z_B$, and $-|Z_A|$, respectively. It is thus necessary to solve an equation of the type of equation 14-12 for only one ion to obtain at once the distribution across the membrane of any other diffusible ion.

If more than one type of cation and anion is present, equation 14-12 could, of course, in general, become very complicated because the concentrations of the other ions would enter into the equation for electrical neutrality, equation 14-1. This difficulty can be circumvented, however, by having one cation and one anion present in great excess, so that other ions make a negligible contribution to the total number of charges on each side.

As an example of the application of equation 14-15 we may refer back to the case previously discussed, with the ions M^+ and X^- the only diffusible ions added. If water is the solvent, as was pointed out above, hydrogen and hydroxyl ions must also be present, though usually not in sufficient quantity to affect equation 14-1. The relative concentrations of these ions may now be given at once without further calculation, since $a_{H^+}/a_{H^+}' = a_+/a_+' = a_{OH^-}'/a_{OH^-} = a_-'/a_-$. This points out the well-known fact that pH values on two sides of a membrane are generally not equal if a non-diffusible ion is confined to one side.

The previous conclusion that the difference in the concentrations of an ion on the two sides may be minimized if the ion is present in excess can now be expanded. This difference will be minimized if *any electrolyte* is present in excess. For example, the pH difference may be effectively eliminated by the presence of an excess of NaCl.

14d. Osmotic pressure of solutions containing macro-ions. In the preceding section we examined the consequence of the equilibrium condition (equation 14-4), $\mu_3' = \mu_3''$. In this section we shall take up the condition $\mu_1' = \mu_1''$, which leads to the osmotic pressure relation given by equation 14-5. We shall be interested in assessing the special effect of the unequal distribution of ions upon the osmotic pressure and, therefore, for simplicity, shall assume all other causes of non-ideality to be absent. (We shall discuss other causes later; cf. Table 14-2.) We may thus use equation 11-14 for the chemical potential of the solvent and write

$$\Pi = \frac{\mathscr{R}T}{\bar{V}_1} \ln \frac{X_1'}{X_1} \tag{14-16}$$

On the side of the membrane which contains no macro-ions we have present, for each $1000/M_1$ moles of solvent (i.e., for each kilogram of solvent) m_+' and m_-' moles of the two kinds of salt ions, so that

$$X_1' = \frac{1000}{1000 + M_1(m_+' + m_-')} \tag{14-17}$$

whereas, on the side containing macro-ions, for each $1000/M_1$ moles of solvent, there are m_+ and m_- moles of salt ions and $m_p = m_2$ moles of macro-ions, so that

$$X_1 = \frac{1000}{1000 + M_1(m_+ + m_- + m_p)} \tag{14-18}$$

Hence

$$\Pi = \frac{\mathscr{R}T}{\bar{V}_1} \ln \frac{1 + M_1(m_+ + m_- + m_p)/1000}{1 + M_1(m_+' + m_-')/1000} \tag{14-19}$$

We shall normally be interested only in dilute solutions for which $c_2 < 0.01$ gram/cc. For $M_2 = 10^4$ or greater, this makes m_2, i.e., m_p, less than 0.001. There is also usually no reason to make m_+ or m_- greater than about 0.1. Thus, with M_1 usually less than 100 (H_2O is the solvent in virtually all experiments with ions), the second term of both denominator and numerator of the logarithm in equation 14–19 becomes very small. We may therefore expand $\ln (1 + x) = x - x^2/2 \cdots$, and retain only the first term, so that

$$\Pi = (\mathscr{R}TM_1/1000\bar{V}_1)(m_p + m_+ - m_+' + m_- - m_-') \tag{14-20}$$

(Actually equation 14–20 is not limited to m_+ or m_- less than 0.1, for the terms in the expansion all involve the differences between m_+ and m_+' and between m_- and m_-'. These differences, as Table 14–1 shows, do not increase as the ion concentration increases.)

To evaluate the right-hand side of equation 14–20 we shall let m_2 and m_3 represent the *equilibrium* concentrations of macromolecular component and of MX on the side of the membrane which contains the macro-ions. The components shall be defined by the Scatchard method, so that equation 14–3 applies. In terms of m_2 and m_3 we can then use equation 14–12 to evaluate the concentrations of M^+ and X^- on the side which does not contain macro-ions. Assuming that $\gamma_\pm = \gamma_\pm'$ (an assumption which will always be good if there is an excess of salt ions over macro-ions and which improves with dilution of the macromolecular component even if the salt concentration is low), we have

$$(m_+')^2 = (m_-')^2 = m_3{}^2 - \tfrac{1}{4}Z^2m_2{}^2 \tag{14-21}$$

It should be noted that neither m_2 nor m_3 represent the initial, pre-equilibrium concentrations on the macromolecular side of the membrane. They must be measured on an aliquot of solution after equilibrium is attained.

Equation 14–21 may be rewritten, after binomial expansion, as

$$m_+' = m_-' = m_3\left(1 - \frac{\tfrac{1}{4}Z^2m_2{}^2}{m_3{}^2}\right)^{1/2}$$

$$= m_3\left(1 - \frac{\tfrac{1}{8}Z^2m_2{}^2}{m_3{}^2} - \frac{\tfrac{1}{128}Z^4m_2{}^4}{m_3{}^4} - \cdots\right) \tag{14-22}$$

Using the expressions for m_+ and m_- in terms of m_3, as given by equation 14-3, we get that

$$m_+ - m_+' + m_- - m_-' = \frac{\frac{1}{4}Z^2 m_2^2}{m_3} + \frac{\frac{1}{64}Z^4 m_2^4}{m_3^3} + \cdots \quad (14\text{-}23)$$

For application to dilute macromolecular solutions we can assume that the total volume containing 1 kg of water is $1000\bar{V}_1/M_1 + m_3\bar{V}_3 \simeq 1000\bar{V}_1/M_1$. (For water, for example, $1000/M_1 = 55.5$. Even if m_3 is 0.1 to 0.5, which would be a very high salt concentration, $m_3\bar{V}_3 \ll 1000\bar{V}_1/M_1$.) This is the volume containing $m_2 M_2$ grams of macromolecular solute; i.e., c_2, in grams per milliliter, is $m_2 M_2 M_1/1000\bar{V}_1$. Thus $m_2 = (1000\bar{V}_1/M_2 M_1)c_2$. Making this substitution and combining equation 14-23 with equation 14-20, we get

$$\frac{\Pi}{c_2} = \mathscr{R}T\left(\frac{1}{M_2} + \frac{1000Z^2 v_1}{4m_3 M_2^2} c_2 + \text{terms in } c_2^3 \text{ and higher}\right) \quad (14\text{-}24)$$

where v_1, the specific volume of solvent, has been written as an approximation for $\bar{V}_1/M_1$, as we have done throughout this chapter for expressions used for limiting behavior in dilute solution.

This equation is of the form of equation 13-8. We see that the limit of Π/c_2 at zero concentration reduces, as for non-electrolyte solutes, to van't Hoff's limiting law, equation 13-5, and that osmotic pressure can therefore be used to determine the molecular weights of macro-ions as well as of neutral macromolecules.

Equation 14-24 is obtained no matter which of the alternative methods of section 14a is used to define the components of the solution. However, the parameters M_2, c_2, etc., have a different meaning, depending on the way in which the components are defined. By the Scatchard definition of components, M_2 is the molecular weight of the macro-ion plus $Z/2$ times the *difference* between the molecular weights of X^- and M^+. If the components are defined as in equations 14-2, then M_2 is the molecular weight of the macro-ion plus Z times the molecular weight of X^-.

In Table 14-2 we have calculated some values of B by equation 14-24 and, for comparison, have included values of B which would be expected on the basis of the excluded volume effect, calculated in section 12. We see at once that the contribution of the Donnan equilibrium to B, as given by equation 14-24, is generally much larger than the contribution of excluded volume. Even if the charge on the macro-ion is as low as one per 5000 molecular weight, the Donnan term far exceeds the excluded volume term at low salt concentration. As the charge is increased, it exceeds it even when m_3 is as high as 0.1 or higher. Only at low values of Z and at high salt concentration should the excluded volume term become the important one.

TABLE 14-2. Second Virial Coefficient for Macro-Ions[a]

| M_2 | Z | B by Equation 14-24 | | | Contribution due to Excluded Volume[b] | |
		$m_3 = 0.001$	$m_3 = 0.01$	$m_3 = 0.1$	Spheres (Eq. 12-11)	Rods ($L/d = 20$) (Eq. 12-13)
10,000	2	0.01	0.001	0.0001	0.0004	0.0020
	20	1.0	0.10	0.01	0.0004	0.0020
100,000	20	0.01	0.001	0.0001	0.00004	0.00020
	200	1.0	0.10	0.01	0.00004	0.00020

[a] Using $v_1 = v_2 = 1$ gram/ml.

[b] We have not included in this table a comparison with flexible polymers, because flexible macro-ions undergo marked configurational changes if charged. Thus the orders of magnitude for B given in Table 12-1 are not applicable to flexible macro-ions. These ions will be discussed in Chapter 7.

Application of the present considerations to proteins is discussed in section 14g. Flexible polyelectrolytes will be taken up in Chapter 7.

14e. Complete expressions for the chemical potential. Scatchard[31] has derived a rigorous expression, applicable to dilute solutions, which takes into account all factors which might influence the chemical potential of a solution of macro-ions. For dilute solutions his result is given in the form of equation 11-21, with specific expressions for the second and third[33] virial coefficients. The expression for B, in the units here used, is

$$B = \frac{1000v_1}{M_2^2} \left(\frac{Z^2}{4m_3} + \frac{\beta_{22}}{2} - \frac{\beta_{23}^2 m_3}{4 + 2\beta_{33}m_3} \right) \tag{14-25}$$

where the β's are derivatives of activity coefficients: $\beta_{22} = \partial \ln \gamma_2/\partial m_2$, $\beta_{23} = \partial \ln \gamma_2/\partial m_3$, $\beta_{33} = \partial \ln \gamma_3/\partial m_3$. The first term in this expression is seen to be the same as the value of B obtained from equation 14-24. The other terms represent the various interactions between the ions of the solution. Included in β_{22} would be the excluded volume effect and the interaction between charges on different macro-ions, whereas β_{23} involves interaction between macro-ions and salt ions, including the actual binding of such ions. The parameter β_{33}, of course, involves interaction between salt ions alone. Experimental values of all three parameters can be obtained by studying the variation of B with the concentration m_3 of the diffusible component in a macro-ion solution, the most complete study of this kind being the work of Scatchard et al.[28] on the protein serum albumin.

The activity coefficient derivatives which appear in equation 14–25 are complex quantities, and at present there is no simple statistical mechanical theory which can be used to predict their values. Theoretical work leading to a solution of this very difficult problem is currently being carried out by Hill.[54] As was pointed out earlier, an essential prerequisite for a solution of the problem is the evaluation of the electrostatic potential at all points in a mixture of macro-ions, small ions, and solvent. The calculation of this potential, in some simple situations, will be described in Chapter 7, but situations sufficiently complex to allow for determination of the chemical potential as a function of macro-ion concentration will not be considered.

14f. The contribution of charge fluctuations to the chemical potential of protein ions. The discussion so far has been limited to solutions containing macro-ions of identical charge Z. It has been pointed out on p. 113, however, that solutions of proteins and other weak electrolytes will generally contain ions which differ in their charge, because the charge of each individual ion will fluctuate about a mean value as a result of interchange of H^+ ions with the solvent. The effect which this has on the thermodynamic properties of a macro-ion solution has been considered by Kirkwood and Shumaker.[34] The general theory proposed by these authors cannot be considered without an understanding of the content of Chapter 7, but the most striking result obtained by them can be simply derived by means of the Debye-Hückel theory and is presented here.

We take up the special case of a solution of macro-ions of average net charge, $\bar{Z}$, zero, in the absence of all other ions. As explained on p. 113, such a solution will contain ions of net charge ± 1, ± 2, etc., as well as ions of actual net charge zero. If C_i is the concentration (moles per liter) of ions of charge Z_i, we have $\bar{Z} = \sum_i C_i Z_i / \sum_i C_i = 0$, whereas $\overline{Z^2}$, given by,

$$\overline{Z^2} = \sum_i C_i Z_i^2 / \sum_i C_i \qquad (14\text{–}26)$$

will be different from zero.

A solution of this kind will resemble solutions of salts such as KCl and $CaCl_2$ in containing both positive and negative ions of about the same size. As in solutions of such simple ions, there will therefore be a net attractive force between the ions, leading to a *negative* contribution to the chemical potential which can be evaluated in a straightforward way by the Debye-Hückel theory, at least in the limit of very low concentrations, where the individual ions will be relatively far apart. Under these conditions the fact that an ion of net charge Z_i may contain a large number of individual positive and negative charges, rather than a charge Z_i concentrated at one point, will not be important.

The Debye-Hückel expression for the activity coefficient of an ion of charge Z_i, at the limit of zero concentration,* is

$$\ln y_i = -Z_i^2 \left(\frac{\epsilon^2}{DkT}\right)^{3/2} \left(\frac{2\pi \mathcal{N}}{1000}\right)^{1/2} I^{1/2} \tag{14-27}$$

where ϵ is the protonic charge, D the dielectric constant, k Boltzmann's constant, and I the ionic strength, $I = \frac{1}{2} \sum_i C_i Z_i^2$. Since small ions are assumed absent, only macro-ions contribute to I, and, by equation 14–26, $I = \frac{1}{2}\overline{Z^2}C_2$, where $C_2 = \sum_i C_i$. The average value of $\ln y_2$ for all the ions in solution may thus be written

$$\ln y_2 = -\left(\frac{\overline{Z^2}\epsilon^2}{DkT}\right)^{3/2} \left(\frac{\pi \mathcal{N}}{M_2}\right)^{1/2} c_2^{1/2} \tag{14-28}$$

where $c_2 = M_2 C_2/1000$ is the solute concentration in grams per cubic centimeter.

This result is readily converted into an expression for the chemical potential of the solvent. By equation 11–20, with $a_2 = y_2 C_2$,

$$\left(\frac{\partial \mu_2}{\partial c_2}\right)_{T,P} = \frac{\mathscr{R}T}{c_2}\left(1 + c_2 \frac{\partial \ln y_2}{\partial c_2}\right)$$

This expression may be combined with the Gibbs-Duhem equation (equation 11–6), which, for dilute solutions of a single solute component, may be written

$$\left(\frac{\partial \mu_1}{\partial c_2}\right)_{T,P} = -\frac{\overline{V_1^0}c_2}{M_2}\left(\frac{\partial \mu_2}{\partial c_2}\right)_{T,P}$$

where the ratio of moles, n_2/n_1, has been expressed in terms of c_2 on the assumption that the total volume of the solution differs insignificantly from the volume $n_1 \overline{V_1^0}$ of solvent alone. The result obtained is that

$$\left(\frac{\partial \mu_1}{\partial c_2}\right)_{T,P} = -\frac{\mathscr{R}T\overline{V_1^0}}{M_2}\left[1 - \frac{1}{2}\left(\frac{\pi \mathcal{N}}{M_2}\right)^{1/2}\left(\frac{\overline{Z^2}\epsilon^2}{DkT}\right)^{3/2}c_2^{1/2}\right] \tag{14-29}$$

Two novel features are inherent in equation 14–29. It shows that the chemical potential in salt-free isoelectric protein solutions can no longer be described by a general equation such as equation 11–21, involving expansion in *integral* powers of c_2, and it predicts negative rather than positive deviations from ideality in properties such as osmotic pressure and

---

* The complete expression for $\ln y$ is derived in section 26. Equation 14–27 should, however, be already familiar to most readers.

light scattering which depend on the chemical potential. These predictions are confirmed by experiment, as we shall show in section 17.

The treatment of charge fluctuations by Kirkwood and Shumaker[34] is more general than that given here, and it is in principle applicable to solutions containing macro-ions of charge other than $\bar{Z} = 0$, in the presence as well as absence of small ions. The preceding equations emerge from it as a limiting result. Kirkwood and Shumaker show that the negative contribution of charge fluctuations to the chemical potential, when $\bar{Z} = 0$, diminishes rapidly when small ions are added, and, at an ionic strength of 0.1, it should be surpassed by the excluded volume effect, leading to positive deviations from ideality. The charge fluctuation effect will presumably also be masked when $\bar{Z}$ differs from zero, because of the large positive second virial coefficient (Table 14–2) produced by the Donnan effect.

14g. The osmotic pressure of protein solutions. Proteins are macromolecules, which have both weakly acidic and weakly basic side chains. At very low pH they have a *positive* charge, which may be as high as $+10$ to $+20$ for each 10,000 of molecular weight. As the pH is increased the weakly acid groups acquire a negative charge, and the net charge falls. Further increase in pH causes dissociation of protons from imidazole and amino groups with a further decrease in the positive charge. The final state at very high pH is one of a net *negative* charge, which again may reach about -10 or -20 for each 10,000 of molecular weight. At some point in between, most commonly somewhere within the range of pH 3 to 11, the net charge becomes zero, and the protein molecule is said to be *isoelectric*. (For further discussion and for the influence of interaction with salt or buffer ions, see sections 24 and 30.) The majority of proteins, as far as we are able to judge, are rigid molecules, often closely approximated by a sphere or long rod. Probably the same over-all shape and size is often maintained over a considerable region of pH, within which the charge may vary from positive through zero to negative.

The preceding discussion has shown that osmotic pressure measurements should give molecular weights of proteins. It is also clear that best values are to be obtained at such a pH that Z is close to zero and the salt concentration high, for then the slopes of extrapolation plots according to equation 13–8 are minimized. Some values obtained in this way are listed in Table 14–3.

The experimental conditions under which the data of Table 14–3 were obtained are those under which the contribution of the Donnan term, $1000Z^2 v_1 / 4 m_3 M_2^2$, to B should be negligibly small. Because of the high ionic strength, the negative contribution of charge fluctuations should also

TABLE 14–3. Molecular Weights and Virial Coefficients of
Proteins from Osmotic Pressures at High Salt Concentration
near $Z = 0$

	Molecular Weight	BM_2 cm³/gram	L/d for Equivalent Rod
β-Lactoglobulin[35]	39,000	3.2	—
Ovalbumin[36]	45,000	1.4	—
Hemoglobin[37]	67,000	3.7	—
Bovine serum albumin[28]	69,000	1.5	—
Hemocyanin (octopus)[38]	(710,000)[a]	(2)[a]	—
Myosin[39]	840,000[b]	75	100
Hemocyanin (helix)[38]	(1,800,000)[a]	(7)[a]	—

[a] These figures are from early data of poor precision. They are included here primarily to show that the high value of BM_2 for myosin is not a general effect of high molecular weight.

[b] More recent work on myosin by other methods, some of which will be discussed in later chapters, indicates that the molecule here studied was a dimer of the true myosin molecule. The value of L/d deduced is therefore that of the dimeric molecule. It will be seen, however, that the true myosin molecule also behaves, as does the dimer, like a long, thin rod.

be small, and B should thus be closely represented by the excluded volume term given by equation 12–11 or 12–13. For spheres, independent of molecular weight, we should have

$$BM_2 = 4v_2 \simeq 3.0$$

whereas for long rods, in terms of L/d,

$$BM_2 = (L/d)v_2 \simeq 0.75(L/d)$$

the numerical values being obtained by setting $v_2 = 0.75$, a value based on the fact that the density of protein material is generally believed to be about 1.33. (Actually, tightly bound water should be included within the excluded volume, as is fully discussed in section 20. However, the experimental precision of osmotic pressure data is not sufficient to warrant such refinements.)

Table 14–3 shows, indeed, that for a number of proteins under the required conditions, BM_2 lies within a factor of 2 of the theoretical value of about 3.0, and this must be accepted as very good evidence that these molecules are in fact close to spherical in shape. It is probably not worthwhile to speculate about possible reasons for the deviation of the observed values of BM_2 from 3.0 (the low values observed for ovalbumin and bovine serum albumin could be due to intermolecular attraction as discussed in section 12e) for the experimental precision is low. A value of $BM_2 = 3$ implies that the value of Π/c_2 at a concentration of 0.01 gram protein per milliliter differs from the extrapolated value by 3%. This is not very much greater than the experimental error; in practice values of B are therefore obtained by carrying out experiments to protein concentrations of 0.05 gram/ml and

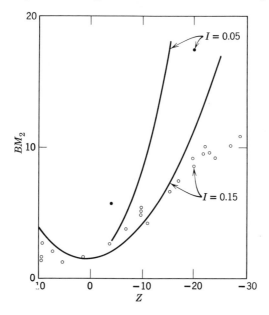

Fig. 14–1. The second virial coefficient in solutions of serum albumin in aqueous NaCl. The experimental data are those of Scatchard et al.[28] The curves are drawn according to equation 14–30.

above. These are high concentrations, and it is perhaps unreasonable to expect that an accurate *limiting slope* is to be obtained in this way. (Note also the very large discrepancy between various experimenters, as shown for β-lactoglobulin by Christensen.[35] Compare the molecular weight for this protein with that given in Table 13–1.)

One protein for which BM_2 far exceeds the expected value for spheres is myosin,[39] for which a value of 75 is obtained near the isoelectric point in 0.5M salt. This, however, is in agreement with other data concerning this molecule, which all suggest is much better approximated by a long rod than by a sphere. (Compare, for example, p. 397.)

Figure 14–1 shows data obtained by Scatchard and coworkers[28,33] for the virial coefficient B of bovine serum albumin at values of Z other than zero. (The values of Z are calculated from titration curves and chloride binding data.) The values are seen to rise considerably above the value for $Z = 0$, as is to be expected. According to equation 14–24 (and including the excluded volume contribution to B, which for serum albumin as given in Table 14–3 is $BM_2 = 1.5$), we should have

$$BM_2 = 1.5 + \frac{1000Z^2 v_1}{4m_3 M_2} \tag{14–30}$$

The curves of Fig. 14–1 were drawn according to this equation and are seen to represent reasonably well the actual variation in BM_2, at least between $Z = +10$ and $Z = -20$. Better agreement between experiment and theory is not to be expected, since equation 14–30 is virtually an ideal

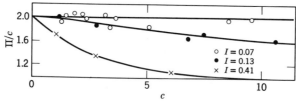

Fig. 14–2. Plots of Π/c versus c for insulin at pH 2.6. The upper curve (open circles) is at ionic strength 0.07, the middle one (filled circles) at ionic strength 0.13, the lowest one (crosses) at ionic strength 0.41. Π is here measured in centimeters of water and c in grams per liter. (Data of Gutfreund[40].)

equation ignoring all electrostatic interactions apart from that imposed by the Donnan effect. The latter, as we have seen, is simply a consequence of the condition that the solutions on both sides of the membrane be electrically neutral.

One further interesting result may be shown, that for insulin given in Fig. 14–2. The study was made by Gutfreund[40] at pH 2.65, where insulin has a positive charge of about $+11$ per 12,000 grams. It is therefore surprising to observe, at a low salt concentration, a slope essentially equal to zero. At higher ionic strength a considerable *negative* slope is observed. This fact strongly suggests that insulin participates in an association-dissociation equilibrium, as discussed in section 12e. That such association actually takes place in insulin has been amply confirmed by other experimental methods. The horizontal plot of reduced osmotic pressure versus concentration at low ionic strength and pH is thus a fortuitous cancellation of the effect of association, leading to a negative slope, and the Donnan effect, leading to a positive slope.

14h. Membrane potentials. If electrodes reversible with respect to one of the ions in the solution (e.g., H^+, X^-) are placed on the two sides of the membrane of Figure 13–1, in the presence of a macro-ion on side II, a potential difference of zero will be observed since the system is at equilibrium. If, however, the membrane is removed and a saturated KCl salt bridge is substituted, with both solutions now at atmospheric pressure, the system is no longer at equilibrium. An electromotive force will be measured since the activities of individual ions on sides I and II are unequal because of the Donnan effect. Its value will be

$$E = \frac{\mathscr{R}T}{z_i \mathscr{F}} \ln \frac{a_i}{a_i'} + E_{\text{KCl}} \tag{14–31}$$

where z_i is the algebraic charge on the ion whose activity the electrode measures and E_{KCl} is the algebraic sum of the liquid junction potentials at the two sides of the salt bridge. As is well known, E_{KCl} is small when a saturated salt bridge is used.

The electromotive force of a galvanic cell is the sum of a number of potentials, one of which exists at each boundary between homogeneous regions. In the cell containing the membrane, the boundaries between the potentiometer leads and the electrodes and between the electrodes and the two solutions are the same as in the cell containing the salt bridge. The difference between the emf's of the two cells must therefore be ascribed to the difference between the potential, E_{KCl}, and the corresponding potential, E_M, existing at the membrane in the cell with zero potential. Thus equation 14–31 represents the quantity $E_{\text{KCl}} - E_M$, so that

$$E_M = -\frac{\mathscr{R}T}{z_i \mathscr{F}} \ln \frac{a_i}{a_i'} \tag{14–32}$$

It is clear from equation 14–15 that the value of E_M, which is known as the *membrane potential*, is independent of the ion chosen to indicate its presence, as, indeed, it must be if it is to have any physical significance.

The membrane potential is a thermodynamic expression of the fact that, at equilibrium, the membrane in Figure 13–1 must be polarized to counteract the ion concentration gradients within the membrane.

(Membranes abound in living matter, and potential differences across them are believed to be important, for example, in the conduction of nerve impulses. It is improbable, however, that the type of equilibrium potential here discussed is primarily responsible for the membrane potentials in living systems. The latter are more likely to be *kinetic potentials* arising from different rates of diffusion of ions across the membranes.)

14i. Osmotic pressure and Donnan effect in the absence of membranes. The phenomena discussed in this section will be observed whenever a solution containing macromolecules is separated from a similar solution without macromolecules, even if no membrane is present at the boundary. For example, in the study of sedimentation, diffusion, or electrophoresis the experimental technique requires such a separation of solutions. The Donnan effect may play an important role in these studies unless it is minimized by the presence of a high salt concentration.

Under certain circumstances even a single macromolecular ion may give rise to a Donnan effect. If such an ion is spread over a large volume, as, for example, a linear polyelectrolyte coil might be, encompassing a considerable amount of solvent, then an imaginary boundary may be drawn around this volume. The solvent and low-molecular-weight ions in the solution can pass freely across this boundary. The charges fixed on the macro-ion, however, cannot, and the imaginary boundary thus behaves much like a semi-permeable membrane. In particular, the concentrations of free ions within the boundary will be different from the corresponding concentrations outside it. (This picture probably does not apply to relatively small macro-ions of low charge density, such as most proteins. For such ions the volume within the boundary is sufficiently small so that neutrality is not required. An ion atmosphere around the ion can compensate for the charge within it. See Chapter 7.)

15. PHASE EQUILIBRIA

The discussion so far presented in this chapter has been confined to systems in which the macromolecules are confined to a single homogeneous liquid phase. The present section will give a brief account of the equilibrium distribution of macromolecules between two contiguous separate phases.

15a. Solubility and freezing point. If the chemical potential μ_i of any component in a solution exceeds the chemical potential $(\mu_i{}^0)_s$ which that component would have in some solid crystalline phase, the solution becomes unstable and component i will precipitate from the solution, forming the crystalline phase. The value of μ_i in the solution then falls, until $\mu_i = (\mu_i{}^0)_s$, at which stage equilibrium is reached. The concentration of component i which corresponds to the equilibrium value of μ_i is known as its *solubility*.

If the substance which precipitates from the solution is the pure solvent, instead of a crystalline phase consisting principally of a solute, then the process of precipitation is the process of freezing. In this case $(\mu_1^{0})_s = (F_1^{0})_s$ and the temperature at which $\mu_1 = (\mu_1^{0})_s$ is the freezing point.

15b. Melting points of crystalline polymers. True melting points do not occur in the field of macromolecular chemistry. The only macromolecular substances which form truly crystalline solid phases, i.e., proteins and viruses, are irreversibly altered on heating, so that their melting points cannot be observed.

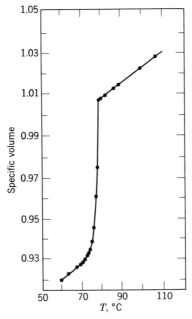

Fig. 15–1. Melting of crystalline regions of poly-(decamethylene adipate). The observed specific volume is plotted as a function of temperature. (Mandelkern, Garrett, and Flory.[44])

Something resembling a true melting point is found, however, in solid polymers with a high degree of crystallinity. Such polymers, as was explained in section 4a, contain numerous crystallites, a few hundred Ångstrom units in extent, embedded in an amorphous medium consisting of entangled polymer chains. When such polymers are heated, melting of the crystallites will occur. The process is accompanied by changes in physical properties, including, for example, specific volume (cf. Fig. 15–1). The most direct result, of course, is a loss of the sharp x-ray diffraction pattern characteristic of highly crystalline polymers.

Melting points of crystalline polymers are not as sharp as the melting points of true crystals. Instead a melting range which may extend over 10°C or more is commonly observed. It should be noted, however, that the extent of conversion from the crystalline to the amorphous state is not spread uniformly over this temperature range. As the example chosen for illustration (Fig. 15–1) shows, the melting range is covered during the conversion of only about half the crystallites. The remainder melt sharply at what is essentially a single temperature.

This phenomenon may be explained by the fact that the crystalline and non-crystalline regions of the solid polymer are intimately connected. As was noted in section 4a, individual crystallites are formed from portions of molecules only. Parts of the same molecules are contained in adjacent amorphous regions and in other neighboring crystallites. When the crystalline regions are few in number and relatively far apart this fact imposes no serious restraint on the system, and, therefore, the final stages of melting occur sharply. When there are many crystalline regions close together, however, the amorphous regions must have less than their normal freedom of internal motion because most of the molecules within these regions will form part of one or more of the neighboring crystallites. Thus the entropy change which accompanies the melting process is larger when there are relatively many crystallites in the solid, since it involves a positive entropy change for the amorphous regions as well as the crystalline ones. If the heat of melting remains unchanged, a lowered melting point will result, since, at the melting point, ΔH for the fusion process is equal to $T \Delta S$.

Several authors[45,46] have developed theories for melting which are much more refined than the simple qualitative explanation here given. These theories also explain the effect of dilution of the polymeric material with additives (depression of freezing point) and associated effects.

The transition from the crystalline to the amorphous state is not the only kind of transition which solid polymers can undergo. Amorphous polymers can undergo a *glass transition*, during which they are converted from a brittle glasslike form to a flexible rubbery form. This is not a true phase change but a change involving local freedom of motion. The phenomenon has been discussed by Boyer and Spencer.[47]

15c. Solubility of crystalline proteins. Much study has been devoted to the solubility of proteins because the solubility characteristics of different proteins are the basis for their separation from one another and subsequent purification. A detailed review of fundamental work in this field is given in Cohn and Edsall's treatise.[8]

Different proteins are found to possess very different solubilities, a fact reflecting differences in their standard chemical potentials, both in the liquid and crystalline phase. It is customary in solubility studies to express the chemical potential of proteins in solution by equation 11–20. Thus the activity of a saturated solution, obtained by setting μ equal to $(\mu^0)_s$, is

$$a_{\text{sat}} = yC_{\text{sat}} = e^{-\Delta\mu^0/\mathscr{R}T} = K \qquad (15\text{--}1)$$

where $y = a/C$ is the activity coefficient in the saturated solution, C_{sat} is the solubility in moles per liter (any other unit of concentration could be used equally well), $\Delta\mu^0 = \mu^0 - (\mu^0)_s$, and K is a constant defined by the equation as $\exp(-\Delta\mu^0/\mathscr{R}T)$. Differences in solubility thus reflect differences in $\Delta\mu^0$. Such differences are purely empirical quantities, and no theory exists by which we can compute lattice energies in the crystalline state and other factors which would be needed for an absolute calculation of $\Delta\mu^0$.

The same statement can be made for the effect of temperature on solubility. The effect of temperature is related to the heat of solution, $d\ln a_{\text{sat}}/dT = d\ln K/dT = \Delta\bar{H}^0/\mathscr{R}T^2$, where $\Delta\bar{H}^0$ is the difference between $\bar{H}^0$ in the solution and $(\bar{H}^0)_s$ in the crystalline state. Like $\Delta\mu^0$, $\Delta\bar{H}^0$ is not susceptible to calculation. Observed values in water range from $+8000$ to $-18{,}000$ cal/mole.

Although absolute solubilities of proteins are not predictable, considerable success has been achieved in rationalizing relative solubilities. Especially consistent are the data on the effect of ionic strength on solubility in water. Since proteins in solution are macro-ions they are, like all ions, stabilized by the addition of low concentrations of inorganic salts. In other words, their activity coefficients are decreased. Thus, by equation 15–1, if C_{sat}^0 is the solubility in pure water and C_{sat} that in a dilute salt solution and if y^0 and y are the corresponding activity coefficients,

$$y^0C_{\text{sat}}^0 = yC_{\text{sat}} = K \qquad (15\text{--}2)$$

and

$$C_{\text{sat}}/C_{\text{sat}}^0 = y^0/y > 1$$

since y is decreased by the addition of salt.

The theory of electrostatic effects on the chemical potential of protein ions is discussed in section 26. The activity coefficient is related to the difference between W_{el} for an isolated protein ion in pure water and W_{el} for a similar ion in an actual solution. For compact protein ions, impenetrable to the solvent, $kT\ln y$ is equal to the last term in equation 26–52. This term depends primarily on the *net* charge Z per ion, and, as a first approximation,

$$kT\ln y = -Z^2\epsilon^2\kappa/2D(1 + \kappa a) \qquad (15\text{--}3)$$

where ϵ is the electronic charge, D the dielectric constant of the solvent, and a the sum of the radii of the protein ion and average electrolyte ions in the solution, and κ is a parameter defined by equation 26–32, which is proportional to the square root of the ionic strength. Equation 15–3 is, of course, just the Debye-Hückel expression for the activity coefficient of any spherical ion of charge Z.

Even protein molecules in an isoelectric solution, i.e., in a solution in which the average value of Z for all protein molecules is zero, have an average negative value for $\ln y$, although equation 15–3 would indicate that $\ln y$ should vanish. There are two reasons for this. In the first place

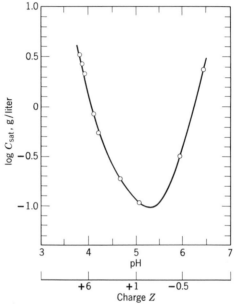

Fig. 15–2. The solubility of insulin in $0.1N$ NaCl as a function of pH. The charge Z is the average charge per 12,000 grams of insulin at the pH values indicated. (Fredericq and Neurath.[48])

(cf. section 14f) a solution containing protein molecules with average net charge zero will actually contain not only neutral molecules but also ions with charge ± 1, ± 2, etc. Thus, though $\bar{Z}$ is zero, $\overline{Z^2}$ is not. Second, equation 15–3 is only a first approximation for $\ln y$. The full expression for this term differs from the Debye-Hückel equation for $\ln y$ and contains terms which depend on the dipole moment, quadrupole moment, etc., which do not vanish when $Z = 0$. These terms, too, are negative in sign, so that $\ln y$ is negative even for those molecules with $Z = 0$.

The fact that the most important term in the expression for $\ln y$ is proportional to Z^2 suggests that the solubility of proteins at constant (low) ionic strength should be smallest at the isoelectric point and should increase with increasing charge. This is ordinarily found to be the case, as shown by the typical data of Fig. 15-2, unless the crystalline phase of the protein is a salt of the type $P^Z X_Z$. Such crystals will have their minimum solubility when the protein ions in solution bear a charge Z.

Whereas the activity coefficient of an ion or of a dipolar or multipolar molecule decreases with ionic strength at low ionic strengths, it increases again in concentrated salt solutions, eventually becoming greater than unity, especially in solutions of salts containing divalent or polyvalent ions. This effect is thought to be primarily due to the fact that the added salt ions are necessarily hydrated and, if present in high concentration, therefore reduce the activity of water in the solution. The activity of solutes is thereby increased and their solubility correspondingly decreased. With $y > y^0$ equation 15-2 clearly predicts that $C_{sat}/C_{sat}^0 < 1$. This property is often used to precipitate ions or polar molecules from a solution by the addition of a highly soluble salt. The process is generally called *salting out*.

Typical solubility data, showing both the increase in solubility at low ionic strength and the salting out effect at high ionic strength, are presented in Fig. 15-3. Comparable data are shown for the amino acid cystine, bearing out the applicability of the principles here discussed to large and small molecules alike.

15d. Solubility as a criterion of purity. A unique solubility may be used as a criterion for purity in the same way as a sharp melting point is used for ordinary organic compounds. The concentration of solute in equilibrium with a pure crystalline solid should be independent of the amount of solid phase present. Thus, if a pure crystalline material is added gradually to a liquid which acts as solvent for it, all the material added should dissolve until the concentration in the liquid phase is exactly equal to C_{sat}, and from that point on no further dissolution should occur. Analytical data for the total solute content should then have the appearance of the data for chymotrypsinogen shown in Fig. 15-4. If, on the other hand, a relatively insoluble impurity is present, we should expect a corresponding plot to have the appearance of curve B in this figure. If there is a small amount of relatively soluble impurity then further dissolution will occur, as in curve C of Fig. 15-4, after C_{sat} for the principal component has been reached.

This criterion, however, is strictly valid only under two alternative conditions. Either the solid phase must contain only a single component, in which case the solubility criterion is valid regardless of the number of components in the liquid phase; or, alternatively, if the solid phase consists of two components (one of them being the solvent), the liquid phase must contain no additional components. These conditions are not satisfied in the kind of experiments usually performed to test the purity of

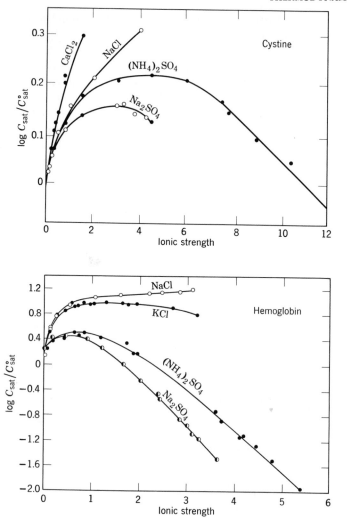

Fig. 15–3. The effect of inorganic salts on solubility. The data for cystine are taken from McMeekin, Cohn, and Blanchard,[50] those for hemoglobin are taken from Cohn.[49] (Both figures from Cohn.[49])

crystalline proteins by means of solubility. Such studies usually involve as solvent an aqueous solution containing a high concentration of an inorganic salt to reduce solubility, as well as a buffer to maintain constant pH. Furthermore, protein crystals invariably contain water. Thus two things can happen to prevent observation of a unique solubility even with a pure protein. If the activity of water in the crystals differs from that of the liquid phase, water may be transferred from one phase to the other, altering the composition of the liquid phase, and, hence, the value of y, and altering the composition of the solid phase, and, therefore, $(\mu^0)_s$. Similar changes can occur as a result of absorption of the inorganic salt ions by the solid phase. Thus C_{sat} may depend on the amount of solid phase present, and failure to observe the theoretical curve of Fig. 15–4 is thus not necessarily a sign of impurity.

The same considerations, incidentally, may enter into the solubility studies described in the preceding section. These studies were discussed with the assumption that the solid phase remains unchanged as pH and ionic strength are varied. This may not always be true.

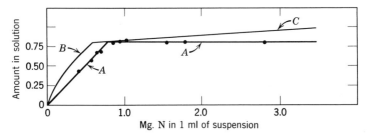

Fig. 15–4. Solubility as a criterion of purity. Curve A shows the behavior of pure chymotrypsinogen dissolved in concentrated $MgSO_4$ (Butler[51]). Curve B shows the expected effect of a relativ_ly insoluble impurity. Curve C shows the effect of a small amount of very soluble impurity.

15e. Miscibility of two-component liquid systems. A more complicated situation, akin to solubility, may arise in mixtures of two liquids. Consider one mole of such a mixture, i.e., $n_1 + n_2 = 1$, and $n_1 = X_1$ and $n_2 = X_2$. Equation 11–22 may then be written as

$$\Delta F_{mix} = (\mu_1 - \mu_1^0) + X_2[(\mu_2 - \mu_2^0) - (\mu_1 - \mu_1^0)] \qquad (15\text{–}4)$$

ΔF_{mix} can be calculated, as a function of composition, if either $\mu_1 - \mu_1^0$ or $\mu_2 - \mu_2^0$ is known as a function of composition, for one of them can always be computed from the other by the Gibbs-Duhem equation, equation 11–6. Some possible graphs of ΔF_{mix} as a function of composition are shown in Fig. 15–5.

The slope of any of the curves of Fig. 15–5, at any value of X_2, is obtained by taking the derivative of equation 15–4. With the Gibbs-Duhem equation, which allows us to place $X_1(\partial\mu_1/\partial X_2) + X_2(\partial\mu_2/\partial X_2) = 0$, we get

$$[\partial(\Delta F_{mix})/\partial X_2]_{T,P} = (\mu_2 - \mu_2^0) - (\mu_1 - \mu_1^0) \qquad (15\text{–}5)$$

This answer is different from equation 11–24 because the derivative with respect to n_2 is taken, not with n_1 constant but with $n_1 + n_2$ constant.

Equation 15–5 leads to a simple geometrical computation of the chemical potentials of the individual components at any composition X_2'. For the tangent at X_2' must have a slope given by equation 15–5 with μ_2 and μ_1 equal to the values μ_2' and μ_1' which are appropriate to the composition X_2'. Moreover, the ordinate at the point of tangency, $X_2 = X_2'$, must be

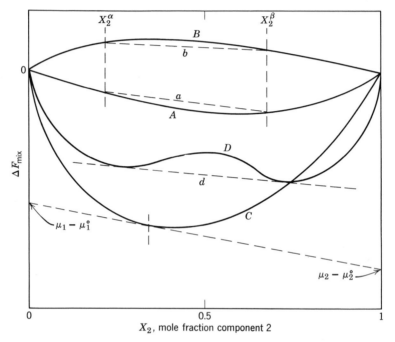

Fig. 15–5. ΔF_{mix} curves for two-component systems. Curves A and C represent complete miscibility; curve B represents complete immiscibility; curve D represents limited mutual solubility.

equal to ΔF_{mix} at that composition, as given by equation 15–4. The equation for the tangent is thus

$$y = \mu_1' - \mu_1{}^0 + X_2[(\mu_2' - \mu_2{}^0) - (\mu_1' - \mu_1{}^0)] \qquad (15\text{–}6)$$

where y is the value of the ordinate. It follows at once that $y = \mu_1' - \mu_1{}^0$ when $X_2 = 0$ and $y = \mu_2' - \mu_2{}^0$ when $X_2 = 1$. Thus the intercepts of the tangent (at $X_2 = 0$ and $X_2 = 1$) give the values of $\mu_1 - \mu_1{}^0$ and $\mu_2 - \mu_2{}^0$ appropriate to the composition at the point of tangency.

We are interested here in the miscibility of the two components. Will they form a homogeneous solution, or will they exist as separate phases? It is easy to show that the answer depends on the *curvature* of curves such as those of Fig. 15–5. Let two possible separate phases be designated as α

and β, and let the corresponding compositions be X_2^α and X_2^β. One of them must be larger and the other smaller than the over-all composition, X_2, of the system as a whole. (What the actual values of X_2^α and X_2^β are does not matter for the moment.) Now the total free energy of any mixture of the two phases, and therefore also ΔF_{mix}, must be a linear function of X_2 between X_2^α and X_2^β.

This statement follows at once from the fact that X_2^α, X_2^β and the corresponding free energies of mixing, $\Delta F_{\text{mix}}^\alpha$ and $\Delta F_{\text{mix}}^\beta$, are numbers independent of X_2. The variables are n_α, the total number of moles in phase α, and n_β, the corresponding number for phase β. These numbers are determined by the conditions that $n_\alpha + n_\beta = 1$ and $n_\alpha X_2^\alpha + n_\beta X_2^\beta = X_2$. Thus both n_α and n_β are linear functions of X_2. Moreover, ΔF_{mix} for the system of separate phases is $n_\alpha \Delta F_{\text{mix}}^\alpha + n_\beta \Delta F_{\text{mix}}^\beta$, and this is therefore also a linear function of X_2.

ΔF_{mix} for the mixture of phases must also coincide with the ΔF_{mix} curve for the homogeneous phase when $X_2 = X_2^\alpha$ (i.e., $n_\alpha = 1$) or when $X_2 = X_2^\beta$ (i.e., $n_\beta = 1$). Thus the ΔF_{mix} value at any composition must lie on straight lines, such as lines a and b of Fig. 15–5. It will always be *above* the ΔF_{mix} curve of the homogeneous phase when the latter is concave up (curves A and a), and the separate phases are then always *unstable* with respect to the homogeneous phase. On the other hand, the ΔF_{mix} value for the separated phases will always lie *below* the ΔF_{mix} curve of the homogeneous phase when the curve is concave down (curves B and b), and the separate phases are then *stable* with respect to the homogeneous phase. In Fig. 15–5, therefore, curves A and C represent a situation where the two components are miscible in all proportions. Curve B, however, represents a case of complete immiscibility, the two pure unmixed components being the only possible stable forms. Curve D, finally, represents the case of partial miscibility. The homogeneous phase is stable when (approximately) $X_2 > 0.8$ and $X_2 < 0.2$. In between it is unstable and will separate into two phases.

It remains to compute the compositions of the two stable phases α and β (i.e., to compute X_2^α and X_2^β) in a case such as that illustrated by the region $0.2 < X_2 < 0.8$ for curve D of Fig. 15–5. The two phases must clearly be such that $\mu_1^\alpha = \mu_1^\beta$ and $\mu_2^\alpha = \mu_2^\beta$ or that $\mu_1^\alpha - \mu_1^0 = \mu_1^\beta - \mu_1^0$ and $\mu_2^\alpha - \mu_2^0 = \mu_2^\beta - \mu_2^0$. We have shown, however, that $\mu_1 - \mu_1^0$ and $\mu_2 - \mu_2^0$, at any composition, are represented by the intercepts of the straight lines tangent to the ΔF_{mix} curve at that composition. The two phases α and β must therefore lie at points of the ΔF_{mix} curve which have a common tangent, as illustrated by curve d of Fig. 15–5.

Whenever partial miscibility of a two-component liquid system is encountered it is found to be strongly temperature dependent, as illustrated schematically by Fig. 15–6. With increasing temperature the region of immiscibility decreases in extent, and, at sufficiently high temperature (T_3

in Fig. 15–6), complete miscibility occurs. It is therefore possible to define a critical temperature, T_{cr}, below which phase separation occurs, while the components are completely miscible above it.

The value of T_{cr} may be expressed in terms of the chemical potential. Below T_{cr} the ΔF_{mix} curve is characterized, in the region of immiscibility, by two minima, a maximum, and two intervening points of inflection. At

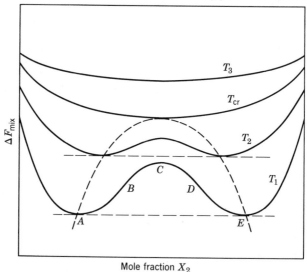

Fig. 15–6. Schematic diagram illustrating the transition, at a critical temperature T_{cr}, from complete to partial miscibility.

points A and E in Fig. 15–6, therefore $\partial(\Delta F_{mix})/\partial X_2 = 0$ and $\partial^2(\Delta F_{mix})/\partial X_2^2 > 0$. At points B and D, $\partial^2(\Delta F_{mix})/\partial X_2^2 = 0$. At point C, $\partial(\Delta F_{mix})/\partial X_2 = 0$ and $\partial^2(\Delta F_{mix})/\partial X_2^2 < 0$. Furthermore, between points B and D, $\partial^2(\Delta F_{mix})/\partial X_2^2$ must pass through a minimum, so that $\partial^3(\Delta F_{mix})/\partial X_2^3 = 0$. As T increases these points come closer together, until, at T_{cr}, they all coalesce into a single point. The ΔF_{mix} curve at T_{cr} is therefore characterized by possession of a single point at which

$$\partial(\Delta F_{mix})/\partial X_2 = \partial^2(\Delta F_{mix})/\partial X_2^2 = \partial^3(\Delta F_{mix})/\partial X_2^3 = 0 \quad (15\text{–}7)$$

Since $\partial(\Delta F_{mix})/\partial X_2 = 0$ it also follows, from equation 15–5, that

$$\mu_2 - \mu_2^0 = \mu_1 - \mu_1^0 \quad (15\text{–}8)$$

Combining all these equations it follows at once that the critical point is that at which

$$\partial\mu_1/\partial X_2 = \partial\mu_2/\partial X_2 = \partial^2\mu_1/\partial X_2^2 = \partial^2\mu_2/\partial X_2^2 = 0 \quad (15\text{–}9)$$

all derivatives being at constant T and P, with $X_1 + X_2 = 1$.

15f. Stability of polymer solutions. To investigate the stability of polymer solutions we use equation 12–40 for the chemical potential, realizing, of course, that it must be regarded as an approximate relation. The condition for incipient instability, as given by equation 15–9, means that $\partial \mu_1 / \partial \phi_1$ and $\partial^2 \mu_1 / \partial \phi_1^2$ must both be zero, i.e., since $\phi_1 + \phi_2 = 1$ and $d\phi_1 = -d\phi_2$, that $\partial \mu_1 / \partial \phi_2$ and $\partial^2 \mu_1 / \partial \phi_2^2$ must both be zero. Thus

$$1/(1 - \phi_2) - (1 - 1/\sigma) - 2\phi_2 \, \Delta\epsilon / \mathscr{R}T = 0$$

$$1/(1 - \phi_2)^2 - 2\Delta\epsilon / \mathscr{R}T = 0$$

Solving these equations for the critical values of $\Delta\epsilon$ and ϕ_2, we get

$$(\phi_2)_{\mathrm{cr}} = 1/(1 + \sigma^{\frac{1}{2}}) \tag{15–10}$$

$$(\Delta\epsilon)_{\mathrm{cr}} = (1 + \sigma^{\frac{1}{2}})^2/2\sigma \tag{15–11}$$

In all situations of interest σ will be much larger than 1, so that

$$(\phi_2)_{\mathrm{cr}} \simeq 1/\sigma^{\frac{1}{2}} \tag{15–12}$$

$$(\Delta\epsilon)_{\mathrm{cr}} \simeq \tfrac{1}{2} + 1/\sigma^{\frac{1}{2}} \tag{15–13}$$

Upon substitution of equation 12–41 for $\Delta\epsilon$, we find that the critical value of $\psi(1 - \Theta/T)$ must be approximately equal to $-1/\sigma^{\frac{1}{2}}$. Since ψ and Θ are constants for a particular solute-solvent system, independent of σ or T, the only variable in the expression for $\Delta\epsilon$ is the temperature. The critical temperature T_{cr} for phase separation is therefore given by

$$\psi(1 - \Theta/T_{\mathrm{cr}}) = -(1/\sigma^{\frac{1}{2}} + 2/\sigma) \simeq -1/\sigma^{\frac{1}{2}} \tag{15–14}$$

It will be recalled that ψ is an entropy function which is generally positive, reducing to the value of $+\tfrac{1}{2}$ for solutions with zero heat of mixing, and that Θ is an energy function which is negative for a good solvent and positive for a poor solvent. It can therefore be concluded at once that equation 15–14 can never be satisfied in a good solvent. The left-hand side of the equation will always be positive in such a solvent. Thus no phase separation will occur in good solvents.

In poor solvents, on the other hand, where Θ is positive, phase separation will occur when Θ/T is slightly greater than unity, i.e., when T_{cr} is approximately equal to Θ, and it becomes equal to Θ at infinite molecular weight, when $1/\sigma = 0$. Thus the critical temperature for phase separation is close to the temperature at which the solution behaves ideally (Θ having been identified with this temperature on p. 202) and, at the limit of infinite molecular weight, becomes equal to it.

It may further be seen from equation 15–12 that the value of ϕ_2 at the critical point is very small.

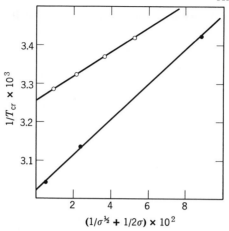

Fig. 15–7. Plot of the reciprocal of the critical temperature against the molecular size function of equation 15–14, for polystyrene fractions in cyclohexane (open circles) and for polyisobutylene fractions in diisobutyl ketone (filled circles). (From Shultz and Flory.[52])

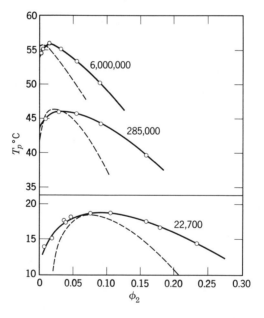

Fig. 15–8. Temperature-composition diagram for incipient phase separation at temperatures below T_{cr}. Diagrams for three polyisobutylene fractions (of indicated molecular weights) in diisobutyl ketone are shown. Solid curves are drawn through the experimental points, dashed curves are calculated. (From Shultz and Flory.[52])

These predictions are more or less confirmed by experimental studies of Shultz and Flory.[52] Figure 15-7 shows, for example, plots of $1/T_{cr}$ for two polymer-solvent systems (both poor solvents, of course) versus $1/\sigma^{1/2} + 2/\sigma$, and, as predicted by equation 15-14, linear plots of positive slope are obtained. From the data it is possible to get values of ψ and Θ for each system, for the intercept at $\sigma = \infty$ (i.e., $1/\sigma^{1/2} + 2/\sigma = 0$) gives $T_{cr} = \Theta$, and the slope is equal to $1/\psi\Theta$.

These values of ψ and Θ may be used to compute the temperature T_p at which phase separation will occur at compositions other than the critical point, i.e., the points of intersection of the dashed line of Fig. 15-6 with the ΔF_{mix} curves of that figure. Equation 12-41 is used to compute $\Delta\epsilon$ at any temperature, and equation 12-40 is used to calculate $\mu_1 - \mu_1^0$ as a function of composition. The Gibbs-Duhem equation is then employed to calculate $\mu_2 - \mu_2^0$ (note that $\mu_1 - \mu_1^0 = \mu_2 - \mu_2^0$ at the critical point; cf. equation 15-8). The compositions ϕ_2 and ϕ_2' are then obtained as those compositions at which $\mu_1 - \mu_1^0 = \mu_1' - \mu_1^0$ and $\mu_2 - \mu_2^0 = \mu_2' - \mu_2^0$. The results of such a calculation, compared with corresponding experiments, are shown in Fig. 15-8. Agreement is seen to be poor, although the general shape of the curves is as predicted. This presumably reflects the approximations involved in the use of equation 12-40 for the chemical potential.

Another consequence of the theory is that the value of Θ obtained from Fig. 15-7 should be identical with that obtained as the ideal temperature from osmotic pressure experiments (cf. Fig. 13-3). In the case of polystyrene and cyclohexane Krigbaum obtained $\Theta = 306.5°$ K as the temperature at which the second virial coefficient of the osmotic pressure plot vanishes. From extrapolation of Fig. 15-7, on the other hand, we obtain $\Theta = 307.2°$ K, essentially in agreement with prediction.

15g. Polymer fractionation. A solution of an unfractionated polymer sample is not a two-component system, and the equations of the preceding section cannot be applied to evaluate theoretical conditions for phase separation. It is, however, reasonable to expect that phase separation will again occur in poor solvents, at low concentrations of polymer and at sufficiently low temperature, as indicated by Figs. 15-6 and 15-8. This expectation is confirmed experimentally. It is found that the two separated phases formed in this process are unequal in composition, a fact of importance in polymer fractionation.

The general theory of this effect can be worked out without precise knowledge of the conditions under which phase separation occurs, merely by stipulating that the chemical potential μ_x of each separate species of degree of polymerization x must be the same in the two phases which are formed. Where primed symbols designate one phase and unprimed symbols the other, we have for every possible value of x,

$$\mu_x = \mu_x' \qquad (15\text{--}15)$$

To apply equation 15–14 we need an expression for μ_x in an unfractionated solution. Applying either the Hildebrand or the Flory-Huggins theory for the entropy of mixing we get in place of equation 12–32 $\Delta S_{\text{mix}} = -\mathscr{R}n_1 \ln \phi_1 - \mathscr{R}(\sum\limits_i n_i \ln \phi_i)$ where n_i is the number of moles of polymer with i segments, and the summation extends from $i = 1$ to ∞. The volume fractions ϕ_1 and ϕ_i are, analogous to equation 12–34,

$$\phi_i = in_i/(n_1 + \sum_i in_i)$$
$$\phi_1 = n_1/(n_1 + \sum_i in_i) \qquad (15\text{--}16)$$

Similarly, we get, in place of equation 12–37, $\Delta H_{\text{mix}} = n_1 \, \Delta \epsilon \sum\limits_i \phi_i$, and hence, since $\Delta F_{\text{mix}} = \Delta H_{\text{mix}} - T \, \Delta S_{\text{mix}}$,

$$\Delta F_{\text{mix}} = n_1 \, \Delta \epsilon \sum_i \phi_i + \mathscr{R}T(n_1 \ln \phi_1 + \sum_i n_i \ln \phi_i) \qquad (15\text{--}17)$$

To obtain $\mu_x - \mu_x^0$ for any value of x we must evaluate (by equation 11–24) $\partial(\Delta F_{\text{mix}})/\partial n_x$, with n_1 and all n_i other than n_x kept constant. In performing this differentiation it should be kept in mind that $\Delta \epsilon$ is independent of the size of the polymer molecule. (We shall again suppose $\Delta \epsilon$ to be a composite term including an energy and an entropy contribution.)

We note from equation 15–16 that $\partial \phi_1/\partial n_x = -xn_1/(n_1 + \sum\limits_i in_i)^2$, that $\partial \phi_x/\partial n_x = x/(n_1 + \sum\limits_i in_i) - xin_i/(n_1 + \sum\limits_i n_i)^2$, and that $\partial \phi_i/\partial n_x$ (for all values of i other than $i = x$) is equal to $-xin_i/(n_1 + \sum\limits_i in_i)^2$. With these relations we get

$$\mu_x - \mu_x^0 = [\partial(\Delta F_{\text{mix}})/\partial n_x]_{T,P,n_1,n_i \atop (i \neq x)} = x \, \Delta\epsilon(\phi_1 - \phi_1 \sum_i \phi_i)$$
$$+ \mathscr{R}T(\ln \phi_x + 1 - x\phi_1 - x \sum_i \phi_i/i) \qquad (15\text{--}18)$$

It is not necessary to simplify this expression but sufficient to note that we can write

$$\mu_x - \mu_x^0 = \mathscr{R}T(\ln \phi_x + 1 + Ax) \qquad (15\text{--}19)$$

where A is a parameter which depends on the over-all compositions of the phase considered, as expressed by the parameters ϕ_1, $\sum\limits_i \phi_i$, and $\sum\limits_i \phi_i/i$, but is *independent of* x.

Equation 15–15 then becomes simply $\ln \phi_x + Ax = \ln \phi_x' + A'x$, or

$$\phi_x'/\phi_x = e^{Bx} \qquad (15\text{--}20)$$

where $B = A - A'$. Clearly, whatever the values of A and A', ϕ_x/ϕ_x' will have a different value for each value of x. This difference leads to the conclusion that phase separation will result in fractionation; i.e., one phase will be enriched in polymer molecules with large values of x, the other in molecules with small values of x. Let V be the total volume of one phase and V' the total volume of the other. Then the total volume of x-mer in one phase is $\phi_x V$, that in the other $\phi_x' V'$. The fractions of x-mer molecules in the two phases are therefore

$$f_x = \phi_x V/(\phi_x V + \phi_x' V')$$
$$f_x' = 1 - f_x = \phi_x' V'/(\phi_x V + \phi_x' V') \tag{15-21}$$

Replacing V'/V by R we get, by equation 15–20,

$$f_x = 1/(1 + Re^{Bx})$$
$$f_x' = Re^{Bx}/(1 + Re^{Bx}) \tag{15-22}$$

Consider now a phase diagram such as that of Fig. 15–6. Suppose a very dilute polymer solution is chosen, such that ϕ_2 is much smaller than the critical value (i.e., that corresponding to T_{cr}). As this solution is cooled phase separation will occur, with formation of a more concentrated second phase. Equation 15–20 shows that B must be positive if the primed phase is to be the more concentrated one and that all x-mers, regardless of the value of x, will be present in greater concentration in this phase. However, the ratio f_x'/f_x will increase with x according to equation 15–22, and the more concentrated phase will become enriched, relative to the more dilute phase, in the larger x-mers.

It is especially useful to make V' much smaller than V, i.e., to keep the volume of the concentrated phase small. Then $R \ll 1$ and $f_x \ll f_x'$ for small values of x. However, for large values of x, Re^{Bx} will become > 1, so that the concentrated phase, though small in volume, will contain the major portion of the longer chains.

15h. Three-component liquid systems. The discussion of phase equilibria in liquid systems can be extended to systems containing a single macromolecular component and *two* low-molecular-weight liquid components. The simplest case is that in which one of the latter is a solvent for the macromolecular component whereas the other is not. Systems of this kind are of interest in particular as an aid to polymer fractionation. There are now *two* variables (temperature and mole fraction of nonsolvent) which can be adjusted for optimum separation of a polymer sample into low- and high-molecular-weight fractions. A valuable summary of work on this subject has been given by Tompa.[5]

16. SEDIMENTATION EQUILIBRIUM[9]

16a. Extension of Gibbs' equilibrium conditions to include the influence of an applied force.[2,55,56] The conditions of equilibrium enumerated in section 11c are readily modified so as to include the influence of an applied force. The modifications appropriate to the influence of gravity, for example, were specifically derived by Gibbs.[2]

We shall confine ourselves here to forces which act only on the mass of any part of a system and are directly proportional to it. Moreover, we shall confine ourselves to forces acting in one specified direction only, which we shall call the x direction. Let the force acting on any mass m, in the positive x direction, be

$$\text{force} = G(x)m \qquad (16\text{-}1)$$

where $G(x)$ is an arbitrary parameter representing the force per unit mass at any point x. The potential energy of the mass m thus becomes a function of x. Relative to an arbitrary reference point at $x = 0$

$$\text{potential energy} = -m \int_{x=0}^{x} G(x)\,dx = H(x)m \qquad (16\text{-}2)$$

where

$$H(x) = -\int_{x=0}^{x} G(x)\,dx \qquad (16\text{-}3)$$
$$d\,H(x)/dx = -G(x)$$

$H(x)$ is the potential energy per unit mass at any point x. The negative sign in equations 16-2 and 16-3 represents the fact that the potential energy is decreased by motion in the direction of the force.

We shall consider a solution which is homogeneous in the absence of the applied force, with temperature and pressure everywhere the same, and we can show at once that one effect of the applied force is to make the pressure a continuously varying function of x. Consider a cylindrical volume element lying in the x direction, of cross-section dA (normal to x) and of length dx. Three forces act on this element: the force $P\,dA$ in the positive x direction at one face; the force $-[P + (dP/dx)\,dx]\,dA$ at the other face, the negative sign indicating that it acts in the opposite direction; and the force given by equation 16-1. The last force is equal to $G(x)\rho\,dA\,dx$, where ρ is the density of the element. The sum of these forces must be zero at equilibrium, so that

$$dP/dx = \rho\,G(x) \qquad (16\text{-}4)$$

and P must be a continuously varying function of x.

One consequence of equation 16–4 is that the originally homogeneous solution can no longer be homogeneous in the field of force; i.e., each region of width dx at any value of x must be considered a separate microscopic phase, and all properties may be expected to vary continuously with x, subject to the conditions of equilibrium which will be derived. One of these conditions is that the temperature must remain the same at all points in the solution, for a small amount of heat may still be transferred from any point x to another point x', without other change. The total entropy change (cf. equation 11–11) must still be zero, and T must therefore be the same at all values of x.

The condition of greatest interest is that obtained by transfer of dn_i moles of a component i from position x to another position x' without other change. As in section 11, we consider the solution of interest to be part of a larger, isolated system at constant volume, so that if the system is to be at equilibrium the *total energy* change attending this process must be zero. This energy change is made up of two parts, that given by equation 11–12 for the internal energy without consideration for the potential energy in the field of force and that given by equation 16–2 for the change in potential energy. The first part is

$$dE = (\mu_{ix'} - \mu_{ix})\, dn_i \qquad (16\text{–}5)$$

where μ_{ix} represents the chemical potential of component i at location x. The second part is

$$dE = [H(x') - H(x)]\, dm \qquad (16\text{–}6)$$

where dm is the mass of dn_i moles of component i; i.e., $dm = M_i\, dn_i$ where M_i is the molecular weight of component i. Combining these equations and setting the total dE equal to zero, we get

$$[\mu_{ix'} - \mu_{ix} + M_i\, H(x') - M_i\, H(x)]\, dn_i = 0$$

and, since dn_i is arbitrary,

$$\mu_{ix'} + M_i\, H(x') = \mu_{ix} + M_i\, H(x) = \text{constant} \qquad (16\text{–}7)$$

It is seen at once that the chemical potential is no longer the same in all parts of the system. Instead it is the sum of the chemical potential and the molar potential energy in the field of force which remains constant. In applying equation 16–7 we must recall that the pressure will be different at x and x' and take this into account in the calculation of the chemical potentials.

Equation 16–7 is often written as

$$(\mu_i)_{\text{total}} = \text{constant} \qquad (16\text{–}8)$$

where $(\mu_i)_{\text{total}}$, the *total potential*, is defined as the sum of the intrinsic chemical potential and the molar potential energy in the field of force. In this form the equation applies to forces other than those which act only on the mass, e.g., to electrical forces.

Another useful form of equation 16–7 is obtained by differentiation; i.e., with equation 16–3 for $d\,H(x)/dx$,

$$\frac{d\mu_i}{dx} + M_i \frac{d\,H(x)}{dx} = \frac{d\mu_i}{dx} - M_i\,G(x) = 0 \tag{16–9}$$

In place of $d\mu_i/dx$ we may use any of the relations of section 11e which express the variation of chemical potential with composition *at constant pressure*, plus the relation

$$\left(\frac{\partial\mu_i}{\partial P}\right)_{T,\,\text{composition}} = \bar{V}_i = M_i\bar{v}_i \tag{16–10}$$

where $\bar{V}_i$ is the partial molal volume of the ith component, and $\bar{v}_i$ its partial specific volume. In so doing we must recognize that a solution of s components will possess $s - 1$ variables of composition, so that the number of independent variables depends on the number of components. For a two-component system we may designate one component as solvent and use the concentration C_2 of the other as the only composition variable. We obtain, for either component (solvent or solute)

$$d\mu_i = \left(\frac{\partial\mu_i}{\partial P}\right)_{T,C_2} dP + \left(\frac{\partial\mu_i}{\partial C_2}\right)_{T,P} dC_2 \tag{16–11}$$

Combining equations 16–9 and 16–11 and using equations 16–4 and 16–10 for dP/dx and $\partial\mu_i/\partial P$, we get for the solute component

$$\left(\frac{\partial\mu_2}{\partial C_2}\right)_{T,P} \frac{dC_2}{dx} = M_2\,G(x)\,(1 - \bar{v}_2\rho) \tag{16–12}$$

This equation is clearly independent of the concentration units employed. If C_2 is in moles per liter, we can use the general equation 11–20 for the chemical potential. Writing $a_2 = y_2 C_2$, where y_2 is the activity coefficient on the moles per liter scale, we get

$$\left(\frac{\partial\mu_2}{\partial C_2}\right)_{T,P} = \frac{\mathscr{R}T}{C_2} + \mathscr{R}T\left(\frac{\partial \ln y_2}{\partial C_2}\right)_{T,P} \tag{16–13}$$

We may then rearrange equation 16–12 as

$$\frac{\mathscr{R}T}{C_2}\frac{dC_2}{dx}\left[1 + C_2\left(\frac{\partial \ln y_2}{\partial C_2}\right)_{T,P}\right] = M_2\,G(x)(1 - \bar{v}_2\rho) \tag{16–14}$$

This equation is still independent of the concentration units, and C_2 may thus be replaced by the concentration in any desired units.

By applying equation 16–13 to the chemical potential of the solvent and using equation 11–21 for μ_1 we easily obtain the alternative relation useful for dilute solutions

$$\frac{\mathscr{R}T}{c_2}\frac{dc_2}{dx}(1 + 2BM_2c_2 + \cdots) = M_2\, G(x)(1 - \bar{v}_2\rho) \qquad (16\text{–}15)$$

where c_2 is now expressed in grams per cubic centimeter. Both equations predict that the concentration will vary with x and provide a means for evaluating this variation. Both equations reduce to the same relation when c_2 is zero or for ideal solutions, and they differ only in the manner in which non-ideality is expressed, this being done by means of an activity coefficient in equation 16–13 and by means of the virial coefficients in equation 16–14.

When more than two components are present the foregoing equations become much more complex. In a multi-component system each solute may be expected to vary in concentration with position, and a series of derivatives dC_i/dx will be needed to express this variation. Moreover, the concentration of any component will affect the chemical potential (or the activity coefficient) of any other component, necessitating the introduction of terms of the type $\partial\mu_i/\partial C_j$ or $\partial \ln y_i/\partial C_j$. The general form of equation 16–12 for a system of s components would become

$$\left(\frac{\partial\mu_i}{\partial C_i}\right)_{T,P,C^*}\frac{dC_i}{dx} = M_i\, G(x)(1 - \bar{v}_i\rho) - \sum_{k=2}^{s}\left(\frac{\partial\mu_i}{\partial C_k}\right)_{T,P,C^*}\frac{dC_k}{dx}$$

$$(16\text{–}16)$$

where the subscript C^* means that all concentration variables, except the one C_i or C_k which appears in the derivative, remain constant. We have separated, on the left-hand side of the equation, the term giving the variation of the chemical potential of the ith component with its *own* concentration from the terms expressing its variation with the concentrations of other components.

16b. Equilibrium under the influence of gravity. The equations of the preceding section could be used to calculate the state of equilibrium under the influence of gravity. Where x is the distance away from the earth's surface, the force of gravity is simply $-gm$, where g is the gravitational constant, the minus sign indicating that the force is in the direction of negative x. Thus, by equation 16–1, $G(x) = -g$, and, by equation 16–3, $H(x) = gx$. Such a calculation would show that the components of a solution will not be uniformly distributed; those for which $\bar{v}_i\rho < 1$ would ordinarily be more concentrated at the bottom of the vessel, those for which $\bar{v}_i\rho > 1$ would ordinarily be more concentrated at the top. However, the concentration gradient would be exceedingly small for all but the very heaviest molecules, and its measurement would be difficult. We shall therefore not proceed with the analysis of this problem but turn at once to discussion of equilibrium under the influence of centrifugal forces which can be made much greater than the force of gravity.

A remark is in order concerning the important influence of the term $\bar{v}_i\rho$ mentioned in the last paragraph. Recalling that $\bar{v}_i$ is the change in solution volume per added gram of component i we note that it would represent the reciprocal of the density of component i if this component did not interact with the solvent; i.e., $\bar{v}_i\rho < 1$ would then mean that the solute is denser than the solution as a whole, and $\bar{v}_i\rho > 1$ would mean the opposite. The fact that the former solute will tend to fall and the latter to rise in the solution is therefore just the obvious consequence of Archimedes' principle. Thus the term $1 - \bar{v}_i\rho$ in equations 16–12, etc., is seen to be the thermodynamic equivalent of this principle for solutions in which interaction between solvent and solute occurs.

16c. The sedimenting force in the ultracentrifuge. The ultracentrifuge was developed by Svedberg[55] about thirty years ago to provide specifically for a sedimenting force greater than that of gravity. It consists of a rotor,

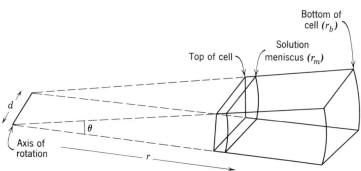

Fig. 16–1. Schematic diagram of an ultracentrifuge cell and of its position relative to the axis of rotation. The figure shows that the cross-section normal to r is curved, with area $d\theta r$. The bottom of the cell is actually flat but can be made curved (as shown) by means of a drop of heavy oil or grease placed on the flat surface.

rotating about an axis with a high and constant velocity. Placed in this rotor, some distance from the axis of rotation, is a small cell into which the solution to be studied is placed. The centrifugal force acting on the solution is directed radially outward, and the walls of the cell are similarly directed, as shown in the diagram of Fig. 16–1.

The centrifugal force acting on any mass m at a distance r from the center of rotation depends on the speed of rotation. The speed of rotation may be expressed in terms of ν, the number of revolutions per second; ω, the angular velocity; or U, the tangential velocity at distance r. These three velocities are related as follows

$$U = \omega r = 2\pi \nu r$$

and the corresponding expressions for the centrifugal force are

$$F = mU^2/r = m\omega^2 r = 4\pi^2 m\nu^2 r \qquad (16\text{--}17)$$

Ultracentrifuges capable of speeds up to 100,000 rpm have been designed.[55] A commercial ultracentrifuge in common use has a top speed of 60,000 rpm, with a cell placed 6.5 cm from the center of rotation. This corresponds to a force about 250,000 times that of gravity.

We shall always express the speed of rotation in terms of the angular velocity ω. Thus the equations of section 16a apply with

$$G(r) = \omega^2 r, \qquad H(r) = -\tfrac{1}{2}\omega^2 r^2 \qquad (16\text{--}18)$$

where r has been used in preference for x since we are dealing with a radial distance. $G(r)$ is positive since the force acts in the positive r direction.

16d. Sedimentation equilibrium in a two-component system. The measurable quantity in sedimentation equilibrium is the variation in composition with position. For a two-component system, ordinarily a dilute solution of a macromolecular solute in a suitable solvent, the variation in composition is given by equation 16–14 or 16–15. Using the former, with c_2 in grams per cubic centimeter in place of C_2 and with equation 16–18 for $G(r)$, we get

$$\frac{1}{c_2}\frac{dc_2}{dr} = \frac{M_2(1 - \bar{v}_2\rho)\omega^2 r}{\mathscr{R}T\left[1 + c_2\left(\dfrac{\partial \ln y_2}{\partial c_2}\right)_{T,P}\right]} \qquad (16\text{--}19)$$

This equation is the fundamental equation for sedimentation equilibrium in a two-component system. It shows that the measurement of the state of equilibrium in the ultracentrifugal field yields the same type of information as the measurement of the state of equilibrium across a semi-permeable membrane (i.e., osmotic pressure). Extrapolation of the left-hand side of equation 16–19 to $c_2 = 0$ gives a value for the molecular weight M_2, and the effect of increasing c_2 gives a value for the deviation from ideal behavior, as expressed by the activity coefficient y_2 or by the virial coefficients of equation 16–15.

An entirely different derivation of equation 16–19, based on opposition of the forces of sedimentation and diffusion, will be presented in section 22.

In applying equation 16–19 to actual measurements it is convenient to define an apparent molecular weight,

$$M_2^{(app)} = M_2 \Big/ \left[1 + c_2 \left(\frac{\partial \ln y_2}{\partial c_2} \right)_{T,P} \right] = M_2/(1 + 2BM_2c_2 + \cdots)$$

$$(16\text{–}20)$$

which approaches M_2 as c_2 approaches zero. By equation 16–19,

$$M_2^{(app)} = \frac{\mathscr{R}T}{(1 - \bar{v}_2\rho)\omega^2} \frac{1}{rc_2} \frac{dc_2}{dr} \qquad (16\text{–}21)$$

If suitable experimental arrangements are made to measure c_2 as a function of r, e.g., by light absorption, then the quantity $(dc_2/dr)/rc_2$ can be measured at various points in the sedimentation cell. Several values of $M_2^{(app)}$ are thus obtained at a given concentration, and the determination of such values at several concentrations permits extrapolation to obtain M_2. Alternatively, equation 16–21 may be integrated, giving

$$\ln c_2 = \frac{(1 - \bar{v}_2\rho)\omega^2}{2\mathscr{R}T} M_2^{(app)} r^2 + \text{constant} \qquad (16\text{–}22)$$

and $M_2^{(app)}$ is obtained from the slope of a plot of $\ln c_2$ versus r^2.

Equation 16–22 may be used to compute the speed of centrifugation which is required to obtain a conveniently measurable concentration gradient in a sedimentation experiment. A convenient gradient is one which has c_2 about five times as large at the bottom of the cell as at the top. Since the usual sedimentation cell has a depth of about 1 cm and is located about 6.5 cm from the center of rotation, this means that it is desirable to have $\ln c_2 (r = 7 \text{ cm}) - \ln c_2 (r = 6 \text{ cm}) \simeq \ln 5$. For a typical macromolecular solution, with $M_2 \sim 10^5$ and $\bar{v}_2 \sim 0.75$, the required value of ω at room temperature would be about 500 sec^{-1}, corresponding to a rotor speed of 4800 rpm, i.e., considerably less than the maximum speed. If the usual maximum speed (60,000 rpm) is used, c_2 falls by a factor of 5 over a distance of 0.006 cm; i.e., the equilibrium state would be one in which virtually all the macromolecules would be packed into the bottom 0.01 cm of the cell, and the top 0.99 cm would consist essentially of free solvent. This situation is of interest in connection with the measurement of sedimentation velocity, as described in section 22.

It is well known that the determination of c_2 as a function of r over the short distance of 1 cm or so which is available in a sedimentation cell is not easily performed with high precision. In the majority of sedimentation equilibrium measurements use has been made of the schlieren optical system or similar techniques which measure *refractive index gradient*, $d\tilde{n}/dr$. This quantity is proportional to the concentration gradient dc_2/dr; i.e., these measurements determine $K (dc_2/dr)$ versus r, where K is a constant.

As is explained in Appendix C at the end of this book, it is possible to use suitable integration procedures to evaluate from such plots auxiliary quantities such as Kc_2^0, where c_2^0 is the initial uniform concentration everywhere in the cell at the commencement of sedimentation, as well as quantities such as $K(c_b - c_m)$ and $K(c_r - c_a)$, where c_b is the concentration at the bottom of the cell (see Fig. 16–1), c_m that at the meniscus, c_r that at any point r in the cell, and c_a that at any suitable reference point. The parameter K is determinable, but it is considered preferable to express all equations in terms of ratios of concentrations, since such ratios, e.g., $(dc_2/dr)/c_2^0$, are clearly equal to the measured ratios without the necessity that K be known.

$M_2^{(\text{app})}$ may now be evaluated in a number of ways from such data.[57] For instance, the total solute content of the cell is $\int c_2 A \, dr$, where A is the cross-sectional area of the cell at any point r, and integration is from the meniscus to the bottom of the cell, i.e., from r_m to r_b in Fig. 16–1. Figure 16–1 shows that A is proportional to r. With $A = \alpha r$, the total solute content becomes $\alpha \int c_2 r \, dr$. This must, however, be equal to the total solute content at the beginning of the experiment, when $c_2 = c_2^0$ everywhere. With α canceling,

$$\int_{r_m}^{r_b} c_2 r \, dr = c_2^0 \int_{r_m}^{r_b} r \, dr = \tfrac{1}{2} c_2^0 (r_b^2 - r_m^2) \qquad (16\text{–}23)$$

Using equation 16–21 for $c_2 r \, dr$ and integrating gives

$$\frac{2\mathscr{R}T}{M_2^{(\text{app})} \omega^2 (1 - \bar{v}_2 \rho)} = \frac{c_2^0}{c_{2b} - c_{2m}} (r_b^2 - r_m^2) \qquad (16\text{–}24)$$

where c_{2b} is the equilibrium value of c_2 at r_b, etc. The right-hand side of equation 16–24 is now in such form as to be determinable from refractive index gradient data, and $M_2^{(\text{app})}$ is then at once obtained.

An alternative procedure is to use equation 16–21 directly. Although c_2 cannot be evaluated from refractive index gradient data, it is possible to evaluate c_2 relative to its value c_{2a} at some arbitrary reference point, e.g., at the meniscus if desired. Equation 16–21 may be rearranged as

$$\frac{1}{r} \frac{dc_2}{dr} = \frac{M_2^{(\text{app})} (1 - \bar{v}_2 \rho) \omega^2}{\mathscr{R}T} [(c_2 - c_{2a}) + c_{2a}] \qquad (16\text{–}25)$$

and $M_2^{(\text{app})}$ is obtained from the slope of a plot of $K(1/r)(dc_2/dr)$ versus $K(c_2 - c_{2a})$. An example of such a plot is shown in Fig. 16–2.

It is implicit in the derivations of equations 16–22 and 16–24, both of which involve integrations, that $M_2^{(\text{app})}$ and $\bar{v}_2 \rho$ are constants independent of r. This is, of course, a false assumption. The difference between $M_2^{(\text{app})}$ and M_2 is due purely to a concentration effect; since c_2 varies with r,

$M_2^{(\mathrm{app})}$ must also vary with r. Similarly $\bar{v}_2\rho$ must depend on concentration and also on pressure, although the effect of this variation is very small, usually small enough to be ignored. Constancy of $M_2^{(\mathrm{app})}$ and $\bar{v}_2\rho$ is not assumed in equation 16–25, but an appreciable concentration or pressure effect would in this case result in a curved plot rather than a linear plot.

In practice the variations in $M_2^{(\mathrm{app})}$ and $\bar{v}_2\rho$ present no problem. The effect of concentration or pressure on $\bar{v}_2\rho$ can be measured and taken

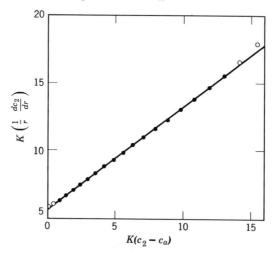

Fig. 16–2. A graph of $K\left(\dfrac{1}{r}\dfrac{dc_2}{dr}\right)$ versus $K(c_2 - c_a)$ for ribonuclease in an aqueous solution. The value of $K(dc_2/dr)$ is obtained directly from the refractive index gradient pattern. The abscissa is simply the integral of $K(dc_2/dr)$ between r and some suitable reference point. (Van Holde and Baldwin.[57])

into account where necessary. The dependence of $M_2^{(\mathrm{app})}$ on r will not be serious for solutions of compact molecules (B small, cf. Table 12–1) nor in solutions of flexible polymers in poor solvents. (Note that there is virtually no curvature in Fig. 16–2.) Any error which enters into the calculation of $M_2^{(\mathrm{app})}$ will decrease as the initial concentration c_2^0 decreases, and it will disappear as part of the extrapolation to zero concentration. If, on the other hand, the deviation from ideality is expected to be large, an estimate of the virial coefficients can be made and an expression of the form $M_2^{(\mathrm{app})} = M_2'/(1 + 2BM_2'c_2 + \cdots)$ can be explicitly introduced into any of the above equations. With suitable approximations this will give, at each concentration, an M_2' value quite close to M_2. Extrapolation to zero concentration is then used to obtain the latter. Methods of this type are discussed by van Holde and Baldwin.[57]

16e. Sedimentation equilibrium in multi-component systems.[9,57,58,64] The treatment of the preceding section can be extended in two ways, by allowing for the presence of more than one low-molecular-weight component and by allowing for the presence of several macromolecular components.

Considering the presence of additional low-molecular-weight components first, we see at once that the equilibrium conditions become much more complex. Supposing that there is just a single additional component (component 3), we get, using equation 16-16,

$$\frac{\mathscr{R}T}{c_2}\frac{dc_2}{dr}\left[1 + c_2\left(\frac{\partial \ln y_2}{\partial c_2}\right)_{T,P,c_3}\right] = M_2(1 - \bar{v}_2\rho)\omega^2 r - \mathscr{R}T\left(\frac{\partial \mu_2}{\partial c_3}\right)_{T,P,c_2}\frac{dc_3}{dr}$$

$$(16\text{--}26)$$

$$\frac{\mathscr{R}T}{c_3}\frac{dc_3}{dr}\left[1 + c_3\left(\frac{\partial \ln y_3}{\partial c_3}\right)_{T,P,c_2}\right] = M_3(1 - \bar{v}_3\rho)\omega^2 r - \mathscr{R}T\left(\frac{\partial \mu_3}{\partial c_2}\right)_{T,P,c_3}\frac{dc_2}{dr}$$

$$(16\text{--}27)$$

where $\partial \mu_i/\partial c_k$ is equal to $\partial \ln y_i/\partial c_k$ (equation 11-20). The difficulty in these equations lies in the last term of the equation for c_2, which does not become zero as c_2 approaches zero. Thus if $M_2^{(app)}$ is defined by equation 16-21, it will not reduce to M_2 when extrapolated to $c_2 = 0$. The sedimentation equilibrium method will thus be generally adaptable to systems of this kind only if $\partial \ln y_2/\partial c_3 = 0$, so that the last term of equation 16-26 vanishes, or if $M_3(1 - \bar{v}_3\rho) \simeq 0$, so that dc_3/dr vanishes when c_2 and dc_2/dr become zero. In this respect osmotic pressure is a simpler tool than sedimentation equilibrium, for a second low-molecular-weight component will always distribute itself equally on both sides of the membrane in an osmometer when c_2 approaches zero.

The presence of several macromolecular components is of greater interest. This situation is likely to arise only when all the macromolecular components are of the same chemical nature (e.g., a heterogeneous polymer mixture) so that the partial specific volume and the refractive index increment become the same for each component. If it is possible to find for such a system a solvent in which all solutes behave ideally (e.g., a Θ solvent in the case of a synthetic polymer mixture), the activity coefficient terms vanish, and we may write for each component the simplified form of equation 16-19,

$$\frac{\mathscr{R}T}{c_i}\frac{dc_i}{dr} = M_i(1 - \bar{v}_2\rho)\omega^2 r \qquad (16\text{--}28)$$

where $\bar{v}_2$ is the partial specific volume common to each solute component. Moreover, because of the equality of the refractive index gradient for each component, the optical systems which measure this quantity will now measure $K(dc_2/dr)$, $Kc_2{}^0$, etc., where $c_2 = \sum_i c_i$.

Equation 16–23 must now apply separately to each component. Introducing equation 16–28 and integrating, we get for each component

$$\frac{2\mathscr{R}T(c_{ib} - c_{im})}{\omega^2(1 - \bar{v}_2\rho)} = c_i{}^0 M_i(r_b{}^2 - r_m{}^2) \qquad (16\text{–}29)$$

where $c_i{}^0$ is the initial concentration and c_{ib} and c_{im} are the equilibrium concentrations at r_b and r_m for each component. Summing over all components and dividing both sides by $\sum_i c_i{}^0$, we get

$$\frac{2\mathscr{R}T}{\omega^2(1 - \bar{v}_2\rho)} \frac{\sum_i (c_{ib} - c_{im})}{\sum_i c_i{}^0} = \frac{\sum_i c_i{}^0 M_i}{\sum_i c_i{}^0} (r_b{}^2 - r_m{}^2) \qquad (16\text{–}30)$$

or, with equation 8–12,

$$\frac{2\mathscr{R}T}{\omega^2(1 - \bar{v}_2\rho)} \frac{c_{2b} - c_{2m}}{c_2{}^0} = \bar{M}_w{}^0(r_b{}^2 - r_m{}^2) \qquad (16\text{–}31)$$

where $c_{2b} = \sum_i c_{ib}$, etc., and where $\bar{M}_w{}^0$ is the weight-average molecular weight, based on the concentrations $c_i{}^0$ of the original polymer sample. The concentration terms on the left-hand side of equation 16–31 are those which are readily obtained from refractive index gradients, and we see then that equation 16–24 is applicable to heterogeneous macromolecular solutes, yielding the *weight-average molecular weight of the sample*.

The analog of equation 16–25 for a heterogeneous sample is more complicated. When equation 16–28 is rearranged and summed over all components we get

$$\frac{\mathscr{R}T}{r} \sum_i \frac{dc_i}{dr} = (1 - \bar{v}_2\rho)\omega^2 \frac{\sum_i M_i c_i}{\sum_i c_i} \sum_i c_i \qquad (16\text{–}32)$$

There is a formal correspondence with equation 16–25, in that we may now plot $K(1/r)(dc_2/dr)$ versus $K(c_2 - c_{2a})$ and obtain a slope equal to a weight-average molecular weight. The weight average, however, is based on the equilibrium concentrations *at the point* r and is clearly one which will have a different value at each point. A plot of the type of Fig. 16–2 will thus be non-linear. Moreover the desired molecular weight of the original sample, $\sum_i M_i c_i{}^0 / \sum_i c_i{}^0$, is not directly obtainable.

A relation in terms of the c_i^0 can, however, be obtained from the straight line joining the terminal points of such a plot, i.e., from evaluation of

$$\left[\frac{1}{r_b}\sum_i\left(\frac{dc_i}{dr}\right)_{r_b} - \frac{1}{r_m}\sum_i\left(\frac{dc_i}{dr}\right)_{r_m}\right] \Big/ \left[\sum_i c_{ib} - \sum_i c_{im}\right] \tag{16-33}$$

(which is one of the quantities obtained directly from refractive index gradient data). Using equation 16–28 for each dc_i/dr and then summing we obtain for the numerator of 16–33

$$\frac{(1 - \bar{v}_2\rho)\omega^2}{\mathscr{R}T} \sum_i M_i(c_{ib} - c_{im})$$

which, with equation 16–29, for each $c_{ib} - c_{im}$ becomes

$$\frac{1}{2}\left[\frac{(1 - \bar{v}_2\rho)\omega^2}{\mathscr{R}T}\right]^2 (r_b^2 - r_m^2) \sum_i M_i^2 c_i^0$$

Similarly, the denominator becomes

$$\frac{(1 - \bar{v}_2\rho)\omega^2}{2\mathscr{R}T} (r_b^2 - r_m^2) \sum_i M_i c_i^0$$

The experimental quantity of equation 16–33 is thus clearly proportional to $\sum_i M_i^2 c_i^0 / \sum_i M_i c_i^0$, which, by equation 8–18, is equal to $\bar{M}_z$.

The fact that $\bar{M}_w$ and $\bar{M}_z$ can be obtained from the same experiments is of great importance, for the difference between these quantities affords a measure of the extent of heterogeneity of a polymer sample. Actually, we can go further than this. Wales, Williams, and coworkers[59] have shown that it is possible to obtain higher averages $\bar{M}_{z+1}$, etc., from the data, that it is often possible to evaluate $\bar{M}_n$ also, and that a distribution function for molecular weight can be computed if the mathematical analysis is carried sufficiently far. They calculated all possible data for a sample of unfractionated polystyrene, with the results shown in Fig. 16–3 and Table 16–1. They found that all the data were in fairly good accord with the simple statistical distribution functions given by equations 8–3 and 8–4.

The development of Wales and coworkers takes non-ideality into account by retaining the term $2BM_2c_2$ of equation 16–15, which, of course has been set equal to zero here. However, they assume B to be independent of molecular weight, which we have shown in section 12 to be incorrect. Their results are probably applicable to ideal solutions only.

16f. Solutions containing macro-ions. As has already been mentioned (section 14) it is usually desirable to carry out the study of macro-ions in solutions containing an inorganic salt. It will be shown presently that this

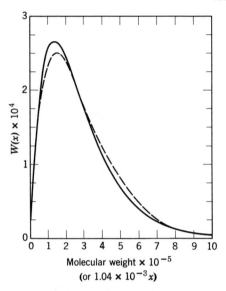

Fig. 16–3. The molecular weight distribution function for an unfractionated sample of polystyrene, as determined by sedimentation equilibrium. The dashed curve shows the distribution predicted by equation 8–3. (Wales and coworkers.[59])

TABLE 16–1. Molecular Weight Distribution of a Polystyrene Sample (Wales and coworkers[59])[a]

	From Sedimentation Equilibrium		
	From Eqs. 16–31, 16–33, etc.	From Exptl. Curve of Fig. 16–3	Other Methods
$\bar{M}_n$	150,000	146,000	144,000 (osm. press.)
$\bar{M}_w$	287,000	298,000	299,000 (light scat.)
$\bar{M}_z$	495,000	464,000	—
$\bar{M}_{z+1}$	785,000	705,000	—
	Sedimentation Equilibrium	Theoretical (Eq. 8–3 or 8–4)	
$\bar{M}_w/\bar{M}_n$	1.97	2.00	
$\bar{M}_z/\bar{M}_w$	1.62	1.50	
$\bar{M}_{z+1}/\bar{M}_z$	1.53	1.33	

[a] These data were obtained from experiments with a non-ideal solution of the polymer. This fact was taken into account in the equations used, but, as mentioned in the text, the assumptions made were not entirely correct. Whether the deviations between experiment and equations 8–3 and 8–4 are due to this error, or whether they are real, is not certain.

is particularly desirable in the study of sedimentation equilibrium, so that we necessarily deal with solutions containing three components. The present discussion of this problem is based on that given by Johnson, Kraus, and Scatchard.[60]

Suppose that the macro-ion P has a positive charge Z and that we express its concentration in terms of the neutral component PX_Z where X is a univalent anion. We shall take as component 3 a neutral salt BX, in which B is a univalent cation. We shall use equation 11–20 for the chemical potential, with the activity expressed as the product of the corresponding ion activities. For each ion activity we shall write $a_i = m_i \gamma_i$ where m_i is the molality and γ_i the activity coefficient. We then have $m_p = m_2$, $m_{B^+} = m_3$, $m_{X^-} = m_3 + Zm_2$, and

$$\mu_2 = \mu_2^0 + \mathscr{R}T \ln m_2 + Z\mathscr{R}T \ln(m_3 + Zm_2) + \mathscr{R}T \ln \gamma_p \gamma_{X^-}{}^Z \quad (16\text{–}34)$$

$$\mu_3 = \mu_3^0 + \mathscr{R}T \ln m_3 + \mathscr{R}T \ln(m_3 + Zm_2) + \mathscr{R}T \ln \gamma_{B^+} \gamma_{X^-} \quad (16\text{–}35)$$

$$\frac{\partial \mu_2}{\partial m_2} = \frac{\mathscr{R}T}{m_2} \left(1 + \frac{m_2 Z^2}{m_3 + Zm_2} + m_2 \frac{\partial \ln \gamma_p \gamma_{X^-}{}^Z}{\partial m_2} \right) \quad (16\text{–}36)$$

$$\frac{\partial \mu_2}{\partial m_3} = \frac{Z\mathscr{R}T}{m_3 + Zm_2} + \mathscr{R}T \frac{\partial \ln \gamma_p \gamma_{X^-}{}^Z}{\partial m_3} \quad (16\text{–}37)$$

$$\frac{\partial \mu_3}{\partial m_3} = \frac{\mathscr{R}T}{m_3} \left(2 - \frac{Zm_2}{m_3 + Zm_2} + m_3 \frac{\partial \ln \gamma_{B^+} \gamma_{X^-}}{\partial m_3} \right) \quad (16\text{–}38)$$

$$\frac{\partial \mu_3}{\partial m_2} = \frac{Z\mathscr{R}T}{m_3 + Zm_2} + \mathscr{R}T \frac{\partial \ln \gamma_{B^+} \gamma_{X^-}}{\partial m_2} \quad (16\text{–}39)$$

where all partial derivatives are at constant T and P, with m_2 or m_3 held constant.

The state of equilibrium is now defined by writing down equation 16–16 for each component, with $G(r) = \omega^2 r$. For component 2 we get

$$\frac{\mathscr{R}T}{m_2} \frac{dm_2}{dr} \left(1 + \frac{m_2 Z^2}{m_3 + Zm_2} + m_2 \frac{\partial \ln \gamma_p \gamma_{X^-}{}^Z}{\partial m_2} \right) = M_2 (1 - \bar{v}_2 \rho) \omega^2 r$$
$$- \frac{dm_3}{dr} \left(\frac{Z\mathscr{R}T}{m_3 + Zm_2} + \mathscr{R}T \frac{\partial \ln \gamma_p \gamma_{X^-}{}^Z}{\partial m_3} \right) \quad (16\text{–}40)$$

Consider this equation with $m_3 = 0$, i.e., for a two-component system containing only PX_Z and solvent. The last term of equation 16–40 now vanishes, and the term $m_2 Z^2 / (m_3 + Zm_2)$ becomes just Z, and thus independent of m_2. We thus get, after rearranging,

$$\frac{\mathscr{R}T}{m_2} \frac{dm_2}{dr} \left(1 + \frac{m_2}{Z+1} \frac{\partial \ln \gamma_p \gamma_{X^-}{}^Z}{\partial m_2} \right) = \frac{M_2}{Z+1} (1 - \bar{v}_2 \rho) \omega^2 r \quad (16\text{–}41)$$

This equation is identical in form with equation 16–19, with a non-ideality term which vanishes when results are extrapolated to $m_2 = 0$, but with the important difference that $M_2/(Z + 1)$ appears in place of M_2. This means that we can use all the methods applicable to non-electrolyte two-component systems, but we shall obtain from the experiments not the molecular weight M_2, but $M_2/(Z + 1)$. It would not ordinarily be possible to obtain M_2 from this quantity because Z is generally unknown, especially in solutions of low ionic strength, because of the tendency of the macro-ion to associate with its counterions; i.e., any real macro-electrolyte is likely to be incompletely dissociated so that the true charge Z' is unknown even if the stoichiometric parameter Z in the formula PX_Z is known.

When m_3 is not zero the last term of equation 16–40 must be included. A counterbalancing gain, however, is that the term $m_2 Z^2/(m_3 + Z m_2)$ now becomes zero when m_2 becomes zero. Moreover, in most ionic solutions, unless there is some strong specific interaction, the derivatives $\partial \ln \gamma_i / \partial m_j$ become negligibly small when m_3 is moderately large. All such terms may therefore be neglected, and equation 16–40 becomes

$$\frac{\mathscr{R}T}{m_2} \frac{dm_2}{dr} \left(1 + \frac{m_2 Z^2}{m_3 + Z m_2}\right) = M_2(1 - \bar{v}_2 \rho)\omega^2 r - \frac{Z \mathscr{R}T}{m_3 + Z m_2} \frac{dm_3}{dr}$$

$$(16\text{–}42)$$

Equation 16–16 for component 3 may now be written, with derivatives of the activity coefficients again neglected, as

$$\frac{\mathscr{R}T}{m_3} \frac{dm_3}{dr} \left(2 - \frac{Z m_2}{m_3 + Z m_2}\right) = M_3(1 - \bar{v}_3 \rho)\omega^2 r - \mathscr{R}T \left(\frac{\partial \mu_3}{\partial m_2}\right) \frac{dm_2}{dr} \quad (16\text{–}43)$$

This equation may be used for dm_3/dr in equation 16–42. We are particularly interested in the limiting form of equation 16–42 at $m_2 = 0$. Setting both m_2 and $dm_2/dr = 0$ we obtain

$$\frac{\mathscr{R}T}{m_2} \frac{dm_2}{dr} = M_2(1 - \bar{v}_2 \rho)\omega^2 r - \tfrac{1}{2} Z M_3(1 - \bar{v}_3 \rho)\omega^2 r \quad (16\text{–}44)$$

The same equation may be used at finite concentrations m_2, provided that the value of M_2 is regarded as $M_2^{(app)}$, similar to that defined in equation 16–21, the true value of M_2 being obtained by extrapolation of $M_2^{(app)}$ to the limit, $m_2 = 0$.

Equation 16–44 again points out the important difference between sedimentation equilibrium and osmotic pressure. The latter yields the true molecular weight in the present system as m_2 approaches zero (equation 14–24). Equation 16–44, on the other hand, is like the two-component equation for a non-ionic solute (equations 16–19 or 16–21), except that M_2 has been replaced by

$$M_2 - \tfrac{1}{2}ZM_3(1 - \bar{v}_3\rho)/(1 - \bar{v}_2\rho) \qquad (16\text{–}45)$$

so that the standard analysis does *not* give the true molecular weight. The difference from the true molecular weight may, however, become small. For typical inorganic salts (e.g., NaCl, KCl) in water, $\bar{v}_3 \sim 0.3$ and $M_3 \sim 65$. Thus the second term of equation 16–45 becomes (with $\bar{v}_2 \sim 0.75$, typical of proteins in water) approximately $90Z$. If Z is fairly small (say two charges per 10,000 grams), equation 16–45 lies within 2% of M_2. In the case of a protein, which can be made essentially electrically neutral by proper adjustment of pH, it should be possible to come even closer to M_2.

TABLE 16–2. Molecular Weights from Sedimentation Equilibrium

	Mol. Wt.
Sucrose (true Mol. Wt. = 342.3)[57]	341
Ribonuclease[57]	13,700
β-Lactoglobulin[61]	38,000
Ovalbumin[55]	43,500
Serum albumin[62]	68,000

It is concluded then that the sedimentation equilibrium method for a two-component system, as outlined earlier, may be applied to a macro-molecular electrolyte such as a protein, in the presence of an inorganic salt, provided that the salt concentration is moderately high and provided that Z is low, i.e., in the case of proteins, provided that the pH is close to that of the isoelectric point.

If a state of low charge cannot be attained (e.g., if a strong polyelectrolyte is considered or if a solute is insoluble unless Z is large), the method can still be applied if the components are defined according to the Scatchard procedure outlined in section 14a. As was pointed out on p. 224, the use of this procedure for expressing the composition of a solution involves use of a μ_2 much closer to μ_p than the μ_2 defined by equation 16–34. As a result M_2 becomes much closer to M_p also. To obtain the correct choice of components for the Scatchard procedure requires, however, that Z be known, and, as pointed out earlier, this may not be easy. Johnson, Kraus, and Scatchard[60] have however developed an iterative procedure which will work even when Z is only approximately known.

The sedimentation equilibrium method has been applied to a number of proteins in aqueous salt solutions. Some results are given in Table 16–2.

In considering these results, it is worth noting that some of the proteins listed have also had their molecular weights measured by osmotic pressure. In each case the results of Table 16–2 agree within experimental error with molecular weights obtained by osmotic pressure, listed in Tables 13–1 and 14–3. If these proteins were substances containing molecules of a variety of molecular weights, the osmotic pressure would yield the number-average molecular weight, whereas sedimentation equilibrium would yield the weight or z-average, depending on the procedure used. These averages would be different (see, for instance, the polystyrene data of Table 16–1). The fact that the same values are obtained is thus proof that these proteins are essentially homogeneous with respect to molecular weight. (The same conclusion also follows from the near linearity of plots of the type of Fig. 16–2.) This subject will be discussed further in connection with the use of light scattering to determine molecular weights (see Table 17–1).

16g. Archibald's method. We have so far made no mention of the major drawback to the sedimentation equilibrium method, which is the fact that a long time is required to reach equilibrium. Although this time has been reduced by modern techniques, such as the use of a short solution column,[57] the time required is still greater than a day for all but the smallest macromolecules. The necessity to keep an ultracentrifuge in operation, at constant speed and temperature, for such a length of time is a distinct disadvantage.

It is for this reason of interest to mention here that the basic equation for sedimentation equilibrium, equation 16–19, applies to two points of a sedimentation cell (at the meniscus and at the bottom) at all stages of a sedimentation experiment, even within a few minutes of the beginning of an experiment. This fact was first pointed out by Archibald,[63] and it may be used to obtain molecular weights which have essentially the same validity as those determined at equilibrium, though not as good an accuracy. Archibald's method is considered in detail in connection with the discussion of sedimentation velocity in section 22, and results obtained by its use are given in Table 22–3.

16h. Sedimentation equilibrium in a density gradient. If a solution of a low molecular weight solute is placed in a centrifugal field of sufficient strength, an appreciable concentration gradient will result. This will lead to a gradient of density. For example, if 7.7 molal CsCl in water is centrifuged at 45,000 r.p.m. in a cell placed 6.5 cm. from the center of rotation, a density gradient of 0.12 g/cc./cm. will be established at equilibrium: ρ will range from about 1.64 to 1.76 g./cc. if the liquid column is 1 cm. high.

We suppose now that the density of the solution is at some point r_0 equal to the reciprocal of the partial specific volume of some macromolecular substance which has been added to the solution. Then no net force will act on the macromolecular particles at r_0. At $r < r_0$, ρ will be less than $1/\bar{v}_2$ (component 2 is the macromolecular solute), $\rho\bar{v}_2 < 1$, and the particles will sediment towards the bottom of the cell. At $r > r_0$, on the other hand, $\rho\bar{v}_2 > 1$ and the particles will rise. At equilibrium all macromolecules will be distributed in a narrow band about r_0.

Meselson, Stahl, and Vinograd[65] have calculated the equilibrium distribution, with the assumption that the equilibration of the low molecular weight solute and the macromolecular particles proceed independently, i.e., their analysis is based on equation 16–21 rather than the correct relation for three-component systems, equation 16–26. (The statements of the preceding paragraph are also based on equation 16–21; otherwise the sedimenting force would not vanish *exactly* when $\rho\bar{v}_2 = 1$.) They limit their derivation to the situation in which the macromolecules exist at equilibrium within a narrow range of r about r_0, so that only small values of $r - r_0$ are of interest and ρ may be written as

$$\rho = 1/\bar{v}_2 + (d\rho/dr)(r - r_0)$$

with $d\rho/dr$ a constant. Equation 16–21 may then be rewritten as

$$\frac{d \ln c_2}{d(r - r_0)} = - \frac{M_2^{(\mathrm{app})}\omega^2\bar{v}_2(d\rho/dr)}{\mathscr{R}T}[r_0 + (r - r_0)](r - r_0)$$

We may neglect $r - r_0$ in comparison with r_0 and obtain, on integration,

$$c_2/(c_2)_{r=r_0} = e^{-(r-r_0)^2/2\sigma^2} \tag{16–46}$$

where

$$\sigma^2 = \mathscr{R}T/\bar{v}_2(d\rho/dr)\omega^2 r_0 M_2^{(\mathrm{app})} \tag{16–47}$$

Equation 16–46 represents a Gaussian distribution about $r = r_0$. Equation 16–47 shows that $M_2^{(\mathrm{app})}$ can be evaluated from σ. The value of $\bar{v}_2$ is obtained directly as $1/r_0$, and $d\rho/dr$ is determinable in a number of ways, most of which are based on the supposition that its value is not appreciably affected by the presence of the macromolecules.

The density gradient method does not possess an advantage over ordinary sedimentation equilibrium in the determination of molecular weight, especially as any slight heterogeneity among the sedimenting molecules will result in broadening of the distribution which cannot be readily distinguished from broadening due to a large value of σ. The method is, however, a powerful one for distinguishing between macromolecules with markedly different values of $\bar{v}_2$. Of particular importance is the fact that a mixture of macromolecules characterized by different values of $\bar{v}_2$, i.e., by different effective densities, will in the absence of strong interactions separate into discrete bands, each located at a value of r at which $\rho = 1/\bar{v}_i$.

The density gradient technique has found its most important application in the study of nucleic acids. It has led to the discovery that DNA preparations from different bacteria have different effective densities and that these densities may be correlated with differences in the base-content of the DNA molecules. DNA molecules with a high content of guanine and cytosine (see p. 9) have a relatively high density, whereas those with a high content of adenine and thymine have a low density.[66,67] The method has also been used to separate nucleic acids labeled with the N^{15} isotope of nitrogen from normal N^{14} nucleic acids.[68]

In connection with the experiments discussed in this section, it should be noted that nucleic acids have a very high intensity of absorption for ultraviolet light, so that a light absorption method, yielding a measure of concentration directly, is usually used to analyze the equilibrium distribution in the sedimentation cell.

General References

1. F. H. MacDougall, *Thermodynamics and Chemistry*, 3rd. ed., John Wiley and Sons, New York, 1939; E. A. Guggenheim, *Thermodynamics*, Interscience Publishers, New York, 1950; W. J. Moore, *Physical Chemistry*, 2nd. ed., Prentice-Hall, Englewood Cliffs, N.J., 1955; etc.
2. *The Collected Works of J. Willard Gibbs*, Longmans, Green and Co., New York, 1928.
3. E. A. Guggenheim, *Mixtures*, Oxford University Press, 1952.
4. J. H. Hildebrand and R. L. Scott, *The Solubility of Non-electrolytes*, 3rd. ed., Reinhold Publishing Corp., New York, 1950.
5. H. Tompa, *Polymer Solutions*, Butterworths Scientific Publications, London, 1956,
6. P. J. Flory, *Principles of Polymer Chemistry*, Cornell University Press, Ithaca, 1953. Chs. XII and XIII; P. J. Flory and W. R. Krigbaum, "Thermodynamics of High Polymer Solutions," *Ann. Rev. Phys. Chem.*, **2**, 383 (1951).
7. R. O. Bonnar, M. Dimbat and F. H. Stross, *Number Average Molecular Weights*, Interscience Publishers, New York, 1958.
8. E. J. Cohn and J. T. Edsall, *Proteins, Amino Acids and Peptides*, Reinhold Publishing Corp., New York, 1943, Chs. 3, 8, 9, 23, 24.
9. J. W. Williams, K. E. Van Holde, R. L. Baldwin, and H. Fujita, "The Theory of Sedimentation Analysis," *Chem. Revs.*, **58**, 715 (1958).

Specific References

10. W. G. McMillan and J. E. Mayer, *J. Chem. Phys.*, **13**, 276 (1945).
11. J. H. Hildebrand, *J. Chem. Phys.*, **15**, 225 (1947).
12. P. J. Flory, *J. Chem. Phys.*, **10**, 51 (1942).

13. M. L. Huggins, *J. Phys. Chem.*, **46**, 151 (1942); *Ann. N.Y. Acad. Sci.*, **41**, 1 (1942); *J. Am. Chem. Soc.*, **64**, 1712 (1942).
14. J. J. van Laar, *Sechs Vorträge über das thermodynamische Potential*, Braunschweig, 1906; J. J. van Laar and R. Lorenz, *Z. anorg. Chem.*, **146**, 42 (1925).
15. B. H. Zimm, *J. Chem. Phys.*, **14**, 164 (1946).
16. L. Onsager, *Ann. N.Y. Acad. Sci.*, **51**, 627 (1949).
17. G. V. Schulz, *Z. Naturforsch.*, **2a**, 348 (1947).
18. P. J. Flory and W. R. Krigbaum, *J. Chem. Phys.*, **18**, 1086 (1950).
19. E. A. Guggenheim, *Proc. Roy. Soc.*, A **183**, 213 (1944).
20. B. H. Zimm, W. H. Stockmayer, and M. Fixman, *J. Chem. Phys.*, **21**, 1716 (1953).
21. T. A. Orofino and P. J. Flory, *J. Chem. Phys.*, **26**, 1067 (1957).
22. M. J. Newing, *Trans. Faraday Soc.*, **46**, 613 (1950).
23. J. Ferry, G. Gee and L. R. G. Treolar, *Trans. Faraday Soc.*, **41**, 340 (1945).
24. G. Gee and W. J. C. Orr, *Trans. Faraday Soc.*, **42**, 507 (1946).
25. W. H. Stockmayer and E. F. Casassa, *J. Chem. Phys.*, **20**, 1560 (1952).
26. R. H. Wagner in A. Weissberger (ed.), *Physical Methods of Organic Chemistry*, 2nd. ed., part I, Interscience Publishing Co., New York, 1949, p. 487–550.
27. G. Gee, *Advances in Colloid Sci.*, **2**, 145 (1946).
28. G. Scatchard, A. C. Batchelder, and A. Brown, *J. Am. Chem. Soc.*, **68**, 2320 (1946).
29. W. R. Krigbaum and P. J. Flory, *J. Am. Chem. Soc.*, **75**, 1775, 5254 (1953).
30. W. R. Krigbaum, *J. Am. Chem. Soc.*, **76**, 3758 (1954).
31. G. Scatchard, *J. Am Chem. Soc.*, **68**, 2315 (1946).
32. F. G. Donnan, *Z. Elektrochem.*, **17**, 572 (1911).
33. G. Scatchard, A. C. Batchelder, A. Brown, and M. Zosa, *J. Am. Chem. Soc.*, **68**, 2610 (1946).
34. J. G. Kirkwood and J. B. Shumaker, *Proc. Natl. Acad. Sci., U.S.*, **38**, 863 (1952).
35. L. K. Christensen, *Compt. rend. trav. lab. Carlsberg, ser. chim.*, **28**, 37 (1952).
36. A. V. Güntelberg and K. Linderstrøm-Lang. *Compt. rend. trav. lab. Carlsberg, ser. chim.*, **27**, 1 (1949).
37. G. S. Adair, *Proc. Roy. Soc.* A **120**, 573 (1928).
38. J. Roche, A. Roche, G. S. Adair and M. E. Adair, *Biochem. J.*, **29**, 2576 (1935).
39. H. Portzehl, *Z. Naturforsch.*, **5b**, 75 (1950).
40. H. Gutfreund, *Biochem. J.*, **42**, 544 (1948); **50**, 564 (1952).
41. A. Munster, *J. Polymer Sci.*, **8**, 633 (1952).
42. C. T. Greenwood, *Advances in Carbohydrate Chem.*, **7**, 289 (1952); **11**, 335 (1956).
43. H. B. Bull and B. T. Currie, *J. Am. Chem. Soc.*, **68**, 742 (1946).
44. L. Mandelkern, R. R. Garrett and P. J. Flory, *J. Am. Chem. Soc.*, **74**, 3949 (1952).
45. E. M. Frith and R. F. Tuckett, *Trans. Faraday Soc.*, **40**, 251 (1944); R. B. Richards, *ibid.*, **41**, 127 (1945).
46. P. J. Flory, *J. Chem. Phys.*, **17**, 223 (1949).
47. R. F. Boyer and R. S. Spencer, *Advances in Colloid Sci.*, **2**, 1 (1946).
48. E. Fredericq and H. Neurath, *J. Am. Chem. Soc.*, **72**, 2684 (1950).
49. E. J. Cohn, *Chem. Revs.*, **19**, 241 (1936).
50. T. L. McMeekin, E. J. Cohn and M. H. Blanchard, *J. Am. Chem. Soc.*, **59**, 2717 (1937).
51. J. A. V. Butler, *J. Gen. Physiol.*, **24**, 189 (1940).
52. A. R. Shultz and P. J. Flory, *J. Am. Chem. Soc.*, **74**, 4760 (1952).
53. H. P. Frank and H. F. Mark, *J. Polymer Sci.*, **17**, 1 (1955).

54. T. L. Hill, *J. Chem. Phys.*, **23**, 623, 2270 (1955); *Discussions, Faraday Soc.*, No. 21, 31 (1956); *J. Am. Chem. Soc.*, **80**, 2923 (1958); D. Stigter and T. L. Hill, *J. Phys. Chem.*, **63**, 551 (1959).
55. T. Svedberg and K. O. Pedersen, *The Ultracentrifuge*, Oxford University Press, 1940.
56. R. J. Goldberg, *J. Phys. Chem.*, **57**, 194 (1953).
57. K. E. van Holde and R. L. Baldwin, *J. Phys. Chem.*, **62**, 734 (1958).
58. W. D. Lansing and E. O. Kraemer, *J. Am. Chem. Soc.*, **57**, 1369 (1935).
59. M. Wales, *J. Phys. and Colloid Chem.*, **52**, 235 (1948); M. Wales, J. W. Williams, J. O. Thompson and R. H. Eward, *ibid.*, **52**, 983 (1948); M. Wales, F. T. Adler and K. E. van Holde, *ibid.*, **55**, 145 (1951); M. Wales, *ibid.*, **55**, 282 (1951).
60. J. S. Johnson, K. A. Kraus, and G. Scatchard, *J. Phys. Chem.*, **58**, 1034 (1954).
61. K. O. Pedersen, *Biochem. J.*, **30**, 961 (1936).
62. T. Svedberg and B. Sjogren, *J. Am. Chem. Soc.*, **50**, 3318 (1928).
63. W. J. Archibald, *J. Phys. Chem.*, **51**; 1204 (1947).
64. L. Mandelkern, L. C. Williams, and S. G. Weissberg, *J. Phys. Chem.*, **61**, 271 (1957).
65. M. Meselson, F. W. Stahl, and J. Vinograd, *Proc. Natl. Acad. Sci., U.S..*, **43**, 581 (1957).
66. N. Sueoka, J. Marmur and P. Doty, *Nature*, **183**, 1429 (1959).
67. R. Rolfe and M. Meselson, *Proc. Natl. Acad. Sci., U.S.*, **45**, 1039 (1959).
68. M. Meselson and F. W. Stahl, *Proc. Natl. Acad. Sci., U.S.*, **44**, 671 (1958).

5

LIGHT SCATTERING

The scattering of electromagnetic radiation by matter is an important topic in the classical theory of electromagnetism. One aspect of this topic, the scattering of x-rays by *ordered* solids, has already been considered in sections 3 and 4. The present chapter is primarily concerned with a very different aspect: the scattering of light from gases and liquid solutions, an essential feature being that scattering particles in which we are interested are *randomly* located, so that each is an independent source of scattered radiation.

The discussion will make use of relations from the general theory of electromagnetism, reference to which may be found in a number of standard textbooks.[1,2] Advanced treatments of the subject are contained in the treatises of Born[3] and Stratton.[4] A more elementary but comprehensive treatment, dealing with all aspects of light scattering except its applications to macromolecular chemistry, has been presented by van de Hulst.[5] The application of light scattering to chemical problems, including macromolecules, has been reviewed in a number of places, e.g., by Oster,[6] Doty and Edsall,[7] and Stacey.[8] A recent review by Geiduschek and Holtzer[9] provides an authoritative review of the use of light scattering in the determination of macromolecular properties.

17. LIGHT SCATTERING BY PARTICLES SMALL COMPARED TO THE WAVELENGTH OF LIGHT

17a. An outline of the derivation of Rayleigh's equation for scattering by dilute gases. If any particle in space is subjected to an electric field of strength $\mathscr{E}$, its constituent electrons become subject to a force in one direction and its constituent nuclei to a force in the opposite direction. Thus a dipole moment p is induced in the particle, which, if the particle is optically isotropic, will be parallel in direction to the electric field. The

275

magnitude of the dipole moment is proportional to the electric field strength, the proportionality constant, α, being known as the *polarizability* of the particle; i.e.,

$$p = \alpha\mathscr{E} \qquad (17\text{-}1)$$

We shall consider here an ideal gas, i.e., a large number of independent particles in vacuum, in the path of a plane-polarized beam of light. The general equation for the electric field of such a light wave is

$$\mathscr{E} = \mathscr{E}_0 \cos 2\pi(\nu t - x/\lambda) \qquad (17\text{-}2)$$

where $\mathscr{E}_0$ is the maximum amplitude; ν the frequency; λ the wavelength, measured in the medium of propagation (here a vacuum); t the time; and x the location along the line of propagation. The direction of the field is always the same: at right angles to the direction of propagation and in the plane of polarization. The electric field is of course periodic, the same value of $\mathscr{E}$ being obtained for all integral values of $\nu t - x/\lambda$, i.e., at time intervals of $1/\nu$ and at intervals of distance λ.

We are confining ourselves in the present section to particles much smaller than λ. Thus each individual particle may be represented by a single value of x in equation 17–2; the electric field at each point of the particle, at any instant, will be identical. We shall treat in section 18 the extension to larger particles, for which different values of x have to be assigned to different portions, so that phase differences exist between the light striking different portions at any instant.

Combining equations 17–1 and 17–2 we see at once that the incident light beam will induce an *oscillating* dipole in any particle in its path, such that

$$p = \alpha\mathscr{E}_0 \cos 2\pi(\nu t - x/\lambda) \qquad (17\text{-}3)$$

An oscillating dipole, however, is itself a source of electromagnetic radiation (e.g., see Frank[1]). This new radiation is what we mean by scattered radiation. Its field strength is proportional to d^2p/dt^2. (The first derivative dp/dt clearly represents an electric current, which, if steady, would give rise to a constant magnetic field. The oscillating dipole produces an oscillating magnetic field, which, in turn, is accompanied by an oscillating electric field. The combination constitutes an electromagnetic wave.) The scattered radiation is a spherical wave, extending in all directions, but the field strength depends on the direction. At large distances from the dipole, where the distance r from the observer to all parts of the dipole will be almost equal, the field due to the positive end of the dipole will almost cancel that due to the negative end. The residual field is perpendicular to the line from the observer to the dipole (as it must be if the disturbance emanating from the dipole is a spherical electromagnetic wave) and its strength, at any given value of r, is proportional to $\sin \theta_1$, where θ_1 is the angle between the dipole axis and the line joining the point of observation to the dipole.

Another property of the scattered radiation is that its field strength must vary as $1/r$. This follows from the law of conservation of energy; the total energy flux through any spherical shell about the dipole must be constant. Thus the intensity (energy flux per square centimeter) must vary as $1/r^2$, and, since the intensity is proportional to $\mathscr{E}^2$ (see below), $\mathscr{E}$ must vary as $1/r$.

Differentiating equation 17–3 to obtain d^2p/dt^2, introducing the factor $\sin \theta_1/r$, and dividing by the square of the velocity of light, $\tilde{c}^2$, for dimensional correctness, we thus obtain

$$\mathscr{E}_s = \frac{4\pi^2 \nu^2 \alpha \mathscr{E}_0 \sin \theta_1}{\tilde{c}^2 r} \cos 2\pi(\nu t - x/\lambda) \qquad (17\text{–}4)$$

where $\mathscr{E}_s$ is used to designate the field strength of the scattered radiation. Equation 17–4 shows that the scattered radiation has *the same frequency as the incident light*, with an amplitude varying both with distance and angle between the observer and the scattering point.

We have assumed, of course, that the frequency chosen for the incident light is not one that is absorbed by the scattering medium. We have also ignored the possibility of interaction between the incident light and the scattering particle which would lead to absorption of part of the light energy and to a scattered wave of different frequency from the incident radiation. Such interaction occurs, of course, and is known as the Raman effect. The intensity of light scattered in this way is, however, so small that it is quite negligible if the measurement with which we are concerned is a measurement of total scattered intensity, as is true in the case under discussion here. The Raman effect is observed only if measurements are made as a function of frequency and then only after long exposure, when lines will be observed at frequencies other than that of the incident radiation.

The experimental measure of the energy in a light wave is the *intensity*, i.e., the energy which falls on 1 cm² of area per second. This quantity, by Poynting's theorem,[1–5] is proportional to the value of $\mathscr{E}^2$, averaged over a period of the vibration (i.e., from $t = 0$ to $t = 1/\nu$). By use of equation 17–2 we thus obtain the intensity I_0 of the incident light; by use of equation 17–4 we obtain the intensity i_s of the scattered light. It is the ratio of the two which is of interest, this ratio being

$$\frac{i_s}{I_0} = \frac{16\pi^4 \alpha^2 \sin^2 \theta_1}{\lambda^4 r^2} \qquad (17\text{–}5)$$

where the wavelength λ (in vacuo) has been written in place of $\tilde{c}/\nu$.

Equation 17–5 was first derived by Lord Rayleigh[10] in 1871. Its most significant feature is the inverse fourth-power dependence on the wavelength of the incident light. The equation accounts for a number of well-known phenomena, e.g., the fact that sunlight scattered by the earth's atmosphere is greatly enriched in the blue wavelengths.

The polarizability α, which occurs in equation 17–5, is, of course, equivalent to the quantities α_d or α_∞ which were defined in section 6b. It may be related to the dielectric constant (at the frequency ν) by equation 6–10, which is applicable here because the electric field with which we are concerned is a local field, rather than a field applied across the plates of a condenser. The dielectric constant, in turn, may be related to the refractive index $\tilde{n}$ by equation 6–12, so that

$$\tilde{n}^2 - 1 = 4\pi N\alpha \tag{17-6}$$

Finally, in a dilute gas, the refractive index will be close to unity (the value of $\tilde{n}$ for a vacuum) so that $\tilde{n}$ may be expanded in a Taylor series with neglect of all but the first two terms; i.e., we may write

$$\tilde{n} = 1 + (d\tilde{n}/dc)c \tag{17-7}$$

or

$$\tilde{n}^2 = 1 + 2(d\tilde{n}/dc)c \tag{17-8}$$

where c is the concentration of particles in the gas, in units of grams per cubic centimeter. Combining equation 17–8 with equation 17–6, we get

$$\alpha = \frac{c(d\tilde{n}/dc)}{2\pi N} = \frac{M(d\tilde{n}/dc)}{2\pi \mathcal{N}} \tag{17-9}$$

where M is the molecular weight of the particles, N their number per cubic centimeter, and $\mathcal{N}$ Avogadro's number, so that $M/\mathcal{N} = c/N$ is equal to the mass per particle.

Equation 17–9 may now be combined with equation 17–5 to give the intensity of light scattered from a single particle,

$$\frac{i_s}{I_0} = \frac{4\pi^2 M^2 \sin^2 \theta_1 (d\tilde{n}/dc)^2}{\mathcal{N}^2 \lambda^4 r^2} \tag{17-10}$$

This equation gives us the second important feature of the intensity of scattered light: its proportionality to the square of the molecular weight of the scattering particles, coupled with proportionality to $(d\tilde{n}/dc)^2$.

A more useful result is the intensity of scattering from unit volume, assuming that the particles are randomly located so that the total intensity is the sum of that of the individual particles. With $N = \mathcal{N}c/M$ particles per cubic centimeter,

$$\frac{i_s}{I_0} = \frac{4\pi^2 \sin^2 \theta_1 (d\tilde{n}/dc)^2 Mc}{\mathcal{N} \lambda^4 r^2} \tag{17-11}$$

The angular dependence of scattering, as given by the term $\sin^2 \theta_1$, is shown in Fig. 17–1. It can be seen that, if vertically polarized light is used and observations are made in the horizontal plane, the scattered intensity will be independent of angle and will be given by equation 17–11 with $\sin^2 \theta_1 = 1$.

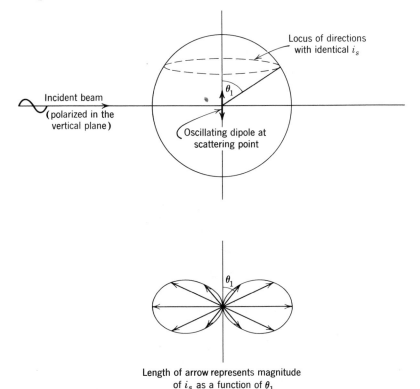

Length of arrow represents magnitude
of i_s as a function of θ_1

Fig. 17–1. Angular dependence of the intensity of scattered light, for polarized incident light. In the lower figure, the length of the arrows represents the magnitude of i_s as a function of the angle θ_1.

Incident light used for scattering experiments is more often unpolarized rather than polarized. An unpolarized light beam is equivalent to the superposition of two plane-polarized beams, independent in phase and of equal intensity, with their planes of polarization perpendicular to one another. The intensity of scattering, which will be called i_θ when the incident light is unpolarized, is therefore the sum of two terms of the form of equation 17–11. Each term represents the scattering from half the incident intensity; i.e., the left side of equation 17–11 will be $i_s/\frac{1}{2}I_0$. The two terms will be identical, except that one contains $\sin^2 \theta_1$ and the other $\sin^2 \theta_2$, where, if the direction of the incident beam is designated as the x axis of a rectangular coordinate system, θ_1 and θ_2 represent the angles made by the line of observation with the y and z axes respectively. The resultant of $\sin^2 \theta_1 + \sin^2 \theta_2$ is then $1 + \cos^2 \theta$, where θ is the angle

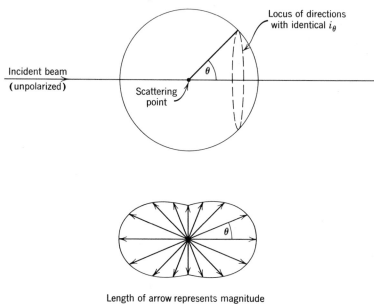

Length of arrow represents magnitude
of i_θ as a function of θ

Fig. 17–2. Angular dependence of the intensity of scattered light, for an unpolarized incident beam. In the lower figure, the length of the arrows represents the magnitude of i_θ as a function of the angle θ.

between the line of observation and the x axis, i.e., between the line of observation and the direction of the incident light. Thus, per unit volume,

$$\frac{i_\theta}{I_0} = \frac{2\pi^2(1 + \cos^2 \theta)(d\tilde{n}/dc)^2 Mc}{\mathcal{N} \lambda^4 r^2} \qquad (17\text{–}12)$$

The angular dependence of scattering of unpolarized incident light, as given by equation 17–12, is shown in Fig. 17–2.

17b. Scattering from transparent crystals and from pure liquids. In the previous section the scattering points were considered entirely independent of one another; in particular, random location with respect to one another was implicit in the discussion. In a crystalline solid the opposite situation prevails; individual scattering particles are rigidly fixed in space relative to one another. The result of this is that there is destructive interference between light scattered from individual particles. Where the wavelength of light is much greater than the distance between particles, the interference is complete, and no scattered light is observed at all. The reason for this is that all the particles in the crystal can always be paired off in such a way that the light paths from the two particles in each pair to an observer at any particular viewing angle θ will differ by essentially exactly one-half wavelength.

By Bragg's law constructive interference occurs only at incident angles θ' given by the equation $\sin \theta' = \lambda/2d$ where d is a regularly repeated distance between neighboring identical scattering points; i.e., its magnitude is of the order of molecular dimensions. It is at once obvious that this equation has no solutions if $\lambda \gg d$. (See section 3.)

Pure liquids are intermediate between crystals and gases. They are ordered but not completely ordered, and, as a result, they give rise to some scattering. The intensity is, however, very much less than it would be for a gas of equal density. The problem has been treated by the method of fluctuations by Smoluchowski.[11] The principle of the method can be seen if we consider two small equal volume elements of liquid (dimensions $\ll \lambda$) separated by just the right distance so that the light paths to an observer differ by half a wavelength. If they were volume elements of a crystal, they would each possess the same number of scattering particles, and complete destruction of the scattered light would occur. In a liquid, however, the number of particles in each volume element is the same only when averaged over a period of time. At any given instant the partially random motion of the particles in the liquid gives rise to fluctuations in the density at any particular point so that the number of particles in one volume element differs from that in the other volume element. (These fluctuations may be considered as fluctuations in local pressure and temperature.) There is thus an excess scattering of light from one element which is not canceled by that from the other element. We shall not go into the details here (for a review see Oster[6]). In principle the solution of the problem resembles that given for concentration fluctuations in a solution in the following section.

17c. Scattering from macromolecular solutions (two-component systems). The interest of macromolecular chemists in light scattering arises from the fact that the scattering intensity from a single particle, by equation 17–10, is proportional to the square of the molecular weight. Thus light scattering appears ideally suited to the study of macromolecules in the presence of smaller molecules. The contribution of the latter should be relatively small.

In order to obtain an equation for the intensity of light scattered by macromolecules in a two-component solution (over and above the small amount scattered by solvent) we might first of all suppose that the solution is ideal, i.e., that the macromolecules are quite independent. In that case the same derivation can be followed as was used for calculating the scattering from a gas, except that the scattering molecules are now immersed in a medium of refractive index $\tilde{n}_0$ (the refractive index of the solvent) instead of free space for which the refractive index is unity. Thus $\tilde{n}^2 - 1$ (equation 17–6) becomes a measure of *total* polarizability per unit volume, including that of the solvent, and $\tilde{n}^2 - \tilde{n}_0^2$ becomes the appropriate equivalent for $4\pi N\alpha$ if N and α are to represent the number of solute molecules per cubic centimeter and their molecular polarizability. (Strictly speaking, α is the difference between the polarizability of a solute molecule and the solvent which it replaces.) In place of equation 17–8, we thus have $\tilde{n}^2 - \tilde{n}_0^2 = 2\tilde{n}_0(d\tilde{n}/dc)c$, and the final equation replacing equation 17–12 becomes

$$\frac{i_\theta}{I_0} = \frac{2\pi^2(1 + \cos^2\theta)\tilde{n}_0^2(d\tilde{n}/dc)^2 Mc}{\mathcal{N}\lambda^4 r^2} \qquad (17\text{--}13)$$

where M is the molecular weight of the macromolecular solute and c its concentration in grams per cubic centimeter.

Macromolecular solutions are not ideal, however, and the preceding derivation is in any event not rigorous. A more illuminating approach to the problem must thus be sought, and it is provided by the fluctuation method discussed above in connection with the scattering of light by pure liquids. We present below the adaptation of this method to solutions, as developed by Einstein[12] and Debye.[13]

We consider the solution made up of small elements of volume ψ. The dimensions of the volume elements are to be much smaller than the wavelength of light, so that the entire element can be considered a single scattering source. The elements, however, are large enough to contain many solvent molecules and a few solute molecules.

A two-component system requires just a single variable of composition, for which we choose the concentration of macromolecular substance in grams per cubic centimeter. The average value of this quantity, over the entire solution, is designated by c', the fluctuating concentration in a volume element being designated by $c = c' + \delta c$, where δc, which may be positive or negative, is the concentration fluctuation. The fluctuations in concentration occur, of course, in a completely random fashion. Corresponding to the fluctuations in concentration there will be fluctuations in polarizability and refractive index. (Fluctuations in these quantities arise also from the fluctuations in local pressure and temperature. We have chosen to ignore them, however, assuming them to be identical to the density fluctuations occurring in pure solvent.) Thus, if α' is the average polarizability of a volume element, $\alpha = \alpha' + \delta\alpha$ represents the actual polarizability of a given element at any instant, $\delta\alpha$ being the contribution due to the concentration fluctuation, δc.

By equation 17–5, then, the light scattered from any one volume element at a particular instant is

$$\frac{i_s}{I_0} = \frac{16\pi^4(\alpha' + \delta\alpha)^2 \sin^2 \theta_1}{\lambda^4 r^2} \qquad (17\text{–}14)$$

We wish to know the time average of i_s/I_0 for N volume elements per cubic centimeter, where N clearly equals $1/\psi$, ψ being the volume of each element. Writing $(\alpha' + \delta\alpha)^2 = \alpha'^2 + 2\alpha'\,\delta\alpha + (\delta\alpha)^2$ we note that the contribution due to α'^2 is the same for each element and thus cancels for the same reason which makes the scattering from a perfect crystal equal to zero. The average value of $N\,\delta\alpha$ must also vanish since positive and negative values of $\delta\alpha$ are equally likely. We are thus left, for the scattering per unit volume containing $N = 1/\psi$ volume elements, with

$$\frac{i_s}{I_0} = \frac{16\pi^4\overline{(\delta\alpha)^2} \sin^2 \theta_1}{\lambda^4 r^2 \psi} \qquad (17\text{–}15)$$

where $\overline{(\delta\alpha)^2}$ is the time average of $(\delta\alpha)^2$ for a single volume element or, what amounts to the same thing, the average value for a large number of elements at any instant.

If we let temperature, pressure, and solute concentration be the independent variables affecting α in a volume element, the fluctuation $\delta\alpha$ may be expressed in terms of fluctuations in these variables; i.e.,

$$\delta\alpha = \left(\frac{\partial\alpha}{\partial P}\right)_{T,c} \delta P + \left(\frac{\partial\alpha}{\partial T}\right)_{P,c} \delta T + \left(\frac{\partial\alpha}{\partial c}\right)_{T,P} \delta c \qquad (17\text{–}16)$$

The first two terms on the right-hand side of this equation are assumed for dilute solutions to be the same as the corresponding terms leading to scattering from pure solvent and are thus ignored. (Experimentally, the scattering of the solvent is deducted from that measured for a solution. The light scattering intensity as here discussed is often called the *excess* scattering intensity for that reason.)

By equation 17–6 we can express $\partial\alpha/\partial c$ in terms of the readily measurable quantity $\partial\tilde{n}/\partial c$. Substituting $N = 1/\psi$, we get

$$\left(\frac{\partial\alpha}{\partial c}\right)_{T,P} = \frac{\psi\tilde{n}}{2\pi}\left(\frac{\partial\tilde{n}}{\partial c}\right)_{T,P} \qquad (17\text{–}17)$$

so that, combining with equations 17–15 and 17–16,

$$\frac{i_s}{I_0} = \frac{4\pi^2\psi\tilde{n}^2(\partial\tilde{n}/\partial c)^2 \sin^2 \theta_1}{\lambda^4 r^2} \overline{(\delta c)^2} \qquad (17\text{–}18)$$

where $\overline{(\delta c)^2}$ is the *average* value of $(\delta c)^2$. Converting to unpolarized light, as in going from equation 17–11 to equation 17–12,

$$\frac{i_\theta}{I_0} = \frac{2\pi^2 \psi \tilde{n}^2 (\partial \tilde{n}/\partial c)^2 (1 + \cos^2 \theta)}{\lambda^4 r^2} \overline{(\delta c)^2} \tag{17-19}$$

It remains to compute $\overline{(\delta c)^2}$. This quantity will be found to depend on the way in which the free energy of the macromolecular solution varies with concentration, and, as a result, the final expression for the scattered intensity will be related to the colligative properties of the solution.

We calculate first the probability of a given fluctuation δc. This may be done in several ways. One method (which avoids the use of a statistical-mechanical treatment of fluctuations) is to consider all volume elements with a given value of δc as an individual chemical species. The abundance of elements of a given kind, relative to elements with the average concentration c', then becomes a problem in chemical equilibrium. The relative abundance has the form of an equilibrium constant which in turn is equal to $\exp(-\Delta F^\circ/kT)$, where ΔF° is the difference in the free energy (per element, not per mole of elements, since k rather than $\mathscr{R}$ appears in the denominator) between one element and the other in some suitable standard state. The standard state can be chosen as the state of unit mole fraction of elements, so that ΔF° becomes just the difference between the free energy (per element) in a solution of concentration $c' + c$ and that in a solution of concentration c'. This quantity would ordinarily be designated δF, so that the probability of a given value of δc becomes proportional to $\exp(-\delta F/kT)$.

We do not anticipate large fluctuations in concentration nor, hence, large values of δF. We may thus expand δF in a Taylor series and retain only the first two terms,

$$\delta F = \left(\frac{\partial F}{\partial c}\right)_{T,P} \delta c + \frac{1}{2!}\left(\frac{\partial^2 F}{\partial c^2}\right)_{T,P} (\delta c)^2 \tag{17-20}$$

But the average concentration about which fluctuations are occurring is the equilibrium concentration at constant pressure, so that $\partial F/\partial c = 0$. Hence the probability of a given value of δc becomes

$$e^{-\delta F/kT} = e^{-(\partial^2 F/\partial c^2)(\delta c)^2/2kT}$$

which, it may be noted, makes positive and negative values of δc equally probable.

The average value of $(\delta c)^2$ can now be written at once,

$$\overline{(\delta c)^2} = \frac{\displaystyle\int_{\delta c=0}^{\infty} (\delta c)^2 e^{-(\partial^2 F/\partial c^2)(\delta c)^2/2kT}\, d(\delta c)}{\displaystyle\int_{\delta c=0}^{\infty} e^{-(\partial^2 F/\partial c^2)(\delta c)^2/2kT}\, d(\delta c)} \tag{17-21}$$

Both of these integrals are well known (Appendix A), and we obtain on integration

$$\overline{(\delta c)^2} = kT/(\partial^2 F/\partial c^2)_{T,P} \qquad (17\text{--}22)$$

It remains to calculate $(\partial^2 F/\partial c^2)_{T,P}$. Let n_1 and n_2 be the number of moles of solvent and solute, respectively, in a volume element. These quantities are not independent but must be so picked as to make the volume of an element equal to ψ; i.e., $n_1 \bar{V}_1 + n_2 \bar{V}_2 = \psi$, or $dn_1 = -(\bar{V}_2/\bar{V}_1)\, dn_2$, where $\bar{V}_1$ and $\bar{V}_2$ are the partial molal volumes of solvent and solute at the prevailing temperature and the average concentration c'.

The free energy change accompanying any change in c at constant temperature and pressure is by definition $dF = \mu_1\, dn_1 + \mu_2\, dn_2 = [-(\bar{V}_2/\bar{V}_1)\mu_1 + \mu_2]\, dn_2$, where μ_1 and μ_2 are the chemical potentials of solvent and solute, respectively. The number of moles of solute per cubic centimeter is $n_2/\psi = c/M$, so that $dn_2 = (\psi/M)\, dc$ and

$$\left(\frac{\partial F}{\partial c}\right)_{T,P} = \left(\mu_2 - \frac{\bar{V}_2}{\bar{V}_1}\mu_1\right)\frac{\psi}{M}$$

Differentiation with respect to c then gives

$$\left(\frac{\partial^2 F}{\partial c^2}\right)_{T,P} = \frac{\psi}{M}\left[\left(\frac{\partial \mu_2}{\partial c}\right)_{T,P} - \frac{\bar{V}_2}{\bar{V}_1}\left(\frac{\partial \mu_1}{\partial c}\right)_{T,P}\right] \qquad (17\text{--}23)$$

But $d\mu_1$ and $d\mu_2$ are related by the Gibbs-Duhem equation, equation 11–6, $n_1\, d\mu_1 + n_2\, d\mu_2 = 0$, so that equation 17–23 becomes

$$\left(\frac{\partial^2 F}{\partial c^2}\right)_{T,P} = -\frac{\psi}{M}\left(\frac{n_1 \bar{V}_1 + n_2 \bar{V}_2}{n_2 \bar{V}_1}\right)\left(\frac{\partial \mu_1}{\partial c}\right)_{T,P}$$

or, since $n_2 M/(n_1 \bar{V}_1 + n_2 \bar{V}_2) = c$,

$$\left(\frac{\partial^2 F}{\partial c^2}\right)_{T,P} = -\frac{\psi}{c \bar{V}_1}\left(\frac{\partial \mu_1}{\partial c}\right)_{T,P} \qquad (17\text{--}24)$$

Introducing this result in equation 17–22, and, hence, in equation 17–19, we get

$$\frac{i_\theta}{I_0} = \frac{2\pi^2 \tilde{n}^2 (\partial \tilde{n}/\partial c)^2 (1 + \cos^2 \theta)c}{\lambda^4 r^2 [-(1/V_1 kT)(\partial \mu_1/\partial c)_{T,P}]} \qquad (17\text{--}25)$$

The result is seen to be independent of the volume ψ, as it must be.

The dependence of the chemical potential of the solvent on the concentration of solute may be expressed in terms of equation 11–21. By differentiation of the latter (with $\bar{V}_1 \equiv V_1{}^0$ in a very dilute solution)

$$-\frac{1}{\bar{V}_1 kT}\left(\frac{\partial \mu_1}{\partial c}\right)_{T,P} = \mathcal{N}\left(\frac{1}{M} + 2Bc + 3Cc^2 + \cdots\right)$$

where M is the molecular weight of the solute, so that equation 17–25 becomes

$$\frac{i_\theta}{I_0} = \frac{2\pi^2 \tilde{n}^2 (\partial \tilde{n}/\partial c)^2 (1 + \cos^2 \theta)c}{\mathcal{N} \lambda^4 r^2 (1/M + 2Bc + 3Cc^2 + \cdots)} \qquad (17\text{–}26)$$

For dilute solutions the difference between $\tilde{n}$ and the refractive index, $\tilde{n}_0$, of the solvent becomes negligible, so that $\tilde{n}^2$ may generally be replaced by $\tilde{n}_0{}^2$. At the limit, $c = 0$, equation 17–26 thus reduces to the ideal equation, equation 17–13, derived earlier.

Equation 17–26 is the fundamental equation for the scattering of unpolarized light by dissolved particles much smaller than the wavelength of light. The equivalent equation for polarized incident light would differ only in containing the factor $2 \sin^2 \theta_1$ in place of $1 + \cos^2 \theta$. In either case we see that the molecular weight and the virial coefficients (of the equation for the chemical potential) may be determined.

Equation 17–26 may be obtained by means of an exact molecular theory,[58] as well as by the method given here.

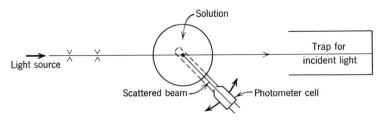

Fig. 17–3. Schematic diagram of the apparatus for light scattering measurements. The photometer cell may be rotated about an axis directly below the center of the vessel containing the solution. The photometer output is amplified and recorded on a galvanometer or recorder.

17d. Experimental determination of light scattering. Detailed descriptions of apparatus for making light scattering measurements are given by Debye,[14] by Zimm,[15] and by Brice, Halwer, and Speiser.[16] These papers, and papers by Carr and Zimm[17] and by Brice, Nutting, and Halwer,[18] discuss also the determination of constants of the instrument and correction for fluorescence and other factors which may interfere with the measurements. Figure 17–3 shows a diagrammatic sketch of a typical apparatus. The light source is usually a mercury arc, from which, by use of filters, we may obtain monochromatic radiation of wavelength 4358 or 5461 Å (these being the ones most commonly employed). The detecting device is a photomultiplier tube supported on an arm which may be rotated about the cell containing the scattering solution. The scattered intensity may thus be measured directly at any angle. The corresponding intensity of light scattered by solvent alone must also be measured and must be subtracted from that observed for the solution, since all our equations represent the difference in scattering between solution and solvent.

The incident intensity, I_0, will usually be several magnitudes higher than the scattered intensity. It may be measured on the same device, either by reduction of the intensity by a passage through a series of calibrated filters or by adjustment of the amplification factor of the detecting device.

Of the factors occurring in equation 17–26, the distance r from the photomultiplier tube to the cell containing the scattering solution is a constant of the instrument, and the volume of the solution from which the scattering occurs is known. The important factor $(\partial \tilde{n}/\partial c)_{T,P}$ is determined on a differential refractometer, using the same wavelength of light as is employed for the light scattering measurements. Its value will differ for each solute-solvent pair. Its concentration dependence in dilute solutions is usually quite negligible. (For recent discussion of the effect of temperature and wavelength, see, for example, Perlman and Longsworth.[19])

The final quantity determined is *Rayleigh's ratio*, $r^2 i_0/I_0$, or more conveniently the ratio

$$R_\theta \equiv \frac{r^2 i_\theta}{I_0(1 + \cos^2 \theta)} \qquad (17\text{–}27)$$

which is independent of the scattering angle and given by

$$R_\theta = \frac{\mathscr{K} c}{1/M + 2Bc + 3Cc^2 + \cdots} \qquad (17\text{–}28)$$

$\mathscr{K}$ being an optical constant given by

$$\mathscr{K} = 2\pi^2 \tilde{n}_0{}^2 (\partial \tilde{n}/\partial c)^2 / \mathscr{N} \lambda^4 \qquad (17\text{–}29)$$

The ordinary procedure (for relatively small solute particles) is to determine i_θ as a function of θ only in so far as it is necessary to prove that R_θ as given in equation 17–27 is indeed independent of θ. Subsequent measurements are made at just a single angle. To determine molecular parameters we rewrite equation 17–28 as

$$\frac{\mathscr{K} c}{R_\theta} = \frac{1}{M} + 2Bc + 3Cc^2 + \cdots \qquad (17\text{–}30)$$

and plot $\mathscr{K} c/R_\theta$ versus c. The information to be obtained from such a plot is essentially the same as that obtained from osmotic pressure measurements on two-component systems (equation 13–8, Fig. 13–2), the intercept giving the molecular weight, the limiting slope being a measure of the second virial coefficient.

The derivation here given limits equation 17–30 to two-component systems and to solute particles much smaller than λ. It will be seen in section 18, however, that it applies to larger solute particles as well, provided that we use values of $\mathscr{H}c/R_\theta$ which have been extrapolated to $\theta = 0°$. The extension to multi-component systems is briefly considered below.

An alternative experimental procedure is to measure the diminution of intensity of the incident beam, i.e., the loss of light intensity as the beam passes through unit volume of solution. Since the intensity is the energy crossing unit area per second, we may obtain the total scattering by integrating over a sphere of any radius r. For any value of θ we may construct an element of surface having an area $2\pi r^2 \sin \theta \, d\theta$ over which i_θ/I_0 will be constant; i.e., the total scattered intensity is

$$\int_{\theta=0}^{\pi} 2\pi r^2 i_\theta \sin \theta \, d\theta = (16\pi/3)R_\theta I_0 = I_0 - I$$

By analogy with absorption spectroscopy we may define the *turbidity* τ as $-\ln(I/I_0)$ where I is the intensity after passing through a volume of solution in a 1 cm cube. Since I is very close to I_0, the fraction of light scattered being very small, we may write $\ln(I/I_0) = \ln [1 - (I_0 - I)/I_0] = -(I_0/I)/I_0$, so that $\tau = (16\pi/3)R_\theta$. We thus obtain, analogous to equation 17–30,

$$\frac{Hc}{\tau} = \frac{\mathscr{H}c}{R_\theta} = \frac{1}{M} + 2Bc + 3Cc^2 + \cdots \tag{17–31}$$

where $H = 16\pi\mathscr{H}/3$. The quantity τ is ordinarily too small to be measured, but experimental results are frequently reported in the form of Hc/τ, even though $\mathscr{H}c/R_\theta$ is the measured quantity.

17e. Optical anisotropy of the scattering particle.

An inherent assumption in the preceding treatment has been that the scattering particle is optically isotropic. If it is not, there arises the possibility of additional scattering due to fluctuations in the *orientation* of the scattering particles.

The existence of this effect may be detected by measuring the ratio of the intensity of the horizontally polarized component of light scattered at 90°, in the horizontal plane, to the intensity of the vertically polarized component at the same viewing position. Unpolarized incident light is used. If the scattering particle is completely isotropic, then the dipole induced in it is parallel to the electric vector of the incident light. If the light is unpolarized, there will be two independent oscillating dipoles, both perpendicular to the direction of incident light, one at 90° in the horizontal plane and one at 90° in the vertical plane. The former component clearly will contribute nothing to radiation in the direction $\theta = 90°$ in the horizontal plane. The light viewed along this line will therefore be completely polarized in the vertical plane, and the ratio of horizontally to vertically polarized scattered light will be zero.

In anisotropic particles, however, the induced moment is not (except at special orientations) parallel to the electric vector of the incident light. The ratio of horizontally to vertically polarized light will therefore not be zero. This ratio is known as the *depolarization ratio*, ρ_u (the subscript u refers to the fact that the incident light is unpolarized) and is easily measured by inserting an analyzer capable of passing only polarized light in front of the photomultiplier tube in Fig. 17–3. (The contribution of the solvent to both components is, of course, to be subtracted.)

Cabannes[20] has shown that the excess scattering due to anisotropy may be related to the depolarization ratio. Thus the value of R_θ due to concentration fluctuations alone is less than the observed value, the correction factor being $(6 - 7\rho_u)/(6 + 6\rho_u)$. The highest value of ρ_u observed for macromolecules has been 0.040 (for polystyrene in methyl ethyl ketone[21]), giving rise to a correction factor of 0.92. Usually the correction factor is considerably closer to unity. If an accuracy of the order of 1 % is desired in the determination of molecular weights, the correction must in principle always be applied.

A note of caution should be inserted here. Depolarization may arise from secondary scattering, as well as from anisotropy. This factor would not affect the intensity of scattering appreciably. In most cases where nonzero values of ρ_u have been reported it is not at all certain whether the measured values of ρ_u really reflect anisotropy at all (see Geiduschek[22]).

17f. Light scattering in multi-component systems. The theory of light scattering for multi-component systems has been developed independently by Kirkwood and Goldberg[23] and by Stockmayer.[24] The same principles are involved as in the discussion of sedimentation equilibrium in multi-component systems, given in section 16. We shall not discuss the theory here but simply state the results obtained.

The most important result is that the equations derived earlier in this section will not apply if more than one solvent component is present, unless the refractive index of the solvent mixture is independent of its composition. The reason for this is that a macromolecular solute will in general show a preference for one of the solvent components in its immediate environment, so that local values of $\partial \tilde{n}/\partial c_2$ (where c_2 is the concentration of macromolecular component) will not be the same as those measured on a large volume of solution.

Fortunately the corrections to the theory which this entails are insignificant for proteins dissolved in aqueous salt solutions of reasonably high concentration (Doty and Edsall[7]). For such solutions the equations presented earlier are valid, with the virial coefficient B given, as in the case of osmotic pressure, by equation 14–24 or 14–25.

The presence of several macromolecular components, in constant proportion so that each concentration goes to zero as the total solute concentration goes to zero, presents no problem if $\partial \tilde{n}/\partial c$ has the same value for each component. This restriction is not serious since in practice we study systems of this kind only when all the components (e.g., the components of a heterogeneous polymer mixture) are chemically identical. The observed scattering in this case is simply the sum of the contributions from each solute component, just as the osmotic pressure of a solution containing several macromolecular solutes is simply the sum of the individual contributions.

There is a most important difference, however, in the molecular weight average which we obtain by light scattering and osmotic pressure studies of solutions of this kind. By equation 17–28, with $\mathscr{K}$ the same for all components (since $\partial \tilde{n}/\partial c$ is the same), the limiting value of R_θ as c tends to zero is

$$\mathscr{L}_{c=0} R_\theta = \mathscr{K} \sum_i c_i M_i \qquad (17\text{–}32)$$

where c_i is the concentration in grams per cubic centimeter of each dissolved species, and M_i the corresponding molecular weight. Where $\sum_i c_i = c$ represents the total macromolecular concentration,

$$\mathscr{L}_{c=0} \frac{\mathscr{K}c}{R_\theta} = \frac{\sum_i c_i}{\sum_i c_i M_i} = \frac{1}{\bar{M}_w} \qquad (17\text{–}33)$$

i.e., the intercept of light scattering plots gives the *weight-average molecular weight* (equation 8–12) rather than the number average obtained under similar circumstances from osmotic pressure determinations. It can be shown in the same way that the value of the limiting slope gives the z-average of B, whereas osmotic pressure yields a weight average.

17g. Experimental results of light scattering studies. It has been shown that light scattering can be used to determine the molecular weight and thermodynamic interaction parameters in macromolecular solutions which contain only a single solvent component, or a multi-component solvent with refractive index independent of composition, or a solvent consisting of water plus a reasonably high concentration of an inorganic salt. For a homogeneous solute the results should be identical with those determined from osmotic pressure; but for a heterogeneous solute light scattering determines averages of molecular weight and virial coefficient different from those determined by osmotic pressure. Thus light scattering becomes an important tool, not only for the determination of molecular weight and thermodynamic interaction parameters but also for the estimation of heterogeneity, the latter being achieved by comparison with corresponding osmotic pressure data.

An additional restriction inherent in the treatment of this section is that the dimensions of the solute molecules be much smaller than the wavelength of light. When larger particles are taken up in section 18 it will be found that the effects of increased size are a function of the angle θ and disappear when $\theta = 0$. Thus larger particles may actually be treated by the methods of this section, if R_{0° is used when equation 17–30 is applied. The extrapolation to zero scattering angle will be disussed in the following section, but results for some large molecules will be discussed here together with data for smaller molecules.

Table 17–1 shows typical results obtained when light scattering and osmotic pressure molecular weights are compared. It is seen that proteins and viruses show essentially no difference between $\bar{M}_n$ and $\bar{M}_w$, and this

TABLE 17–1. Weight- and Number-Average Molecular Weights

	$\bar{M}_n$ Osmotic Pressure	$\bar{M}_w$ Light Scattering	$\dfrac{\bar{M}_w}{\bar{M}_n}$
β-Lactoglobulin[25,26]	39,000	36,000	1.0
Ovalbumin[26,27]	45,000	46,000	1.0
Serum albumin[28,29]	69,000	70,000	1.0
Tobacco mosaic virus[46,47]	49,000,000[a]	39,000,000	(1.0)[a]
Unfractionated polystyrene[30]	785,000	1,550,000	2.0
Fraction B5 of same[30]	330,000	372,000	1.13
Unfractionated cellulose trinitrate[37]	94,000	273,000	3.7
Cellulose trinitrate fractions[48]	35,000	41,000	1.17
	89,000	128,000	1.44
	257,000	573,000	2.23
Starch amylopectin[49]	300,000	80,000,000	267.0

[a] $\bar{M}_n$ was obtained by particle counting in the field of an electron microscope. $\bar{M}_w/\bar{M}_n$ has been assigned a value of 1.0 since values smaller than this cannot occur.

must mean that these molecules are essentially homogeneous with respect to molecular weight. This fact was already inferred in section 4 from the fact that these substances can form true crystals, but Table 17–1 is a direct experimental proof. (Similar proof was provided by sedimentation equilibrium data discussed in section 16.)

Synthetic polymers and polysaccharides quite generally show a large difference between $\bar{M}_n$ and $\bar{M}_w$, indicative of molecular weight heterogeneity. Unfractionated polystyrene is seen to have $\bar{M}_w/\bar{M}_n = 2$, in accord with predictions based upon a purely statistical distribution of chain lengths, as discussed on p. 147. The same ratio was found for polystyrene from sedimentation equilibrium data shown in Table 16–1.

Especially noteworthy in Table 17–1 are the data of Stacy and Foster[49] for starch amylopectin, which lead to an $\bar{M}_w/\bar{M}_n$ ratio of nearly 300, indicative of an extremely wide and skewed molecular weight distribution. Such a result turns out to be theoretically expected for molecules which contain randomly placed branching points (Erlander and French[50]). That such branching points occur in amylopectin is known from chemical studies.

The data for polystyrene in Table 17–1 show that fractionation of synthetic polymers can lead to a marked reduction in $\bar{M}_w/\bar{M}_n$, i.e., to samples with a narrower molecular weight distribution than the original unfractionated preparation. On the other hand, ordinary fractionation procedures do not necessarily lead to homogeneous samples, as the data for cellulose nitrate show. Some fractionated samples of this substance still show a high degree of heterogeneity, although they are presumably less heterogeneous than an unfractionated preparation would be.

In considering the experimental data here presented it should be borne in mind that the experimental techniques required to obtain really precise results have not yet been completely established. It is not uncommon for different investigators to report molecular weight values for the same protein, by the same method, which differ by 10%, as is shown, for instance, by the osmotic pressure molecular weights listed for β-lactoglobulin in Tables 13–1 and 14–3. In the case of synthetic polymers similar differences of 50% or more have been reported for measurements on identical samples.[31] In the case of light scattering such differences may often result from the presence of dust particles, which exert a disproportionately great effect when a weight-average molecular weight is being measured.

Hermans and Hermans[61] have used light scattering to determine the molecular weights of forty-one different preparations of deoxyribose nucleic acid (DNA), almost all of them from the same source, calf thymus glands. Their results range from $\bar{M}_w = 2 \times 10^6$ to 36×10^6. It is probable that some of the preparations studied by Hermans and Hermans were not entirely freed from the protein with which DNA is associated in nature. Nevertheless, some lack of agreement is typical of molecular-weight determinations of DNA in different laboratories, and it is likely that most preparations examined have been to some extent degraded.

One exception may be noted. Sinsheimer[62] has found that the virus bacteriophage ϕX 174 has a molecular weight (by light scattering) of 6.2×10^6. This virus contains 25% by weight of DNA, and when the DNA was extracted it was found by Sinsheimer to have a molecular weight of 1.7×10^6. This can only mean that all the DNA in the virus particle is a single molecule and that it was extracted without degradation.

Figures 17–4 and 17–5 and Table 17–2 show a comparison between second virial coefficients determined by osmotic pressure and by light scattering. In the case of bovine serum albumin no difference is observed, nor is any expected. In the case of polystyrene (Table 17–2) a lower value is expected from light scattering in a good solvent because B decreases with increasing molecular weight (Table 13–3) and because light scattering measures a higher molecular-weight average of B than does osmotic pressure. The large experimental error obscures the difference in the data shown. Figure 17–5 shows the striking effect of temperature on the behavior of polystyrene fractions in a poor solvent (cf. Fig. 13–3).

TABLE 17–2. Second Virial Coefficient for Polystyrene
Standard Sample V[a]

	Average Values of $B \times 10^4$ Obtained by Several Investigators[31]	
	In Toluene (good solvent)	In Methyl Ethyl Ketone (poorer solvent)
Osmotic pressure	4.7 ± 0.6	1.3 ± 0.7
Light scattering	4.2 ± 0.8	1.31 ± 0.28

[a] $\bar{M}_n$ for this sample is about 460,000; $\bar{M}_w$ about 550,000

A special kind of interaction, leading to a negative second virial coefficient, is that of association between solute molecules, as discussed on p. 203. The osmotic data for insulin shown in Fig. 14–2 indicate that such association occurs in solutions of this protein. Light scattering data of Doty and Myers,[34] given in Fig. 17–6, strikingly confirm this. Doty and Myers, assuming $B = 0$, were able to treat the data in terms of an equilibrium between molecules of molecular weight 12,000 and 24,000. (If NaCl was used in place of phosphate, a third species of molecular weight 36,000 had to be assumed present.) It should be noted that none of these species represents the smallest possible insulin molecule (p. 7), which has a molecular weight of about 6000.

Another kind of interaction leading to negative deviations from ideality results from charge fluctuations in macro-ion solutions. This effect was discussed in section 14f, and specific equations were developed for isoelectric solutions (average charge $\bar{Z}$ of the macro-ions equal to zero) containing no salt ions. It was predicted that equation 11–21 should no longer be applicable to such solutions, and, hence, equation 17–30 should also be invalid. To obtain the correct equation for the concentration

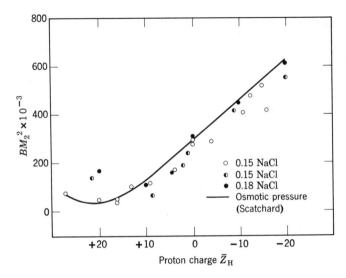

Fig. 17–4. Comparison of the second virial coefficients of serum albumin, obtained by osmotic pressure and by light scattering. The abscissa is the average charge due to protons alone, which is more positive by about ten charges than the true net charge because this protein binds chloride ions. The points shown were obtained by light scattering. The curve is from the osmotic pressure data of Scatchard et al., shown in Fig. 14–1. (From Edsall et al.[51])

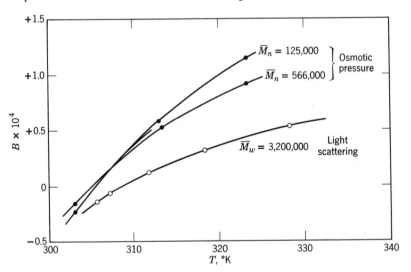

Fig. 17–5. Second virial coefficients for polystyrene in cyclohexane (a poor solvent). Each curve represents a separate fraction, and all three curves show the increase in B which always accompanies an increase in the temperature of polymer solutions in poor solvents. (Data from Krigbaum[32] and Krigbaum and Carpenter.[33])

dependence of light scattering in this case, we must use equation 14–29 for $\partial \mu_1/\partial c$ in equation 17–25. The result is that

$$\frac{\mathscr{H}c}{R_\theta} = \frac{1}{M}\left[1 - \frac{1}{2}\left(\frac{\pi \mathscr{N}}{M}\right)^{\frac{1}{2}}\left(\frac{\overline{Z^2}\epsilon^2}{Dk\,T}\right)^{\frac{3}{2}}c^{\frac{1}{2}}\right] \qquad (17\text{--}34)$$

with the understanding that this is a limiting expression, applicable only near $c = 0$.

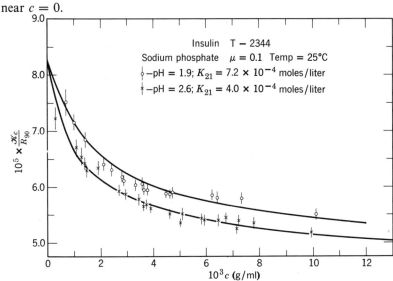

Fig. 17–6. Light scattering data for insulin at low pH. K_{21} is the equilibrium constant for dissociation of an aggregate (assumed mol. wt. 24,000) into two identical units of mol. wt. 12,000. (Doty and Myers.[34])

The validity of equation 17–34 is fully confirmed by light scattering data determined by Timasheff et al.[59] for essentially isoelectric solutions of serum albumin. Figure 17–7 shows their data as plotted according to equation 17–30, and it is seen that $\mathscr{H}c/R_\theta$ cannot be represented as a linear function of c. Figure 17–8 shows the same data plotted versus $c^{\frac{1}{2}}$, and we see that a linear relation is obtained, as predicted by equation 17–34. The straight line (as c approaches zero), which is drawn through the data, is actually a calculated one, with $\overline{Z^2}$ evaluated from experimental titration data by the method given in section 30.

It should be noted that it is actually impossible to obtain a truly isoelectric protein solution, in the absence of all other ions, in a solvent such as water, because of the presence of H^+ and OH^- ions. We shall postpone discussion of this problem until section 30, but we mention it here to indicate that a small correction will often be required in applying equation 17–34 to actual experimental data. The correction is discussed by Kirkwood and Timasheff.[60]

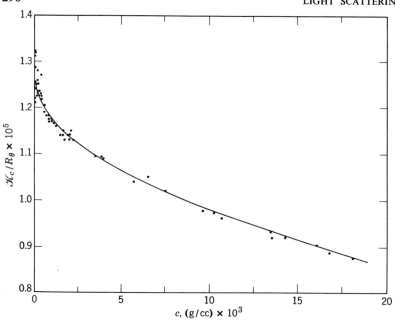

Fig. 17–7. Light scattering data for serum albumin, essentially at its isoelectric point, in de-ionized water. The data are plotted according to equation 17–30. (Timasheff et al.[59])

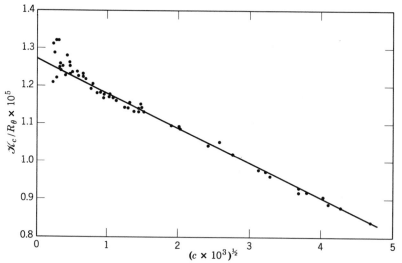

Fig. 17–8. The same data as in Fig. 17–7, plotted according to equation 17–34. The straight line representing the approach to $c = 0$ is calculated according to the equation. (Timasheff et al.[59])

18. LIGHT SCATTERING BY LARGER PARTICLES

The discussion of light scattering in section 17 was confined to particles much smaller than the wavelength of light. In practice, this means that all the molecular dimensions must be less than about $\lambda/20$, i.e., if the mercury arc is used as the incident light source, less than about 200 to 270 Å. Most macromolecules, other than the smaller globular proteins, have an extension in at least one dimension which exceeds this figure.

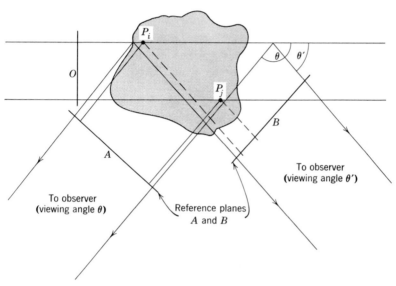

Fig. 18–1. Scattering of light from a particle with dimensions greater than $\lambda/20$. There is destructive interference between light scattered from P_i and P_j because the distances OP_iA and OP_jA (or OP_iB and OP_jB) are different.

The effect which this has is illustrated in Fig. 18–1, where two points, P_i and P_j, of a large scattering particle are shown. A reference plane O is drawn, where all of the incident light is in phase. A second reference plane A is drawn perpendicular to the scattered beam at viewing angle θ. Light which is in phase at the plane A will remain in phase when it reaches the observer, phase differences at A will likewise be maintained. The distance OP_iA is clearly considerably less than the distance OP_jA, so that light scattered from P_j will reach the observer out of phase with that scattered from P_i. This phase difference results in interference and diminution of the scattered intensity, as shown in Fig. 3–1*b* on p. 17.

Figure 18–1 also shows a beam scattered at an angle θ' smaller than θ, with a reference plane B perpendicular to it. It is clear that the difference between the distances OP_iB and OP_jB is less than the difference between OP_iA and OP_jA. There will thus be less interference and less diminution of intensity at the viewing angle θ' than at the angle θ.

Finally, the path difference disappears altogether when the viewing angle becomes zero. This result is quite independent of the locations of P_i and P_j.

Quite generally, the effect of large size may be described by a function $P(\theta)$,

$$P(\theta) \equiv \frac{\text{scattered intensity for large particle}}{\text{scattered intensity without interference}} \qquad (18\text{–}1)$$

which may be much less than 1 when θ is relatively large but which will increase as θ becomes smaller and become unity when $\theta = 0°$. At this point the equations of section 17 would apply. In particular, equation 17–30 is applicable to measured values of $\mathscr{K}c/R_{0°}$. All the conclusions of the preceding section thus apply to large particles, if data are obtained at several angles θ and extrapolated to $\theta = 0°$.

If P_i and P_j (Fig. 18–1) are sufficiently far apart so that the difference between OP_iA and OP_jA may exceed $\lambda/2$, complete destruction of the scattered light is possible, and a further increase in angle may result in an increase in scattered intensity. We shall find that this situation is not of interest to us here. We shall confine ourselves, for all particles, to scattering angles sufficiently small so that path differences greater than $\lambda/2$ make little or no contribution to the scattering intensity.

18a. The general equation for $P(\theta)$. We have shown that light scattered from large particles at $0°$ will obey the same equations as light scattered from small particles at the same angle. At angles of $\theta > 0°$ the light scattered by large particles is diminished by a factor designated $P(\theta)$. At first sight this effect may appear to be a nuisance, requiring an extrapolation to zero angle to obtain the information ordinarily sought from light scattering data. Further consideration, however, shows this to be far from true. In fact, we shall find that the interference effect can provide us with perhaps more useful information than any other obtainable from light scattering. Accordingly, we shall derive here a general relation between $P(\theta)$ and molecular configuration. The derivation will assume that the particles in solution are all alike and that they are sufficiently far apart to be independent; i.e., it will apply only to data at very low concentrations. Since light scattering data must in any event be extrapolated to zero concentration if the molecular weight is to be obtained, this imposes no new requirement on experimental procedure.

Figure 18–2 represents a three-dimensional analog of Fig. 18–1. It represents a region of space filled by a large molecule. The point P_i represents a single scattering element of the large molecule. We are to imagine that the entire volume contains a very large number of such points.

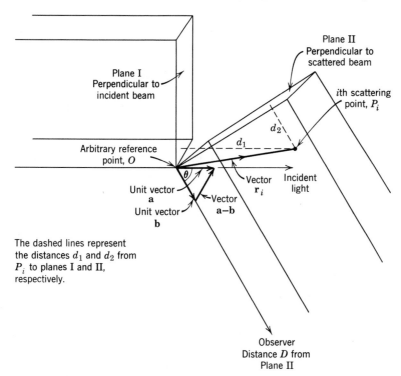

Fig. 18–2. Representation of the various parameters used in the derivation of an expression for $P(\theta)$. The entire figure represents space within the domain of a large molecule, only one point of which, P_i, is shown explicitly.

For a solid molecule the scattering elements are contiguous and fill the entire space; for a flexible polymer molecule the natural choice is to have each individual segment represent a scattering element.

It may be noted at this point that the definition of the "segments" of a flexible polymer molecule (in section 9) was primarily concerned with the *lengths* of the successive bonds which join one end of the chain to the other. It was shown that the detailed distribution of *mass* with respect to the location of the bonds was not important in the consideration of dimensional parameters, so that the total mass of the molecule could usually be taken as located in the manner most convenient for calculation of the radius of gyration. The usual choice was to locate the mass at the points of junction between the bonds.

The same latitude is permitted in the present context. The scattering intensity from any point of a molecule depends on the effective electron density, but, within distances up to 100 Å or more (for visible light), it does not matter exactly where the electron density is located since all light scattered from within short distances is essentially in phase. Thus the electron density may be taken as proportional to the mass and distributed in the same way.

A reference point O is indicated in the figure, and through it are drawn planes perpendicular to the incident ray and to the ray scattered at an arbitrary angle θ. Unit vectors **a** and **b** are drawn from O in the direction of the incident and scattered beams. We shall find shortly that the difference, **a** − **b**, between these vectors will enter into our calculations. The magnitude of this vector is the base of an isosceles triangle with unit sides and included angle θ; i.e., it is clearly $2 \sin (\theta/2)$. Its direction is at an angle of $90° - \theta/2$ to the direction of the incident light and at an angle of $90° + \theta/2$ to that of the scattered light. We may designate by **c** a unit vector along this direction, so that

$$\mathbf{a} - \mathbf{b} = 2\mathbf{c} \sin (\theta/2) \tag{18-2}$$

Consider next the length of path from reference plane I to the observer. From Fig. 18–2 this is equal to $d_1 + (D - d_2)$, these quantities being defined in the figure. Where $\mathbf{r}_i$ is a vector from O to P_i, d_1 is clearly the projection of $\mathbf{r}_i$ along the direction of the unit vector **a** and d_2 is the projection along the direction of **b**; i.e., in vector notation $d_1 = (\mathbf{r}_i \cdot \mathbf{a})$, $d_2 = (\mathbf{r}_i \cdot \mathbf{b})$, hence the distance traveled to the observer will be $D + \mathbf{r}_i \cdot (\mathbf{a} - \mathbf{b})$. We shall call this distance Δ_i; i.e., by equation 18–2,

$$\Delta_i = D + 2(\mathbf{r}_i \cdot \mathbf{c}) \sin (\theta/2) \tag{18-3}$$

This quantity will have a different value for each of the scattering points of which the scattering particle is composed. The difference between the values of Δ_i for different points P_i will produce phase differences at the point of observation.

Using equation 17–4, we can evaluate the electric field, from the ith scattering point, as seen by the observer. If the origin of the distance parameter ($x = 0$) is located at the point of observation,

$$\mathscr{E}_i = (\text{constant}) \cos 2\pi(\nu t - \Delta_i/\lambda) \tag{18-4}$$

where the wavelength λ is again measured in the scattering medium. If there are σ scattering points in all, the total scattered radiation is represented by

$$\mathscr{E}_s = \sum_{i=1}^{\sigma} \mathscr{E}_i = (\text{constant}) \sum_{i=1}^{\sigma} \cos 2\pi(\nu t - \Delta_i/\lambda) \tag{18-5}$$

It is not necessary to write down the constant coefficient in equation 18–4 and 18–5 because we are interested only in the phase differences due to differences in the Δ_i and the resulting ratio of the scattered intensity to that which would be observed in the absence of such phase differences.

We obtain the scattered intensity, as before (p. 277), by averaging the value of $\mathscr{E}^2$ over a period $1/\nu$, so that

$$i_\theta = (\text{constant})\, \nu \int_{t=0}^{1/\nu} \left(\sum_{i=1}^{\sigma} \cos A_i\right)^2 dt \qquad (18\text{--}6)$$

where $A_i = 2\pi(\nu t - \Delta_i/\lambda)$. By straightforward algebra

$$\left(\sum_{i=1}^{\sigma} \cos A_i\right)^2 = \left(\sum_{i=1}^{\sigma} \cos A_i\right)\left(\sum_{j=1}^{\sigma} \cos A_j\right)$$

$$= \sum_{i=1}^{\sigma} \sum_{j=1}^{\sigma} \cos A_i \cos A_j$$

$$= \sum_{i=1}^{\sigma} \sum_{j=1}^{\sigma} \left[\tfrac{1}{2} \cos (A_i - A_j) + \tfrac{1}{2} \cos (A_i + A_j)\right] \quad (18\text{--}7)$$

Thus

$$i_\theta = (\text{constant})\, \nu \int_{t=0}^{1/\nu} \sum_{i=1}^{\sigma} \sum_{j=1}^{\sigma} \{\tfrac{1}{2} \cos 2\pi(\Delta_j - \Delta_i)/\lambda$$

$$+ \tfrac{1}{2} \cos 2\pi[2\nu t - (\Delta_i + \Delta_j)/\lambda]\} \, dt$$

$$= (\text{constant}) \sum_{i=1}^{\sigma} \sum_{j=1}^{\sigma} \cos 2\pi(\Delta_j - \Delta_i)/\lambda \qquad (18\text{--}8)$$

for the integral of the time-dependent term is zero.

If $\Delta_j - \Delta_i$ is much smaller than λ for all values of i and j, i.e., if we have a small scattering particle, then each cosine in equation 18–8 becomes equal to unity. Under these conditions, however, i_θ must reduce to the scattered intensity in the absence of interference, which we may call i_θ'. Thus, since $\sum_{i=1}^{\sigma} \sum_{j=1}^{\sigma} 1 = \sigma^2$, $i_\theta' = (\text{constant})\sigma^2$, and using the definition of $P(\theta)$ given by equation 18–1 gives

$$P(\theta) = \frac{i_\theta}{i_\theta'} = \frac{1}{\sigma^2} \sum_{i=1}^{\sigma} \sum_{j=1}^{\sigma} \cos 2\pi(\Delta_j - \Delta_i)/\lambda \qquad (18\text{--}9)$$

Substituting for $\Delta_j - \Delta_i$ by equation 18–3 and noting that the vector $\mathbf{r}_j - \mathbf{r}_i$ is merely the vector $\mathbf{r}_{ij}$ leading from the ith to the jth scattering point, we get

$$P(\theta) = \frac{1}{\sigma^2} \sum_{i=1}^{\sigma} \sum_{j=1}^{\sigma} \cos \mu[(\mathbf{r}_j - \mathbf{r}_i) \cdot \mathbf{c}]$$

$$= \frac{1}{\sigma^2} \sum_{i=1}^{\sigma} \sum_{j=1}^{\sigma} \cos \mu(\mathbf{r}_{ij} \cdot \mathbf{c}) \qquad (18\text{--}10)$$

where, to avoid repetition of a term which will continue to appear in all subsequent discussion, we write

$$\mu \equiv \frac{4\pi}{\lambda} \sin \frac{\theta}{2} \qquad (18\text{--}11)$$

Equation 18–10 represents the value of $P(\theta)$ for a scattering particle *fixed in space*. We are interested, however, in a particle free to orient itself in any direction with respect to the vector $\mathbf{c}$. Thus we seek the average value of $P(\theta)$ over all possible orientations. This may be obtained with the aid of Fig. 6–6. When α is the angle between $\mathbf{c}$ and $\mathbf{r}_{ij}$, $\mathbf{r}_{ij} \cdot \mathbf{c}$ becomes $r_{ij} \cos \alpha$. The probability of a given value of α between α and $\alpha + d\alpha$ is proportional to $2\pi r_{ij} \sin \alpha \, d\alpha$, and, therefore

$$\overline{\cos \mu(\mathbf{r}_{ij} \cdot \mathbf{c})} = \frac{\displaystyle\int_{\alpha=0}^{\pi} \cos(\mu r_{ij} \cos \alpha) \sin \alpha \, d\alpha}{\displaystyle\int_{\alpha=0}^{\pi} \sin \alpha \, d\alpha}$$

$$= \frac{\sin \mu r_{ij}}{\mu r_{ij}} \qquad (18\text{--}12)$$

Substitution in equation 18–10 thus leads to the value of $P(\theta)$ based on random orientation of the scattering particle,

$$P(\theta) = \frac{1}{\sigma^2} \sum_{i=1}^{\sigma} \sum_{j=1}^{\sigma} \frac{\sin \mu r_{ij}}{\mu r_{ij}} \qquad (18\text{--}13)$$

This relation was first obtained by Debye,[35] in connection with the problem of x-ray scattering. The equation may be used both for polarized and unpolarized incident light.

18b. The limiting form of $P(\theta)$ and its relation to the radius of gyration. The form taken by the function $P(\theta)$ will depend on the geometric shape of the scattering particle, or, for flexible particles, on the average shape. The derivation of appropriate forms is discussed in section 18d.

A more important result of equation 18–13, however, is that $P(\theta)$ becomes independent of particle shape as θ approaches zero and, under these limiting conditions, becomes a measure of the radius of gyration of the scattering particle. This result, first recognized by Guinier,[36] is unique, for no other physical measurement provides a measure of the dimensions of a macromolecular particle without any assumption regarding its general form (i.e., whether it is an ellipsoid, a random coil, etc.).

The required relation is obtained simply by expanding each term under the summation sign of equation 18–13 in terms of a power series. Since $\sin x = x - x^3/3! + x^5/5! - \cdots$, we get

$$P(\theta) = \frac{1}{\sigma^2} \sum_{i=1}^{\sigma} \sum_{j=1}^{\sigma} (1 - \mu^2 r_{ij}^2/3! + \mu^4 r_{ij}^4/5! - \cdots) \qquad (18\text{–}14)$$

For a limiting expression we retain only the first two terms of each expansion, and, since $\sum_{i=1}^{\sigma} \sum_{j=1}^{\sigma} 1 = \sigma^2$, we get

$$\mathscr{L}_{\theta \to 0^\circ} P(\theta) = 1 - (\mu^2/3! \, \sigma^2) \sum_{i=1}^{\sigma} \sum_{j=1}^{\sigma} r_{ij}^2 \qquad (18\text{–}15)$$

In section 9j, equations 9–33 or 9–34, we proved, however, that $\sum_i \sum_j r_{ij}^2$ is just $2\sigma^2$ times the square of the radius of gyration or, for flexible molecules, that $\sum_i \sum_j \overline{r_{ij}^2}$ is $2\sigma^2$ times the square of the average radius of gyration. Furthermore, although the objective of section 9j was to evaluate R_G for linear polymers, the limitation to linear polymers (which takes the form of stating that r_{ij} depends only on the separation between the mass units i and j) was not introduced until after equation 9–34 was derived. Thus equations 9–33 and 9–34 are perfectly general and applicable to every kind of dissolved particle. With complete generality, therefore,

$$\mathscr{L}_{\theta \to 0} P(\theta) = 1 - \mu^2 R_G^2/3 \qquad (18\text{–}16)$$

where R_G, in general, is the root-mean-square average radius of gyration. Substituting the expression for μ given by equation 18–11, we get

$$\mathscr{L}_{\theta \to 0} P(\theta) = 1 - \frac{16\pi^2}{3\lambda^2} R_G^2 \sin^2 \frac{\theta}{2} \qquad (18\text{–}17)$$

Or, since $1/(1 - x)$, for small x, is equal to $1 + x$,

$$\mathscr{L}_{\theta \to 0} 1/P(\theta) = 1 + \frac{16\pi^2}{3\lambda^2} R_G^2 \sin^2 \frac{\theta}{2} \qquad (18\text{–}18)$$

In view of the fact that the equation for $P(\theta)$ is itself valid only at very high dilution it is necessary to extrapolate the data at each angle to zero concentration before we can perform the extrapolation indicated by equation 18–17 or 18–18. At the same time, it is desirable to extrapolate the data at each concentration to zero angle, so that $\bar{M}_w$ and B can be determined according to equation 17–30, which, for large molecules, is valid only at zero angle.

Both these objectives can be achieved by the method of Zimm,[15] which consists of plotting $\mathscr{K} c/R_\theta$ versus $\sin^2 (\theta/2) + kc$, where k is an arbitrary constant chosen so as to provide a convenient spread of the data on a piece of graph paper.

An example of such a plot is shown in Fig. 18–3. From it we obtain, as seen, the two types of limiting plot. Extrapolating to $\theta = 0°$, we obtain a plot of $\mathcal{K}c/R_{0°}$ versus kc, which, by equation 17–30, gives $1/M$ as intercept

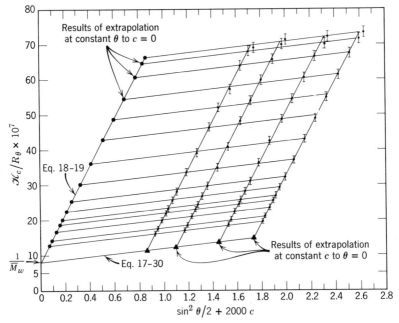

Fig. 18–3. Zimm plot for extrapolation of light scattering data. The data are for a cellulose nitrate fraction in acetone (Benoit, Holtzer, and Doty.[37])

and $2B/k$ as limiting slope. Extrapolating to $c = 0$, we obtain a plot of $\mathcal{K}c/R_\theta$ versus $\sin^2 (\theta/2)$. Since R_θ is directly proportional to i_θ (equation 17–27), we get, by combining equations 17–30, 18–1, and 18–18,

$$\mathcal{L}_{c=0} \frac{\mathcal{K}c}{R_\theta} = \frac{1}{M P(\theta)} = \frac{1}{M}\left(1 + \frac{16\pi^2}{3\lambda^2} R_G{}^2 \sin^2 \frac{\theta}{2} + \cdots\right)$$

$$(18–19)$$

Thus the intercept of this plot is $1/M$, the same as the intercept of the $\theta = 0°$ plot, and the limiting slope is immediately a measure of R_G; i.e., limiting slope/intercept $= (16\pi^2/3\lambda^2)R_G{}^2$. In using this relation, it must be remembered that the value of λ to be used is that measured in the scattering medium, or, where λ_0 is the wavelength in vacuum and $\tilde{n}$ the refractive index of the medium, $\lambda = \lambda_0/\tilde{n}$.

We have noted previously that the molecular weight obtained from these measurements for a heterogeneous mixture is the weight average. The value of $R_G{}^2$, on the other hand, is a z-average. By equation 18–19, for extrapolated data at $c = 0$,

$$R_\theta = \mathscr{K} \sum_i M_i \, P_i(\theta) c_i$$

where the subscript i refers to each component of a mixture which contributes to R_θ. With equation 18–17, relating each $P_i(\theta)$ to the corresponding radius of gyration, $(R_G)_i$, and with $\mathscr{K} c = \mathscr{K} \sum_i c_i$, we get

$$\frac{\mathscr{K} c}{R_\theta} = \sum_i c_i \Big/ \left\{ \sum_i M_i c_i - \frac{16\pi^2}{3\lambda^2} \sin^2 \frac{\theta}{2} \sum_i M_i (R_G{}^2)_i c_i \right\}$$

For small values of $\sin^2 \theta/2$, as in going from equation 18–17 to equation 18–18, this gives

$$\frac{\mathscr{K} c}{R_\theta} = \frac{1}{\bar{M}_w} \left(1 + \frac{16\pi^2 \langle R_G{}^2 \rangle_z}{3\lambda^2} \sin^2 \frac{\theta}{2} \right)$$

where
$$\langle R_G{}^2 \rangle_z = \frac{\sum_i M_i c_i (R_G{}^2)_i}{\sum_i M_i c_i} = \frac{\sum_i N_i M_i{}^2 (R_G{}^2)_i}{\sum_i N_i M_i{}^2}$$

and N_i is the concentration of molecules of kind i.

Section 9k should be consulted for definition of the various possible averages. It may be noted that $\langle R_G{}^2 \rangle_z$ is not equal to $(\langle R_G \rangle_z)^2$ and that neither average gives the radius which characterizes molecules which have $M = \bar{M}_z$.

It should be noted also that the extrapolation method here given is applicable only over a limited range of dimensions, $0.05 \leqslant R_G/\lambda \leqslant 0.5$. If R_G falls below the lower limit given, $P(\theta)$ becomes too small for accurate estimation even at obtuse angles; if it is above the upper limit it becomes impossible to attain values of θ sufficiently small to reach the limiting form as given by equation 18–17, owing to the fact that we cannot make scattering measurements too close to $\theta = 0°$ because of interference of the incident beam which is transmitted through the scattering cell. With the mercury arc the permissible range thus lies between about 200 and 2000 Å. Except for the small globular proteins, this is very often the range of greatest interest. For molecules with smaller dimensions we can however obtain the same kind of information by means of x-ray scattering (section 18f); for those with larger dimensions we can use the complete expressions for $P(\theta)$ given in section 18d.

18c. Experimental radii of gyration and their interpretation. The ability to obtain an experimental value for the radius of gyration provides an extraordinarily powerful tool in the determination of molecular structure, for macromolecules of different configurational types (e.g., solid spheres, long rods, random coils) have radii of gyration of very different magnitudes.

Assuming all mass elements to have the same mass (see p. 299) we can use equation 9–4 as the definition. For solid spheres of radius R the center of mass is at the center of the sphere, and the number of mass elements at a

TABLE 18–1. Calculated Radius of Gyration for Model
Molecule, $M = 500,000$[a]

	R_G, Å
Solid sphere	45
Flexible coil (poor solvent, $\alpha = 1$)	188
(good solvent)	295
Long rod (diameter 25 Å)	410
(diameter 15 Å)	1150
(diameter 10 Å)	2550

[a] For the solid sphere and long rod the specific volume has been taken as unity. For the flexible coil a vinyl polymer has been taken, with M_0 (per segment of half a monomer unit) $= 50$, $\beta = 4.6$ Å per carbon–carbon bond, and α as given by equation 9–22.

distance between r and $r + dr$ from the center is proportional to $4\pi r^2 \, dr$, so that

$$R_G{}^2 = \int_{r=0}^{R} 4\pi r^4 \, dr \Big/ \int_{r=0}^{R} 4\pi r^2 \, dr = \tfrac{3}{5}R^2 \qquad (18\text{–}20)$$

For a long straight rod of length L the center of mass is at the center of the rod, and the number of mass elements at a distance between r and $r + dr$ is just proportional to dr, so that (the maximum value of r being $L/2$)

$$R_G{}^2 = \int_{r=0}^{L/2} r^2 \, dr \Big/ \int_{r=0}^{L/2} dr = L^2/12 \qquad (18\text{–}21)$$

For a flexible coil $R_G{}^2$ is given by equation 9–40.

Suppose now that we have a macromolecule of known molecular weight and specific volume. We can then immediately determine what its radius would be if it were a sphere or what its length would be if it were a cylindrical rod of any desired cross-section. By equations 18–20 or 18–21 we can then compute its radius of gyration. If we suppose the molecule to be a flexible polymer, we can likewise estimate the effective length per segment, β, and compute R_G by equation 9–40. Table 18–1 shows the kind of result which is obtained; R_G is seen to be very small for a sphere, several times larger for a random coil, and up to a hundred times greater for a long rod. The table, of course, does not by any means exhaust the possible variety of models, so that an experimental determination of R_G does not permit the immediate assignment of configuration. It does, however, greatly limit the choice of model structure, and, as we shall show, auxiliary measure-

TABLE 18–2. Radii of Gyration from Light Scattering or X-Ray Scattering

	Mol. Wt.	Calculated Result, Unsolvated Sphere		Experimental Result
		Density[a] grams/cc	R_G, Å	R_G, Å
Serum albumin[45]	66,000	1.33	21	29.8[b]
Catalase[52]	225,000	1.37	31	39.8[b]
Myosin[53]	493,000	1.37	41	468
Polystyrene fraction[33]	3.2×10^6	1.0	84	494[c]
DNA[43]	4×10^6	1.8	74	1170
Bushy stunt virus[54]	10.6×10^6	1.35	113	120[b]
Tobacco mosaic virus[55]	39×10^6	1.33	175	924

[a] Reasonable values have been assumed. The calculation is, of course, quite insensitive to the density chosen.

[b] Obtained by x-ray scattering.

[c] Value depends on nature of the solvent, as shown in Table 18–3. The value given is the smallest observed, in a very poor solvent.

ments can then be made to determine what the actual configuration must be. (The value of R_G given in Table 18–1 for a sphere is, of course, too small to be determined by light scattering. X-ray scattering would have to be employed.)

Some experimental radii of gyration are listed in Table 18–2. The results are compared with radii of gyration calculated with the assumption that the molecules are dry, solid spheres. It is at once apparent that only three of the substances listed in the table are at all close to being solid spheres of this kind (the two globular proteins, serum albumin and catalase, and the bushy stunt virus). The relatively small differences between the observed and calculated R_G values for these molecules are presumably due to incorporation of solvent within the scattering particle, extending its dimensions, and to relatively minor deviations from spherical shape. The R_G value for serum albumin, for example, can be accounted for[45] if the molecule is a rectangular box of dimensions $80 \times 60 \times 30$ Å.

The other molecules listed in Table 18–2 must all have very much greater extension in space, and there are then a variety of methods by which a choice among possible models (the simplest of which would be a rigid rod or a flexible coil) can be made.

One of the methods is to see whether a given model is compatible with several different types of measurement. This method will be applied to myosin in section 23, where viscosity data are compared with the light scattering radius of gyration. (The result will be that the molecule behaves more nearly like a rod than a flexible coil.) It may also be applied to

TABLE 18–3. Effect of Temperature on the Radius of
Gyration of Polystyrene in Cyclohexane[a]

$T, \degree K$	R_G, Å
305.7	494
307.2	518
311.2	576
318.2	625
328.2	665
333.2	690

[a] The molecular weight of this sample was $\overline{M}_w = 3.2 \times 10^6$. The values of B for the same sample are given in Fig. 17–5.

tobacco mosaic virus, which, by electron microscopy (section 6), appears to be a rod-shaped particle of length about 3000 Å. If the R_G value listed in Table 18–2 is that of a rod, the length of the rod can be determined by equation 18–21. The result is $L = 3200$ Å, in excellent agreement with the electron microscope value.

Yet another method is to investigate the angular dependence of light scattering at higher angles, i.e., beyond the limiting angles to which equation 18–17 applies. This method will be applied to DNA in section 18e.

If a molecule is thought to be a flexible coil, this conclusion can be tested by observing its contraction in a poor solvent and its expansion on being heated in a poor solvent. This property of flexible polymers has already been considered in terms of the second virial coefficient, and pertinent data from osmotic pressure and from light scattering (at $\theta = 0\degree$) have been presented in Figs. 13–3 and 17–5. A more direct method of observing the same phenomenon is by measuring the radius of gyration in a poor solvent. Typical results obtained by Krigbaum and Carpenter[33] are shown in Table 18–3, and they strikingly confirm the interpretation given to the thermodynamic data.

For polymers which are available in a series of fractions of different molecular weight, immediate evidence concerning structure can be obtained from the molecular-weight dependence of R_G. This is shown, for example, by the data of Table 18–4. For poly-γ-benzyl-L-glutamate in chloroform-formamide as solvent, R_G is seen to be proportional to the first power of the molecular weight. This is incontrovertible evidence that the molecule is a rigid rod (cf. section 9i), for R_G varies as $M^{0.5}$ to $M^{0.6}$ for all randomly

TABLE 18–4. The Dependence of Radius of Gyration on Molecular Weight

Mol. Wt.	R_G, Å	$\dfrac{R_G}{M} \times 10^3$	$\dfrac{R_G}{M^{1/2}}$	$\dfrac{R_G}{M^{0.55}}$
(a)	Poly-γ-benzyl-L-glutamate in chloroform-formamide, 25° C. (Doty, Bradbury, and Holtzer.[39])			
262,000	528	2.02	1.03	—
208,000	408	1.96	0.89	—
130,000	263	2.02	0.73	—
(b)	Polystyrene in butanone, 22° C. (Outer, Carr, and Zimm.[40])			
1,770,000	437	0.25	0.33	0.160
1,630,000	414	0.25	0.32	0.158
1,320,000	367	0.28	0.32	0.158
940,000	306	0.33	0.32	0.159
524,000	222	0.42	0.31	0.159
230,000	163	0.71	0.34	0.183

coiled molecules. The data in Table 18–4 for polystyrene, on the other hand, immediately confirm that it is of the randomly coiled type, since R_G is very close to being proportional to the square root of M. (The data at the five highest molecular weights fit closely the relation $R_G \propto M^{0.55}$, suggesting that the interaction parameter α is proportional to about $M^{0.05}$.)

It was shown in section 4 that x-ray diffraction data indicate that poly-γ-benzyl-L-glutamate exists in the solid state as a helical rod (Fig. 4–7), the length of which is 1.50 Å per monomer unit. The data of Doty, Bradbury, and Holtzer[39] given in Table 18–4 indicate that it exists in solution (in the particular solvent used) as a rod of radius of gyration 0.0020 Å per gram of molecular weight. The monomer unit of poly-γ-benzyl-L-glutamate has a molecular weight M_0 of 219, so that the length per monomer unit becomes 1.51 Å. The agreement between this figure and the solid state length is truly remarkable, and it must be concluded that the molecules in solution have the identical helical configuration which they possess in the solid state. (It will be seen in the next chapter that this conclusion applies only to certain solvents. In some solvents the molecule becomes a flexible coil.)

18d. Radius of gyration and the dimensions of flexible polymer coils. Figure 17–5 shows a plot of the second virial coefficient B versus temperature for the same polystyrene sample used to obtain the data of Table 18–3. It is seen that B is negative at the lowest temperatures ($\alpha < 1$) and becomes positive at higher temperatures. At 309° K, $B = 0$ and this temperature is therefore the ideal temperature at which $\alpha = 1$. The radius of gyration (equation 9–40) is then simply

$$R_G = \beta \sigma^{\frac{1}{2}}/2.45$$

and the effective bond length β can be determined. The degree of polymerization of polystyrene of molecular weight 3.2×10^6 is 30,750, and the molecular chains in this fraction thus contain 61,500 carbon–carbon bonds. Using this value for σ we obtain $\beta = 5.40$ Å as the effective length of a carbon–carbon bond. The true average length of such a bond (l_{av}) is 1.54 Å, so that $\beta = 3.5 l_{av}$. This value is considerably larger than the value $\beta = 2^{\frac{1}{2}} l_{av}$, which would be predicted on the basis of free rotation (section 9d), and thus provides experimental confirmation for the assertion made in section 9g that rotation about single bonds of polymer chains is generally quite restricted, leading to β of the order $3 l_{av}$.

The radius of gyration in a good solvent, e.g., the data for polystyrene given in Table 18–4, can be used in a similar way to determine values for the non-ideality parameter α as a function of molecular weight and solvent system, so that a complete compilation of the dimensional parameters of flexible polymer coils can be obtained. (We shall show in section 23 that the same information can be obtained from the intrinsic viscosity of solutions of flexible polymer molecules. Since intrinsic viscosity is much easier to measure than the angular dependence of light scattering, it becomes the preferred tool for this kind of analysis, and detailed tabulations of values of β and α will thus appear in section 23, where viscosity will be discussed in detail.

18e. The complete expression for $P(\theta)$. As was mentioned earlier, the complete expression for $P(\theta)$ depends on the geometric shape of the scattering particle. The simplest to evaluate is that for long thin rods, which may be considered as composed of $\sigma + 1$ scattering units, in a straight line, adjacent units being separated by small distances l. The total length of the rod is thus $L = \sigma l$.

The determination of the sum of equation 18–13 then becomes similar to that encountered in section 9d, where a sum of the form $\sum_i \sum_j a_i a_j$ was evaluated. There will be 2σ terms for which $r_{ij} = 0$, $2(\sigma - 1)$ terms for which $r_{ij} = l$, $2(\sigma - 2)$ terms for which $r_{ij} = 2l$, and, in general, $2(\sigma - k)$ terms for which $r_{ij} = kl$. Thus, by equation 18–13,

$$P(\theta) = \frac{1}{(\sigma + 1)^2} \sum_{k=0}^{\sigma} 2(\sigma - k) \frac{\sin \mu kl}{\mu kl} \qquad (18\text{--}22)$$

The scattering elements may be chosen sufficiently small so that $\sigma + 1$ may be replaced by σ, and the sum of equation 18–22 by an integral. Thus

$$P(\theta) = \frac{2}{\sigma} \int_{k=0}^{\sigma} \frac{\sin \mu kl}{\mu kl} \, dk - \frac{2}{\mu l \sigma^2} \int_{k=0}^{\sigma} (\sin \mu kl) \, dk \qquad (18\text{--}23)$$

The integration of the second term is straightforward. By placing $\mu kl \equiv u$ and recalling that $\sigma l = L$, we obtain the final result

$$P(\theta) = \frac{2}{\mu L} \int_{u=0}^{\mu L} \frac{\sin u}{u} \, du - \left[\frac{\sin (\mu L/2)}{\mu L/2} \right]^2 \qquad (18\text{--}24)$$

The first integral is a simple series,

$$\int \frac{\sin u}{u} \, du = u - \frac{u^3}{3.3!} + \frac{u^5}{5.5!} - \cdots$$

which obviously need be evaluated only for $u = \mu L$. Equation 18–24 was first derived by Neugebauer.[41]

A similar equation, for large solid spheres, was derived by Rayleigh[42] in 1910, and $P(\theta)$ has also been evaluated for solid ellipsoids of revolution, cylinders, and flat discs. References to these are given by Geiduschek and Holtzer.[9]

The simplest relation for flexible molecules is that of Debye.[13] It applies to polymer chains in which the separation between any segments i and j follows a Gaussian distribution; i.e., the probability of a given separation h_{ij} is proportional to $h_{ij}^2 \exp(-\text{constant} \times h_{ij}^2)$. (The complete equation for the separation between the first and the terminal segment, i.e., the distribution function for end-to-end distances, is given by equation 10–19.) Debye's result is

$$P(\theta) = (2/w^2)(e^{-w} + w - 1) \qquad (18\text{--}25)$$

where

$$w = \mu^2 R_G^2$$

This equation applies when all the polymer molecules have the same molecular weight. The modification required for a heterogeneous sample is given by Benoit.[38]

An extension of equation 18–25, applicable to *stiff* chains, to which equation 10–19 cannot be expected to apply, has been derived by Peterlin.[57]

An important paper by Benoit and Goldstein[56] has shown that the function $P(\theta)$ which we obtain by assuming a *spherically symmetrical* distribution of segments about the center of mass, as given by equation 10–26, is quite different from the function given by equation 18–25, which, as we have said, rests on a Gaussian distribution of separation between segments. This result throws some doubt on the statement made on p. 178 (and used in several places in this book) that equation 10–26 can be used as a suitable starting point for theories of the behavior of all flexible polymer molecules which obey equation 10–19.

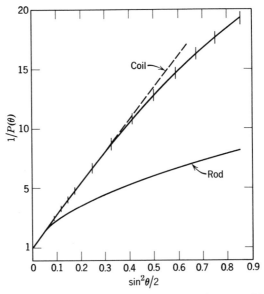

Fig. 18–4. The angular dependence of light scattering from a DNA solution. Experimental values of $1/P(\theta)$ are compared with theoretical curves for a flexible coil (equation 18–25) and for a rigid rod (equation 18–24). (Doty and Bunce.[43])

An example of the application of the complete equation for $P(\theta)$ to deduce information concerning molecular configuration is provided by the data of Fig. 18–4 for deoxypentose nucleic acid (DNA). It shows an experimental plot of $1/P(\theta)$ for this substance, as compared to theoretical curves according to equations 18–24 and 18–25. Clearly the nucleic acid molecule resembles a random coil more closely than a rod. (The Peterlin equation for a stiff coil provides an even better fit.[9]) If the double-stranded helical structure assigned to DNA on the basis of x-ray data (Fig. 4–13) is maintained in solution, and there is much chemical evidence that it is,

Fig. 18–4 indicates that occasional points of flexibility must interrupt this structure.

It might be expected that there would be some difficulty in constructing an optically acceptable scattering cell capable of being used for measurements at all possible values of the angle θ. On the other hand, there should be no difficulty in constructing a cell with planar windows at specified angles of θ. In particular, a semi-octagonal cell has often been used which permits measurements only at $\theta = 45$, 90, and 135°. If this kind of cell is used, the angular dependence is usually expressed in terms of the dissymmetry, z, defined as the ratio, $i_{45°}/i_{135°}$. Since $\cos^2 \theta$ has the same value for both 45 and 135°, $\mathcal{K}$ of equation 17–29 also has the same value for both angles, so that the intensity which would be obtained in the absence of internal interference would be the same, and $z = 1$. When internal interference is present,

$$z = \frac{i_{45°}}{i_{135°}} = \frac{R_{45°}}{R_{135°}} = \frac{P(45°)}{P(135°)} \tag{18–26}$$

The value of z is measured at several concentrations and extrapolated to zero concentration, so that the preceding equations for $P(\theta)$ can be applied to evaluate $P(45°)/P(135°)$.

If we now have previous knowledge of the configuration of the scattering particle, we can obtain its dimensions by equations 18–24 and 18–25, or similar equations for other shapes. For a long rod, for instance, z becomes a measure of the length L, the relation between z and L following from equation 18–24 with the substitution $\mu L = 1.5307\pi L/\lambda$ for $P(45°)$ and $\mu L = 3.6955\pi L/\lambda$ for $P(135°)$.

This method is obviously a far more limited one than that employing measurements over a wide range of the angle θ. Since, in actual fact, there appears to be no real difficulty in constructing a scattering cell for measurements at all possible values of θ (as the experimental data cited earlier in this section amply testify), the dissymmetry method leaves little to recommend it.

18f. The Mie equation for large spheres. In conclusion we should point out again that we have omitted virtually all reference to particles with dimensions greater than $\lambda/2$. For such particles the scattered intensity diminishes very rapidly with increasing θ, and $P(\theta)$ may become an oscillating function of θ. Another important factor, however, arises. It is that for large particles the difference between the refractive index of the particles and that of the solvent produces a distortion of the electric field of the incident radiation. The complete treatment then required is exceedingly complicated. The problem was solved for spheres by Mie[44] fifty years ago, and a solution for oriented cylinders has also been obtained.[5]

18g. X-ray scattering. In view of the fact that the internal interference of light scattered from large particles provides us with a powerful tool for estimating their shape and dimensions, it would be desirable to extend the method to smaller particles, by use of a shorter wavelength. A wavelength of about 100 Å, for example, would provide the same kind of information for the small globular proteins as the wavelengths of the mercury arc provide for the more extended molecules discussed earlier. Unfortunately, we know of no material which transmits a wavelength of 100 Å; i.e., this wavelength is experimentally inaccessible. It is necessary to go to the range of 0.7 to 2 Å, i.e., the x-ray range, to find another experimentally accessible source of radiation. Even for the smallest proteins this corresponds to a ratio, $R_G/\lambda \sim 10$.

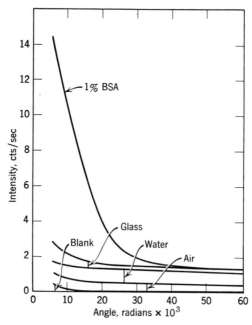

Fig. 18–5. Low angle x-ray scattering by bovine serum albumin (BSA). The figure compares the intensity of scattering by the protein solution with the background intensity due to air, water, and the containing vessel. (Anderegg et al.[45])

From the preceding discussion it is clear that this value of R_G/λ is higher than desirable. As a result, almost complete interference is observed at values of θ as low as 2 or 3 Å. A compensating factor, however, is that measurements of x-ray scattering can be made at values of θ much lower than can be reached in the scattering of visible light. Thus we can in fact determine the radius of gyration by suitable extrapolation.

Some data obtained in this way have been included in Table 18–2, and an experimental scattering curve is shown in Fig. 18–5. The experimental difficulties are greater than for light scattering, and careful corrections have to be made for scattering by the glass walls of the vessel containing the solution, etc.

General References

1. N. H. Frank, *Introduction to Electricity and Optics*, 2nd. ed., McGraw-Hill Book Co., New York, 1950; J. C. Slater and N. H. Frank, *Introduction to Theoretical Physics*, McGraw-Hill Book Co., New York, 1933.
2. M. Abraham and R. Becker, *The Classical Theory of Electricity and Magnetism* (J. Dougall, trans.), Blackie and Son, London, 1937.
3. M. Born, *Optik*, Julius Springer, Berlin, 1933.
4. J. A. Stratton, *Electromagnetic Theory*, McGraw-Hill Book Co., New York, 1941.
5. H. C. van de Hulst, *Light Scattering by Small Particles*, John Wiley and Sons, New York, 1957.
6. G. Oster, "The Scattering of Light and its Applications to Chemistry," *Chem. Revs.*, **43**, 319 (1948).
7. P. Doty and J. T. Edsall, "Light Scattering in Protein Solutions," *Advances in Protein Chem.*, **6**, 35 (1951).
8. K. A. Stacey, *Light-Scattering in Physical Chemistry*, Academic Press, New York, 1956.
9. E. P. Geiduschek and A. Holtzer, "Application of Light Scattering to Biological Systems," *Advances in Biol. and Med. Phys.*, **6**, 431 (1958).

Specific References

10. J. W. Strutt (Lord Rayleigh), *Phil. Mag.*, [4], **41**, 107, 447 (1871).
11. M. Smoluchowski, *Ann. Physik*, [4], **25**, 205 (1908); *Phil. Mag.*, [6] **23**, 165 (1912).
12. A. Einstein, *Ann. Physik*, [4], **33**, 1275 (1910).
13. P. Debye, *J. Applied Phys.*, **15**, 338 (1944); *J. Phys. and Colloid Chem.*, **51**, 18 (1947).
14. P. P. Debye, *J. Applied Phys.*, **17**, 392 (1946).
15. B. H. Zimm, *J. Chem. Phys.*, **16**, 1099 (1948).
16. B. A. Brice, M. Halwer, and R. Speiser, *J. Opt. Soc. Am.*, **40**, 768 (1950).
17. C. I. Carr and B. H. Zimm, *J. Chem. Phys.*, **18**, 1616 (1950).
18. B. A. Brice, G. C. Nutting, and M. Halwer, *J. Am. Chem. Soc.*, **75**, 824 (1953).
19. G. E. Perlman and L. G. Longsworth, *J. Am. Chem. Soc.*, **70**, 2719 (1948).
20. J. Cabannes, *La diffusion moléculaire de la lumière*, Presses universitaires de France, Paris, 1929.
21. P. Doty and S. J. Stein, *J. Polymer Sci.*, **3**, 763 (1948).
22. E. P. Geiduschek, *J. Polymer Sci.*, **13**, 408 (1954).
23. J. G. Kirkwood and R. J. Goldberg, *J. Chem. Phys.*, **18**, 54 (1950).
24. W. H. Stockmayer, *J. Chem. Phys.*, **18**, 58 (1950).
25. L. K. Christensen, *Compt. rend. trav. lab. Carlsberg*, *Sér. chim.*, **28**, 37 (1952).
26. M. Halwer, G. C. Nutting, and B. A. Brice, *J. Am. Chem. Soc.*, **73**, 2786 (1951).
27. A. V. Güntelberg and K. Linderstrøm-Lang, *Compt. rend. trav. lab. Carlsberg*, *Sér. chim.*, **27**, 1 (1949).
28. G. Scatchard, A. C. Batchelder, and A. Brown, *J. Am. Chem. Soc.*, **68**, 2320 (1946).
29. W. B. Dandliker, *J. Am. Chem. Soc.*, **76**, 6036 (1954).
30. J. Oth and V. Desreux, *Bull. soc. chim. Belges.*, **63**, 285 (1954).
31. H. P. Frank and H. F. Mark, *J. Polymer Sci.*, **17**, 1 (1955).
32. W. R. Krigbaum, *J. Am. Chem. Soc.*, **76**, 3758 (1954).
33. W. R. Krigbaum and D. K. Carpenter, *J. Phys. Chem.*, **59**, 1166 (1955).

34. P. Doty and G. E. Myers, *Discussions Faraday Soc.*, 1953, No. 13, 51.
35. P. Debye, *Ann. Physik*, [4], **46**, 809 (1915).
36. A. Guinier, *Ann. Phys.*, [11], **12**, 161 (1939).
37. H. Benoit, A. M. Holtzer, and P. Doty, *J. Phys. Chem.*, **58**, 635, (1954).
38. H. Benoit, *J. Polymer Sci.*, **11**, 507 (1953).
39. P. Doty, J. H. Bradbury, and A. M. Holtzer, *J. Am. Chem. Soc.*, **78**, 947 (1956).
40. P. Outer, C. I. Carr, and B. H. Zimm, *J. Chem. Phys.*, **18**, 830 (1950).
41. T. Neugebauer, *Ann. Physik*, [5], **42**, 509 (1943).
42. Lord Rayleigh, *Proc. Roy. Soc.*, A **84**, 25 (1910).
43. P. Doty and B. H. Bunce, *J. Am. Chem. Soc.*, **74**, 5029 (1952).
44. G. Mie, *Ann. Physik*, [4], **25**, 377 (1908).
45. J. W. Anderegg, W. W. Beeman, S. Shulman, and P. Kaesberg, *J. Am. Chem. Soc.*, **77**, 2927 (1955).
46. R. C. Williams, R. C. Backus, and R. L. Steere, *J. Am. Chem. Soc.*, **73**, 2062 (1951).
47. H. Boedtker and N. S. Simmons, *J. Am. Chem. Soc.*, **80**, 2550 (1958).
48. M. L. Hunt, S. Newman, H. A. Scheraga, and P. J. Flory, *J. Phys. Chem.* **60**, 1278 (1956).
49. C. J. Stacy and J. F. Foster, *J. Polymer Sci.*, **20**, 57 (1956); **25**, 39 (1957).
50. S. Erlander and D. French, *J. Polymer Sci.*, **20**, 7 (1956).
51. J. T. Edsall. H. Edelhoch, R. Lontie, and P. R. Morrison, *J. Am. Chem. Soc.*, **72**, 4641 (1950).
52. A. G. Malmon, *Biochim. et Biophys. Acta*, **26**, 233 (1957).
53. A. L. Holtzer and S. Lowey, *J. Am. Chem. Soc.*, **78**, 5954 (1956), **81**, 1370 (1959).
54. B. R. Leonard, Jr., J. W. Anderegg, S. Shulman, P. Kaesberg, and W. W. Beeman, *Biochim. et Biophys. Acta*, **12**, 499 (1953).
55. H. Boedtker and N. S. Simmons, *J. Am. Chem. Soc.*, **80**, 2550 (1958).
56. H. Benoit and M. Goldstein, *J. Chem. Phys.*, **21**, 947 (1953).
57. A. Peterlin, *Makromol. Chem.*, **9**, 244 (1953); *J. Polymer Sci.* **10**, 425 (1953).
58. M. Fixman, *J. Chem. Phys*, **23**, 2074 (1955).
59. S. N. Timasheff, H. M. Dintzis, J. G. Kirkwood, and B. D. Coleman, *Proc. Natl. Acad. Sci., U.S.*, **41**, 710 (1955).
60. J. G. Kirkwood and S. N. Timasheff, *Arch. Biochem. Biophys.*, **65**, 50 (1956).
61. J. Hermans, Jr. and J. J. Hermans, *J. Phys. Chem.*, **63**, 170, 175 (1959).
62. R. L. Sinsheimer, *J. Mol. Biol.*, **1**, 37 (1959).

6

TRANSPORT PROCESSES. VISCOSITY

19. FLOW PROCESSES IN VISCOUS FLUIDS

19a. Thermodynamics of irreversible processes.[3] The laws of thermodynamics are sweeping generalizations which characterize the state of equilibrium of a system, regardless of any other considerations. They may be adapted to any conceivable kind of equilibrium, as was shown, for instance, in section 16, where they were applied to equilibrium in the presence of an external force. They are sufficiently general to be independent of the structure of a chemical system; i.e., we know that they apply to macromolecular solutions even though our knowledge of the structure of such solutions is incomplete and our concept of this structure may undergo changes.

Thermodynamics tells us that a system not at equilibrium will always move in the direction of equilibrium, so that processes not at equilibrium are always irreversible, but no generalization as powerful as the laws of thermodynamics exists to govern the rate with which such irreversible processes occur. In recent years, however, it has become generally accepted that certain laws can be formulated for irreversible processes in systems subject to small forces, i.e., in systems not far removed from the equilibrium state. These laws comprise what is called *thermodynamics of irreversible processes*. The most important of these laws, from the point of view of this chapter, are the so-called *phenomenological equations*, which state that the flow of matter is a linear function of the forces which cause the flow.

The forces which are involved in the processes to be described in this chapter are always exceedingly small, so that the phenomenological equations may be assumed applicable. We can, furthermore, usually design our experiments so that all such forces act along a single direction, so that the phenomenological equations may be expressed in terms of a single dimension x. The equations then take the form

$$J_i = \sum_k L_{ik} X_k \tag{19-1}$$

where J_i is the flow (or flux) per unit area, of component i, across any plane perpendicular to x, and the X_k are the forces which act in the direction of x. The flows J_i may be expressed in any convenient units, such as moles per square centimeter per second, molecules per square centimeter per second, or grams per square centimeter per second, and the X_k may likewise represent force per molecule, force per gram, etc. The L_{ik} of equation 19-1 are called *phenomenological coefficients*. Like thermodynamic variables, they are functions of temperature, pressure, and composition. Their values will also depend on the units in which the flows and forces are expressed. The important point is that they are independent of the magnitudes of the X_k as long as the X_k remain small.

Equation 19-1 applies not only to the flow of matter, i.e., to the flow of the various chemical components of a system, but also to the flow of heat, electricity, etc. An important aspect of the equation is that every kind of flow may in general arise from every kind of force. For example, a temperature gradient is a force which we think of naturally as leading to a flow of heat. The same force, however, will also produce a flow of the various chemical components, leading to the process of thermal diffusion. Conversely, a centrifugal force, which acts primarily on the elements of mass, as described in section 16, will also produce a flow of heat.

We shall be interested in applying equation 19-1 to flow processes in dilute liquid solutions. A separate equation will then apply to the flow of each component i of such a solution, but it will be sufficient to consider explicitly the flow of only $q - 1$ components of any system of q components. The reason for this is that we always study flow phenomena in a fluid which is at rest with respect to a cell which contains it. This condition automatically determines the flow of the qth component if $q - 1$ flows have been specified. In dilute solutions the natural choice is to consider explicitly the flow of solutes, leaving the flow of solvent to be determined, if needed, from the condition that the fluid as a whole is at rest. The same procedure is normally followed in specifying the composition of dilute solutions: only $q - 1$ composition variables are needed (cf. section 11). Since the phenomenological coefficients L_{ik} are functions of composition, they too depend on $q - 1$ composition variables, these being normally the solute concentrations.

It is implicit to the above discussion that the dimension x be chosen with reference to the cell in which flow is being studied, and this procedure will be followed here. This is, however, not always the logical way to define x, and the reader will find, in the literature of irreversible thermodynamics, that x is often chosen with the center of mass of the fluid as reference point. Another point of minor importance here is that flow in *compressible* fluids may be accompanied by a volume change, which would have to be taken into account when the condition that the fluid as a whole is at rest is specifically introduced. This factor need not be considered for macromolecular solutions which can nearly always be taken as being incompressible.

The phenomenological equations are not the only general relations of irreversible thermodynamics. Another general relation is Onsager's equation for the rate of entropy increase[111]; another is Onsager's reciprocal relation which states that each $L_{ik} = L_{ki}$ if the frame of reference is properly chosen. The reader is referred to de Groot[3] for discussion of these relations. We shall not make use of them here.

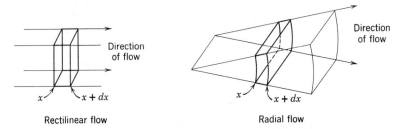

Fig. 19-1. Volume elements for derivation of the equation of continuity. The cross-sectional area is independent of x for rectilinear flow, but proportional to x for radial flow.

Finally, it is of course necessary that conservation of energy and conservation of mass apply to any flow situation. In the absence of chemical reactions conservation of mass must apply separately to each component.

Conservation of mass is frequently expressed in terms of the *equation of continuity*. Consider any volume element (Fig. 19-1) with parallel boundaries at x and $x + dx$, perpendicular to x. Let the cross-sectional areas at x and $x + dx$ be A and $A + dA$, respectively; let the flow of any component at x be J_i in the positive x direction, and let it be $J_i + dJ_i$ at $x + dx$. If J_i is expressed in moles per square centimeter per second, the number of moles accumulating in the volume element per second is $J_iA - (J_i + dJ_i)(A + dA)$. Ignoring products of infinitesimals, we see that this becomes $-d(AJ_i)$ or $-[\partial(AJ_i)/\partial x]_t\, dx$. This number divided by the volume of the element ($A\, dx$) is, however, the change in concentration per second; i.e., where C_i represents moles per cubic centimeter,

$$\left(\frac{\partial C_i}{\partial t}\right)_x = -\frac{1}{A}\left[\frac{\partial(AJ_i)}{\partial x}\right]_t \tag{19-2}$$

(For rectilinear flow A would be a constant and would cancel.) In the absence of chemical reactions which can convert one component into another, equation 19–2 must apply separately to each component. (The equation is applicable, of course, with C_i and J_i expressed in any consistent units.)

It is sometimes of interest to consider problems in which chemical reactions occur simultaneously with flow. In that case conservation of mass for each separate component would have to include terms for the rate of interconversion of components. Equation 19–2 would hold without such terms only for $\sum_i (\partial c_i/\partial t)_x$ expressed in grams per cubic centimeter per second.

In concluding this brief introduction to the thermodynamics of irreversible processes, it should be pointed out that this subject can give no information concerning the magnitude of the coefficients L_{ik} of equation 19–1, just as equilibrium thermodynamics can give no information about the magnitude of chemical potentials and similar thermodynamic variables. To calculate these quantities requires the use of specific mechanical models (e.g., those used in section 12 to compute chemical potentials), and it is with such specific models that this chapter is primarily concerned. The value of the general approach is that it provides a criterion of acceptability for mechanical theories which can be proposed. If they are not consistent with equation 19–1, they cannot be true. The mechanical theories, if consistent with equation 19–1, will provide numerical values for the phenomenological coefficients L_{ik}.

19b. Chemical potential gradient as a force. The state of equilibrium in a system may be characterized by the statement that the sum of all forces acting on any component is zero. On the other hand, a purely thermodynamic statement of equilibrium, given by equation 16–8, is that the total chemical potential must be independent of x, or that $(\partial \mu_{\text{total}}/\partial x)_t = 0$. The total potential μ_{total} was defined as the sum of the potential energy in the field of an externally applied force and the chemical potential; i.e., where $\mathscr{V}_i$ is the potential energy per mole of component i in an externally applied field of force, and μ_i its chemical potential,

$$-(\partial \mathscr{V}_i/\partial x)_t - (\partial \mu_i/\partial x)_t = 0 \qquad (19\text{–}3)$$

neither of the two terms of equation 19–3 by itself being zero. Now $-(\partial \mathscr{V}_i/\partial x)$ is just the applied force in the positive x direction (cf. equation 16–3), so that, since equation 19–3 must be compatible with the statement that all forces vanish at equilibrium, the term $-(\partial \mu_i/\partial x)$ must also represent a force in the positive x direction. It is for this reason that μ_i, which, by definition, is the partial molal free energy of component *i*, is called the *chemical potential*. Equation 19–3 indicates that it clearly represents a property in all respects similar to gravitational potential, electrostatic potential, etc. It will become clear in section 21 that $-(\partial \mu_i / \partial x)$ is in fact the force which results in diffusion and, mechanically, represents the force due to the thermal motion of molecules.

It will be convenient to separate this force from externally applied forces. We must suppose, in doing so, that each force acts on each component, i.e., that each component will be subject to flow as a result of gradients in the chemical potentials of other components. We thus write

$$J_i = \sum_k L_{ik} F_k - \sum_j L_{ij} \left(\frac{\partial \mu_j}{\partial x} \right)_t \qquad (19\text{--}4)$$

where the F_k represent the various external forces, and the second sum includes the chemical potentials of all but one of the components, i.e., in general, all components other than the solvent.

It may be pointed out here that we shall be interested in isothermal flow only, so that no thermal diffusion forces need be considered.

19c. Viscosity.[1,2] A viscous fluid is a fluid in which there are attractive forces between neighboring portions of the fluid. Any motion of one part of the fluid, relative to another, is opposed by these attractive forces. It is clear that all liquids must be viscous fluids, for the existence of attractive forces is a prerequisite for the existence of the liquid state. Real gases are also viscous, but their viscosity is much lower than that of liquids.

In the theoretical treatment of viscosity a liquid is thought of as a structureless continuous fluid. Consider two adjacent volume elements of such a liquid. If the two elements are at rest with respect to one another, the net effect of the forces between them must be zero; if it were not, the elements would move under the influence of the force. If, however, one of the volume elements is in motion relative to the other, under the influence of an external force, the local forces will oppose such motion, striving to return the volume elements to their equilibrium positions. This opposing force is called the *frictional force*.

It should not be difficult for the reader, if he so desires, to visualize a mechanism for friction in terms of some suitable picture for liquid structure. At rest, the molecules are in equilibrium positions with respect to one another. Any motion of one part of the liquid with respect to a neighboring portion will require that the molecules be pushed away from their equilibrium positions, and this is opposed, of course, by a restoring force in the direction of equilibrium. When the molecules have been removed a sufficient distance from the original equilibrium positions this force will vanish, but they will by then presumably be in new equilibrium positions, displacement from which will again be opposed by a restoring force. Compare, in this connection, the liquid model of Eyring and coworkers,[119] in which successive equilibrium positions are pictured as separated by potential energy barriers.

To define viscosity quantitatively, we may refer to Fig. 19–2. Suppose that one of the volume elements there pictured is moving with a velocity du relative to the second element. It is to be expected that the frictional force will be proportional to the relative velocity du and to the contact area dA between the adjacent volume elements. It should be inversely proportional to the distance dx between the centers of the elements. The proportionality constant relating the force to these variables is known as the *coefficient of viscosity*, or, more simply, as the *viscosity*, η. Calling the frictional force F_f, we have

$$F_f = \eta(du/dx)\, dA \qquad\qquad (19\text{–}5)$$

This definition of viscosity is originally due to Newton. (This definition is, of course, a microscopic one, not in terms of measurable quantities. See section 19h for its relation to quantities which can be measured.)

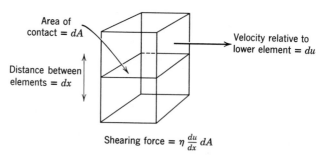

Area of contact $= dA$

Velocity relative to lower element $= du$

Distance between elements $= dx$

Shearing force $= \eta \dfrac{du}{dx}\, dA$

Fig. 19–2. Definition of viscosity coefficient, η. The shearing force between the two elements is set equal to $\eta(du/dx)\, dA$, thus defining the value of η.

If a liquid is in motion, one of the results of the viscous force is to prevent any very abrupt change in velocity with distance. In terms of equation 19–5, the larger du/dx the greater the force tending to decrease it. This has two important consequences:

(1) If a liquid flows along a tube, the wall of the tube is by definition stationary. Since attractive forces exist between the material in the wall of the tube and the liquid, elements of volume immediately adjacent to the wall can have but a very low velocity. Those slightly further removed may have a slightly higher velocity, and so forth. There will thus be velocity gradients everywhere in the tube. The maximum velocity will be at the center of the tube. In calculating the total resistance to flow, as we shall do in section 19g, for example, we must take into account the frictional forces existing (owing to a non-zero du/dx) at all points in the tube. We shall use equation 19–5, and η will represent throughout the coefficient applicable to the friction between volume elements identical

in composition (i.e., it will be the viscosity of the liquid), except at the wall of the tube, where η would represent the coefficient applicable to the friction between the liquid and the material of which the wall is composed. The force at the wall is, however, a negligibly small portion of the whole, so that we may without error suppose that η there, too, is the viscosity of the liquid.

(2) If instead we consider a solid particle moving through a stationary liquid, the same situation prevails. Elements of fluid immediately adjacent to the particle will move with almost the same velocity as the particle, those somewhat further away with a slightly lower velocity, etc. Liquid will be in motion up to a considerable distance away from the particle, and the total frictional force will involve again largely the friction between adjacent elements of liquid. The resistance to motion thus is independent of the material of which the particle is made; it depends only on the viscosity of the liquid.

19d. The frictional coefficient. The simplest problem in mechanics is that of a single particle moving under an applied force F. In free space the motion is described by the equation $ma = F$ where m is the mass of the particle and $a = du/dt$ is its acceleration. Thus the velocity of the particle increases continuously. In a viscous fluid, however, an opposing frictional force is set up as soon as the velocity becomes finite. If the velocity is not too large, this frictional force will be directly proportional to the velocity u, so that the equation of motion becomes

$$m\, du/dt = F - fu \tag{19-6}$$

where f is called the frictional coefficient. Thus du/dt decreases as u increases until fu becomes equal to F when du/dt becomes zero, and no further change in velocity occurs. The final steady-state velocity which will be maintained henceforth is given by setting $du/dt = 0$; i.e.,

$$u = F/f \tag{19-7}$$

If F/f is large, high velocities are reached before a steady state is attained. If F/f is small, however, the terminal velocity is low and thus attained almost immediately upon application of the force. Transport processes in macromolecular solutions involve only very small applied forces (relative to f), so that equation 19-7 will apply throughout except at the very instant when a force is first applied.

It is to be noted that equation 19-7 is consistent with the phenomenological equation, equation 19-1. The "particles" of the mechanical description of the system are one of the "components" of the thermodynamic description. The flow of particles across unit cross-section (perpendicular to x) in time dt is simply the number of particles in a volume $u\,dt$. Since the force F of equation 19–7 is the force per particle, J_i must be expressed in terms of particles per square centimeter per second. If N_i is the number of particles per cubic centimeter, the flow in time dt is $uN_i\,dt$ or, per second,

$$J_i = uN_i = N_iF/f_i \tag{19-8}$$

The coefficient L_{ik} of equation 19–1 is seen to be equal to N_i/f_i.

We shall more often express J_i in terms of moles per square centimeter per second and F as force per mole. With these units and with f_i still the frictional coefficient per particle, L_{ik} becomes smaller by $\mathcal{N}^2$, where $\mathcal{N}$ is Avogadro's number; i.e.,

$$L_{ik} = N_i/\mathcal{N}^2f_i = C_i/\mathcal{N}f_i \tag{19-9}$$

where C_i represents concentration in moles per cubic centimeter. Equation 19–9 will be applicable whenever a mechanical force k is acting on a component i which can be considered to consist of discrete particles of only one kind.

19e. Frictional coefficients of solid spheres and ellipsoids. As was indicated in section 19c, the resisting force, and, therefore, the frictional coefficient, will be proportional to the viscosity of the liquid and independent of the material of which the moving particle is made. Its calculation is a formidable problem involving a computation of the pattern of liquid flow in the entire region around the particle, followed by integration of equation 19–5 over all volume elements in such a region. The problem was solved for spherical particles by Stokes (in 1856)[19] and for ellipsoids of revolution by Perrin.[20] (See also Herzog, Illig, and Kudar.[21]) A derivation of Stokes' equation is given by Page.[2]

The problem is necessarily one which requires a three-dimensional equation of motion for fluid flow. This equation may be written quite generally as

$$\rho(\partial \mathbf{u}/\partial t) + \rho \mathbf{u} \cdot \nabla \mathbf{u} = \rho \mathbf{F} - \nabla p + \tfrac{1}{3}\eta \nabla \nabla \cdot \mathbf{u} + \eta \nabla \cdot \nabla \mathbf{u} \tag{19-10}$$

Here $\mathbf{u}$ is the velocity of flow at any time t at any point (x, y, z), which may be thought of as composed of components u_x, u_y, u_z, such that $\mathbf{u} = u_x\mathbf{i} + u_y\mathbf{j} + u_z\mathbf{k}$, where $\mathbf{i}$, $\mathbf{j}$, and $\mathbf{k}$ are unit vectors along the x, y, and z directions. The term $\mathbf{F}$ represents an external applied force per unit mass; ρ and p are, respectively, the density and hydrostatic pressure at any point (x, y, z); ∇ is the vector differential operator, $\nabla = \mathbf{i}(\partial/\partial x) + \mathbf{j}(\partial/\partial y) + \mathbf{k}(\partial/\partial z)$; and η is the coefficient of viscosity. The fluid is assumed isotropic so that a single viscosity coefficient η applies regardless of position or direction of contiguous surface elements.

Equation 19–10 represents a general equation of the type mass ×
acceleration = force, per unit volume of fluid. The left-hand side represents the total derivative $\rho\, du/dt$, split into two parts, the first of which is
the change in **u** with time at a given position, and the second the change
in **u** which would result from changes in position alone. The right-hand
side is the sum of the acting forces, the first term representing external
forces, the second hydrostatic pressure, the third and fourth the frictional
forces.

The equation of motion may be considerably simplified for application
to the present problem. In the first place we are interested in the flow of
only incompressible liquids, so that ρ becomes the same everywhere. This
condition requires that $\nabla \cdot \mathbf{u} = 0$, so that the third term on the right-hand
side vanishes. Further, we are interested in only the steady-state solution,
i.e., the case where the flow pattern becomes independent of time. Thus
$\partial \mathbf{u}/\partial t = 0$. Next, we shall limit the applications to very small velocities,
so that the term $\mathbf{u} \cdot \nabla \mathbf{u}$, which depends on the square of the magnitude of
u, may be neglected. Finally, in the problem here under consideration,
no *external* forces are acting on any part of the fluid. (The external force
which acts on the *particle* immersed in the fluid is by definition one which
does not act on the fluid itself, its purpose being to move the particle
through the fluid. The fact that this force exists expresses itself as a
boundary condition, namely that the layer of fluid in direct contact with
the particle must have the same velocity as the particle itself.) Thus **F**
may be set equal to zero, and equation 19–10 becomes

$$\eta \nabla \cdot \nabla \mathbf{u} = \nabla p \qquad (19\text{–}11)$$

We require the solution of equation 19–11 with the condition that the
spherical particle have a constant velocity u_0 with respect to the liquid
(this being the velocity u of equation 19–7). However, an entirely equivalent solution may be obtained by supposing that the liquid moves with a
velocity u_0 with respect to the particle, which is taken to be stationary;
that is, the mathematical boundary conditions may be stated as $\mathbf{u} = 0$
at the surface of the particle and $\mathbf{u} = -u_0$ in a particular direction at all
points far removed from the particle.

With these boundary conditions equation 19–11 gives **u** and p as a
function of position. (The hydrostatic pressure, of course, is not a constant, being greater upstream along the direction of flow.)

The solution to equation 19–11 may be used to calculate the total force
on the immersed particle. This force consists of two parts: (*a*) the force
due to the variation of p with position, which is normal to the surface, and
(*b*) the frictional force, which is parallel to the surface. When integrated
over the entire surface of the sphere, the net effect of both forces turns out

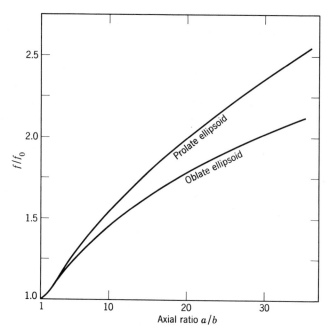

Fig. 19–3. Perrin's factor for the frictional coefficient of ellipsoids for relatively low values of the axial ratio, a/b.

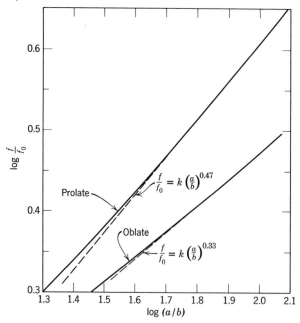

Fig. 19–4. A logarithmic plot of Perrin's factor for large values of a/b. The equations are those for the straight lines which are indicated in the figure.

(as is to be expected) to be in the direction of fluid flow. Both factors are proportional to the coefficient of viscosity and to the radius R of the particle. The contributions of factors (a) and (b) are, respectively, $-2\pi\eta Ru_0$ and $-4\pi\eta Ru_0$, so that the total force on the particle is $-6\pi\eta Ru_0$. To maintain the particle stationary with respect to the moving liquid requires a force which exactly counterbalances the force exerted by the liquid; i.e.,

$$F = 6\pi\eta Ru_0 \qquad (19\text{--}12)$$

(It may at first sight be surprising that the force due to hydrostatic pressure is, like the shearing force, proportional to η. It should be recalled, however, that, were it not for viscosity, no pressure gradient would be required to maintain liquid flow.)

The force given by equation 19–12 is, of course, exactly the same force which is needed to give a velocity u_0 to a particle moving through a stationary liquid, so that, by comparison with equation 19–7, we get Stokes' equation for the frictional coefficient,

$$f_0 = 6\pi\eta R \qquad (19\text{--}13)$$

where f_0 is to be used to designate the frictional coefficient of a sphere.

Perrin[20] and Herzog, Illig, and Kudar[21] independently extended Stokes' equation to ellipsoids of revolution. Their equations are most conveniently expressed in terms of the ratio of the frictional coefficient, f, to that which would be observed for a sphere of the same volume, the latter, f_0, being given by equation 19–13. Different relations are required for the two kinds of ellipsoids illustrated by Fig. 7–2. We obtain for *prolate ellipsoids* (semi-axes a, b, b)

$$\frac{f}{f_0} = \frac{f}{6\pi\eta R_0} = \frac{(1 - b^2/a^2)^{\frac{1}{2}}}{(b/a)^{\frac{2}{3}} \ln \dfrac{1 + (1 - b^2/a^2)^{\frac{1}{2}}}{b/a}} \qquad (19\text{--}14)$$

and for *oblate ellipsoids* (semi-axes a, a, b)

$$\frac{f}{f_0} = \frac{f}{6\pi\eta R_0} = \frac{(a^2/b^2 - 1)^{\frac{1}{2}}}{(a/b)^{\frac{2}{3}} \tan^{-1}(a^2/b^2 - 1)^{\frac{1}{2}}} \qquad (19\text{--}15)$$

where R_0 is the radius of a sphere of equal volume to the ellipsoid, i.e., $\frac{4}{3}\pi R_0^3 = \frac{4}{3}\pi ab^2$ (prolate ellipsoid) or $\frac{4}{3}\pi a^2 b$ (oblate ellipsoid).

A graphical representation of these functions for low values of a/b is shown in Fig. 19–3. A logarithmic plot for high values of a/b is shown in Fig. 19–4. It is seen that $\log (f/f_0)$ approaches a linear dependence on $\log (a/b)$. Where $a/b \simeq 100$, f/f_0 varies roughly as $(a/b)^{0.48}$ for prolate ellipsoids or as $(a/b)^{0.33}$ for oblate ellipsoids.

The frictional force and the frictional coefficient of asymmetric particles must depend on their orientation with respect to the direction of flow; e.g., a rod encounters less resistance when moving lengthwise than when moving broadside. The equations here derived are based on random orientation. Random orientation will prevail when the flow velocity is sufficiently small. As was already indicated above, the flow velocity of particles of the size of macromolecules is likely to be very small under the influence of any applied force; i.e., the equations here given are the appropriate ones for macromolecular solutions.

Frictional coefficients of particles in the macromolecular size range cannot, of course, be measured directly by application of equation 19–7 because such particles are too small for direct observation. We shall show in sections 20 and 21, however, that frictional coefficients are simply related to diffusion and sedimentation coefficients and can thus be measured indirectly. Stokes' equation has been confirmed by measurement of the sedimentation coefficients of spherical polystyrene particles whose radii could be measured by electron microscopy.[22]

19f. Newtonian flow. Laminar flow. The coefficient of viscosity defined by equation 19–5 has been defined as a proportionality *constant* and thus independent of the actual velocity of flow. This situation does not necessarily prevail. For some liquids the phenomenon of flow itself gives rise to orientation of the constituent molecules, which in turn alters the retarding force between adjoining elements. This orientation effect may be expected to occur particularly in a liquid containing very asymmetric molecules, or easily deformed molecules, and at high velocity gradients. Liquids showing such orientation effects are said to be *non-Newtonian*; flow which may be described by a viscosity independent of velocity gradient is called *Newtonian*.

In the following sections we shall discuss the measurement and meaning of the viscosity of macromolecular solutions. We shall assume that Newtonian flow prevails. As a test of this η may be measured at several velocity gradients. If it increases with decreasing velocity gradient, then it is non-Newtonian (orientation at high velocity gradient will be such as to decrease the resistance to flow). At sufficiently low velocity gradient Newtonian behavior may be expected for all macromolecular solutions, for the thermal motion of macromolecules ensures random orientation when the shearing forces become small. Thus even when non-Newtonian behavior is observed, we may evaluate the Newtonian viscosity by extrapolation to zero velocity gradient.

When a fluid flows slowly along a tube without obstacles, the flow pattern observed experimentally is *streamline* or *laminar* flow. It is as if the tube were constructed of numerous parallel microscopic tubes along the direction of flow, the fluid in any one microscopic tube remaining always within this tube. At sufficiently high velocities, or near obstacles, this pattern of flow is disturbed. Vortices may appear, and, in general, large masses of fluid will move as a unit, possessing both velocity in the direction of motion and rotational velocity. This disturbed flow is called *turbulent flow*. In cylindrical tubes the transition from laminar to turbulent flow generally occurs when the *Reynolds' number* (a dimensionless quantity given by $2\rho a u/\eta$, where ρ, u, and η are the density, mean velocity, and viscosity of fluid in a tube of radius a) exceeds 2200.

It is not necessary to have a tube to show the transition from laminar to turbulent flow. If a lighted cigarette is placed on an ashtray smoke rising from it will be seen to exhibit laminar flow for a considerable distance. Transition to turbulent flow occurs as the smoke spreads out some distance above the cigarette or if an obstacle is placed in the laminar stream.

If large-scale flow experiments are to be used as a measure of viscosity the pattern of flow must be known; i.e., we must be able to determine the velocity of any volume element with respect to its neighbor (cf. equation 19–5). In general, this requires that the flow be laminar, since the flow pattern for turbulent flow is exceedingly complex. Fluid flow in narrow capillaries or between slowly rotating cylinders placed close together is laminar at all reasonable speeds of flow, and these form the basis for the measurement of viscosity, as discussed below.

19g. Poiseuille's law for capillary flow.[23] Consider a uniform capillary of radius a and length l through which a liquid flows under the influence of a uniform pressure P. The pressure is to be sufficiently small so that the flow is laminar. We may imagine the moment at which the pressure P is first applied. The liquid in the capillary will be accelerated. The frictional force will oppose the acceleration, and a steady state will soon be reached. We seek the flow velocity at any point after the steady state is reached. This velocity will be zero at the capillary wall and a maximum at the center. Symmetry considerations indicate that at intermediate points it must be a function only of the distance r from the center. Since liquids have negligible compressibility, the flow velocity must be independent of distance along the cylinder axis, for the same volume must cross any cross-section per second.

Consider (Fig. 19–5) a volume element contained between two concentric cylinders, a distance dr apart, with the inner cylinder located r centimeters from the center of the capillary. The volume element runs the whole length (l) of the capillary. Since the flow is laminar, the liquid flowing down this volume element does not mix with the liquid in adjacent volume elements.

The volume element we are considering is adjacent to two other volume elements, one on the outside with an area of contact $2\pi(r + dr)l$, and one on the inside with an area of contact $2\pi rl$. The fluid in the outside adjacent element is moving more slowly, that in the inside adjacent element more rapidly. Two frictional forces are therefore acting in opposite directions.

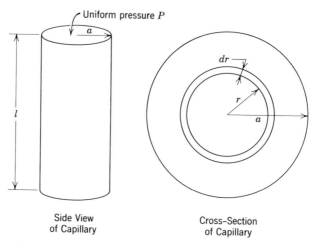

Uniform pressure P

Side View
of Capillary

Cross-Section
of Capillary

Fig. 19–5. Model of capillary used in the derivation of Poiseuille's law.

When du/dr has reached its equilibrium value, the algebraic sum of these forces must be equal in magnitude and opposite in direction to the applied force. The latter is given as the product of pressure and cross-sectional area, i.e., $P(2\pi r\, dr)$. The frictional force is given by equation 19–5. Thus

$$2\pi Pr\, dr = -2\pi(r + dr)l\eta \left(\frac{du}{dr}\right)_{r=r+dr} + 2\pi rl\eta \left(\frac{du}{dr}\right)_{r=r} \quad (19\text{--}16)$$

where η is the viscosity of the liquid. The sign on the right side of equation 19–16 is determined so that the frictional force on the outside, which is a retarding force, is reckoned positive. Since du/dr is negative, a negative sign is thus required in front of the first term on the right-hand side.

By Taylor's expansion du/dr (where $r = r + dr$) is equal to du/dr (where $r = r$) plus $(d^2u/dr^2)\, dr$. Neglecting terms in $(dr)^2$ and dividing each side by $2\pi\, dr$, we get for equation 19–16

$$-Pr/l\eta = \frac{du}{dr} + r\frac{d^2u}{dr^2} = \frac{d}{dr}\left(r\frac{du}{dr}\right)$$

where du/dr is now the value of the velocity gradient r centimeters from the center, and the subscript $r = r$ may be dropped.

Integration leads to

$$r \frac{du}{dr} = -\frac{Pr^2}{2l\eta} + A_1 \qquad (19\text{--}17)$$

where A_1 is an integration constant. Dividing by r and integrating again gives

$$u = -\frac{Pr^2}{4l\eta} + A_1 \ln r + A_2 \qquad (19\text{--}18)$$

where A_2 is another integration constant. Since the velocity is certainly finite when r is zero, A_1 must be set equal to zero. Since the velocity is zero when $r = a$, A_2 must be equal to $Pa^2/4l\eta$, so that

$$u = \frac{P}{4l\eta}(a^2 - r^2) \qquad (19\text{--}19)$$

This equation gives the well-known parabolic velocity profile in a capillary and is shown graphically in Fig. 19–6.

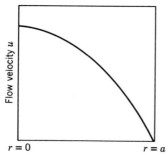

Fig. 19–6. Flow velocity for laminar flow in a capillary tube, as given by equation 19–19.

A more convenient property than the linear flow velocity is the volume rate of flow. The volume crossing any cross-section per second will have a different value for each volume element (Fig. 19–5), given by u multiplied by the cross-sectional area, i.e., by $2\pi u r \, dr$. The total volume flow per second is obtained by integration over all volume elements. Introducing equation 19–19, we have

$$U = \frac{\pi P}{2l\eta} \int_{r=0}^{a} (a^2 - r^2) r \, dr = \frac{\pi P a^4}{8\eta l} \qquad (19\text{--}20)$$

where U is the rate of flow in cubic centimeters per second. Equation 19–20 is known as Poiseuille's law.

It may be noted that Poiseuille's law is again of the general phenomen-ological type, the flow being proportional to the force, $\pi P a^2$. In this case we shall not try to interpret the phenomenological coefficient, $a^2/8\eta l$, in terms of a mechanical model of a liquid, but we shall make such an inter-pretation for the change in this coefficient which occurs when suspended particles are added to an initially homogeneous liquid. This will be done in section 19j, after the measurement of viscosity is discussed.

19h. Experimental measurement of the viscosity of liquids. Equation 19–20 forms the basis for one common method for the determination of viscosity, by use of a capillary viscometer.[24] The quantity measured is the time t required for a given volume to flow through the capillary, the driving force being provided by gravity. The initial and final liquid levels, h_1 and h_2, are fixed. The volume rate of flow varies as h drops from h_1 to h_2, and P drops correspondingly from $\rho g h_1$ to $\rho g h_2$, where g is the gravitational constant and ρ the density of the liquid. The volume dv flowing through the capillary in time dt, at an instant when the liquid level is h, is given by equation 19–20 as $dv = U\,dt$, so that the total time required for the liquid level to fall from h_1 to h_2 is

$$t = \int_{h_1}^{h_2} \frac{dv}{U} = \frac{8\eta l}{\pi g \rho a^4} \int_{h_1}^{h_2} \frac{dv}{h} \qquad (19\text{–}21)$$

The integral on the right-hand side of equation 19–21 is a constant of the viscometer, as are all other parameters except η and ρ. The product ρt is clearly a measure of the viscosity of the liquid, the product of constant terms being determined by measuring the flow time of a liquid of known ρ and η in the same viscometer. (It may also be noted that the flow time alone measures the quantity η/ρ, which is known as *kinematic viscosity*.)

This derivation has ignored the fact that the liquid issuing from the capillary has greater kinetic energy than that which enters the capillary. Thus the effective pressure difference driving the liquid through the capillary is actually less than $g\rho h$, for some of the pressure difference is used to accelerate the liquid. A correction for this difference must be applied for viscometers with short flow times (i.e., large velocities.)

Another type of viscometer is the Couette viscometer,[25] which consists of two concentric cylinders, one rotating and one at rest. The flow lines are circular, with u zero at the surface of the resting cylinder and a maximum at the surface of the rotating one. The viscosity of the liquid is linearly related to the torque exerted on the resting cylinder at a given speed of rotation. One other method worthy of mention is the "falling sphere" method, which, in essence, uses equations 19–7 and 19–13 to determine the viscosity.

19i. Viscosity and the dissipation of energy. In the preceding section were listed several experimental procedures which can be used as a macroscopic measure of viscosity. All of these methods have one thing in common: they measure the work which must be done to maintain steady-state flow; i.e., they measure the energy dissipated by the frictional forces inherent to viscous flow.

Consider, for example, a capillary tube such as that of section 19g (Fig. 19–5). The work done per second to maintain the steady-state flow in any volume element of cross-sectional area $2\pi r\, dr$, traveling down the capillary at a velocity of u centimeters per second, is clearly $2\pi Pur\, dr$, P being the pressure which drives the liquid through the capillary. The total work done in maintaining capillary flow for 1 sec is therefore the integral of this quantity over all volume elements. With equation 19–19 for u this gives

$$W = (\pi P^2/2l\eta) \int_{r=0}^{a} (a^2 - r^2)r\, dr = \pi a^4 P^2/8\eta l. \tag{19–22}$$

Using equation 19–20 we can express W in terms of the desired rate of flow U,

$$W = 8\eta l U^2/\pi a^4 \tag{19–23}$$

Thus the viscosity is clearly a measure of the energy dissipated (as heat) in maintaining a desired rate of flow in a capillary of given dimensions. Or, more generally, viscosity is a measure of the energy dissipated in maintaining flow in any apparatus of suitable design.

19j. The Einstein-Simha equation for the viscosity of suspensions. When solid particles much larger than solvent molecules, but still much smaller than the dimensions of experimental apparatus (such as capillary tubes), are suspended in a liquid, the viscosity on a *microscopic* scale remains unchanged; i.e., the viscosity of the solvent still governs the frictional force between adjacent volume elements. If such a suspension is allowed to flow through a capillary tube, however, or if any of the other methods for macroscopic measurement of viscosity are used, a change in the measured quantity is observed. This arises from the fact that a suspended particle considerably larger than a solvent molecule must stretch across a number of flow lines and must distort the pattern of flow.

To calculate the effect produced we must again use the three-dimensional equation of flow as given by equation 19–10. We again need take no explicit account of any externally applied force but can introduce the existence of such a force implicitly by specifying appropriate boundary conditions. The other simplifications applied to equation 19–10 in outlining the derivation of Stokes' equation also apply, so that equation 19–10 reduces to equation 19–11.

The effect produced by a spherical particle was computed by Einstein[6],[26] in 1906. He assumed no special pattern of flow but simply specified that the flow velocity **u** at all points far removed from the suspended particle must have an assigned value. This is, of course, equivalent to assigning a value to the total rate of flow. In the particular case of flow through a capillary, for instance, conservation of mass requires that the rate of flow through any cross-section of the capillary be equal to U. Thus a cross-section far removed from the suspended particle may be used to assign a value to the total rate of flow.

The assigned **u** far from the particle represents one of the boundary conditions needed to solve equation 19–11. The other boundary condition is again the one that **u**, at any point at the surface of the suspended particle, must be identical in direction and magnitude to the velocity of motion of that point of the particle. However, in contrast to the situation prevailing in the case of Stokes' law, the particle is not stationary but is carried along with the liquid. No external forces are acting on the particle other than the viscous forces implied by the equivalence of fluid and particle velocity at all points on the surface.

A solution of equation 19–11, with these boundary conditions, is now obtained for a region about the suspended particle. From the computed flow pattern the energy dissipated by viscous forces is calculated and compared with the corresponding energy dissipated in the absence of the suspended particle. It is found that the suspended particle increases W, so that, if the coefficient of viscosity is defined in terms of W, as in effect it is in all macroscopic experiments used to measure viscosity, this coefficient also increases. The result obtained by Einstein, for any number of suspended particles, far enough apart so as to prevent overlap of the distortion of the flow lines produced by each individual particle, is that the macroscopic viscosity η' of the suspension is

$$\eta' = \eta(1 + 2.5\phi) \tag{19-24}$$

where η is the viscosity of the solvent and ϕ the volume fraction of the particles in the total volume of the suspension. An important aspect of the equation is that η' for spheres is independent of the size of the spheres.

Extension of Einstein's treatment to asymmetric particles leads to the result that the formal viscosity depends on the orientation of the particles. If they are oriented parallel to the streamlines, as they would be for large velocity gradients, a smaller viscosity increment would be anticipated than if the particles were oriented at random, so that some of them would lie perpendicular to the streamlines. In other words, such suspensions show non-Newtonian behavior even though the solvent is Newtonian. At sufficiently low velocity gradients the orientation will become negligible, and for this situation Simha[28] has extended the Einstein treatment to particles which are ellipsoids of revolution. The result obtained is that

$$\eta' = \eta(1 + \nu\phi) \tag{19–25}$$

where ν, which by equation 19–24 is equal to 2.5 for suspended spheres, is invariably larger than 2.5 for ellipsoids. For prolate (rod-shaped) ellipsoids

$$\nu = \frac{J^2}{15(\ln 2J - \frac{3}{2})} = \frac{J^2}{5(\ln 2J - \frac{1}{2})} + \frac{14}{15} \tag{19–26}$$

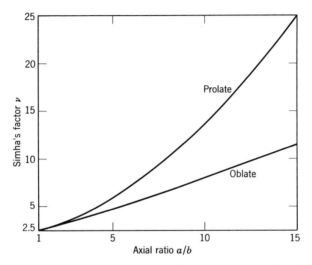

Fig. 19–7. Simha's factor for the viscosity increment of ellipsoids, for relatively low values of the axial ratio, a/b.

and for disc-shaped (oblate) ellipsoids

$$\nu = \frac{16}{15} \frac{J}{\tan^{-1} J} \tag{19–27}$$

where J is the axial ratio, a/b. It should be noted that these equations apply only to large values of J ($J > 10$). The more complete equations applicable to all values of J are given by Simha.[28]

Values of ν (calculated by Simha's complete equations) are shown plotted in Fig. 19–7 for low values of the axial ratio. A logarithmic plot for large values of a/b is shown in Fig. 19–8. At large values of a/b, $\log \nu$ becomes close to linear in $\log a/b$, such that ν varies roughly as $(a/b)^{1.8}$ for prolate ellipsoids and as $(a/b)^{1.0}$ for oblate ellipsoids. These approximate relations will be found useful in detecting the formation of rigid rods (rather than the more usual flexible coils) by synthetic polymer molecules, as shown for instance in the discussion of the data of Table 23–8.

It is important to note that Einstein and Simha's equations (like Stokes' law and Perrin's equations) apply only to very low flow velocities, because of neglect of the second term of equation 19–10 and, in the case of the equations for ellipsoids, because of the assumption of random orientation.

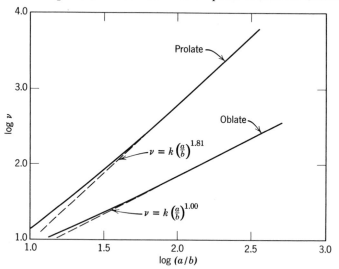

Fig. 19–8. A logarithmic plot of Simha's factor for large values of a/b. The equations correspond to the straight lines shown in the figure.

They also apply only to infinitely dilute solutions, because it has been assumed that the distortion of flow produced by one suspended particle does not interfere with that produced by other particles.

Equation 19–24 has been confirmed experimentally by measurement of the (macroscopic) viscosity of small suspended spheres of known diameter (Perrin[27] and Cheng and Schachman[22]).

20. MACROMOLECULES AS HYDRODYNAMIC "PARTICLES"

The preceding section considered two general aspects of flow processes and of viscosity: *thermodynamic* considerations expressed in terms of phenomenological coefficients, and *mechanical* considerations which allowed calculation of these coefficients for solid particles. If the mechanical considerations are to apply to macromolecular solutions, models of the solutions must be constructed in which the macromolecules appear as solid particles. It is the objective of the present section to construct appropriate models.

It is important in this connection to recall that the mechanical theories of the preceding section are based on the disturbance of the flow pattern about an *individual particle* on the assumption that no other particles are close enough to have any effect. In making the transition to macromolecular solutions it is clearly necessary, therefore, that these solutions be studied at high dilution, or, best of all, at a series of concentrations, followed by extrapolation to zero concentration. In viscosity determinations extrapolation to zero rate of flow may also be necessary for very asymmetric or for easily deformed macromolecules, so that the equations derived for random orientation may apply.

20a. Rigid macromolecules. Rigid macromolecules are those which possess internal bonding which allows little or no latitude in the location of one part of the molecule relative to another. For such molecules the assumption that they behave as solid particles would seem to be a good one, with the understanding that these particles may have any conceivable shape; they may be spheres, ellipsoids, rods, etc., or may have shapes close to such geometric figures. On the other hand, completely irregular shapes are equally possible.

It is likely that a certain amount of solvent will often be an inherent part of the macromolecule. Protein molecules, for example, possess ionic groups and polar groups, which, in a polar solvent, will generally be associated with one or more solvent molecules. In addition to this inherent solvation, some solvent is likely to be part of the macromolecule *for the purpose of hydrodynamic experiments*, for any spaces which exist within such molecules, such as the hollow core in tobacco mosaic virus (Fig. 4–15) or any sharp indentations on the surface, will, naturally, contain solvent, and, even if no inherent attraction (other than weak van der Waals' forces) exists between this solvent and the macromolecular substance, this "trapped" solvent will travel with the same velocity as the adjoining macromolecular substance because the frictional forces will prevent a large value of du/dx.

This solvation must be taken into account when the mass and volume of the hydrodynamic particle formed by a macromolecule are computed. We shall introduce for this purpose the parameter δ_i, signifying the number of grams of a solvent component associated with 1 gram of the unsolvated macromolecular substance. Thus, in a two-component system (solvent designated by subscript 1), the mass m_h of each hydrodynamic particle is

$$m_h = \frac{M(1 + \delta_1)}{\mathcal{N}} \qquad (20\text{--}1)$$

where M is the molecular weight of unsolvated macromolecules. If the solvent consists of more than one component,

$$m_h = \frac{M(1 + \sum_i \delta_i)}{\mathcal{N}} \tag{20-2}$$

where the summation extends over all solvent components.

The experimental determination of the factor δ_1, or the factors δ_i, is an exceedingly difficult problem. The same problem arises in the chemistry of small molecules and has received particular attention in the case of ions in aqueous solution. Recent reviews by Bockris and Conway[29] bring out eloquently the nature of the latter problem and provide a measure of the certainty with which at least the inherent solvation may be measured. For alkali and halide ions in water, for instance, we find from one to six H_2O molecules per ion, with a probable error of from one to two H_2O molecules. The problem with which we are concerned is an even more difficult one, for inherent solvations is only a part of the solvation we wish to determine.

One fairly straightforward method for the direct determination of solvation in macromolecules has been suggested by Wang.[30] He determines the rate of self-diffusion of the solvent, i.e., the rate of diffusion of an isotopic form (e.g., H_2O^{18} into H_2O). The result is clearly related to the volume unavailable to diffusing solvent molecules. According to Wang it is only slightly dependent on the shape of the regions of unavailable volume. Since the regions of unavailable volume should be essentially the same thing as the corresponding hydrodynamic particles, this provides a method for the determination of the volume of the hydrodynamic particles. By the relation to be given in the following section this volume is directly related to δ_1. Using this method, Wang has estimated that for ovalbumin in water $\delta_1 = 0.18 \pm 0.01$ gram/gram of dry protein, whereas, for deoxyribose nucleic acid (DNA), δ_1 was found to be about 0.35 gram/gram dry DNA.

It might be noted that ovalbumin has a molecular weight of 45,000 and, at its isoionic pH, the pH used by Wang, possesses about forty-one positive charges (cationic nitrogen) and an equal number of negative charges (carboxylate groups). If the tightly bound water for these groups is the same as that for 41 moles of guanidine hydrochloride or alkyl ammonium chlorides (given by Conway and Bockris[29]), about 2 ± 1 moles of water would be tightly bound for each pair of groups. This would lead to a total of 82 ± 41 moles of water or to $\delta_1 = 0.032 \pm 0.016$. That the experimental value is about six times as great is reasonable in view of the fact that δ_1 actually includes more than the tightly bound solvent.

The procedure of Wang has so far not been used except for the two macromolecules cited above. Furthermore, it is not certain whether the δ_1 value determined by his method is *exactly* that which is required for other applications. At the present time, therefore, the solvation factor is generally regarded as an unknown parameter.

20b. Expression for the hydrodynamic volume in terms of the partial specific volume. The relations obtained in section 19 for the hydrodynamic properties of suspended particles are expressed in terms of the volume of such particles. When applied to dissolved macromolecules these equations will therefore involve the volume v_h of the particles which these molecules form, including the volume of solvent which is incorporated within them. It is clear from the preceding section that v_h, like m_h, will in general not

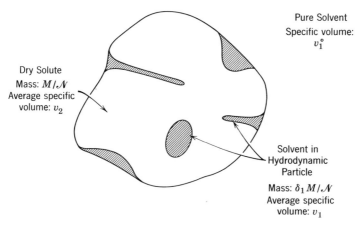

Fig. 20–1. Schematic diagram of a solvated macromolecular particle of arbitrary shape.

have a predictable value. It is convenient, however, to express v_h in terms of the same solvation factor δ_1 (or δ_i) used in discussing m_h.

Consider first a two-component system, containing g_1 grams of solvent and g_2 grams of an electrically neutral macromolecular solute. Figure 20–1 shows a diagrammatic sketch of a hydrodynamic particle in such a solution. It is supposed that the particle contains δ_1 grams of solvent per gram of dry macromolecular material. This solvent may well vary considerably in specific volume from that of pure solvent ($v_1{}^0$), especially if it is tightly bound. Let v_1 represent the *average* specific volume of all of the shaded portions of Fig. 20–1. Let v_2 represent the *average* specific volume of all the macromolecular portions of the particle. (Average values allow for variation from point to point within the particle.) The total volume of the particle is thus

$$v_h = \frac{M}{\mathcal{N}}\,(v_2 + \delta_1 v_1) \qquad (20\text{–}3)$$

Specific combination of solvent with macromolecular substance is likely to change the specific volume. As previously explained, the δ_1 grams of solvent which form part of the hydrodynamic particle contain specifically bound solvent, as well as solvent which is more or less trapped. For the latter solvent we would expect a specific volume not appreciably different from v_1^0. The specific volume v_1 is the average for all solvent within the particle. As regards solvent outside the particle, there is no reason to believe that any of it will have a specific volume differing from v_1^0.

The total volume of a solution containing g_1 grams of solvent and g_2 grams of dry solute may now be calculated, allowing for the fact that $\delta_1 g_2$ grams of solvent incorporated in macromolecular particles has a specific volume v_1 and the remainder has a specific volume equal to that of pure solvent. Where V is the total volume

$$V = g_2 v_2 + g_2 \delta_1 v_1 + (g_1 - g_2 \delta_1) v_1^0 \qquad (20\text{--}4)$$

We are interested, as explained in section 19, only in very dilute solutions where δ_1 must be a constant independent of concentration (i.e., solvent activity is essentially unity at all concentrations). The *partial specific volume* of the solute in the solution may therefore be obtained by differentiation of equation 20–4 as

$$\bar{v}_2 = \left(\frac{\partial V}{\partial g_2}\right)_{T,P,g_1} = v_2 + \delta_1 v_1 - \delta_1 v_1^0 \qquad (20\text{--}5)$$

The desired relation between v_h and δ_1 can then be obtained at once by combining equation 20–5 with equation 20–3,

$$v_h = \frac{M}{\mathcal{N}} (\bar{v}_2 + \delta_1 v_1^0) \qquad (20\text{--}6)$$

The concept of a solvated hydrodynamic "particle," as presented here, is due primarily to Oncley.[16] It is somewhat of an oversimplification, since it is unrealistic to suppose that solvent molecules in the domain of a dissolved macromolecule can be sharply divided into those which are an immovable part of the molecular particle and those which are completely free. The parameter δ_1 thus represents an *effective* solvation, and its value may depend to some extent on the experiment which is being carried out; i.e., the macromolecule may appear in one experiment as a "particle" with a value of δ_1 different from that which it appears to have in another experiment, even though both experiments were performed in the same solvent. The value for δ_1 for rigid particles must, however, be small (as the experiments to be cited in subsequent sections will show), and variations in δ_1 from one experiment to another are thus generally considered as negligible. In any event, the fact that δ_1 represents an effective solvation in no way limits the generality of equation 20–6 for a two-component system, since this equation simply consists of the replacement of one unknown property of the "particle," v_h, by another, δ_1.

An extreme point of view concerning the foregoing question has been taken by Scheraga and Mandelkern.[106] These authors have taken the position that even rigid macromolecules in solution can *never* be assumed to behave as particles in the sense here required. The hydrodynamic particle of volume v_h, which we need to call into existence if the equations of Perrin, Simha, etc., are to be applied to interpret measured values of the frictional coefficient or of viscosity, becomes, in their view, an artificial device not necessarily related to the real molecule at all. A consequence of this position is that Scheraga and Mandelkern object to the use of the parameter δ_1 and of equation 20–6. A further consequence would seem to be to prohibit any comparison between molecular dimensions which we compute from transport processes and viscosity with the dimensions of the *real* molecule which we obtain, for instance, from x-ray diffraction or from the angular dependence of light scattering. In the view taken here such a comparison is valid, provided we keep in mind the fact that we shall have to approximate the shape of the molecular particle as an ellipsoid of revolution because appropriate equations are not available for other shapes. In other words, hydrodynamic properties will lead us to the best ellipsoid which represents the *real* molecular particle.

20c. Systems with more than one solvent component. Macro-ions. Equation 20–6 applies to two-component systems: solvent and electrically neutral macromolecular solute. It may be extended to systems which contain a single electrically neutral macromolecular solute but a number of solvent components, giving

$$v_h = \frac{M}{\mathcal{N}}\left(\bar{v}_2 + \sum_i \delta_i \bar{v}_i{}^0\right) \tag{20–7}$$

where the summation extends over all solvent components and $\bar{v}_i{}^0$ designates the partial specific volume of each component in the solvent mixture The value of $\bar{v}_2$, of course, is that measured on solutions of the macromolecular substance in the mixed solvent. Each of the δ_i in this equation becomes an independent variable since the dissolved macromolecule must in general be assumed to have a different affinity for each individual solvent component.

In many experimental studies the solvent consists of a dilute solution of salts or buffers in water. The concentration of the salts or buffers in these solutions is normally so low that their contribution to $\sum_i \delta_i \bar{v}_i{}^0$ is negligible unless specific binding of the salt or buffer to the macromolecule occurs. Thus equation 20–6 is applicable (component 1 referring to water), with the understanding that the salts or buffers employed are always substances for which the macromolecules are known to have no particularly strong affinity.

If the solute of interest is a macro-ion, there is an additional complication in that v_h and $\bar{v}_2$ must both refer to the volume of the neutral macromolecular *component*, which may consist of small ions in addition to the macro-ion. Equation 20–6 would then represent the sum of the volumes of several particles, the number and nature of which are determined by the manner in which these components are defined (section 14a). If the Scatchard definition of components is used, the sum of simple electrolyte ions included in the component is zero, and v_h of equation 20–7 may be taken to represent the volume of only the macro-ion. For macro-ions of low charge, e.g., proteins near their isoelectric points, this is true no matter how the components are defined, since the volume of a small number of inorganic ions is negligible in comparison with the volume of the macro-ion particle.

In the remainder of this chapter we shall generally use equation 20–6 for the volume of the hydrodynamic particle formed by macro-ions in dilute salt solutions. The foregoing assumptions should be kept in mind when this is done.

20d. Long rod-shaped molecules. The preceding discussion has been limited to rigid macromolecules but has not been restricted to any particular shape. However, the equations for hydrodynamic properties of suspensions, which were presented in section 19, are available only for spherical or ellipsoidal particles. We are thus in practice compelled to approximate the shape of all particles to be considered by choosing an appropriate equivalent ellipsoid or sphere.

In view of the fact that equations for thermodynamic excluded volume (section 12) and for the angular dependence of light scattering (section 18) are available for particles which are *rod shaped*, it is worthwhile to consider specifically what ellipsoid might be chosen which would most closely approximate a long cylindrical rod, of length L and diameter d. It is generally regarded that a prolate ellipsoid (semi-axes a, b, b; Fig. 7–2) of the same length and volume as the rod provides the best approximation. (All dimensions refer to the over-all particle, of course.) The equivalence of length and volume may be stated as $L = 2a$ and $4\pi ab^2/3 = \pi L d^2/4$. Solving these equations for a and b, we find that

$$\frac{a}{b} = \left(\frac{2}{3}\right)^{\frac{1}{4}} \frac{L}{d} \qquad\qquad (20\text{–}8)$$

Thus a rod of length L and diameter d is considered equivalent to a prolate ellipsoid of the same volume, with a/b given by equation 20–8.

20e. Flexible polymers. In considering flexible polymers it is essential to inquire whether the solvent within the domain of the coiled molecule is free to travel at its own velocity or whether it is forced to travel with the velocity of the polymer molecule. In the former case the polymer molecule is said to be *free draining*, and there would be little resemblance to a solid suspended particle. However, if the latter is true, the polymer molecule, with all of the solvent within its coils, would be well approximated as a solid particle.

In section 10c we discussed the density of polymeric material within the space occupied by a flexible polymer chain. It was concluded that the density is greatest at the center of mass and decreases as we proceed outward. More than half the polymer segments must lie within the radius of gyration, and the average density within the radius of gyration must be such that only about 10^{-2} of the volume is occupied by polymeric material, the exact value depending on the degree of polymerization. Knowing this density, we may compute a rough value for the average distance between nearest neighbor segments of a dissolved polymer molecule. If we imagine each segment at the center of a cube, we see that the ratio of the volume of a segment to that of each cube is the volume fraction of polymeric material, i.e., about 10^{-2}. The volume of each cube is thus on the average 100 times the segment volume, or, when taking cube roots, the side of each cube is about five times an average linear dimension of each segment. The side of the cube, however, represents also the average distance between nearest neighbor segments. This distance is thus likely to be of the order of five times an average linear dimension (per segment), i.e., not more than 25 Å, usually. This means that all the solvent within the interior (say, within the radius of gyration) is likely to lie within 25 Å or less of *several* polymer segments. In view of the restriction on du/dx imposed by the viscous forces, the velocity of this solvent relative to the polymer molecule must therefore be very small. From the hydrodynamic point of view this solvent is trapped and essentially indistinguishable from solvent which might be inherently combined with the polymer chain. Solvent which lies further away from the center of the polymer coil (outside the radius of gyration, perhaps) will become progressively less restricted as the segment density decreases.

The concept of a hydrodynamic solvated "particle" is thus applicable to flexible coils as well as to rigid macromolecules, and equations 20–1, 20–6, etc., are applicable. The amount of solvent in the hydrodynamic particle must, however, be very large; depending on exactly where the dividing line is which separates restricted from unrestricted solvent, the solvent may exceed the amount of polymer substance by a factor of 100 or more. If equation 20–3 or 20–6 is used for v_h, the term $\delta_1 v_1{}^0$ becomes by far the predominant term. Under these circumstances equation 20–3 or 20–6 is no longer useful. It becomes imperative to evaluate δ_1 in terms of other parameters.

The use of an equivalent sphere in the treatment of the thermodynamically excluded volume of a polymer molecule (section 12) immediately suggests that the same device be used here. We thus introduce an *equivalent hydrodynamic sphere*, which we visualize as a solid sphere of radius R_e, to represent the polymer molecule. We may suppose, as before, that R_e is related to the radius of gyration; i.e., we write

$$R_e = \xi R_G \tag{20–9}$$

where the parameter ξ remains to be discussed.

It should be noted that the equivalent hydrodynamic sphere represents a much simpler concept than the equivalent sphere used for the purpose of calculation of the excluded volume. The excluded volume is a measure of the interaction between two polymer molecules. It is a measure of the volume occupied as modified by intermolecular attractions. When the attractive forces become large, as they do in a poor solvent, they play the dominant role, and the thermodynamic excluded volume can then become zero or even negative. Hydrodynamic measurements, however, are made in (or extrapolated to) very dilute solutions, in which intermolecular interaction plays no part at all. The hydrodynamic volume is always a measure of the volume occupied in the solution by a single particle. In calculating this volume, we have in mind a picture of a polymer particle, somewhat analogous to a gel. The water content of a gel may be as much as 99%, yet the liquid is incapable of independent flow.

20f. Approach to an exact theory for flexible polymer chains. An exact theory for the behavior of flexible polymer molecules in solution requires that the idea of a rigid solid impermeable particle be rejected. In its place we must use an appropriate distribution of mass elements. The density of mass elements will be everywhere small (cf. Fig. 10–2), and the fact that they are connected can be expressed by giving them all the same velocity of flow. We can now follow the same procedure as that used in obtaining the Stokes and Einstein equations; i.e., equation 19–10 or 19–11 is solved with appropriate boundary conditions, with the difference that the single particle used by Stokes and Einstein is replaced by the assembly of mass elements which represent the polymer chain. The flow velocity of solvent is now determined everywhere, including the regions within the molecular domain. The frictional coefficient and the viscosity increment are then computed as before. The result of such a calculation is qualitatively the same as that given in the preceding section; most of the solvent deep within the molecular domain moves with essentially the same velocity as the macromolecule, whereas that nearer the periphery has greater freedom of motion. The models used so far (see below) have resulted in equations which are identical with equation 20–9; i.e., the use of an equivalent sphere is validated, with the difference that numerical values for ξ are obtained.

The simplest model of this kind is that used by Debye and Bueche,[33] which consists of a sphere with a *uniform* distribution of mass elements, the elements being treated as spherical beads. A more realistic model is that of Kirkwood and Riseman,[32] which treats the polymer chain as a connected chain of spherical beads, with bond lengths and bond angles treated as in the derivation of the equations for end-to-end distance and radius of gyration in section 9. The final result is averaged over all possible configurations of the polymer chain.

The result of Kirkwood and Riseman's treatment is that the frictional coefficient of a flexible chain molecule is equal to

$$f = \frac{(3\pi^{\frac{1}{2}}/8)6\pi\eta R_G}{1 + 9\pi^{\frac{3}{2}}\eta R_G/4\sigma\zeta} \tag{20-10}$$

where σ is the number of mass elements, ζ is the frictional coefficient of each mass element, and η is the viscosity of the solvent. This result is seen to be equivalent to equation 20–9 (combined with Stokes' equation, equation 19–13), except that ξ appears to be a complicated function of the size and shape of the mass elements, as well as the size of the polymer molecule, because of the second term in the denominator. It is easily seen, however, that the second term becomes very small as σ becomes large. For by equation 9–40 R_G is equal to $\alpha\beta\sigma^{\frac{1}{2}}/6^{\frac{1}{2}}$, where α is of order unity and β of order $3l_{av}$, l_{av} being the average bond length, i.e., distance between mass elements. At the same time each mass element may be taken as a sphere of radius $\sim l_{av}$, so that its frictional coefficient becomes $6\pi\eta l_{av}$. Thus

$$\frac{9\pi^{\frac{3}{2}}\eta R_G}{4\sigma\zeta} \sim \frac{1}{2}\left(\frac{\pi}{\sigma}\right)^{\frac{1}{2}}$$

and this becomes much smaller than unity when σ exceeds 1000 or so; i.e., for all practical purposes we obtain

$$f = 6\pi\eta R_e \tag{20-11}$$

$$R_e = (3\pi^{\frac{1}{2}}/8)R_G = 0.665R_G \tag{20-12}$$

A similar result is obtained for the viscosity increment. For sufficiently large σ equation 19–24 applies, with the volume of each polymer particle equal to $\frac{4}{3}\pi R_e^3$ and

$$R_e = 0.875R_G \tag{20-13}$$

It is seen that the same equivalent sphere cannot be used for calculating both the frictional coefficient and the viscosity increment.

The major deficiency of the Kirkwood-Riseman theory is that it is based on a polymer chain with ideal dimensions ($R_G \propto \sigma^{1/2}$), so that it cannot be taken as exact. In writing equations 20–12 and 20–13, without restriction to ideal solutions, we assume that the relation between R_e and R_G will remain unaffected by the nature of the solvent, but this assumption is dubious. The lowered segment density in good solvents (cf. Fig. 10–2) might be expected to provide greater freedom of motion for the solvent and, hence, to decrease ξ. (Such an effect is, in fact, observed experimentally. See Fig. 23–2.)

In the subsequent sections we shall therefore treat ξ as an empirical factor. We shall, however, take into account specifically the theoretically expected difference between the values of ξ for calculation of the frictional coefficient and the viscosity, writing ξ_f for the former and ξ without subscript for the latter.

It is important to point out that the preceding treatment of the Kirkwood-Riseman theory incorporates within it ideas which are due to Flory.[5] In the original theory it was assumed that R_G can be given by the ideal expression, i.e., equation 9–40 with $\alpha = 1$. The original theory predicts that ξ will be a constant for relatively high values of σ only ($\sigma > 1000$). Experimentally there exists no evidence for such a limitation. The constant proportionality between the radius of the equivalent sphere and the radius of gyration will necessarily fail when σ becomes sufficiently small, but it is applicable to much lower values of σ than the theory predicts.

It should also be noted that the conclusion reached on p. 343 that the solvent within the domain of a flexible polymer molecule is largely immobile justifying the use of an equivalent sphere model in the first place, is likely to fail for exceedingly stiff chains and that for such chains a free-draining model of the polymer chain may be more applicable than the model we have employed. Such a model is described by Flory[5] (who gives references to the original derivations based on it). We shall not describe it here since no application of it is made in the subsequent discussion.

21. DIFFUSION[7]

Whenever a concentration gradient exists in a solution, a flow of matter is observed which tends to equalize the concentration everywhere. This is the process of diffusion. In general it would be a process occurring in three dimensions, but by the simple device of allowing a concentration gradient to exist in one direction only (e.g., along the length of a tube) it can be reduced to a one-dimensional problem and will be treated as such here.

A basic law for one-dimensional diffusion was proposed by Fick[34] more than a hundred years ago. Reasoning that the flow of matter along a concentration gradient should be governed by a law analogous to the law for heat flow along a temperature gradient, Fick deduced that the flow of particles per second, J, across unit area of a plane perpendicular to the direction x of the concentration gradient should be proportional to the concentration gradient; i.e.,

$$J = -D\left(\frac{\partial C}{\partial x}\right)_t \qquad (21\text{--}1)$$

where C is in any suitable concentration units (e.g., moles or grams per cubic centimeter) and J has corresponding units (e.g., moles per square centimeter per second). The negative sign of equation 21–1 takes cognizance of the fact that particles will flow in the direction of decreasing concentration. The coefficient D is known as the *diffusion coefficient*.

An equation known as Fick's second law may be obtained by combining equation 21–1 with the equation of continuity (equation 19–2). For diffusion in a tube of constant cross-sectional area A, we obtain

$$\left(\frac{\partial C}{\partial t}\right)_x = \left[\frac{\partial}{\partial x}\left(D\frac{\partial C}{\partial x}\right)\right]_t \qquad (21\text{--}2)$$

which, if D is independent of concentration, reduces to

$$\left(\frac{\partial C}{\partial t}\right)_x = D\left(\frac{\partial^2 C}{\partial x^2}\right)_t \qquad (21\text{--}3)$$

Equations 21–2 and 21–3 may be used with any desired concentration units.

The objective of this section will be to examine the extent of validity of this very early phenomenological approach to diffusion, to identify the force which produces the flow of diffusion, and to show how the measurement of the diffusion coefficients of macromolecules can lead to useful information about them.

21a. Diffusion in two-component systems. Diffusion occurs in a system which is completely at equilibrium (e.g., the pressure is everywhere the same) except for the existence of a concentration gradient. A concentration gradient implies a gradient of chemical potential. The general phenomenological equation (equation 19–4) for diffusion of the solute in a two-component system is thus

$$J = -L\left(\frac{\partial \mu}{\partial x}\right)_t \qquad (21\text{--}4)$$

where both J and the force are in *molar* units because μ is the chemical potential per mole.

If the macromolecular component is a non-electrolyte, then its molecules may be considered "particles" in the sense of section 20, and the coefficient L may be identified with the frictional coefficient by equation 19–8 or 19–9. Where J and F are in molar units the latter equation applies, so that

$$J = -\frac{C}{\mathcal{N}f}\left(\frac{\partial \mu}{\partial x}\right)_t \tag{21-5}$$

where C represents the solute concentration in moles per cubic centimeter. The use of equation 19–9 implies that the force $\partial \mu/\partial x$ represents a mechanical force and that this is indeed the case will be shown in the following section.

The subscript i has been omitted in writing equations 21–4 and 21–5 because in a two-component system we are concerned with only a single flow equation, that of the solute. The flow of solvent could be calculated from equation 21–4, if we so desired, by use of the condition that the solution as a whole remains stationary with respect to the coordinate x, which represents a dimension of the cell containing the fluid.

Any of the general expressions of section 11e may now be used for the chemical potential of the solute. We shall here use equation 11–20, with the activity replaced by the product of molar concentration and an activity coefficient y; i.e., $a = 1000Cy$, where C is in moles per cubic centimeter. (The factor 1000 is introduced so that y may represent the customary activity coefficient referred to the concentration in moles per liter.) Thus

$$\left(\frac{\partial \mu}{\partial x}\right)_t = \frac{\mathcal{R}T}{C}\left[1 + C\left(\frac{\partial \ln y}{\partial C}\right)_{T,P}\right]\left(\frac{\partial C}{\partial x}\right)_t \tag{21-6}$$

and equation 21–5 becomes

$$J = -\frac{\mathcal{R}T}{\mathcal{N}f}\left[1 + C\left(\frac{\partial \ln y}{\partial C}\right)_{T,P}\right]\left(\frac{\partial C}{\partial x}\right)_t \tag{21-7}$$

This equation is clearly equivalent to equation 21–1 and thus proves that Fick's law is valid for a two-component system. Moreover, it identifies the diffusion coefficient as being simply related to the frictional coefficient; i.e.,

$$D = \frac{\mathcal{R}T}{\mathcal{N}f}\left[1 + C\left(\frac{\partial \ln y}{\partial C}\right)_{T,P}\right] \tag{21-8}$$

Equation 21–8 shows that the diffusion coefficient in a two-component system is not a constant but depends on the concentration C, not only because of the term in brackets in equation 21–8 but also because the mechanical picture of frictional motion, given in section 19, clearly indicates that f must vary with concentration. If the diffusion coefficient can be measured at several concentrations, however, and extrapolated to $C = 0$, then the second term in brackets becomes zero, and f reduces to the frictional coefficient for an isolated particle which was discussed in sections 19 and 20. Thus, where D^0 represents the extrapolated value of D,

$$D^0 = \mathscr{R}T/\mathscr{N}f = kT/f \qquad (21\text{-}9)$$

where $k = \mathscr{R}/\mathscr{N}$ is Boltzmann's constant.

This derivation of the relation between D^0 and f has been somewhat abstract. The same result can be obtained more simply if we are merely interested in the behavior of a two-component system approaching infinite dilution. Such a system will be an ideal system, and the equilibrium distribution of particles under the influence of an arbitrary force (such as that of equation 16-1) can be evaluated by Boltzmann's equation; the concentration C at any value of x will depend on exp $[-m\,H(x)/kT]$, where $H(x)$ is the potential energy as given by equation 16-2. Thus

$$\ln C = \text{constant} + m \int_{x=0}^{x} G(x)\,dx/kT$$

The same distribution may be evaluated in another way by invoking the condition that, at equilibrium, the flow of particles due to the applied force must be exactly counterbalanced by the flow due to diffusion. Thus the sum of J as given by equation 19-8 $[J = Cm\,G(x)/f]$, and J as given by equation 21-1 must be zero. The resulting equation may be integrated to give

$$\ln C = \text{constant} + m \int_{x=0}^{x} G(x)\,dx/fD$$

and comparison with the Boltzmann distribution at once leads to equation 21-9.

21b. Kinetic molecular theory of diffusion. In the preceding section Fick's law of diffusion was derived on the assumption that a formal force, equal to the gradient of chemical potential, provides the driving force for the diffusion process. The concept of frictional resistance to motion was then used to compute the rate of flow. In view of the strictly mechanical picture of frictional resistance which was provided in section 19 this approach to the diffusion problem lacks conviction unless it can be shown that the formal force of equation 21-4 or 21-5 is in fact an actual mechanical force. That this is indeed so is shown by Einstein's classic papers on the diffusion process.[6,46] The mechanical force is that provided by the random collisions of diffusing molecules with solvent molecules (and, at higher concentrations, with one another). This force leads to random motion of the diffusing molecules, with equal likelihood that a given solute molecule will move towards a region of higher concentration or towards a region of lower concentration. The probability that a given molecule will move a distance x in time t is the same for all molecules and is independent of the direction of motion. This random motion must, however, lead to a net transfer of solute from a region of high concentration to one of low concentration because there are more dissolved particles at the place of high concentration to which the equal probability of moving a given distance applies.

We first focus our attention on a single particle and consider its motion over a small period of time, during which no appreciable concentration changes have occurred but during which the particle will have undergone many collisions. (The number of collisions suffered by a particle in 1 sec is very large, ordinarily about 10^{15} per second). It is clear that the motion of the particle must represent a random flight process of the kind discussed in sections 9 and 10. In particular, the average value of the square of the displacement x after σ collisions must be proportional to σ, i.e., $\overline{x^2} = $ (constant) σ, just as the average square of the distance between the beginning and end of a polymer chain with σ links is proportional to σ (cf. equation 9–7). Since the number of collisions suffered by a particle in time t is proportional to t, it follows that $\overline{x^2}$, too, is proportional to t. Thus, if x is the distance through which a particle is displaced in time t, we have, on the average,

$$\overline{x^2}/t = \text{constant} \tag{21–10}$$

In the calculation of end-to-end distances of linear polymer molecules it was necessary to correct the corresponding equation, equation 9–7, for the fact that a polymer chain cannot cross its path. This restriction, of course, does not apply in the present case, and equation 21–10 is therefore exact, except in so far as composition of the solution may influence the constant of equation 21–10.

We consider now a tube of cross-sectional area A, with a concentration gradient along the length of the tube only. We let $N(x, t)$ represent the number of particles per cubic centimeter, at any point x, at any time t, and consider the number of particles in a volume element of width dx about the point $x = 0$, after a small time interval t has elapsed from an arbitrary zero time. At $t = 0$ all of these particles must have been either in the same or in some other volume element. Now let $W(x)$ be the probability* that a particle will have undergone in time t a displacement of just the right magnitude to move it from a volume element at $x = x$ to the element at $x = 0$. The total number of particles going in time t from a volume element at $x = x$ to that at $x = 0$ is thus the total number of particles in the element at $x = x$ at $t = 0$ (i.e., $N(x, 0)A\, dx$) multiplied by $W(x)\, dx$. The number of particles in the element at $x = 0$ at time t is the sum (i.e., integral) of these products over all possible volume elements; i.e.,

$$N(0, t)A\, dx = \int_{-\infty}^{\infty} N(x, 0)A\, dx\, W(x)\, dx \tag{21–11}$$

Equation 21–11, of course, contains a term under the integral, where $x = 0$, for particles which suffered no displacement at all, i.e., which were in the same volume element at $t = 0$. Dividing both sides by $A\, dx$, we get for equation 21–11

$$N(0, t) = \int_{-\infty}^{\infty} N(x, 0)\, W(x)\, dx \tag{21–12}$$

Our treatment is confined to small intervals of time, since our objective is to obtain a

---

* An exact formulation of $W(x)$ is not required for this derivation. However, $W(x)$ must clearly have a form analogous to equation 10–14.

differential equation for $\partial N/\partial t$ at a given instant. Thus $N(0, t)$ will differ only little from $N(0, 0)$ and may be expressed in terms of $N(0, 0)$ by a Taylor series,

$$N(0, t) = N(0, 0) + \left(\frac{\partial N}{\partial t}\right)_x t + \frac{1}{2}\left(\frac{\partial^2 N}{\partial t^2}\right)_x t^2 + \cdots \tag{21–13}$$

Furthermore, in a small interval of time, $W(x)$ will have a value different from zero only if x is quite small. An appreciable contribution to the integral of equation 21–11 will be made only within a range of x lying very close to $x = 0$. We may thus express $N(x, 0)$ also in terms of $N(0, 0)$; i.e.,

$$N(x, 0) = N(0, 0) + \left(\frac{\partial N}{\partial x}\right)_t x + \frac{1}{2}\left(\frac{\partial^2 N}{\partial x^2}\right)_t x^2 + \cdots \tag{21–14}$$

Since every particle must go somewhere, $\displaystyle\int_{-\infty}^{\infty} W(x)\, dx = 1$, and

$$\int_{-\infty}^{\infty} N(0, 0)\, W(x)\, dx = N(0, 0) \tag{21–15}$$

Furthermore,

$$\int_{-\infty}^{\infty} \left(\frac{\partial N}{\partial x}\right)_t x\, W(x)\, dx = \left(\frac{\partial N}{\partial x}\right)_t \bar{x} = 0 \tag{21–16}$$

Here we have taken $\partial N/\partial x$ to be a constant, since we are considering only a small time interval, and have placed $\bar{x} = 0$, since positive and negative displacements are equally probable. Finally,

$$\int_{-\infty}^{\infty} \left(\frac{\partial^2 N}{\partial x^2}\right)_t x^2\, W(x)\, dx = \left(\frac{\partial^2 N}{\partial x^2}\right)_t \overline{x^2} \tag{21–17}$$

Combining equations 21–13 to 21–17 with equation 21–12 and neglecting the term in t^2 in comparison with that in t and that in x^3 in comparison with that in x^2, we get

$$\left(\frac{\partial N}{\partial t}\right)_x t = \frac{1}{2}\left(\frac{\partial^2 N}{\partial x^2}\right)_t \overline{x^2}$$

But we have already shown in equation 21–10 that $\overline{x^2}/t$ is a constant, so that

$$\left(\frac{\partial N}{\partial t}\right)_x = \tfrac{1}{2}(\text{constant})\left(\frac{\partial^2 N}{\partial x^2}\right)_t \tag{21–18}$$

which result is, of course, identical with the form of Fick's second law, equation 21–3, which was obtained when D was assumed independent of concentration. We see by comparing equation 21–18 with equation 21–3 that the "constant" of equation 21–10 is in fact just twice the diffusion coefficient. (The derivation here given was first used by Einstein in 1905.[46])

21c. Multi-component systems. Macro-ions.
We have shown that Fick's equations are completely valid for the diffusion of a non-electrolyte solute in a two-component system. This general validity cannot be demonstrated for systems of more than two components. Taking a three-component

system as the simplest example of such a system, we need to specify the flow of two of the components. Designating them as components 2 and 3, we find that equation 19–4 gives

$$J_2 = -L_{22}\left(\frac{\partial\mu_2}{\partial x}\right)_t - L_{23}\left(\frac{\partial\mu_3}{\partial x}\right)_t$$

$$J_3 = -L_{33}\left(\frac{\partial\mu_3}{\partial x}\right)_t - L_{32}\left(\frac{\partial\mu_2}{\partial x}\right)_t$$

(21–19)

The equation for $\partial\mu/\partial x$ also becomes more complicated, because each component affects the activity of the other, so that we get, in place of equation 21–6,

$$\frac{\partial\mu_2}{\partial x} = \frac{\mathscr{R}T}{C_2}\left(1 + C_2\frac{\partial\ln y_2}{\partial C_2}\right)\frac{\partial C_2}{\partial x} + \mathscr{R}T\frac{\partial\ln y_2}{\partial C_3}\frac{\partial C_3}{\partial x} \qquad (21\text{–}20)$$

with a similar equation for $\partial\mu_3/\partial x$. Combination of equations 21–19 and 21–20 clearly leads to diffusion equations of the type

$$J_2 = -D_{22}\left(\frac{\partial C_2}{\partial x}\right)_t - D_{23}\left(\frac{\partial C_3}{\partial x}\right)_t$$

$$J_3 = -D_{33}\left(\frac{\partial C_3}{\partial x}\right)_t - D_{32}\left(\frac{\partial C_2}{\partial x}\right)_t$$

(21–21)

where D_{22} and D_{33} are the diffusion coefficients which would obtain for diffusion of each component in the absence of the other, and D_{23} and D_{32} are coefficients which depend upon the interaction between the flows of component 2 and 3 when both are present together. Equations 21–21 are clearly not identical with Fick's law, so that it cannot be generally valid for systems of more than two components.

The difference between equation 21–21 (for J_2) and equation 21–1 would be unimportant if these equations became identical whenever the concentration of component 3 is initially everywhere uniform. But this is not in general the case, for flow of component 3 will be expected to occur as a result of the term $D_{32}(\partial C_2/\partial x)$, so that C_3 will not remain uniform as diffusion proceeds. That this actually occurs has been demonstrated by Dunlop[36] in low-molecular-weight systems.

The situation becomes even more complicated in solutions which contain macro-ions. For the reasons given in section 14 (p. 238) macro-ions are ordinarily studied in the presence of a low-molecular-weight electrolyte, so that three-component systems are involved. In addition, however, the flow of the macro-ion will be subject to modification by gradients in the electrostatic potential ψ. Although ψ is initially the same everywhere, it will not remain so as diffusion proceeds; i.e., a potential gradient will be set up in the same way as the initially uniform concentration of a third component can be disturbed.

It has been customary for many years to ignore these difficulties in the study of diffusion of macromolecules, and of macro-ions of low charge, in aqueous salt or buffer solutions in which the concentration of the salt or buffer component is moderately high (say 0.1 molar) and initially uniform. For the latter component substances have been chosen that are expected to interact only weakly or not at all with the macromolecular component, so that interactions leading to flow of the third component have been assumed insignificant. Moreover, a moderately high concentration of salt has been assumed to prevent the appearance of any appreciable gradient of electrostatic potential, by analogy with the fact that it largely eliminates the Donnan effect (p. 226) and the effect of charge on the apparent molecular weight which is obtained from measurement of sedimentation equilibrium (p. 268).

The validity of these assumptions has recently been questioned by Gosting,[7] but no quantitative estimate of the probable error has been made. We shall assume that the error is small in situations where only a single species is originally present at non-uniform concentration and that the diffusion of this species follows Fick's laws as if all other components were a single-component solvent. The reader should, however, keep in mind the possibility that future research on this subject may demonstrate that the error introduced in doing so is larger than is at present believed.

21d. Experimental determination of diffusion coefficients. In all practical applications of diffusion to macromolecular solutions, the macromolecules (or macro-ions) are assumed to act as if they were solute particles in a two-component system, so that Fick's law applies. Studies are usually carried out at low concentrations, so that the second term of equation 21–8 becomes very small and D becomes virtually independent of x. Equation 21–3 then becomes the differential equation describing the diffusion process. The particular way in which a diffusion experiment is set up provides initial and boundary conditions with the aid of which equation 21–3 may be integrated to yield expressions for C, as a function of the time t and the position x, in terms of the diffusion coefficient. This coefficient may then be evaluated from experimental curves of C versus x at various times. Results are obtained at several concentrations and extrapolated to $C = 0$ to yield a value of D^0, which is then related to the frictional coefficient by equation 21–9.

The most precise experimental method is the *free diffusion* method. In this method a sharp boundary is initially set up between a solution of uniform concentration C_0 and pure solvent. The solution to equation 21–3 under these conditions is

$$C = \frac{C_0}{2}\left[1 - \frac{2}{\pi^{1/2}}\int_0^{x/2(Dt)^{1/2}} e^{-y^2}\,dy\right] \qquad (21\text{–}22)$$

The integral in equation 21–22 is known as the probability integral and is, of course, a function of $x/2(Dt)^{1/2}$. Tabulated values of this integral, which vary in value from 0 to $\frac{1}{2}$ as $x/2(Dt)^{1/2}$ varies from 0 to ∞, are available in many handbooks.[35] A graphical representation of equation 21–22 is shown in Fig. 21–1.

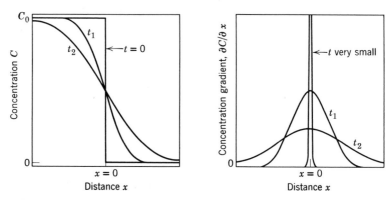

Fig. 21–1. Progress of a diffusion experiment with initially sharp boundary at $x = 0$.

Experimental curves of C versus x can be obtained at various time intervals, using, for example, records of light absorption at a wavelength corresponding to an absorption peak of the diffusing solute. These curves can be compared with theoretical curves calculated by equation 21–22 for a series of values of Dt. From the curves a value of D at each time t can be computed.

More often, however, the progress of diffusion is followed by measuring refractive index gradient, by one of the methods given in Appendix C. This quantity is proportional to $\partial C/\partial x$, so that the derivative of equation 21–22 is needed for analysis;

$$\left(\frac{\partial C}{\partial x}\right)_t = -\frac{C_0}{2(\pi Dt)^{1/2}}\,e^{-x^2/4Dt} \qquad (21\text{–}23)$$

This function is represented graphically in Fig. 21–1.

Equation 21–23 shows that the maximum value of $|\partial C/\partial x|$ will occur at $x = 0$, and its value will be $C_0/2(\pi Dt)^{1/2}$. The maximum height (h_{max}) of the corresponding refractive index gradient curve (Appendix C) will thus be

$KC_0/2(\pi Dt)^{1/2}$, where K is a constant. At the same time, the area A under the refractive index gradient curve will be KC_0. Thus

$$A/h_{max} = (4\pi Dt)^{1/2} \qquad (21\text{–}24)$$

and this provides a convenient way of determining D. (See Gosting[7] for alternative procedures.)

Because of the concentration dependence of D, the experimental curves for $\partial C/\partial x$ will generally differ slightly in shape from the theoretical curves of Fig. 21–1. The diffusion coefficient determined, as, for instance, by the height-area method of the preceding paragraph, will therefore be an apparent diffusion coefficient rather than the true value of D at a particular concentration. However, if D is determined in this way at several concentrations and extrapolated to $C = 0$, the error so introduced may be

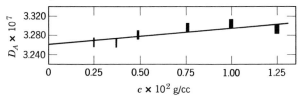

Fig. 21–2. Dependence of the diffusion coefficient of bovine serum albumin on concentration. The symbol D_A is used to indicate that D was determined by the height-area method, equation 21–24. (Wagner and Scheraga.[41])

expected to disappear, and the limiting value will be equal to D^0. A typical extrapolation of this kind is shown in Fig. 21–2.

Rigorous treatment of the diffusion process requires, of course, that equation 21–2 rather than equation 21–3 be used as the basic differential equation, with the concentration dependence of D specifically provided for before integration. This procedure is rarely used in practice, though suitable equations have been derived (see Gosting[7]).

Equation 21–24 may be used to determine an average apparent diffusion coefficient in systems containing several diffusing components, i.e., several components initially present on only one side of the boundary, although the over-all curve of $\partial C/\partial x$ versus x would in this case be quite different from that given by equation 21–23. (This situation would arise, for example, in the diffusion of an unfractionated synthetic polymer sample.) This would lead to an unnatural average of D, however, even if the problem of interacting flows could be ignored, which in this situation is unlikely. If no interaction between flows occurs, the weight average of D, $\bar{D}_w$, can be evaluated as

$$\bar{D}_w = \mathscr{L}_{c=0} \frac{\int_{-\infty}^{\infty} x^2 (\partial c/\partial x)\, dx}{2tc_0} \qquad (21\text{–}25)$$

(where c refers to concentration in grams per cubic centimeter) which in turn is easily obtained from refractive index gradients if each component has the same refractive index increment.

21e. Diffusion coefficients of proteins. Protein molecules in solution bear electrostatic charges. Even at the isoelectric point, where the average net charge is zero, there will be molecules with net charge ± 1, ± 2, etc., and even those molecules which actually possess zero net charge are multi-polar ions; i.e., they contain many positive and negative charges, in equal numbers. The diffusion of proteins is therefore undoubtedly influenced by electrostatic forces. It is ordinarily believed that a moderately large concentration of a low-molecular-weight electrolyte, such as KCl, will eliminate any such influence, and this contention is borne out by the fact that diffusion coefficients at moderately high ionic strengths become essentially independent of the net charge (e.g., Creeth[42]). Introducing an electrolyte of course means that we are no longer dealing with a two-component system. As was pointed out on p. 353, however, this factor is believed to be unimportant if the third component (i.e., the added electrolyte) is initially present at the same concentration everywhere. In any event, proteins dissolved in salt solutions are generally regarded as two-component systems, and D^0 values are obtained for them by extrapolation of apparent diffusion coefficients to zero concentration.

To interpret diffusion data so obtained the protein molecules or ions are regarded as particles in the sense discussed in section 20. Equation 20–6 (or, where necessary, equation 20–7) gives the volume of such a particle in terms of an empirical degree of solvation, δ_1. The radius of a sphere with this volume ($v_h = 4\pi R_0^3/3$) would be

$$R_0 = \left[\frac{3M}{4\pi \mathcal{N}} (\bar{v}_2 + \delta_1 v_1^0) \right]^{1/3}$$

and the frictional coefficient of this hypothetical sphere is given by equation 19–13 as $f_0 = 6\pi\eta R_0$. The actual frictional coefficient of the particle may be expressed in terms of f_0 by means of the ratio f/f_0; i.e., $f = (f/f_0)6\pi\eta R_0$. The ratio f/f_0 will be unity if the particle actually is a sphere but, in general, will be greater than unity.

Combining these relations with equation 21–9 we can now write for the diffusion coefficient extrapolated to zero concentration

$$f = \frac{kT}{D^0} = 6\pi\eta \frac{f}{f_0} \left[\frac{3M(\bar{v}_2 + \delta_1 v_1^0)}{4\pi \mathcal{N}} \right]^{1/3} \tag{21-26}$$

It is seen by equation 21–26 that the diffusion coefficient depends on the molecular weight and partial specific volume of the protein, both of which are measurable quantities. It depends also, however, on *two* unknown factors: f/f_0, representing the deviation of the shape of the protein hydrodynamic particle from a sphere, and δ_1, representing the solvation. We clearly cannot determine both these factors from a single measurement of the diffusion coefficient. Thus there is an ambiguity in the interpretation of diffusion coefficients, as the discussion of the data below will show.

To obtain a function containing only these unknown parameters we calculate the *minimum* possible frictional coefficient, f_{min}, or the corresponding maximum diffusion coefficient, D^0_{max}, compatible with the known molecular weight and $\bar{v}_2$. This is obtained by setting $f/f_0 = 1$ and $\delta_1 = 0$ in equation 21–26, giving

$$f_{min} = \frac{kT}{D^0_{max}} = 6\pi\eta\left(\frac{3M\bar{v}_2}{4\pi\mathcal{N}}\right)^{\frac{1}{3}} \tag{21–27}$$

The ratio of the observed diffusion coefficient to this calculated value is then the desired quantity,

$$\frac{f}{f_{min}} = \frac{D^0_{max}}{D^0} = \frac{f}{f_0}\left(\frac{\bar{v}_2 + \delta_1 v_1^0}{\bar{v}_2}\right)^{\frac{1}{3}} \tag{21–28}$$

(In many literature compilations this ratio is referred to as f/f_0. We have introduced the new symbol f/f_{min} since we have used f/f_0 to represent the effect of shape alone on the frictional coefficient.)

It should be noted at this point that diffusion coefficients of proteins are conventionally corrected to "standard conditions," the standard solute being pure water at 20° C. The values of D^0 so corrected are those which the protein solutions in pure water at 20° C would have if the molecular parameters M, $\bar{v}_2$, f/f_0, and δ_1 were all to remain constant, independent of the solvent. If this were so, then, with D^0 representing the diffusion coefficient measured at temperature T in a solvent (e.g., a salt solution) of viscosity η and T_s representing the "standard" temperature and η_s the viscosity of the "standard" solvent at that temperature, the "standard" diffusion coefficient, D_s^0, is given by equation 21–26 as

$$D_s^0 = D^0 T_s \eta / T \eta_s \tag{21–29}$$

With $T_s = 293.16°$ K and $\eta_s = 0.01002$ poise we get the usual standard conditions, the corresponding value of D^0 being often designated $D^0_{20,w}$. It should be noted that this correction has no effect on the experimental ratio f/f_{min} given by equation 21–28.

It is important to observe that the computation of $D^0_{20,w}$ by equation 21–29 is purely an algebraic procedure. The experimental value of D^0 which would actually be obtained in pure water at 20° C would often be quite different from $D^0_{20,w}$, for the molecular parameters such as M and f/f_0 often change with even minor alterations in temperature or electrolyte concentration.

Typical experimental data for a number of proteins are shown in Table 21–1, and the ratio f/f_{min} has been calculated for each example. The most significant feature of the table is that it indicates that proteins may be divided into two classes, those which have f/f_{min} quite close to unity, indicating hydrodynamic behavior not very different from the hypothetical spheres for which f_{min} is calculated, and those which have much larger values of f/f_{min}, indicating hydrodynamic behavior which differs greatly from that of unsolvated spheres. The proteins which fall into the first

TABLE 21–1. Diffusion Coefficients of Proteins in
Aqueous Solutions

	M	$\bar{v}_2$	$D^0_{20,w} \times 10^7$	f/f_{min}
Ribonuclease[47]	13,683	0.728	11.9	1.14
Lysozyme[48]	14,100	0.688	10.4	1.32
Chymotrypsinogen[37]	23,200	0.721	9.5	1.20
β-Lactoglobulin[51]	35,000	0.751	7.82	1.25
Ovalbumin[49]	45,000	0.748	7.76	1.17
Serum albumin[38,41]	65,000	0.734	5.94	1.35
Hemoglobin[49]	68,000	0.749	6.9	1.14
Catalase[39]	250,000	0.73	4.1	1.25
Urease[50]	480,000	0.73	3.46	1.20
Tropomyosin[40]	93,000	0.71	2.24	3.22
Fibrinogen[109]	330,000	0.710	2.02	2.34
Collagen[43]	345,000	0.695	0.69[a]	6.8
Myosin[72]	493,000	0.728	1.16[a]	3.53

[a] This value was calculated from the sedimentation coefficient by equation 22–23 using a molecular weight determined by light scattering.

group are the globular proteins, which may be crystallized and whose crystal unit cells (section 4) indicate that they are, in the solid state, compact and symmetrical. It is clear from Table 21–1 that they retain such a configuration in aqueous solution.

It should be noted that these conclusions agree completely with those reached from evaluation of radii of gyration, as given in Table 18–2, although the latter quantity has been determined for relatively few proteins. Proteins with f/f_{min} close to unity also have R_G close to the value calculated for unsolvated spheres, and vice versa.

Equation 21–28 shows that the ratio f/f_{min} depends on two factors, solvation and asymmetry, which can be resolved only if additional physical properties are determined which depend on the same variables. We shall postpone discussion of this subject until later in this chapter (p. 396), but in the meantime it is of interest to discover the range of possible values which these variables might possess. At one extreme we can assume that the difference between the observed value of $f/f_{\min}$ and the "ideal" value of 1.0 for unsolvated spheres is to be ascribed only to solvation. This

TABLE 21–2. Solvation and/or Asymmetry Calculated from the Diffusion Coefficients of Table 21–1[a]

	Maximum Solvation $(f/f_0 = 1)$			Maximum Asymmetry $(\delta_1 = 0,$ $f/f_0 = f/f_{\min})$	Compromise $(\delta_1 = 0.2)$	
	δ_1 grams/gram	R_e Å	a/b of prolate ellipsoid		$\dfrac{f}{f_0}$	a/b of prolate ellipsoid
Ribonuclease	0.35	18.0	3.4		1.05	2.1
Lysozyme	0.89	20.6	6.1		1.21	4.3
Chymotrypsinogen	0.52	22.5	4.2		1.11	3.0
β-Lactoglobulin	0.72	27.4	4.9		1.16	3.7
Ovalbumin	0.45	27.6	3.8		1.08	2.5
Serum albumin	1.07	36.1	6.5		1.25	4.9
Hemoglobin	0.36	31.0	3.4		1.05	2.1
Catalase	0.70	52.2	4.9		1.15	3.6
Urease	0.53	61.9	4.2		1.11	3.0
Tropomyosin	23.0	96	62			
Fibrinogen	8.4	106	31			
Collagen	218	310	300			
Myosin	49	215	100			

[a] This general method of analysis is primarily due to Oncley.[16]

assumption leads to the maximum possible value of δ_1, or, in other words, the value of δ_1 required to produce a sphere with the observed diffusion coefficient. (The radius R_e of this sphere is obtained from equation 21–9, together with Stokes' equation, $f = 6\pi\eta R_e$.) At the other extreme, we may ascribe the value of $f/f_{\min}$ entirely to asymmetry, assuming $\delta_1 = 0$. We express the magnitude of the asymmetry in terms of the axial ratio, a/b, which an ellipsoid would have to produce the observed value of $f/f_{\min} = f/f_0$, using the data of Figs. 19–3 and 19–4. This calculation leads to the maximum possible value of a/b for such an ellipsoid.

Both types of calculation are shown in Table 21–2, the value of a/b which is given being that for a prolate ellipsoid. Somewhat smaller values would have been obtained if the oblate ellipsoidal model had been used.

Table 21–2 shows that the globular proteins can be neither highly solvated nor highly asymmetric. The maximum solvation never exceeds about 1 gram of solvent per gram of protein and the maximum asymmetry is that of a prolate ellipsoid with a/b about 6. Since it is highly unlikely that a protein hydrodynamic particle is a perfect sphere and impossible that it should be unsolvated (cf. section 20a), a compromise calculation such as is shown in the last two columns of Table 21–2 probably affords a reasonably accurate picture of these molecules in water solution. In this calculation δ_1 has been assumed to be 0.2 gram/gram of protein and f/f_0 has been computed from equation 21–28. The axial ratios of equivalent ellipsoids were then obtained from Fig. 19–3. It should be noted, of course, that these molecules are not to be thought of as *actually* forming ellipsoidal hydrodynamic particles. Their over-all shape may well be quite irregular. We have given the axial ratios of corresponding ellipsoids only because this is the only type of body for which equations are available by means of which f/f_0 may be interpreted.

It is worth noting that really high precision in the evaluation of diffusion measurements is a quite recent achievement and that the probable accuracy of most of the diffusion coefficients of Table 21–1 is only about $\pm 5\%$. Moreover, all the data of Table 21–1 were obtained in aqueous salt solutions, with the assumption that Fick's equations are applicable. As was stated on p. 353, this assumption is unlikely to be entirely correct, contributing an additional small uncertainty to the interpretation put upon the data. These uncertainties do not invalidate the differentiation between "compact" and "non-compact" proteins, for the differences in the corresponding f/f_{min} values exceed 100%. They do indicate, however, that it would be dangerous to place any quantitative interpretation on the small differences between individual globular proteins of Tables 21–1 and 21–2. Differences between them certainly exist, but it should not be considered an established fact that, for example, serum albumin is necessarily the most asymmetric or most highly hydrated molecule in the group.

It should also be pointed out that the data of Tables 21–1 and 21–2, showing that globular proteins have compact and relatively symmetric configurations, were obtained in aqueous salt solutions near the isoelectric pH values of the proteins. Experiments conducted under other conditions show that the compact configurations are easily lost, as indicated by marked changes in the diffusion coefficient. Two examples of such changes will be cited in Figs. 22–8 and 22–9 in conjunction with corresponding changes in sedimentation coefficients.

When we consider those proteins which do not fall into the class of globular proteins, e.g., tropomyosin, fibrinogen, collagen, and myosin in Tables 21–1 and 21–2, the question of deciding whether the observed large values of f/f_{min} are due to solvation or asymmetry becomes a crucial one. If the observed effect is principally due to solvation, the values of δ_1

computed in Table 21–2 become so large that it may be presumed that the molecules are behaving much like randomly coiled molecules, which, as was pointed out in section 20e, trap the large amount of solvent within their coils so that most of it forms an inherent part of the hydrodynamic particle. On the other hand, if the observed effect is to be ascribed primarily to asymmetry, the prolate ellipsoids required to represent the behavior of the molecular particles have such large axial ratios, a/b, so as to be essentially equivalent to long thin rods. To decide between these possibilities on the basis of diffusion coefficients alone is, of course, not possible. For this reason a discussion of this topic is best postponed until after viscosity measurements on these molecules have been considered, since all available data should be taken together in arriving at a decision. The problem will be taken up in section 23.

21f. Diffusion coefficients of macromolecules other than proteins. Most macromolecules, apart from proteins, do not lend themselves particularly

TABLE 21–3. Diffusion Coefficients of Macromolecules Other than Proteins

	Reference State	M	$\bar{v}_2$	$D^0 \times 10^7$	f/f_{min}
Bushy stunt virus[45]	Water, 20°	10,700,000	0.74	1.15	1.3
Tobacco mosaic virus[64]	Water, 20°	50,000,000	0.73	0.3	2.9
DNA[a,c]	Water, 20°	6,000,000	0.53	0.13	15
Amylose acetate (Corn)[66]	Methyl acetate, 25°	108,000		6.8[b]	2.7
(Potato)[66]	Methyl acetate, 25°	69,000		8.0[b]	2.7
PMA[a,55]	Acetone, 20°	410,000	0.63	3.95	3.9
PMA[a,55]	Acetone, 20°	1,000,000	0.63	2.25	5.0

[a] DNA = deoxyribose nucleic acid; PMA = polymethyl methacrylate.
[b] The amylose samples used in this work were probably highly heterogeneous with respect to both molecular weight and diffusion coefficient. This factor was not taken into consideration when the measurements were made.
[c] D^0 evaluated from the sedimentation coefficient (ref. 68) by equation 22–23.

well to diffusion measurements. The principal reason is that these molecules tend to show a greater effect of concentration, so that extrapolation, from the region of concentration required to get measurable data to zero concentration, becomes quite uncertain. A selection of available data are shown in Table 21–3; their reliability is lower than that for the corresponding data on proteins.

Table 21–3 shows only one substance, bushy stunt virus, which appears to have a high degree of compactness and symmetry, as indicated by a low value of f/f_{min}. A few other viruses (but not tobacco mosaic virus) fall into this class. No other macromolecules, however, have this type of configuration. This result is, of course, in accord with the structural information on these molecules which was given in Chapter 2. Apart from the globular proteins there are only a few viruses (among them bushy stunt virus) which form single crystals with the large symmetrical unit cells which we would expect from the packing of large compact molecules. Similarly, among molecules large enough to be examined by electron microscopy, only the bushy stunt virus and a few other viruses appear as close to spherical particles. The same conclusion was also reached from the radii of gyration listed in Table 18–2.

In considering the data of Table 21–3 for tobacco mosaic virus, DNA, and amylose acetate we are again confronted with the fact that the observed high value of f/f_{min} might be due to effective solvation or to asymmetry, and we shall postpone further discussion until section 23. In the case of polymethyl methacrylate, however, this difficulty can be at least partly resolved because this substance can be prepared with any desired molecular weight. We can therefore gain information not only from the value of f/f_{min}, which tells us that the molecule is not a compact sphere, but also from the molecular weight dependence of this quantity or of the diffusion coefficient itself.

Our expectation (section 7) is, of course, that the polymer will be randomly coiled. Under these circumstances, as indicated in section 20e, it becomes convenient to treat the hydrodynamic particle resulting from a polymer molecule as a sphere; this equivalent sphere has a radius $R_e = \xi_f R_G$, where ξ_f, according to the best theory at present available, is a universal constant equal to 0.665. The frictional coefficient of such a sphere, by equation 19–13, is $6\pi\eta\xi_f R_G$, and, hence,

$$D^0 = kT/f = kT/6\pi\eta\xi_f R_G \qquad (21\text{--}30)$$

To obtain the dependence on molecular weight we substitute equation 9–40 for R_G into equation 21–30, obtaining the relation

$$D^0 = kT/f = kT M_0^{\frac{1}{2}}/6^{\frac{1}{2}}\pi\eta\alpha\beta\xi_f M^{\frac{1}{2}} \qquad (21\text{--}31)$$

Recalling that β is independent of molecular weight and ξ_f probably so, equation 21–31 would lead us to predict that D^0 should vary as $1/M^{\frac{1}{2}}$ in a poor solvent and as $1/M^{\sim0.55}$ in an average good solvent, since α varies roughly as $M^{0.05}$ in such a solvent. This prediction is entirely confirmed by the data of Meyerhoff and Schulz,[55] on polymethyl methacrylate in acetone, which are plotted logarithmically in Fig. 21–3. The straight line of this plot has a slope of -0.56, i.e., $\log D^0$ varies as 0.56 $\log M$, in accord with the predicted result in a good solvent.

If polymethyl methacrylate had been a thin rod-shaped molecule, with constant diameter, d, and with molecular length, L, directly proportional to the molecular weight, then, by equation 20–8, the axial ratio, a/b, of the corresponding prolate ellipsoid would vary directly with molecular weight. Furthermore, by Fig. 19–4, f/f_0 would vary roughly as $(a/b)^{0.48}$, i.e., as $M^{0.48}$. Assuming that the high value of f/f_{min} is entirely due to asymmetry,

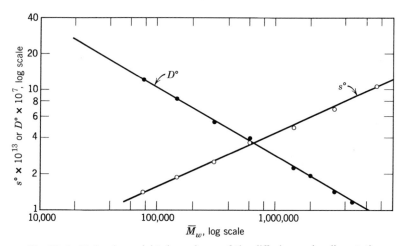

Fig. 21–3. Molecular weight dependence of the diffusion and sedimentation coefficients of polymethyl methacrylate. (Meyerhoff and Schulz.[55])

we then conclude that f/f_{min} would vary as $M^{0.48}$. By equation 21–27, D^0_{max} is proportional to $1/M^{0.33}$, so that D^0 would have to be proportional to $1/M^{0.81}$, a result very different from that actually observed.

Meyerhoff and Schulz used both equation 21–24 and 21–25 to determine their diffusion coefficients and found that each gave the same value of D^0 upon extrapolation to zero concentration. Since the two methods should give different averages, this suggests that the polymer fractions they examined were quite homogeneous. Molecular weight determinations did not support this conclusion, the ratio $\bar{M}_w/\bar{M}_n$ being between 1.3 and 1.4 for most of the fractions. We have assumed that the D^0 values represent weight averages and have plotted them against $\bar{M}_w$. The slope of such a plot should not differ appreciably from the slope of a plot of D^0 versus M for truly homogeneous fractions.

It should be noted, finally, that equation 21–30 could be used, in principle, to test the Kirkwood-Riseman theory and to evaluate the parameter ξ_f experimentally. This could be done by comparing measured values of D^0 with values of R_G determined from the angular dependence of light scattering (section 18). As was previously pointed out, however, there are considerable obstacles in the way of an accurate determination of diffusion constants of flexible polymers, and such a direct comparison has, in fact not been made. (It has been made for sedimentation coefficients, which are also a function of the frictional coefficient. See section 22.)

22. SEDIMENTATION IN THE ULTRACENTRIFUGE[8,9,10]

It was explained in section 16 that an ultracentrifuge is a device capable of subjecting a solution to a centrifugal force more than 100,000 times as great as the force of gravity. The application of such a force causes macromolecules to sediment outward, i.e., towards the bottom of a cell of the type shown in Fig. 16–1. A calculation made on p. 260 indicated that the final equilibrium state would depend on the speed of rotation. At relatively low speed macromolecules will, at equilibrium, still be distributed over the entire cell, with concentration increasing gradually from top to bottom. At high speed, however, the equilibrium concentration gradient would become much greater, so that virtually all the macromolecules will eventually become packed into a small region near the bottom of the cell.

This section will be primarily concerned with high speed operation, and we shall be interested in the velocity with which macromolecules move to the bottom of the cell. In section 22i, however, we shall return to the problem of low speed operation and events occurring during the approach to the equilibrium state.

22a. Sedimentation velocity in a two-component system. Mechanical description. We consider first the sedimentation of a single kind of suspended particle in a uniform solvent, with the understanding that non-electrolyte macromolecules in a two-component system will behave like such particles. As was discussed in section 20, the particles formed by macromolecules will contain solvent, and their mass and volume (m_h and v_h) will in general be unknown. Equations for m_h and v_h in terms of solvent content are given by equations 20–1 and 20–6.

The centrifugal force per unit mass is given by $G(r)$ of equation 16–18; i.e., the force per particle is $m_h\omega^2 r$, where ω is the angular velocity of the rotor and r the distance from the center of rotation. Countering this force will be the buoyant force exerted by the solvent, which is numerically the same as the centrifugal force exerted on a volume of solvent equal to the volume of the sedimenting particle; i.e., the buoyant force is $v_h\rho_0\omega^2 r$, where ρ_0 is the density of the solvent. The net force per particle is thus

$$F = \omega^2 r(m_h - \rho_0 v_h) \tag{22-1}$$

It was shown on p. 254 that the sedimentation process produces a pressure gradient in the sedimentation cell, pressure increasing with r. The buoyant force is simply the force exerted on the sedimenting particle by this pressure gradient and could have been computed with the aid of equation 16–4.

Equations 20–1 and 20–6 are now introduced for m_h and v_h. Since ρ_0, the solvent density, is the reciprocal of v_1^0, the solvent specific volume, the unknown solvation number δ_1 is seen to cancel, and equation 22–1 becomes

$$F = (M/\mathcal{N})\omega^2 r(1 - \bar{v}_2 \rho_0) \tag{22-2}$$

where M is the molecular weight of dry unsolvated macromoles, $\mathcal{N}$ is Avogadro's number, and $\bar{v}_2$ is the thermodynamic partial specific volume of the solute in the solution. (In discussing two-component systems we shall in general omit subscripts except where, as in the partial specific volume, there may be a question as to whether solvent or solute is being referred to.)

The force given by equation 22–2 is a simple mechanical force which will impart a constant velocity u to each particle. This velocity is given by equation 19–7 as

$$u = F/f = (M/\mathcal{N}f)\omega^2 r(1 - \bar{v}_2 \rho_0) \tag{22-3}$$

where f is the frictional coefficient of the sedimenting particle.

It was pointed out in section 19 that a mechanical equation for transport processes cannot be considered valid unless it is consistent with the phenomenological equations (equation 19–1). This test will be applied to equation 22–3 in the following section, and we shall see that equation 22–3 is indeed valid for two-component systems at infinite dilution, the equation for finite concentrations having ρ in place of ρ_0; i.e., for any concentration,

$$u = (M/\mathcal{N}f)\omega^2 r(1 - \bar{v}_2 \rho) \tag{22-4}$$

That ρ_0 should appear in equation 22–3 is in accord with the fact that the mechanical description of the sedimentation process can be strictly valid only at infinite dilution.

The flow velocity u, like the sedimenting force, depends on rotor speed and on the distance r from the center of rotation. It is convenient, therefore, to define a new quantity, the *sedimentation coefficient*, s,

$$s = \frac{u}{\omega^2 r} = \frac{M(1 - \bar{v}_2 \rho)}{\mathcal{N}f} \tag{22-5}$$

which depends only on molecular parameters. This is the quantity determined by experiment.

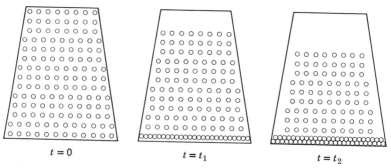

Fig. 22–1. Schematic representation of the progress of a sedimentation experiment at high velocity, with neglect of the effect of diffusion. Each point represents an individual sedimenting particle.

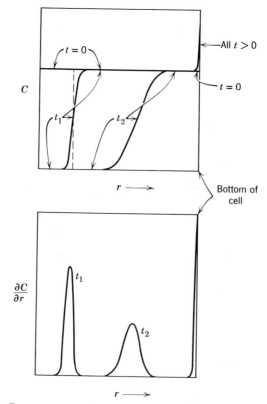

Fig. 22–2. Concentration and concentration gradient during a sedimentation run. The dashed line at t_1 shows an infinitely sharp boundary as is envisaged in Fig. 22–1. The solid lines show the boundary as it actually exists, broadened by diffusion.

Discussing next the actual measurement of sedimentation coefficients, in a solution in which all the macromolecules are identical, we note first that the sedimenting force at any value of r must be the same for all the dissolved particles, so that they will all move outward with the same velocity. A sharp boundary will therefore be formed as shown in Fig. 22–1. This boundary will move towards the bottom of the cell with the velocity given by equation 22–4, i.e., where r_H designates the boundary position

$$u = \frac{dr_H}{dt} \qquad s = \frac{1}{r_H \omega^2} \frac{dr_H}{dt} \qquad (22\text{–}6)$$

so that s can be determined directly from a plot of log r_H versus time. The boundary position at any time can be determined by any of the procedures already discussed in connection with sedimentation equilibrium and diffusion measurements; e.g., it would be the position at which the refractive index gradient $d\tilde{n}/dr$ is a maximum (i.e., infinite in the situation here under discussion).

In an actual experiment occupying a finite length of time the effect of diffusion will necessarily be superimposed upon that of sedimentation. This effect will lead to a broadening of the otherwise sharp boundary between the macromolecular solution and the solvent it is leaving behind, as shown schematically by the plots of Fig. 22–2. It is physically reasonable, however, to refer the diffusion process to the boundary between solution and solvent and to consider it independent of the motion of the boundary with respect to the cell. The diffusion process should therefore not affect the rate with which the boundary itself moves towards the bottom of the cell. Thus equation 22–6 remains correct, with r_H representing the position of maximum $d\tilde{n}/dr$ as before. (This conclusion will be slightly modified below.)

22b. Derivation from the phenomenological equation. The phenomenological equations are written in terms of the thermodynamic components of a system and not in terms of physical particles. Discussion of the sedimentation process in terms of these equations therefore eliminates the need to introduce solvated particles of unknown mass and volume. The sedimenting force per mole of solute is simply the force per M grams of solute, where M is the molecular weight; i.e., it is $M\omega^2 r$. For a two-component system we need write only a single flow equation (equation 19–1), that for the flow of solute, which now becomes

$$J = LM\omega^2 r - L'(\partial\mu/\partial r)_t \qquad (22\text{–}7)$$

where J is the flow across any cross-section in the units of moles per square centimeter per second and μ is the chemical potential per mole. Since both of the forces involved are mechanical forces and since, in a non-electrolyte two-component system, a one-to-one identification between solute molecules and corresponding particles is always possible, we may use equation 19–9 for the phenomenological coefficients L and L'; i.e., with both J and the forces in molar units,

$$L = L' = C/\mathcal{N}f \tag{22-8}$$

where C is the solute concentration in moles per cubic centimeter and f is the frictional coefficient of an individual hydrodynamic particle.

The chemical potential μ varies with r for two reasons: because the pressure varies with r and because the concentration varies with r. Using equation 16–11, we have

$$\left(\frac{\partial \mu}{\partial r}\right)_t = \left(\frac{\partial \mu}{\partial P}\right)_C \left(\frac{\partial P}{\partial r}\right)_t + \left(\frac{\partial \mu}{\partial C}\right)_P \left(\frac{\partial C}{\partial r}\right)_t$$

Or, with equation 16–10 for $\partial \mu / \partial P$, equation 16–13 for $\partial \mu / \partial C$, and equations 16–4 and 16–18 for $\partial P / \partial r$,

$$\left(\frac{\partial \mu}{\partial r}\right)_t = M\bar{v}_2 \rho \omega^2 r + \frac{\mathscr{R}T}{C}\left(1 + C\frac{\partial \ln y_2}{\partial C}\right)\left(\frac{\partial C}{\partial r}\right)_t$$

where $\bar{v}_2$ is the partial specific volume and y_2 the activity coefficient (molar scale) of solute, and ρ is the solution density.

Combining these equations with equation 22–7 gives the final flow equation

$$J = \frac{CM\omega^2 r}{\mathcal{N}f}(1 - \bar{v}_2 \rho) - \frac{\mathscr{R}T}{\mathcal{N}f}\left(1 + C\frac{\partial \ln y_2}{\partial C}\right)\left(\frac{\partial C}{\partial r}\right)_t \tag{22-9}$$

The second term on the right-hand side of this equation is identical with equation 21–7 and represents the flow due to the force of diffusion. The first term clearly represents the flow due to the sedimenting force.

It is the first term of equation 22–9 which is needed for comparison with the equations of the preceding section. To evaluate the flow velocity u due to the sedimentation force alone we proceed as on p. 324 (equation 19–8). With J in moles per square centimeter per second and C in moles per cubic centimeter, $J = uC$. The result is equation 22–4, which has already been discussed.

The fact that the present treatment leads to the same flow equation as the mechanical treatment is, of course, not sufficient information to enable us to *measure* the sedimentation coefficient; i.e., it is necessary to justify the use of equation 22–6 or to provide an alternative means for determining u or s. To consider this question we must obtain the complete differential equation for the flow process. With the right-hand side of equation 22–5 as a definition of sedimentation coefficient and with equation 21–8 as a definition of the diffusion coefficient, equation 22–9 becomes

$$J = sC\omega^2 r - D\left(\frac{\partial C}{\partial r}\right)_t \qquad (22\text{–}10)$$

This equation is combined with the equation of continuity, equation 19–2. Noting that the cross-sectional area A of the sector-shaped sedimentation cell (Fig. 16–1) is proportional to r, we get

$$\partial C/\partial t = -(1/r)[\partial(Jr)/\partial r]$$

or

$$r\left(\frac{\partial C}{\partial t}\right)_r = -\left[\frac{\partial}{\partial r}\left(s\omega^2 r^2 C - Dr\frac{\partial C}{\partial r}\right)\right]_t \qquad (22\text{–}11)$$

This equation was first derived by Lamm.[53] Like the differential equation for diffusion (equation 21–2) equation 22–11 is independent of the concentration units; i.e., C may now be taken to represent any desired units.

The exact solution of equation 22–11 is a formidable infinite series (Archibald[65]) in terms of integrals which can be computed only by numerical integration. A number of approximate solutions have been obtained, all of which are reviewed by Williams et al.[9] The simplest is the solution of Faxén,[52] which applies whenever sedimentation is rapid in comparison with diffusion, i.e., whenever there are regions on both sides of the boundary where (as in Fig. 22–2) $\partial C/\partial r = 0$. It is also assumed that s and D are independent of concentration. Faxén's solution is

$$\frac{C}{C_0} = \tfrac{1}{2}e^{-2\omega^2 st}\left(1 - \frac{2}{\pi^{1/2}}\int_0^y e^{-x^2}\,dx\right) + \text{higher terms}$$

where C_0 is the initial uniform concentration and

$$y = (r_m - re^{-\omega^2 st})\left[\frac{\omega^2 s}{2D(1 - e^{-2\omega^2 st})}\right]^{1/2}$$

with r_m giving the position of the meniscus. If all but the first term of Faxén's equation are omitted, the validity of equation 22–6 can be immediately established. This result is, however, of limited value, and a much more satisfactory treatment of the problem of measuring s is the treatment of Goldberg,[54] which does not require a general solution of the sedimentation equation.

The Goldberg treatment is valid for any sedimentation experiment in which there is a region between the boundary and the bottom of the cell where $\partial C/\partial r = 0$, as in Fig. 22–2. This region is called the *plateau region*, and within it equations 22–10 and 22–11 take on a much simpler form. Designating the variables in this region by the subscript p, we have

$$J_p = s_p C_p \omega^2 r_p \qquad (22\text{–}12)$$

$$dC_p/dt = -2s_p \omega^2 C_p \qquad (22\text{–}13)$$

the time derivative of C_p being a total derivative since C_p is independent of r. (It should be noted that C_p decreases with time, the physical reason for this being that the sedimenting particles are continually moving toward the bottom of the ultracentrifuge cell, i.e., in the direction of increasing cross-sectional area. Since s must in general depend on concentration, s_p will also depend on time.)

We consider now a cross-sectional plane at a fixed value of r_p. The area of such a plane (by Fig. 16–1) is proportional to r_p and we shall call it αr_p, where α is a constant. The total number of moles of solute crossing the plane per second is $\alpha r_p J_p$.

The number of moles crossing the plane at r_p in unit time must be equal to the decrease in unit time in the total number of moles which lie between the meniscus and the plane at r_p. Since the number of moles between r and $r + dr$ at any time is $\alpha r C \, dr$, we thus have

$$\alpha J_p r_p = -\frac{\partial}{\partial t} \left(\int_{r_m}^{r_p} \alpha r C \, dr \right) \qquad (22\text{–}14)$$

We now define a new variable r_z as that value of r which marks off a volume behind r_p which, if it were to contain solute at the plateau concentration C_p, would at any time contain just the amount of solute actually present between r_m and r_p; i.e., we define r_z so that, at any time t,

$$\int_{r_z}^{r_p} \alpha r C_p \, dr = \int_{r_m}^{r_p} \alpha r C \, dr \qquad (22\text{–}15)$$

The value of r_z defined in this way would, of course, denote the position of the sedimentation boundary if that boundary were infinitely sharp as in the dashed line of the upper part of Fig. 22–1.

The left-hand side of equation 22–15 is just $\frac{1}{2}\alpha C_p(r_p{}^2 - r_z{}^2)$. This quantity may be substituted in place of the integral of equation 22–14. Noting that r_p is a fixed coordinate, independent of time, we get,

$$J_p r_p = -\tfrac{1}{2}(r_p{}^2 - r_z{}^2)(dC_p/dt) + C_p r_z(dr_z/dt)$$

Introducing equations 22–12 and 22–13 for J_p and dC_p/dt, we get

$$s_p = \frac{1}{r_z \omega^2} \frac{dr_z}{dt} \qquad (22\text{–}16)$$

and we see at once that r_z is the quantity which must be measured, as a function of time, to determine the rate of sedimentation. Equation 22–6 will be true only if $r_H = r_z$.

To obtain an experimental expression which will allow r_z to be determined, we note that it can be obtained directly from equation 22–15 if an experimental curve of C versus r is available. More usually, however, we determine $K(\partial C/\partial r)$, where K is an optical constant (Appendix C), so that r_z will be expressed in terms of $\partial C/\partial r$. To achieve this, we first integrate the right-hand side of equation 22–15 by parts, obtaining

$$\alpha \int_{r_m}^{r_p} rC \, dr = \tfrac{1}{2}\alpha(C_p r_p{}^2 - C_m r_m{}^2) - \tfrac{1}{2}\alpha \int_{r_m}^{r_p} r^2 \left(\frac{\partial C}{\partial r}\right)_t dr \quad (22\text{–}17)$$

where C_m is the concentration at the meniscus. If, as is usually the case, the sedimentation boundary has completely separated away from the meniscus before any measurements are made, C_m may be set equal to zero.

Furthermore C_p may be replaced by $\int_{r_m}^{r_p} (\partial C/\partial r) \, dr$, so that

$$r_z{}^2 = \frac{\displaystyle\int_{r_m}^{r_p} r^2 (\partial C/\partial r)_t \, dr}{\displaystyle\int_{r_m}^{r_p} (\partial C/\partial r)_t \, dr} \quad (22\text{–}18)$$

If a sedimentation boundary is perfectly symmetrical about its peak r_H, then $\partial C/\partial r$ has identical values at $r = r_H - x$ and $r = r_H + x$. Under these conditions the difference between r_z and r_H becomes sufficiently small so that equation 22–6 (though not rigorously true) may be considered valid for most practical purposes. For boundaries which are markedly skewed, however, the use of equation 22–6 introduces a large error, and it is necessary that equation 22–18 be used to determine a value of r_z from each individual photograph of the sedimentation run, the sedimentation coefficient being determined from a plot of $\log r_z$ versus t, according to equation 22–16.

The derivation just given is due to Goldberg.[54] Its importance lies in the fact that it depends essentially only on the principle of conservation of mass. It is, moreover, independent of the shape of the sedimentation boundary or of the manner in which s depends on concentration. It not only gives a criterion by which we can decide whether equation 22–6 is valid but also shows how s must be determined when this criterion is not satisfied. Finally, the treatment shows that the sedimentation coefficient which is determined is that appropriate to the concentration C_p in the plateau region, and not to the concentration at r_z.

Another relation which is readily obtained from the preceding treatment is the radial dilution law of Trautman and Schumaker,[164] which gives C_p

as a function of r_z. Combining equations 22–13 and 22–16 and integrating between $t = 0$ (at which time $r_z = r_m$) and $t = t$, we have

$$C_p/C_0 = r_m{}^2/r_z{}^2 \qquad\qquad (22\text{–}19)$$

where C_0 is the initial uniform concentration.

Further details of the experimental determination of sedimentation coefficients are described in the detailed treatise of Schachman.[10] A rigorous discussion of the complete solution of equation 22–11 is given by Williams et al.[9]

22c. Effect of concentration. Secondary effects of pressure. It is evident from equation 22–5 that the sedimentation coefficient must vary with concentration, since the frictional coefficient on which it depends varies with concentration. As has been pointed out before, we do not have a general theory for the concentration dependence of f, and the concentration dependence of the sedimentation coefficient is therefore also unknown. For fairly compact solute particles the sedimentation coefficient may usually be taken as varying linearly with concentration; i.e., s may be written as

$$s = s^0 - kc \qquad\qquad (22\text{–}20)$$

whereas for synthetic polymers and other less compact or less symmetric particles a better empirically linear relation has the form

$$1/s = 1/s^0 + kc \qquad\qquad (22\text{–}21)$$

In both these relations s^0 represents the sedimentation coefficient at the limit of zero concentration and c the concentration in grams per unit volume. In both relations k is always positive; i.e., s always decreases with increasing concentration. The magnitude of the concentration effect is far greater than in the case of diffusion coefficients because the latter depend on the ratio of a thermodynamic term to f (cf. equation 21–8), both of which terms increase with concentration, whereas the sedimentation coefficient depends on $1/f$ alone. Typical plots representing the variation in s with concentration are shown in Fig. 22–3. A discussion of the theory has been presented by Schachman.[10]

Experimental values of sedimentation coefficients are always extrapolated to zero concentration for the same reason that applies to all hydrodynamic measurements, namely, that mechanical interpretations (in this case involving the frictional coefficient) are only valid under those conditions.

An important consequence of the effect of concentration on sedimentation coefficients is that the spreading of the sedimenting boundary which

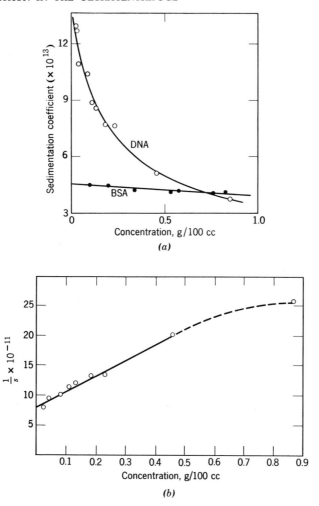

Fig. 22–3. (a) The effect of concentration on the sedimentation coefficient of bovine serum albumin (BSA; Kegeles and Gutter[60]) and of deoxyribose nucleic acid (DNA; Kahler[62]). (b) Kahler's data plotted as $1/s$ versus c.

is caused by diffusion, as discussed above, is partially and at times completely counteracted. The effect of diffusion is to produce a gradual rather than an abrupt change in concentration at the boundary. The region of lowest concentration, however, is that which is farthest from the bottom of the cell. Since the sedimentation coefficient in this region will be higher than that in the more concentrated region at the advancing front of the boundary, the molecules in it will tend to catch up with the more concentrated region, thus diminishing or canceling the rate at which the boundary spreads. One result of this is that the measurement of diffusion coefficients from the rate of boundary spreading during a sedimentation run is not a practical procedure, although the diffusion coefficient can be determined in this way if the dependence of s on concentration is known.[61]

We must also consider the influence of the high pressures which may exist at the bottom of a sedimentation cell. (These pressures may amount to several hundred atmospheres.) The *direct* effect of this pressure, in opposing the flow of molecules to the bottom, has already been taken into account in deriving the general equations for the sedimentation process, but there is an additional *indirect* effect in that the high pressures may alter the values of $\bar{v}_2$, ρ, or f, in which case the sedimentation coefficient becomes pressure dependent. This effect has been discussed in several recent papers[57,58,59] and is usually negligible, except in situations where $\bar{v}_2\rho$ is close to unity, where even small changes in these variables may have a large effect on the sedimentation rate.[58]

22d. Multi-component systems. Macro-ions. When equation 22–7 is rewritten for a system containing several components, or for systems containing macro-ions, the same difficulties will clearly be encountered as in the corresponding analysis of the diffusion process in section 21c. However, we shall again ignore these difficulties in systems which contain a single kind of macromolecular solute, or a macro-ion solute of low charge, in an aqueous salt or buffer solution. As a matter of fact the majority of sedimentation studies in the literature have been carried out in such solutions.

The effects of electrostatic charge and of the concentration of added salts have been investigated experimentally by Pedersen.[63] The major effect (called *primary* charge effect) is due to the electrostatic field which is set up when a charged macro-ion sediments away from its counter-ions. This field opposes the centrifugal force and, in the absence of added electrolyte, may easily decrease the sedimentation velocity to half its original velocity. This effect is, however, almost entirely removed when a moderate concentration of electrolyte is added. As far as the primary charge effect is concerned, therefore, Pedersen's results confirm the assumption of the previous paragraph.

Two additional effects may, however, arise in macro-ion solutions containing an added electrolyte. One of them (called the *secondary* charge effect) occurs when one of the ions of the added electrolyte has a tendency to sediment away from its counter-ion. Here the presence of the salt alone sets up an electric field, which will increase the sedimentation velocity if the heavier salt ion has a charge opposite in sign to that of the macro-ion and will decrease it if the opposite is true. Thus Pedersen has found that the presence of $0.2M$ CsCl increases the sedimentation coefficient of serum albumin when the macro-ion charge is negative, whereas $0.2M$ LiI decreases it. This is, of course, the expected result since Cs^+ sediments more rapidly than Cl^-, whereas I^- sediments more rapidly than Li^+. The magnitude of the effect (with the macro-ion charge $Z \simeq -30$) was less than 5%, however, so that it is reasonable to assume that the secondary charge effect will be quite small when salts with a smaller difference in the mass of the ions are used.

The other effect arises when one of the salt ions is strongly bound to the protein ion. This effect will increase the sedimentation coefficient if the bound ion is denser than the solvent. The effect has been discussed on p. 341 in terms of the resulting changes in the mass and volume of the hydrodynamic particle and is generally avoided by choosing for the electro-lyte needed to eliminate charge effects one whose ions do not react with the macro-ion being studied.

Pedersen's results tend to confirm the assumption that a solution of macro-ions, in an aqueous solution of an inert salt such as NaCl, may be treated as a two-component system with relatively little error. But there has been no rigorous theoretical study nor an experimental study aiming at really high precision. The comment made in discussing the influence of these same effects on diffusion (p. 353) thus applies also here; i.e., the effects may well be considered more serious in the future, when greater precision of sedimentation data will undoubtedly be attained.

It should be noted in conclusion that the presence of a second non-macromolecular component affects the density and the viscosity of the solution and, hence the frictional coefficient and the sedimentation coefficient (see equation 22–5 or equation 22–24). These effects are, however, readily calculated. The effects discussed above are those observed in addition to these normal effects.

Of greater interest than the effect of more than one solvent component is a discussion of what occurs in a system which contains several macro-molecular components. Figure 22–4 is a schematic representation of what occurs (ideally) when a solution containing just a few such solutes is subjected to the sedimenting force. Each type of macromolecule should move with its own velocity, according to equation 22–4, and, if these velocities are sufficiently different, the single sharp boundary of Fig. 22–2 should separate into several distinct individual boundaries, each of which should yield the sedimentation coefficient of one of the solutes. Further-more, the area under each peak in a concentration gradient diagram of the type shown in Fig. 22–5 should be equal to the over-all change in

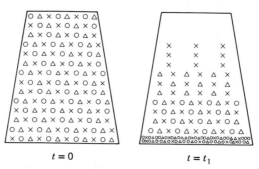

Fig. 22–4. Schematic representation of a sedimentation experiment involving three different kinds of molecules, each with a different sedimentation coefficient. (The effect of diffusion is neglected.)

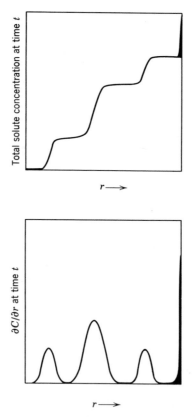

Fig. 22–5. Concentration and concentration gradient at a particular instant during sedimentation of a mixture of three kinds of sedimenting molecules.

concentration across the boundary represented by that peak (Appendix C); i.e., the area under each peak should be a measure of the concentration of the corresponding component. If the solution contains a mixture of macromolecules covering a range of molecular weights (e.g., a heterogeneous polymer mixture), separate boundaries are, of course, not to be expected, and, instead, the appearance of a very broad sedimentation boundary, with continuous variation of composition, would be predicted.

As we shall see, the concentration dependence of sedimentation coefficients will require some modification of these predictions. However, the ideal picture happens to be qualitatively correct for globular proteins (whose sedimentation coefficients vary less markedly with concentration than do those of less compact macromolecules) and has been of the greatest historical importance in the characterization of such proteins. Before the advent of the ultracentrifuge it was still quite generally believed that proteins were colloidal aggregates rather than true molecules (cf. section 1). They were thought to be heterogeneous association products of some as yet undiscovered smaller molecule. This belief was shattered by the work of Svedberg and coworkers. "The most astonishing result obtained with the ultracentrifuge was undoubtedly the fact that the native soluble proteins are either mono-disperse or paucidisperse in solution, i.e., they consist either of a single or of a few molecular species of well-defined molecular mass and shape." (Svedberg and Pedersen.[8]) The conviction that globular proteins are substances whose molecules are all essentially identical (which, in this book, was first reached from x-ray diffraction studies in section 4) thus had its historical inception from the fact that sedimentation diagrams of the type of Fig. 22–5, or, in some cases Fig. 22–2, were obtained from their examination in the ultracentrifuge.

The way in which the concentration dependence of sedimentation coefficients interferes with the simple picture of Fig. 22–5 was first brought out by Johnston and Ogston.[56] Suppose that there are two components, one of which sediments much faster than the other. Then the faster component will sediment in the usual way (its value of s, however, being determined by the *sum* of the concentrations of the two components). The slower component, however, will have a lower sedimentation coefficient ahead of the first boundary, where both components are present, than behind this boundary, where only the second component is present. This component will therefore tend to accumulate behind the faster boundary as shown diagrammatically in Fig. 22–6. The areas under the concentration gradient curves will now no longer be a measure of the concentrations of the two components. The total change in refractive index in crossing the first boundary is not that produced by a change in concentration of the faster component from zero to c_0 but is the sum of

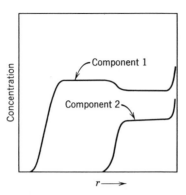

Fig. 22–6. The Johnston-Ogston effect. The concentration of the slower component changes at the boundary of the faster one because of the effect of the latter on its sedimentation coefficient. (After Johnston and Ogston.[56])

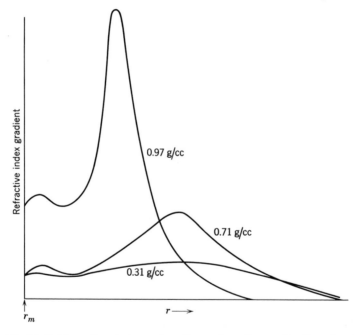

Fig. 22–7. The sedimentation pattern for amylopectin as a function of concentration. The boundary is sharpened at high concentrations because of the concentration dependence of the sedimentation coefficient. The small peak near the meniscus is ascribed to an impurity. (Data of Stacy and Foster,[70] provided through the courtesy of Dr. J. F. Foster.)

that change and the change produced by the decrease in concentration of the slower component across the fast boundary.

An even more spectacular effect may be produced in heterogeneous synthetic polymer mixture, or in other mixtures containing molecules with highly concentration-dependent sedimentation coefficients. Here the effect of concentration on sedimentation coefficients may be so great that the sharpening of the sedimentation boundary described earlier may outweigh all other factors. A single hypersharp sedimentation peak may be obtained in place of the broad peak predicted by the ideal theory. This phenomenon is well illustrated by Fig. 22–7 which shows sedimentation diagrams for starch amylopectin obtained by Stacy and Foster.[70] At a concentration of 0.007 gram/cc the sedimentation peak is sharp, suggesting that amylopectin is relatively homogeneous with respect to molecular weight (i.e., sedimentation coefficients for all molecules are close to the same). As the concentration is reduced, however, so that all s values approach closer to s^0 and the changes in s which occur as concentration changes along the cell become smaller, the sedimentation peak progressively broadens. The lowest concentration used clearly indicates that amylopectin is in fact extremely heterogeneous, a conclusion borne out by other studies on this substance, such as the $\bar{M}_w/\bar{M}_n$ ratio shown in Table 17–1. A distribution function, giving the fraction of molecules with any given range of sedimentation coefficient, can be obtained by suitable extrapolation of data such as those of Fig. 22–7 to zero concentration.[70,71]

An important result of the effects discussed in this section is that a single sedimentation velocity experiment cannot be used as an infallible guide to the semi-quantitative or quantitative composition of a macromolecular mixture, as had at one time been believed. Such information can be obtained only upon careful extrapolation to zero concentration. (See discussion by Williams, Baldwin, and coworkers.[9,71] These authors show that distribution functions $g(s)\,ds$ which give the fraction of solute material possessing sedimentation coefficients between any value of s and $s + ds$ can be obtained if interacting flow and diffusion are neglected. If diffusion cannot be neglected, its effects can be removed by performing experiments at different times and extrapolating to $t = \infty$. This technique is successful because boundary spreading due to heterogeneity increases with time more rapidly than boundary spreading due to diffusion.)

22e. Molecular weights from sedimentation and diffusion. The most important use of sedimentation during the past thirty years has been in the determination of molecular weights by the combined measurement of sedimentation and diffusion coefficients. If s and D are measured in solutions of identical composition, the same frictional coefficient will apply to both, so that, for a two-component system, or for systems which behave like two-component systems, equations 22–5 and 21–8 may be combined to yield

$$\frac{s}{D} = \frac{M(1 - \bar{v}_2\rho)}{\mathcal{R}T[1 + C(\partial \ln y/\partial C)]} \tag{22–22}$$

Equation 22–22 shows that s/D is a thermodynamic quantity resembling the apparent molecular weight determinable from sedimentation equilibrium (equation 16–20). The true molecular weight can be determined by extrapolation of s/D to zero solute concentration, and values of y or the equivalent virial coefficients can be evaluated from the concentration dependence.

An alternative procedure for determining molecular weight is to extrapolate s and D to zero concentration separately, giving extrapolated values s^0 and D^0 from which M is evaluated as

$$M = \frac{s^0 \mathcal{R}T}{D^0(1 - \bar{v}_2\rho_0)} \tag{22–23}$$

where ρ_0 is the density of the solvent.

If equation 22–23 is to be applied to heterogeneous polymer mixtures, the molecular weight average which we obtain depends on the manner in which we choose to determine s and D from sedimentation and diffusion data.[74] If equations 22–16 and 22–18 form the basis for the evaluation of s, then the weight average of the sedimentation coefficient is obtained. This may be combined with the weight average of the diffusion coefficient (p. 355), but the ratio of these quantities will not ordinarily yield the weight-average molecular weight. It is generally considered preferable to use osmotic pressure, light scattering, sedimentation equilibrium, or Archibald's method for heterogeneous mixtures, because exact mathematical treatments for these procedures are available. For this reason we shall not consider here the molecular weights which have been obtained for heterogeneous samples by use of the sedimentation-diffusion method.

Proteins and viruses are the commonest macromolecular substances which are homogeneous with respect to molecular weights and it is for these substances that the sedimentation-diffusion method has been primarily used. Typical results are shown in Table 22–1, and they are seen to be in excellent agreement with molecular weights for the same substances obtained by other methods, as listed elsewhere in this book. (All the data shown were obtained in aqueous salt solutions, near room temperature and near the isoelectric points of the proteins. Under these conditions, as indicated above, the probable error due to the effects of electric charge and the presence of the third component should be quite small.)

It should be noted that sedimentation coefficients of proteins, like diffusion coefficients (p. 357), depend on the properties of the solvent and are normally reported as reduced to standard conditions. By using

TABLE 22–1. Sedimentation Coefficients and Molecular Weights of Some Proteins and Viruses

	$s^0_{20,w} \times 10^{13}$	$D^0_{20,w} \times 10^7$	$\bar{v}_2$	M
Ribonuclease[47]	1.64	11.9	0.728	12,400
Lysozyme[48]	1.87	10.4	0.688	14,100
Chymotrypsinogen[37]	2.54	9.5	0.721	23,200
β-Lactoglobulin[51]	2.83	7.82	0.751	35,000
Ovalbumin[49]	3.55	7.76	0.748	45,000
Serum albumin[67]	4.31	5.94	0.734	66,000
Hemoglobin[60]	4.31	6.9	0.749	60,000
Catalase[39]	11.3	4.1	0.73	250,000
Fibrinogen[69]	7.9	2.02	0.706	330,000
Urease[50]	18.6	3.46	0.73	480,000
Myosin[72]	6.4	1.0	0.728	570,000
Bushy stunt virus[45]	132	1.15	0.74	10,700,000
Tobacco mosaic virus[64]	170	0.3	0.73	50,000,000

equation 22–23 we can express s^0 in terms of D^0 and can use equation 21–29 to compute the "standard" sedimentation coefficient, $s_s{}^0$. Where symbols without subscripts refer to the solvent in which measurements were made and the subscript s refers to the "standard" solvent,

$$s_s{}^0 = s^0 \eta (1 - \bar{v}_2 \rho_0)_s / \eta_s (1 - \bar{v}_2 \rho_0) \qquad (22\text{–}24)$$

with the same underlying conditions as those applied to equation 21–29. If the values of η_s and $(\rho_0)_s$ are those of water at 20° C, we obtain the standard solvent usually employed. Sedimentation coefficients corrected to this solvent are designated $s^0_{20,w}$.

22f. Frictional coefficients from sedimentation coefficients. If the molecular weight of a homogeneous macromolecular substance is known, then, as equation 22–5 shows, the sedimentation coefficient is a measure of the frictional coefficient. The sedimentation coefficient at zero concentration, s^0, can then be interpreted exactly as frictional coefficients determined from diffusion were interpreted in Tables 21–1 to 21–3.

Some of the frictional coefficients listed in Tables 21–1 and 21–3 were in fact obtained from sedimentation rather than from diffusion coefficients. The majority are the result of simultaneous determination of s^0 and D^0, yielding values for the molecular weight by equation 22–23 and for the frictional coefficient by equation 21–9 or 22–5.

22g. Sedimentation coefficients of flexible polymers. It has been pointed out that sedimentation velocity studies of unfractionated heterogeneous polymer mixtures yield results which are difficult to interpret. The study of well-fractionated samples which may be regarded as essentially homogeneous with respect to molecular weight, is, however, of considerable interest. The average frictional coefficient should in this case be that of the equivalent sphere discussed in section 20e. The radius R_e of this sphere is $\xi_f R_G$, where R_G is the radius of gyration and ξ_f a coefficient whose

TABLE 22-2. Sedimentation Coefficients for Polyisobutylene
Fractions in Cyclohexane[a] (Mandelkern et al.[73])

M	$[\eta]$ cc/gram	R_G, Å (eq. 23-5)	$s^0 \times 10^{13}$	ξ_f (eq. 22-25)
30,000	34.2	71	0.925	0.63
86,000	70.6	128	1.49	0.61
172,000	112	187	1.94	0.64
672,000	287	403	3.33	0.67
1,420,000	489	618	4.45	0.69
			Average	0.65

[a] At 20° C with $\bar{v}_2 = 1.091$ cc/gram, $\rho_0 = 0.779$ gram/cc and $\eta = 0.00985$ poise.

value is given as 0.665 by the best quantitative theory at present available. With equation 19–13 we get $f = 6\pi\eta\xi_f R_G$, or, from equation 22–5,

$$s^0 = \frac{M(1 - \bar{v}_2\rho_0)}{6\pi\eta\mathcal{N}\xi_f R_G} \qquad (22\text{–}25)$$

where s^0 is the sedimentation coefficient extrapolated to zero concentration and ρ_0 the density of the solvent.

If equation 9–40 is introduced for the dependence of R_G on molecular weight, we get

$$s^0 = \frac{M^{\frac{1}{2}}(1 - \bar{v}_2\rho_0)}{(6/M_0)^{\frac{1}{2}}\pi\eta\mathcal{N}\alpha\beta\xi_f} \qquad (22\text{–}26)$$

leading to the prediction that, for a series of polymer fractions varying in molecular weight, s^0 should vary as $M^{\frac{1}{2}}$ in a poor solvent and roughly as $M^{0.45}$ in a better solvent. This result is confirmed by the few studies which have been made; for example, by the data on polymethyl methacrylate shown in Fig. 21–3, which show that log s^0 varies as 0.44 log M.

Equation 22–25 can be used directly to evaluate the parameter ξ_f, by measuring M, s^0, and R_G for the same fractions of a given polymer preparation. R_G can be measured by light scattering (section 18), or it can be obtained from viscosity measurements by means of equation 23–5, which will be discussed in the following section. The result of one such calculation is shown in Table 22–2. The average value of ξ_f is 0.65, in excellent agreement with the value of 0.665 predicted by the Kirkwood-Riseman theory (p. 345).

For a number of other polymers[73] values of ξ_f ranging from about 0.55 to about 0.75 have been observed. In each case fairly good solvents were employed. It can not yet be said to be experimentally established whether ξ_f (like ξ) varies with the nature of the solvent. Nor is it really certain at this time that it is completely independent of molecular weight. Certainly a trend with molecular weight is observed in the data of Table 22–2.

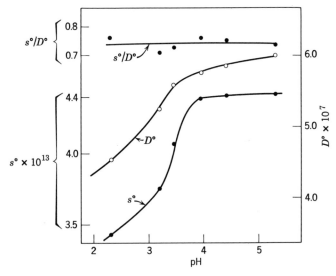

Fig. 22–8. Sedimentation and diffusion coefficients of serum albumin as a function of pH. (Harrington, Johnson, and Ottewill.[126])

22h. Sedimentation coefficients as a measure of configurational change in proteins. It has been clearly demonstrated by all the experimental methods discussed in this book that protein molecules, examined at or near their isoelectric points, have a unique molecular weight and a unique compact configuration, the uniqueness being inferred from the x-ray diffraction patterns which are obtained from crystals of these proteins. It has also been mentioned, however, that both molecular weight and configuration may easily change under the influence of altered external conditions. Such changes are, of course, reflected in changes in the value of the sedimentation coefficient and are often first detected in this way. Two examples of this kind of process are shown in Figs. 22–8 and 22–9. In both examples there is a decrease in sedimentation coefficient as the pH is reduced. In the case of serum albumin (Fig. 22–8) the decrease in sedimentation coefficient is accompanied by a parallel decrease in the diffusion coefficient, while the ratio s^0/D^0 remains constant. The process occurring is thus clearly one involving an increase in frictional coefficient without change in molecular weight. On the other hand, in the case of hemoglobin (Fig. 22–9), the diffusion coefficient *increases* slightly as the sedimentation coefficient falls, and the ratio of s/D falls. The process here clearly represents a drop in average molecular weight without appreciable change in configuration.

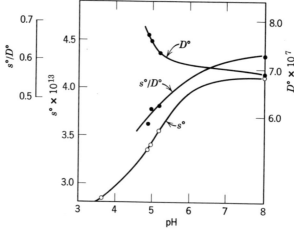

Fig. 22–9. Sedimentation and diffusion coefficients of hemoglobin as a function of pH. (Field and O'Brien.[127])

22i. Approach to sedimentation equilibrium. To conclude this section we return to a discussion of slow-speed operation of the ultracentrifuge, which leads to equilibrium distribution of the solutes as described in section 16. Figure 22–10 shows the manner in which the equilibrium state is approached during a sedimentation run, by showing solute distribution curves at various times. In this section we shall consider the distribution curves, for a single solute, at early stages of the experiment, when the central portion of the distribution curve is still independent of r.

To do so we consider the flow of solute across any cross-section of the sedimentation cell, as produced by the combined effects of sedimentation and diffusion. The number of moles, J, crossing a unit cross-sectional area in unit time was already evaluated in section 22b, and it is given by equation 22–10. We shall confine ourselves to considering cross-sections of the cell across which *no net flow occurs*. For such cross-sections J must be equal to zero, and equation 22–10 becomes

$$\frac{\mathscr{R}T}{C}\left(\frac{\partial C}{\partial r}\right)_t = \frac{M(1 - \bar{v}_2\rho)\omega^2 r}{1 + C(\partial \ln y/\partial C)_{T,P}} \tag{22–27}$$

It should be obvious that equation 22–27 represents an alternative derivation of the equation for sedimentation equilibrium in a two-component system, for the state of equilibrium may be defined as that state in which no net flow occurs anywhere in the cell. Another aspect of equilibrium is that C at any r becomes independent of time; i.e., if we wish to

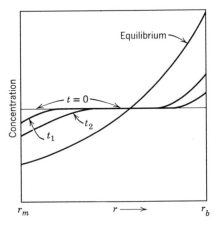

Fig. 22–10. Sedimentation equilibrium and the early stages in the approach to equilibrium. The distances r_m and r_b (from the axis of rotation) represent the positions of the meniscus and of the bottom of the sedimentation cell.

use equation 22–27 to describe the state of equilibrium, then $(\partial C/\partial r)_t$ may be replaced by dC/dr. With this replacement equation 22–27 becomes identical with equation 16–19, as it must.

As Archibald[130] was the first to point out, however, equation 22–27 must also apply *at any time* during the approach to equilibrium at any cross-section of the cell across which no flow occurs. There are clearly two such cross-sections, one at the meniscus separating the sedimenting solution from the air above it, and the other at the bottom of the sedimenting cell. The information ordinarily obtained from sedimentation equilibrium can thus be obtained without the necessity of waiting for equilibrium to occur if the factor $(1/rC)(\partial C/\partial r)_t$, at the meniscus or at the bottom of the cell, can be evaluated. We shall show below that this factor is indeed readily obtained, at any value of r, from refractive index gradient measurements. The values at the meniscus and at the bottom cell are then

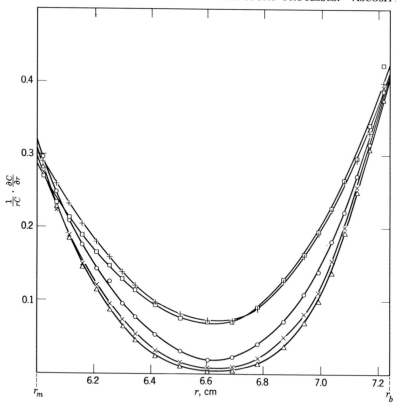

Fig. 22–11. Application of Archibald's method to determination of the molecular weight of lysozyme. From photographs taken 427 ($\triangle$), 491 ($\times$), 619 ($\bigcirc$), 1061 ($+$), and 1084 ($\square$) minutes after attaining full rotor speed. The method of extrapolation used here is slightly different from that suggested in the text, where $\partial C/\partial r$ is extrapolated first and then combined with values of C_m and r_m or C_b and r_b. (Smith, Wood, and Charlwood.[129])

determined by extrapolation, as is shown by the example of Fig. 22–11. By equation 22–27 these extrapolated values give the apparent molecular weight

$$M_{app} = \frac{M}{1 + C(\partial \ln y/\partial C)_{T,P}} \tag{22–28}$$

M_{app} can be determined at several concentrations, yielding the true molecular weight on extrapolation to $C = 0$, i.e., at $C = 0$

$$\frac{1}{r_m}\left[\frac{1}{C}\left(\frac{\partial C}{\partial r}\right)\right]_{t,\,r=r_m} = \frac{1}{r_b}\left[\frac{1}{C}\left(\frac{\partial C}{\partial r}\right)\right]_{t,\,r=r_b} = \frac{M(1 - \bar{v}_2\rho_0)\omega^2}{\mathscr{R}T} \tag{22–29}$$

where r_m and r_b are the values of r at the meniscus and at the bottom of the cell, respectively.

It will be noted from Fig. 22–11 that the values of $(1/rC)\,(\partial C/\partial r)$ at the meniscus and at the bottom of the cell are independent of time, as they must be. However, exactly the same values are not obtained from the top and bottom of the cell. This is presumably due to the fact that Fig. 22–11 represents an experiment at a rather high concentration. Since the concentration at the bottom of the cell must be different from that at the mensicus, the activity coefficient term in equation 22–28 will have a different value at the two positions.

It remains to describe the method for evaluating $(1/C)(\partial C/\partial r)$. For the early stages of an experiment, in which a central region of the cell exists which has C independent of r (cf. Fig. 22–1), the necessary relation can be derived according to the following method, due to Baldwin,[131] Gutfreund and Ogston[110] and Kegeles and coworkers.[112]

We begin with Lamm's general differential equation for the ultracentrifuge, equation 22–11. In the plateau region, where $\partial C/\partial r = 0$, the diffusion term in this equation vanishes, so that equations 22–12 and 22–13 again replace equations 22–10 and 22–11. If an arbitrary plane $(r = r_p)$ in this region is chosen as a reference plane, we may (as on p. 370) set the number of moles of solute which traverse this plane in unit time equal to $\alpha r_p J_p$. The total number of moles crossing this plane between the beginning of an experiment $(t = 0)$ and any subsequent time is the integral of this quantity; i.e., it is given by

$$\alpha r_p \int_0^t J_p\, dt = \alpha r_p{}^2 \int_0^t C_p s_p \omega^2\, dt \qquad (22\text{–}30)$$

With the substitution of equation 22–13 the integral on the right-hand side of equation 22–30 becomes equal to $-\dfrac{1}{2}\displaystyle\int_0^t dC_p$, and this, in turn, is just one-half the difference between the value of C_p at $t = 0$ and C_p at time t. At $t = 0$, however, the concentration of the molecules is the same everywhere in the cell, equal to the concentration C_0 in the initially homogeneous solution. Thus equation 22–30 becomes

$$\alpha r_p \int_0^t J_p\, dt = \tfrac{1}{2}\alpha r_p{}^2(C_0 - C_p) \qquad (22\text{–}31)$$

We consider next the total number of moles of solute which lie between the meniscus of the liquid in the cell $(r = r_m)$ and the arbitrary reference plane at $r = r_p$. This quantity was considered in section 22b, and it is given by equation 22–17. At $t = 0$, however, C is everywhere C_0 and the total number of moles between r_m and r_p is just $\tfrac{1}{2}\alpha C_0(r_p{}^2 - r_m{}^2)$. The difference between this number and that given by equation 22–17 must be the number of moles which have crossed the plane r_p in the time t, as given by equation 22–31. Thus, with α canceling,

$$C_m = C_0 - \frac{1}{r_m^2} \int_{r_m}^{r_p} r^2 \left(\frac{\partial C}{\partial r}\right)_t dr \qquad (22\text{–}32)$$

A similar relation may be obtained for the bottom of the ultracentrifuge cell,

$$C_b = C_0 + \frac{1}{r_b^2} \int_{r_p}^{r_b} r^2 \left(\frac{\partial C}{\partial r}\right)_t dr \qquad (22\text{–}33)$$

The quantities necessary for evaluation of the left-hand side of equation 22–27 can now be obtained from refractive index gradient measurements. As Appendix C shows, these measurements provide values of $K(\partial C/\partial r)$ at any point in the sedimentation cell, where K is a constant which depends on the value of the refractive index increment, $d\tilde{n}/dC$, of the substance under investigation. These measurements may be extrapolated to $r = r_m$ or $r = r_b$ to yield the limiting values required. They may also be multiplied by r^2 and integrated to give the integrals on the right-hand side of equations 22–32 and 22–33. Finally, we can determine the value of KC_0 from a separate experiment on the same solution in which a boundary between the solution and solvent is formed, either by a layering technique or by a sedimentation velocity experiment. (The latter would determine KC_p rather than KC_0; KC_0 can then be obtained by equation 22–19).

Combining equations 22–27 and 22–32 and multiplying by the optical constant K, we obtain

$$K\left(\frac{\partial C}{\partial r}\right)_{r=r_m} = \frac{M_{app}(1 - \bar{v}_2\rho)\omega^2 r_m}{\mathscr{R}T}\left[KC_0 - \frac{K}{r_m^2}\int_{r_m}^{r_p} r^2 \left(\frac{\partial C}{\partial r}\right)_t dr\right] \qquad (22\text{–}34)$$

and a similar relation applicable to the bottom of the cell. The value of M_{app} is thus obtained directly, with the true molecular weight determined by extrapolation to zero concentration.

An alternative procedure, due to Trautman,[165] is to evaluate $(\partial C/\partial r)_{r=r_m}$ and the integral on the right-hand side at several different times in the initial stages of the sedimentation process. By plotting $K(\partial C/\partial r)_{r=r_m}$ versus the value of the integral we obtain M_{app} from the slope. This method has the advantage of eliminating the necessity for a separate determination of KC_0.

Some molecular weights which have been obtained by the Archibald method are given in Table 22–3. It will be noted that relatively few examples are provided, as compared with the large number given in

Table 22–1. The sole reason for this is that the Archibald method has come into use only quite recently. It will probably be used more frequently in the future than the sedimentation-diffusion method.

The discussion so far presented has been concerned only with systems containing a single sedimenting solute. Equation 22–29 may, however, be readily applied to a mixture of sedimenting components. A weight-average molecular weight is obtained, as the derivation below shows. (All non-ideality terms are ignored in this derivation, it being assumed that they are eliminated by extrapolation to zero concentration. See Kegeles et al.[112] for an exact treatment.)

TABLE 22–3. Molecular Weights by Archibald's Method

Raffinose (true Mol. Wt. = 504.5)[128]	495
Ribonuclease[128,166]	14,000
Ribonuclease[167]	13,500
Lysozyme[129]	14,100
Ovalbumin[167]	46,300
Serum albumin[128]	70,000

The quantity determined at the meniscus (after extrapolation to zero concentration) is

$$\frac{1}{r_m}\left[\frac{1}{c}\left(\frac{\partial c}{\partial r}\right)_t\right]_{r=r_m} = \frac{1}{r_m}\left[\frac{1}{\sum_i c_i}\sum_i\left(\frac{\partial c_i}{\partial r}\right)_t\right]_{r=r_m}$$

where c_i is the concentration of the ith component, all concentrations being expressed in grams per cubic centimeter. For the simple case in which $\bar{v}_2$ is the same for all components (e.g., a heterogeneous polymer preparation)

$$\frac{1}{r_m}\left[\frac{1}{c_i}\left(\frac{\partial c_i}{\partial r}\right)_t\right]_{r=r_m} = M_i\frac{(1-\bar{v}_2\rho)\omega^2}{\mathscr{R}T}$$

so that

$$\frac{1}{r_m}\left[\frac{1}{c}\left(\frac{\partial c}{\partial r}\right)_t\right]_{r=r_m} = \frac{(1-\bar{v}_2\rho)\omega^2}{\mathscr{R}T}\left(\frac{\sum_i c_i M_i}{\sum_i c_i}\right)_{r=r_m} = \frac{(\bar{M}_w)_{r=r_m}(1-\bar{v}_2\rho)\omega^2}{\mathscr{R}T}$$

The Archibald method thus measures $\bar{M}_w$ for the mixture at the meniscus at the time of measurement or, by the corresponding equation at r_b, the value of $\bar{M}_w$ at the bottom of the cell. These values will be different. However, both may be evaluated at various times, and extrapolation to

$t = 0$ will yield $\bar{M}_w$ for the original mixture. (If Trautman's graphical method is used, a heterogeneous mixture will yield a non-linear plot, from which $\bar{M}_w$ can be evaluated by determining the slope corresponding to zero time.[168])

23. VISCOSITY[14]

To measure and interpret the viscosity of macromolecular solutions is easier than to measure diffusion or sedimentation coefficients. First, the apparatus required is much simpler (section 19h). Second, an unequivocal numerical value is obtained directly from measured flow times and densities; the measurement is made as easily for a heterogenous polymer mixture as for a homogeneous preparation. Finally, the viscosity is measured on a homogeneous solution *without concentration gradients*. Thus no complications arise from the use of multi-component solvents. These solvents may of course alter the composition of the hydrodynamic particle in viscosity as well as in sedimentation and diffusion. But in sedimentation and diffusion they also lead to the creation of new forces in the form of chemical potential gradients, which interfere with the desired measurement of flow resulting from an applied force of known magnitude. When viscosity measurements are made, we determine in effect the additional local energy dissipation produced when large particles are introduced into a solvent consisting of small molecules. Whether or not these small molecules are all alike does not affect the validity of the measured quantity.

23a. The intrinsic viscosity. In section 19j we described the viscosity of suspensions of particles in terms of the difference between the macroscopic viscosity η' of the suspension and the viscosity η of the corresponding pure solvent. It is convenient to express experimental data in terms of the *specific viscosity*,

$$\eta_{\text{sp}} \equiv \frac{\eta' - \eta}{\eta} \qquad (23\text{-}1)$$

Equations 19–24 and 19–25 predict that this quantity, in the limit of infinite dilution, should be proportional to the number of suspended particles per unit volume. Considering, as we have done throughout this chapter, that a macromolecular solution is equivalent to a suspension of particles, we would expect η_{sp} to be proportional to the concentration, c, most conveniently measured in grams per cubic centimeter. Thus the quantity η_{sp}/c (often called *reduced viscosity*) should ideally be independent of concentration, and it becomes so at the limit of zero concentration. This limiting value of η_{sp}/c is called the *intrinsic viscosity*, $[\eta]$,

$$[\eta] \equiv \mathscr{L}_{c=0} \frac{\eta_{sp}}{c} \equiv \mathscr{L}_{c=0} \frac{\eta' - \eta}{\eta c} \qquad (23\text{-}2)$$

The intrinsic viscosity is the quantity of primary interest for macromolecular solutions. It is determined by measuring $(\eta' - \eta)/\eta c$ at various concentrations and extrapolating to $c = 0$.

An equivalent result is obtained by measuring $(1/c) \ln (\eta'/\eta)$ and extrapolating to zero concentration. For, $\ln(\eta'/\eta) = \ln [1 + (\eta' - \eta)/\eta]$. As the limit of zero concentration is approached $(\eta' - \eta)/\eta$ becomes very small and the preceding logarithm may be replaced by $(\eta' - \eta)/\eta$; i.e.,

$$\mathscr{L}_{c=0} \frac{1}{c} \ln \frac{\eta'}{\eta} = \mathscr{L}_{c=0} \frac{\eta' - \eta}{\eta c}$$

A general relation between reduced viscosity and molecular parameters, valid at infinite dilution, is given by equation 19–25. The factor ν in this equation was shown to be 2.5 for spheres and larger for more asymmetric particles. Values were given for ellipsoids of revolution. They are not available for particles of other shape.

Equation 19–25 contains the volume fraction ϕ of suspended particles. This is not an experimentally determinable quantity and must be replaced by the weight concentration c. If v_h is the hydrodynamic volume of a dissolved macromolecule, and c the macromolecular concentration in grams per cubic centimeter, then $\mathscr{N} c/M$ is clearly the number of particles per cubic centimeter, and $\phi = \mathscr{N} c v_h/M$. Thus equation 19–25, combined with equation 23–2, becomes

$$[\eta] = \nu \frac{\mathscr{N} v_h}{M} \qquad (23\text{-}3)$$

It remains only to introduce the relations for v_h given in section 20. For rigid particles we use equation 20–6, with the understanding that this limits the investigation to solutions containing a single solvent component, or, if macro-ions are present, to solutions of reasonably high ionic strength. Thus, for rigid macromolecules,

$$[\eta] = \nu(\bar{v}_2 + \delta_1 v_1^0) \qquad (23\text{-}4)$$

Flexible polymers, on the other hand, behave as spheres of radius $R_e = \xi R_G$. Thus $v_h = \frac{4}{3}\pi \xi^3 R_G^3$. With equation 9–40 for R_G, and giving ν the value 2.5 appropriate to spheres, we find that equation 23–3 becomes

$$[\eta] = \frac{10\pi \mathscr{N}}{3M} \xi^3 R_G^3 = \frac{10\pi \mathscr{N} \alpha^3 \beta^3 \xi^3 M^{\frac{1}{2}}}{3(6M_0)^{\frac{3}{2}}} \qquad (23\text{-}5)$$

It will be noted that the molecular weight does not appear in equation 23–4, owing to the fact that v_h for rigid molecules varies as M to the first power, thus canceling the $1/M$ of equation 23–3. For flexible molecules, however, v_h depends on $\alpha^3 M^{\frac{3}{2}}$, leaving the residual molecular weight dependence given by equation 23–5.

It is also worth noting that the intrinsic viscosity of rigid macromolecules should be essentially independent of temperature or of the nature of the solvent as long as the molecular parameters v, δ_1, and $\bar{v}_2$ remain unchanged. This is in contrast to diffusion and sedimentation coefficients which are sensitive to both solvent and temperature through their direct dependence on T and on solvent viscosity. On the other hand, as we shall see in section 23f, the intrinsic viscosity of flexible polymers is markedly influenced by solvent and temperature, principally because of their influence on the solute-solvent interaction parameter α.

23b. The effect of concentration on reduced viscosity. It has been emphasized throughout this chapter that hydrodynamic properties reflect the behavior of individual molecules only at infinite dilution, so that experimental data obtained at finite concentrations must in general be extrapolated to zero concentration. In the case of viscosity measurements this necessity is incorporated in the very definition of the intrinsic viscosity, which is the quantity usually measured.

The quantity η_{sp}/c, or reduced viscosity, which is extrapolated to give the intrinsic viscosity, may be strongly concentration dependent. The concentration dependence is often expressed in terms of the relation

$$\eta_{sp}/c = [\eta] + k[\eta]^2 c \qquad (23\text{--}6)$$

where k is a constant known as the Huggins constant. For solid uncharged spheres k is approximately 2.0, both in theory[99] and practice.[100] For flexible polymer molecules[101] in good solvents k is often near 0.35. Somewhat higher values occur in poor solvents. These figures are such that η_{sp}/c at a concentration $c = 0.01$ may be only 5% larger than $[\eta]$ for uncharged spheres whereas it may be twice as large as $[\eta]$ for randomly coiled molecules. The magnitude of the effect of concentration on η_{sp}/c is thus of the same order as for the effect of concentration on sedimentation coefficients (cf. Fig. 22–3).

23c. The effect of rate of shear. It was pointed out in section 19f that very asymmetric molecules or easily deformable molecules would tend to be oriented or deformed by high velocity gradients. Such behavior would lead to a reduction in the disturbance of the flow pattern at high velocity gradients and hence to viscosities below those characteristic of unoriented or undeformed molecules. This effect is an important one for molecules with a very high intrinsic viscosity, and for such molecules it is essential that experimental determinations be made at several different values of the mean velocity gradient and that intrinsic viscosities be extrapolated to zero velocity gradient. An example of the effect produced is shown by

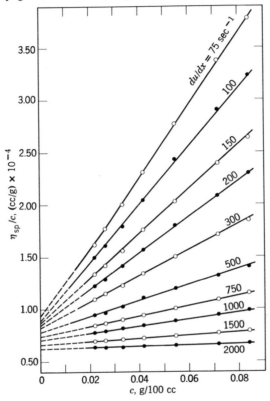

Fig. 23–1. The viscosity of a sample of cellulose nitrate as a function of the rate of shear. (Timell.[103])

data for cellulose nitrate in Fig. 23–1. It should be noted that the mean velocity gradient in a viscometer, $(du/dx)_{av}$, has the dimensions of sec^{-1} and is ordinarily called the *rate of shear*.

For molecules of relatively low intrinsic viscosity the rate of shear has generally no appreciable effect on $[\eta]$, though the factor k of equation 23–6 may be altered.

Convenient apparatus for determining the effect of shear has been described by several investigators, e.g., Fox, Fox, and Flory[88] and Timell.[103]

23d. The intrinsic viscosity of proteins. Dissolved proteins exist as macro-
ions of considerable electrostatic charge at most pH values. As a result,
an ionic atmosphere is maintained around them, and this causes a further
disturbance of the flow pattern in viscous flow and a corresponding
increase in viscosity. It was believed at one time that this effect is quite
large, even at relatively high ionic strength, but a recent treatment of the
problem by Booth[82] has shown that this is not true. His equation shows
that the effect of charge is quite small in most cases, even at an ionic

TABLE 23–1. Intrinsic Viscosities of Proteins in
Aqueous Salt Solutions[a]

	M	$\bar{v}_2$ cc/gram	$[\eta]$ cc/gram	f/f_{min} (Table 21–1)
Ribonuclease[75]	13,683	0.728	3.30	1.14
β-Lactoglobulin[113]	35,000	0.751	3.4	1.25
Serum albumin[76]	65,000	0.734	3.7	1.31
Hemoglobin[77]	68,000	0.749	3.6	1.14
Catalase[114]	250,000	0.73	3.9	1.25
Tropomyosin[79]	93,000	0.74	52	3.22
Fibrinogen[79,109]	330,000	0.710	27	2.34
Collagen[43]	345,000	0.695	1150	6.8
Myosin[72]	493,000	0.728	217	3.53

[a] Literature values of $[\eta]$ are sometimes given in units of deciliters per gram. In these
units $[\eta]$ is $1/100$ of the values here given.

strength of 0.01. Experimental studies have amply confirmed this conclusion. Intrinsic viscosities of proteins may therefore be treated in terms
of the relations of section 23a, which make no allowance for electrostatic
effects, in the same way as diffusion and sedimentation coefficients, at
reasonably high ionic strength, can be treated without allowance for such
effects. (It should be noted, however, that there is a large effect of electrostatic charge on the concentration dependence of reduced viscosity, i.e.,
on the factor k of equation 23–6.[75,76]).
    Experimental values of the intrinsic viscosity of a number of proteins
are listed in Table 23–1. Most of the results were obtained at 20 or 25° C,
near the isoelectric points of the protein, in aqueous solutions containing
reasonably high concentrations of neutral salt. The results obtained are
seen to be comparable to those obtained from diffusion studies (Table
21–1). One group of proteins, the globular proteins, is characterized by
values of $[\eta]$ between 3.3 and 4.0. The same proteins have f/f_{min} not very

different from unity. The second group of proteins, by contrast, have very much larger values of $[\eta]$, and the same proteins also have values of $f/f_{\min}$ much larger than unity.

Equation 23–4 shows that $[\eta]$, like $f/f_{\min}$, depends on two factors, the factor ν, representative of the shape of the hydrodynamic particle, and the factor δ_1, which is a measure of solvation. As in the case of $f/f_{\min}$ these two factors cannot be resolved by considering viscosity measurements alone.

TABLE 23–2. Molecular Dimensions of Proteins from Intrinsic Viscosities

	Maximum Solvation ($\nu = 2.5$)		Maximum Asymmetry ($\delta_1 = 0$)		Compromise ($\delta_1 = 0.2$)	
	δ_1 grams/gram	R_e Å	ν	a/b, Prolate Ellipsoid	ν	a/b, Prolate Ellipsoid
Ribonuclease	0.59	19.3	4.5	3.9	3.6	2.9
β-Lactoglobulin	0.61	26.6	4.5	3.9	3.6	2.9
Serum albumin	0.75	33.7	5.0	4.4	4.0	3.3
Hemoglobin	0.69	34	4.8	4.1	3.8	3.1
Catalase						
Tropomyosin	20	91	70	29		
Fibrinogen	10.1	112	38	20		
Collagen	460	400	1660	175		
Myosin	86	257	298	68		

However, we can again make the same kind of calculations as those made in Table 21–2. We can calculate the maximum value of δ_1 by assigning to ν its minimum value of 2.5, i.e., by assuming the hydrodynamic particle to be spherical (with R_e obtained by setting v_h in equation 23–3 equal to $\frac{4}{3}\pi R_e^3$). Alternatively we can calculate the maximum value of ν, i.e., the maximum deviation from spherical shape, by setting $\delta_1 = 0$. In the latter case the observed value of ν can be interpreted in terms of the ellipsoidal model of Simha, using Fig. 19–7 or 19–8. The prolate ellipsoidal model has been used in Table 23–2, giving the axial ratio a/b of the most asymmetric ellipsoid which could be used to represent the hydrodynamic particle.

The calculations of Table 23–2 yield essentially the same information as those of Table 21–2. The globular proteins can be neither highly solvated nor very asymmetric. It is probable that their molecules (in water) contain some hydration and that they deviate somewhat from spherical shape. The compromise calculation shown in the last two columns of Table 23–2 probably provides a reasonably accurate picture of the hydrodynamic particles which these molecules form in solution, with the understanding that asymmetry is described in terms of an ellipsoidal model simply because this is the only one for which quantitative relations are available and not because the hydrodynamic particle is actually believed to be ellipsoidal in shape. Within the accuracy of the measurements, the description of globular proteins in aqueous solution provided by Table 23–2 is identical with that provided by Table 21–2.

It should again be noted that the globular proteins maintain the relatively compact and symmetrical configurations indicated in Table 23–2 only over a limited range of conditions. The protein ribonuclease, for instance, remains compact in aqueous solution at acid pH (in contrast to serum albumin; cf. Fig. 22–8), as shown[75] by intrinsic viscosities of 3.4 and 3.5, respectively, at pH 4 and pH 3. The same protein, however, acquires an intrinsic viscosity of 8.9 cc/gram in $8M$ aqueous urea[115] and an intrinsic viscosity of 6.0 cc/gram in the solvent 2-chloroethanol.[116]

Turning next to the proteins with high values of $[\eta]$, it is clear that these proteins form either highly asymmetric particles or else highly solvated (i.e., probably randomly coiled) particles, even in aqueous solution near their isoelectric points. This again corresponds to the conclusion reached from frictional coefficients in section 21.

23e. Structural information from comparison of hydrodynamic and light scattering data. It remains to decide between the two possible extreme structures allowed by Tables 21–2 and 23–2 for molecules with high values of $[\eta]$ and of $f/f_{\min}$, i.e., to decide whether these molecules, in solution, more nearly resemble randomly coiled chains or long thin rods. To achieve this end it is necessary to examine the result of applying *several* experimental techniques to the same molecule. Four suitable experimental quantities have been discussed so far in this book: the thermodynamic second virial coefficient B (section 12), the radius of gyration R_G from light scattering or low angle x-ray scattering (section 18), the frictional coefficient (i.e., D^0 or s^0) (section 21), and the intrinsic viscosity. (An additional quantity is the rotary diffusion coefficient to be described in section 25.) Of these parameters the second virial coefficient B is too sensitive to solvent-solute interaction to be a reliable measure of configuration alone, so that it is generally used only as a last resort. Deferring discussion of the rotary diffusion coefficient to section 25 we shall here consider the information to be derived from a comparison of the light scattering radius of gyration with the measurements of viscosity and of the diffusion or sedimentation coefficient.

To do so we shall compute the radius of gyration to be anticipated from the hydrodynamic measurements and compare it with the value of R_G actually obtained by light scattering. Two procedures may be followed.

(1) If the molecule is a flexibly coiled molecule, equation 23–5 can be used to determine R_G directly from $[\eta]$, equation 22–25 can be used to determine R_G from s^0, and equation 21–30 can be used to determine R_G from D^0. There is some question about the value of the parameter ξ or ξ_f (p. 383), but the resulting uncertainty is only about 10%.

(2) If the molecule is considered to be rod shaped, the length of the rod $(L = 2a$, where a is the major semi-axis of the corresponding prolate ellipsoid) may be computed by combining the hydrodynamic volume, $v_h = \frac{4}{3}\pi ab^2$, as given by equation 20–6, with the ratio a/b as computed in Tables 21–2 or 23–2. The radius of gyration is then obtained by equation 18–21.

TABLE 23–3. Radius of Gyration of Myosin and Collagen

	Myosin[72]	Collagen[43]
Assuming random coil		
From $[\eta]$, $\xi = 0.875$	295 Å	450 Å
From $[\eta]$, $\xi = 0.775$	330 Å	510 Å
From s^0 or D^0, $\xi_f = 0.66$	280 Å	480 Å
Assuming rigid rod		
From $[\eta]$	495 Å	830 Å
From s^0 or D^0	516 Å	1180 Å
Experimental value		
From light scattering	468 Å	870 Å

The values of a/b given in Tables 21–2 and 23–2 are *maximum* values, corresponding to $\delta_1 = 0$. It is reasonable to suppose, however, that protein molecules will always bind some solvent quite strongly, regardless of configuration, so that a value of δ_1 somewhat greater than zero would be a better choice. This would lead to smaller values of a/b, but it would also give a larger value to v_h. The combined effect is to leave the calculated value of R_G virtually unchanged; changing δ_1 from zero to 0.5 gram/gram protein alters the calculated R_G by only 5%. The use of the present data to distinguish between flexible coils and long rods does not, therefore, depend on assumptions concerning the extent of solvation of the latter.

In Table 23–3, both procedures are applied to two of the non-compact protein molecules cited in Table 23–2. The data show conclusively that neither of these molecules can be flexibly coiled.

In the case of collagen the radius of gyration calculated from the intrinsic viscosity with the assumption that the molecule is a rigid rod is in very good agreement with the experimental value. The value of R_G computed from the sedimentation coefficient is not in quite such good agreement, but, as we have pointed out, the determination of sedimentation coefficients for molecules as asymmetric as collagen is subject to considerable error because of the pronounced concentration dependence. In any event, Boedtker and Doty,[43] who obtained the data cited in Table 23–3 concluded that collagen is best represented as a long rod with a diameter of 13.6 Å and a length of 3000 Å. These dimensions agree well with those obtained by electron microscopy. The diameter corresponds closely to that of the structural unit in solid collagen fibers, as determined by x-ray diffraction.

In the case of myosin the data are about equally strong in suggesting that a rigid rod is a closer representation of the actual structure than a random coil. On the basis of information concerning some smaller molecules which can be formed from myosin, Rice and Holtzer[104] have suggested that the actual molecule may be a rodlike rigid structure with non-uniform thickness.

One other technique should be mentioned which provides information on the configuration of protein molecules which are not compact and very symmetrical. It will be shown in section 27 that flexible molecules which bear an electrostatic charge become very much more extended as the charge is increased and that this process is progressively reversed when an inorganic salt is added. Rigid molecules, of course, should be unaffected by charge or ionic strength. This procedure has not been used on myosin and collagen, but it is frequently employed in the investigation of the globular proteins in their denatured state, i.e., under conditions where these proteins are no longer compact and symmetrical. In nearly every instance the denatured state is found to be one in which the molecules have become flexible, at least to some extent, for sizable changes in sedimentation coefficient and intrinsic viscosity (in the predicted direction) result from alterations in electrostatic charge and ionic strength.[105]

It may be observed that a combination of measured values of $[\eta]$ and f, without other measurements, does not provide a sensitive tool for the determination of configuration. Combining equations 23–4, 22–5, and 21–26 and eliminating δ_1, we get[106]

$$\frac{\mathcal{N} s[\eta]^{1/3} \eta}{1 - \bar{v}_2 \rho_0} = (100)^{1/3} \beta' M^{2/3}$$

with the constant β' proportional to $v^{1/3} f_0 / f$. (A similar relation involving D and $[\eta]$ would be obtained from equations 23–4 and 21–26 alone.) It happens that $v^{1/3}$ and f/f_0 depend on configuration in much the same way, so that the parameter β' is markedly insensitive to configuration. For example, β' lies within the limits of 2.12 to 2.50 $\times$ 10^6 for solid spheres, flexible coils, oblate ellipsoids of any axial ratio, and prolate ellipsoids of axial ratio up to $a/b = 13$. Only for very long rods does β' become significantly larger.

(This insensitivity of β' to configuration may in fact be utilized to advantage. With assignment of a reasonable value of β', the equation here given may be used for the evaluation of an approximate molecular weight.)

23f. Intrinsic viscosity and the dimensions of flexible polymers. It has already been pointed out that whereas diffusion, sedimentation, and viscosity measurements are all used to provide information concerning protein solutions, diffusion and sedimentation are difficult techniques to apply to flexible polymer molecules. For such polymers, therefore, viscosity measurements are especially important. A great deal of thought has been devoted to their interpretation. Definitive in this respect is the analysis of the problem by Flory and Fox,[83] which forms the basis of the discussion here given.

It should be pointed out that there is considerable difference between the notation here used and that given in the original papers by Flory and Fox, and in other experimental studies based upon them. The notation used here has a distinct advantage in that each symbol has a definite physical significance; i.e., β is an effective bond length, ξ a dimensionless quantity relating the radius of the equivalent hydrodynamic sphere to the radius of gyration, etc. Since virtually all experimental studies of flexible polymers give results in the Flory-Fox notation we give a detailed comparison of the two notations, so that the reader can readily adapt experimental results to the present notation, and vice versa.

One difference between the present notation and that of Flory and Fox is that our intrinsic viscosities are expressed in normal cgs units (cubic centimeter per gram) whereas those of Flory and Fox are expressed in deciliters per gram. Our values of $[\eta]$ are therefore 100 times greater.

Second, the present treatment is centered explicitly about an equivalent sphere, the radius of which is ξ times R_G. Grouping together all numerical constants of equation 23–5 we have

$$[\eta] = 6.308 \times 10^{24} \xi^3 R_G^3/M = 4.291 \times 10^{23} \frac{\alpha^3 \beta^3 \xi^3}{M_0^{3/2}} M^{1/2} \tag{23-7}$$

In the Flory-Fox treatment the concept of an equivalent sphere enters implicitly rather than explicitly, and they write

$$[\eta] \text{ (in dl/gram)} = \Phi h_{av}^3/M = K\alpha^3 M^{1/2} \tag{23-8}$$

with the additional difference that they write $(\overline{r^2})^{3/2}$ in place of h_{av}^3, r representing end-to-end distance in their notation. Since $h_{av} = 6^{1/2} R_G$, Φ is clearly proportional to ξ^3, whereas K is proportional to $\beta^3 \xi^3/M_0^{3/2}$.

The difference in notation has been summarized in Table 23–4, which also lists appropriate numerical conversion factors.

We shall now show that it is possible to use equation 23–7 to evaluate, in turn, each of the parameters α, β, and ξ. Since they are the parameters which determine completely the configuration of flexible polymer molecules in solution, we are in effect using viscosity measurements, together with other data, to obtain experimental evidence concerning this configuration.

TABLE 23–4. The Notation of Flory and Fox

Symbols for Polymer Dimensions

This Chapter	Flory and Fox
R_G	$(\overline{s^2})^{\frac{1}{2}}$
h, h_0, h_{av}	$r, r_0, (\overline{r^2})^{\frac{1}{2}}$
α	α
β	$(\overline{r_0^2}/M)^{\frac{1}{2}}$

Equations for Intrinsic Viscosity

This Chapter, $[\eta]$ in cc/gram	Flory and Fox, $[\eta]$ in dl/gram
$[\eta] = (10/3)\pi \mathcal{N} \xi^3 R_G^3/M$	$[\eta] = \Phi(\overline{r^2})^{\frac{3}{2}}/M$
$[\eta] = 4.291 \times 10^{23}\,(\alpha^3\beta^3\xi^3/M_0^{\frac{3}{2}})M^{\frac{1}{2}}$	$[\eta] = K\alpha^3 M^{\frac{1}{2}}$

Numerical Conversion Factors

$$[\eta] \text{ in cc/gram} = 100 \; ([\eta] \text{ in dl/gram})$$
$$\overline{h^2} = \overline{r^2} = 6R_G^2 = 6\overline{s^2}$$
$$\Phi = 4.291 \times 10^{21}\xi^3$$
$$K = 4.291 \times 10^{21}\,\beta^3\xi^3/M_0^{\frac{3}{2}}$$

We first consider the parameter ξ (or the equivalent Flory-Fox parameter Φ). Since the radius of gyration can be directly determined from the angular dependence of light scattering, as discussed in section 18, it is necessary only to determine R_G and $[\eta]$ in the same solution. We can then solve for ξ directly by equation 23–7.

It was pointed out in section 20f that the Kirkwood-Riseman theory predicts that ξ should be a universal constant at sufficiently high molecular weight, its value being 0.875. The first results obtained from experiment appeared to verify this conclusion, and Φ was considered a universal constant in the original Flory-Fox formulation. More recent data, however, indicate that this is not true. It appears certain that ξ varies somewhat, depending on the nature of the polymer-solvent interaction. In a poor solvent the radius of gyration is relatively small and the density of polymer segments correspondingly relatively large. Thus solvent is effectively trapped to a larger distance from the center of mass than in good solvents.

A number of results of various authors have been collected by Krigbaum and Carpenter,[84] and they are shown in Fig. 23-2. The nature of

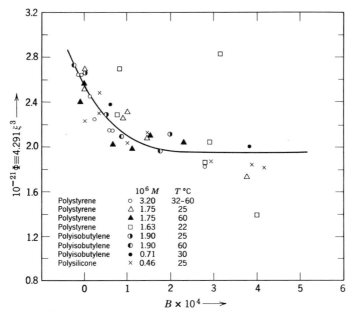

Fig. 23-2. Dependence of the parameter Φ or ξ on the thermodynamic virial coefficient B. (Kirgbaum and Carpenter.[84])

the polymer-solvent interaction is expressed by the second virial coefficient B (cf. sections 12,13, and 17), which is positive for good solvents, zero for ideal solvents, and negative for very poor solvents. The corresponding values of the Flory-Fox "constant" Φ are seen to vary from about 2.0×10^{21} to about 2.8×10^{21}. By Table 23-4 the corresponding values of ξ vary from 0.775 to 0.86, as compared to the theoretical value of 0.875.

The actual values of ξ or Φ here given are tentative. All our equations apply strictly only to homogeneous polymer fractions. Even the best fractions used in practice retain, however, some heterogeneity. Different values of ξ or Φ are then obtained, depending on which average of M and R_G is employed. This is a factor which most authors have not yet considered with sufficient care. It should also be mentioned that it is not impossible that more careful work in the future will show some trend in ξ

or Φ with molecular weights, as is in fact predicted at low molecular weights by the Kirkwood-Riseman theory. No such trend has yet been detected experimentally.

The next step in the analysis of viscosity data for flexible polymers is to examine the relation between $[\eta]$ and molecular weight in ideal solvents. They are solvents for which the virial coefficient B is equal to zero. In practice, any poor solvent will act as an ideal solvent at an appropriate temperature, designated as the Flory temperature, Θ (cf. Fig. 13–3). For an ideal solvent $\alpha = 1$ and equation 23–7 thus reduces to

$$[\eta] = 4.291 \times 10^{23} \beta^3 \xi^3 M^{\frac{1}{2}} / M_0^{\frac{3}{2}} \qquad (23\text{–}9)$$

Representative plots[85] of $[\eta]$ versus M in such solvents are shown in Fig. 23–3, and the square root dependence predicted by equation 23–9 appears

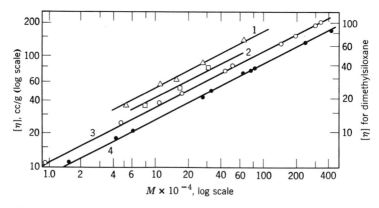

Fig. 23–3. Intrinsic viscosities of polymer solutions at the ideal temperature Θ. Curve 1 represents poly(dimethylsiloxane) in methyl ethyl ketone at 20°C (with $[\eta]$ given by right-hand ordinate). Curve 2 is cellulose tricaprylate in γ-phenylpropanol at 48°C; curve 3 is polyisobutylene in benzene at 24°C; curve 4 is polystyrene in cyclohexane at 34°C. (From Flory.[5])

to be entirely confirmed. Using the value of ξ appropriate to ideal solvents ($\xi = 0.835$) these data can be used to obtain values for the important dimensional parameter β, and a few values so obtained are shown in Table 23–5.

For polystyrene and polyisobutylene the C—C bond has been considered as the basic unit from which the polymer chain is constructed. Since each styrene or isobutylene monomer unit contains two such bonds, the value of M_0 used becomes half the molecular weight of such a monomer unit, i.e., $M_0 = 52$ for polystyrene and 28 for polyisobutylene. For cellulose tricaprylate the calculations are based on the dimensions given in Fig. 9–3. Each cellulose unit is considered composed of three successive bonds,

$a = 1.43\,\text{Å}$, $b = 2.95\,\text{Å}$, $a = 1.43\,\text{Å}$, so that the average (root-mean-square) bond length is $l_{av} = 2.06\,\text{Å}$, and M_0 is one-third of the molecular weight of a monomer unit, i.e., $M_0 = 180$.

The effective bond lengths listed in Table 23–5 should be compared with those calculated in sections 9d and 9f, based on the idea that there is free rotation about all simply connected single bonds. For polymethylene

TABLE 23–5. Dimensions of Flexible Polymer Molecules
in Ideal Solvents[a]

Polymer	Solvent	Ideal Temperature °K	Eff. Bond Length β, Å	β/l_{av} Observed	β/l_{av} Calculated, Free Rotation
Polystyrene[86]	Cyclohexane–CCl$_4$	288	5.04	3.27	
	Cyclohexane	307	4.96	3.22	1.41
	Ethylcyclohexane	343	4.83	3.14	
Polyisobutylene[81]	Benzene	297	4.00	2.60	1.41
	Anisole	378.5	3.78	2.45	
Cellulose tricaprylate[87]	γ-Phenyl propanol	321	10.75	5.22	2.66
	Dimethyl formamide	413	10.55	5.12	

[a] The statistical unit is the C—C bond ($l_{av} = 1.54\,\text{Å}$) in polystyrene and in polyisobutylene. In the cellulose ester it is the root-mean-square average ($l_{av} = 2.06\,\text{Å}$) of the bonds a, b, a shown in Fig. 9–3.

chains this calculation led to the result, $\beta = 2^{1/2}l_{av}$, and the observed values are seen to be very much larger than this. This was already predicted in section 9 and is due to the fact that rotation about single bonds in polymers is not free, but considerably hindered. It is of interest that the phenyl side chain in polystyrene exerts a greater hinderance to free rotation (i.e., β is larger) than do the two methyl side chains in polyisobutylene.

In the case of cellulose tricaprylate the ratio β/l_{av} is already 2.66 (equation 9–15) if free rotation about simply connected single bonds is permitted because no rotation can occur in any event about the glucose ring (bond b in Fig. 9–3). The observed value of β is about twice as big as this, however, and this indicates that the cellulose chain is an exceedingly stiff chain.

It was mentioned in section 9g that β should be a constant for a given polymer, independent of the solvent in which it is dissolved. Table 23–5, however, shows some variation in β with solvent. It should be noted, however, that the studies in each solvent were made at a different temperature, because each solvent becomes ideal at a different temperature. In each case the variation of β is towards lower values at higher temperature. Such variation is easily explained, for an increase in temperature is expected to lead to greater freedom of rotation and, hence, to a closer approach of β to the value calculated on the basis of completely free rotation. The variation in β is small, and there would seem to be no reason to seek a more complicated cause for it.

TABLE 23–6. The Parameter α for Polystyrene
in Benzene at 20° C

M	$[\eta]$	α^3	α
44,500	26.8	1.88	1.23
65,500	35.6	2.05	1.27
262,000	107	3.09	1.46
694,000	207	3.67	1.54
2,550,000	554	5.12	1.72
6,270,000	1175	6.93	1.91

For subsequent calculation we may plot β versus temperature, regardless of the solvents used in the experimental studies. Such plots give the values of β to be used at each temperature in any solvent whatever, ideal or non-ideal.

Only the interaction parameter α remains to be calculated. In the ideal solvents used to obtain a value for β, α has the value unity. Its value in other solvents can be obtained directly by application of equation 23–7 or 23–8; i.e.,

$$\alpha^3 = \frac{[\eta]M_0^{3/2}}{4.291 \times 10^{23}\beta^3\xi^3M^{1/2}} = \frac{[\eta] \text{ (in dl/gram) } M_0^{3/2}}{\Phi\beta^3M^{1/2}} \qquad (23\text{–}10)$$

In using this relation it is necessary to use a value of ξ or Φ appropriate to the type of solvent in which the viscosity measurements were made, and a value of β^3 appropriate to the temperature of the measurements.

As an example, Table 23–6 shows the calculation of α for a series of polystyrene fractions in benzene at 20° C. The solvent is a good solvent (cf. Table 13–3), and we have used for Φ the value 2.0×10^{21}. Table 23–5 suggests a value of about 5.02 Å for β. The most significant result of

Table 23–6 is that it shows α to increase with molecular weight, as predicted by the theoretical considerations of section 9h. Within probable error the values of α obey the relation

$$\alpha = 0.492M^{0.086} \tag{23–11}$$

Relations of this type appear to be applicable to all polymer-solvent systems. It should be noted that the exponent of M in equation 23–11, but not the coefficient 0.483, could be obtained merely by plotting log $[\eta]$ versus log M (cf. section 23i). An alternative expression equivalent to equation 23–11 would be

$$\alpha = 0.734\sigma^{0.086}$$

where σ is the number of monomer units in a chain.

The parameters α and β occur not only in the equation for intrinsic viscosity but also in equations 12–9 and 12–21, for the second virial coefficient B of the chemical potential equation. Since B can be evaluated experimentally from osmotic pressure or light scattering measurements, we can in principle evaluate α and β (once ξ is known) by simultaneous solution of equations 12–9, 12–21, and 23–7. However, equation 12–21 depends on the correctness of the statistical theory of Flory and Krigbaum, as it applies to both intermolecular and to intramolecular forces. That this theory is entirely free from approximations is not yet established, and α and β are therefore ordinarily not evaluated from thermodynamic data.

It is worth noting that in section 13c values of α, computed from viscosity measurements alone, were used to predict the molecular weight dependence of B for polyisobutylene, using equation 12–21. The calculation was found to agree only moderately well with experiment.

Since the value of α is a measure of how good a solvent is for the polymer being studied, we should expect that the intrinsic viscosity of flexible polymers should normally vary considerably with the nature of the solvent and that, for poor solvents, it should vary with temperature, $[\eta]$ increasing with increasing temperature as the solvent is improved. These conclusions are spectacularly confirmed, for example, by the work of Fox and Flory[81] on a polyisobutylene sample of molecular weight 1,460,000. The data are shown in Fig. 23–4.

23g. The effect of chain branching. There is no detailed theory as yet concerning the effect of chain branching on the physical properties of flexible polymers. Qualitatively we would predict that the major effect on the viscosity would be the result of the reduction in the radius of gyration discussed in section 9l. Partially compensating for this reduction might be a small increase in ξ, resulting from the increase in density within the particle. In any event, the net result should be a decrease in $[\eta]$. This prediction has been confirmed in a study of Thurmond and Zimm[102] on a polystyrene sample containing a few branching points.

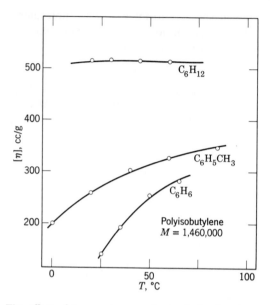

Fig. 23–4. The effect of temperature on the intrinsic viscosity of polyisobutylene in a good solvent (cyclohexane) and in two poor solvents (toluene and benzene). (Fox and Flory.[81])

TABLE 23–7. Intrinsic Viscosities of Various Macromolecules[a]

	M	$[\eta]$ cc/gram
Bushy stunt virus[78]	10,700,000	3.44
Tobacco mosaic virus[93]	39,000,000	36.7
DNA[117]	6,000,000	5000
Natural rubber (typical fractions)[91]	1,130,000	530
	190,000	152
	42,000	61
Cellulose acetate (typical fractions)[92]	184,000	290
	66,000	180
	11,600	38
Amylose (typical fractions)[169]	2,220,000	173
	847,000	115
	270,000	60
Amylopectin (unfractionated)[70]	90,000,000[b]	127

[a] Natural rubber in toluene; cellulose acetate in acetone. The others are aqueous solutions.

[b] Weight-average molecular weight.

23h. Intrinsic viscosities of some other macromolecules. Intrinsic viscosities of some macromolecules other than proteins and flexible polymers are listed in Table 23–7. Among the molecules listed, only bushy stunt virus is found to have the low intrinsic viscosity characteristic of a nearly spherical essentially unsolvated particle. All the other molecules listed have larger values of $[\eta]$. This result is clearly in agreement with other studies of these molecules, as discussed in earlier parts of this book.

Of the molecules listed in the table, one, tobacco mosaic virus, is a rod-like molecule. With $\delta_1 = 0.2$ we get $a/b = 20$, so that, by equation 20–8, $L/d = 24.5$. This result is in good accord with measurements cited elsewhere. Amylopectin, as pointed out earlier (p. 12) is a branched polysaccharide, and it probably possesses a flexible configuration. The intrinsic viscosity is very low for so high a molecular weight, a result due to the chain branching. DNA, as pointed out on p. 312, is probably a very stiff flexible chain. The intrinsic viscosity listed is that of a particular preparation; considerable variations in properties, including intrinsic viscosity, have been reported for different samples of DNA (see p. 292).

The other molecules listed have all been studied as fractionated samples of different molecular weight. All of them are flexible coils, as the following section will show.

23i. Intrinsic viscosity and molecular weight. For molecules which can exist with a variety of molecular weights, the relation between $[\eta]$ and M is one of the most important properties. By equation 23–4, $[\eta]$ will be independent of M if the molecular configuration is close to an unsolvated sphere. For flexible linear chains, however, equation 23–5 shows that $[\eta]$ increases with molecular weight, and the same is true for molecules which are long rods of constant diameter, since the length of such rods, and, hence, the ratio L/d, will increase with molecular weight.

Considering flexible chains first (equation 23–5), we note that, in a given solvent and at a given temperature, β and ξ should be constants independent of molecular weight. (Possibly a slight trend in ξ might occur; within present experimental accuracy it has not been detected.) Thus, since α can be expressed as a function of the form AM^x, we obtain for $[\eta]$ the relation

$$[\eta] = K'M^a \tag{23–12}$$

where both K' and a are constants whose values depend on the nature of the polymer and solvent and on the temperature, The variation in the exponent a must, however, be quite limited, for, according to the Flory-Krigbaum theory (section 12), the exponent x of the relation $\alpha = AM^x$ can never be greater than 0.10, so that a cannot exceed 0.80. In addition, polymer solutions become ideal when $\alpha = 1$, independent of molecular weight, and they become unstable when α falls appreciably below this value (i.e., phase separation occurs, as explained in section 15). Thus $x = 0$ and $a = 0.50$ become essentially minimum values. For flexible linear polymers which behave in accord with the theories which we have

TABLE 23–8. Intrinsic Viscosity and Molecular
Weight of Polymers[a]
$$[\eta] = K'M^a$$

Polymer	Solvent	K'	a	Coefficient x in Relation, $\alpha = AM^x$
Polystyrene[86,89,90]	Benzene, 25°	0.0095	0.74	0.080
	Toluene, 25°	0.017	0.69	0.063
	Butanone, 25°	0.039	0.58	0.027
	Cyclohexane, 34°[b]	0.081	0.50	0
Polyisobutylene[81]	Cyclohexane, 30°	0.026	0.70	0.067
	Benzene, 60°	0.026	0.66	0.053
	Benzene, 24°[b]	0.083	0.50	0
Natural rubber[91]	Toluene, 25°	0.050	0.67	0.057
Cellulose acetate[92]	Acetone, 25°	0.0090	0.90	0.133
Amylose[169]	0.33N KCl (aq.), 25°	0.113	0.50	0
	Dimethyl sulfoxide, 25°	0.0306	0.64	0.047
Poly-γ-benzyl-L-glutamate[94]	CHCl₂COOH, 25°	0.00278	0.87	0.123
	Dimethyl formamide,[c] 25°	1.4×10^{-7}	1.75	—

[a] $[\eta]$ is expressed in cubic centimeters per gram. The more usual experimental units are deciliters per gram. In these units K' would be smaller than here given by a factor of 100.

[b] Ideal solvent; cf. Table 23–5.

[c] Identical data were obtained in the mixed solvent, chloroform–formamide.

used throughout this book, we are thus limited to values of a between 0.50 and 0.80.

Turning next to rod-shaped molecules, with constant diameter and with length proportional to molecular weight, we predict from equation 20–8 that they can be represented as ellipsoids, with a/b directly proportional to molecular weight. From Figs. 19–7 and 19–8 we note further that the Einstein-Simha coefficient ν is a complex function of a/b. For sufficiently large molecular weights, however, $\log \nu$ becomes very close to a linear function of $\log a/b$ such that ν, and, therefore, by equation 23–4, also the intrinsic viscosity, varies closely as $(a/b)^{1.8}$, i.e., as $M^{1.8}$. Thus rod-shaped molecules should also tend to obey equation 23–12, with the exponent $a \simeq 1.8$ rather than 0.5 to 0.8.

A selection of values of K' and of a, for various macromolecules in a variety of solvents, is given in Table 23–8. For most of the synthetic polymers, as well as for amylose, a is seen to lie between 0.5 and 0.8, as predicted for flexible linear chains. The larger values then indicate that the solvent being used is a good solvent; the smaller values indicate a poor solvent.

The fact that amylose behaves as a flexible chain is of particular interest, since the complex formed between amylose and iodine was shown in section 5 to have a helical configuration. Evidently this configuration is not maintained in the absence of iodine.

One of the viscosity–molecular weight relations of Table 23–8, that for poly-γ-benzyl-L-glutamate in dimethyl formamide, clearly agrees with the expected behavior of rodlike molecules. The actual data of Doty et al.,[94] given in Fig. 23–5, show that log $[\eta]$ is not quite linear with respect to log M, just as log v in Fig. 19–8 is not quite linear with respect to log (a/b). Combining equation 19–26 with equation 23–4, assuming $\bar{v}_2 = 0.76$ and $\delta_1 = 0$, they obtained for the ellipsoidal semi-axis b an average value of 9.1 Å. By the treatment of section 20d, since $L = 2a$, the diameter d of the equivalent cylindrical rod becomes $d = (2/3)^{1/2}(2b) = 14.9$ Å. This is almost exactly the value to be expected for the cylinder resulting from the Pauling-Corey-Branson α-helix (section 4), which leads to $d = 15.3$ Å. That poly-γ-benzyl-L-glutamate *in the solid state* forms an α-helix has been established by a variety of methods. Here we have proof that it retains essentially this helical form in certain solvents. The data for the same polymer in dichloroacetic acid, on the other hand, indicate that there must be another group of solvents in which the molecular configuration must instead be quite close to that of a flexibly coiled chain. (The agreement between the viscosity data here reported, and the light scattering data of Table 18–4 should also be noted.)

The results for cellulose acetate and for poly-γ-benzyl-L-glutamate in dichloroacetic acid represent a deviation from the expected behavior of flexible coils which is much too small to be explained on the basis of the formation of rigid rods. What it most likely indicates is the presence of flexible coils for which R_G increases with molecular weight somewhat more markedly than usual. As was indicated in section 9i (equation 9–25) this kind of deviation probably results from unusual stiffness (i.e., resistance to rotation) of the polymer. If this explanation is correct, the exponent a should slowly decrease with increasing molecular weight, as was discussed on p. 165. To detect this would require more homogeneous fractions than have so far been employed in the examination of these substances.

A somewhat different result of interest here is that obtained for the polysaccharide *dextran* by Senti and coworkers.[108] This polysaccharide is a branched polyglucose obtained from the fermentation of sucrose by

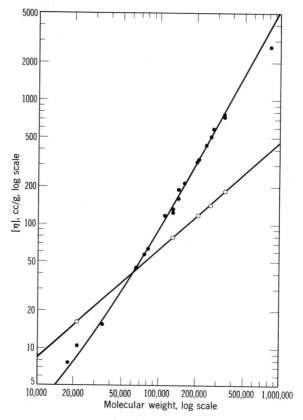

Fig. 23–5. The intrinsic viscosity of poly-γ-benzyl-L-glutamate in various solvents. Open circles represent data obtained in dichloroacetic acid, filled circles data obtained in dimethylformamide or in a mixture of chloroform and and formamide. (Doty, Bradbury, and Holtzer.[94])

certain bacteria. The particular samples studied by Senti and coworkers were separated into fractions ranging in molecular weight from 18,000 to 10,000,000. It was found that $[\eta]$ was proportional to $M^{0.5}$ for the fractions of low molecular weight. With increasing molecular weight, however, $[\eta]$ increases less markedly, until, when M approaches 10,000,000, $[\eta]$ varies roughly as $M^{0.11}$. The only reasonable explanation of this observation is that the extent of chain branching is a function of molecular weight. Since $[\eta]$ becomes smaller with increased branching, the results suggest that the extent of branching was much greater in the high-molecular-weight fractions.

23j. Viscosity as a measure of molecular weight. Application to heterogeneous samples. Whenever equation 23–12 is applicable it may be used as a convenient practical method for determining the molecular weight of a given polymer preparation. To employ the method it is necessary to have only two well-fractionated samples of the polymer, the molecular weights of which are known. These two samples can then be dissolved in a series of solvents at a series of temperatures, and K' and a for each set of conditions can be determined. Equation 23–12 then serves as a measure of molecular weight for all subsequent samples. This use of viscosity measurements is, in fact, much older than the theoretical analysis which has formed the basis of our discussion. Staudinger[95] suggested the relation $[\eta] = K'M$, i.e., equation 23–12 with $a = 1.0$, as early as 1930; and equation 23–12, with a variable, was first suggested as an empirical equation by Mark[96] and Houwink[97] about 1940.

Most unfractionated polymer samples are heterogeneous with respect to molecular weight. If the viscosity method (calibrated by use of homogeneous fractions) is employed for such samples, we obtain a rather complex average of the molecular weight. By equation 23–2 a concentration c_i grams per cubic centimeter of a polymer of molecular weight M_i makes a contribution to $(\eta' - \eta)/\eta$ given (in the limit of infinite dilution) by

$$\frac{\eta' - \eta}{\eta} = [\eta]_i c_i = K' M_i{}^a c_i \qquad (23\text{–}13)$$

where $[\eta]_i$ is the intrinsic viscosity of the species of molecular weight M_i, as given by equation 23–12. For a heterogeneous sample of a given polymer, K' and a, being by definition independent of molecular weight, will have the same value for all the species in a given solution. Thus

$$[\eta] = \mathscr{L}_{c=0} \frac{\eta' - \eta}{\eta c} = K' \frac{\sum\limits_i M_i{}^a c_i}{\sum\limits_i c_i} \qquad (23\text{–}14)$$

where $c = \sum\limits_i c_i$. If we now interpret the measured intrinsic viscosity in terms of equation 23–12 and call the average molecular weight obtained the *viscosity average*, $\bar{M}_v$, i.e., we place

$$[\eta] = K'(\bar{M}_v)^a \qquad (23\text{–}15)$$

then clearly

$$\bar{M}_v = \left(\frac{\sum\limits_i M_i{}^a c_i}{\sum\limits_i c_i}\right)^{1/a} \tag{23-16}$$

This average is equal to the weight-average molecular weight (section 8d) if $a = 1$. For the fractional values of a ordinarily observed, however, it is an average falling between $\bar{M}_n$ and $\bar{M}_w$. For a polymer whose distribution function is given by equation 8-3, for example,[98]

$$\bar{M}_n:\bar{M}_v:\bar{M}_w = 1:\{(1 + a)\,\Gamma(1 + a)\}^{1/a}:2$$

where $\Gamma(1 + a)$ is the Γ function of $1 + a$.

It is to be noted that in the case of the distribution function just cited, as well as in certain other distribution functions, the ratios $\bar{M}_v/\bar{M}_n$ and $\bar{M}_v/\bar{M}_w$ are constants independent of molecular weight. Thus equation 23-15 may be written as

$$[\eta] = K_1(\bar{M}_n)^a = K_2(\bar{M}_w)^a \tag{23-17}$$

where K_1 and K_2 are different constants. In this case $[\eta]$ can be used to measure $\bar{M}_n$ or $\bar{M}_w$ if the calibration is carried out, not with homogeneous fractions, but with heterogeneous samples of known $\bar{M}_n$ or $\bar{M}_w$ which can be expected to follow the same molecular weight distribution function as the samples to which equation 23-17 is to be applied.

24. TRANSPORT IN AN ELECTRIC FIELD[11,12,13]

If an electric field of strength $\mathscr{E}$ acts upon an isolated charged particle of charge q suspended in an insulating medium, it will produce a force $\mathscr{E}q$ in the direction of the field. By equation 19-7, the particle will then acquire a steady-state velocity u in the same direction, given by

$$u = \mathscr{E}q/f \tag{24-1}$$

where f is the frictional coefficient. For macromolecular particles q is usually expressed in terms of the number Z of unit charges. Where ϵ is a unit charge (4.80×10^{-10} esu), $q = Z\epsilon$.

If the particle under consideration is a sphere of radius R, the frictional coefficient is given by equation 19-13, and we get

$$u = \frac{\mathscr{E}Z\epsilon}{6\pi\eta R} \tag{24-2}$$

where η is the viscosity of the medium. It is customary to assign a positive sign to the direction along which $\mathscr{E}$ is acting, so that u will be positive if q or Z is positive and negative if q or Z is negative.

The experimental study of macromolecular ions in an electric field cannot ordinarily be carried out under conditions to which equations 24–1 or 24–2 are applicable. Macro-ions are necessarily studied in neutral solutions which contain ions of opposite charge. Moreover, as has been repeatedly pointed out, it is usually desirable to have an excess of some low-molecular-weight electrolyte present in such a solution, so that there are usually present small ions of both positive and negative sign, with those of sign opposite to that of the macro-ion (called counterions or gegenions) in excess. The excess of gegenions will not be randomly distributed through the solution but will be localized in the vicinity of the macro-ions. The electric field which drives the macro-ions will act on these counterions also, but in the *opposite direction*. Since the moving counterions drag solvent along with them and the solvent in turn acts on the macro-ion, the net effect is a secondary force on the macro-ion, opposite in direction to the primary force. The result is that the velocity of motion in the direction of the field may be reduced well below that predicted by equation 24–1 or 24–2.

The retardation described in the preceding paragraph is called the *electrophoretic effect*. Its calculation is one of the major problems of physical chemistry. Its solution is required for a basic understanding of conductance as well as for the problem under consideration in this section. It is closely related to other phenomena involving a connection between flow and an electric field, called *electrokinetic phenomena*. (No general discussion of the latter will be given here. A detailed account is given by Kruyt.[132])

The foregoing introductory statement indicates that an analysis of the transport of macro-ions in an electric field requires a knowledge of the interaction between macro-ions and surrounding small ions. The reader may therefore wish to postpone consideration of the remainder of this section until section 26 is taken up. However, only a few of the results of section 26 will be used, and, if the reader is willing to accept them without proof, no difficulty in following the argument will be encountered.

24a. Electrophoretic mobility. The movement of ions in an electric field is generally called *electrophoresis*, and the velocity u with which they are transported is called the electrophoretic velocity. This quantity is ordinarily assumed to be proportional to the electric field strength. The corresponding quantity which depends on only molecular parameters (analogous to the sedimentation coefficient defined by equation 22–5) is the *electrophoretic mobility* U, defined as

$$U = u/\mathscr{E} \qquad (24\text{–}3)$$

It is this quantity which is ordinarily determined from experimental data. Experimental methods for its determination are described in section 24e.

24b. Theory of the electrophoretic mobility. It was pointed out at the beginning of this chapter that any theory of flow processes should be consistent with the phenomenological equations of the thermodynamics of irreversible processes. The flow of macro-ions in a system containing other ions should thus be of the form (equation 19–4)

$$J = L_0\mathscr{E} + \sum_i L_i(\partial\mu_i/\partial x) \qquad\qquad (24\text{–}4)$$

where $\mathscr{E}$, the electric field strength, is a measure of the electrical force and $\partial\mu_i/\partial x$ represents the force due to the concentration gradient of any ionic or neutral species in the solution, the summation extending over all such species except the solvent. The phenomenological coefficients, L_0 and the various L_i, depend on molecular parameters and on the composition of the solution, but not on the magnitude of the forces. It should be recalled (equation 19–8) that the flow J and the velocity u are directly proportional to one another, so that equation 24–4 may be considered an equation for the transport velocity u.

By extending the summation of equation 24–4 over all species, including ionic species, we are including more terms than equation 19–4 requires, the terms required being one for each independent *component*. Ordinarily the number of ion species exceeds the number of corresponding neutral components by one. There would thus be one excessive term in the sum of equation 24–4. It would be removed by application of the condition that neutrality is everywhere maintained.

No existing theory of electrophoretic mobility is consistent with equation 24–4. It is always assumed, as in equation 24–3, that u or J is proportional to the electric field strength alone. Subsequent analysis will show that concentration gradients of the various ionic species inevitably arise in all experimental studies of electrophoresis, so that existing theories are *a priori* not completely rigorous. It will be seen, however, that existing theories are inexact for several other reasons, and it is probable that the error made in ignoring the forces due to chemical potential gradients is a relatively unimportant one.

It should be noted that one of the terms in the sum of equation 24–4 is simply the term describing the diffusion of macro-ions. As was true in the case of sedimentation (p. 367) this term should not have an effect on the average rate of motion which results from the applied force. The other terms represent interaction between the flow of macro-ions and the flows of other constituents of the solution. To assume that these terms can be neglected is an approximation, but it is the same approximation which has been made already in considering the sedimentation and diffusion of macro-ions in aqueous salt solutions (see p. 353).

With the foregoing assumption, the most searching analysis is that of F. Booth.[133] The procedure is similar to that already used in section 19 for the movement of particles under the influence of *any* applied force. As in section 19, we do not use the direct approach of calculating the particle velocity u (resulting from the applied force) in a stationary fluid. We solve instead the equivalent problem of calculating the force required to keep the particle stationary in a fluid with velocity $-u$. The advantage of this procedure is that the force on the particle does not enter into the problem directly, for what is done is to compute the force which the moving fluid exerts on the particle. The applied force enters into the problem (as far as it effects the central particle) only in the final step, where it is set equal to the force exerted by the fluid.

The analysis begins with equation 19–10, with the usual condition that the fluid is stationary at the boundary of the particle and that its velocity far away from the particle is equal in magnitude but opposite in direction to the electrophoretic velocity of the particle, which is the quantity we eventually seek to compute. Since the flow velocity of fluid will not be in the direction of the field in the vicinity of the particle, we must use vector notation; i.e., the symbol **u** represents flow velocity at any point, and the boundary condition far from the particle takes the form of assigning both a direction and a magnitude to **u**.

If the same assumptions could be made that were made in deriving Stokes' equation for the frictional coefficient, equation 19–10 would reduce to equation 19–11. The solution (for spheres) would be equation 19–13, and this is equivalent to equation 24–2 for the electrophoretic velocity of spheres, or to equation 24–1 for particles of any shape. All the assumptions that were made in the derivation of Stokes' equation are, however, not applicable to the present problem because the applied electric field acts not only on the central particle but also on the ions in the surrounding fluid, as explained above. Thus the term ρF of equation 19–10, which could be discarded in the derivation of Stokes' equation because it was specified that the applied force acted solely on the central particle, has to be retained. All other assumptions made in deriving Stokes' equation (i.e., incompressibility of the fluid, low flow velocity, etc.) may be assumed to hold true.

In equation 19–10 **F** represents a force per unit mass and ρ the density, so that ρF is force per unit volume. In the present context it is preferable to express **F** as the force *per unit charge*, for if we do so we can replace **F** by the gradient of electrostatic potential ψ (i.e., by $\nabla \cdot \psi$). If we let ρ be the charge density at any point in the fluid (i.e., the density of charge due to small ions near the central macro-ion), $\rho\mathbf{F} = \rho\nabla\psi$ will again have the desired units of force per unit volume. The resulting equation is

$$\eta\nabla\cdot\nabla\mathbf{u} = \nabla p + \rho\nabla\psi \qquad (24\text{-}5)$$

where $\mathbf{u}$ is the vectorial flow velocity at any point, p the pressure, and η the viscosity, which is assumed to be the same near the central macro-ion as elsewhere in the solution.

Booth obtained a solution for equation 24–5 which is applicable to spherical, non-conducting, charged particles. As expected, the force on the stationary particle is in the same direction as the fluid flow velocity.

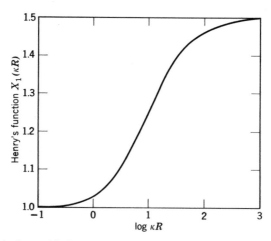

Fig. 24–1. A graphical representation of Henry's function (equations 24–6 and 24–7).

In equating this force to the required applied force we may therefore drop the vector notation, with the final result that

$$U = \frac{u}{\mathscr{E}} = \frac{Z\epsilon}{6\pi\eta R}\frac{X_1(\kappa R)}{1+\kappa R} + \sum_{j=2}^{\infty}\frac{Z^j\epsilon^{2j-1}}{6\pi\eta R^j(DkT)^{j-1}}(X_j + Y_j + W_j) \quad (24\text{-}6)$$

In this equation R is the radius of the sphere, D the dielectric constant of the fluid (which, like the viscosity, is assumed independent of position), k is Boltzmann's constant, T the temperature, and κ the well-known Debye-Hückel constant (given by equation 26–32) which depends on the ionic strength of the solution.

The function $X_1(\kappa R)$ which occurs in equation 24–6 is a complicated function (first given by D. C. Henry, see below), which, however, always lies between 1.0 and 1.5. A graphical representation of this function is shown in Fig. 24–1. The terms X_j and Y_j also represent functions of the concentrations and charges of the mobile ions of the fluid, and the W_j represent functions of these variables and also of the corresponding ion mobilities (i.e., the W_j depend on the variables which determine the conductance of the fluid). The functions X_2, Y_2, and W_2 vanish for symmetrical electrolytes (i.e., for the case where all mobile ions have the same valence), and the same is true of X_4 and Y_4. The functions X_3, Y_3, W_3, and W_4 are shown in graphical form in Booth's paper. Higher terms of the sum in equation 24–6 have not been computed.

Booth's equation is clearly a very elaborate equation, and calculation of mobilities from the equation is a laborious procedure. A more practical equation for estimation of *approximate* values for the mobilities of spherical macro-ions is obtained by using only the first term of equation 24–6; i.e., the equation is

$$U = \frac{u}{\mathscr{E}} = \frac{Z\epsilon}{6\pi\eta R} \frac{X_1(\kappa R)}{1 + \kappa R} \tag{24-7}$$

Equation 24–7 is one form of Henry's equation,[134] a more general form of which will be considered later. The equation is a relatively crude approximation because it is based on a solution of equation 19–10 in which the distribution of mobile ions in the fluid about the macro-ion is assumed to be that which exists at equilibrium in the absence of the applied field. In actual fact the field must distort the mobile ion distribution, and this distortion is described by the terms Y_j and W_j of Booth's equation. The terms Y_j describe that part of the distortion which would occur even if the macro-ion and the surrounding fluid were at rest with respect to one another; the terms W_j describe the additional distortion due the disturbance of equilibrium by the flow. (The terms X_j in Booth's equation represent the fact that Booth used a higher approximation in computing the electrostatic potential than that which is used to obtain equation 24–7.)

Booth's equation, despite its elaborate and painstaking analysis of the problem, is of only limited applicability to real macro-ions. Earlier sections of this chapter have shown that such ions cannot be represented as solid non-conducting spheres. Even if they could be represented as spheres, their charge distributions would not be expected to be spherically symmetrical (see p. 475). More important, perhaps, is the fact that the macro-ion charges are mobile. For, almost all macro-ions are weak electrolytes, and we are almost always interested in studying them under conditions where the net charge Z is less than the maximum possible charge. In other words, a typical macro-ion will contain, for example, both —COOH and —COO$^-$ groups, or both —NH$_3^+$ and —NH$_2$ groups, on or near its surface. There will be a continuous rapid exchange of protons among such sites, which is another way of saying that the macro-ion surface is a *conducting* surface. This in turn means that the applied field would disturb the charge distribution on the surface. This effect has been considered theoretically by Booth[151], and it has been shown to have no effect at the level of approximation represented by Henry's equation, provided that the interior of the macro-ion is non-conducting.

A further omission in equation 24–6 has been pointed out by Gorin[136]. It stems from the fact, to be considered in section 26, that there are really two radii characteristic of a macro-ion in solution. One of them is the actual particle radius R; the other is the radius of exclusion, a, which represents the closest distance of approach of the mobile ions of the fluid to the macro-ion. The value of a exceeds R by 2 to 3 Å, depending on our estimate of the radius of a typical mobile electrolyte ion. Booth's treatment assumes that the difference between R and a is negligible, so that his solution is strictly applicable only to particles which are somewhat larger than most common macro-ions. A correction for this effect at the level of Henry's equation (i.e., to the first term of Booth's equation) has been made by Gorin. His result is[136]

$$U = \frac{u}{\mathscr{E}} = \frac{Z\epsilon}{6\pi\eta R} \frac{1 + \kappa r_i}{1 + \kappa a} X_1(\kappa R) \tag{24–8}$$

where r_i is the radius of a typical mobile ion and $a = R + r_i$. Equation 24–8 predicts mobilities which are about 20% higher than those predicted by Henry's equation at ionic strengths of 0.1 to 0.2. At lower ionic strengths the difference between them is less.

In conclusion, then, it is clear that no existing relation between electrophoretic mobility and molecular parameters is rigorously applicable to macro-ions. Equation 24–7 may be considered adequate to arrive at semi-quantititive information about molecular charge, size, or shape, but it cannot be expected to have the same degree of validity as the equations for the sedimentation coefficient or the diffusion coefficient of macro-ions, even though the latter are themselves approximate to the extent that they neglect the forces due to interaction between the flows of the various components and of electric charge.

The discussion of this section has been confined to a single theoretical approach to the problem of electrophoresis. Numerous other theoretical studies will be found in the literature (reviewed by Overbeek[12]). Only one of them, that of Overbeek,[135] is of the same order of exactness as Booth's theory. It should also be noted that the problem of electrophoresis is closely connected to the problem of the conductance of electrolytes in general. The definitive treatment of this subject is that of Onsager and Fuoss.[137]

24c. Zeta potential. Relaxation effect. To conclude the theoretical part of this discussion, mention must be made of two concepts which do not enter directly into Booth's treatment of electrophoresis but which have been the subject of much discussion in the literature of electrophoresis. One of these is the *relaxation effect*, which is usually cited as a retardation of the mobility distinct from the electrophoretic effect. Such a distinction, however, arises only when the electrophoretic effect is computed (as in equation 24–7) on the basis of the mobile ion distribution which would apply in the absence of an electric field. The calculation must in that case be corrected for the fact that the atmosphere of mobile ions, in the presence of the field, will always lag behind the central ion in its movement through the solution. The relaxation effect is represented in Booth's equation by the functions Y_j and W_j.

The *zeta potential*, ζ, is the potential ψ at the surface of shear of a charged particle; i.e., for macro-ions it is the potential at the surface of the hydrodynamic particles (section 20) which such an ion forms. This quantity occurs in the literature of electrophoresis because an exceedingly simple relation may be obtained between electrophoretic mobility and zeta potential. This relation, due to Smoluchowski,[138] is

$$U = \frac{u}{\mathscr{E}} = \frac{D\zeta}{4\pi\eta} \qquad (24\text{–}9)$$

This relation is of only borderline interest in the physical chemistry of particles of molecular size, for two reasons: (1) Because it applies only to particles much larger than macromolecular particles. It applies to such particles regardless of shape (provided that ζ has the same value at all points of the surface), but this is clearly irrelevant here. (2) Because the charge Z, and not the zeta potential, is the molecular charge parameter by which macro-ions are characterized. This would be a trivial objection if ζ and Z were easily related, but such a relation can only be made if we know the exact distribution of mobile ions about the central macromolecular ion. But this is just what we do not know. The problem is a difficult one even in the absence of a field (see section 26). To calculate the distribution in the presence of the field, with the solvent and central ion in motion with respect to one another, is an intimate part of the problem of calculating the complete state of the system, including the electrophoretic mobility; i.e., to calculate the relation between ζ and Z is no easier than to calculate U from Z directly.

It should be noted that the most general form of Henry's equation[134] is also in terms of the zeta potential, i.e.,

$$U = \frac{u}{\mathscr{E}} = \frac{D\zeta X_1(\kappa R)}{6\pi\eta} \qquad (24\text{–}10)$$

and it applies in this form to spherical ions of any size. Although more general than equation 24–7, equation 24–10 is still inexact, and, of course, objection 2 of the preceding paragraph applies to it as well as to Smoluchowski's equation.

One other term commonly encountered in the literature of electrophoresis is the term *double layer*. This term arises from the fact that equations such as equation 24–9 were derived before the advent of the Debye-Hückel theory. It was generally supposed that charged particles carry their charge on a surface layer and that a second enveloping layer carries the counterions which neutralize the charge. If the second layer is regarded as diffuse, with the charge distribution calculated by means of the Debye-Hückel theory or related theories, then this picture is, of course, correct, and many authors continue to use the term *double layer* to describe this situation. This does not therefore imply any difference from the models used here and in section 26, where this term is not employed.

24d. Electrophoresis as a tool in protein chemistry. It is clear from the preceding discussion that an exact equation relating electrophoretic mobility to molecular parameters is not available. Within the approximation implied by ignoring all but the first term on the right side of the phenomenological equation (equation 24–4), an approximation not different from that used in analyzing sedimentation and diffusion data of macromolecular electrolytes in salt solution, two definite statements can, however, be made. (1) The mobility U is always directly proportional to the charge Z of the macro-ion. (2) The mobility is always inversely proportional to the frictional coefficient, as indicated in equations 24–6, 24–7, and 24–8, all of which apply only to spherical ions, by the term $6\pi\eta R$ in the denominator. These properties make electrophoresis a powerful semi-quantitative tool, one which has been of the greatest importance in protein chemistry. Many of the applications of this tool are analytical in nature and fall outside the scope of this book, but others provide useful information about molecular properties, and a brief account of them will be given here. The discussion is limited to soluble proteins because the bulk of all work in this field has been done on proteins. (An example of electrophoresis of a synthetic polyelectrolyte will be shown in section 27.)

24e. The moving boundary method for a single macro-ion species.[11] Macromolecular electrolytes are ordinarily studied in solution in the presence of an excess of an inert electrolyte of low molecular weight. The reason for this is the requirement of electrical neutrality everywhere in a solution. If a macromolecular electrolyte, consisting of macro-ions of charge Z and Z

univalent ions of opposite charge is dissolved in a solvent containing no other ions, then the macro-ion cannot move independently of its counter-ions, and all properties of the macro-ion which we wish to observe are obscured. The addition of an excess of other low-molecular-weight ions largely eliminates this difficulty because neutrality can be maintained by relatively tiny movements by a large number of these ions.

Proteins and virtually all other common macromolecular electrolytes are *weak* electrolytes, owing their charges to weakly acidic or basic groups such as $-COOH \leftrightharpoons -COO^- + H^+$. The charge Z of macro-ions which

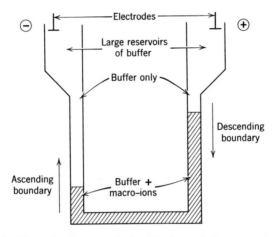

Fig. 24–2. Schematic diagram of the Tiselius cell for measuring electro-phoretic mobility by the moving boundary method. The macro-ions are assumed to have a positive charge; for negatively charged ions the electrodes would be reversed.

are formed by these electrolytes depends strongly on pH, so that a buffered solution is required if we desire to study macro-ions whose charge will not vary from place to place in the apparatus. The necessity for buffering and for an added electrolyte is usually, though not always, satisfied by addition of a single system, consisting of a low-molecular-weight acid or base and the corresponding salt, e.g., acetic acid and sodium acetate. In the sub-sequent discussion we shall always refer to the added low-molecular-weight solute as a "buffer," with the understanding that simple electrolytes such as KCl may sometimes be present in addition to or instead of the buffer system.

The usual apparatus for determining mobility is a U-shaped cell of the type shown in Fig. 24–2. Two sharp boundaries are formed between the solution being studied and a corresponding solution containing the same solvent and buffer but devoid of macro-ions. (The two solutions are usually dialyzed against one another before electrophoresis is begun, so that no gradient in the chemical potentials of buffer components exist at the beginning of the experiment.) A potential difference is then applied across the two electrodes shown in Fig. 24–2, as a result of which the ions of the solution will move, each with its appropriate velocity. This velocity will depend on the mobility of the ion and on the electric field strength (equation 24–3).

In order to obtain convenient rates of flow a rather high field strength is required. This in turn produces a high current, mostly due to flow of buffer ions, and a resulting heating of the solution. The difficulties caused by this heating (principally the formation of convection currents) were not overcome for several years. The Tiselius apparatus[139] is the only one which has been successful in this regard.

It follows from Ohm's law that the electric field strength at any point in the electrophoresis cell depends on the local conductivity. The two limbs of the electrophoresis cell are constructed with a cross-sectional area A which is everywhere constant. Consider an element of volume $A\,dx$ in either limb, across which there is a potential difference $d\psi$. By Ohm's law $d\psi$ is the product of the current I and the resistance dR of the element, or, where $i = I/A$ is the *current density*,

$$d\psi = -iA\,dR$$

the negative sign indicating that the flow of electricity is in the direction of decreasing potential. The resistance dR may be replaced by $r\,dx/A$ or by $dx/\kappa A$, where r is the specific resistance and κ the specific conductance of the element. The potential difference may be related to the electric field strength $\mathscr{E}$ (i.e., the electric force), since $\mathscr{E} = -d\psi/dx$. With these relations Ohm's law may be restated as

$$\mathscr{E} = i/\kappa \qquad\qquad (24\text{–}11)$$

Conservation of electricity requires that i be the same at all points in the two arms (since A is constant). The specific conductance κ of the macro-ion solution will, however, in general differ from that of the buffer solution, so that the field strength $\mathscr{E}$ will not be the same at all points. As a result the transport velocities u of the various ions (equation 24–3) will also be different, and it is seen at once that the flow processes occurring in the cell may become quite complex.

It will be noted that the symbol κ used here for specific conductance is the same as the symbol used for the Debye-Hückel ionic-strength parameter earlier. This dual use of the same symbol follows standard practice. No ambiguity should arise because the specific conductance will not occur explicitly in the same context as any of the theoretical equations for mobility.

It is possible to visualize a simple ideal situation in which no complexity arises. If the macro-ion solution contains a large excess of buffer, the buffer ion concentrations on the two sides of the boundary will be essentially the same (see Table 14–1 for equilibrium concentrations for a macro-ion solution dialyzed against a solution of diffusible electrolyte). Moreover, the specific conductance will be almost entirely that of the small ions present; the slow-moving macro-ions will make an exceedingly small contribution to it. Under these circumstances it becomes an excellent approximation to consider κ a constant, and, as a result, $\mathscr{E}$ will be constant and so will the characteristic velocity u of each of the constituent ions.

If $\mathscr{E}$ is indeed constant everywhere, the concentration of buffer ions will also remain constant everywhere, since flow of these ions simply consists of transfer from one reservoir through the cell into the other reservoir. The movement of the macro-ions, all of which are of the same kind in the present discussion, will formally be similar to that which occurs in sedimentation (cf. Figs. 22–1 and 22–2). Each individual macro-ion will move towards the appropriate electrode with the identical velocity, given by equation 24–3, and, as a result, the sharp boundaries themselves will move with the same velocity. As in the case of sedimentation, the boundaries will be broadened by diffusion, and, a given time after the potential difference has been applied, a concentration pattern such as that of Fig. 24–3 will be observed. Again, as was true in the case of sedimentation, the concentration gradient is more readily measured than the concentration gradient is more readily measured than the concentration itself (Appendix C).

The ideal situation envisaged above cannot be achieved in practice. Even for a large excess of buffer ions, a small difference in concentrations must exist on the two sides of the initial boundary. Moreover, when the solutions on the two sides are dialyzed so as to equate the chemical potentials of the buffer components, a small difference in conductivity must result. In other words, we can arrange to have *any one* of the variables (chemical potential of a neutral component, concentration of an ion species, or conductivity) the same on both sides of the initial boundary, but we cannot make more than one of them the same simultaneously.

The resulting complications have been rigorously analyzed by Dole,[140] for systems in which the macro-ion mobility (but not, of course, its velocity) is constant. In protein solutions there is the additional complication that changes in buffer ion concentrations across a boundary lead to changes in pH, which in turn lead to changes in charge, with an accompanying change in mobility. We shall not describe the theoretical approaches to this problem here (they have been extensively reviewed, e.g.,

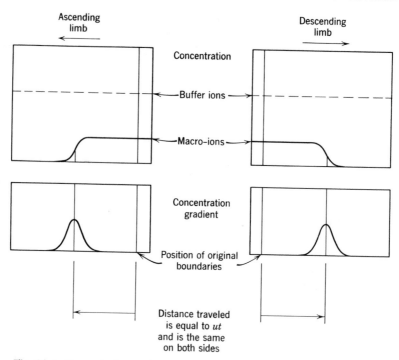

Fig. 24–3. Concentration and concentration gradient (and, hence, refractive index gradient) as a function of position in the cell. The patterns are for hypothetical ideal electrophoresis of a single macro-ion species in a solution containing an excess of buffer. The areas of both peaks are equal and proportional to the concentration of macro-ions in the original solution (Appendix C).

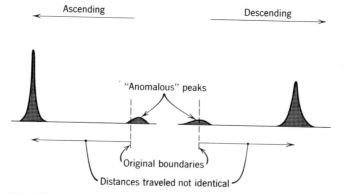

Fig. 24–4. Electrophoresis of crystalline bovine serum albumin in 0.10 ionic strength diethylbarbiturate buffer at pH 8.6. The molecular charge at this pH is about -25. (After Alberty.[11]).

by Longsworth[141]), but merely point out that the theory accounts in a general way for the observed deviations from ideal behavior which actual systems exhibit. These deviations, illustrated by Fig. 24–4, are:

(1) The appearance of essentially stationary boundaries at the positions of the initial boundaries.

(2) The moving boundaries in the two limbs do not move with exactly the same velocities. (Only the descending boundary can be used to determine mobility, the field strength being computed from equation 24–11 by measurement of the current density and of the conductance of the macro-ion solution.)

(3) The ascending boundary is sharper, the descending boundary less sharp, than we would expect on the basis of diffusion alone.

(4) The areas under the two peaks are not equal and do not represent macro-ion concentrations because buffer ion concentrations are different on the two sides of each boundary (Appendix C shows that the area is a measure of the sum of the concentration changes across the boundary).

24f. The isoelectric point and the charge of protein ions. Protein molecules owe their ionic character to the presence of weakly acidic or basic side chains of the type

$$—COOH \rightleftharpoons —COO^- + H^+$$

$$—NH_3^+ \rightleftharpoons —NH_2 + H^+$$

$$—C_6H_5OH \rightleftharpoons —C_6H_5O^- + H^+$$

as will be discussed more fully in section 30. At sufficiently low pH all side chains are in their acidic form, and the net charge Z per molecule will be positive; at sufficiently high pH all the groups will be in their basic form, and Z will be negative. The *isoelectric point* is defined as that pH at which $Z = 0$. It is one property of protein molecules which can be measured unequivocally by means of electrophoresis, for at this point U becomes zero, so that the uncertainty concerning the proportionality constant between U and Z is not important. To determine the isoelectric point we simply measure the mobility at a series of pH values and determine by interpolation the pH at which U becomes zero. An example is provided by Fig. 24–8.

At a given pH in a protein solution there is an equilibrium between molecules having different value of Z (over a narrow range), so that the value of Z referred to in this section should strictly be regarded as an average value.

As a first approximation, the isoelectric point may be considered to be the pH at which the number of positive charges due to cationic groups (e.g., $—NH_3^+$) is equal to the number of negative charges due to anionic

groups (e.g., —COO⁻), and it is thus determined by the relative number of
the various kinds of acidic and basic groups and by the pH range in
which the dissociation of H^+ ions from these groups occurs. For instance
a protein molecule which possesses many more —COOH groups than
basic groups such as —NH_3^+ will have a low isoelectric point because
—COOH groups dissociate in the range of pH 1 to pH 5, and, if they are
in excess, only a fraction need become dissociated for Z to become zero.
On the other hand, a protein molecule with an excess of basic groups will
have a high isoelectric point. In these proteins dissociation of every
—COOH group will not produce a sufficient number of negative charges
to make $Z = 0$. It is necessary to dissociate —NH_3^+ and/or —C_6H_5OH
and/or —SH groups to bring Z to zero, and these dissociations occur
characteristically in the range of pH 9 to pH 12. Experimentally we
observe isoelectric points which range from below pH 2 to above pH 11,
so that different proteins clearly possess very different distributions of side
chains.

The explanation of isoelectric points in terms of the dissociation of
acidic and basic groups is, however, only a first approximation. Isoelectric points of individual proteins are found to depend to some extent
on the nature and concentration of buffers used. This indicates that
complex formation with buffer ions (or with ions such as Na^+ or Cl^-)
occurs, and these ions of course contribute to the net charge Z. That such
ion binding in fact occurs can be confirmed by direct measurement, as
discussed in section 31.

Once the pH at which the net charge of a protein molecule is zero has
been determined, we would like to be able to measure the actual charge
at other pH's, although, as already stated, this can be done only approximately, because of the uncertainty in the relation between mobility and
charge. We should, in principle, be able to ascertain how large this
uncertainty is, by comparing values of Z determined by electrophoresis
with values of Z obtained from measurements of ion binding, but, as the
following examples will show, the necessary ion binding data are at present
not generally available.

Figure 24–5 shows values of Z for the protein ovalbumin, as a function
of pH, determined by Longsworth[142] from electrophoretic mobilities,
which were interpreted according to equation 24–7. The values of Z were
compared with values calculated from the titration curve of ovalbumin
(cf. section 30), and the agreement is seen to be poor. Part of the discrepancy can be ascribed to the fact that the titration curve does not give
a value for the *total* charge but only for that part due to the presence of
charged acidic and basic groups, i.e., to the binding and dissociation of H^+
ions. It is established that ovalbumin binds other ions also (in the present

case it would be Cl^- and/or K^+ ions, since the supporting electrolyte which was used was KCl). Data on binding of such ions are sparse, but what data are available indicate that ion binding cannot be the sole reason for the discrepancy.

As was pointed out on p. 418, Gorin's equation (equation 24–8) is to be preferred to Henry's equation for small macro-ions such as ovalbumin. When this equation is used, the discrepancy of Fig. 24–5 becomes worse

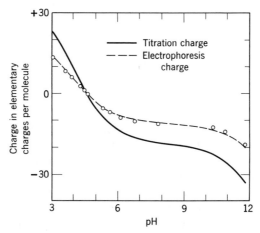

Fig. 24–5. The charge of ovalbumin as a function of pH as determined from electrophoretic mobilities using equation 24–7. The charge is compared with the charge due to acid-base equilibria as determined from the titration curve. Part of the discrepancy can be explained by the fact that charges due to acidic and basic groups on the molecule do not constitute the total charge on the protein ion, but part of it must also be ascribed to inadequacy of equation 24–7. The ionic strength in these experiments was 0.10. (Data of Longsworth[142]. The figure is taken from Overbeek.[12]).

rather than better. Booth's equation has not been applied because of the complex calculation required, but Overbeek[12] has applied his more or less equivalent theory. It changes the Z values calculated by Henry's equation by less than 5%. We are left only with intangible factors such as shape, charge distribution, and surface conduction. One other possibility is that the ovalbumin molecule changes size or shape as the pH is changed, so that the frictional coefficient varies continuously. However, measurements of diffusion and sedimentation coefficients (or of viscosity which gives equivalent information) indicate that this does not occur in the pH range covered by Fig. 24–5.

Similar data for the protein lysozyme are contained in a paper by Beychok and Warner.[143] There is an even larger discrepancy between Z

values determined from electrophoresis and from the titration curve. In this instance actual binding data for chloride ion were available. The electrophoretic experiments were, however, carried out in buffer solutions, containing $H_2PO_4^-$, $HCOO^-$, or CH_3COO^- ions. Assuming that these ions are bound to the same extent as chloride ion, reasonable agreement (i.e., to within about 10%) between the two sets of Z values could be obtained. However, it is not at all reasonable to expect that anions as different as those here cited should all be bound to the same degree.

Both the instances just discussed illustrate one of the major difficulties in deciding just how good an approximation equation 24–7 or even equation 24–6 might represent. This difficulty is the absence of accurate data on the binding of buffer ions, so that a reliable evaluation of the true charge Z is not available.

24g. The moving boundary method for a mixture of macro-ions.

Mixtures of proteins are generally studied in buffers of such pH that all the protein ions possess a charge of the same sign. A mixture of proteins with iso-electric points at pH 4.5, 6.0, 7.5, and 8.0, for example, is studied at pH < 4.5, where all the ions would have a positive charge, or at pH > 8.0, where they would all have a negative charge. The hypothetical ideal behavior of such a mixture ($\mathscr{E}$ constant) would again be similar to the ideal behavior of a mixture in sedimentation (cf. Fig. 22–5). The refractive index gradient diagram would be like Fig. 24–3 except that there would now be a separate peak for each component. Each peak would move with a constant velocity characteristic of the component in question; the area under each peak would be a measure of the concentration of that component. Identical diagrams would be obtained from the ascending and descending limbs. Actual diagrams (e.g., Fig. 24–6) differ from the ideal in the same way as Fig. 24–4 differs from Fig. 24–3. An anomalous peak occurs on each side, approximately at the location of the inital boundary. The peaks due to macro-ion components are sharper in the ascending than in the descending limb; they do not move with velocities which can be simply related to their mobilities; the areas under them do not exactly correspond to their concentrations. However, the ideal picture is *qualitatively* correct in that the number of peaks, apart from the two stationary ones, is equal to the number of components (cf. Dole[140]) and the areas are an approximate measure of the amount of each which is present. Electrophoresis thus becomes a powerful tool for the qualitative or semi-quantitative analysis of mixtures of proteins.

The most celebrated example of the electrophoretic analysis of a protein mixture is that illustrated by Fig. 24–6. It shows the refractive index gradient pattern obtained for human blood plasma (i.e., the fluid obtained after red and white cells are removed) and indicates the presence of at least six distinct proteins. Actually there are many more than this, since some of the peaks represent mixtures of proteins with similar mobilities. When this pattern was obtained for the first time[144] only three components of blood plasma were known (these being albumin, globulin, and fibrinogen), and the discovery that the globulin component could be separated into

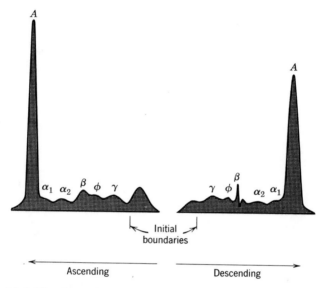

Fig. 24–6. The electrophoretic refractive index gradient pattern for normal human blood plasma, measured at pH 8.6, where all the proteins are negatively charged. The peak labeled A represents serum albumin; α_1, α_2, β, and γ are mixtures of serum globulins, and ϕ represents fibrinogen. The two unlabeled peaks are the stationary peaks near the positions of the initial boundary. (Alberty.[11])

three fractions (α, β, and γ; the α_1 peak was not resolved in the original work) provided a far-reaching stimulus for the isolation and purification of individual proteins, not only from blood but also from other fluids and tissues of living systems.

Sedimentation in the ultracentrifuge provides a second method by which protein mixtures may be separated into a number of component groups. However, the basis for separation is quite different, being the ratio of molecular weight to frictional coefficient. Combination of the two methods leads to identification of at least twelve separate proteins in human serum.[145]

24h. Heterogeneity of "pure" proteins. An important result obtained from electrophoretic analysis of protein solutions has been the discovery that many proteins which are "pure" by other criteria in fact consist of more than one kind of molecule. A striking example of this is provided by ovalbumin (Fig. 24–7). This protein is crystalline (implying that all molecules are close to identical) and appears homogeneous in the ultracentrifuge (implying that all molecules have the same molecular weight

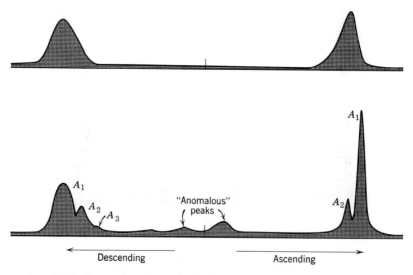

Fig. 24–7. Electrophoretic analysis of crystalline ovalbumin. The upper pattern (ionic strength 0.3) indicates that only a single component is present. At an ionic strength of 0.05 (lower pattern), however, where the lower specific conductance creates a much higher field strength for a given applied potential, resolution into three components is observed. (Cann.[146])

and frictional coefficient). It also appears homogeneous in electrophoresis experiments under a variety of conditions, such as those used by Longsworth to obtain the data of Fig. 24–5 (see also Fig. 24–7, upper pattern). If, on the other hand, high field strengths and runs of long period are used, then conditions are favorable for the separation of components with closely similar mobilities. Under these conditions, as Fig. 24–7 shows, the "pure" protein is found to consist of three separate components. It is believed[147] that the three components are identical except in the amount of phosphate which they contain, there being two, one, and no phosphate ions, respectively, in the three components. The phosphate is linked to the protein through a hydroxyl side chain, and each phosphate group may thus change Z by up to two units, depending on pH.

A closely related discovery is that the same protein of a given animal species may exist in two or more slightly different forms. An individual possesses only one of these forms, the choice being genetically determined. The most striking example of this phenomenon has been the discovery that persons who suffer from a disease called sickle cell anemia possess a form of hemoglobin which differs from normal hemoglobin. The dif-

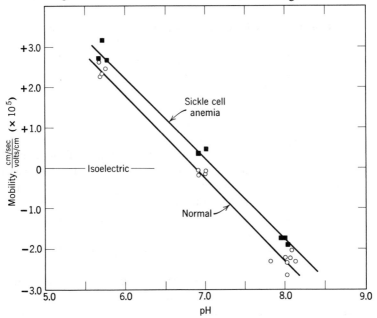

Fig. 24–8. Electrophoretic mobility of normal and sickle cell hemoglobin as a function of pH. The isoelectric point of the sickle cell protein is seen to be about 0.2 pH units higher than that of the normal protein. There is a more or less constant difference in mobility at all pH values shown, the magnitude of which suggests that Z for normal hemoglobin is more negative by about two charges throughout this pH range. (Pauling et al.[148])

ference was first observed in determination of electrophoretic mobility, as shown by Fig. 24–8.[148] Sickle cell hemoglobin is seen to have a slightly higher isoelectric point than normal hemoglobin, and its mobility is seen to be uniformly more positive at other pH's. The results indicate that U/Z is the same for both proteins but that Z at a given pH is more positive (by two charges) in sickle cell hemoglobin. This conclusion has been confirmed by partial determination of the amino acid residue sequence.[149] Both hemoglobins appear to have the identical number and sequence of monomer units except that an uncharged side chain ($-CH(CH_3)_2$) appears in one position in the normal form where sickle cell hemoglobin has a negatively charged side chain ($-CH-CH_2COO^-$). The hemoglobin molecule is a dimer with two symmetrical halves, so that a charge difference of two per molecule results.

These examples demonstrate the utility of electrophoresis as a tool in protein chemistry. Only one other technique, that of chromatography, would have been capable of leading to the discoveries here cited.

24i. Other methods and applications. The electrophoretic motion of bacteria, red blood cells, glass particles, oil droplets, and other large particles may be observed directly under a microscope. It was at one time believed that proteins adsorbed on such particles would exhibit the same electrophoretic mobility as in the free state. (If the Smoluchowski equation were valid, both for the large particles and the free protein ions and if "adsorption" were a purely physical phenomenon which left charge distributions unaltered, then this result would actually be theoretically expected.) If this were true, microscopic measurement of the mobility of proteins adsorbed on such large particles would be equivalent to a measurement of the mobility of the free protein ion, at any pH, ionic strength, etc. It has been conclusively demonstrated,[150] however, that the two mobilities are in fact not equivalent, so that the microscopic method can give no information about the behavior of free protein ions.

Another method is to use a porous medium, such as filter paper, which may be saturated with a solution containing macro-ions. The movement of these ions under an applied electric field can then be followed by a suitable staining technique. This method is obviously a most difficult one from which to determine absolute mobilities. On the other hand it is much easier experimentally than the moving boundary method, and it has become the preferred method for the majority of analytical applications which do not require a knowledge of absolute mobilities. (See Bier.[13])

It should be mentioned, finally, that electrophoresis is readily adapted to the physical separation of components of protein mixtures. Techniques used for this purpose are described in Bier's monograph.[13]

25. ROTARY MOTION[15]

In previous sections we have dealt with several examples of the *translational* motion of macromolecular particles under the influence of various forces. In this section we shall take up their *rotational* motion. Since the general principles which are involved are similar to those underlying translational motion, only a brief treatment of the subject will be presented.

25a. Rotary motion and rotary diffusion in one dimension. The rotary motion of a particle may be described as one-dimensional if the motion of every point of the particle is a simple rotation about a single axis, drawn in space, which passes through the center of gravity of the particle. Motion of this type can be described in terms of a single angle ϕ between a suitable reference axis *of the particle* and a suitable reference axis *in space*, both axes passing through the center of gravity at right angles to the rotation axis. The change in ϕ with time is expressed in terms of the angular velocity, $\omega = d\phi/dt$.

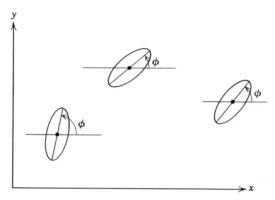

Fig. 25–1. One-dimensional rotation. Each ellipse represents a reference plane of a particular particle, and it is supposed that all such reference planes must lie parallel to the xy plane. Rotation can occur about the z axis, which lies perpendicular to the xy plane and passes through the center of gravity of each particle. The angle ϕ represents the coordinate describing the rotational position.

A very dilute solution of macromolecules may be taken as equivalent to a suspension of independent particles (section 20). The rotation of all the particles may be considered together, if we imagine all the centers of gravity placed at a single spot (or, what amounts to the same thing, we can suppose a replica of our coordinate system centered at the center of gravity of each particle). The rotation of these particles will be one dimensional if the particles are planar and lie parallel to a single plane or if some reference plane of a three-dimensional particle is so situated. As Fig. 25–1 shows there is then a single parameter ϕ by which the rotational coordinate of each particle may be indicated.

A variable $\rho(\phi)$ is now introduced, such that the number of particles, per cubic centimeter of solution, with orientation between ϕ and $\phi + d\phi$, is $\rho(\phi)\,d\phi$. If the particles are subject to no forces other than random collisions with solvent molecules, the equilibrium state of the system will be one in which all values of ϕ are equally probable; $\rho(\phi)$ will then be a constant ($1/2\pi$ times the concentration) independent of ϕ. Under the influence of a torque, however, some orientations become more probable than others, and $\rho(\phi)$ becomes dependent on ϕ. Finally, if the torque producing an oriented system is removed, the system will no longer be at equilibrium, and a gradual redistribution of orientations will occur until $\rho(\phi)$ again becomes constant. This process is known as *rotary diffusion*. In short, $\rho(\phi)$ is a variable of rotary motion which is analogous to the volume concentration, $C(x)$, in one-dimensional translational motion.

The theoretical treatment of one-dimensional rotary diffusion is similar to the theory of one-dimensional translational diffusion. If $J(\phi)\, dt$ is the net number of particles per cubic centimeter which, in time dt, traverse the orientation angle ϕ in the direction of positive ϕ, then a phenomenological law analogous to Fick's first law (equation 21–1), may be written as

$$J(\phi) = -\Theta[\partial \rho(\phi)/\partial\phi]_t \qquad (25\text{–}1)$$

where Θ is called the *rotary diffusion coefficient*. By an appropriate analogue of the equation of continuity (equation 19–2) a second law analogous to equation 21–2 or 21–3 is obtained,

$$\frac{\partial \rho(\phi)}{\partial t} = \Theta \frac{\partial^2 \rho(\phi)}{\partial\phi^2} \qquad (25\text{–}2)$$

It should be noted that no approximation is involved in the present situation in replacing $\partial[\Theta\, \partial\rho/\partial\phi]/\partial\phi$ by $\Theta\partial^2\rho/\partial\phi^2$. In translational diffusion particles passing from one concentration to another undergo a change in physical environment, with a resulting change in diffusion coefficient. In rotary motion this does not occur; the environment at any orientation is exactly the same. Similarly, equations 25–1 and 25–2 unlike the corresponding equations for translational diffusion, are in no way restricted to two-component systems. The composition of the solution will have an effect on the value of Θ, but no factor analogous to the chemical potential gradients of other components is encountered during the rotation process.

Continuing the analogy with translational motion, we may define a rotary frictional coefficient ζ. A torque $\mathscr{T}$ acting in the plane of Fig. 25–1, in the direction of positive ϕ, on any particle, will lead to an angular acceleration of the particle. This acceleration is opposed by friction between the particle and the viscous solvent in which it moves, and, at low angular velocities, the opposing frictional torque will be proportional to the angular velocity ω. The net torque is then $\mathscr{T} - \zeta\omega$, and the angular acceleration will decrease as ω increases. A steady state is reached when $\mathscr{T} - \zeta\omega$ becomes zero, and thereafter $d\omega/dt = 0$ and

$$\omega = \mathscr{T}/\zeta \qquad (25\text{-}3)$$

equation 25-3 being the analogue of equation 19-7.

The frictional coefficient applies to the rotation which results from thermal motion as well as to any other, and a relation may thus be derived between the rotary frictional and diffusion coefficients. Since the *chemical* potential is independent of ϕ, we may use the method of p. 349 (applicable only to *ideal* solutions in the case of translational diffusion). A torque $\mathscr{T}$ is imagined as acting on each particle. Rotation through an angle $d\phi$ will then require an amount of work $- \mathscr{T}d\phi$, the work being positive if the rotation opposes $\mathscr{T}$. Thus the potential energy of a particle with orientation ϕ becomes $-\int_0^\phi \mathscr{T}d\phi$, location at $\phi = 0$ being designated as the state of zero energy. Application of the Boltzmann distribution law then gives the equilibrium distribution

$$\rho(\phi) = (\text{constant})\ e^{\int_0^\phi \mathscr{T}d\phi/kT} \qquad (25\text{-}4)$$

or, upon differentiation,

$$\frac{d\,\rho(\phi)}{d\phi} = \frac{\mathscr{T}\rho(\phi)}{kT} \qquad (25\text{-}5)$$

The equilibrium distribution is also obtained by equating the flow due to the applied torque alone to the flow (in the opposite direction) due to rotary diffusion as given by equation 25-1. The condition that these cancel one another is

$$\frac{d\,\rho(\phi)}{d\phi} = \frac{\mathscr{T}\rho(\phi)}{\zeta\Theta} \qquad (25\text{-}6)$$

and it follows at once that

$$\Theta = kT/\zeta \qquad (25\text{-}7)$$

a relation similar to equation 21-9.

Rotary frictional coefficients depend on the shape and size of the particle and on the location of the rotation axis with respect to the axes of the particle. They may be computed theoretically by methods similar to those discussed on p. 324 for the evaluation of translational frictional coefficients. Stokes,[118] for instance, has evaluated ζ for rotation of a sphere of radius R about any axis through its center, obtaining

$$\zeta = 8\pi\eta R^3 \qquad (25\text{-}8)$$

where η is the viscosity of the solvent. Edwardes[152] has evaluated ζ for ellipsoids, obtaining separate values for rotation about each of the three axes. For ellipsoids of revolution (Fig. 7–2) two of the axes are identical so that only two separate frictional coefficients are required. Of particular importance is the rotary frictional coefficient for a long prolate ellipsoid with semi-axes of length a and b ($a > 5b$), for rotation about one of the b axes, which is given by Perrin[153] as

$$\zeta = \frac{16\pi\eta a^3}{3[-1 + 2 \ln (2a/b)]} \tag{25-9}$$

This relation is a useful one because it relates the effective length of an elongated particle (i.e., $2a$ in the ellipsoidal model) to an observable parameter, ζ, with very little dependence on any other dimensions of the particle; i.e., the value of a which corresponds to any observed value of ζ depends only in a minor way on the choice of a value for a/b.

25b. Rotary motion in three dimensions. Relaxation time. The rotation of a molecule in space must in general involve not one but three independent modes of rotation of the kind discussed in the preceding paragraph. Moreover, the three modes cannot be described in terms of the usual rectangular coordinate system but require a special set of coordinates, the ones usually chosen being Euler's angles. The necessary analysis is considered in most introductory textbooks of mechanics or theoretical physics and will not be described here. It suffices to say that to specify the orientation of a molecular particle in general requires three parameters and that rotary diffusion must ordinarily be described in terms of three diffusion coefficients, each related to a frictional coefficient by equation 25–7.

Many three-dimensional problems are more conveniently considered in terms of relaxation times, which were defined on p. 114 in connection with the ability of dipoles to adjust their orientations to a rapidly changing electric field. The definition on p. 114 was in terms of the average value of the cosine of the angle between the dipole and the direction of the electric field. In the present case the definition is in terms of the average cosines of the orientation angles of the three rotation axes. That is, an initial ordered state may be imagined in which the orientations of the three rotation axes of a group of molecules are not randomly distributed with respect to suitable axes in space. The average values of the cosines between the particle axes and the space axes may be assigned values $\overline{(\cos \phi_1)_0}$, $\overline{(\cos \phi_2)_0}$, $\overline{(\cos \phi_3)_0}$, the subscripts 1, 2, and 3 referring to the three axes, the subscript 0 to the fact that these averages describe the initial state. If the forces which produced the ordered state are relaxed, the usual random motions will destroy the order and lead to random orientation, with each $\overline{\cos \phi} = 0$. In analogy with equation 6–18, writing $(\overline{\cos \phi})_\infty = 0$, we have, at any time t,

$$\frac{\overline{\cos \phi}}{(\overline{\cos \phi})_0} = e^{-t/\tau} \tag{25–10}$$

where τ is the relaxation time. In general there will be three relaxation times, one for the orientation of each axis, and each of them is related to the ease of rotation about the *other two* axes. Thus

$$\tau_1 = \frac{1}{\Theta_2 + \Theta_3} = \frac{\zeta_2 \zeta_3}{kT(\zeta_2 + \zeta_3)} \tag{25–11}$$

with symmetrical relations for τ_2 and τ_3. For a sphere, of course, all three τ's are equal, and we have

$$\tau = 1/2\Theta = \zeta/2kT \tag{25–12}$$

For an ellipsoid of revolution two of the τ's are equal. A single relaxation time will suffice for *long prolate* ellipsoids if we are interested in the orientation of the long axis only.

25c. Flow birefringence.[17,18] The most widely used experimental method for observation of rotary diffusion is the method of flow birefringence. It is a method which utilizes the shearing force produced in a flowing liquid in order to orient asymmetric particles. This orientation is opposed by the process of rotary diffusion, and the dependence of the extent of orientation on the shearing force is a measure of Θ. To determine the extent of orientation, we make use of the fact that a solution containing oriented asymmetric particles exhibits birefringence, as was discussed in section 6.

The apparatus consists of two concentric cylinders, one of which rotates while the other is fixed. The annular space between the cylinders contains the solution under study. The motion of the cylinders sets the solution in motion; the layer adjacent to the moving cylinder will move with the same velocity as that cylinder, the layer adjacent to the stationary cylinder will be stationary (cf. p. 322). A velocity gradient is thus set up in the fluid. Since the annular space is very narrow, the gradient will be a constant independent of position (i.e., the fluid motion is equivalent to that produced by an infinite plate moving with respect to a similar parallel plate). Any region which is small compared to the dimensions of the cylinder may be described in terms of a rectangular coordinate system, in which all flow lines are parallel. We shall use such a system, as in Fig. 25–2, with x designating the direction of the flow lines, y the direction

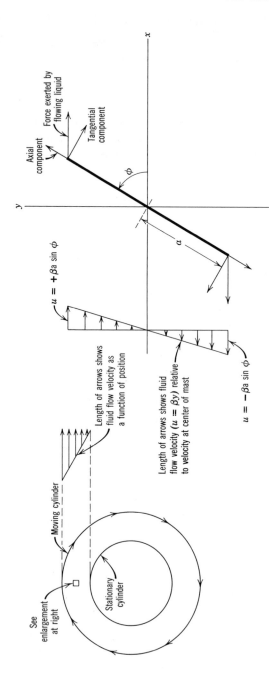

Fig. 25–2. Cross-section through flow birefringence apparatus and enlarged view of a rod-shaped particle lying in the plane of the cross-section. This diagram shows the outer cylinder as the rotating cylinder, with the inner cylinder stationary. The reverse situation can be used equally well. (See Cerf and Scheraga.[18]) The thickness of the annular space containing the fluid is exaggerated in the figure.

perpendicular to the cylinder surfaces, and z the vertical direction. The flow velocity will depend on y but will be independent of x and z.

The flow of fluid will cause all suspended particles to be carried along with it. The center of gravity will move with the velocity u_0 appropriate to the fluid at the position where it is located. The motion of interest here, however, is the rotary motion of the particle relative to its center of gravity. It is easily described for thin rod-shaped particles located in the xy plane, as shown in Fig. 25–2. The two ends of each particle will clearly have y coordinates of $+a \sin \phi$ and $-a \sin \phi$ relative to the center of mass, where $2a$ is the length of each rod, and ϕ the angle of orientation relative to the x direction. If $\beta = du/dy$ is used to designate the velocity gradient, the two ends of the rod will move with velocities $\pm \beta a \sin \phi$ relative to u_0. The result is clearly a rotation about an axis parallel to the z direction.

To calculate the angular velocity $\omega = d\phi/dt$ we need only know the tangential velocities of the ends of the rod. They will be the tangential components of the fluid flow velocities; i.e., they will be equal to $\pm \beta a \sin^2 \phi$. The tangential velocity is equal to $a\omega$, so that

$$\omega = d\phi/dt = \beta \sin^2 \phi \qquad (25\text{–}13)$$

When a solution containing many particles is considered, the rotation of the individual particles will lead to alteration in the distribution function $\rho(\phi)$. The number of particles per cubic centimeter which, in time dt, traverse the orientation angle ϕ in the direction of positive ϕ is just the number which have initial orientations between $\phi = \phi - \omega \, dt$ and $\phi = \phi$; i.e., it is equal to $\rho(\phi)\omega \, dt$. The number (per cubic centimeter) which traverse the orientation angle $\phi + d\phi$ in the same direction is

$$[\rho + (\partial \rho/\partial \phi)_t \, d\phi][\omega + (d\omega/d\phi) \, d\phi] \, dt$$

The difference between these quantities is the net change (in time dt) in the concentration $\rho \, d\phi$ of particles with orientation between ϕ and $\phi + d\phi$. Equation 25–13 gives $d\omega/d\phi = 2\beta \sin \phi \cos \phi = \beta \sin 2\phi$, so that

$$\left(\frac{\partial \rho}{\partial t}\right)_\phi = -\beta \sin^2 \phi \left(\frac{\partial \rho}{\partial \phi}\right)_t - \beta \rho \sin 2\phi \qquad (25\text{–}14)$$

At the same time, ρ is changing because of rotary diffusion, with $\partial \rho/\partial t$ given by equation 25–2. The combination of these effects will lead to a steady state in which all particles rotate in such a way that no further change in $\rho(\phi)$ takes place. Adding equations 25–2 and 25–14, setting the sum equal to zero, and replacing partial derivatives by total derivatives because ρ is now independent of time, we get

$$\frac{d^2\rho}{d\phi^2} + \frac{\beta}{\Theta}\left(\sin^2 \phi \frac{d\rho}{d\phi} + \rho \sin 2\phi\right) = 0 \qquad (25\text{–}15)$$

It is seen that ρ depends on the ratio of the shearing force produced by the fluid flow, and here represented by the velocity gradient β, to the rotary diffusion coefficient.

Equation 25–15 was first obtained by Boeder.[155] A solution in closed form cannot be obtained, but Boeder has given a series solution applicable to small values of the ratio β/Θ, i.e., to conditions under which the velocity gradient is just beginning to be large enough to perturb a completely random distribution of orientations. His result is

$$\rho = (\text{constant})\left[1 + \frac{\beta}{\Theta}\frac{\sin 2\phi}{4} + \left(\frac{\beta}{\Theta}\right)^2\left(\frac{\cos 2\phi}{16} - \frac{\cos 4\phi}{64}\right) + \cdots\right] \quad (25\text{–}16)$$

the constant being determined by the condition that $\int \rho\, d\phi$ over all possible values of ϕ must be equal to the total concentration.

Boeder's solution is, of course, not applicable to real macromolecular solutions because it is limited to rods of infinitesimal cross-sectional area lying entirely in the xy plane. Both these restrictions are removed in the three-dimensional analysis of Peterlin and Stuart,[17,156] in which the particles are treated as rigid ellipsoids of revolution (semi-axes a and b) and in which the orientation (angle θ) with respect to the z axis is allowed to vary. Their limiting expression for small values of β/Θ is

$$\rho(\phi, \theta) = (\text{constant})\left[1 + \frac{\beta}{\Theta}\frac{A\sin 2\phi\sin^2\theta}{4} + \left(\frac{\beta}{\Theta}\right)^2\right.$$
$$\left. \times \left(\frac{A\cos 2\phi\sin^2\theta}{24} + \frac{A^2\cos 4\phi\sin^4\theta}{64} - \frac{A^2\sin^4\theta}{64} + \frac{A^2}{120}\right) + \cdots\right] \quad (25\text{–}17)$$

where $A = (a^2 - b^2)/(a^2 + b^2)$. Series solutions extending to higher values of β/Θ are given by Scheraga et al.[157]

The next phase of the problem concerns the dependence of the optical behavior of the solution on the distribution function ρ. It is clear from the considerations of section 6 that a solution in which ρ is not independent of ϕ and θ will be birefringent. Equations 6–20 and 6–21 cannot be used to evaluate the birefringence because they apply to solutions in which all particles have identical orientations, but modifications of these relations exist which could be used. We are, however, not interested in the birefringence as such but in the orientation, and this factor can be approached simply by considering the location of the optic axes. If the solution is examined by means of light which enters in the vertical direction, all the electric vibrations are in the xy plane, so that only a single optic axis needs to be considered.

We first suppose that the particles are thin rods, all lying in the xy plane and all oriented at the same angle ϕ_1 with respect to the flow lines (Fig.

25–3). This direction is then the optic axis. Polarized light is now passed through the solution in a vertical direction, the plane of polarization being everywhere the same, as shown in Fig. 25–3. In general, the plane of

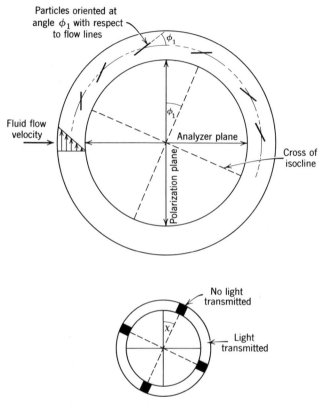

Fig. 25-3. Experimental manifestation of flow birefringence when all solute particles have the same orientation angle ϕ_1 with respect to the flow lines. The upper diagram shows the *cross of isocline*, which points to the four locations where the optic axes of solute particles are exactly parallel to the analyzer or polarizer plane. The lower diagram shows the observable result in terms of transmission of light through the annular space between the two cylinders of the apparatus. It is to be noted that the extinction angle χ, which is the angle between the cross of isocline and the polarization planes, is here equal to the angle ϕ_1.

polarization will be neither parallel nor perpendicular to the optic axis, so that the light will split into two components, whose polarization planes are, respectively, parallel and perpendicular to the axis. These components travel through the solution with different velocities, so that the emerging rays will no longer recombine to produce the original polarized beam.

The emerging light will then not be stopped by a second (analyzer) Nicol prism set at right angles to the polarizing prism; i.e., the two perpendicular polarizing prisms, which would ordinarily pass no light at all, will now do so.

There are, however, four positions where the situation described by the previous paragraph does not hold, these being the positions where the optic axis is either parallel or perpendicular to the plane of polarization. Light passing through the solution at these positions will travel as a single polarized beam, and it will be completely stopped by the analyzer prism. Thus four dark areas will be seen in the emerging light. They will be at the corners of a cross called the *cross of isocline*. This makes an angle χ with the cross formed by the planes of polarization of the two Nicol prisms, this angle being called the *extinction angle*. As Fig. 25–3, shows, the angle χ is identical with the orientation angle ϕ_1, so that the latter may be directly determined.

When the solution contains particles distributed over a range of orientations, the cross of isocline is still observed, with χ related to an appropriate average of the orientation angle. This is given by Peterlin and Stuart[17,156] as

$$\tan 2\chi = - \frac{\langle \sin 2\phi \sin^2 \theta \rangle_{av}}{\langle \cos 2\phi \sin^2 \theta \rangle_{av}} \tag{25–18}$$

With the distribution function given by equation 25–17 for small values of β/Θ, equation 25–18 may be converted to the form

$$\chi = 45° - \frac{1}{12} \frac{\beta}{\Theta} + \left(\frac{1}{1296} + \frac{A^2}{1890} \right) \left(\frac{\beta}{\Theta} \right)^3 + \cdots \tag{25–19}$$

Numerical values for larger values of β/Θ have been computed by Scheraga et al.[157]

Equation 25–19 shows that for rigid solute particles χ will always approach 45° as the orienting force, β, approaches zero. The relation between χ and β should be linear for low values of β, with a slope equal to $-\beta/12\Theta$, so that Θ can be directly determined. It will be noted that the limiting slope does not depend on the shape factor a/b, as contained in the factor A.

When the velocity gradient becomes very large, all particles will tend to become oriented along the stream lines. The value of χ will thus approach 0° as β tends to infinity. This condition is usually not experimentally attainable because the flow pattern of the fluid ceases to be laminar at very high values of β.

An example of the results of an experimental determination of the extinction angle in flow birefringence of a solution containing rod-shaped solute particles is shown in Fig. 25–4. This particular example was chosen because it shows the concentration dependence of flow birefringence. The equations which have been given in this section are, of course, applicable to zero concentration only, being based on the idea that the solution is an assembly of *independently* rotating particles.

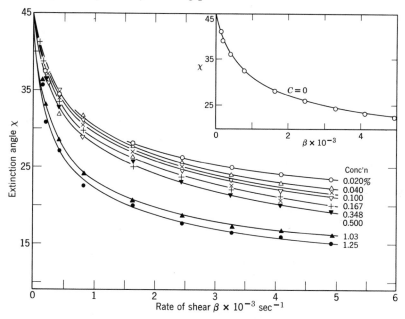

Fig. 25–4. A plot of extinction angle χ versus velocity gradient β for poly-γ-benzyl-L-glutamate in *m*-cresol, which is a solvent in which the polypeptide exists as a helical rod. The figure shows experimental data at various concentrations and, in the insert, an extrapolated curve for infinite dilution. The sample used in these experiments was heterogeneous ($\bar{M}_w/\bar{M}_n = 1.3$), so that the extrapolated curve does not obey equation 25–19. The curve has therefore not been used to evaluate a length for the polypeptide particles. (Yang.[154])

It should be noted that the equations of the present section also assume that all solute particles are identical. If a sample is heterogeneous, consisting of solute particles with a variety of rotary diffusion coefficients, then equation 25–19 will not be obeyed as it stands. The modifications required are discussed in the review of Cerf and Scheraga.[18] If equation 25–19 is applied to limiting slopes obtained with heterogeneous samples (such as the data of Fig. 25–4), too small a value of Θ will be obtained. This is because the most asymmetric molecules will retain their orientations to the lowest values of β.

Flow birefringence is employed primarily in the investigation of macromolecules which are believed to form rigid rod-shaped particles in solution.

TABLE 25–1. Lengths of "Particles" from Rotary Diffusion Measurements

Molecule	Mol. Wt.	Method and Temperature	Θ,[a] sec^{-1}	From Θ, Eqs. 25–7, 25–9	Length, Å By Other Methods[b]
Tobacco mosaic Virus[120,121]	39×10^6	Flow biref., 25°	370	3,400	3,000[EM]
		Elec. biref., 25°	333	3,500	3,200[LS]
Fibrinogen[80,122]	330,000	Flow biref., 20°	39,400	670	820[FR]
		Elec. biref., 20°	36,000	700	710[VIS]
Poly-γ-benzyl-L-glutamate[123]	208,000[d]	Non-Newt. visc., 25°	9,000	1,430	1,430[LS]
Serum albumin[124,125,170]	67,000	Flow biref., 21°	0.84×10^6	190[c]	168[FR]
		Dielec. disp., 25°	1.39×10^6	180[e]	129[VIS]
		Elec. biref., 25°	0.83×10^6	200	

[a] The experimental data were obtained in aqueous solution, with two exceptions: the work on poly-γ-BLG was carried out in m-cresol, and the flow birefringence of serum albumin was studied in 88.5% aqueous glycerol. The values of Θ in these two cases have been corrected by equations 25–7 and 25–9 to what they would have been in a solvent with the viscosity of water, with a and b unchanged. The observed values of Θ were 500 sec^{-1} for poly-γ-BLG and 4800 sec^{-1} for serum albumin.

[b] EM = electron microscopy. LS = light scattering, using the measured radius of gyration and equation 18–21. FR = frictional coefficient, interpreted in terms of a prolate ellipsoid of revolution (length = $2a$), with $\delta_1 = 0.2$. VIS = intrinsic viscosity interpreted in the same way.

[c] The experimental data were not extrapolated to zero concentration. The resulting error has been discussed by Yang[154], and he estimates that the length obtained from extrapolated data would have been 170 Å.

[d] The sample used was not homogeneous, having $\bar{M}_w = 208,000$, $\bar{M}_w/\bar{M}_n = 1.3$. Experiments were carried out in m-cresol, a solvent in which the molecular configuration is that of a helical rod (see Table 18–4 and Fig. 23–5).

[e] Two critical frequencies were obtained, from which two values of Θ may be computed, one for rotation about the long axis of the assumed ellipsoidal model, and one for rotation about either short axis. It is the latter which is used in the table.

There are two reasons for this: (1) Other things being equal, rod-shaped particles have the smallest rotary diffusion coefficients, and they are thus readily oriented at low velocity gradients. (2) The rotary diffusion coefficient is an especially useful property to evaluate for such particles because it provides, by equation 25–7 and 25–9, a measure of the *length* of the particles without the necessity that the molecular weight or other dimensions be known with high degree of accuracy. Some results obtained for such particles are given in Table 25–1, together with similar data from other measurements of rotary diffusion by methods which will be briefly described subsequently.

The data of Table 25–1 provide conclusive corroboration for the rod-shaped configuration already assigned to tobacco mosaic virus on the basis of electron microscopy (Fig. 6–2) and to poly-γ-benzyl-L-glutamate (in certain solvents, such as m-cresol) because of the dependence of the radius of gyration and intrinsic viscosity on molecular weight (Table 18–4, Fig. 23–5). The molecular lengths which we obtain from the rotary diffusion coefficient and other methods are not in perfect agreement with one another, at least in the case of tobacco mosaic virus, but closer agreement could hardly have been expected in view of the theoretical simplifications which have been made, such as the assumption that a long rod may be considered equivalent to a prolate ellipsoid of the same length (section 20d).

As regards fibrinogen, viscosity and diffusion data presented previously are compatible with either rod-shaped particles (large a/b) or highly solvated flexible coils (large δ_1). The birefringence data of Table 25–1 unequivocally exclude the flexible structure because it would not give rise to appreciable flow birefringence at the low rates of shear here employed (see below). Moreover, there is reasonably good agreement between the rod lengths deduced from frictional coefficients, from viscosity, and from the two kinds of birefringence measurements.

The molecular lengths calculated for fibrinogen from the frictional coefficient and viscosity depend on the choice of the solvation number δ_1 (i.e., on the choice of molecular volume). The values listed in Table 25–1 were obtained with $\delta_1 = 0.2$ gram of water per gram of protein, this figure having been chosen because it represents a reasonable estimate of the extent of hydration of *compact* protein molecules in aqueous solution. An extended rigid protein molecule would not be expected to differ greatly in its hydration properties from a compact one. If other reasonable values of δ_1 are chosen, e.g., $\delta_1 = 0.5$ gram/gram, then the molecular lengths derived from viscosity and frictional coefficients can change by about 10%. It should be noted that really large values of δ_1, while mathematically allowed, are physically prohibited in the present analysis because they correspond to a flexible coil model, whereas we are treating the viscosity and frictional coefficients on the basis of a rigid model structure.

Considering, finally, the data of Table 25–1 for serum albumin, we must note that serum albumin is by all criteria (e.g., Tables 21–1 and 23–1) a compact molecule. The diffusion coefficient of serum albumin (Table 21–1) indicates that its hydrodynamic "particle" could be represented as a prolate ellipsoid with $\delta_1 = 0.2$ and $a/b = 4.9$, the corresponding dimensions being $2a = 168$ Å, $2b = 34$ Å. The viscosity data of Table 23–1

indicate, on the other hand, that a suitable ellipsoid with $\delta_1 = 0.2$ would
have $a/b = 3.3$, i.e., $2a = 129$ Å, $2b = 39$ Å. (The lack of agreement
between the two sets of dimensions would persist regardless of the choice
of δ_1.) It was pointed out, in discussing these data, that the choice of the
value 0.2 for the parameter δ_1 was made arbitrarily, and the possibility
that the serum albumin hydrodynamic particle is, for instance, actually a
sphere, with a considerably larger value of δ_1, could not be excluded. In
the present measurements a spherical shape is immediately excluded, since
for a sphere no birefringence would be observed. Moreover, the lengths
derived from the rotary diffusion coefficients depend relatively little on the
chosen value of a/b. The three values given were calculated with the
assumption that $a/b = 4$ (using not equation 25–9, but the more complete
form of Perrin's equation, applicable to $a/b < 5$), and they lead, as
shown, to a length of about 190 Å. Had we assumed $a/b = 2$, this figure
would have become 155 Å; had we assumed the maximum value of a/b
permitted by the diffusion coefficient ($a/b = 6.5$, see Table 21–2), it
would have become 210 Å. Any of these lengths are clearly incompatible
with the diffusion coefficient and intrinsic viscosity unless the latter is
interpreted in terms of a sparingly hydrated ellipsoid with a/b in the
range of 3 to 6, so that the present measurements confirm in a general way
the compromise solution adopted in the interpretation of viscosity and
diffusion data of the compact proteins.

The data for fibrinogen discussed above are particularly significant
because Hall and Slayter[171] have obtained electron micrographs for this
protein (Fig. 6–2), with which the hydrodynamic measurements may be
compared. The electron micrographs indicate (Fig. 6–3) that the molecule
consists of three more or less spherical beads of diameter 65, 50, and 65 Å,
respectively, joined by thin threads of diameter 15 Å or less. The over-all
length is 475 Å. The hydrodynamic data, on the other hand, are all
compatible with a prolate ellipsoid of length about 700 Å and maximum
diameter 38 Å.

In comparing these data we must allow for the possibility that the
electron micrograph is misleading. It may be that the frozen and dried
molecule which the electron microscope sees differs from the molecule in
solution. It is also possible that the threads which join the beads (and
which are poorly resolved) extend beyond the outside beads, so that the
true molecular length exceeds the 475 Å distance measured between the
external edges of these beads. These doubts concerning the validity of
the electron microscope data are, however, irrelevant to the main point
which the comparison between the two representations brings out. The
main point is that the dimensions of Fig. 6–3 are entirely possible and
not incompatible with the hydrodynamic data. Since the equations for

interpreting hydrodynamic data are limited, necessitating the use of a prolate ellipsoidal model for *all* rigid extended particles, it follows that we can deduce from hydrodynamic data only the dimensions of the ellipsoid most closely resembling the actual molecule. The difference between the "best" ellipsoid for fibrinogen and the shape given by Fig. 6–3 must be regarded as a measure of the uncertainty which applies to all the shapes deduced from hydrodynamic measurements, even when they are combined with light scattering data.

The discussion of flow birefringence has so far been limited to rigid rod-shaped particles, and it remains to consider the behavior of the other kinds of solute particles which are amenable to mathematical analysis, i.e., rigid spheres, rigid discs (oblate ellipsoids), and flexible chains. The first two present no problem. Rigid spheres are not oriented by a streaming fluid at all and thus exhibit no birefringence. Oblate ellipsoids may be treated by the same theory as prolate ellipsoids, with the result that, for a given value of the axial ratio, higher velocity gradients will in general be needed to produce birefringence for oblate ellipsoids than for prolate ellipsoids.

Flexible chains present a more complicated problem. It might at first be expected that they will show no birefringence at all, since they may be represented hydrodynamically as swollen spheres. It will be recalled, however, that such a representation can hold true only if the polymer chain is not subject to stress. Under the influence of a shearing stress, deformation will occur, and the particles will become oriented, as was pointed out in the discussion of viscosity on p. 392. Birefringence will therefore be observed. However, the extent of orientation and the resulting birefringence are much smaller than for asymmetric rigid particles, and a relatively large velocity gradient is required if appreciable birefringence is to be observed.

There has been considerable work in recent years on the subject of flow birefringence of flexible polymers, for experimental data on this phenomenon may be expected to yield information on the *internal* freedom of motion of polymer chains, as reflected in the ease with which they may be deformed. Work up to 1952 has been reviewed by Cerf and Scheraga.[18]

25d. Non-Newtonian viscosity. It was pointed out in section 23 that the viscosity of a macromolecular solution may depend on the velocity gradient under which the viscosity is measured and that this effect is due to either orientation of rigid asymmetric particles or deformation of flexible molecules, The underlying cause of this effect is clearly the same as that of flow birefringence, and the effect can thus be used to obtain the same information. The theory for rigid ellipsoids has been developed by Saito and Sugita[158] and by Scheraga[159], reference to earlier papers being given by these authors.

The experimental manifestation of non-Newtonian viscosity is a decrease in the intrinsic viscosity with increasing β. The kind of results obtained are shown in Fig. 25–5.

Figure 25–5, incidentally, provides a fine illustration of the point made earlier, in connection with the discussion of flow birefringence, that the

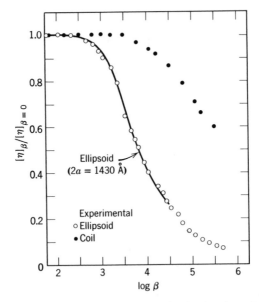

Fig. 25–5. Non-Newtonian viscosity of a sample of poly-γ-benzyl-L-glutamate in *m*-cresol (open circles), where the polymer behaves as a rigid rod, and in dichloroacetic acid (filled circles), where the polymer behaves as a flexible coil. The data are expressed as the ratio of $[\eta]$ at a given value of β to the value of $[\eta]$ when $\beta = 0$. The solid line through the open circles is a theoretical curve for a rigid ellipsoid of length $2a = 1430$ Å. (Yang.[123])

orientation effect produced by the distortion of a flexible coil requires a much greater rate of shear than the effect due to orientation of a rigid rod. The data of the figure are for the *same* sample of poly-γ-benzyl-L-glutamate, the configuration of which is a rigid rod in one solvent and a flexible coil in the other. A much larger value of β is seen to be necessary to produce a decrease in $[\eta]$ for the coiled form.

25e. Dielectric dispersion.[16] It was shown in section 6c that the critical frequencies at which the dielectric constant of a macromolecular solution undergoes marked changes may be related to corresponding relaxation times. If the change in dielectric constant which is observed is the loss of the contribution due to permanent dipole moments of the dissolved macromolecules, the corresponding relaxation time is that for the rotation of the macromolecular particle; i.e., it is the relaxation time related to rotary diffusion coefficients by equation 25–11.

Oncley[16,125] has studied the dielectric dispersion of several globular proteins. He has found that the dispersion curves (Fig. 6–6) which correspond to the longest relaxation time, i.e., those which are presumably associated with rotation of macromolecular particles rather than solvent molecules, cannot usually be described in terms of a single relaxation time but require a minimum of two values of τ. From them, two values of Θ can be computed by equation 25–11, and they may be taken to be the two rotary diffusion coefficients of a prolate or an oblate ellipsoid of revolution. The dimensions of the ellipsoid can then be calculated. A single calculation of this kind is included in Table 25–1. A detailed summary of all such studies is contained in two reviews[16,125]

It was mentioned in section 6 that the low frequency dielectric increment of protein solutions can often be accounted for on the basis of dipole moment fluctuations alone, with the assumption that the average moment, $\bar{\mu}$, of all solute molecules is zero. Whether or not this affects the interpretation of dielectric *dispersion* data depends on the rate at which dipole moment fluctuations occur. No critical analysis of this problem has been made.

25f. Electric birefringence. The birefringence produced by the action of an electric field on a macromolecular solution is the result of dipole orientation. (See Section 6c.) By using a sinusoidal alternating field, or a rectangular electrical pulse of short duration, we can measure the relaxation time associated with this process, and this quantity may again be related to the rotary diffusion coefficient. When a rectangular pulse is used, two independent properties can be observed: the growth of birefringence following establishment of the applied field, and the decay of birefringence after the pulse is complete. According to Tinoco,[161] comparison of these two phenomena permits the unambiguous determination of the relaxation time due to rotary diffusion, because the effects of the permanent and fluctuating dipole moments can in principle be distinguished. The theory and measurement of electric birefringence are fully discussed by Benoit[160] and Tinoco.[161] Three rotary diffusion coefficients obtained by this method are included in Table 25–1.

25g. Depolarization of fluorescence.[162] Fluorescence occurs when a molecule absorbs light of frequency ν_1, loses part of the absorbed energy by a radiationless transition, and emits the remainder as light of frequency ν_2 (where, of course, $\nu_2 < \nu_1$). The exciting frequency ν_1 is usually in the ultraviolet region of the spectrum. An important aspect of the phenomenon is that it does not occur immeasurably fast; there is a time lapse, typically about 10^{-8} sec, between absorption and the subsequent emission of light.

The absorption of the exciting radiation produces a change in electronic state at a particular site on the absorbing molecule, and this in turn produces a local change in dipole moment in a specified direction. (See section 5d for discussion of the corresponding phenomenon in infrared absorption.) If the exciting radiation is polarized, the probability of absorption depends on the orientation of the absorbing sites, being a maximum if the direction of the dipole moment change is parallel to the direction of oscillation of the $\mathscr{E}$ vector of the radiation and reducing to a minimum of zero if these directions are perpendicular. On the other hand, the radiation emitted (as fluorescence) from a given site will always be polarized parallel to the orientation of the dipole moment change. If, therefore, all absorbing sites in a particular sample are immobile, so that their orientations do not change between absorption and emission, then the fluorescent radiation resulting from excitation by polarized light will be partially depolarized, the degree of depolarization being readily calculated from the relation between original orientation and the absorption probability.

In a *solution* of molecules with absorbing sites, however, rotational movement occurs. Since this is a random process, it leads to randomization of orientation of the fluorescing sites in the time which elapses between absorption and emission, and this in turn leads to a depolarization of the fluorescent radiation, which is greater than that calculated for the process described in the last paragraph. The observed depolarization is thus a measure of the rotational freedom of the site on the molecule which is responsible for the fluorescence. If the molecule is rigid and incapable of internal rotations, depolarization clearly becomes a measure of the rotary diffusion coefficients. A detailed analysis of its use for this purpose has been presented by Weber.[162] It should be noted that the method may be applied not only to macromolecules which are naturally fluorescent but also to those which may be rendered fluorescent by chemical combination with or adsorption of a fluorescent molecule of small size.

It must be noted that the use of fluorescence depolarization to measure rotary diffusion coefficients can never be entirely unambiguous because we cannot in general arbitrarily dismiss the possibility that a fluorescent site on a macromolecule is capable of rotation by itself, quite apart from the rotation of the molecule as a whole. As a result, it is likely that the most fruitful application of the method will be to use it together with other techniques for determining macromolecular configuration (i.e., radius of gyration, viscosity, flow birefringence, etc.). The fluorescence depolarization will then not be a measure of over-all shape or configuration, but rather a measure of the freedom of *internal* rotation.

An example is provided by measurements of Weber[163] on serum albumin. Depolarization measurements made near the isoelectric pH of the protein yield rotary diffusion coefficients which agree with those determined by other methods (Table 25–1), which implies that the fluorescing site (in this case a conjugated dye molecule) is incapable of independent motion. When the pH is reduced, a sharp *increase* in the extent of depolarization is observed; i.e., the rotational motion of the fluorescing site becomes easier. At the same time, as was shown in Fig. 22–8, there is an increase in the translational frictional coefficient, which, without other information, could be interpreted as meaning either that serum albumin takes on a new elongated rigid configuration or that its structure has become a looser, swollen structure, approaching that of a flexible coil. In either event there would be an accompanying increase in the rotary frictional coefficient of the molecule as a whole. The observed increase in freedom of rotation of the fluorescing site must therefore correspond to acquisition of internal rotational freedom; i.e., it suggests that the new configuration of serum albumin is a loose, flexible structure.

General References

1. H. Lamb, *Hydrodynamics*, 6th ed., Cambridge University Press, 1932.
2. L. Page, *Introduction to Theoretical Physics*, 3rd ed., D. Van Nostrand Co., Princeton, N.J., 1952, Ch. VI.
3. S. R. de Groot, *Thermodynamics of Irreversible Processes*, Interscience Publishers, New York, 1951.
4. J. T. Edsall, "The Size, Shape and Hydration of Protein Molecules" in H. Neurath and K. Bailey (ed.), *The Proteins*, vol. IB, Academic Press, New York, 1953, Ch. 7.
5. P. J. Flory, "Configurational and Frictional Properties of the Polymer Molecule in Dilute Solution" in *Principles of Polymer Chemistry*, Cornell University Press, Ithaca, 1953, Ch. XIV.
6. A. Einstein, *Investigations on the Theory of the Brownian Movement*, Dover Publications, New York.
7. L. J. Gosting, "Measurement and Interpretation of Diffusion Coefficients of Proteins," *Advances in Protein Chem.*, **11**, 429, (1956).
8. T. Svedberg and K. O. Pedersen, *The Ultracentrifuge*, Oxford University Press, 1940.
9. J. W. Williams, K. E. van Holde, R. L. Baldwin, and H. Fujita, "The Theory of Sedimentation Analysis," *Chem. Revs.*, **58**, 715 (1958).
10. H. K. Schachman, *Ultracentrifugation in Biochemistry*, Academic Press, New York, 1959.

11. R. A. Alberty, "An Introduction to Electrophoresis," *J. Chem. Educ.*, **25**, 426, 619 (1948).
12. J. Th. G. Overbeek, "Quantitative Interpretation of the Electrophoretic Velocity of Colloids," *Advances in Colloid Sci.*, **3**, 97 (1950).
13. M. Bier (ed.), *Electrophoresis*, Academic Press, New York, 1959.
14. H. L. Frisch and R. Simha, "The Viscosity of Colloidal Suspensions and Macromolecular Solutions" in F. R. Eirich, (ed.), *Rheology*, Academic Press, New York, 1956, vol. 1, Ch. 14.
15. J. T. Edsall, "Rotary Brownian Movement. The Shape of Protein Molecules as Determined from Viscosity and Double Refraction of Flow" in E. J. Cohn and J. T. Edsall, *Proteins, Amino Acids and Peptides*, Reinhold Publishing Corp., New York, 1943, Ch. 21.
16. J. L. Oncley, "The Electric Moments and the Relaxation Times of Proteins as Measured from their Influence upon the Dielectric Constants of Solutions" in E. J. Cohn and J. T. Edsall, *Proteins, Amino Acids and Peptides*, Reinhold Publishing Corp., New York, 1943, Ch. 22.
17. A. Peterlin and H. A. Stuart, "Birefringence, Especially Artificial Birefringence" in *Hand u. Jahrb. chem. Phys.*, vol., 8, Pt. 1B, Becker and Erler, Leipzig 1943.
18. R. Cerf and H. A. Scheraga, "Flow Birefringence in Solutions of Macromolecules", *Chem. Revs.*, **51**, 185 (1952).

Specific References

19. Sir G. Stokes, *Trans. Cambridge Phil. Soc.*, **8**, 287 (1847), **9**, 8 (1851).
20. F. Perrin, *J. phys. radium*, [7], **7**, 1 (1936).
21. R. O. Herzog, R. Illig and H. Kudar, *Z. phys. Chem.*, A **167**, 329 (1934).
22. P. Y. Cheng and H. K. Schachman, *J. Polymer Sci.*, **16**, 19 (1955).
23. J. L. M. Poiseuille, *Mémoirés présentés par divers savants à l'académie royale des sciences de l'institut de France*, **9**, 433 (1846); cf. *Rheological Memoirs*, **1**, No. 1 (1940).
24. L. Ubbelohde, *Ind. Eng. Chem., Anal. Ed.*, **9**, 85 (1937); M. R. Cannon and M. R. Fenske, *ibid.*, **10**, 297 (1938).
25. M. Couette, *Ann. chim. phys.*, [6], **21**, 433 (1890).
26. A. Einstein, *Ann. Physik*, [4], **19**, 289 (1906); **34**, 591 (1911).
27. M. Perrin, quoted in Refs. 6 and 26.
28. R. Simha, *J. Phys. Chem.*, **44**, 25 (1940); J. W. Mehl, J. L. Oncley and R. Simha, *Science*, **92**, 132 (1940).
29. J. O'M. Bockris, *Quart. Rev. (London)*, **3**, 173 (1949); B. E. Conway and J. O'M. Bockris, in J. O'M. Bockris, ed., *Modern Aspects of Electrochemistry*, Academic Press, New York, 1954, Ch. 2.
30. J. H. Wang, *J. Am. Chem. Soc.*, **76**, 4755 (1954), **77**, 258 (1955).
31. H. S. Harned and B. B. Owen, *Physical Chemistry of Electrolytic Solutions*, 2nd. ed., Reinhold Publishing Corp., New York, 1950, pp. 158, 253.
32. J. G. Kirkwood and J. Riseman, *J. Chem. Phys.*, **16**, 565 (1948), J. G. Kirkwood, R. W. Zwanzig, and R. J. Plock, *ibid*, **23**, 213 (1955); P. L. Auer and C. S. Gardner, *ibid.*, **23**, 1546 (1955).
33. P. Debye and A. M. Bueche, *J. Chem. Phys.*, **16**, 573 (1948).
34. A. Fick, *Ann. Physik.* **94**, 59 (1855).
35. E.g., U.S. National Bureau of Standards, *Tables of Probability Functions*, vol. 1, Washington, D.C., 1941.

36. P. J. Dunlop, *J. Phys. Chem.*, **61**, 994, 1619 (1957).
37. G. W. Schwert, *J. Biol. Chem.*, **190**, 799 (1951).
38. R. L. Baldwin, L. J. Gosting, J. W. Williams, and R. A. Alberty, *Discussions Faraday Soc.*, No. 20, 13 (1955).
39. J. B. Sumner and N. Gralen, *Science*, **87**, 284 (1938); *J. Biol. Chem.*, **125**, 33 (1938).
40. K. Bailey, H. Gutfreund, and A. G. Ogston, *Biochem J.* **43**, 279 (1948).
41. M. L. Wagner and H. A. Scheraga, *J. Phys. Chem.* **60**, 1066 (1956).
42. J. M. Creeth, *Biochem. J.*, **51**, 10 (1952).
43. H. Boedtker and P. Doty, *J. Am. Chem. Soc.*, **78**, 4267 (1956).
44. K. Laki and W. R. Carroll, *Nature*, **175**, 389 (1955).
45. H. Neurath and G. R. Cooper, *J. Biol. Chem.*, **135**, 455 (1940).
46. A. Einstein, *Ann. Physik*, [4], **17**, 549 (1905); **19**, 371 (1906).
47. A. Rothen, *J. Gen. Physiol.* **24**, 203 (1940).
48. J. R. Colvin, *Can. J. Chem.*, **30**, 831 (1952).
49. O. Lamm and A. Polson, *Biochem. J.* **30**, 528 (1936).
50. J. B. Sumner, N. Gralen, and I.-B. Eriksson-Quensel, *J. Biol. Chem.*, **125**, 37 (1938).
51. R. Cecil and A. G. Ogston, *Biochem., J.*, **44**, 33 (1949); *ibid.*, **43**, 592 (1948).
52. H. Faxén, *Arkiv. Mat. Astron. Fysik*, **21B**, No. 3 (1929).
53. O. Lamm, *Arkiv. Mat. Astron. Fysik*, **21B**, No. 2 (1929).
54. R. J. Goldberg, *J. Phys. Chem.*, **57**, 194 (1953).
55. G. Meyerhoff and G. V. Schulz, *Markromol. Chem.*, **7**, 294 (1952).
56. J. P. Johnston and A. G. Ogston, *Trans. Faraday Soc.*, **42**, 789 (1946).
57. H. Fujita, *J. Am. Chem. Soc.*, **78**, 3598 (1956).
58. P. Y. Cheng and H. K. Schachman, *J. Am. Chem. Soc.*, **77**, 1498 (1955).
59. J. Oth and V. Desreux, *Bull. soc. chim. Belges*, **63**, 133 (1954).
60. G. Kegeles and F. J. Gutter, *J. Am. Chem. Soc.*, **73**, 3770 (1951).
61. R. L. Baldwin, *Biochem. J.*, **65**, 490, 503 (1957).
62. H. Kahler, *J. Phys. and Colloid Chem.*, **52**, 676 (1948).
63. K. O. Pedersen, *J. Phys. Chem.*, **62**, 1282 (1958); Ref. 8, p. 27.
64. H. K. Schachman, *J. Am. Chem. Soc.*, **73**, 4808 (1951).
65. W. J. Archibald, *Ann. N.Y. Acad. Sci.*, **43**, 211 (1942).
66. C. O. Beckman and J. L. Rosenberg, *Ann. N.Y. Acad. Sci.*, **46**, 329 (1945).
67. S. Shulman, *Arch. Biochem. Biophys.*, **44**, 230 (1953).
68. M. E. Reichmann, S. A. Rice, C. A. Thomas, and P. Doty, *J. Am. Chem. Soc.*, **76**, 3047 (1954).
69. P. Ehrlich, S. Shulman, and J. D. Ferry, *J. Am. Chem. Soc.*, **74**, 2258 (1952).
70. C. J. Stacy and J. F. Foster, *J. Polymer Sci.*, **25**, 39 (1957).
71. J. W. Williams, R. L. Baldwin, W. M. Saunders, and P. G. Squire, *J. Am. Chem. Soc.*, **74**, 1542 (1952); R. L. Baldwin, *ibid.*, **76**, 402 (1954).
72. A. Holtzer and S. Lowey, *J. Am. Chem. Soc.*, **78**, 5954 (1956); **81**, 1370 (1959).
73. L. Mandelkern, W. R. Krigbaum, H. A. Scheraga, and P. J. Flory, *J. Chem. Phys.*, **20**, 1392 (1952).
74. P.-O. Kinell and B. G. Ranby, *Advances in Colloid Sci.*, **3**, 161 (1950).
75. J. G. Buzzell and C. Tanford, *J. Phys. Chem.*, **60**, 1204 (1956).
76. C. Tanford and J. G. Buzzell, *J. Phys. Chem.*, **60**, 225 (1956).
77. E. J. Cohn and A. M. Prentiss, *J. Gen. Physiol.*, **8**, 619 (1927).
78. R. Markham, *Progress in Biophysics*, **3**, 61 (1953).
79. T-C. Tsao, K. Bailey, and G. S. Adair, *Biochem. J.*, **49**, 27 (1951).

454

80. L. Nanninga, *Arch. néerl. physiol.*, **28**, 241 (1947); C. S. Hocking, M. Laskowski, Jr., and H. A. Scheraga, *J. Am. Chem. Soc.*, **74**, 775 (1952).
81. T. G. Fox, Jr., and P. J. Flory, *J. Phys. and Colloid Chem.*, **53**, 197 (1949); *J. Am. Chem. Soc.*, **73**, 1909 (1951).
82. F. Booth, *Proc. Roy. Soc.*, A **203**, 533 (1950).
83. P. J. Flory and T. G. Fox, Jr., *J. Am. Chem. Soc.*, **73**, 1904 (1951).
84. W. R. Krigbaum and D. K. Carpenter, *J. Phys. Chem.*, **59**, 1166 (1955).
85. P. J. Flory, Ref. 5, p. 613.
86. A. R. Shultz and P. J. Flory, *J. Polymer Sci.*, **15**, 231 (1955).
87. L. Mandelkern and P. J. Flory, *J. Am. Chem. Soc.*, **74**, 2517 (1952).
88. T. G. Fox, Jr., J. C. Fox, and P. J. Flory, *J. Am. Chem. Soc.*, **73**, 1901 (1951).
89. W. R. Krigbaum and P. J. Flory, *J. Polymer Sci.*, **11**, 37 (1953).
90. P. Outer, C. I. Carr, and B. H. Zimm, *J. Chem. Phys.*, **18**, 830 (1950).
91. W. C. Carter, R. L. Scott, and M. Magat, *J. Am. Chem. Soc.*, **68**, 1480 (1946).
92. H. J. Philipp and C. F. Bjork, *J. Polymer Sci.*, **6**, 383, 549 (1951); see also P. M. Doty and H. M. Spurlin in E. Ott, H. M. Spurlin, and M. W. Grafflin (ed.), *Cellulose and Cellulose Derivatives*, 2nd ed., Interscience Publishers, New York, 1955, p. 1164.
93. H. Boedtker and N. S. Simmons, *J. Am. Chem. Soc.*, **80**, 2550 (1958).
94. P. Doty, J. H. Bradbury, and A. M. Holtzer, *J. Am. Chem. Soc.*, **78**, 947 (1956).
95. H. Staudinger and W. Heuer, *Ber.*, **63**, 222 (1930); H. Staudinger and R. Nodzu, *ibid.*, **63**, 721 (1930).
96. H. Mark, *Der feste Körper*, Hirzel, Leipzig, 1938, p. 103.
97. R. Houwink, *J. prakt. Chem.*, **157**, 15 (1941).
98. J. R. Schaefgen and P. J. Flory, *J. Am. Chem. Soc.*, **70**, 2709 (1948).
99. E. Guth and O. Gold, *Phys. Revs.*, **53**, 322ᶦ (1938); R. Simha, *J. Appl. Phys.*, **23**, 1020 (1952).
100. P. Y. Cheng and H. K. Schachman, *J. Polymer Sci.*, **16**, 19 (1955).
101. M. L. Huggins, *J. Am. Chem. Soc.*, **64**, 2716 (1942).
102. C. D. Thurmond and B. H. Zimm, *J. Polymer Sci.*, **8**, 477 (1952).
103. T. E. Timell, *Svensk Papperstidn*, **57**, 777 (1954).
104. A. Holtzer and S. A. Rice, *J. Am. Chem. Soc.*, **79**, 4847 (1957).
105. C. Tanford, *Symposium on Protein Structure* (A. Neuberger, ed.), Methuen and Co., London, 1958.
106. H. A. Scheraga and L. Mandelkern, *J. Am. Chem. Soc.*, **75**, 179 (1953).
107. B. A. Dombrow and C. O. Beckman, *J. Phys. & Colloid. Chem.*, **51**, 107 (1947).
108. F. R. Senti, N. N. Hellman, N. H. Ludwig, G. E. Babcock, R. Tobin, C. A. Glass, and B. L. Lamberts, *J. Polymer Sci.*, **17**, 527 (1955).
109. S. Shulman, *J. Am. Chem. Soc.*, **75**, 5846 (1953).
110. H. Gutfreund and A. G. Ogston, *Biochem. J.*, **44**, 163 (1949).
111. L. Onsager, *Phys. Revs.*, **37**, 405 (1931); **38**, 2265 (1931).
112. G. Kegeles, S. M. Klainer and W. J. Salem, *J. Phys. Chem.*, **61**, 1286 (1957).
113. L. G. Bunville, Ph.D. Thesis, State U. of Iowa, 1959.
114. R. E. Lovrien, Ph.D. Thesis, State U. of Iowa, 1958.
115. W. F. Harrington and J. A. Schellman, *Compt. rend. trav. lab. Carlsberg, Sér. chim.*, **30**, 21 (1956).
116. R. E. Weber and C. Tanford, *J. Am. Chem. Soc.*, **81**, 3255 (1959).
117. M. E. Reichmann, S. A. Rice, C. A. Thomas, and P. Doty, *J. Am. Chem. Soc.*, **76**, 3047 (1954).
118. Sir G. Stokes, *Mathematical and Physical Papers*, Cambridge University Press, 1880.

119. S. Glasstone, K. J. Laidler, and H. Eyring, *The Theory of Rate Processes*, McGraw-Hill Book Co., New York, 1941, Ch. 9.
120. H. Boedtker and N. S. Simmons, *J. Am. Chem. Soc.*, **80**, 2550 (1958).
121. C. T. O'Konski and A. J. Haltner, *J. Am. Chem. Soc.*, **78**, 3604 (1956).
122. I. Tinoco, Jr., *J. Am. Chem. Soc.*, **77**, 3476 (1955).
123. J. T. Yang, *J. Am. Chem. Soc.*, **80**, 1783 (1958).
124. J. T. Edsall and J. F. Foster, *J. Am. Chem. Soc.*, **70**, 1860 (1948).
125. J. L. Oncley, *Chem. Revs.*, **30**, 433 (1942).
126. W. F. Harrington, P. Johnson and R. H. Ottewill, *Biochem. J.* **62**, 569 (1956).
127. E. O. Field and J. R. P. O'Brien, *Biochem. J.*, **60**, 656 (1955).
128. S. M. Klainer and G. Kegeles, *J. Phys. Chem.*, **59**, 952 (1955); *Arch. Biochem. Biophys.*, **63**, 247 (1956).
129. D. B. Smith, G. C. Wood, and P. A. Charlwood, *Can. J. Chem.*, **34**, 364 (1956).
130. W. J. Archibald, *J. Phys. Chem.* **51**, 1204 (1947).
131. R. L. Baldwin, *Biochem. J.*, **55**, 644 (1953).
132. H. R. Kruyt (ed.), *Colloid Science*, vol. I, Elsevier Publishing Co., Amsterdam, 1952.
133. F. Booth, *Proc. Roy. Soc.*, **A 203**, 514 (1950).
134. D. C. Henry, *Proc. Roy. Soc.*, **A 133**, 106 (1931).
135. J. Th. G. Overbeek, *Kolloid Beih.*, **54**, 287 (1943).
136. M. H. Gorin, *J. Chem. Phys.*, **7**, 405 (1939).
137. L. Onsager, *Physik. Z.*, **27**, 388 (1926); **28**, 277 (1927); L. Onsager and R. M. Fuoss, *J. Phys. Chem.*, **36**, 2689 (1932); R. M. Fuoss and L. Onsager, *ibid.*, **61**, 668 (1957).
138. M. Smoluchowski, *Z. physik. Chem.*, **92**, 129 (1918).
139. A. Tiselius, *Trans. Faraday Soc.*, **33**, 524 (1937).
140. V. P. Dole, *J. Am. Chem. Soc.*, **67**, 1119 (1945).
141. L. G. Longsworth, in Ref. 13, Chs. 3 and 4.
142. L. G. Longsworth, *Ann. N.Y. Acad. Sci.*, **41**, 267 (1941).
143. S. Beychok and R. C. Warner, *J. Am. Chem. Soc.*, **81**, 1892 (1959).
144. A. Tiselius, *Biochem. J.*, **31**, 1464 (1937).
145. G. Wallenius, R. Trautman, H. G. Kunkel, and E. C. Franklin, *J. Biol. Chem.*, **225**, 253 (1957).
146. J. R. Cann, *J. Am. Chem. Soc.*, **71**, 907 (1949).
147. G. E. Perlmann, *Advances in Protein Chem.*, **10**, 1 (1955).
148. L. Pauling, H. A. Itano, S. J. Singer, and I. C. Wells, *Science*, **110**, 543 (1949).
149. J. A. Hunt and V. M. Ingram in A. Neuberger (ed.), *Symposium on Protein Structure*, Methuen and Co., London, 1958.
150. H. B. Bull, *J. Am. Chem. Soc.*, **80**, 1901 (1958); D. K. Chattoraj and H. B. Bull, *ibid.*, **81**, 5128 (1959).
151. F. Booth, *J. Colloid Sci.*, **6**, 549 (1951).
152. D. Edwardes, *Quart. J. Pure and App. Math.*, **26**, 70 (1893).
153. F. Perrin, *J. Phys. radium*, [7] **5**, 497 (1934).
154. J. T. Yang, *J. Am. Chem. Soc.*, **80**, 5139 (1958).
155. P. Boeder, *Z. Physik*, **75**, 258 (1932).
156. A. Peterlin, *Z. Physik*, **111**, 232 (1938); A. Peterlin and H. A. Stuart, *ibid.*, **112**, 1, 129 (1939).
157. H. A. Scheraga, J. T. Edsall, and J. O. Gadd, Jr., *J. Chem. Phys.*, **19**, 1101 (1951).
158. N. Saito, *J. Phys. Soc., Japan*, **6**, 297 (1951); N. Saito and M. Sugita, *ibid.*, **7**, 554 (1952).

159. H. A. Scheraga, *J. Chem. Phys.*, **23**, 1526 (1955).

160. H. Benoit, *Ann. Phys.*, [12], **6**, 561 (1951); *J. chim. phys.*, **48**, 612 (1951).

161. I. Tinoco, Jr., *J. Am. Chem. Soc.*, **77**, 4486 (1955).

162. G. Weber, *Advances in Protein Chem.*, **8**, 415 (1953).

163. G. Weber, *Biochem. J.*, **51**, 155 (1952). See also Ref. 126.

164. R. Trautman and V. N. Schumaker, *J. Chem. Phys.*, **22**, 551 (1954).

165. R. Trautman, *J. Phys. Chem.*, **60**, 1211 (1956).

166. R. Trautman and C. F. Crampton, *J. Am. Chem. Soc.*, **81**, 4036 (1959).

167. S. R. Erlander and J. F. Foster, *J. Polymer Sci.*, **37**, 103 (1959).

168. D. A. Yphantis, *J. Phys. Chem.*, **63**, 1742 (1959).

169. W. W. Everett and J. F. Foster, *J. Am. Chem. Soc.*, **81**, 3464 (1959).

170. S. Krause and C. T. O'Konski, *J. Am. Chem. Soc.*, **81**, 5082 (1959).

171. C. E. Hall and H. S. Slayter, *J. Biophys. Biochem. Cytol.*, **5**, 11 (1959).

7

ELECTROSTATIC
FREE ENERGY AND
POLYELECTROLYTES

It has already been pointed out that many macromolecules exist in aqueous solution (or in other polar solvents) in the form of macro-ions. Included in this group, for example, are the proteins, the nucleic acids, and synthetic polyelectrolytes. Some discussion of such macro-ions has already been given in section 14, where it was shown, by purely thermodynamic arguments, that solutions of such ions tend to show deviations from ideality simply because of the necessity to maintain neutrality. This source of non-ideality, it was shown, could be suppressed by the addition of a neutral salt such as NaCl. It also vanishes at the limit of zero concentration of the macro-ion.

A little consideration shows, however, that the effect just discussed cannot be the only nor even the most important consequence of the presence of electrostatic charges on a macromolecule. There must be strong electrostatic forces between the charges on an individual macro-ion, and they clearly do not vanish when the macro-ion concentration is reduced. Furthermore, macro-ion solutions always contain small ions. There must be at least enough of them to neutralize the charge on the macro-ions. The electrostatic interaction between the macro-ion and the small ions will also not vanish at infinite dilution of the macro-ion.

These electrostatic forces have important effects on the configuration of macro-ions and on their chemical combination with other ions, including their acidic and basic properties. A fairly detailed discussion of their influence is therefore called for.

It will be found that the most straightforward approach to this problem involves the evaluation of the electrostatic potential field generated by the macro-ion charges, and the calculation of the electrostatic contribution to the free energy, which is equivalent to the work of placing the charges on a macro-ion. Accordingly we describe in the first section of this chapter the general procedures used to accomplish this.

26. CALCULATION OF THE ELECTROSTATIC POTENTIAL AND FREE ENERGY

Since the effects in which we are principally interested here persist at the limit of zero concentration of macro-ions we shall confine our discussion in this section largely to this limiting situation. The calculations will then be simplified because the properties of the macro-ions in such a solution are simply additive, and we shall need to calculate only the electrostatic free energy of a single macro-ion. In some situations the extension of the results to solutions of finite concentration will be discussed (for instance, sections 26d and 26j), but in general the problem at finite concentration is more complicated, not only because of the effects already discussed in section 14 but also because of electrostatic forces between the macro-ions in the solution.

Before proceeding to actual calculations, it should be recalled that, at constant temperature and pressure, a free energy change is a direct measure of work done on a system (excluding work of expansion, which, in solutions is in any case negligible), so that the electrostatic contribution to the free energy of a macro-ion is simply the work done in placing charges on it. This electrostatic work, in turn, is the product of potential and charge. If an increment of charge dq is to be placed at a position of electrostatic potential ψ, the infinitesimal amount of work done is

$$dW_{el} = \psi \, dq \qquad (26\text{--}1)$$

The electrostatic free energy, W_{el}, is defined with respect to a hypothetical discharged state of the macro-ion (although we shall have occasion to modify this definition to exclude self-energies of discrete charges where they occur), i.e.,

$$W_{el} = \int \psi \, dq \qquad (26\text{--}2)$$

where the integral extends over all places on the macro-ion where charges occur, and from $q = 0$ to its final value.

26a. General procedure for calculating the potential in a system of fixed and mobile charges.
To obtain W_{el}, as expressed by the integral of equation 26–2, it is necessary to know ψ as a function of location and of the positions

of charges already in place. The general procedure by which this knowledge may be obtained is a well-known development of nineteenth century classical physics, and a convenient discussion of it is given, for instance, by Stratton.[1]

In any region devoid of charges Laplace's equation applies,

$$\nabla^2 \psi = 0 \tag{26-3}$$

where ∇^2 is the *Laplacian operator*, the precise expression for which depends on the choice of coordinate system. In rectangular coordinates (x, y, z),

$$\nabla^2 \psi \equiv \frac{\partial^2 \psi}{\partial x^2} + \frac{\partial^2 \psi}{\partial y^2} + \frac{\partial^2 \psi}{\partial z^2} \tag{26-4}$$

In spherical coordinates (r, θ, ϕ),

$$\nabla^2 \psi \equiv \frac{1}{r^2} \frac{\partial}{\partial r}\left(r^2 \frac{\partial \psi}{\partial r}\right) + \frac{1}{r^2 \sin \theta} \frac{\partial}{\partial \theta}\left(\sin \theta \frac{\partial \psi}{\partial \theta}\right) + \frac{1}{r^2 \sin \theta} \frac{\partial^2 \psi}{\partial \phi^2} \tag{26-5}$$

For systems with spherical symmetry, where ψ is independent of θ and ϕ, equation 26–5 reduces to the simpler expression

$$\nabla^2 \psi \equiv \frac{d^2 \psi}{dr^2} + \frac{2}{r}\frac{d\psi}{dr} \equiv \frac{1}{r}\frac{d^2(r\psi)}{dr^2} \tag{26-6}$$

Equations for ∇^2 in a number of other coordinate systems are given by Margenau and Murphy.[2]

Laplace's equation applies to regions of space devoid of charges. In regions of space containing a *continuous* distribution of charge, of density ρ (which will in general be a function of position), the potential is obtained from Poisson's equation

$$\nabla^2 \psi = -\frac{4\pi\rho}{D} \tag{26-7}$$

where D is the dielectric constant of the region. In virtually all our applications we shall be concerned with regions of space of constant dielectric constant.

When the potential in a solution is calculated, the origin of the co-ordinate system is normally located on one of the ions of the solution (in our case ordinarily the center of mass of the macro-ion). The charges on that ion are therefore *fixed charges*; i.e., their positions are indicated by a fixed set of coordinates. All other charges, however, are *mobile* with respect to the coordinate system chosen, and they may therefore be treated as a continuous distribution of charge with the same approximation that allows the solvent to be treated as a continuum of given dielectric constant.

Equation 26–7 therefore applies to regions containing mobile charges, with ρ representing the time-average charge density at any position.

At times it may be convenient to consider fixed charges also as distributed continuously over a region of space. This is especially true if there are many of them and we do not know their exact locations. At other times, however, fixed charges will be treated as *point charges* of fixed location. In that event the positions at which they are located will be positions of infinite potential, since the potential near a fixed charge q is q/Dr' where r' is the distance from the charge. These positions must then be treated mathematically as *singularities*, which means that equation 26–3 or 26–7 (depending on whether the region of space considered also contains mobile ions) applies to the region of space considered at all points except those occupied by fixed charges. At a point occupied by a particular fixed charge q_i, ψ must become infinite as q_i/Dr'. If $\mathbf{r}$ is a vector from the center of the coordinate system to any position in space and $\mathbf{r}_i$ is the vector to the location of the charge, then at or very near this location

$$\psi = \frac{q_i}{D|\mathbf{r} - \mathbf{r}_i|} \tag{26–8}$$

$|\mathbf{r} - \mathbf{r}_i|$ designating the absolute value of the vector $\mathbf{r} - \mathbf{r}_i$, which is of course the same as the distance r'.

Yet another possibility is that it may be appropriate to consider the fixed charges distributed over a surface. This situation is treated mathematically in terms of a boundary condition (equation 26–11).

The procedure just discussed clearly permits us to set up differential equations for the potential for any given distribution of physical boundaries and of fixed or mobile charges. In general, it will be found necessary to divide the solution into several distinct regions in each of which Laplace's or Poisson's equation applies. The solution to these equations, like the solution to any differential equation, then depends on specification of pertinent parameters at the boundaries of the regions to which the equations apply.

One of the pertinent conditions is that the potential ψ is continuous everywhere except at singularities. At a boundary between two different regions ψ must therefore have the same value on both sides. Where subscripts 1 and 2 refer to different regions, therefore, at the boundary,

$$\psi_1 = \psi_2 \tag{26–9}$$

An independent condition is that the gradient of potential normal to the boundary obeys the condition

$$D_2\left(\frac{\partial \psi}{\partial \mathbf{n}}\right)_2 = D_1\left(\frac{\partial \psi}{\partial \mathbf{n}}\right)_1 \tag{26–10}$$

when there is no charge on the surface representing the boundary. If this surface contains a charge of density σ, then equation 26–10 is replaced by

$$D_2 \left(\frac{\partial \psi}{\partial \mathbf{n}}\right)_2 - D_1 \left(\frac{\partial \psi}{\partial \mathbf{n}}\right)_1 = -4\pi\sigma \qquad (26\text{–}11)$$

In equations 26–10 and 26–11, D_1 and D_2 represent the dielectric constants of the adjacent regions, and $\mathbf{n}$ is a normal to the surface directed from region 1 to 2. For spherical boundaries with center at the origin $\partial \psi / \partial \mathbf{n}$ becomes $\partial \psi / \partial r$. Finally, ψ cannot become infinite except at singularities and, in particular, must become zero at an infinite distance from fixed charges.

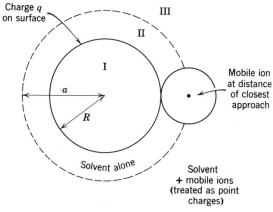

Fig. 26–1. Model used in derivation of the Debye-Hückel theory. Region I is a fixed ion, and its center is the origin of the coordinate system.

26b. The Debye-Hückel theory for spherical, impenetrable ions. The most important advance in our understanding of ionic solutions resulted from the calculation by Debye and Hückel, in 1923, of the electrostatic free energy of small spherical ions.[6] Since their method is the basis of all the calculations in this section, it is given here in some detail.

A definitive account of the theory, and of older theories which led up to it, is given by Falkenhagen.[3] Another good discussion is given by Harned and Owen.[7] There have been a number of critical examinations of the foundations of the theory, one of the most recent being by Kirkwood and Poirier.[8] References to earlier work may be found in their paper.

The model for the Debye-Hückel calculation is shown in Fig. 26–1. The ion whose electrostatic free energy we wish to determine is taken to be spherical with a radius R. It is placed at the center of the coordinate system. In the original theory this ion represents an ordinary ion, such as Na^+, but there is nothing in the development to prevent it from being a macro-ion of arbitrary charge. It is assumed that the charge q of the ion is evenly distributed over the surface of the sphere, though we shall see that any spherically symmetrical location of the charge would give the same result for the activity coefficients and other properties of small ions which the Debye-Hückle theory is normally used to compute.

The ion is immersed in a solvent of dielectric constant D (the dielectric constant inside the ion need not be specified, since it does not enter into the result), which contains mobile charges. We shall assume here that all mobile ions are univalent; the extension to higher valence types is well known and leads to the same result in terms of ionic strength (cf. equation 26–30 and the subsequent discussion).

The univalent mobile ions are treated as point charges and enter into the calculation only in that they determine the charge density in the solution around the central ion. It is obviously not necessary to specify the chemical nature of the mobile ions—they need be divided only into positive and negative ions with charge $+\epsilon$ and $-\epsilon$, respectively, where ϵ is the protonic charge, 4.80×10^{-10} esu. That the mobile ions are in fact of finite size need not concern us when we calculate a charge density from the concentration of ions. Their finite size will, however, limit how closely they can approach the central ion. As Fig. 26–1 shows, the center of a mobile ion can approach no closer than a distance a to the center of the coordinate system, where a is the sum of the radii of the central and the mobile ion. A sphere of radius a is therefore drawn about the ion, as shown in Fig. 26–1, representing the distance within which there can be no mobile charges. Since there are always at least two kinds of mobile ion the distance a is in fact an average distance of closest approach.

The electrostatic problem now resolves itself naturally into three separate regions of space, as shown by Fig. 26–1. Laplace's equation applies in region I, within the central ion, and in region II, which consists of solvent of dielectric constant D but no mobile ions. The surface between these regions has a uniform charge density, $\sigma = q/4\pi R^2$, where q is the charge on the central ion. Region III contains the mobile ions, and Poisson's equation therefore applies to it. It is necessary to determine the distribution of the mobile ions, i.e., to determine ρ as a function of position, before this equation can be solved. This distribution will clearly be non-uniform. At distances far removed from the central ion the time-average charge density and the potential will both be zero and the concentrations of positive and negative ions will be equal. These concentrations will be the bulk concentrations of these ions in the solution, which we shall designate by N ions per cubic centimeter. Near the central ion, however, the concentration N_+ of positively charged mobile ions will differ from the concentration N_- of negatively charged ions, the population of the former being greater if the central ion bears a negative charge and smaller if it bears a positive charge.

The cornerstone of the Debye-Hückel theory is that this concentration difference may be computed by the Boltzmann distribution law, i.e., that the ratio of the concentration of one kind of ion near the central ion to its concentration far from the central ion is equal to $e^{-W_i/kT}$, where W_i is the work which must be done to move the ion from infinity ($\psi = 0$) to the potential ψ prevailing at the point at which the concentration is to be computed. Clearly $W_i = \epsilon\psi$ for a positive ion of charge $+\epsilon$ and $W_i = -\epsilon\psi$ for a negative ion of charge $-\epsilon$. The Boltzmann distribution law thus becomes

$$N_+ = Ne^{-\epsilon\psi/kT}$$
$$N_- = Ne^{+\epsilon\psi/kT}$$

(26-12)

Equation 26-12 makes $N_+ > N_-$ near a negative central ion (ψ negative) and vice versa, whereas $N_+ = N_- = N$ far from the central ion where $\psi = 0$.

The charge density ρ at any point is now given by

$$\rho = N_+\epsilon - N_-\epsilon = N\epsilon(e^{-\epsilon\psi/kT} - e^{\epsilon\psi/kT})$$

which, on expansion of the exponentials, becomes

$$\rho = -2N\epsilon\left[\frac{\epsilon\psi}{kT} + \frac{1}{3!}\left(\frac{\epsilon\psi}{kT}\right)^3 + \frac{1}{5!}\left(\frac{\epsilon\psi}{kT}\right)^5 + \cdots\right]$$

(26-13)

One further assumption characterizes the Debye-Hückel theory, namely, that the potential is sufficiently small at all points where Poisson's equation is to be used so that $\epsilon\psi/kT \ll 1$. If this is true, all but the first term in the expansion of equation 26-13 can be neglected, so that

$$\rho = -\frac{2N\epsilon^2\psi}{kT}$$

(26-14)

This assumption is justified for central ions of sufficiently low charge density and for external solutions with sufficiently low concentration of mobile ions. (The assumption will generally not be true *immediately* adjacent to any real central ion, but at low concentrations of mobile ions there will be a very low probability that any such ions are immediately adjacent, and the potential will be largely determined by the distribution of more distant mobile ions.)

What the *practical* limitations resulting from the introduction of equation 26-14 are cannot be definitely stated. The Debye-Hückel theory, with the distance parameter *a* as adjustable variable, is reasonably successful in predicting the thermodynamic behavior of ordinary univalent ions up to an electrolyte concentration of the order of 0.1 molar.

For ordinary ions of higher valence it works less well, to an ionic strength closer to 0.01 than to 0.1. When the results of the theory are applied to proteins, which are macro-ions of relatively low charge density, they again seem applicable (in so far as they are applicable at all) to solutions containing *univalent* mobile ions up to an ionic strength of about 0.1 molar. As regards fully charged synthetic polyelectrolytes, which ordinarily have a relatively high charge density, it is sometimes asserted that equation 26–14 cannot be applied at all, but whether or not this is true has not yet been established.

With the introduction of equation 26–14, Poisson's equation, equation 26–7, becomes

$$\nabla^2\psi = \frac{8\pi N\epsilon^2}{DkT}\,\psi = \kappa^2\psi \qquad (26\text{–}15)$$

where κ is the well-known Debye-Hückel parameter

$$\kappa = \left(\frac{8\pi\epsilon^2}{DkT}\right)^{\frac{1}{2}} N^{\frac{1}{2}} \qquad (26\text{–}16)$$

which is proportional to the square root of the mobile ion concentration and is a constant as far as integration of equation 26–15 is concerned. Equation 26–15 is known as the Poisson-Boltzmann equation.

Because the charge distribution of the central ion is spherically symmetrical, the potential field generated by it must also have spherical symmetry, so that equation 26–6 may be used for $\nabla^2\psi$. Thus equation 26–15 becomes

$$\frac{d^2(r\psi)}{dr^2} = \kappa^2 r\psi \qquad (26\text{–}17)$$

the general solution of which is $r\psi = A_1 e^{-\kappa r} + A_2 e^{\kappa r}$, or

$$\psi = A_1 e^{-\kappa r}/r + A_2 e^{\kappa r}/r \qquad (26\text{–}18)$$

where A_1 and A_2 are integration constants depending on the boundary conditions. One of them, as previously pointed out, is that ψ must become zero as r approaches infinity. Since $e^{\kappa r}/r$ becomes infinite at this limit it is necessary that A_2 of equation 26–18 be set equal to zero, so that the final expression for the potential in region III is

$$\psi = A_1 e^{-\kappa r}/r \qquad (26\text{–}19)$$

In regions I and II Laplace's equation applies, i.e., for spherical symmetry,

$$\frac{d^2(r\psi)}{dr^2} = 0 \qquad (26\text{–}20)$$

with the general solution

$$\psi = A_3 + A_4/r \qquad (26\text{–}21)$$

Equation 26–21 may be taken as the solution for region II, with the constants A_3 and A_4 to be determined. Region I, however, contains the point $r = 0$ and, since ψ must remain finite there, cannot have a term in $1/r$ in its equation for ψ. Hence ψ in region I must be constant,

$$\psi = A_5 \tag{26-22}$$

It remains to solve for the integration constants A_1, A_3, A_4 and A_5 from the boundary conditions given by equations 26–9, 26–10, and 26–11, applied to the two boundaries at $r = R$ and $r = a$. We obtain, where $r = R$,

$$A_5 = A_3 + A_4/R \tag{26-23}$$

$$DA_4 = q \tag{26-24}$$

(Note that equation 26–24, arising from application of equation 26–11, does not contain a term involving the dielectric constant inside the sphere because ψ is constant inside the sphere and $\partial\psi/\partial r$ therefore is zero. This is why the dielectric constant inside the sphere does not have to be specified in the solution of this particular problem.) Similarly where $r = a$

$$A_1 e^{-\kappa a}/a = A_3 + A_4/a \tag{26-25}$$

$$A_1 e^{-\kappa a}(1 + \kappa a) = A_4 \tag{26-26}$$

Using equations 26–23 to 26–26 to solve for the constants and introducing these into equations 26–19, 26–21, and 26–22, we get for region III

$$\psi = \frac{q}{D}\frac{e^{\kappa a}}{1 + \kappa a}\frac{e^{-\kappa r}}{r} \tag{26-27}$$

for region II

$$\psi = \frac{q}{Dr}\left(1 - \frac{\kappa r}{1 + \kappa a}\right) \tag{26-28}$$

and for region I, including the surface at $r = R$,

$$\psi = \frac{q}{DR}\left(1 - \frac{\kappa R}{1 + \kappa a}\right) \tag{26-29}$$

Equation 26–29 gives the potential at the surface where the charge is located and is therefore the one to be used in computing W_{el} by equation 26–2. If the final charge on the central ion is to be $Z\epsilon$, then

$$W_{el} = \int_{q=0}^{Z\epsilon} \frac{q}{DR}\left(1 - \frac{\kappa R}{1 + \kappa a}\right)dq = \frac{Z^2\epsilon^2}{2DR}\left(1 - \frac{\kappa R}{1 + \kappa a}\right) \tag{26-30}$$

This is the result we sought. As here derived it is applicable only to the situation where the mobile ions are univalent. As is well-known, however, and shown in any elementary discussion of the Debye-Hückel theory, equation 26-30 applies equally well for mobile ions of higher valence type, provided that the concentration of mobile ions (as it occurs in the parameter κ) is expressed in terms of the *ionic strength*

$$I = \tfrac{1}{2} \sum_i C_i z_i^2 \tag{26-31}$$

where C_i is the molar concentration of any mobile ion and z_i its charge, the summation extending over all kinds of mobile ion present. If, therefore, we express κ, as given by equation 26-16, in terms of the ionic strength, equation 26-30 is applicable for mobile ions of all charge types.

In the model used here, with two kinds of mobile ion, each having $z_i^2 = 1$ and $C_i = 1000N/\mathcal{N}$, where $\mathcal{N}$ is Avogadro's number,

$$I = 1000N/\mathcal{N}$$

so that, by equation 26-16,

$$\kappa = \left(\frac{8\pi \mathcal{N} \epsilon^2}{1000\, DkT} \right)^{\!\!\tfrac{1}{2}} I^{\tfrac{1}{2}} \tag{26-32}$$

This is then the required expression for κ.

The Debye-Hückel theory is ordinarily applied to solutions of ionized salts, in which the central ion is just one of many similar ions (e.g., Na^+) also included among the mobile ions; the central ion is differentiated from the others like it only by the fact that it has been made the origin of the coordinate system. It is clear then that W_{el} for one mole of such ions is just $\mathcal{N}$ times the result given by equation 26-30. This is by definition the electrostatic contribution to the chemical potential of this particular kind of ion.

The most familiar result derived from equation 26-30 is the expression for the activity coefficient of a small ion. By definition $\mathscr{R}T \ln \gamma$ is just the difference between the electrostatic contribution to the chemical potential at any ionic strength and the same contribution calculated at infinite dilution of all ions ($I = \kappa = 0$); i.e.,

$$kT \ln \gamma = -\frac{Z^2 \epsilon^2}{2D} \frac{\kappa}{1 + \kappa a} \tag{26-33}$$

The calculations of the present section have been based on the assumption that the charge on the central ion is evenly distributed over the surface of the ion. It should be noted, however, that some of the conclusions are independent of this assumption. Thus equation 26-27, giving the potential for $r > a$, will remain unchanged for *any spherically symmetrical distribution* of a charge q within the sphere $r = a$. The ionic-strength-dependent portions of equations 26-28, 26-29, and 26-30 are similarly independent of the location of the charge, provided that it is spherically symmetrical. An important result of this is that equation 26-33, for the activity coefficient, is independent of the location of the charge, and its success in predicting experimental deviations from ideality does not therefore permit us to draw any conclusion concerning the location of the charge on small ions.

26c. Approximate expression for the electrostatic free energy of globular proteins. As has been shown, for instance in Chapter 6, many proteins in aqueous solution are close to spherical in shape and so compactly folded as to include only a small amount of solvent within their domain. A portion of this solvent is specifically bound to ionic sites (section 20), and it is likely that the remainder may also be molecularly dispersed, e.g., hydrogen bonded to polar groups. It is therefore a reasonable first approximation to consider such a protein ion as an impenetrable sphere

TABLE 26-1. Electrostatic Free Energy of Impenetrable
Spherical Protein Ion of Radius 25 Å
(Mol. Wt. $\sim$ 40,000) in Aqueous Solution at 25° C

Net Charge	W_{el}, cal/mole			
	Ionic Strength (Molar)			
Z	0.001	0.01	0.05	0.15
10	6,700	4,800	3,300	2,500
20	27,000	19,200	13,200	9,900
30	60,500	43,100	29,700	22,200
40	108,000	76,800	52,800	39,600

and to calculate its electrostatic free energy by equation 26–30. Indeed, such an approximation was suggested by Linderstrøm-Lang[9] immediately after publication of the Debye-Hückel theory, and it has been used with considerable success. One application of this approximation has already been discussed in section 24.

In estimating the radius R of the protein ion for such calculations, we use equation 20–6 for the volume, i.e.,

$$\tfrac{4}{3}\pi R^3 = \frac{M}{\mathscr{N}} (\bar{v}_2 + \delta_1 v_1^0) \qquad (26\text{--}34)$$

thus including within the sphere the incorporated solvent as defined by δ_1. From the discussion of section 20 we must conclude that δ_1 for this purpose is not necessarily the same as the δ_1 used in computing hydrodynamic volume. However, the difference cannot be great, and it need certainly not concern us when equation 26–30 is used, since this is in any case an approximate equation. In the absence of other information we normally assign to δ_1 a value near 0.2, as in the calculation of hydrodynamic properties.

In order to obtain an idea of the magnitude of W_{el}, we show, in Table 26–1, some calculations by equation 26–30 for a spherical protein ion of radius 25 Å. For a typical protein such a radius would correspond to a molecular weight of about 40,000. A protein of this molecular weight would ordinarily have a maximum charge Z of the order of ± 40. Table 26–1 shows that W_{el} can attain considerable magnitude and that it can be reduced, but by no means eliminated, by an increase in ionic strength. From the magnitude of the figures of Table 26–1 it is apparent that the electrostatic free energy can be expected to play an important role in determining the properties of protein ions.

The application of equation 26–30 to proteins is an approximation, principally for the two following reasons:

(1) A protein ion of net charge Z does not have a charge $Z\epsilon$ spread evenly over its surface. Instead, the charge exists as discrete unit charges ($-NH_3^+$, $-COO^-$, etc.) at fixed locations. The total number of charges is generally greater than the net charge, which is the difference between the number of positive and negative charges. The calculation of W_{el} for a more realistic model involving discrete charges will be discussed in section 26g. It might be mentioned here, however, that the locations of the fixed charges are ordinarily not known, so that the improved equation for W_{el} has at present only limited practical applicability.

(2) Chapter 6 has shown that an ellipsoid of revolution of axial ratio $a/b \simeq 3$ to 4 is probably a better representation of many of the compact protein ions than a sphere. It will be shown in section 26h that this is likely to reduce W_{el}, but only by a small amount.

In the remaining sections of this book many instances will arise which require a value of W_{el} for globular proteins. Ordinarily, in the absence of exact information regarding shape and the location of charges, we shall use equation 26–30 for this purpose.

26d. The effect of concentration. Protein ions in the absence of added electrolyte. When equation 26–30 is applied to protein ions it should, strictly speaking, be used only when they are at a concentration approaching zero. It might be supposed that a higher concentration could be accounted for by including protein ions among the mobile ions of the solution. However, a protein ion is so large, and its charges are so far apart that the various individual charges must ordinarily be at locations of different potential. To include the protein ion in the ionic strength as an ion of charge Z is therefore unrealistic, except when Z is quite small and when the total ionic strength is also small so that the potential changes relatively slowly with distance.

An alternative procedure to extend the validity of equation 26–30 to higher concentrations is to treat a protein ion, in its contribution to the ionic strength, as Z univalent ions. This procedure, however, is more reasonable for long chain polyelectrolytes, the charges of which may be expected to be capable of independent motion, than it is for the more rigid protein ions.

An exact solution of this problem is clearly difficult, and it will not be discussed here. It should be noted, however, that the problem is essentially the same as that of the calculation of the electrostatic contribution to the second virial coefficient of the chemical potential, to which reference was made in section 14. W_{el} is the electrostatic contribution to the chemical potential of an ion, and what we have done here is to evaluate it (for a particular way of distributing charges) at infinite dilution of the macro-ion. When we ask how this quantity behaves as the macro-ion concentration increases, we are in effect asking how electrostatic forces influence the concentration dependence of the chemical potential. (For a general solution the excellent recent papers of Hill and Stigter[55] should be consulted.)

A quantitative solution of the problem of the interaction between protein ions, under conditions where the average net charge of the ions is zero and where no small ions are present at all, was given in section 14 on p. 231. This is a very special situation, however, in which the difficulties of the general problem all disappear.

It is possible to prepare solutions which contain, besides protein ions, no other ions except the concentration of counterions sufficient to maintain neutrality, and the ions resulting from ionization of the solvent. Such solutions cannot be expected to obey the equations derived for isolated ions in the presence of a reasonably high concentration of mobile ions. The derivations given above can be applied to experimental data only if we can approach conditions under which the presence of ions of opposite charge reduces the potential to zero at a distance less than the average separation between macro-ions. This condition will be approached if the dilution of macro-ions is carried out in the presence of a constant, reasonably high concentration of added low-molecular-weight electrolyte. It cannot be attained if no electrolyte is added, for then the concentration of counterions decreases along with the concentration of macro-ions, so that the distance required to attain zero potential increases parallel with the increase in macro-ion separation. For this reason experimental studies of proteins have virtually always been made in the presence of a low-molecular-weight electrolyte, such as NaCl of KCl.

26e. The effect of solvent penetration on the electrostatic free energy of protein ions.[10] Although the Debye-Hückel theory in its original form provides a reasonable approximation for the calculation of W_{el} for many compact protein ions in aqueous solution, it cannot be applied to proteins which are rod shaped, nor to proteins which are not compactly folded (for instance, most proteins in urea solution), nor to flexible polyelectrolytes, which must be expected to resemble flexible polymers in being spread (loosely coiled) over a large volume of solution. This and the following sections will provide models appropriate to the calculation of W_{el} for such macro-ions.

Figure 26-2 shows a spherical model permitting penetration of solvent, including small mobile ions, into the domain of the macro-ion. To keep the model general, we allow a central core of radius R_1 to remain impenetrable to solvent (R_1 can be set equal to zero to allow for the special case of complete penetration). Between R_1 and R there is a mixture of the solvent and of the macro-ion substance. We suppose that all the charge of the macro-ion is distributed throughout this region with uniform density ρ_m. If q is the total charge of the macro-ion, then

$$\rho_m = \frac{3q}{4\pi(R^3 - R_1^3)} \tag{26-35}$$

We let α^2 represent the fraction of the volume of region II accessible to small ions. To calculate α^2 we calculate the total volume of protein substance by equation 26-34, calling this volume $\frac{4}{3}\pi R_0^3$. In doing so we include the usual small amount of tightly bound solvent (i.e., $\delta_1 \simeq 0.2$) in the material impenetrable to the mobile ions. (The result will turn out to be quite insensitive to the choice of R_0 within reasonable limits.) In terms of R_0, α^2 is clearly given as

$$\alpha^2 = \frac{R^3 - R_0^3}{R^3 - R_1^3} \tag{26-36}$$

We again assume that the mobile ions are all univalent and make the same assumptions as in the Debye-Hückel theory concerning their distribution. In region III (Fig. 26-2) therefore the charge density is again given by equation 26-14. In region II the same equation applies to any element of volume accessible to the mobile ions. Since only a fraction α^2 is accessible to these ions, the actual density ρ_i of mobile ions in region II becomes, $\rho_i = -2\alpha^2 N\epsilon^2\psi/kT$, and the total charge density in this region, including that of fixed charges as given by equation 26-35, becomes

$$\rho = \rho_m + \rho_i = \frac{3q}{4\pi(R^3 - R_1^3)} - \frac{2\alpha^2 N\epsilon^2\psi}{kT} \tag{26-37}$$

Poisson's equation for this region thus becomes

$$\frac{1}{r}\frac{d^2(r\psi)}{dr^2} = \alpha^2\kappa^2\psi - \frac{3q}{D(R^3 - R_1^3)} \tag{26-38}$$

where κ is again given, for univalent mobile ions, by equation 26-16 and where the result can again be extended to mobile ions of higher valence by the substitution of equation 26–32. (The dielectric constant in region II has been assigned the same value as in region III. This can be true only if α^2 is close to unity. The present treatment is thus limited to the situation where the macro-ion material in region II is highly diffuse. We could treat the intermediate situation between this and a compact sphere by assigning a dielectric constant $D' = \alpha^2 D$ to region II.)

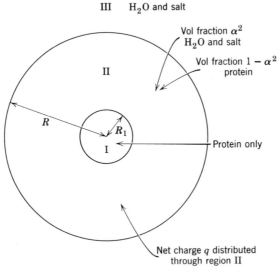

Fig. 26–2. Model for solvent-permeable macro-ions.

The general solution of equation 26–38 is

$$\psi = \frac{B_1 e^{-\alpha\kappa r}}{r} + \frac{B_2 e^{\alpha\kappa r}}{r} + \frac{3q}{\alpha^2\kappa^2 D(R^3 - R_1^{\,3})} \qquad (26\text{–}39)$$

where B_1 and B_2 are integration constants. In regions I and III the same differential equations apply as in the Debye-Hückel theory, and the potential is again given by equations 26–22 and 26–19, respectively.

The constants A_1, A_5, B_1, and B_2 are again obtained by application of the boundary conditions, as expressed by equations 26–9 and 26–10 (this time there is no surface bearing charges, so that equation 26–11 is not used), and the final solution for region II is

$$\psi = \frac{3q(1+\kappa R)}{\kappa^3 D(R^3 - R_0^{\,3})}\left\{\frac{\kappa}{1+\kappa R} - \frac{\dfrac{1+\alpha\kappa R_1}{1-\alpha\kappa R_1}e^{\alpha\kappa(r+R-2R_1)} - e^{\alpha\kappa(R-r)}}{r\left[(1+\alpha)\dfrac{1+\alpha\kappa R_1}{1-\alpha\kappa R_1}e^{2\alpha\kappa(R-R_1)} - (1-\alpha)\right]}\right\}$$

$$(26\text{–}40)$$

To calculate W_{el} we divide region II into spherical layers of thickness dr. For each increment dq in total charge, there will be an increment $4\pi r^2\, dr\, dq / \tfrac{4}{3}\pi(R^3 - R_1^3)$ in each layer. The total work of charging, to a final charge $Z\epsilon$, is obtained by integrating equation 26-2 from $r = R_1$ to $r = R$ and from $q = 0$ to $q = Z\epsilon$, i.e.,

$$W_{el} = \frac{3}{R^3 - R_1^{\,3}} \int_{q=0}^{Z\epsilon} \int_{r=R_1}^{R} r^2 \psi\, dq\, dr \qquad (26\text{--}41)$$

which, by use of equation 26-40, becomes

$$W_{el} = \frac{9Z^2\epsilon^2}{2\kappa^2 D(R^3 - R_0^{\,3})}$$

$$\times \left\{ \frac{1}{3} + \frac{1+\kappa R}{\kappa^3(R^3 - R_0^{\,3})} \frac{(1-\alpha\kappa R)\dfrac{1+\alpha\kappa R_1}{1-\alpha\kappa R_1} e^{2\alpha\kappa(R-R_1)} - (1+\alpha\kappa R)}{(1+\alpha)\dfrac{1+\alpha\kappa R_1}{1-\alpha\kappa R_1} e^{2\alpha\kappa(R-R_1)} - (1-\alpha)} \right\}$$

$$(26\text{--}42)$$

Equation 26-42 gives W_{el} for a single ion. To obtain the electrostatic free energy per mole, equation 26-42 must be multiplied by Avogadro's number.

TABLE 26–2. Electrostatic Free Energy of a Solvent-Permeated Spherical Protein Ion of Radius 50 Å ($R_0 = 25$ Å, $M.W. \simeq 40{,}000$) in Aqueous Solution at 25° C

Net Charge	W_{el}, cal/mole			
	Ionic Strength (Molar)			
Z	0.001	0.01	0.05	0.15
10	4,200	1,600	630	270
20	16,800	6,500	2,500	1,100
30	37,800	14,600	5,700	2,400
40	67,200	25,900	10,100	4,300

Table 26-2 illustrates a typical calculation using equation 26-42. We have chosen a sphere with $R_0 = 25$ Å (corresponding, for normal values of $\bar{v}_2$ and δ_1, to a molecular weight of 40,000) and have set the expanded radius R equal to 50 Å. The value of R_1 turns out to be unimportant; i.e., any value between the allowed limits of 0 and 25 Å leads to essentially the same value for W_{el}.

The values of W_{el} shown in the table are seen to be much smaller than those calculated for an impenetrable protein of the same molecular weight in Table 26-1. This is due in part to the fact that the fixed charges of the ion are spread over a larger volume in the model of Fig. 26–2. At higher ionic strength, however, the decrease in W_{el} is primarily a result of the much greater effectiveness of the mobile ions of the present model in shielding the fixed charges from one another. It will be noted that a relatively high ionic strength can almost eliminate W_{el} for a penetrable ion, whereas W_{el} is always quite large, even at ionic strength 0.15, if the mobile ions cannot penetrate within the protein sphere.

A model similar to that of Fig. 26–2, but with charges confined to the surface at R_1, has also been treated.[10] The results are of the same order of magnitude as those of Table 26–1.

An impenetrable sphere with charge distributed throughout its volume, instead of being confined to the surface, has been treated by Hill.[11] As would be expected, W_{el} for this model is *higher* than that predicted by the Debye-Hückel treatment.

26f. The approximate electrostatic free energy of long rod-shaped ions. Using exactly the same techniques employed in the preceding sections, Hill[11] has computed W_{el} for long cylindrical ions. He has assumed that a cylindrical ion of length L can be considered to have the same electrostatic free energy as a section of length L of an infinite cylinder; i.e., he has neglected the fact that the electrostatic field at the ends of real cylinders must be somewhat different from that nearer the middle. For long cylinders this should not lead to appreciable error.

He has considered two kinds of cylinders. One is an impenetrable cylinder of radius R and radius of exclusion a (cross-section shown in Fig. 26–3a), with a charge Z spread uniformly over the surface. He obtains

$$W_{el} = \frac{Z^2 \epsilon^2}{DL} \left[\frac{K_0(\kappa a)}{\kappa a \, K_1(\kappa a)} + \ln \frac{a}{R} \right] \tag{26-43}$$

where $K_0(x)$ and $K_1(x)$ represent modified Bessel functions of the second kind.[12]

The second kind of cylinder used by Hill is illustrated by the cross-section in Fig. 26–3b. This cylinder has a cavity of radius R_1 down its center, which contains solvent. There is a layer from which ions are excluded both on the inside and outside; i.e., there are mobile ions only where $r < d$ and $r > a$, as shown in the figure. For this model, with Z charges spread uniformly over the *outside* of the cylinder ($r = R$), Hill obtains

$$W_{el} = \frac{Z^2 \epsilon^2}{DL} \frac{\left[\dfrac{K_0(\kappa a)}{\kappa a \, K_1(\kappa a)} + \ln \dfrac{a}{R} \right] \left[\dfrac{D_i \, I_0(\kappa d)}{D\kappa d \, I_1(\kappa d)} + \dfrac{D_i}{D} \ln \dfrac{R_1}{d} + \ln \dfrac{R}{R_1} \right]}{\left[\dfrac{D_i \, I_0(\kappa d)}{D\kappa d \, I_1(\kappa d)} + \dfrac{D_i \, K_0(\kappa a)}{D\kappa a \, K_1(\kappa a)} + \dfrac{D_i}{D} \ln \dfrac{aR_1}{dR} + \ln \dfrac{R}{R_1} \right]}$$

$$\tag{26-44}$$

whereas, if the charge is on the *inside* surface of the cylinder,

$$W_{el} = \frac{Z^2 \epsilon^2}{DL} \frac{\left[\dfrac{I_0(\kappa d)}{\kappa d \, I_1(\kappa d)} + \ln \dfrac{R_1}{d}\right]\left[\dfrac{D_i \, K_0(\kappa a)}{D\kappa a \, K_1(\kappa a)} + \dfrac{D_i}{D} \ln \dfrac{a}{R} + \ln \dfrac{R}{R_1}\right]}{\left[\dfrac{D_i \, I_0(\kappa d)}{D\kappa d \, I_1(\kappa d)} + \dfrac{D_i \, K_0(\kappa a)}{D\kappa a \, K_1(\kappa a)} + \dfrac{D_i}{D} \ln \dfrac{aR_1}{dR} + \ln \dfrac{R}{R_1}\right]}$$

(26–45)

In equations 26–44 and 26–45 D_i is the dielectric constant between $r = R_1$
and $r = R$, i.e., the dielectric constant of the macro-ion material; the

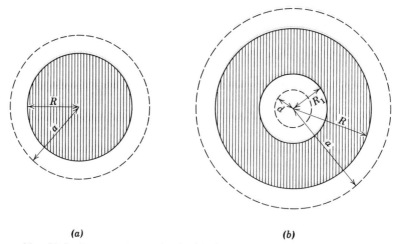

(a)   (b)

Fig. 26–3. Cross-sections of cylindrical macro-ions to which equations
26–43 to 26–45 apply. The shaded portion represents macro-ion substance,
assumed impenetrable to solvent. The unshaded portions represent solvent.
The dashed lines represent the distance of closest approach of salt ions to the
macro-ion cylinder.

K functions are again modified Bessel functions of the second kind, and
the I functions modified Bessel functions of the first kind.[12]

Equation 26–43 should provide a good approximation for the electrostatic free energy of rod-shaped proteins such as myosin or proto-collagen.
Equations 26–44 and 26–45 may be applicable to nucleic acids, if the
structure proposed for these substances by Crick and Watson (section 4)
is correct.

As in the case of spherical ions, the equations of this section are not
applicable to appreciably high concentrations nor to solutions containing
no added electrolyte.

It is believed that flexible polyelectrolytes may become close to fully
extended in the absence of added electrolyte, in which case they would

resemble long rod-shaped particles. A discussion of appropriate equations is given in section 26j.

26g. More rigorous calculation of the electrostatic free energy of impenetrable protein ions.[14] It was pointed out in section 26c that an accurate representation of a protein ion must take into account the fact that its charges are not uniformly distributed over a given surface or volume but that they exist instead as discrete unit charges of both positive and negative sign. Accordingly, the present section presents a calculation of W_{el} for the more realistic model illustrated by Fig. 26–4. The protein ion is

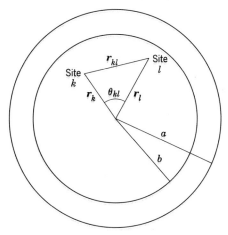

Fig. 26–4. Protein ion with discrete charges. The points k and l represent the locations of two of the charges.

represented by an impenetrable sphere of radius R. The internal dielectric constant must be specified for this calculation and is designated by D_i. The charges are located at p fixed points, each located at the end of a vector $\mathbf{r}_k$ $(k = 1, 2, \ldots, p)$ drawn from the center of the sphere. The charge at each point is $\zeta_k \epsilon$, where ζ_k is $+1$ for a positive charge (e.g., NH_4^+) and -1 for a negative charge (e.g., COO^-). The finite radius of the mobile ions in the solvent is again taken into account by drawing a sphere of radius a to represent the surface from within which mobile ions are excluded.

In principle, the calculation of W_{el} presents no new difficulties. The fixed charges are treated as singularities, as explained in section 26a, so that Laplace's equation applies in region I of Fig. 26–4, the discrete charges being excluded from the region. Laplace's equation also applies in region II, and, with the same assumptions that are made in the Debye-Hückel theory, the Poisson-Boltzmann equation (equation 26–15), applies to region III. The charges, however, are now no longer distributed with spherical symmetry, so that the potential field will also be asymmetric, and hence (for the spherical coordinate system required for the model) equation 26–5 must be used for $\nabla^2 \psi$.

Under these conditions the solutions to Laplace's equation and to the Poisson-Boltzmann equation naturally become quite complicated. One way to write them is as a superposition of the solutions for the individual charges; i.e., for Laplace's equation (cf. equation 26–21)

$$\psi = A + \sum_{k=1}^{p} \frac{B_k}{|\mathbf{r} - \mathbf{r}_k|} \qquad (26\text{--}46)$$

and for the Poisson-Boltzmann equation (cf. equation 26–18)

$$\psi = \sum_{k=1}^{p} \left(\frac{\mathscr{A}_k e^{-\kappa|\mathbf{r}-\mathbf{r}_k|}}{|\mathbf{r} - \mathbf{r}_k|} + \frac{\mathscr{B}_k e^{\kappa|\mathbf{r}-\mathbf{r}_k|}}{|\mathbf{r} - \mathbf{r}_k|} \right) \qquad (26\text{--}47)$$

However, these equations obviously cannot be used if there are boundary conditions like those of the present problem, since the boundaries cannot be described in a simple way in terms of the parameters $|\mathbf{r} - \mathbf{r}_k|$. Instead it becomes necessary to expand $|\mathbf{r} - \mathbf{r}_k|$ and its functions in terms of infinite series, most conveniently in terms of Legendre polynomials. We get for Laplace's equation the general solution

$$\psi = \sum_{n=0}^{\infty} \sum_{m=-n}^{n} \left\{ \frac{G_{nm}}{r^{n+1}} + H_{nm}r^n \right\} P_n{}^m(\cos \theta) \, e^{im\phi} \qquad (26\text{--}48)$$

where $P_n{}^m(x)$ represents an associated Legendre polynomial of the first kind, i is the square root of -1, and the G_{nm} and H_{nm} are integration constants. This solution applies directly to region II of Fig. 26–4. In region I the terms in $1/r^{n+1}$ must vanish in order that ψ may remain finite at the origin. To the solution in region I must also be added the condition that ψ becomes infinite at the positions $\mathbf{r}_k$, which is done by introduction of equation 26–8, so that

$$\psi = \sum_{n=0}^{\infty} \sum_{m=-n}^{n} E_{nm}r^n \, P_n{}^m(\cos \theta) \, e^{im\phi} + \sum_{k=1}^{p} \frac{\zeta_k \epsilon}{D_i|\mathbf{r} - \mathbf{r}_k|} \qquad (26\text{--}49)$$

The solution of the Poisson-Boltzmann equation for region III has been given by Kirkwood[13] as

$$\psi = \sum_{n=0}^{\infty} \sum_{m=-n}^{n} \frac{I_{nm}}{r^{n+1}} e^{-\kappa r} \chi_n(\kappa r) \, P_n{}^m(\cos \theta) \, e^{im\phi} \qquad (26\text{--}50)$$

where the $\chi_n(x)$ are functions given by

$$\chi_n(x) = \sum_{s=0}^{n} \frac{2^s \, n! \, (2n - s)!}{s! \, (2n)! \, (n - s)!} x^s \qquad (26\text{--}51)$$

The usual boundary conditions are now applied (separately to the coefficients of each term in the expansions), and the work of charging is calculated by equation 26–2. The result, first obtained by Kirkwood,[13] may be converted into a form amenable to calculation by making use of suitable properties of Legendre polynomials. In this form

$$W_{\text{el}} = \frac{\epsilon^2}{2R} \sum_{\substack{k=1 \\ k \neq l}}^{p} \sum_{=1}^{p} \zeta_k \zeta_l (A_{kl} - B_{kl}) - \frac{\epsilon^2}{2a} \sum_{k=1}^{p} \sum_{l=1}^{p} \zeta_k \zeta_l C_{kl} \qquad (26\text{–}52)$$

where

$$A_{kl} = R/D_i r_{kl}$$

$$B_{kl} = \frac{1 - 2\delta}{D_i (1 - 2\rho_{kl} \cos \theta_{kl} + \rho_{kl}^2)^{\frac{1}{2}}} + \frac{1}{D\rho_{kl}}$$

$$\times \ln \left[\frac{(1 - 2\rho_{kl} \cos \theta_{kl} + \rho_{kl}^2)^{\frac{1}{2}} + \rho_{kl} - \cos \theta_{kl}}{1 - \cos \theta_{kl}} \right] \qquad (26\text{–}53)$$

with

$$\delta = D_i/D \quad \text{and} \quad \rho_{kl} = r_k r_l/R^2$$

and C_{kl} is a complicated function which, however, at low ionic strength reduces simply to $(1/D)\kappa a/(1 + \kappa a)$, so that at low ionic strength, since

$$\sum_k \sum_l \zeta_k \zeta_l = (\sum_k \zeta_k)(\sum_l \zeta_l) = Z^2,$$

$$W_{\text{el}} = \frac{\epsilon^2}{2R} \sum_{\substack{k=1 \\ k \neq l}}^{p} \sum_{l=1}^{p} \zeta_k \zeta_l (A_{kl} - B_{kl}) - \frac{Z^2 \epsilon^2}{2D} \frac{\kappa}{1 + \kappa a} \qquad (26\text{–}54)$$

Tables of values of $A_{kl} - B_{kl}$ useful for most calculations are given by Tanford and Kirkwood,[14] who also give the full expression for C_{kl}.

In the first term of equation 26–54, the terms with $k = l$, which represent the work of creating the p individual point charges and which are necessarily infinite, have been omitted. These terms (called the *self-energies* of the charges) cancel in any application of the equation since charges are not destroyed in any process of interest.

It should be noted that the ionic-strength-dependent term of equation 26–54 is identical with that of equation 26–30. It was pointed out in connection with equation 26–30 that this term is independent of the location of charges on the macro-ion, provided that the distribution is spherically symmetrical. What the actual calculations of the last term of equation 26–52 prove is that at low ionic strength most of the mobile ions tend to be sufficiently far from the macro-ion so that even deviations from a symmetrical distribution do not appreciably affect this term.

A further complication arises when equation 26–52 or 26–54 is applied to real protein ions or models representing them. This complication is that such ions virtually always have numerous possible configurations, differing in the distribution of charge. Consider, for instance, the very simple model illustrated by Fig. 26–5a: it represents a spherical molecule which has six carboxyl groups, C, and two amino groups, N, located at the corners of a cube within the molecule. Suppose we are interested in the form of this molecule which is an ion of charge $Z = -2$. Because of the relatively strong basicity of amino groups we may assume that both of the amino groups of the model have a positive charge when $Z = -2$, so that the ion must have four of its six carboxyl groups in the ionized state, i.e., negatively charged. There are, however, fifteen ways of choosing four

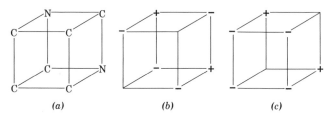

$$(a)\qquad\qquad\qquad (b)\qquad\qquad\qquad (c)$$

Fig. 26–5. A hypothetical ion with six —COOH groups and two —NH₂ groups, located as shown in (a). Two possible configurations, both representing a net charge of − 2, are shown in (b) and (c). Uncharged corners in (b) and (c) represent undissociated −COOH groups. Charged corners represent —NH₃⁺ and —COO⁻.

carboxyl groups out of six, and, therefore, fifteen possible configurations of the ion, two of which are shown in Fig. 26–5b and c. Each such configuration will usually have a different value of W_{el}. For instance, if the cube of Fig. 26–5 is considered to be concentric with the molecular sphere, to which we assign $R = 8$ Å and $D_i = 2$, and if we suppose that the corners of the cube are 0.8 Å below the surface, then, in water, $W_{el} = -2500$ cal/mole for the configuration of Fig. 26–5b, and $W_{el} = +700$ cal/mole for the configuration of Fig. 26–5c. If we wish, therefore, to calculate an observable value of W_{el} for such an ion, we must obtain an average over all configurations, keeping in mind that all configurations are not equally probable, the relative probability of a given configuration being $e^{-W_{el}/kT}$.

It should be pointed out that, in addition, a real solution containing ions of charge, say, $Z = -2$, will in general also contain ions with $Z = 0$, $-1, -3, -4$, etc., since all such ions are in equilibrium with one another (cf. section 30). This aspect of the problem, however, need not concern us at the present time since no comparisons with experiment are being made, but it will have to be considered subsequently.

In view of the fact that different configurations involve different sites for the interacting charges, it becomes convenient to extend the summation of equation 26–54 over all sites which *may* bear charges (i.e., p becomes the number of sites rather than the number of charges). The fact that a particular site k is uncharged in a given configuration can then be introduced by setting ζ_k for that site equal to zero. This procedure is advantageous because $A_{kl} - B_{kl}$ is clearly independent of configuration (the sites being fixed), so that, by summing over all pairs of sites rather than pairs of charges, we use the same set of $A_{kl} - B_{kl}$ values for each configuration, the configurations differing only in that different ζ_k are set equal to zero.

When equation 26–52 or 26–54 is applied to simple models[14] (such as that of Fig. 26–5) it is found, after averaging in each case over all possible configurations, that W_{el} depends on the locations of the various sites and especially on their depth below the surface. (This would not be true if D_i were equal to D. It is generally believed, however, that the dielectric constant in the interior of organic molecules is about 2). Almost any desired value of W_{el} can be obtained, even without appreciably altering the location of sites relative to one another, by adjusting the depth below the surface (i.e., r_k/R in Fig. 26–4). The dependence of W_{el} on this parameter is particularly strong when the sites are very close to the surface, and, since this is where charged sites on impenetrable proteins are probably located, it constitutes a serious obstacle to the use of equation 26–52 or 26–54.

To get around this obstacle it is reasonable to make the assumption that the charges on protein ions come just as close to the surface as do charges on small organic molecules. The interaction between charges on such molecules can be studied experimentally,[16] and from such studies it can be concluded that charges tend to be close to 1.0 Å from the surface.[15] This figure is presumably related to the inadequacy of a model which pictures the ion as a cavity of low dielectric constant, at the surface of which there is a jump in dielectric constant to that of the solvent (in this case water). In actual fact solvent molecules immediately adjacent to a charged site must be largely frozen in position owing to charge-dipole interaction. As a result the dielectric constant of the solvent does not attain its full value until some distance from the charges, so that, if the model calls for a sharp transition in dielectric constant *at* the surface, the charges appear to be below the surface by an amount which represents the effective thickness of a solvent molecule and which varies little from one ion to another.

This argument demonstrates why it is reasonable that charges on macroions should have the same *minimum* distance of approach to the solvent. The minimum distance is, however, the only feasible distance, for the self-energy of a charged site increases tremendously with increasing depth in a cavity of low dielectric constant; i.e., any configuration which requires an isolated charge to be anywhere other than at its closest possible distance from the solvent cannot have stability.

For protein ions in water (at 25° C) we shall thus always assume that $R - r_k = 1.0$ Å for every charge. With this assumption we have made calculations of W_{el} for the two simple models illustrated by Fig. 26–6. Both of them contain six —COOH groups (C) capable of existing as uncharged or negative sites, three imidazole groups (I) capable of existing as uncharged or positive sites, and three amino groups (N) with the same alternatives. The net charge Z may thus vary from $+6$ to -6.

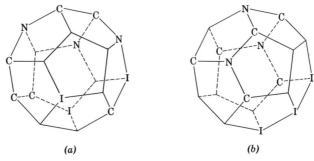

(a)   (b)

Fig. 26–6. Models used for calculating the electrostatic free energies of Table 26–3. The letters C, I, and N represent, respectively, carboxyl, imidazole, and amino side chains. The dodecahedra are inscribed in spheres of radius 10 Å, with vertices 1 Å below the surface. The distance between nearest neighboring vertices is 6.5 Å.

In going from $Z = +6$ to -6, in aqueous solution, the —COOH groups will lose their protons first; the imidazole groups will follow, and the process $-NH_3^+ \rightarrow -NH_2$ will be the final one, the reason for this being that this is the order of the acid dissociation constants of the groups involved (see section 30). When $Z = 0$ there will thus ordinarily be twelve charges, six positive and six negative.

The calculations shown in Table 26–3 indicate that W_{el} is positive when all or nearly all charges have the same sign but that W_{el} is negative near $Z = 0$ when charges of both signs are present in about equal proportion. The actual values of W_{el} will, moreover, depend on the actual positions of the charges, as shown by the considerable difference between the values for Model A and Model B. Finally, it should be noted, the effect of ionic strength is relatively small, in agreement with the result obtained by means of the Debye-Hückel equation in Table 26–1.

The last two columns of Table 26–3 serve as a comparison between the equations of this section and the simple Debye-Hückel theory (equation 26–30). One of the things it shows is, of course, that equation 26–30 cannot lead to negative values of W_{el}. This failing can be corrected, however, if account is taken of the fact that equation 26–30 includes the self-energies of all of the charges, whereas equation 26–54 does not. The self-energies represent the work done in placing each individual charge on the ion in the absence of other charges. In the Debye-Hückel model the self-energies are thus given by equation 26–30 with $Z = 1$ and $\kappa = 0$, so that, if they are subtracted out, equation 26–30 becomes

$$W_{el} = \frac{Z^2 \epsilon^2}{2DR}\left(1 - \frac{\kappa R}{1 + \kappa a}\right) - (v_+ + v_-)\frac{\epsilon^2}{2DR} \qquad (26\text{–}55)$$

where v_+ and v_- represent, respectively, the number of positive and negative charges on the macro-ion.

TABLE 26–3. Electrostatic Free Energy of Model Ions
Shown in Fig. 26–6

W_{el}, cal/mole

| Net Charge Z | Discrete Charge Treatment (Eq. 26–54) | | | | Debye-Hückel | |
	Model A $I = 0$	Model A $I = 0.015$	Model A $I = 0.060$	Model B $I = 0$	Eq. 26–30 $I = 0$	Eq. 26–55 $I = 0$
+6	+7530	+5880	+5110	+8510	+7600	+6330
+4	+380	−30	−170	+250	+3380	+1690
+2	−4260	−3880	−3610	−6170	+840	−1270
0	−6390	−5650	−5200	−10770	0	−2530
−2	−4060	−3670	−3400	−11110	+840	−1270
−4	+1100	+700	+550	−4350	+3380	+1690

Table 26–3 shows that equation 26–55 gives a reasonably reliable representation of W_{el}. It can never, of course, allow for the effects of variation in the positions of individual charges. However, since these positions are usually unknown, we shall make use of the equation with the understanding that it can be regarded only as an approximation to the true value of W_{el}. It should be noted finally, as remarked before, that we are usually interested in using the equations of this section to calculate the change in W_{el} in processes during which the self-energies remain unchanged, so that it makes no difference whether we use equation 26–30 or 26–55 for the calculation. We shall ordinarily use equation 26–30.

26h. Ellipsoidal protein ions. It has been pointed out that many proteins are better represented by ellipsoids of low eccentricity ($a/b \sim 3$ or 4) than by spheres. Equations for the calculation of W_{el} for discrete charges on an ellipsoid are available[17,18] but have never been used for numerical calculations. Ellipsoids with uniform charge distributions have also not been discussed in the literature.

It is likely that the principal effect in going from a sphere to a relatively symmetrical ellipsoid of equal volume is the resulting increase in surface area, hence in the average distance between charges, if they are located near the surface. For an ellipsoid with $a/b \simeq 3$ or 4 the distance between charges will increase by about 10%, and this will result in a corresponding *decrease* in W_{el} by about the same amount.

For proteins like myosin, which are well represented by prolate ellipsoids with high values of a/b, the best available equation for the calculation of W_{el} is that for cylindrical sections given in section 26f.

26i. Flexible linear polyelectrolytes.

It was shown in section 10 that flexible polymers in solution are spread over a large volume, of which the actual polymer segments occupy only a very small fraction. The density of segments relative to the center of mass could be described, on the average by a distribution function with spherical symmetry (equation 10–26).

If this picture is extended to a polyelectrolyte, i.e., a polymer bearing charges on some or all of its segments, then W_{el} should be calculable by the equations of section 26e, or simple modifications thereof. If, for instance, we take equation 26–42 and set $R_1 = 0$, since polyelectrolytes would not possess an impenetrable core, and set $\alpha^2 = 1$, since all but a very small fraction of the volume containing the polymer should be accessible to mobile ions, we get

$$W_{el} = \frac{3Z^2\epsilon^2}{2DR_e}\left\{\frac{1}{\kappa^2R_e^2} - \frac{3}{2\kappa^5R_e^5}\left[\kappa^2R_e^2 - 1 + (1 + \kappa R_e)^2 e^{-2\kappa R_e}\right]\right\} \quad (26\text{–}56)$$

where R_e is the radius of a sphere equivalent to the polyelectrolyte ion; i.e., it is a quantity of the same order of magnitude as the radius of gyration R_G. Equation 26–56 was first derived by Hermans and Overbeek,[19] and it has been frequently applied to calculate properties of polyelectrolyte ions.

The model underlying equation 26–56 (section 26e) assumes a uniform distribution of the fixed charges within the sphere of radius R_e, whereas the average segment distribution as given by equation 10–26 decreases exponentially with the distance from the center. Hermans and Overbeek[19] have shown, however, that this has only a negligibly small effect on the calculation of W_{el}.

We have shown in the case of proteins, however, that the substitution of a continuous charge distribution for the discrete fixed charges leads generally to an incorrect evaluation of the electrostatic free energy, though it may predict correctly the effect of ionic strength. It is therefore probable that the Hermans-Overbeek theory is also only a crude approximation to W_{el}. The error involved here may, in fact, be more serious than in the case of compact proteins, for the segments of the polyion occupy only a very small portion of the equivalent sphere which represents the polyion, so that the resemblance between the actual distribution of charge and any

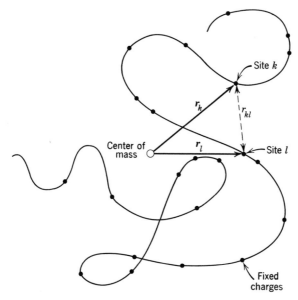

Fig. 26-7. Model for calculation of the electrostatic free energy of a flexible polyelectrolyte ion.

continuous distribution must be quite small. In any event, the Hermans-Overbeek equation fails to provide more than qualitative agreement with experimental behavior (Table 27-2).

Wall and Berkowitz[20] have shown that another cause for failure of the Hermans-Overbeek treatment is that the linearization assumption of the Debye-Hückel theory, i.e., neglect of the higher terms in the expansion of equation 26-13, cannot be valid for their model. Whether this same criticism can be applied to the discrete charge model discussed below is at present unknown.

A considerably more rigorous treatment of the electrostatic problem of polyelectrolyte ions has been given by Harris and Rice,[21] and we shall present here a modification of their procedure, applicable to typical polyions at infinite dilution.

We consider a polyelectrolyte ion at infinite dilution, *in a particular spatial configuration*, such as the one illustrated by Fig. 26-7. The configuration chosen may be described by a set of Z fixed charges, each of charge q. As before q will be a variable during the charging process used to compute W_{el}. In the fully charged polyion each of the fixed charges q is equal to $+\epsilon$, the protonic charge, for a cationic polymer and equal to $-\epsilon$ for an anionic one. No assumption need be made at first as to how these charges are joined to one another; i.e., the calculation will apply in principle to cross-linked polyelectrolytes and to solvent-permeated proteins, as well as to linear polyelectrolytes. However, we shall assume that the polymer segments occupy negligible volume and that the dielectric constant in the domain of the polymeric ion is that of the solvent, an assumption which in general will apply only to linear polyelectrolytes. (This assumption is not made in the original papers of Harris and Rice[21] so that their treatment is capable of more general applications than that given here.)

Since we are confining ourselves to the treatment of a single polyion, we have only a single region of space to consider, exclusive of the singularities at the locations of charges. Poisson's equation applies, which, with the same assumptions used in the Debye-Hückel theory, may again be written in the form of the Poisson-Boltzmann equation, equation 26–15, with κ given by equation 26–16 or 26–32.

Since there are no boundaries to consider we may simply use the solution to the Poisson-Boltzmann equation given by equation 26–47. Since ψ must tend to zero at a large distance from the polyion, where all $|\mathbf{r} - \mathbf{r}_k|$ tend to infinity, each $\mathscr{B}_k$ must be set equal to zero. At each singularity ψ must approach the value given by equation 26–8 and this condition requires that each $\mathscr{A}_k$ be set equal to q. The final result is then simply

$$\psi = \sum_{k=1}^{Z} \frac{q}{D|\mathbf{r} - \mathbf{r}_k|} e^{-\kappa|\mathbf{r}-\mathbf{r}_k|}$$

At any site l, the distance $|\mathbf{r} - \mathbf{r}_k|$ is just the distance r_{kl} between sites k and l (Fig. 26–7). The work done in placing a charge $\pm\epsilon$ at each site, $l = 1, 2, \cdots, Z$, is therefore (equation 26–2) given by

$$W_{\text{el}} = \sum_{l=1}^{Z} \int_{q=0}^{\pm\epsilon} q \, dq \sum_{\substack{k=1 \\ k \neq l}}^{Z} \frac{e^{-\kappa r_{kl}}}{D r_{kl}} = \frac{\epsilon^2}{2D} \sum_{k=1}^{Z} \sum_{\substack{l=1 \\ k \neq l}}^{Z} \frac{e^{-\kappa r_{kl}}}{r_{kl}} \qquad (26\text{–}57)$$

Just as in the case of equation 26–52, we have omitted the self-energies of the individual charges, as given by the terms of equation 26–57 with $k = l$.

Equation 26–57 can now be applied to any desired *spatial* configuration of a polyion by specification of the vector coordinates $\mathbf{r}_k$ of its charges, from which, as shown by Fig. 26–7, the r_{kl} at once follow. Each given spatial configuration may, as in the case of proteins (section 26f), itself exist in numerous subconfigurations arising from the fact that alternative positions for the Z charges may exist. Equation 26–57 may be written in a form similar to equation 26–52 for this purpose; i.e., we can sum over all sites which *can* bear charges, and we write

$$W_{el} = \frac{\epsilon^2}{2D} \sum_{\substack{k=1 \\ k \neq l}}^{p} \sum_{l=1}^{p} \zeta_k \zeta_l \frac{e^{-\kappa r_{kl}}}{r_{kl}} \qquad (26\text{–}58)$$

where the summation now extends over all sites, of number p, ζ_k or ζ_l being unity for a charged site and zero for an uncharged one. Again W_{el} as given by equation 26–58 must be averaged over all subconfigurations to give the average value for a given spatial configuration.

It will appear in section 27c that some charged sites of polyelectrolytes may enter into specific combination with mobile ions of opposite charge. If this occurs, the site in question must be regarded as an uncharged site in equation 26–57 or 26–58.

It is instructive to compare equation 26–58 with the corresponding equations for impenetrable proteins, equations 26–52 and 26–54. At first sight the latter appear to be more complicated in view of the complexity of the functions A_{kl}–B_{kl}. In actual fact, however, A_{kl}–B_{kl} is not much harder to calculate than the corresponding exponential function in equation 26–58, and the equation for proteins thus actually becomes easier to apply, first because the number of charges is generally much smaller and second because the locations of the potentially charged sites is fixed. Thus after the sum of equation 26–52 or 26–54 is evaluated and averaged over all possible ways of distributing the charges over the available sites, the calculation of W_{el} is complete. When the same steps have been completed for polyelectrolytes, however, we have evaluated W_{el} for only one of many possible spatial configurations, and a further averaging process is required to obtain the average value of W_{el} for all polyelectrolyte ions of a given charge. Finally, it should be pointed out that the use of the expression $e^{-\kappa r_{kl}}/r_{kl}$ in equation 26–58 is an approximation which neglects the fact that polyelectrolyte ions, just like proteins, are composed of material of low dielectric constant. This is justified for many calculations since most of the pairs of interacting charges are separated by large distances, the intervening material being largely solvent. Suppose, however, that we are specifically interested in the interaction between charges on adjacent monomer segments of the polyion. Then this neglect of the regions of low dielectric constant is no longer justified, and we would have to use, as in the case of proteins, a more complicated function than the simple $e^{-\kappa r_{kl}}/r_{kl}$ to describe the interaction between the charges.

We shall make no attempt here to calculate an exact value for W_{el} but confine ourselves to a simple calculation which will enable us to obtain later an estimate of the average polyelectrolyte ion's configuration. Following Harris and Rice[21] we represent the polymer chain in terms of an equivalent "statistical" chain of σ_K elements (Kuhn chain) as discussed in section 9g. The number of elements is smaller than the number of monomer segments, and the length l_K of each element is accordingly many times that

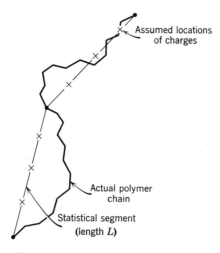

Fig. 26–8. Two adjacent statistical segments of a chain with three charges per segment. (In the example chosen each statistical segment replaces twelve monomer units of the actual polymer chain.)

of a monomer segment. The equivalent statistical chain is completely unrestricted (in the absence of charges) as to the angle between adjacent elements. This condition enables us, from a measurement or estimate of the mean square end-to-end distance, $\overline{h^2}$, for the *uncharged* polymer, to determine both the number and length of statistical elements, using the relation $\overline{h^2} = \sigma_K l_K{}^2$ and equation 9–18. Intramolecular interaction of the kind discussed in section 9h is neglected.

We consider next the charged form of the same polymer, describing it in terms of the same σ_K statistical elements of the same length l_K, and we avoid the problem due to the presence of numerous subconfigurations by assigning reasonable *average* positions to the polyion's charges. We suppose that each statistical element has the same number, z, of charges, such that $\sigma_K z = Z$, and, further, we suppose that they are placed at equal intervals along the straight line representing the element, as shown in Fig. 26–8. (This is clearly a crude approximation. It is, however, a very much less serious approximation than to introduce a *continuous* distribution of charge.)

The macro-ion is now completely specified for application of equation 26–57, except for the angles γ (Fig. 26–8) between successive elements, which, in this model, determine the spatial configuration. (It will be recalled that equation 26–57 was designed to apply only to a particular spatial configuration.)

Anticipating the detailed calculation to be described in section 27b we see at once that the presence of charges will result in a configurational change, for, whereas σ_K and l_K were chosen so as to give each possible configuration the same energy in the uncharged state, W_{el} in the charged state is clearly a function of the angles γ, so that different configurations will have different energies.

Proceeding now with the calculation of W_{el} for a particular configuration, using equation 26–57, we note that this equation is a sum representing the interaction of each charged site with every other charged site. We note further that the term $e^{-\kappa r_{kl}}/r_{kl}$ decreases rapidly as r_{kl} increases. Accordingly, and especially in view of the simplifying approximations already made, it is necessary to include in the sum only those terms for which r_{kl} is relatively small. Rice and Harris include, for each charge k, its interaction terms with other charges on the same statistical element and with those on the next nearest neighboring element.

Computing first all interactions between pairs of charges on the *same* element, we note that there will be $z - 1$ pairs for which $r_{kl} = l/z, z - 2$ terms for which $r_{kl} = 2l/z$, etc. Furthermore, each pair appears in the sum twice (once with k and l reversed), so that the sum of these terms, for any one element, is

$$(W_{el})_1 = \frac{\epsilon^2}{Dl} \sum_{j=1}^{z-1} \frac{z(z-j)}{j} e^{-j\kappa l/z} \qquad (26\text{–}59)$$

Equation 26–59 is, of course, independent of configuration.

It might be thought a poor approximation to choose a model in which the terms making the largest contribution to W_{el}, i.e., the terms involving the charges which are closest together, are independent of the spatial configuration. But this actually reflects the true state of the polyion. For instance, for a fully charged polyion bearing charges on each monomer segment, r_{kl} for nearest neighbor segments is fixed by bond distances and angles and that for several next nearest neighbors is also allowed little variation, especially if rotation is sterically hindered. For polyions of relatively low charge, of course, the situation is different, for the charges are likely to be relatively far apart. But this will again be reflected in the model in that there may then be only one or two charges per statistical element, with a resulting diminution of the configuration-independent contribution to W_{el}. (In that case, however, interaction between more than nearest-neighbor elements should probably be taken into account.)

To calculate the interaction between charges on neighboring statistical elements, Rice and Harris assume that the average contribution to W_{el} can be obtained with sufficient accuracy by placing all z charges of an element at its center. Thus r_{kl} becomes $l \cos (\gamma/2)$ for each pair of charges. Since z^2 pairs of charges are involved in the interaction between each pair of adjacent elements and each pair occurs twice in equation 26–57, we get for the contribution of any pair of adjacent elements to W_{el},

$$U(\gamma) = \frac{z^2 \epsilon^2}{Dl} \frac{e^{-\kappa l \cos (\gamma/2)}}{\cos (\gamma/2)} \qquad (26\text{–}60)$$

It should be noted that there are $\sigma_K - 1$ independent angles γ which must be specified to define a given configuration. There will therefore also be $\sigma_K - 1$ independent parameters $U(\gamma)$. If γ_s $(s = 1, 2, \cdots, \sigma_K - 1)$ represents any given angle γ, the over-all electrostatic free energy becomes

$$W_{el} = \frac{\sigma_K \epsilon^2}{Dl} \sum_{j=1}^{z-1} \frac{z(z-j)}{j} e^{-j\kappa l/z} + \sum_{s=1}^{\sigma_K-1} U(\gamma_s) \qquad (26\text{–}61)$$

The derivation of these equations is meaningful only when z is integral. However, if $z = Z/\sigma_K$ is non-integral this merely means that some elements must be assigned the next higher integral value of z and others the next lower one. The non-integral average value of z may be used in equation 26–60, while the sum of equation 26–59 may be given a value interpolated between the values obtained for the higher and lower integral values.

Rice and Harris[22] have extended their calculations to polyampholytes, i.e., polyelectrolytes with both positive and negative charges, but only to the special case in which potentially positive and negative sites are present in equimolar ratio and arranged alternately along the polymer chain or arranged randomly.

26j. Polyelectrolytes in the absence of added salt. We have already noted in section 26d that the simple theoretical treatment which involves only a single macro-ion fails if we try to apply it to protein ions in the absence of a low-molecular-weight electrolyte. The same conclusion applies to polyelectrolyte solutions; as in the case of protein solutions it becomes necessary in calculating the electrostatic potential and free energy to consider the forces between and spatial distribution of the polyions themselves.

Theoretical studies on this subject have proceeded as follows. It has been assumed that highly charged polyelectrolyte ions at very low ionic strength tend to approach a fully extended, rodlike shape, so that they can be treated as sections of cylindrical rods (cf. section 26f). Because such rods would be extremely long, they would, even at the lowest feasible concentration, interfere with one another's freedom of motion. The solution would tend to become highly ordered, with the rods arranged parallel to one another.[23] We can take this into account in solving the electrostatic problem by choosing an average distance r_1 between adjacent rods, setting up Poisson's equation with cylindrical coordinates about a particular rod, and then, instead of imposing the condition that ψ must vanish at infinity, imposing the condition that ψ must pass through a minimum half-way between adjacent rods. In other words, we let $d\psi/dr$ become zero at a distance $r = r_1/2$ from the center of each rod.

The problem just outlined can be solved without the necessity of expanding the exponential $\exp(\pm\epsilon\psi/kT)$, and solutions have been obtained by Fuoss, Katchalsky, and Lifson[24] and by Alfrey, Berg, and Morawetz.[25] The solutions, however, are based on a continuous distribution of charge and, as we have seen, will therefore probably not yield correct values of W_{el}. For this reason we shall not present them here. (The problem was solved primarily to obtain an idea of the distribution of counterions about the polyions. For this purpose, as we have seen, the assumption of a continuous distribution of charge is much less apt to be serious.)

The discrete charge model of Harris and Rice could be adapted to a rodlike configuration as easily as to any other. However, it has not been solved for boundary conditions suitable to the problem.

27. FLEXIBLE LINEAR POLYELECTROLYTES[4,5]

27a. Expansion of flexible polyelectrolytes. As we have indicated repeatedly, an uncharged flexible polymer molecule may, by means of rotation about its single bonds, take on any configuration compatible with its fixed bond lengths and angles and any other steric restrictions which may exist. Furthermore, a given molecule may be expected, in the course of time, to take on different configurations covering the entire range of possible configurations.

In sections 9 and 10 were calculated some properties to be expected for such polymers, based on the assumption that all possible configurations have equal free energy, aside from statistical multiplicity. In particular, it was shown that it would be reasonable to expect the average density of segments about the center of mass to be roughly spherically symmetrical and, moreover, that the segments would fill only a fraction of the order of 0.01 of the space in their immediate vicinity. Experimental data, especially those of Chapters 5 and 6, were seen to confirm this conclusion.

When such a polymer is electrically charged, i.e., when it becomes a flexible polyelectrolyte, nothing is done to interfere with the *ability* of each molecule to take on all possible configurations. However, the free energy of each configuration will no longer be the same, for the repulsion between the molecule's charges will give a relatively high free energy to compact configurations and a relatively low free energy to expanded ones. (We are considering here only polyelectrolytes with just one kind of charge. Polyampholytes will be taken up in section 27e.)

TABLE 27–1. Electrostatic Free Energy of a Hypothetical
Polyelectrolyte with $Z = 1000$

Ionic Strength (Molar)	$R_e = 100$ Å	W_{el}, cal/mole 150 Å	300 Å	1000 Å
0.001	1.12×10^7	5.6×10^6	1.25×10^6	5.0×10^4
0.01	3.43×10^6	1.23×10^6	1.80×10^5	5600
0.05	9.4×10^5	3.00×10^5	4.05×10^4	1150

In Table 27–1, for instance, are shown some calculations, using the Hermans-Overbeek equation, equation 26–56, for an hypothetical polymer of degree of polymerization 2000, with half its segments charged (i.e., $Z = 1000$). Such a polymer in its *uncharged* state might be expected to have an average radius of gyration, $R_G \simeq 150$ Å (the exact value depending, of course, on the nature of the polymer), but Table 27–1 shows that the presence of the charges leads to prohibitively high free energies for dimensions of this order of magnitude. At relatively high ionic strength W_{el} becomes considerably reduced by moderate expansion, and the average configuration may be expected to continue to have spherical symmetry, but with an increased radius of gyration. At low ionic strength, however, considerably greater expansion is required. The Hermans-Overbeek model, used to compute these data, is, of course, an oversimplification, and the figures given should not be assigned more than semi-quantitative significance. However, any other theory predicts essentially the same result, that a polyelectrolyte ion should be more expanded than the uncharged polymer molecule from which it is derived, and all experimental data confirm the prediction.

It is clear that the polymer will have its lowest possible electrostatic free energy when fully stretched. This follows at once from equation 26–57. The values of r_{kl} in this equation, for any pair of charges, attain their maxima, and W_{el} therefore its minimum, when the polyelectrolyte ion is stretched to its full length. This configuration, of course, has a low statistical probability, and it is inconceivable that all the ions in a solution will ever be in this configuration. Nevertheless, at very low ionic strength, where moderate expansion does not suffice to provide a large decrease in W_{el} there should be a *tendency* in this direction, reflected in an over-all shape resembling a long rod rather than a sphere.

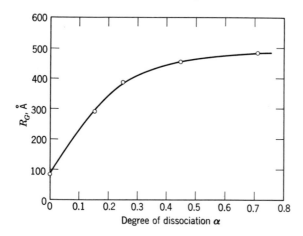

Fig. 27-1. The radius of gyration of a sample of polymethacrylic acid as a function of the degree of dissociation of the carboxyl groups. Data were determined by light scattering from solutions containing no added salts. (Data of Oth and Doty.[26])

We have pointed out that the greatest degree of expansion should occur at very low ionic strength, and the most striking experimental evidence for it may thus be obtained from experiments in the absence of added electrolyte.

Figure 27-1, for instance, shows the effect of charge on the radius of gyration of polymethacrylic acid.[26] This is a polymer whose repeating unit is a weak electrolyte,

$$\begin{array}{c} COOH \\ | \\ -C-CH_2- \\ | \\ CH_3 \end{array}$$

so that the charge per polyelectrolyte ion may be adjusted at will by the addition of NaOH or other suitable base. The R_G values of Fig. 27-1 were obtained from light scattering dissymmetries (equation 18–26), and clearly show that the variation with charge is in accord with expectation.

Since a different amount of NaOH has to be added to achieve each different value of Z and no other electrolyte was added, the ionic strength is not constant over the range of charge considered.

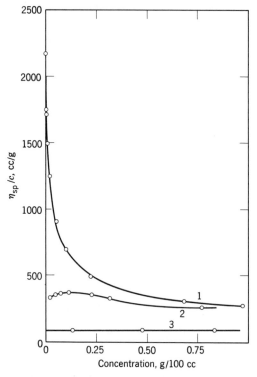

Fig. 27–2. The viscosity of polyvinylbutylpyridinium bromide, dissolved in water (curve 1), $0.001M$ KBr (curve 2), and $0.0335M$ KBr (curve 3). The ionic strength is not constant during the dilution process because the concentration of polymer ions and their accompanying bromide ions decreases without a corresponding increase in the KBr concentration. (Fuoss.[27])

The effect of ionic strength is illustrated by Fig. 27–2, which shows viscosity data[27] for polyvinylbutylpyridinium bromide, which is a strong electrolyte, with the repeating unit

$$C_5H_4N(C_4H_9)^+$$
$$|$$
$$-CH-CH_2-$$

which cannot be converted to an uncharged form. The data of Fig. 27–2 show that the intrinsic viscosity, i.e., the limiting value of η_{sp}/c at zero concentration, decreases sharply with increasing ionic strength, signifying a corresponding decrease in size, again as expected.

An interesting feature of Fig. 27–2 is the anomalous shape of the curve of η_{sp}/c versus concentration in the experiment without added salt. As was pointed out in section 23, η_{sp}/c ordinarily *rises* with increasing concentration. The reason for the anomaly here observed is not hard to find. For, in the absence of added salt, the only mobile ions present are the Z counterions (Br^- in this case) accompanying each macro-ion. The concentration of these counterions, and therefore the ionic strength, falls as the polyelectrolyte concentration is decreased, causing further expansion. Another contributing factor may be that because the polyelectrolyte ions are so greatly expanded in the absence of any added salt they can be expected to interfere with one another, as indicated in section 26j, so that there may be an influence of concentration on configuration. In any event, the ordinary viscosity plot clearly does not permit in this case a meaningful extrapolation of η_{sp}/c to zero concentration. Fuoss and Strauss[27] found that the curve in Fig. 27–2 could be described by an equation of the form

$$\frac{\eta_{sp}}{c} = \frac{A}{1 + B\sqrt{c}} \qquad (27\text{–}1)$$

This equation is purely empirical but holds to the lowest concentration reached, as Fig. 27–3 shows. In the absence of contrary information it is assumed to hold to the experimentally unattainable limit of $c = 0$. In this case A is clearly the limiting value of η_{sp}/c; i.e., $A = [\eta]$. To evaluate A, we plot c/η_{sp} versus $c^{1/2}$, as shown in Fig. 27–3, the intercept at $c^{1/2} = 0$ being $1/[\eta]$.

Equation 27–1, sometimes slightly modified by a small additive term, has been found applicable to viscosity data obtained with a variety of polyelectrolytes. Of particular interest is the work of Strauss and coworkers[40] on sodium polyphosphate. The charges on this polyion are very close to one another (see p. 13), and electrostatic interaction therefore particularly pronounced. Strauss and coworkers obtained intrinsic viscosities in the complete absence of salt, for samples of different molecular weight, by use of equation 27–1. The result, shown in Fig. 27–8, indicates that $[\eta]$ varies approximately as $M^{1.9}$. From the discussion on p. 408 this clearly demonstrates that the polyphosphate ion, in the complete absence of salt, behaves as a rod-shaped rather than a coiled particle. It is presumably close to a fully extended chain, which, as pointed out above, would be the configuration of lowest electrostatic free energy. As will be brought out later, the configuration reverts to that of a flexible coil in the presence of salt.

In a similar experiment (using, however, only two samples differing in molecular weight) Oth and Doty[26] showed that for polymethacrylic acid, in the absence of salt, $[\eta]$ varies as $M^{0.82}$ when the degree of dissociation is 0.001 but as $M^{1.87}$ when the degree of dissociation is 0.7.

The simplest procedure for avoiding the decrease in ionic strength which accompanies dilution in the experiments just noted is to dilute the polyelectrolyte not with pure water but with a salt solution of the same ionic

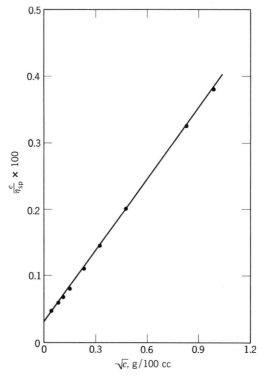

Fig. 27-3. The data of curve 1 of Fig. 27-2, plotted according to equation 27-1. (Fuoss.[27])

strength as the most concentrated polyelectrolyte solution. To do this we must decide how the polyion enters into the expression for the ionic strength. The usual procedure is to assume that the polyelectrolyte charges act as univalent ions, which is reasonable in view of the fact that the individual charges are able to move in the solution more or less independently of other charges. (The same assumption is often made in protein solutions, where, as stated in section 26d, it is less reasonable.) With this assumption the dilution procedure requires merely that the total concentration of mobile counterions be kept constant. The procedure is often called *isoionic dilution*.

Experimentally,[28,29] it is generally found that better data, i.e., more nearly linearly rising plots of η_{sp}/c versus c, are obtained if it is assumed that the concentration of mobile counterions is smaller than the total counterion concentration. This is equivalent to saying that some counterions are fixed to charged sites on the polyelectrolyte. It will be shown in section 27c that this is indeed so. The fraction of counterions which is mobile is determined empirically by trial and error until linear viscosity plots are obtained. The results[28] are in good agreement with the more direct measurements of counterion association to be described in section 27c.

The use of isoionic dilution of course prevents us from reaching the lowest ionic strengths in the examination of polyelectrolytes, and, as a result, the degree of expansion is reduced.

An example of the use of constant ionic strength dilution is shown by the data of Pals and Hermans[29] in Fig. 27-4 on solutions of sodium carboxymethylcellulose. This is a polyelectrolyte obtained from cellulose by substituting $—CH_2COOH$ for one or more of the hydroxyl hydrogen atoms of each glucose unit. Figure 27-4 shows that normal curves of η_{sp}/c versus c are now obtained, which can be extrapolated without difficulty to evaluate intrinsic viscosities. A plot of $[\eta]$ versus ionic strength at constant charge is shown in Fig. 27-5; it shows the sharp change at low ionic strength which was seen less well defined in Fig. 27-2.

We have demonstrated the expansion of polyelectrolytes on the basis of viscosity and light scattering measurements. Similar conclusions could, of course, have been reached from the measurement of other physical properties sensitive to configuration, such as sedimentation rate and flow birefringence.[30]

27b. Calculation of the extent of expansion. A calculation of the extent of expansion may be made from a knowledge of W_{el} as a function of configuration, by assigning to each possible configuration a probability proportional to $\exp(-W_{el}/kT)$. Such a calculation was made by Rice and Harris[21] for sodium carboxymethylcellulose and compared with experimental data obtained by Schneider and Doty.[31]

The sample used by these authors had a weight-average molecular weight of 435,000. The properties for the polyelectrolyte in its hypothetical uncharged state were obtained by extrapolation of light scattering data to infinite ionic strength, where all charge effects should be suppressed (see, however, section 27d), and the polyelectrolyte molecules should behave as if uncharged. Indeed the dimensions found under these conditions are very much like those of uncharged derivatives of cellulose. It was concluded that the molecule in an uncharged state could be represented

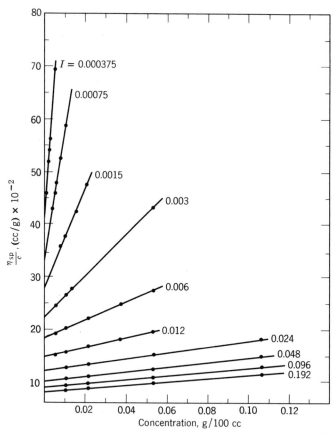

Fig. 27–4. Viscosity data for carboxymethylcellulose, obtained by dilution at constant ionic strength. (Pals and Hermans.[29])

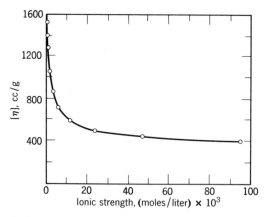

Fig. 27–5. The intrinsic viscosity of carboxymethylcellulose as a function of ionic strength. The data plotted are the intercepts from Fig. 27–4. (Pals and Hermans.[29])

by an equivalent statistical unrestricted chain (section 9g) of, on the average, 39.5 segments each of length 335 Å. (In actual fact the carboxymethyl-cellulose used was highly heterogeneous, so that all the experimental data represent appropriate averages. For simplicity, we shall neglect all discussion of the averaging procedures, presenting the data as if the sample had been homogeneous with respect to molecular weight.) If all configurations of this molecule have identical energy, as is true of the hypothetical uncharged state, then the radius of gyration, as given by equations 9–17 and 9–38, is 860 Å.

The total number of fixed charges per molecule was determined by titration to be about 2000. There are therefore about fifty charges per statistical segment.

W_{el} for any configuration can now be evaluated by equation 26–61. It will be noted that the first term of this equation does not depend on configuration and therefore does not enter into the present calculation. The second term of equation 26–61 is configuration dependent, but it is seen to be a sum of *independent* contributions, depending on the position of each statistical segment relative to the one preceding it, as defined by the successive angles γ of Fig. 26–8.

To obtain the *average* configuration of the polyelectrolyte ion it is necessary to know only the *average* value of each angle γ. Since they are independent of one another and since, by the limitations of the model, the electrostatic forces between each pair of statistical elements is the same, each of the $\sigma_K - 1$ angles will have the same average value.

The uncharged polymer was described in terms of an equivalent chain completely unrestricted as to the bond angle between the statistical chain segments. The average value of γ is 90°, the average of $\cos \gamma$ is zero, and the average end-to-end distance is given by $\overline{h_0}^2 = \sigma_K l_K^2$. In the charged polyion smaller values of γ result in a lower value of $U(\gamma)$ (equation 26–61) than do large values of γ. Thus $\overline{h^2} > \overline{h_0}^2$ in the same way as the polymethylene chain of section 9d is more extended than the unrestricted chain discussed in section 9c. Analogous to equation 9–13 we get

$$\overline{h^2} = \sigma_K l_K^2 \frac{1 + \overline{\cos \gamma}}{1 - \overline{\cos \gamma}} \tag{27-2}$$

where $\overline{\cos \gamma}$ is the average value of $\cos \gamma$ in the charged polyion, given by

$$\overline{\cos \gamma} = \frac{\displaystyle\int_{\gamma=0}^{\pi} \cos \gamma e^{-U(\gamma)/kT} \sin \gamma \, d\gamma}{\displaystyle\int_{\gamma=0}^{\pi} e^{-U(\gamma)/kT} \sin \gamma \, d\gamma} \tag{27-3}$$

Where $(R_G)_0$ and R_G represent, respectively, the radii of gyration in the uncharged and charged forms,

$$\frac{\overline{h^2}}{\overline{h_0^2}} = \frac{R_G^2}{(R_G)_0^2} = \frac{1 + \overline{\cos \gamma}}{1 - \overline{\cos \gamma}} \tag{27-4}$$

The integrals occurring in equation 27–3 were evaluated numerically by Rice and Harris, and the resulting calculations by equation 27–4 are shown in Table 27–2. They show very good agreement with the experimental data of Schneider and Doty. It is significant that only poor

TABLE 27-2. The Expansion of Sodium Carboxymethylcellulose

$$R_G/(R_G)_0$$

Ionic Strength (Molar)	Exptl.[b]	Calcd. by Rice and Harris			Calcd. by Hermans and Overbeek
		No Binding	33% Binding[a]	67% Binding[a]	
0.50	1.09	1.07	1.07	1.06	1.41
0.05	1.16	1.13	1.11	1.09	1.81
0.01	1.39	1.38	1.30	1.19	2.45
0.005	1.59	1.72	1.57	1.31	2.87

[a] Cf. section 27c for a discussion of counterion binding.
[b] The experimental values of R_G are 940, 1000, 1200, and 1360 Å at the four ionic strengths given, as compared with 860 Å for the uncharged polymer.

quantitative agreement with experiment is obtained if W_{el} is computed by the Hermans-Overbeek theory, as the table shows. Even poorer agreement results if the calculations are based on a rodlike model of the kind suggested in section 26j.

It should be noted that the success of the preceding calculation is due in part to the fact that carboxymethylcellulose is, even in the absence of charge, a relatively extended molecule, as indicated by the considerable length (335 Å) of the elements of the corresponding Kuhn chain. A consequence of this great length of the statistical units is that the Rice-Harris calculation, which neglects all interactions other than between charges on nearest-neighboring statistical segments, takes into account the interaction of each charge with a very considerable number of other charges, crude though the assumptions concerning the location of these charges may be. Rice and Harris[21] attempted the same sort of calculation for Oth and Doty's data on polymethacrylic acid[26] and were considerably less successful. The reason is that polymethacrylic acid is a highly flexible polymer (length of Kuhn statistical element only 10 Å), so that the dependence of W_{el} on configuration can not be computed merely from the interactions between nearest-neighbor elements. To obtain good agreement in this case it would presumably be necessary to evaluate the sum of equation 26–58 in greater detail than was actually done.

27c. The distribution of counterions in polyelectrolyte solutions. By combining equation 26–14 with equation 26–27 we can obtain an expression for the average density of mobile charges near a small impenetrable ion of charge $q = Z\epsilon$. By integration we can obtain the total charge q' within

TABLE 27–3. Fraction of Total Counterions $(q'/-\epsilon Z)$ within Distance of $1/\kappa$ and $2/\kappa$ from Central Ion Surface

	$r' = a + 1/\kappa$	$r' = a + 2/\kappa$
$\kappa a \ll 1$	0.26	0.59
$\kappa a = 1$	0.44	0.73
$\kappa a \gg 1$	0.63	0.86

the volume between $r = a$ (the distance of closest approach to the central ion) and any desired value $r = r'$. The result of such a calculation is given by equation 27–5,

$$q' = \int_{r=a}^{r'} 4\pi r^2 \rho \, dr = -\epsilon Z \left[1 - \frac{1 + \kappa r'}{1 + \kappa a} e^{-\kappa(r'-a)} \right] \quad (27\text{–}5)$$

We see at once that $q' = -\epsilon Z$ if $r' = \infty$, a necessary consequence of the fact that the total solution must be electrically neutral.

Of greater interest is the calculation of the charge within much smaller distances of the central ion. For instance, if we set $r' = a + 1/\kappa$,

$$q' = -\epsilon Z \left(1 - \frac{2 + \kappa a}{1 + \kappa a} e^{-1} \right) \quad (27\text{–}6)$$

Some numerical calculations based on equation 27–6, and a similar calculation for $r' = a + 2/\kappa$, are shown in Table 27–3. The result depends somewhat on the value of κa, but it is clear that a majority of the counterions will always be found within a distance $2/\kappa$ from the central ion.

This calculation is of interest because, as we saw in section 26i (equation 26–57), the potential in the domain of a polyelectrolyte ion is just the sum of the contributions arising from the individual charges of the ion. By equation 26–14 the same must be true of the density of mobile charges.

Thus we may expect the results of Table 27–3 to be applicable to the domain of a polyelectrolyte ion as well as to the neighborhood of a small impenetrable ion.

Table 27–4 shows some values of κ and $1/\kappa$ at 25° C, and it is seen that at relatively high ionic strength (already at $I = 0.001$) $1/\kappa$ becomes considerably smaller than the typical dimensions of a polyelectrolyte ion. Under these conditions most of the counterions required to neutralize most of the charges of the polyion must be within the domain of the polyion, and the "equivalent sphere" corresponding to the polyion and including the solvent within a distance of the order of R_G from the center, must have a much smaller charge than that due to the fixed charges of the polyion alone.

TABLE 27–4. Values of the Debye Parameter κ

Ionic Strength (Molar)	κ, cm^{-1}	$1/\kappa$, Å
0.0001	3.28×10^5	295
0.001	1.04×10^6	96.5
0.01	3.28×10^6	29.5
0.05	7.35×10^6	13.6
0.10	1.04×10^7	9.7

Even at lower ionic strength, where $1/\kappa > R_G$, the effect of a finite polyion concentration may produce the same result. Polyelectrolyte ions being in their most expanded state at low ionic strength, their domains may overlap already at quite low concentrations (0.001 gram/cc, for instance). If they do, the requirement for over-all electrical neutrality clearly implies that the domain of each polyion must be close to electrically neutral.

The conclusions just reached, that macro-ions ordinarily carry with them a major portion of the counterions needed to neutralize them, would presumably not be true for a solution of an isolated macro-ion and its counterions in the absence of any added electrolyte, for under these conditions $1/\kappa$ would tend to infinity, while R_G of course cannot increase indefinitely. However, a concentration low enough for this situation to occur is ordinarily too low for experimental measurements, and experimental data at finite concentrations generally fully support the contention that the domain of the polyion is close to electrically neutral.

The simplest and most conclusive experiments supporting this point of view come from study of the interaction *between* polyelectrolyte ions in solution. It was shown in Chapter 4 that the second virial coefficient, B, as determined by either light scattering or osmotic pressure, is a measure of such interaction. Specifically, for an *uncharged* macromolecule B depends only on the *volume* of the molecule and on the nature of solvent-segment interaction. It can be calculated from equation 12–9. For macro-ions, on the other hand, B contains the same factor and in addition a term representing electrostatic repulsion between the ions as given approximately by equation 14–24.

Schneider and Doty[31] made a very thorough study of carboxymethyl-cellulose ($\bar{M}_w = 435,000$; $Z \simeq 2000$) as a function of ionic strength, determining both radii of gyration (given in Table 27–2) and B from light scattering measurements. For comparison with the experimental values,

TABLE 27–5. Second Virial Coefficient for Sodium
Carboxymethylcellulose

Ionic Strength (Molar)	$B \times 10^3$		
	Exptl.	Equation 27–7	$1000Z^2/4M^2m_3$
0.500	1.45	1.25	1.01
0.050	5.70	3.00	10.1
0.010	15.6	9.25	51.0
0.005	26.0	25.0	101.0

B can now be calculated, on the basis of the assumption that the domain of the macro-ion is essentially neutral, by combining equations 12–9 and 12–17, which lead to

$$B = \frac{16\pi\gamma^3 \mathcal{N} R_G{}^3}{3M^2} \qquad (27\text{–}7)$$

The interaction parameter γ^3 which occurs in equation 27–7 can be obtained by comparing B, as determined from extrapolation of light scattering data to infinite ionic strength, with the value of R_G under the same conditions (cited in Table 27–2). The value of γ^3 so obtained is essentially the same as that of uncharged cellulose derivatives, as required by the assumption inherent in this procedure that the polyelectrolyte ion at infinite ionic strength behaves like the corresponding uncharged polymer, retaining none of the effects due to the presence of its charges.

Values of B so calculated are shown in Table 27–5, and they are seen to be in reasonably good agreement with the observed values. From this it can be concluded that, in its interaction with other polyions, the polyion behaves as if it were essentially uncharged.

Table 27–5 also shows values for the Donnan term $1000Z^2/4M^2m_3$ (equation 14–24), which is the principal contribution to B for a polyion with a net charge Z. This term would have to be added to the excluded volume term if the effective charge were equal to Z, and the experimental data clearly allow no room for this. As a matter of fact the Donnan term *alone* is much larger than B at all but the highest ionic strength used.

Another type of experiment confirming that a large fraction of the total counterion concentration is within the polyion domain is a simple transference experiment performed by Huizenga, Grieger, and Wall.[32] A long horizontal tube, divided in the middle by a sintered glass disc, was filled

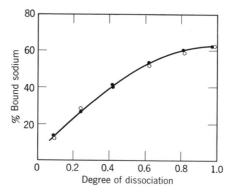

Fig. 27–6. Extent of association between Na^+ ions and the negative charges of polyacrylic acid, as a function of the degree of dissociation of the latter. Open circles represent diffusion experiments; solid circles represent transference measurements. All data were obtained in the absence of added salt. (Huizenga, Grieger, and Wall.[32])

with a dilute solution of polyacrylic acid ($\bar{M}_w \simeq 200,000$). The molecules had on the average 3000 carboxyl groups, various fractions of which were ionized by addition of appropriate amounts of NaOH. The solutions used therefore contained negative polyacrylate ions and Na^+ counterions. If these ions were independent, then, when current passes through the solution, it would be carried (regardless of the electrode reaction) in part by Na^+ ions traveling towards the cathode and in part by polyions traveling towards the anode. The anode side of the sintered glass disc would therefore *accumulate* polymer and be *depleted* of Na^+, the relative amounts of change being determined by the relative mobilities of the ions. Experimentally this result was not observed. When the polymer was highly charged, the amount of both polymer and Na^+ in the cathode compartment *increased*. This result can be accounted for only if we suppose that the polyions carry with them a considerable amount of Na^+, the actual quantities calculated from the data being shown in Fig. 27–6.

Yet another way of showing the same phenomenon is by measurement of the self-diffusion constant of Na^+ in solutions containing sodium polyacrylate. The method is similar to that discussed in section 20a for the determination of water bound to proteins. Results obtained by this method[32] agree with those measured by transference, as Fig. 27–6 shows.

One further question arises upon careful consideration of the above results; it concerns the possibility of chemical combination between specific sites of the polyion (in this case —COO⁻ groups) and the counter-ions. The affinity between free ions such as CH_3COO^- and Na^+ is ordinarily very small, but nevertheless some affinity exists as shown by formation of

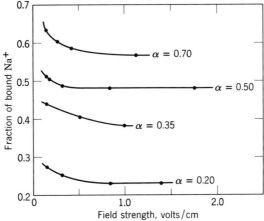

Fig. 27–7. The effect of varying electric field strength on the transference experiments of Fig. 27–6. (Wall, Terayama, and Techakumpuch.[34])

"ion pairs" in non-aqueous solutions and by observed deviations from ideality of salt solutions at high concentrations. (For instance, Harned and Owen[33] have estimated that the equilibrium constants $C_{KNO_3}/C_{K^+}C_{NO_3^-}$ and $C_{CaAc^+}/C_{Ca^{++}}C_{Ac^-}$, in water with concentrations in molar units, are of the order of unity.) In polyelectrolytes this weak attraction would be aided by the electrostatic field,[16] with the result that a considerable fraction of the counterions might be bound to specific sites.

The observations concerning the second virial coefficient can give no information on this point, but those involving the mobility under the influence of an electric field can do so, for the extent to which mobile ions are carried along with a central ion depends on the force exerted, i.e., on the field strength. At high field strength the central ion is torn away from its surrounding ion atmosphere, whereas at low field strength the Boltzmann distribution is maintained. (This phenomenon gives rise to the Wien effect in conductance measurements.[33])

Figure 27–7 accordingly shows a repetition of the experiments of Fig. 27–6 with varying field strength. This experiment clearly shows that the major part of the ions within the polyelectrolyte domain is *not* susceptible to being left behind by an increase in field strength, and the logical conclusion is that these ions are therefore strongly bound to specific sites. Only those sodium ions which appear to be part of the polyion at low fields, but not at high fields, can be regarded as part of the ion atmosphere in the sense of a Debye-Hückel type of calculation.

A similar result was obtained for polyvinylbutylpyridinium bromide by measurement of the Wien effect.[35]

Specific binding of ions is, of course, a property which depends on the exact chemical nature of the macro-ion. For instance, it is likely that the large amount of specific ion binding observed for polyacrylate is not shared by carboxymethylcellulose. Though both of these polymers contain —COO⁻ groups, these groups are 2.5 Å apart in polyacrylate, but 5 Å apart in carboxymethylcellulose. In the former case a very much greater intrinsic affinity for cations would therefore be expected (cf. Kirkwood and Westheimer[16]).

The extent of specific binding has an effect on the calculation of the electrostatic free energy, as explained in section 26i, for charged sites which are chemically combined with cations are treated as neutral sites. The extent of specific binding would thus also effect calculation of the degree of expansion by the method of section 27b. Thus Table 27–2 shows how the calculated expansion of carboxymethylcellulose, as previously discussed, is affected by the degree of specific binding. The columns headed "33% binding" and "67% binding" were computed with the assumption that one-third and two-thirds of the fixed charges are neutralized by specific binding. It is seen that in this case the best agreement with experiment is obtained when a relatively low degree of chemical binding is assumed.

It should be pointed out, finally, that we have here used the term "chemical," or "specific," binding to indicate any kind of binding which *fixes* the counterion to a specific site (presumably at a distance less than that to which mobile ions are excluded). We do not intend to make the implication that electrostatic forces may not be responsible for this type of binding as well as for the non-specific attraction of the Debye-Hückel type.

27d. Configuration of polyelectrolytes at low charge and very high ionic strength and in non-aqueous solvents. It has been taken for granted that the polyelectrolytes which have been discussed in this section, would, in their uncharged form and at very high ionic strength, have the random coil configurations typical of ordinary uncharged polymers such as polystyrene and polyisobutylene. The most striking confirmation of this assumption comes from the work of Strauss and coworkers[40,41] on the effect of NaBr on the behavior of sodium polyphosphates. Intrinsic viscosities for samples of various molecular weights are shown in Fig. 27–8. In the complete absence of salt, as already mentioned, $[\eta]$ varies as $M^{1.9}$, indicating that the polyphosphate ions are essentially fully extended rods.

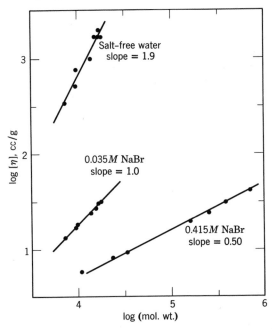

Fig. 27–8. The intrinsic viscosity of sodium polyphosphate. Points for the upper curve (Strauss and coworkers[40]) were obtained from specific viscosities of salt-free solutions, extrapolated by means of the Fuoss-Strauss equation as in Fig. 27–3. The other data (Strauss and Wineman[41]) were obtained by dilution at constant ionic strength, as in Fig. 27–4.

A moderate ionic strength sharply reduces $[\eta]$ and also the dependence of $[\eta]$ on molecular weight. At $0.415M$ NaBr, finally, the polymer solutions actually become ideal, $[\eta]$ varying exactly as $M^{0.50}$. Parallel changes in the second virial coefficient were observed; B is large at low ionic strength, because of the large excluded volume of the expanded polymer chain, but it becomes zero in $0.415M$ NaBr. When the NaBr concentration is increased further, phase separation occurs, exactly as happens (p. 249) when uncharged polymers are dissolved in solvents which would be expected to produce negative second virial coefficients.

Some interesting comparisons have been made between the actual dimensions of polyelectrolytes, under conditions where electrostatic effects are suppressed, and the dimensions of similar non-electrolyte polymers. Sodium carboxymethylcellulose, for example, at high ionic strength, approaches dimensions essentially identical to those of uncharged cellulose derivatives of the same molecular weight. Polyvinylbutylpyridinium bromide,[27] on the other hand, becomes more compact at high ionic strength than the uncharged parent polymer, polyvinylpyridine. This result is presumably due to the presence of counterions within the coils, with a resulting tendency to form $-N^+-Br^--N^+-$ networks, which draw the polymer chains closer together.

This same phenomenon produces unusual results when polyvinyl-butylpyridinium bromide is dissolved in mixed solvents of relatively low dielectric constant. It might at first be expected that the increased inter-action between charges in a medium of low dielectric constant would lead to an increased expansion. When the polyelectrolyte is dissolved in ethanol-water mixtures, however, the reverse result is observed,[27] a *decrease* in intrinsic viscosity being found. In 91% ethanol the intrinsic viscosity of polyvinylbutylpyridinium bromide becomes less than that of polyvinylpyridine in the same solvent even in the absence of other electro-lytes. Both these results are presumably a consequence of the formation of $N^+-Br^--N^+$ bonds within the polymer coils.

No detailed studies have been made of any of the above phenomena, so that the interpretations which have been made of them must be re-garded as speculative.

27e. Linear polyampholytes. Polyampholytes are polyelectrolytes which may bear both positive and negative charges. An example is the copolymer of methacrylic acid and of diethyl (or dimethyl) amino ethyl methacrylate, which contains the monomer units

$$O=C-O-(CH_2)_2-NR_2H^+ \qquad\qquad COO^-$$
$$\underset{\displaystyle CH_3}{\overset{\displaystyle |}{|}} \qquad\qquad$$
$$-C-CH_2- \qquad and \qquad -C-CH_2-$$
$$\underset{\displaystyle CH_3}{|} \qquad\qquad \underset{\displaystyle CH_3}{|}$$

where R may represent ethyl or methyl groups. These polymers have a net positive charge at low pH and a net negative charge at high pH. In between there is, as in proteins, a pH at which the net charge is zero. This may be seen, for instance, from the electrophoretic mobilities of such a polymer shown in Fig. 27-9.

Far from the isoelectric point these polymers would be expected to behave like ordinary polyelectrolytes, i.e., to be highly expanded, with their expansion suppressed by an increase in ionic strength. Near the isoelectric point, however, the attraction between the oppositely charged

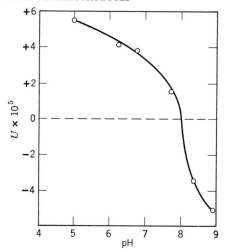

Fig. 27–9. The electrophoretic mobility of a typical polyampholyte as a function of pH. (Alfrey et al.[37])

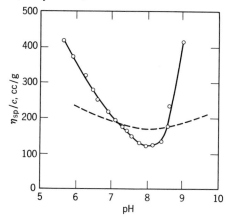

Fig. 27–10. The reduced viscosity of a polyampholyte near its isoelectric point. The data are from Alfrey et al.[37] and were obtained at low ionic strength. The dashed line shows the predicted behavior at high ionic strength. An actual illustration of the reversal of the usual ionic strength effect at the isoelectric pH is provided by Fig. 27–11.

side chains, present in equal numbers, should lead to relatively tight coiling of the polyions. In this case an *increase* in ionic strength should lead to *expansion*. This behavior is confirmed in a qualitative way by the data of Figs. 27–10 and 27–11.

It is of interest[38] that the second virial coefficient of solutions of these polymers in their isoelectric state is negative, indicating that the attraction between the positive and negative charges may be effective intermolecularly as well as intramolecularly. As a matter of fact, many polyampholytes of this type tend to become insoluble at their isoelectric points.[39]

It should also be noted that the electrophoretic mobilities of Fig. 27–9 are far from linear functions of the charge, contrary to what we would

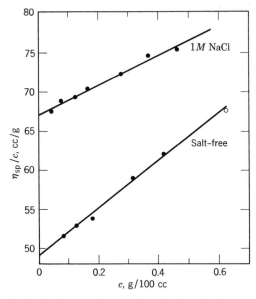

Fig. 27–11. Reduced viscosity for a polyampholyte (similar to that used to obtain the data of Fig. 27–10) at its isoelectric pH, in the absence and presence of added salt. (Ehrlich and Doty.[38])

expect from equations 24–7 or similar relations. This is, of course, a consequence of the expansion accompanying increasing charge which results in a corresponding increase of the frictional coefficient.

28. POLYPEPTIDES, POLYSOAPS, PROTEINS, NUCLEIC ACIDS

The preceding section has dealt with the effect of electrostatic forces on the behavior of *flexible* polyelectrolyte ions or molecules. In a solution of such ions or molecules there is a dynamic equilibrium between all conceivable configurations, and any change in electrostatic forces produces an immediate effect, shifting the equilibrium in the direction of configurations with lower electrostatic free energy.

In this section we shall consider the effect of electrostatic forces on the behavior of those polyelectrolyte ions or molecules which, in the absence of such forces, would have a unique preferred configuration. (Most of the macromolecules assigned to this class in section 7, i.e., the proteins, nucleic acids, and viruses, are, in fact, polyampholytes.) When molecules of this kind acquire an electric charge, they will, like flexible polyelectrolyte ions, possess a lower electrostatic free energy in an expanded configuration, where their charges are further apart, than in a compact configuration, where charges are close together. (This potential reduction in W_{el} may be illustrated, for example, by comparing the W_{el} values given in Table 26–1 for a compact protein ion with the values given in Table 26–2 for an ion of the same molecular weight and charge in an expanded configuration with twice the original radius.) The only exception to this generalization would occur for polyampholytes, at or very near their isoelectric points, where the presence of an equal number of positive and negative charges would of course favor a compact configuration, just as flexible polyampholyte ions, under similar circumstances, tend to contract rather than expand.

The actual configuration of these polyelectrolyte ions in solution must, of course, be the result of a minimization of the total free energy and not of W_{el} alone. Since the existence of a preferred configuration (when the net charge is low) implies the existence of internal bonds, such as hydrogen bonds or hydrophobic bonds (Chapter 2, especially section 7), any expansion of the configuration would clearly be accompanied by a positive free energy change due to the necessity of disrupting the original preferred structure. Whether or not expansion is thermodynamically favored therefore depends on the relative magnitudes of the decrease in W_{el} which would occur and the increase in internal free energy which would be required to achieve it. The study of the behavior of polyelectrolytes with a definite structure, under conditions where expansion is favored because of the electrostatic free energy, thus yields important information on the strength of the internal bonds which maintain the internal structure when the charge is low.

28a. Synthetic polypeptides.[36] Synthetic polypeptides such as poly-γ-benzyl-L-glutamate have been shown to be in a helical, rod-shaped configuration in solid films (p. 51) and in certain organic solvents (p. 410). The principal factor in this configuration is believed to be the formation of hydrogen bonds between —C=O and —NH groups of the peptide links. Many of the synthetic polypeptides are polyelectrolytes, examples being polyglutamic acid and polylysine, with the repeating units

$$-NH-CH-CO- \qquad\qquad -NH-CH-CO-$$
$$\quad | \qquad\qquad\qquad\qquad\qquad | $$
$$(CH_2)_2 \qquad\qquad\qquad\qquad (CH_2)_4$$
$$\quad | \qquad\qquad\qquad\qquad\qquad | $$
$$COO^- \qquad\qquad\qquad\qquad NH_3^+$$
$$(or\ COOH) \qquad\qquad\qquad (or\ NH_2)$$

the side chain groups being charged or uncharged, depending on pH. When these polypeptides, in their uncharged form, are dissolved in water, or in certain other solvents, they take up the same helical configuration typical of non-electrolyte polypeptides. This is illustrated, for instance, by the data in Fig. 28–1, for a poly-L-glutamic acid sample with a degree of polymerization near 280. The specific optical rotation (Na D line) in the uncharged state is seen to be $+5°$. Since the molecular weight per monomer unit is 129, the mean residue rotation (equation 6–23) is $+6°$, a figure which, within probable error, is identical to the value $+10°$ which is characteristic of α-helical polypeptide chains. (A figure of $-100°$ would be typical of a randomly coiled configuration; cf. p. 122.)

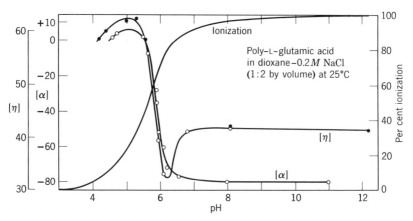

Fig. 28–1. The helix-coil transition accompanying the ionization of poly-L-glutamic acid. (Doty, Wada, Yang, and Blout.[36])

The intrinsic viscosity of about 60 cc/gram is also essentially identical to that which we would calculate for a helical polyglutamic acid chain of degree of polymerization 280, but this result cannot be used as a criterion of configuration because this particular chain length happens to be one where the intrinsic viscosities of the helical and flexible configurations are not very different. In other words, if intrinsic viscosities were available for both configurations at various molecular weights, similar to those shown in Fig. 23–5 for the benzyl ester of polyglutamic acid, we would be operating close to the point where the two curves cross. What is significant about the intrinsic viscosity measurements is the fact that close to 40% of the carboxyl groups can be ionized without appreciable change in $[\eta]$, in sharp contrast to the immediate expansion which accompanies ionization of flexible polyelectrolytes.

Figure 28–1 shows a sharp transition in both viscosity and optical rotation, near pH 6, over a pH range within which the degree of ionization changes from 40 to 80%. The value of $[\alpha]_D$ changes to $-80°$, the corresponding mean residue rotation being $-103°$, exactly as expected for a randomly coiled polypeptide chain. The intrinsic viscosity drops from about 60 to about 40 cc/gram. The fact that the change in viscosity is essentially parallel to the change in optical rotation indicates that both experimental procedures reflect the same configurational change. A quantitative interpretation of the viscosity change is not possible, however, because in the present situation $[\eta]$ for a flexible coil could be either larger or smaller than that for a helical rod, depending on the value of the interaction parameter α of equation 23–7, i.e., depending on whether the solvent for the coil is a good or a poor solvent.

It would clearly be desirable to repeat these experiments for a sample of higher molecular weight, where viscosity measurements could unequivocally distinguish between a randomly coiled and a rod-shaped particle. The optical rotation data by themselves, however, are reasonably convincing and clearly indicate that the hydrogen bonds maintain the helical structure up to about 40% ionization. The corresponding molecular charge, assuming no neutralization by counterions, is 112, and the value of W_{el}, as computed by equation 26–43 is 182,000 cal/mole. (Assumed dimensions of cylinder: length 420 Å, diameter 13.5 Å. The dielectric constant of the dioxane-water mixture is 53.)

To calculate the corresponding value of W_{el} which the ion would have in a randomly coiled configuration, we use the Hermans-Overbeek equation (equation 26–56), which, it will be recalled, can only be regarded as providing a crude approximation to the true value. We assume that R_e in that equation is equivalent to R_G and compute the latter from the intrinsic viscosity (35 cc/gram) which the polyion attains, as shown by Fig. 28–1, when in the flexible form. Using equation 23–5 we get $R_G = 67.5$ Å, and, hence, $W_{el} = 26,000$ cal/mole.

It may at first be surprising that the helix-coil transition should be accompanied by so large a decrease in W_{el}, for in going from a helical rod to a random coil the end-to-end distance will usually be decreased, so that charges which are far apart in the helix may be brought somewhat closer to each other by the transition. It is the interaction between closely spaced charges, however, which makes the major contribution to W_{el}, and the distance between these is increased by the transition.

The potential decrease in W_{el} which would accompany unraveling of the helix is therefore 156,000 cal/mole, and, since unraveling does not occur, the free energy stabilizing the helix must be of this same order of magnitude. Since the helix is maintained by 280 $-C\!\!=\!\!O \cdots H\!\!-\!\!N-$ hydrogen bonds, this means that each such bond must contribute about 500 cal. Since the rupture of such a hydrogen bond is presumably accompanied by the formation of new hydrogen bonds to water, the reaction may be written schematically as

$$-C{=}O \cdots H{-}N{-} + HOH \cdots OH_2 \rightarrow$$
$$-C{=}O \cdots HOH + -N{-}H \cdots OH_2$$

That the free energy change in this reaction, per peptide hydrogen bond, should be about 500 cal is entirely reasonable, especially since the solvent in this case is a water-dioxane mixture, in which ΔF for the reaction would be somewhat greater than in water alone.

The free energy change of 500 cal per bond represents, of course, a complex cooperative phenomenon which involves the entire helix. The free energy change is determined by the relative energies of the different hydrogen bonds, by the difference in entropy between the rigid and flexible polypeptide chain, and by the loss in entropy of water molecules attached to —CO or —NH groups. A variety of theoretical and experimental approaches to this problem all indicate that, for all but very short polypeptide chains, ΔF should be of the order of a few hundred calories per monomer unit and independent of the total length of the chain.

It should be emphasized finally that the preceding calculation must be regarded as a crude approximation. Partial neutralization of the charge by counterions has, for instance, not been taken into account. Equations 26–43 and 26–56 are themselves so approximate, however, that refinements of the calculation are not justified.

28b. Polysoaps.[42] An interesting class of synthetic polymers which have not been mentioned before are the *polysoaps*. They are copolymers containing both ionic groups and long non-polar side chains. The principal example is provided by a series of polymers which have been prepared and studied by Strauss and coworkers.[42] These polymers are copolymers of 4-vinyl-N-ethyl pyridinium ion and 4-vinyl-N-dodecyl pyridinium ion with the repeating units

They are strong electrolytes and were prepared in the form of the bromide salts.

The polysoaps are of interest because they would be expected to have a tendency to form intramolecular hydrophobic bonds in water solution, as discussed on p. 130. Bonding of this kind would arise from the fact that side chains such as the —$C_{12}H_{25}$ chain in the monomer unit B of the present example can be accommodated in an aqueous medium only at the expense of a considerable free energy increase, which results from breaking up the solvent structure to form a hydrocarbon–water interface. Thus the

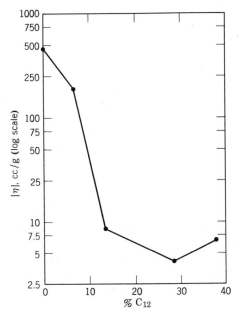

Fig. 28–2. The effect of dodecyl group content on the intrinsic viscosity of copolymers of vinyl-N-ethyl and vinyl-N-dodecyl pyridinium bromide. The solvent is 0.0226M KBr. (Strauss and coworkers.[42])

formation of an intramolecular micelle, from which solvent is excluded, is favored. The positively charged quaternary N^+ atoms would be left on the outside, in contact with solvent, or, if neutralized by formation of N^+Br^- ion pairs, they could be incorporated in the micelle. A fraction of them, sufficient to cover the outside surface, would always have to be on the outside.

Strauss and coworkers have demonstrated convincingly that such intramolecular micelles are in fact formed. The data of Fig. 28–2, for example, show intrinsic viscosities for a series of copolymers, all having an average degree of polymerization of 18,000 but varying in the relative content of monomer units of types A and B. Polyions consisting entirely of A monomers show typical polyelectrolyte behavior, with a high intrinsic viscosity which decreases with increasing ionic strength. (Only a single ionic strength is shown in the figure.) A sufficiently high ionic strength to achieve ideal behavior was not used, but the limiting minimum ideal intrinsic viscosity was estimated from data on similar polymers to be 112 cc/gram, compared to the value of 500 cc/gram observed at ionic strength 0.0226. When monomers of type B are substituted for those of type A, a much more dramatic decrease in viscosity occurs, and, when the content of type B is 28.5%, a minimum value of 4 cc/gram is attained. Comparison with Tables 23–1 and 23–7 shows that this value of $[\eta]$ is of the same order of magnitude as that observed for globular proteins and crystalline viruses. The dissolved ions must be exceedingly compact, containing very little solvent within their coils.

An analysis of the molecular dimensions could be made, based on equation 23–4. The same ambiguity would arise as that observed for globular proteins in Table 23–2; i.e., the dissolved particles could be spheres containing about 0.5 grams of water per gram of polymer, or they could be compact ellipsoids, with less solvation, the maximum possible axial ratio being about 4 : 1. For the purpose of subsequent calculation this ambiguity makes little difference. Since equations for electrostatic work are available only for spheres, a spherical shape will be assumed, but the W_{el} values of the corresponding ellipsoids would certainly not be appreciably different. The radius to be assigned to the spherical particle is obtained from equation 23–3, by setting $\nu = 2.5$ and $v_h = \frac{4}{3}\pi R^3$.

Before calculations of W_{el} can be made it is necessary to decide how many of the pyridine N^+ charges are neutralized by Br^- counterions. Since the charges are quite closely spaced, the fraction neutralized is likely to be much higher than in the case of the polypeptide discussed in the preceding section. The charge separation being about the same as in polyacrylic acid, we can use the data of Figs. 27–6 and 27–7 as an indication of what to expect. Figure 27–6 shows that two-thirds of the charges become neutralized when the degree of ionization of polyacrylic acid approaches 100%. Figure 27–7 shows, moreover, that the major part of this neutralization occurs by means of specific binding and not just by virtue of the presence of counterions in the solvent trapped within the polymer coils. (No such retention of ions could occur in the present case, since the polysoap ion is assumed impenetrable to solvent.) In the present example an even greater degree of neutralization can be anticipated (a) because we are dealing with a compact configuration, whereas the data for polyacrylic acid apply to a randomly coiled configuration, and (b) because the added electrolyte (KBr) provides a higher counterion concentration than was present in the experiments with polyacrylic acid, which were conducted in the absence of added salt. It is also worth noting that the experiments cited on p. 506 (on the behavior of polyvinylbutylpyridinium bromide) indicated that the attraction between pyridine N^+ and Br^- might be unusually large.

It is of interest in this connection that the allowed dimensions of the compact polysoap ion, based on the value of $[\eta]$ regardless of whether we choose an ellipsoidal or a spherical model, are such that the *minimum* dimension must be larger than the length of two fully extended $-C_{12}H_{25}$ groups.[42] There must therefore be N^+ charges within the micelle, and since these charges must necessarily be neutralized, they provide a further limitation on the possible number of free charges.

We shall assume, for the calculation of W_{el}, that two-thirds of the charges are neutralized, i.e., that each polysoap ion (degree of polymerization 18,000) has 6000 free charges. It is highly probable that this is too high an estimate of the net charge, so that the calculation will actually represent a *maximum* value for W_{el}.

Using $Z = 6000$, then, as a maximum figure, we have, for the polyion containing 28.5% monomer of type B, a molecular weight of 4×10^6, including therein the weight of 12,000 Br^- ions. The radius of the corresponding sphere (equation 23–3) is 140 Å. Taking the sphere to be impenetrable, we can compute W_{el} by equation 26–30, obtaining, at an ionic strength of 0.0226, $W_{el} = 7.6 \times 10^7$ cal/mole.

To calculate the value of W_{el} which the same ion would have as an expanded coil, we assume that R_G would have the same value as is actually observed for the polymer containing 100% monomers of type A, i.e., a value based on an intrinsic viscosity of 500 cc/gram. Using this value for R_e in equation 26–56, we get $W_{el} = 4 \times 10^5$ cal/mole, which is negligibly small in comparison with the electrostatic free energy of the compact ion. It is concluded then that the ion could gain 7.6×10^7 cal/mole by expansion, although it should be recalled that this a maximum figure.

To account for the fact that expansion does not occur, it must thus be assumed that 7.6×10^7 cal/mole have been gained in formation of the intramolecular micelle. There being $18,000 \times 0.285 = 5130$ $-C_{12}H_{25}$ side chains, this means that the transfer of each such chain from the aqueous medium to the interior of the micelle must result in a free energy decrease of about 15,000 cal (maximum). This is not an unreasonable figure, for it was estimated on p. 130 that transfer of a CH_2 group from water to a hydrophobic region might result in a free energy decrease of 1000 to 2000 cal, so that the corresponding estimate for a $-C_{12}H_{25}$ group would be 12,000 to 24,000 cal.

28c. Globular proteins. The globular proteins, when examined in aqueous solution, near their isoelectric points, have compact configurations, impenetrable to water. It was pointed out in section 7 that various forces may cooperate in achieving such a configuration, including the hydrophobic and hydrogen bonds which have been discussed in connection with the behavior of polypeptides and polysoaps. It is probable that the exact combination of forces, and the free energy required to disrupt the initial native structure, varies from one protein to another, so that their ability to withstand the effects of an electrostatic charge varies likewise, and this is borne out by actual experimental data.

Some proteins appear to undergo no configurational change at all when charged. Ribonuclease,[43] for example, does not expand in acid solution, at 25° C and ionic strength 0.25, even when it attains its maximum charge of $+19$. At the same temperature, at ionic strength 0.05, it has been taken to a charge of $+16$ without expansion. The electrostatic free energies attained under these conditions may be computed by equation 26–30, using a radius of 19 Å, which is the radius of an equivalent sphere obtained from any of the hydrodynamic measurements cited in Chapter 6. We obtain, for the maximum charge attained both at ionic strength 0.25 and 0.05, $W_{el} = 13,000$ cal/mole.

If expansion to a solvent-permeated configuration were to occur, W_{el} would become negligibly small in comparison; i.e., the free energy inherent in the internal bonds of this molecule must be of the order of 13,000 cal/mole. This is actually a very small figure, for ribonuclease contains sixteen fairly large hydrophobic side chains such as $-C_3H_7$, $-C_4H_9$, and $-CH_2C_6H_5$ and can in addition form a very large number of hydrogen bonds. If the sequence of amino acid residues were such that all hydrophobic side chains could form a micelle and all possible hydrogen bonds (between $-C{=}O$ and $-NH$ groups) could be formed, the free energy of stabilization of a compact configuration could without difficulty exceed 100,000 cal/mole.

An important aspect of the stability of the compact configuration of maximally charged ribonuclease is that it does not persist to temperatures much above 25° C. Extensive unfolding occurs on heating to 40° C, the intrinsic viscosity rising to about 7 cc/gram.[44] The unfolding process is, however, completely reversible, i.e., we can be sure that the stability at 25° C represents true thermodynamic stability. This statement would not be true for all globular proteins. Ovalbumin[45,46] and pepsin,[47,48] for example, are two molecules which can acquire a considerable charge before expanding. When they do expand, the process is an *irreversible* one. This means that the apparent stability which is observed for these proteins may be a kinetic phenomenon rather than a thermodynamic one. Data obtained in the study of the unfolding of these proteins could not form the basis of a calculation of the strength of internal bonds, such as that which was carried out for ribonuclease.

Serum albumin, in contrast to ribonuclease, is a protein which expands very readily. Intrinsic viscosities of this protein, in acid solution at 25° C, are shown in Fig. 28–3. It is seen that the molecule undergoes a preliminary transition, slightly acid to its isoelectric point, forming a new configuration which differs only slightly from the original native form, having an intrinsic viscosity of 4.4 cc/gram instead of the original value of 3.7 cc/gram. This new form undergoes expansion as it acquires increasing

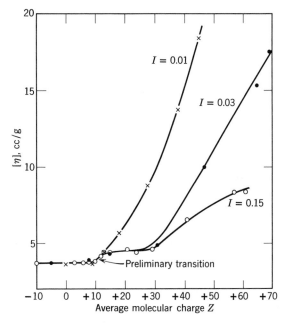

Fig. 28–3. The intrinsic viscosity of serum albumin in acid solution at 25°C. The data at ionic strength 0.01 are from Yang and Foster;[50] those at ionic strengths 0.03 and 0.15, from Tanford et al.[49]

positive charge, and the extent of expansion varies with ionic strength in the same manner as the extent of expansion of a flexible polyelectrolyte ion. Figure 28–3 shows that the charge required to initiate this expansion is only $+15$ at $I = 0.01$ and $+35$ at $I = 0.15$. The corresponding values of W_{el}, as calculated by equation 26–30, are 6500 and 14,000 cal/mole, respectively, using $R = 35.5$ Å, a figure based on the molecular weight of 65,000 and the intrinsic viscosity of 4.4 cc/gram. (If the isoelectric viscosity of 3.7 cc/gram were used, we would obtain $R = 34$ Å, and W_{el} would be changed by less than 10%.)

Serum albumin is a molecule of nearly five times them olecular weight of ribonuclease. It possesses 130 hydrophobic side chains of the size of $-C_3H_7$ or larger, compared to sixteen for ribonuclease, and can form five times as many hydrogen bonds. If the ability to form internal bonds were utilized as well as it is in ribonuclease (and, as we have seen, it is not particularly well utilized by ribonuclease), the protein ion would be expected to attain a W_{el} value of 65,000 cal/mole or more before expansion occurs. The fact that expansion occurs when W_{el} is only 10 to 20% of this value thus indicates that the internal bonding in serum albumin is unusually weak.

The compact configuration of serum albumin is somewhat more stable on the alkaline side than on the acid side of the isoelectric point. A charge of -40 at ionic strength 0.05 can be attained before any expansion sets in. This corresponds, by equation 26–30 to $W_{el} = 32,000$ cal/mole. The reason for the greater stability towards negative charges is not known. It could possibly be an artefact, resulting from the use of equation 26–30 to calculate W_{el}. The charges might well be so located that W_{el} increases more rapidly with charge on the acid side than on the alkaline side. An alternative possibility is that the unfolding process does not initially involve the entire molecule, but only a part of it. In that event unfolding would set in when W_{el} of any such portion exceeds the internal bond energies, and there would be no necessary connection with W_{el} of the molecule as a whole. Finally, it should be kept in mind that the energy of the bonds stabilizing the compact structure may itself be pH dependent. This would be true, for example, if hydrogen bonds involving ionizable side chain groups make a sizable contribution to the free energy of the compact structure.

It should be noted that the conclusions which we have reached here on the basis of viscosity measurements can be corroborated by a variety of other data. For instance, the sedimentation and diffusion data of Fig. 22–8[51] lead to the same conclusion concerning the acid expansion of serum albumin as the data of Fig. 28–3. Other evidence comes from changes in optical rotation[50] and in the depolarization of fluorescence.[51,52]

It can be seen from Fig. 28–3 that the maximum intrinsic viscosity which is attained by serum albumin in its flexible form at ionic strength 0.15 is about 9 cc/gram. At lower ionic strengths, larger values are observed, and an extrapolation by Yang and Foster[50] to zero ionic strength at maximum charge yields a value of 75 cc/gram for the maximum attainable value. The corresponding radius of gyration can be calculated by equation 23–7 and comes out to be about 110 Å, the exact value depending slightly on the choice of the parameter ξ. This radius of gyration represents the dimensions of a fully charged flexible serum albumin ion in the complete absence of salt and is therefore larger than that which would represent a hypothetical *uncharged* serum albumin molecule in the random coil form. However, the extra expansion due to electrostatic repulsion would not be as great as that which occurs for most of the polyelectrolytes discussed in section 27 because a fully charged serum albumin molecule has only about 100 charges on about 600 monomer units, the remainder being incapable of bearing a charge in acid solution. We can only guess what the actual electrostatic effect might be. In Fig. 27–1 it was shown that a degree of dissociation of 1/6 changes R_G for polymethacrylic acid, in the absence of salt, from about 100 to 300 Å, but the side chains in this polymer are closer to one another than the side chains on a polypeptide backbone. It is probable, therefore, that the effect of electrostatic repulsion in the present situation is at most a factor of 2. A hypothetical uncharged, randomly coiled serum albumin molecule would therefore probably have an R_G value somewhere above 55 Å, and certainly less than 110 Å.

This figure may be compared with the radius of gyration of poly-γ-benzyl-L-glutamate in a solvent in which it has the flexible coil configuration. Appropriate data are given in Fig. 23–5. Choosing a molecule of the same degree of polymerization as serum albumin (i.e., about 600 monomer units), we find $[\eta] = 75$ cc/gram and, by equation 23–7, a corresponding radius of gyration of 135 Å. This value of R_G is double the estimated R_G of an uncharged flexible serum albumin molecule. The reason for this is undoubtedly the existence of sixteen disulfide cross-links in the serum albumin molecule. These cross-links must considerably restrict the ability of the molecule to expand.

In conclusion, it should be pointed out that three major problems concerning the configurational stability of serum albumin, ribonuclease, and similar proteins remain unsolved. (1) It is not known why the unfolding is sometimes reversible, as in the examples used in the present section, and sometimes irreversible, as in the case of ovalbumin and pepsin. (2) It is not established whether expansion is a gradual process or an all-or-none unfolding analogous to a phase transition. That is, the increase in $[\eta]$ shown in Fig. 28–3 could represent the simultaneous, continuous expansion of all solute molecules. It could also represent a shift in an equilibrium between compact molecules and completely expanded ones, without the existence of molecules of intermediate structure. (3) The preliminary transition shown in Fig. 28–3 at a charge of $+10$ and similar transitions which occur in other proteins have only recently been discovered, and there is at present no clear understanding of the changes which they represent.

28d. Association of proteins. The various secondary bonds, which, acting intramolecularly, cause proteins to be ordinarily in compactly coiled configurations, may also be effective intermolecularly to produce association between protein molecules. Insulin, for instance, has already been cited in Figs. 14–2 and 17–6 as an example of this effect.

If the protein molecules involved in the reaction are charged ions, such a reaction will always involve the close association of the charges of one ion with those of the other, If the associating ions have opposite charge, the association will be aided; if they have the same charge, it will be hindered. We can make a very crude estimate of this effect by use of equation 26–30. Consider for instance, an association of the type $2A \rightleftharpoons A_2$. If the ion A can be represented as a sphere of radius R and net charge Z, we may suppose that the ion A_2 can be represented as a sphere of twice the volume and charge, i.e., of radius $2^{1/3}R$ and charge $2Z$. The change in electrostatic free energy is then

$$\Delta W_{el} = 2^{2/3} \frac{Z^2 \epsilon^2}{DR} \left(1 - \frac{2^{1/3} \kappa R}{1 + 2^{1/3} \kappa a}\right) - \frac{Z^2 \epsilon^2}{DR} \left(1 - \frac{\kappa R}{1 + \kappa a}\right) \quad (28\text{--}1)$$

which will make a positive contribution to $\Delta F°$ for the association reaction.

For instance, Doty and Myers[53] have calculated from light scattering data the thermodynamic constants for the association of insulin at low pH. They find that $\Delta F°$ for the dimerization reaction (in $0.1M$ phosphate) is -4690 cal at pH 2.6. At pH 1.9, on the other hand, $\Delta F° = -4320$ cal; i.e., it is more positive by 370 cal. The value of Z for the monomer is about 10 at pH 2.6 and about 11 at pH 1.9. If this value of Z is inserted into equation 28–1, with a reasonable value of R, we find ΔW_{el} to be about 600 cal larger at pH 1.9 than at pH 2.6, a value which is in as good agreement with the experimental change in $\Delta F°$ as the very artificial nature of the model can be expected to yield.

As an example of the electrostatic attraction between proteins of opposite charge we may cite the result of Steiner[54] who found that lysozyme (isoelectric at pH 11) and serum albumin (isoelectric at pH 5) form association complexes between their isoelectric points, but not at pH < 5 or pH > 11, where both have a net charge of like sign. As is to be expected, the association is suppressed by an increase in ionic strength.

In examining the effect of pH on association reactions of proteins it should always be kept in mind that a change in net charge is only a part of what happens to a protein molecule when the pH is changed. We have already noted that configuration changes may occur, but even in their absence there will be changes which may affect association equilibria. For instance, Fig. 28–4 shows the pH dependence of the molecular weight of β-lactoglobulin.[56] The molecule normally consists of two loosely joined polypeptide chains, of molecular weight 17,500 each, and dissociates into individual chains only when the maximum positive charge is approached. At 2° C, however, there appears also an aggregate of higher molecular weight, with maximum stability near pH 4.6. The dissociation of this aggregate, acid to pH 4.6, is in accord with expectation and, in fact, has been shown by Townend and Timasheff[56] to agree closely with what would be predicted by equation 28–1. But the dissociation alkaline to pH 4.6 is anomalous, because it occurs while the net charge is decreasing (the isoelectric point being at pH 5.2). The possible explanations for this

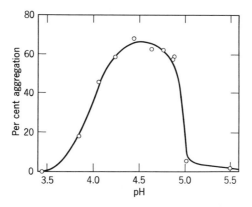

Fig. 28–4. The effect of pH on the aggregation of β-lactoglobulin. The isoelectric point is at pH 5.2. The molecular charge Z for unaggregated molecules (molecular weight 35,000) is approximately $+10$ at pH 4.5 and $+20$ at pH 3.5. (Townend and Timasheff.[56])

anomaly are essentially the same as those cited on p. 518 to account for the lack of symmetry about the isoelectric point which is observed in the stability of a compact configuration for serum albumin.

28e. Nucleic acids. At the present time deoxypentose nucleic acid (DNA) is the best known nucleic acid and the only one for which extensive physico-chemical data are available. (See Doty[57] for a summary of current information.) Its molecular weight is in the vicinity of 6 to 7 million, and its molecules in neutral aqueous solution have the configuration of exceedingly stiff flexible coils. As was explained on p. 312, the stiffness is believed due to the fact that the basic structural unit is not a single molecular chain but the helical structure of Fig. 4–13, consisting of two intertwined chains. This structure would ordinarily lead to a molecule which behaves as a long thin rod, but occasional points of interruption from the helical structure are believed to occur and to give rise to a sufficient number of flexible joints to account for the over-all behavior as a stiff coil.

The helical structure just described is destroyed in either acidic or basic solution, with the formation of coils of much greater flexibility. The experimental manifestation of this process is a sharp decrease in intrinsic viscosity (as shown in Fig. 28–5) and in other properties which reflect the extent of the molecular domain. The breakdown of the helical structure will be accompanied by a decrease in electrostatic interaction, because,

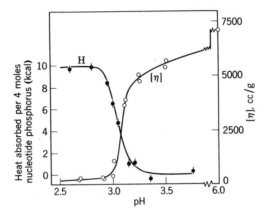

Fig. 28–5. The collapse of the helical structure of DNA in acid solution at 25°C and ionic strength 0.1. The figure shows the sharp drop in intrinsic viscosity which characterizes this process and also shows the heat uptake (measured calorimetrically) which accompanies it. (Sturtevant and Geiduschek.[58])

although the over-all dimensions of the molecule shrink, the separation between the closely spaced charges of the individual rodlike portions is increased, in the same way as the helix-coil transition in polypeptides increases the local distance between charges and, hence, the magnitude of W_{el}. The details of the unfolding process are not as yet fully established, because the process appears to consist of several consecutive stages, most of which are irreversible. Experimental results therefore depend considerably on how low or high the pH is at which the reaction is studied and how long the nucleic acid is exposed to that pH. The fact that unfolding occurs is however beyond question.

An important aspect of the unfolding of DNA is that it appears at first sight to violate the principle that configurational changes should occur in such a direction as to decrease W_{el}. For DNA is a negatively charged ion (one charge per nucleotide) at neutral pH and approaches the isoelectric state, by addition of hydrogen ions to the weakly basic amino groups of its pyrimidine and purine side chains, as the pH is decreased. Thus the tendency to unfold should be greater in neutral solution, contrary to what is observed. The probable explanation of this apparent anomaly lies in the location of the various charges in the double-stranded helical structure (Fig. 4–13). It will be noted that the negative phosphate groups, on which the charge at neutral pH is located, lie on the *outside* of the helical structure and, hence, are relatively far apart. The purine and pyrimidine rings, on the other hand, occupy the *inside* of the helix and lie in close proximity to one another. As a result the work of charging the purine and pyrimidine nitrogen atoms is much higher than the work of charging the phosphate groups, and the magnitude of W_{el} is determined largely by the state of the basic groups, regardless of the total charge. Since the groups in the center of the helix are uncharged at neutral pH and become charged in acid solution, W_{el} rises as the pH is reduced, even though the total charge is decreased.

On the alkaline side of neutrality, the inside of the helix acquires a negative charge because of dissociation of protons from $=\!N\!-\!C(OH)\!-$ groups of guanine and thymine, and unfolding results as before. In this case there is of course no anomaly since the total charge becomes more negative also.

General References

1. J. A. Stratton, *Electromagnetic Theory*, McGraw-Hill Book Co., New York, 1941.
2. H. Margenau and G. M. Murphy, *The Mathematics of Physics and Chemistry*, D. Van Nostrand Co., Princeton, N.J., 1943.
3. H. Falkenhagen, *Electrolytes, Oxford University Press*, 1934.
4. P. Doty and G. Ehrlich, "Polymeric Electrolytes," *Ann. Rev. Phys. Chem.*, **3**, 81 (1952).
5. H. Eisenberg and R. M. Fuoss, "Physical Chemistry of Synthetic Polyelectrolytes" in J. O'M. Bockris, (ed.), *Modern Aspects of Electrochemistry*, Academic Press, New York, 1954.

Specific References

6. P. Debye and E. Hückel, *Physik. Z.*, **24**, 185 (1923).
7. H. S. Harned and B. B. Owen, *Physical Chemistry of Electrolytes*, 2nd ed., Reinhold Publishing Corp., New York, 1950, Chs. 2, 3.
8. J. G. Kirkwood and J. C. Poirier, *J. Phys. Chem.*, **58**, 591 (1954).
9. K. Linderstrøm-Lang, *Compt. rend. lab. Carlsberg*, *Sér. chim.*, **15**, No. 7 (1924).
10. C. Tanford, *J. Phys. Chem.*, **59**, 788 (1955).
11. T. L. Hill, *Arch. Biochem. Biophys.*, **57**, 229 (1955).
12. G. N. Watson, *Bessel Functions*, Cambridge University Press, 1922, p. 79.
13. J. G. Kirkwood, *J. Chem. Phys.*, **2**, 351 (1934).
14. C. Tanford and J. G. Kirkwood, *J. Am. Chem. Soc.*, **79**, 5333, 5340 (1957).
15. C. Tanford, *J. Am. Chem. Soc.*, **79**, 5348 (1957).

16. J. G. Kirkwood and F. H. Westheimer, *J. Chem. Phys.*, **6**, 506, 513 (1938).
17. T. L. Hill, *J. Chem. Phys.*, **12**, 147 (1944).
18. K. Linderstrøm-Lang, *Compt. rend. lab. Carlsberg. Sér. chim.*, **28**, 281 (1953).
19. J. J. Hermans and J. Th. G. Overbeek, *Rec. trav. chim.*, **67**, 761 (1948).
20. F. T. Wall and J. Berkowitz, *J. Chem. Phys.*, **26**, 114 (1957).
21. F. E. Harris and S. A. Rice, *J. Phys. Chem.*, **58**, 725 (1954); *J. Chem. Phys.*, **25**, 955 (1956); S. A. Rice and F. E. Harris, *J. Phys. Chem.*, **58**, 733 (1954).
22. S. A. Rice and F. E. Harris, *J. Chem. Phys.* **24**, 326 (1956).
23. L. Onsager, *Ann. N.Y. Acad. Sci.*, **51**, 627 (1949).
24. R. M. Fuoss, A. Katchalsky, and S. Lifson, *Proc. Natl. Acad. Sci., U.S.* **37**, 579 (1951); S. Lifson and A. Katchalsky, *J. Polymer Sci.*, **13**, 43 (1954).
25. T. Alfrey, Jr., P. W. Berg, and H. Morawetz, *J. Polymer Sci.*, **7**, 543 (1951).
26. A. Oth and P. Doty, *J. Phys. Chem.*, **56**, 43 (1952).
27. R. M. Fuoss and U. P. Strauss, *J. Polymer Sci.*, **3**, 246, 602 (1948); R. M. Fuoss, *Discussions Faraday Soc.*, No. 11, 125 (1951).
28. H. Terayama and F. T. Wall, *J. Polymer Sci.*, **16**, 357 (1955).
29. D. T. F. Pals and J. J. Hermans, *Rec. trav. chim.*, **71**, 433 (1952).
30. R. M. Fuoss and R. Signer, *J. Am. Chem. Soc.*, **73**, 5872 (1951).
31. N. S. Schneider and P. Doty, *J. Phys. Chem.*, **58**, 762 (1954).
32. J. R. Huizenga, P. F. Grieger, and F. T. Wall, *J. Am. Chem. Soc.*, **72**, 2636, 4228 (1950).
33. Ref. 7, Chs. 6 and 7.
34. F. T. Wall, H. Terayama, and S. Techakumpuch, *J. Polymer Sci.*, **20**, 477 (1956).
35. F. E. Bailey, A. Patterson, and R. M. Fuoss, *J. Am. Chem. Soc.*, **74**, 1845 (1952).
36. P. Doty, A. Wada, J. T. Yang, and E. R. Blout, *J. Polymer Sci.*, **23**, 851 (1957); P. Doty, K. Imahori, and E. Klemperer, *Proc. Natl. Acad. Sci., U.S.*, **44**, 424 (1958).
37. T. Alfrey, Jr., R. M. Fuoss, H. Morawetz, and H. Pinner, *J. Am. Chem. Soc.*, **74**, 438 (1952).
38. G. Ehrlich and P. Doty, *J. Am. Chem. Soc.*, **76**, 3764 (1954).
39. A. Katchalsky and I. R. Miller, *J. Polymer Sci.*, **13**, 57 (1954).
40. U. P. Strauss, E. H. Smith, and P. L. Wineman, *J. Am. Chem. Soc.*, **75**, 3935 (1953); U. P. Strauss and E. H. Smith, *ibid.*, **75**, 6186 (1953).
41. U. P. Strauss and P. L. Wineman, *J. Am. Chem. Soc.*, **80**, 2366 (1958).
42. U. P. Strauss and N. L. Gershfeld, *J. Phys. Chem.*, **58**, 747 (1954); U. P. Strauss, N. L. Gershfeld, and E. H. Crook, *ibid.*, **60**, 577 (1956).
43. J. G. Buzzell and C. Tanford, *J. Phys. Chem.*, **60**, 1204 (1956).
44. R. E. Weber, unpublished data.
45. P. S. Lewis, *Biochem. J.*, **20**, 978 (1926).
46. R. J. Gibbs, M. Bier, and F. F. Nord, *Arch. Biochem. Biophys.*, **35**, 216 (1952).
47. J. Steinhardt, *Kgl. Danske Videnskab. Selskab, Mat. fys. Medd.*, **14**, No. 11 (1937).
48. H. Edelhoch, *J. Am. Chem. Soc.*, **78**, 2644 (1956); **80**, 6640 (1958).
49. C. Tanford, J. G. Buzzell, D. G. Rands, and S. A. Swanson, *J. Am. Chem. Soc.*, **77**, 6421 (1955).
50. J. T. Yang and J. F. Foster, *J. Am. Chem. Soc.*, **76**, 1588 (1954).
51. W. F. Harrington, P. Johnson, and R. H. Ottewill, *Biochem. J.*, **62**, 569 (1956).
52. G. Weber, *Biochem. J.*, **51**, 155 (1952); *Discussions Faraday Soc.*, No. 13, 33 (1953).
53. P. Doty and G. E. Myers, *Discussions Faraday Soc.*, No. 13, 51 (1953).
54. R. F. Steiner, *Arch. Biochem. Biophys.*, **47**, 56 (1953).
55. T. L. Hill, *Discussions Faraday Soc.*, No. 21, 31 (1956); *J. Am. Chem. Soc.*, **80**, 2923 (1958); D. Stigter and T. L. Hill, *J. Phys. Chem.*, **63**, 551 (1959).

56. R. Townend and S. N. Timasheff, *Arch. Biochem. Biophys.*, **63,** 482 (1956); *J. Am. Chem. Soc.*, **79,** 3613 (1957), **82,** 3161, 3168 (1960).

57. P. Doty, *Proceedings 3rd. Intl. Congress of Biochemistry*, Academic Press, New York, 1956, p. 135.

58. J. M. Sturtevant and E. P. Geiduschek, *J. Am. Chem. Soc.*, **80,** 2911 (1958).

59. C. Tanford, *Symposium on Protein Structure* (A. Neuberger, ed.), Methuen and Co., London, 1958.

8

MULTIPLE EQUILIBRIA

29. GENERAL THEORY OF MULTIPLE EQUILIBRIA[1-3]

When macromolecules combine reversibly with smaller molecules or ions the usual laws of equilibrium govern the reaction. The number of combining sites on the macromolecule may be large, so that a large number of equilibrium constants may be required, but no new principle is thereby introduced. The combination of a macromolecule P with a smaller molecule or ion A may thus be described by n association constants,

$$k_1 = \frac{(PA)}{(P)C}, \quad k_2 = \frac{(PA_2)}{(PA)C}, \cdots, \quad k_n = \frac{(PA_n)}{(PA_{n-1})C} \qquad (29\text{--}1)$$

where parentheses represent concentrations, C being used in place of the concentration (A) of free A, and n is the total number of combining sites on the macromolecule. Alternatively, if we so prefer, we may describe the reaction in terms of the corresponding *dissociation* constants,

$$K_1 = \frac{(PA_{n-1})C}{(PA_n)}, \quad K_2 = \frac{(PA_{n-2})C}{(PA_{n-1})}, \cdots, \quad K_n = \frac{(P)C}{(PA)} \qquad (29\text{--}2)$$

where $K_1 = 1/k_n$, $K_2 = 1/k_{n-1}, \cdots, K_n = 1/k_1$.

The equilibrium constants $k_1, k_2, \cdots, k_n$, or $K_1, K_2, \cdots, K_n$, are not true constants but depend on the solvent, salt concentration, etc., as do all equilibrium constants which are expressed in terms of *concentrations* rather than *activities*. True thermodynamic constants, expressed in terms of the activities of the combining species, could be obtained from these constants in the usual way, for instance by extrapolation to infinite dilution of all solutes, a condition under which activities become by definition equal to concentrations. The usual practice, however, is not to determine the true thermodynamic constants, because the accuracy of experimental data rarely justifies this refinement.

526

All equilibrium constants, including those used here, may be related to standard free energy changes by the relation $\Delta F^\circ = -\mathscr{R}T \ln k$. The value of ΔF° so defined is the free energy change which accompanies the stoichiometric reaction, all reactants and products being in their standard states. The standard states are states of unit activity if the k's are true thermodynamic constants. In the present case the standard states are unit concentrations in the particular solvent medium being used.

The concentrations of the individual species (P), (PA), $\cdots$, (PA$_n$) are not usually experimentally determinable. Instead, in a study of the combination of a macromolecule with a smaller molecule or ion, it is usually possible to determine only the *average* number of molecules (or ions) A associated with each macromolecule. This number is designated by $\bar{\nu}$ and is measured as

$$\bar{\nu} = \frac{\text{moles combined A}}{\text{total moles P}} \tag{29-3}$$

A summary of some of the experimental methods used to determine $\bar{\nu}$ has been given by Klotz.[3] Ordinarily the method depends on a determination of the number of moles of free, uncombined A in a solution containing known total amounts of A and P. The amount of combined A is then obtained by difference and $\bar{\nu}$ evaluated by equation 29-3.

The methods most frequently employed are *dialysis equilibrium and potentiometry*. In dialysis equilibrium the macromolecular solution is separated by a semi-permeable membrane from a similar solution which contains no macromolecules. The low-molecular-weight substance A passes freely across the membrane, such that its activity on both sides is the same when equilibrium has been reached. The solutions on the two sides are usually made sufficiently similar (e.g., by addition of electrolyte) so that activity coefficients should also be the same on both sides; thus C must be the same on both sides. (If A and P are ionic, we must use a sufficiently high concentration of added salt so as to supress the Donnan effect described in section 14, or else a correction for it must be made.) Standard analytical methods are then used to determine the total amount of A on both sides. The measurement on one side yields the sum of free and combined A for the macromolecular solution; that on the other side is a measure of the concentration C. Sometimes only C is determined experimentally, the amount of combined A being obtained from the total amount originally added and from the volumes of solution on each side of the membrane.

The potentiometric method is more direct. A suitable cell is constructed containing an electrode behaving reversibly with respect to A. The electromotive force of the cell then measures the activity of A in the cell solution. Activity coefficients determined from solutions of known concentration of A are used to convert activity to concentration, and combined A is determined as the difference between the added amount and that observed present in the free state. If the potentiometric method is used, the equilibrium constants of equations 29–1 and 29–2 are often expressed in mixed units, with C replaced by the *activity* of A, while terms of the type (PA_v) continue to represent concentrations. The potentiometric method is applicable to relatively few substances because reversible electrode systems are few in number. The recent development of ion exchanger electrodes promises that the potentiometric method may be extended to virtually all ions.[7,8]

By means of equation 29–1 it is possible to express $\bar{v}$ in terms of the n association constants $k_1, \cdots, k_n$, for the total number of moles of the macromolecular species (per liter of solution or per kilogram of solvent, depending on the concentration units used) is just $(P) + (PA) + (PA_2) + \cdots + (PA_n)$ and the total number of moles of combined A (again per liter of solution or per kilogram of solvent) is $(PA) + 2(PA_2) + 3(PA_3) + \cdots + n(PA_n)$. By means of equation 29–1 we can now write

$$(PA) = k_1 C(P)$$
$$(PA_2) = k_1 k_2 C^2(P)$$
$$\vdots$$
$$(PA_n) = k_1 k_2 \cdots k_n C^n(P)$$

(29–4)

so that the equation for $\bar{v}$ becomes

$$\bar{v} = \frac{k_1 C + 2k_1 k_2 C^2 + \cdots + nk_1 k_2 \cdots k_n C^n}{1 + k_1 C + k_1 k_2 C^2 + \cdots + k_1 k_2 \cdots k_n C^n}$$

(29–5)

A similar equation may again be obtained in terms of dissociation constants. If $\bar{r} = n - \bar{v}$ is the average number of A ions or molecules dissociated from the fully saturated form PA_n, then

$$\bar{r} = \frac{K_1/C + 2K_1 K_2/C^2 + \cdots + nK_1 K_2 \cdots K_n/C^n}{1 + K_1/C + K_1 K_2/C^2 + \cdots + K_1 K_2 \cdots K_n/C^n}$$

(29–6)

It will be noted that the macromolecular concentration does not enter into equations 29–5 and 29–6. This is easily seen to be a consequence of the fact that we have not allowed for complexes between P and A which contain more than one molecule of P (e.g., P_2A). It is found that this simplification can be made in virtually every example of interest, as is easily checked experimentally by the observation that the relation between $\bar{v}$ or $\bar{r}$ and C is in fact independent of the macromolecular concentration and that there are no accompanying changes in average molecular weight.

Another important aspect of any experimental study of equilibrium is a test of whether we are measuring equilibrium binding at all. Such a test usually takes the form of a test of reversibility. A solution for which $\bar{\nu}$ is large is diluted so that C is decreased. The value of $\bar{\nu}$ at the new concentration C should be identical to that obtained at the same concentration by the addition of A to a solution containing P only. If this test fails, the addition of A to P is causing irreversible changes, and the measurements of binding do not represent equilibrium.

A convenient way to represent binding data is to plot $\bar{\nu}$ versus log C. Typical plots are shown in Fig. 29–1. It is seen that the general character of such a plot is quite similar to that of the formation curves of metal complexes, such as the metal ammines, where the same equations apply.[2]

The constants $k_1, k_2, \cdots, k_n$ can be obtained from the binding curve by a series of successive approximations. Several techniques are discussed and applied to metal ammine formation by J. Bjerrum.[2] For macromolecules, where n may be very large, the procedure would be extremely laborious.

For two reasons the representation of multiple equilibria of macromolecules by equations 29–5 or 29–6 is a procedure which has been rarely applied to cases where n is appreciably large. The first reason is that the evaluation of the n equilibrium constants from experimental data is laborious, so that, as a mere description of the data, a graphical representation such as that of Fig. 29–1 is more convenient. The second reason is that, if we wish to use the equilibrium constants as a basis for interpreting the details of the reaction, such as which sites are involved, the n constants defined in equation 29–1 or 29–2 turn out to be badly suited for this purpose.

This difficulty in interpretation occurs when only two combining sites are present, even in a small molecule. Consider, for example, the dissociation of the two protons of an aminothiol, ^+H_3N—CH_2—R—CH_2—SH, where R represents any desired structure. This reaction[11] is completely described thermodynamically in terms of two dissociation constants, K_1 and K_2, which may turn out, for example, to have the values 10^{-8} and 10^{-10}. We intuitively assign them to the amino and sulfhydryl protons, respectively, and are then led to ask which constant refers to which proton. This intuitive argument is, however, wrong. There are in fact *four* meaningful constants involved in the reaction,

$$
\begin{array}{ccc}
 & {}^+H_3N\text{—}CH_2\text{—}R\text{—}CH_2\text{—}S^- + H^+ & \\
 & [{}^+HNS^-] & \\
K_A \nearrow & & \searrow K_C \\
{}^+H_3N\text{—}CH_2\text{—}R\text{—}CH_2\text{—}SH & & H_2N\text{—}CH_2\text{—}R\text{—}CH_2\text{—}S^- + 2H^+ \quad (I) \\
[{}^+HNSH] & & [NS^-] \\
K_B \searrow & & \nearrow K_D \\
 & H_2N\text{—}CH_2\text{—}R\text{—}CH_2\text{—}SH + H^+ & \\
 & [NSH] & \\
\end{array}
$$

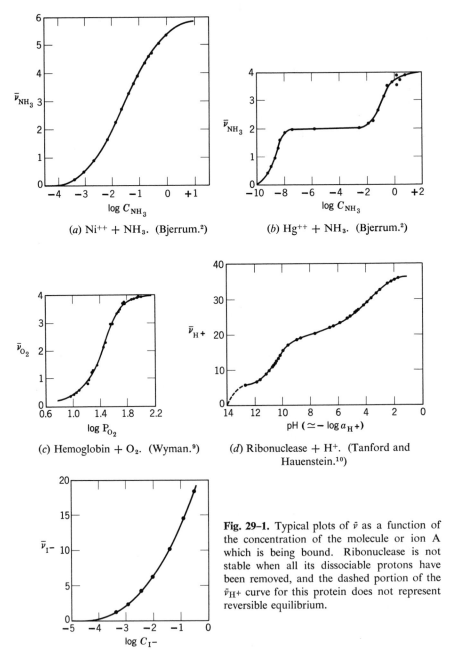

(a) Ni^{++} + NH$_3$. (Bjerrum.[2])

(b) Hg^{++} + NH$_3$. (Bjerrum.[2])

(c) Hemoglobin + O$_2$. (Wyman.[9])

(d) Ribonuclease + H$^+$. (Tanford and Hauenstein.[10])

Fig. 29–1. Typical plots of $\bar{\nu}$ as a function of the concentration of the molecule or ion A which is being bound. Ribonuclease is not stable when all its dissociable protons have been removed, and the dashed portion of the $\bar{\nu}_{H^+}$ curve for this protein does not represent reversible equilibrium.

(e) Serum albumin + I$^-$. (Scatchard et al.[7])

although only three of them are independent since $K_A K_C = K_B K_D$, both of these quantities being equivalent to $(NS^-)(H^+)^2/(^+HNSH)$. The experimental constants K_1 and K_2, as defined by equation 29-2 are easily related to the four constants of equation I,

$$K_1 = \frac{\{(NSH) + (^+HNS^-)\}(H^+)}{(^+HNSH)} = K_A + K_B$$

$$K_2 = \frac{(NS^-)(H^+)}{\{(NSH) + (^+HNS^-)\}} = \frac{K_C K_D}{K_C + K_D}$$

It is thus possible that the dissociation constants, K_A and K_B, of the original aminothiol may actually be about equal, in which case also $K_C \simeq K_D$. What makes K_2 smaller than K_1 might then be the fact that $K_C < K_B$ by virtue of the negative charge on the sulfur atom of $^+HNS^-$ and that $K_D < K_A$ because of the positive charge on ^+HNSH. Whether this hypothesis is correct cannot be decided from the experimental dissociation curves alone, since they yield only the two constants K_1 and K_2.

Two procedures are possible as a way out of this difficulty, and it is worthwhile mentioning them since in principle they resemble what would have to be done on a larger scale if the thermodynamic equilibrium constants of macromolecular reaction are to be interpreted in a corresponding way. The first procedure is to devise an auxiliary measurement which determines specifically the extent of dissociation of one of the two protons. In the present example, for instance, we can use a spectroscopic method[11] which specifically measures the extent of conversion of $-SH$ to S^-. This permits the calculation of an additional experimental constant, which, with the relation $K_A K_C = K_B K_D$, allows all four of these constants to be evaluated.

The second procedure is one involving calculation alone. For example, in the case of an aminothiol, we can assume that the reactive sites are sufficiently well separated so as to have no effect on one another except by virtue of their electrostatic charges, if they have them. Thus K_D can be assumed to have a value close to the dissociation constant of a suitable monobasic acid (e.g., C_2H_5SH) and the relation between K_A and K_D (i.e., the influence of the $-NH_3^+$ group) can be computed from electrostatic theory.[12] And in the same way K_B may be taken as roughly equal to the dissociation constant of $C_2H_5NH_3^+$, and K_C may be calculated from K_B. If K_1 and K_2 calculated in this way are not in agreement with experiment, a faulty assignment of structural parameters (in the calculation of the electrostatic effects, for example) is indicated. They are then adjusted until agreement is obtained; i.e., the calculation leads indirectly to information concerning the structure of the molecule at the same time.

Another disadvantage of the purely thermodynamic representation of the equilibrium in terms of n constants is that this representation fails to distinguish between molecules in which all the combining sites are different and those in which all or some of the sites may be identical. Basically, the problem must clearly be capable of simplification in the latter case. To take again a simple example, the dissociation of hydrogen ions from an acid such as $HOOC—(CH_2)_x—COOH$ is described thermodynamically by two constants, just as in the case of the aminothiol previously referred to. That it is fundamentally a much simpler situation can be seen, however, if we again write the equilibrium in the form of equations I. We see at once that for the dicarboxylic acid $K_A = K_B$, $K_C = K_D$, $K_1 = 2K_A$, and $K_2 = K_C/2$; i.e., in this case the constants K_A, K_B, K_C, and K_D are determinable unambiguously without auxiliary measurements.

If these considerations are extended to a macromolecule with n combining sites, we would have to take into account $2^n - 1$ independent constants of the type K_A, K_B, $\cdots$. Ordinarily this is clearly an impossible task (see however section 30), and it is clearly desirable to seek an alternative representation of the equilibrium between a macromolecule and low-molecular-weight substances. The following sections will be devoted to this purpose, with the object of reducing the number of parameters required to describe the equilibrium where this can reasonably be expected to be possible, i.e., when many of the combining sites are identical.

29a. Identical and completely independent sites. In the simplest possible situation, where the n combining sites are identical and have no effect upon one another whatsoever, the n conventional equilibrium constants can all be related to a single equilibrium constant, k. Suppose that the n identical sites of a macromolecule P could be labeled in some way, so that they could all be distinguished. There would then be n different forms of the species PA, $n(n - 1)/2!$ forms of the species PA_2, and, in general, $n!/i! (n - i)!$ forms of the species PA_i. The forms P and PA_n would be unique. If PA* represents a *particular* species of PA, i.e., a species with A attached to a particular site, then the association constant $(PA*)/(P)C$ would, for identical sites, have the same value k for every possible species. There being n equally probable forms of PA and only one form of P, the first association constant, $(PA)/(P)C$, which is the sum of all the possible $(PA*)/(P)C$, would be equal to nk. There being $n(n - 1)/2!$ forms of PA_2, $k_1 k_2 = (PA_2)/(P)C^2$ would similarly be equal to $n(n - 1)k^2/2!$; i.e., k_2 would be equal to $\frac{1}{2}(n - 1)k$. It is easily seen that the general expression for k_i will be

$$k_i = k\,\frac{n - i + 1}{i} \tag{29-7}$$

and if this is substituted in equation 29–5 we get

$$\bar{\nu} = \frac{\sum\limits_{i=1}^{n} i \left\{ \prod\limits_{j=1}^{i} \left(\frac{n-j+1}{j} \right) \right\} k^i C^i}{1 + \sum\limits_{i=1}^{n} \left\{ \prod\limits_{j=1}^{i} \left(\frac{n-j+1}{j} \right) \right\} k^i C^i} \qquad (29\text{–}8)$$

The product occurring in this equation can be simplified

$$\prod_{j=1}^{i} \left(\frac{n-j+1}{j} \right) = \frac{n!}{(n-i)!\, i!} \qquad (29\text{–}9)$$

so that equation 29–8 becomes

$$\bar{\nu} = \frac{\sum\limits_{i=1}^{n} i \frac{n!}{(n-i)!\, i!} (kC)^i}{1 + \sum\limits_{i=1}^{n} \frac{n!}{(n-i)!\, i!} (kC)^i} \qquad (29\text{–}10)$$

According to the binomial theorem

$$(1 + kC)^n = 1 + \sum_{i=1}^{n} \frac{n!}{(n-i)!\, i!} (kC)^i \qquad (29\text{–}11)$$

Differentiation with respect to kC leads to

$$kC \frac{d(1 + kC)^n}{d(kC)} = kCn(1 + kC)^{n-1} = \sum_{i=1}^{n} i \frac{n!}{(n-i)!\, i!} (kC)^i \qquad (29\text{–}12)$$

Equations 29–11 and 29–12 are seen to be the same as the denominator and numerator, respectively, of equation 29–10, so that the latter becomes

$$\bar{\nu} = \frac{kCn(1 + kC)^{n-1}}{(1 + kC)^n} = \frac{nkC}{1 + kC} \qquad (29\text{–}13)$$

Rearranging equation 29–13 we obtain

$$\frac{\bar{\nu}}{n - \bar{\nu}} = kC \qquad (29\text{–}14)$$

It will now be shown that equation 29–14 can be derived in a much simpler way. The constant k was defined above as the equilibrium constant for the reaction at a single combining site. Let the state of equilibrium at any such site be represented by its *degree of association*, designated by θ. This is defined as the probability that this site on a given molecule is combined with the reacting substance A; in other words, it represents the fraction of such sites (on different molecules) which are combined with A. The uncombined fraction is then $1 - \theta$, and the equilibrium condition becomes

$$\frac{\theta}{1 - \theta} = kC \tag{29–15}$$

If there are now n sites on a macromolecule and these sites are identical and completely independent, the degree of association must be the same at each site and, moreover, must have the same value at each site as it would have if there were only a single site present. It therefore follows that $\bar{\nu} = n\theta$, and equation 29–15 leads at once to equation 29–14. By focusing our attention on an individual site and then multiplying the result by the number of sites, we have been able to avoid the statistical factors which occur, when we focus our attention on the molecule as a whole, as a result of the fact that different combinations of events at individual sites lead to the same over-all state of the molecule as a whole.

The single constant k which occurs in these equations is an association constant for the reaction

$$\text{free site} + \text{A} \rightleftharpoons \text{combined site} \tag{II}$$

and it may be related in the usual way to a standard free energy change for the same reaction (stoichiometrically, from left to right), by the relation $\Delta F^\circ = -\mathscr{R}T \ln k$. The usual standard state, in terms of concentrations, would be the state of unit concentration of sites, and the general expression for the concentration dependence of the chemical potential of either kind of site would be $\mu = \mu^\circ + \mathscr{R}T \ln C_s$, where C_s is the concentration of sites. In the present case, however, both kinds of sites are attached to macromolecules, the concentration (C_M) of which does not change during the reaction. Thus $C_s = \theta C_M$ for the combined sites and $C_s = (1 - \theta)C_M$ for the free sites, and the general expression for the free energy change of reaction II, at any concentration, becomes

$$\Delta F = \Delta F^\circ + \mathscr{R}T \ln \frac{\theta}{(1 - \theta)C} \tag{29–16}$$

where C is the concentration of A. It is important to note that ΔF and ΔF° of this equation represent free energy changes *per mole of combining sites*.

A noteworthy feature of equation 29–16 is that C_M does not appear in it, for the reason that only the *relative* concentrations of the two kinds of sites enter into the expression.

It should also be noted that ΔF for reaction II becomes zero at equilibrium. Equation 29–16, together with the relation $\Delta F^\circ = -\mathscr{R}T \ln k$, then leads at once to equation 29–15.

A treatment similar to that here applied to association may be applied to equilibria viewed in terms of dissociation. If α is the *degree of dissociation* at any site and K the dissociation constant for the reverse of reaction II, then, with $\bar{r}$ defined by equation 29–6,

$$\frac{\alpha}{1 - \alpha} = \frac{\bar{r}}{n - \bar{r}} = \frac{K}{C} \qquad (29\text{–}17)$$

An equation analogous to equation 29–16 may also be written, with $\alpha C/(1 - \alpha)$ replacing $\theta/(1 - \theta)C$.

The advantage of equations 29–14 and 29–17 is that they represent polyequilibria by a single constant, rather than by n separate constants. The assumption made in their derivation, that the combining sites are identical and independent, is not, of course, generally applicable; i.e., these equations rarely, if ever, fit curves like those of Fig. 29–1. The remainder of this chapter will be devoted to methods for taking into account the deviations from these assumptions, while maintaining, as far as possible, the simple form of equations 29–14 and 29–17.

29b. Identical sites with interaction between them. Suppose a macromolecule has n *identical* sites which interact with one another in such a way that binding at any site affects the binding affinity at other sites. It is possible to allow for this effect without in any way altering equations 29–14 to 29–17, simply by allowing k and ΔF° to vary as $\bar{\nu}$ varies. Specifically we first define an *intrinsic association constant*, k_{int}, which will be the limiting value of k when $\bar{\nu} = 0$, i.e., before any interaction due to binding has occurred. Corresponding to it there will be an *intrinsic standard free energy change*, $\Delta F^\circ_{\text{int}}$, applying to reaction II when $\bar{\nu} = 0$. The effect of interaction is then introduced by defining a completely arbitrary function, $\phi(\bar{\nu})$, such that, for $\bar{\nu} > 0$,

$$\Delta F^\circ = \Delta F^\circ_{\text{int}} + \mathscr{R}T\,\phi(\bar{\nu})$$

or $\qquad\qquad\qquad\qquad\qquad\qquad\qquad\qquad\qquad\qquad (29\text{–}18)$

$$k = k_{\text{int}}\, e^{-\phi(\bar{\nu})}$$

If $\phi(\bar{\nu})$ is an increasing function of $\bar{\nu}$, occupation of some sites makes binding at other sites more difficult; if it is a decreasing function of $\bar{\nu}$, there is cooperative interaction between sites; i.e., occupation of some sites facilitates binding at other sites. By definition $\phi(\bar{\nu})$ is zero when $\bar{\nu} = 0$. The fact that $\phi(\bar{\nu})$ is a completely unrestricted function means that any effect whatever can be formally described.

It should be noted that interacting identical sites do not remain identical when binding takes place; sites located nearest an occupied site will be subjected to stronger interaction than those farther away. ΔF° and $\phi(\bar{\nu})$

as here defined are therefore *average* properties, averaged over all sites and all configurations at any value of $\bar{\nu}$.

By combining equations 29–18 and 29–15 we obtain

$$\frac{\bar{\nu}}{n - \bar{\nu}} = \frac{\theta}{1 - \theta} = kC = k_{int}\, e^{-\phi(\bar{\nu})}\, C \qquad (29\text{–}19)$$

If the combining sites in a particular reaction can reasonably be expected to be identical and if their number, n, is known, then this relation can be applied directly to determine the qualitative nature of the interaction

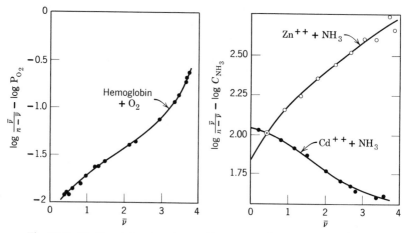

Fig. 29–2. Binding data plotted according to equation 29–20. The data for the hemoglobin-oxygen reaction are from Fig. 29–1. Those for the combination of NH_3 with Zn^{++} and Cd^{++} are from Bjerrum.[2] The number of binding sites, n, is four in each example.

between sites. This is done most conveniently if equation 29–19 is converted to logarithmic form,

$$\log \frac{\bar{\nu}}{n - \bar{\nu}} - \log C = \log k_{int} - 0.434\, \phi(\bar{\nu}) \qquad (29\text{–}20)$$

Figure 29–2, for instance, shows a plot of the left-hand side of equation 29–20 versus $\bar{\nu}$ for the combination of O_2 with hemoglobin. The combining sites here are the four porphyrin-bound iron atoms of the hemoglobin molecule. These sites should have identical intrinsic affinity for O_2. Figure 29–2 shows clearly that $\phi(\bar{\nu})$ is a decreasing function of $\bar{\nu}$, i.e., that the binding of the first molecule of O_2 *increases* the affinity of the remaining sites for further binding. The more usual situation is the reverse of this.

The best examples are provided in reactions in which the interaction is electrostatic in nature, and they will be discussed in more detail below.

The applicability of equation 29-20 is not limited to macromolecular reactions. Figure 29-2 shows the same equation applied to the combination of ammonia with simple metal ions. It is seen that both cooperative and interfering interaction may occur in these simple cases just as they may in macromolecular reactions.

29c. Approximate treatment of electrostatic interaction. In many of the best-known examples of reversible reactions between large molecules and substances of low molecular weight the small combining particles are simple ions. In this case the predominant form of interaction between sites is likely to be electrostatic in nature; i.e., $\phi(\bar{\nu})$ is directly related to the effect of charged sites on the chemical potential of the reacting species.

In view of the non-specific nature of electrostatic forces it becomes convenient to modify the general' treatment of section 29b so as to allow for interaction between all sites which bear charges, i.e., not only the charges resulting from the $\bar{\nu}$ bound ions but also those which may be present elsewhere on the molecule. It becomes appropriate to redefine the intrinsic properties of a site as those which it would have in the absence of *any* charges, rather than when $\bar{\nu} = 0$. Many macromolecules, however, are polyampholytes and possess no form devoid of charges. (The majority of all equilibrium studies have involved such molecules.) For these molecules the most rigorous procedure is to define the intrinsic properties as those which a molecule would possess in a *hypothetical discharged state*. A treatment incorporating this definition is, however, not simple. A brief description will be given in section 30. In the present section we shall confine ourselves to an approximate treatment which assumes that all electrostatic interactions vanish when the average net charge per molecule, $\bar{Z}$, is equal to zero, and we thus define the intrinsic properties of a site as those which it possesses when $\bar{Z} = 0$. For a macromolecule which can possess charges of only one sign, such as a simple synthetic polyelectrolyte, the state $\bar{Z} = 0$ is indeed a discharged state. If both positive and negative charges are present, however, the interaction between them does not in general vanish when $\bar{Z} = 0$, as was shown in section 26, and the intrinsic properties, as here defined, then differ from those for a completely discharged state.

We now write, analogous to equations 29-18

$$\Delta F^\circ = \Delta F^\circ_{\text{int}} + \mathscr{R}T\,\phi(\bar{Z}) \qquad (29\text{-}21)$$

where $\phi(\bar{Z})$ will be a positive function of $\bar{Z}$ for combination with a positive ion and a negative function for combination with a negative ion; i.e., a net (positive) charge will always tend to repel a positive ion and attract a negative one. Similarly, analogous to equation 29–19

$$\frac{\bar{\nu}}{n - \bar{\nu}} = \frac{\theta}{1 - \theta} = kC = k_{\text{int}}\, e^{-\phi(Z)} C \qquad (29\text{–}22)$$

It will now be shown that $\phi(Z)$ can be simply related to the electrostatic free energy, W_{el}, of the macromolecule if we make the simple approximation that the average free energy per macromolecule in a solution can be replaced by the free energy of the average molecule (characterized by the average charge Z).

We consider the process of adding, on the average, $d\bar{\nu}$ ions of charge z_i to a macromolecule which already has an average charge Z. On the average, $d\bar{\nu}/n$ ions will be added to each one of the n identical sites for binding. The resulting free energy change, per mole of sites, will be just $d\bar{\nu}/n$ times the value of ΔF given by equation 29–16; i.e.,

$$dF = \left[\Delta F^{\circ} + \mathscr{R}T \ln \frac{\theta}{(1 - \theta)C}\right]\frac{d\bar{\nu}}{n} \qquad (29\text{–}23)$$

An expression equivalent to equation 29–23 may be obtained by carrying out the same process by steps. We first discharge the macromolecule with charge Z. The resulting free energy change per mole of macromolecules will be $-\mathscr{N} W_{\text{el}}(Z)$, where W_{el} is the electrostatic free energy of a single macro-ion as given by one of the relations of section 26 and $\mathscr{N}$ is Avogadro's number. The corresponding free energy change per mole of sites will be $-\mathscr{N} W_{\text{el}}(Z)/n$. Per mole of sites $d\bar{\nu}/n$ moles of small ions are being added, and we now discharge them also, with a free energy change of $-\mathscr{N} W_i\, d\bar{\nu}/n$, where W_i is the work of charging a single ion of charge z_i.

We next combine the sites on the discharged macromolecule with the discharged ions. Since the dependence of ΔF on θ and on C is entirely independent of whether the combining species are charged, the result will again be $d\bar{\nu}/n$ times the value of ΔF given by equation 29–16, the only difference being that ΔF° now has a different value, since it now represents the standard free energy change for reaction of discharged species, which may be very different from that for reaction of the charged species. Using the symbol ΔF^* for this standard free energy, we then have

$$dF = \Delta F^*\, d\bar{\nu}/n + \mathscr{R}T(d\bar{\nu}/n) \ln \theta/(1 - \theta)C$$

As the final step, we restore the charge previously removed. All of it now goes onto the macro-ion, which now has a charge of $(Z + z_i\, d\bar{\nu})$. The free energy change per mole of sites is $+\mathscr{N} W_{\text{el}}(Z + z_i\, d\bar{\nu})/n$ or $\mathscr{N} W_{\text{el}}(Z)/n + \mathscr{N}(\partial W_{\text{el}}/\partial Z)z_i\, d\bar{\nu}/n$.

The over-all free energy change is obtained by adding all these contributions, with the result

$$dF = \left[\Delta F^* - \mathcal{N}W_i + z_i\mathcal{N}\left(\frac{\partial W_{el}}{\partial Z}\right) + \mathcal{R}T\ln\frac{\theta}{(1-\theta)C} \right]\frac{d\bar{\nu}}{n}$$

This expression must be identical with equation 29–23, and it follows at once that

$$\Delta F^\circ = \Delta F^* - \mathcal{N}W_i + z_i\mathcal{N}(\partial W_{el}/\partial Z) \qquad (29\text{–}24)$$

We now use for W_{el} one of the *approximate* expressions of section 26 (e.g., equations 26–30, 26–42, and 26–56). All of them may be written in the form

$$W_{el} = A(\bar{Z})^2 \qquad (29\text{–}25)$$

(We are essentially committed to using an expression of this kind, rather than a more exact one, by the assumption that W_{el} vanishes when $\bar{Z} = 0$.) Differentiating this expression to obtain $(\partial W_{el}/\partial Z)$ we get

$$\Delta F^\circ = \Delta F^* - \mathcal{N}W_i + 2A\mathcal{N}z_i\bar{Z}$$

By definition, $\Delta F^\circ = \Delta F^\circ_{int}$ when $\bar{Z} = 0$. Thus $\Delta F^* - \mathcal{N}W_i = \Delta F^\circ_{int}$, and we get

$$\Delta F^\circ = \Delta F^\circ_{int} + 2A\mathcal{N}z_i\bar{Z} \qquad (29\text{–}26)$$

Comparing with equation 29–18 we see that

$$\phi(\bar{Z}) = 2A\mathcal{N}z_i\bar{Z}/\mathcal{R}T = 2wz_i\bar{Z} \qquad (29\text{–}27)$$

where

$$w = A\mathcal{N}/\mathcal{R}T \qquad (29\text{–}28)$$

is a constant independent of z_i or $\bar{Z}$. The value of w depends on the model chosen to represent the macromolecule, i.e., on which particular equation is used for W_{el} and on the parameters occurring in these equations, such as radius or length, ionic strength, or temperature. It should be noted that w is independent of the nature of the combining site, a consequence of the fact that we have assumed that W_{el} and its dependence on $\bar{\nu}$ can be expressed in terms of the net average charge.

29d. Sites with different intrinsic affinities. If a macromolecule is known to have sites with different intrinsic affinities for a combining molecule or ion, it is easily introduced in a formal way into the preceding equations. Let there be n_1 sites with intrinsic association constant $k_{int}^{(1)}$, n_2 sites with intrinsic constant $k_{int}^{(2)}$, etc. Then in the absence of interaction between sites, by equation 29–13,

$$\bar{\nu} = \frac{n_1 k_{int}^{(1)}C}{1 + k_{int}^{(1)}C} + \frac{n_2 k_{int}^{(2)}C}{1 + k_{int}^{(2)}C} + \cdots \qquad (29\text{–}29)$$

If there is interaction between sites, we use k's as given by equation 29–18 in place of intrinsic constants; i.e.,

$$\bar{\nu} = \frac{n_1 k_{\mathrm{inte}}^{(1)} e^{-\phi_1(\bar{\nu})} C}{1 + k_{\mathrm{inte}}^{(1)} e^{-\phi_1(\bar{\nu})} C} + \frac{n_2 k_{\mathrm{inte}}^{(2)} e^{-\phi_2(\bar{\nu})} C}{1 + k_{\mathrm{inte}}^{(2)} e^{-\phi_2(\bar{\nu})} C} + \cdots \tag{29–30}$$

Here the $\phi_i(\bar{\nu})$ may or may not be identical functions of $\bar{\nu}$. If the reaction is an ionic one, then k is given by equation 29–22 instead of equation 29–18. Using equation 29–27 for $\phi(Z)$ we get

$$\bar{\nu} = \frac{n_1 k_{\mathrm{inte}}^{(1)} e^{-2wz_i Z} C}{1 + k_{\mathrm{inte}}^{(1)} e^{-2wz_i Z} C} + \frac{n_2 k_{\mathrm{inte}}^{(2)} e^{-2wz_i Z} C}{1 + k_{\mathrm{inte}}^{(2)} e^{-2wz_i Z} C} + \cdots \tag{29–31}$$

Here the interaction parameter is the same for all sites.

Equations 29–30 and 29–31 are useful, of course, only if the number of terms on the right is relatively small, i.e., if many of the combining sites are identical. In general we expect this condition to be satisfied because the same kind of monomer unit occurs repeatedly in a given macromolecule.

It is conceivable, however, that environmental effects will make a set of combining sites similar, but not truly identical. This would result in a statistical distribution of k_{int} values about a suitable mean. Equations for this situation, in the absence of interaction, have been derived by Karush and Sonenberg.[13] It is not known whether the situation to which they apply actually occurs in any real macromolecular reaction.

29e. Procedure in the absence of information concerning the number and nature of combining sites. When we are studying the dissociation of hydrogen ions from polymethacrylic acid of degree of polymerization n, we can with certainty identify the sites involved; they will be the side chain carboxyl groups of the polymer. The sites will all be identical, and their number will be n. On the other hand, when we are studying the combination of methyl orange or of iodide ion with serum albumin, this information is lacking and can be obtained only from an examination of the experimental data on the combination. We shall now discuss how experimental results may be treated to obtain this information. We shall find that in many cases it cannot be obtained at all from equilibrium data alone.

We consider first the determination of the total number n of combining sites. In principle it is easily evaluated, since n is the maximum value of $\bar{\nu}$ obtained at high concentrations of the combining species. Thus in Fig. 29–1a the value of n is clearly 6, in Fig. 29–1c it is 4, in Fig. 29–1d it is 36. Where the combining sites fall into classes of *very* different intrinsic affinity the corresponding numbers n_1, n_2, $\cdots$, can be determined in a similar way since combination with sites of the strongest affinity will be complete, so that $\bar{\nu}$ no longer increases with concentration, before combination with sites of weaker affinity has even begun. Thus Fig. 29–1b clearly indicates four sites (not necessarily identical in intrinsic affinity) which bind ammonia strongly and two which bind it only weakly. Figure 29–1d shows close to six sites with very strong affinity for H^+, about fourteen sites with a somewhat less strong affinity, about five more with a weaker affinity, and, finally a group of about eleven sites with weaker affinity still. (Figure 29–1d is an example of equilibrium with hydrogen ions, as discussed in section 30. It will be seen that sites with different affinity are in this case easily identified.)

Unfortunately it is not possible in many cases of relatively weak affinity to reach a value of $\bar{\nu}$ which approaches its maximum value. For example, we cannot determine from Fig. 29–1e the number of sites on a serum albumin molecule which can combine with iodide ion. [It might be thought that n can be evaluated by extending the data to higher concentrations, but this cannot be done for the reason that $\bar{\nu}$ is determined from the *difference* between the *total* concentration of I^- and the *free* concentration as determined (in this case) from an electromotive force measurement. Suppose we have a protein concentration of about 70 grams/liter, which is already undesirable, being high enough to produce effects due to interaction between protein molecules. In terms of molarity this concentration represents 10^{-3} mole/liter. Suppose that the free iodide concentration is about $1M$. Figure 29–1e indicates that we would then expect $\bar{\nu}$ to be about 20, so that the concentration of bound I^- is about $0.02M$ and the total iodide concentration $1.02M$. Clearly we would have to be able to determine the free concentration C to better than 1 part in 1000 to obtain a meaningful difference between 1.02 and $1.00M$, and this is a degree of precision which is experimentally unattainable.]

If all the combining sites involved in a reaction were identical and if there is no interaction between them, n could be evaluated graphically by extrapolation of the data to infinite concentration. Thus equation 29–13 can be written as $1/\bar{\nu} = 1/n + 1/nkC$. A plot of $1/\bar{\nu}$ versus $1/C$ would be linear with the intercept $1/n$ where $1/C = 0$. Alternatively, from equation 29–14, we see that $\bar{\nu}/C = k(n - \bar{\nu})$ so that $\bar{\nu}/C$ is a linear function of $\bar{\nu}$ which intersects the abscissa where $\bar{\nu} = n$. In most real situations, however, the combining sites will not be identical and will interact with one another so that the corresponding plots are far from linear. This is shown, for instance, by Figs. 29–3 and 29–4 which are based on Scatchard and coworkers' data on the serum albumin–iodide reaction.

The data on the serum albumin–iodide reaction cover a fairly wide range of ionic strength, so that the activity coefficient of iodide ion varies from about 1.0 to 0.7. Under these conditions it becomes desirable to replace the concentration C in all the preceding

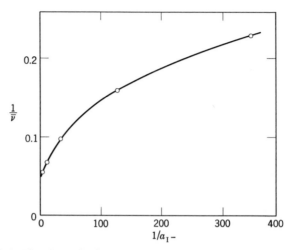

Fig. 29–3. The data of Fig. 29–1e plotted according to equation 29–13. The non-linearity of the plot shows that the combining sites are not identical sites without interaction. The ordinate at $1/a_{I^-} = 0$ is equal to $1/n$, but it cannot be evaluated with any certainty because of the curvature at low values of $1/a_{I^-}$.

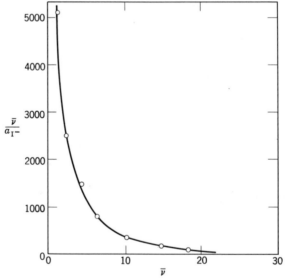

Fig. 29–4. The same data as in Fig. 29–3, plotted according to equation 29–14. The non-linearity again shows that the combining sites are not identical sites without interaction. The total number of sites, n, can be determined as the value of $\bar{\nu}$ at which $\bar{\nu}/a_{I^-} = 0$, but the figure shows that n can have any value above 25, since there is no information on which to base an extrapolation of the curve to higher $\bar{\nu}$.

equations by the activity a_{I^-}, or else a 30% variation in the association constants k or k_{int} must be expected. This procedure has been followed in the analysis of Scatchard and coworkers' data, the activity coefficient of iodide ion being computed by standard techniques (equation 26–33).

It can be shown that the intercepts of Figs. 29–3 and 29–4 still have the same meaning as they do for identical, non-interacting sites. For $\bar{\nu} = n$ when $C = \infty$, no matter what the nature of the combining sites may be. And it then clearly follows that $1/\bar{\nu} = 1/n$ when $1/C = 0$ and also that $\bar{\nu}/C = 0$ when $\bar{\nu} = n$. It is, however, impossible to determine these intercepts because we have no way of predicting whether the curvature of these plots will increase or decrease as we approach the limit of extrapolation. The best we can do is to extrapolate to a reasonable limit to obtain a working value of n for analysis of our data. Thus for the combination of iodide with serum albumin, we might guess from Fig. 29–3 that $1/n \simeq 0.04$ or from Fig. 29–4 that $n = 25$ to 30. We can then say that the data to the highest concentration used can probably be represented by means of about twenty-five combining sites but that there is no guarantee that more sites may not be present. (From general knowledge concerning serum albumin it is reasonable to predict that there are actually about 100 sites on the molecule which can react with iodide ion to some extent.)

When n, or at least a working value for n, has been evaluated we ask next whether the sites are all intrinsically identical or whether it is necessary to postulate the existence of sets of sites with different affinities. This question arises also in a situation like that illustrated by Fig. 29–1d, where a preliminary separation into sets of *greatly* different sites is possible. The sites in any one set may still not be intrinsically identical (and in the example of Fig. 29–1d they are in fact not identical).

It is immediately obvious that this question cannot be answered from experimental equilibrium data alone. For, with $\phi(\bar{\nu})$ an *arbitrary* function of $\bar{\nu}$, each individual term of equation 29–30 is a completely general equation for $\bar{\nu}$ as a function of C. In other words, with $\phi(\bar{\nu})$ an arbitrary function of $\bar{\nu}$, it is always possible to fit experimental data by using only a single term of equation 29–30. If more information is sought concerning the nature of the combining sites, it is necessary that we be able to make a reasonable estimate of the true interaction between occupied sites, so that the functions $\phi_1(\bar{\nu})$, $\phi_2(\bar{\nu})$, etc., of equation 29–30 can be assigned numerical values.

This procedure was used by Scatchard, Coleman, and Shen[7] to analyze the equilibria between serum albumin and various anions, including the data for iodide given earlier. The reaction being ionic, it may be assumed that the interaction between sites is entirely electrostatic. It is known, furthermore, that under the conditions of the experiments the serum albumin molecule is compact, impenetrable, and close to spherical, so that equation 26–30 can be used to evaluate W_{el}. The data should therefore be capable of representation by equation 29–31, with w given by equations 29–28 and 26–30 as

$$w = \frac{\epsilon^2}{DRkT}\left(1 - \frac{\kappa R}{1 + \kappa a}\right) \qquad (29\text{-}32)$$

The radius R can be estimated from the molecular weight and a reasonable density (equation 26–34). (The equation for w so obtained can be tested,

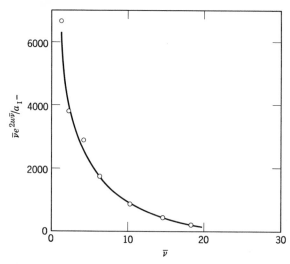

Fig. 29–5. The same data as in Figs. 29–3 and 29–4, plotted according to equation 29–34. This equation includes a term which removes that part of the curvature of Fig. 29–4 which is caused by electrostatic interaction between sites. That the result is still not a linear plot indicates that the binding sites do not belong to a single class with identical k_{int}. (Scatchard and co-workers.[7])

for it should apply to other ionic reactions of serum albumin, such as the reaction with H^+ ion. For this reaction the nature and number of sites is known.)

If now there is only one kind of site, we can represent the equilibrium by using only one term of equation 29–31. Upon rearrangement this gives

$$\frac{\bar{\nu}e^{2wz_iZ}}{C} = k_{int}(n - \bar{\nu}) \qquad (29\text{-}33)$$

In the experiments under consideration the protein molecules have a net charge of essentially zero before any binding has occurred, so that the net charge after binding becomes $Z = \bar{\nu} z_i$. For the iodide ion $z_i = -1$, so that equation 29–33 becomes

$$\frac{\bar{\nu} e^{2w\bar{\nu}}}{C} = k_{\text{int}}(n - \bar{\nu}) \qquad (29\text{–}34)$$

A plot of the left-hand side of equation 29–34 versus $\bar{\nu}$ (again with activity replacing concentration) is shown in Fig. 29–5. It is clearly *not linear*, and the combining sites can therefore not belong to a single class. It should be noted, incidentally, that this test does not require a previous choice of the total number of combining sites.

Continuing to assume that the interaction between sites is entirely electrostatic, we can now proceed to find the number of classes of sites which must be present, how many sites there are in each class, and the corresponding values of k_{int}. Essentially the experimental curves must be fitted by trial and error or by successive approximation. For details of some convenient procedures the reader is referred to Scatchard and co-workers' paper.[7] For the iodide–serum albumin reaction the data could not be fitted with fewer than three classes of sites. The following constants were obtained: $n_1 = 1$, $k_{\text{int}}^{(1)} = 9250$; $n_2 = 8$, $k_{\text{int}}^{(2)} = 385$; $n_3 = 18$, $k_{\text{int}}^{(3)} = 12.7$. It should be pointed out that this solution is not unique. More classes of sites could have been introduced. However, we would always find one or perhaps two combining sites with much stronger affinity than all the rest and always about seven or eight sites with stronger affinity than the remainder.

29f. Identification of combining sites. A study of the equilibrium between a macromolecule and smaller molecules or ions is not complete until the combining sites have been identified and the corresponding values of k_{int} have been shown to agree reasonably well with the expected affinities of those sites. This objective is nearly always achieved in the study of the equilibrium between hydrogen ion and acidic or basic macromolecules, as will be seen in section 30. It is also achieved very strikingly for the equilibrium between serum albumin and Zn^{++}, as described in section 31. Often, however, this final step in the analysis of equilibrium data has not yet been achieved.

For instance, the data of Fig. 29–1c clearly show that hemoglobin has four combining sites for oxygen. These sites are readily identified as being located at the four iron atoms of this molecule, for many metal complexes are known which are capable of binding molecular oxygen. Most of them are complexes of Co^{++}, but some Fe^{++} complexes with this property are also known. However, no Fe^{++} complex is known which combines with O_2 with the same affinity as the Fe^{++} present in hemoglobin.

In the combination of iodide and other anions with serum albumin the situation is even less satisfactory. None of the side chain groups of this molecule can be expected to have any appreciable affinity for simple anions. Most other proteins, although they possess the same kinds of side chain groups, do not combine to an appreciable extent with simple ions. The combining sites of this protein can therefore not be identified at present. It should be noted, however, that several anions besides I^- combine with this protein and that, whenever a detailed study has been made, there appear to be, as for I^-, one or two sites with strongest affinity, about seven or eight with somewhat weaker affinity, and further sites with still less combining power. We therefore expect the pertinent sites to be non-specific to the extent that the same sites seem to bind all anions, and we naturally look to positively charged sites for an answer. However, the sites with positive charges are all known from the hydrogen ion titration curve, and do not fall into classes corresponding to the numbers $n_1 = 1$, $n_2 = 8$, $n_3 = 18$.

It is necessary to conclude, in both the examples here cited, that the particular configuration of the protein molecule, resulting from a particular arrangement of internal bonds, leads to the occurrence of unusually reactive sites.

29g. Intrinsic free energies, heats, and entropies. No study of chemical equilibrium is complete without a determination of the effect of temperature, followed by application of the Gibbs-Helmholtz equation to determine the heat and entropy of reaction. If any intrinsic equilibrium constant has been determined at several temperatures, the corresponding thermodynamic properties are evaluated as follows

$$\Delta F^\circ_{int} = -\mathscr{R}T \ln k_{int}$$

$$\Delta H^\circ_{int} = -\mathscr{R}\frac{d \ln k_{int}}{d(1/T)} \tag{29-35}$$

$$\Delta S^\circ_{int} = \frac{\Delta H^\circ_{int} - \Delta F^\circ_{int}}{T}$$

An approximate value of the heat of reaction can sometimes be obtained even without knowledge of the number of combining sites or their intrinsic equilibrium constants. For instance, if there is only a single kind of combining site in an ionic reaction and it appears reasonable that electrostatic interaction is the only kind of interaction between sites, then, by combination of equations 29–22 and 29–27,

$$\ln \frac{\bar{\nu}}{n - \bar{\nu}} = \ln k_{int} - 2wz_iZ + \ln C \tag{29-36}$$

From representations of experimental data such as those of Fig. 29–1 we can obtain the value of C at any given value of $\bar{\nu}$ as a function of temperature, and, by differentiation, evaluate

$$\left[\frac{\partial \ln C}{\partial (1/T)}\right]_{\bar{\nu}} = -\frac{d \ln k_{int}}{d(1/T)} + 2z_i\left[\frac{\partial (w\bar{Z})}{\partial (1/T)}\right]_{\bar{\nu}} \qquad (29\text{-}37)$$

Since $\bar{Z}$ is determined experimentally as $\bar{\nu}$ is determined we know its variation with temperature; with a single kind of combining site $\bar{Z}$ will be a unique function of $\bar{\nu}$ and will be constant when $\bar{\nu}$ is constant. The

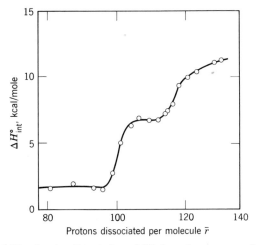

Fig. 29–6. ΔH°_{int} for the dissociation of H^+ ions from serum albumin, as a function of the average number of protons dissociated per molecule. Where ΔH°_{int} is independent of $\bar{r}$ it is likely that a single type of acidic site is involved. (Tanford et al.[14])

derivative $dw/d(1/T)$ can be evaluated from the assumed equation for electrostatic free energy. Electrostatic forces normally involve only a small temperature dependence, and $dw/d(1/T)$ will usually be quite small. Thus, using equation 29–35,

$$\left[\frac{\partial \ln C}{\partial (1/T)}\right]_{\bar{\nu}} = \frac{\Delta H^\circ_{int}}{\mathcal{R}} + 2z_i\bar{Z}\frac{dw}{d(1/T)} \simeq \frac{\Delta H^\circ_{int}}{\mathcal{R}} \qquad (29\text{-}38)$$

For a single type of combining site ΔH°_{int} must be independent of the value of $\bar{\nu}$ at which $\partial \ln C/\partial (1/T)$ is evaluated.

As an example of the application of equation 29–38 we show, in Fig. 29–6, a plot of 2.303 $\mathcal{R}[\partial pH/\partial (1/T)]_{\bar{r}}$ for the dissociation of H^+ ions from bovine serum albumin.[14] According to equation 29–38 this term should be independent of $\bar{r}$ and approximately equal to the heat of ionization if hydrogen ions are dissociated from a single type of site. Figure 29–6 then indicates that for $\bar{r} < 100$ a single type of site is involved ($\Delta H \simeq 1.5$ kcal/mole) and again for $105 < \bar{r} < 115$ a single type of site is involved ($\Delta H \simeq 7$ kcal/mole). In the other regions of the figure the derivative is not independent of $\bar{r}$, and, therefore, more than one term of equation 29–31 (or the corresponding *dissociation* equation, equation 30–1) must be contributing in these regions.

29h. Adsorption. A solid particle in contact with a gas or immersed in a liquid will often combine with molecules in the surrounding medium. This process is known as *adsorption*. Presumably the sites involved are on the particle surface. The principles involved in adsorption are clearly the same as those involved in multiple equilibria of the type here discussed. There will be a large number of combining sites, often with different intrinsic affinity, and there may be interaction between sites. The empirical or semi-empirical *isotherms* which describe adsorption equilibria are merely simple binding equations which are not expressed in the language of this chapter because we do not possess a knowledge of the binding sites on a solid surface comparable to our knowledge of the binding sites on macromolecules. It is worth mentioning that the simplest of all adsorption isotherms is that of Langmuir[41]

$$\frac{\theta}{1 - \theta} = (\text{constant})C$$

It is seen to be identical with equation 29–15, and it thus implies the presence of a set of identical and non-interacting binding sites.

Early work on the binding of small molecules or ions by macromolecules was sometimes called "adsorption" and described in terms of adsorption isotherms. The present-day better understanding of the structure of large molecules has replaced this procedure by the more clearly defined analysis described in the preceding pages.

30. HYDROGEN ION TITRATION CURVES

The equilibrium which exists between hydrogen ions and macromolecules which are acidic, basic, or amphoteric in nature has been studied more extensively than most other equilibria. This is due at least in part to the fact that pH measurement provides an easy means of determining the activity of hydrogen ions over a range from about unit activity to 10^{-14}.

30a. General equations based on approximate treatment of electrostatic interaction. Hydrogen ion equilibria are generally represented in terms of dissociation constants rather than association constants. Furthermore, it is usually supposed that, unless there is definite evidence to the contrary, the interaction between sites is purely electrostatic. As an approximation we may thus describe the interaction by the method of section 29c. It gives, for the average number, $\bar{r}$, of hydrogen ions dissociated per macromolecule, in analogy with equation 29–31,

$$\bar{r} = \frac{n_1 K_{\text{int}}^{(1)} e^{2w\bar{Z}}/a_{\text{H}^+}}{1 + K_{\text{int}}^{(1)} e^{2w\bar{Z}}/a_{\text{H}^+}} + \frac{n_2 K_{\text{int}}^{(2)} e^{2w\bar{Z}}/a_{\text{H}^+}}{1 + K_{\text{int}}^{(2)} e^{2w\bar{Z}}/a_{\text{H}^+}} + \cdots \qquad (30\text{–}1)$$

In equation 30–1 we have used a_{H^+}, the hydrogen ion activity, rather than the concentration C which has occurred in most of the equations of section 29, because the measurement of pH, although it is *not* a measurement of the activity of hydrogen ions as we would usually define it, is closer to it than it is to a measurement of hydrogen ion concentration. (The reader is reminded that the chemical potential and, hence, the activity of an ion are not measurable quantities in any event; as discussed in section 11. If the reader likes he can consider hydrogen ion activity, as used in this section, to be *defined* by the relation $pH \equiv -\log a_{\text{H}^+}$. The difference between this and more usual definitions of a_{H^+} arises as a result of liquid junction potentials which are present in the cell usually employed to determine pH from emf measurements.[4])

The constants K_{int} which occur in equation 30–1 are *intrinsic dissociation constants* for reactions such as $-\text{COOH} \rightleftharpoons -\text{COO}^- + \text{H}^+$ and $-\text{NH}_3^+ \rightleftharpoons -\text{NH}_2 + \text{H}^+$ at single sites. They are defined, like the association constants of section 29c, as the values for the dissociation constants at those sites when $\bar{Z} = 0$. Each K_{int} corresponds to an intrinsic standard free energy change, as before. The exponential term $e^{2w\bar{Z}}$ has a *positive* exponent, instead of the *negative* exponent of equation 29–31, because the net charge *decreases* by z_i when an ion is dissociated; i.e., when equation 29–24 is applied to dissociation rather than association, it contains the term $-z_i \mathcal{N}(\partial W_{\text{el}}/\partial \bar{Z})$. In obtaining equation 30–1 we have placed $z_i = +1$.

30b. Titration curves of synthetic polyelectrolytes. The simplest of all macromolecular titration curves are those of simple linear polyelectrolytes. Here all the dissociable sites are clearly identical. Furthermore, the polyelectrolytes are simple derivatives of well-known acids so that the intrinsic properties of the dissociable sites can be predicted with reasonable certainty. With the approximate treatment for the electrostatic interaction, equation 30–1 should describe the equilibrium, with the simplification that there is only one kind of site. Rearranging equation 30–1, we get, analogous to equation 29–36,

$$\log \frac{\alpha}{1 - \alpha} - \mathrm{pH} = \mathrm{p}K_{\mathrm{int}} - 0.868w\bar{Z} \qquad (30\text{–}2)$$

where $\alpha = \bar{r}/n$ and $\mathrm{p}K_{\mathrm{int}} = -\log K_{\mathrm{int}}$.

Figure 30–1 shows, as an example, titration curves for polymethacrylic acid, obtained by Arnold and Overbeek.[15] The sample used was fractionated

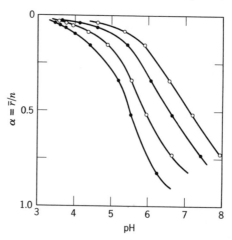

Fig. 30–1. Titration curves of polymethacrylic acid in the presence of various amounts of added KCl. Top curve: $0.001M$ KCl (ionic strength $I = 0.001$ to 0.008 during the course of titration). Second curve: $0.01M$ KCl ($I = 0.010$ to 0.017). Third curve: $0.1M$ KCl ($I = 0.100$ to 0.107). Bottom curve: $1.0M$ KCl ($I = 1.00$). (Data of Arnold and Overbeek.[15])

and had a molecular weight near 139,000. This corresponds to a degree of polymerization of about 1600, hence, to the presence of 1600 dissociable carboxyl groups.

Figure 30–2 shows the same data plotted in logarithmic form. The left-hand side of equation 30–2 is the ordinate and α the abscissa. Since $\bar{Z} = 0$ when $\alpha = 0$, the intercepts at $\alpha = 0$ should give us $\mathrm{p}K_{\mathrm{int}}$. The value obtained is 4.85, which is indeed a very reasonable intrinsic $\mathrm{p}K$ for a carboxyl group (e.g., acetic acid has $\mathrm{p}K = 4.76$).

Actually the intrinsic properties of the carboxyl groups of polymethacrylic acid should resemble those of the carboxyl groups of glutaric acids, $HOOC\!\!-\!\!(CHR)_3\!\!-\!\!COOH$, rather than those of an acid with an isolated carboxyl group. K_{int} should be one-half the first dissociation constant of such an acid, leading to $\mathrm{p}K_{\mathrm{int}} \simeq 4.6$. However, glutaric acids with a methyl group in the α-position have not been studied, and they could well have a somewhat higher $\mathrm{p}K$. Some hydrogen bonding between uncharged carboxyl groups would also raise the $\mathrm{p}K$. (See Fig. 31–3.)

We examine next the dependence of $\{\log[\alpha/(1 - \alpha)] - pH\}$ on α. In so doing we shall make two simplifying assumptions. (1) We assume that the net average charge $\bar{Z}$ is numerically equal to the number of protons dissociated, i.e., that $\bar{Z} = -\alpha n$. This assumption is true only at very low values of α. At higher degrees of dissociation there is, as we saw in section 27c, binding of counterions such that $|\bar{Z}| < \alpha n$. (2) We assume that W_{el} can be computed by the Hermans-Overbeek equation, equation

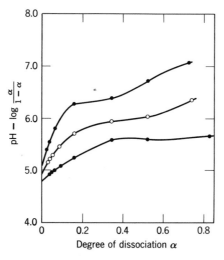

Fig. 30–2. The data of Fig. 30–1 plotted according to equation 30–2. Ionic strength 0.010 to 0.017 for the top curve, 0.10 for the middle curve, 1.0 for the bottom curve.

26–56. We have already seen that this is a relatively poor approximation. With these assumptions

$$w = \frac{3\epsilon^2}{5DR_e}(1 + 0.6\kappa R_e + 0.4\kappa^2 R_e^2)^{-1} \qquad (30\text{–}3)$$

and

$$\log \frac{\alpha}{1 - \alpha} - pH = pK_{int} - 0.868wn\alpha \qquad (30\text{–}4)$$

We note that the curves of Fig. 30–2 are not linear. This can only mean that w is not a constant, which in turn requires that R_e is not constant. This conclusion, however, is entirely in accord with expectation, for we saw in section 27 that polyelectrolytes must expand with increasing charge.

By means of equation 30–4 we can calculate w as a function of α; hence, by equation 30–3, we can obtain the effective radius R_e as a function of α. The result of such a calculation for two of the curves of Fig. 30–2 is shown in Fig. 30–3. Qualitatively the variation of R_e with α agrees with the results obtained for typical polyelectrolytes from light scattering and viscosity measurements, as shown in section 27a.

For a more quantitative comparison, Arnold and Overbeek[15] measured the viscosities of the same solutions used to obtain the titration curve. From them they evaluated η_{sp}/c as a function of α. The solutions had a concentration of 8.5×10^{-4} gram/cc, and η_{sp}/c is therefore a little larger

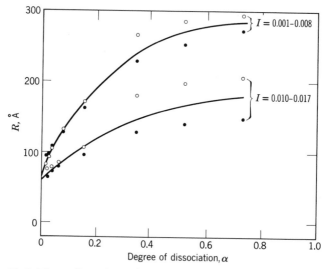

Fig. 30–3. The radius of a sphere equivalent to a polymethacrylic acid molecule, as a function of the extent of ionization of the carboxyl side chains. Open circles represent viscosity measurements; filled circles are based on the titration data of Fig. 30–2. (Original data of Arnold and Overbeek.[15])

than the intrinsic viscosity, but it is not a serious error to assume that η_{sp}/c at this concentration is equal to the intrinsic viscosity. We can then calculate an equivalent hydrodynamic radius by means of equation 23–3, placing the shape factor $\nu = 2.5$; i.e., we have

$$\frac{\eta_{sp}}{c} \simeq [\eta] = 2.5 \frac{\mathcal{N}}{M} (\tfrac{4}{3}\pi R_e^{3}) \qquad (30\text{–}5)$$

The values of R_e so obtained are also shown in Fig. 30–3. Quantitative agreement with the radii obtained from the titration data is poor, except at low values of α. In view of the approximations made in interpreting the titration curves this is not at all unexpected. It is clear, however, that the over-all picture as here presented is semi-quantitatively correct.

30c. Synthetic polyampholytes. There has been no careful study of the hydrogen ion equilibria of synthetic polyelectrolytes which bear both positive and negative charges. The data which are available show clearly, however, the influence of the configuration changes which occur in these polymers.

As was discussed in section 27e, polyampholytes have a very compact configuration near their isoelectric points, owing to the attraction between

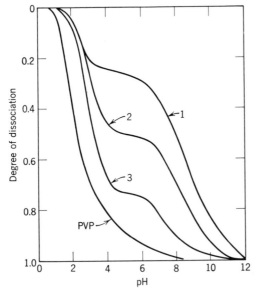

Fig. 30–4. Titration of the pyridine side chains of a copolymer of vinylpyridine and methacrylic acid, as determined from spectroscopic measurements. Three samples are shown containing (1) 6% pyridyl side chains, (2) 27.4% pyridyl side chains, (3) 45% pyridyl side chains. The curve labeled PVP shows the titration of polyvinyl pyridine containing no acidic side chains. (Data of Katchalsky and Miller.[16])

their positive and negative charges. This makes it most difficult either to add or to remove H^+ ions because the initial value of W_{el} is large and negative and $\partial W_{el}/\partial |Z|$ is positive and also large. As a result, the pH may be changed by several pH units in either direction from the isoelectric point, without appreciable titration of either acidic or basic groups.

For instance, Fig. 30–4 shows the titration of the pyridine side chains of a copolymer of vinyl pyridine and methacrylic acid.[16] Initially (near pH 2) the net charge of this polyampholyte is positive. As the pH is increased the net charge decreases, largely as a result of the titration of carboxyl side chains ($—COOH \rightarrow —COO^-$), though some of the pyridine chains are also titrated ($-NH^+ \rightarrow -N$) and contribute to the net charge decrease. As the isoelectric pH is approached it becomes increasingly difficult to titrate either kind of side chain, and there is very little change in net charge over about two pH units. As the net charge slowly becomes negative, however, great expansion occurs, and the titration of subsequent pyridine side chains becomes much easier.

30d. Titration curves of proteins.[4] All proteins contain a variety of acidic and basic sites. In some proteins the state of equilibrium between these sites and H^+ ions can be determined only with difficulty because of irreversible configuration changes ("denaturation") which may sometimes follow the increase in charge which accompanies titration. In many of the small proteins, however, this difficulty does not arise or occurs only at the extremes of the titration curve. In these cases the equilibrium is easily determined experimentally and should approximately obey equation 30–1. If the proteins in question are globular proteins, their molecules may be compact and impenetrable to solvent over a wide range of pH, and, in that event, the electrostatic interaction factor w which occurs in equation 30–1 should be that derived from equation 26–30 for W_{el}, as given by equation 29–32. The intrinsic constants K_{int} should correspond to the dissociation constants of appropriate small molecules containing the same dissociable groups.

We shall consider here in some detail the hydrogen ion equilibria of ribonuclease,[10] in order to determine how accurately the approximate theory accounts for the experimental results. Ribonuclease has been chosen because it is one of the simplest proteins and because its content of amino acids is known exactly,[18] so that the numbers of titratable groups of each kind which occur (i.e., n_1, n_2, $\cdots$) are known quantities. Furthermore, viscosity data have shown that this protein has a compact almost spherical shape at all pH values between pH 2 and 11.

Hydrogen ion dissociation curves of ribonuclease at several ionic strengths are shown in Fig. 30–5. (One of these curves was shown as an *association* curve in Fig. 29–1.) We note at once that the curve may be divided into three distinct regions. In the first region, pH 1 to about 5, eleven H^+ ions are dissociated. In the second region, pH 5 to 8, five H^+ ions are dissociated. About thirteen H^+ ions are dissociated in the third region between pH 8 and 12. However, the titration is not complete at pH 12, so that there are more titratable sites involved in this region. A reasonable extension, assuming an S-shaped curve, leads to a total of perhaps sixteen or seventeen H^+ ions for this region. It should be noted that the titration curve is reversible only up to pH 11.5. The curve beyond this pH therefore does not represent thermodynamic equilibrium.

We have pointed out in section 5 that the titration of phenolic groups can be followed independently by making use of the difference in ultra-violet absorption spectrum between the undissociated and the dissociated phenol. The results of such a titration for ribonuclease have been shown in Fig. 5–8. Six phenolic groups are titrated in all, but we see that they

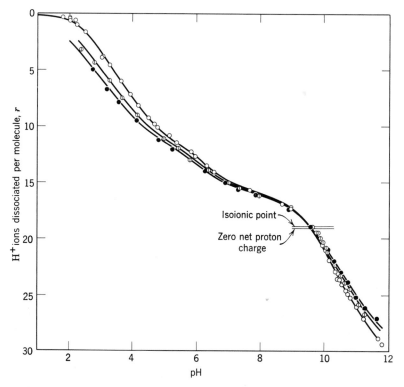

Fig. 30–5. Titration data for ribonuclease at 25°C, ionic strengths 0.01 (●), 0.03 (◑), and 0.15 (○). The figure shows the point of zero net charge (corresponding to $r = 19$) and also the isoionic pH at the particular concentration which was used. The isoionic pH is discussed on p. 561. (Tanford and Hauenstein.[10])

are not identical. Only three of the groups are titrated in the region below pH 11.5, in which the titration curve is reversible. The other three phenolic groups are titrated above pH 12 more or less all at once (the observed degree of dissociation depends on time as much as on pH; at 6° C these groups are hardly titrated at all). This indicates that these three phenolic groups are, in the native state of the protein, in the interior of the molecule and thus inaccessible to the water molecules or OH⁻ ions which are presumable involved in the mechanism of the ionization process. Between pH 11.5 and 12 a configurational change occurs which brings these three phenolic groups to the surface. (In the reverse titration, after the configurational change has occurred, all six phenolic groups are titrated together. In urea, in which solvent the protein is no longer compactly folded, all six groups are titrated together in either direction.)

TABLE 30–1. Titratable Groups of Ribonuclease[10]

	pK_{int} Expected from Data on Small Molecules	Number of Sites from Amino Acid Analysis[18]	Data Obtained from Titration Curve	
			Number of Sites	pK_{int}
α-COOH	3.75	1	(1)[a]	—
Side chain COOH	4.6	10	11 { 10	4.7
Imidazole	7.0	4	4	6.5
α-NH₂	7.8	1	5 { (1)[a]	7.8
Phenolic	9.6	6	6[c]	9.95[c]
Side-chain NH₂	10.2	10	16 { 10	10.2
Guanidyl	>12	4	4[b]	>12

[a] These numbers were assumed correct because of the overwhelming evidence that ribonuclease consists of a single polypeptide chain.

[b] This number is obtained from Fig. 30–5 by the procedure outlined on p. 564.

[c] The data for the phenolic groups are obtained from the spectrophotometric titration. Without this auxiliary measurement the sixteen groups titrated in the alkaline region would not have been separable into amino and phenolic groups. As discussed in the text, only three of the phenolic groups possess the pK_{int} here given.

Returning now to an analysis of Fig. 30–5, we note that the first region must represent the titration of the most acidic sites of the molecule, the second region those of intermediate acidity, etc. This makes it possible to identify these sites at once, for the pK values which we expect to find for the various possible titratable groups are well known, and they are shown in Table 30–1. It is seen that the carboxyl groups should be the most acidic, imidazole and α-amino groups should be intermediate, and phenolic and side chain amino groups should be the least acidic. (Guanidyl groups are also present, but they would not be expected to be titrated until well above pH 12.) That this identification of the groups is correct can be checked in several ways, which need not be discussed here. (One of the ways involves determination of approximate heats of dissociation, as in Fig. 29–6.) It suffices for our purpose that, if we identify the groups in this way and thus conclude that there is a total of eleven carboxyl groups, five imidazole plus α-amino groups, and about sixteen phenolic and side chain amino groups, the observed count of dissociable groups agrees with the number known to be present from the amino acid content of the molecule, as shown in Table 30–1.

It should be noted that the sites titrated in each region are not expected to be identical. One of the carboxyl groups is the terminal α-carboxyl group of the polypeptide chain, and it should be more acidic than the ten side chain carboxyl groups. In the intermediate region the terminal α-amino group is different from the four imidazole side chains, and there are again two kinds of groups in the alkaline region. It should be noted, however, that the single α-carboxyl group makes only a small contribution to the total dissociation in the acidic region. It is possible to assume a reasonable titration curve for this one site (if necessary several alternatives can be tried) and to subtract it from the observed result. This computation will leave us with an experimental curve representing the titration of only the ten side chain carboxyl groups. In the same way a reasonable assumption concerning the titration of the single α-amino group will leave us with a curve representing the course of titration of the four imidazole groups. Finally, in the alkaline region, no assumption need be made at all, for the phenolic groups have been titrated separately, so that the $\bar{r}$ values for side chain amino groups at any pH are determinable by difference.

If we now make the reasonable assumption that all sites of a given chemical variety (Table 30–1) have the same *intrinsic* dissociation constant, making an exception only for the phenolic groups, of which there are clearly two different kinds (Fig. 5–8), then there will be eight terms on the right-hand side of equation 30–1. Four of them, as indicated in the previous paragraph, can be obtained individually from the experimental data. To two terms, each involving only one site, we have assigned assumed values of K_{int}. The two other terms, that for the three phenolic groups buried inside the molecule and that for the four guanidyl groups, make no contribution to the titration curve in the reversible region below pH 11.5.

We can now separately analyze the four terms corresponding to the side chain carboxyl, amino, imidazole, and phenolic groups. Each term can be converted to the logarithmic form as given by equation 30–2. The values of α at any pH are given by the experimental data. The average net charge $\bar{Z}$ is obtained from the titration curve, Fig. 30–5, by merely counting the number of protons bound or dissociated relative to the point of zero net charge. (Section 30e describes how this point is obtained.) In calculating Z in this way we assume that the protein does not combine with any other ions present in the solution, in this case K^+ or Cl^- ions. This assumption can be confirmed experimentally. If other ions are bound, as in serum albumin, for example, the computation of Z at.any pH must include a measured value of the number of such ions bound at that pH.

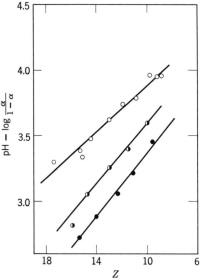

Fig. 30–6. Titration data for the side chain carboxyl groups of ribonuclease, plotted according to equation 30–2. Ionic strength is 0.15 for the top curve, 0.03 for the middle curve, and 0.01 for the lowest curve. (Data of Tanford and Hauenstein.[10])

The left-hand side of equation 30–2 is now plotted against Z. The result for the carboxyl groups of ribonuclease is shown in Fig. 30–6. We see that linear plots are obtained, as predicted by equation 30–2. The slopes, given in Table 30–2, are a little *larger* than predicted by the value

TABLE 30–2. Values of w for Ribonuclease

Ionic Strength (Molar)	w, Calcd. by Eq. 29–32	w, Observed Carboxyl Groups	w, Observed Phenolic Groups
0.03	0.113	0.134	0.093
0.15	0.079	0.102	0.061

of w calculated by equation 29–32. When the lines are extended to $\bar{Z} = 0$ we obtain an intercept of 4.7, in good agreement with the expected value of pK_{int} for carboxyl groups.

A similar plot for the three reversibly titrated phenolic groups of ribonuclease is shown in Fig. 30–7. This time the slopes are a little *lower* than predicted by the value of w calculated by equation 29–32. The intercept at $\bar{Z} = 0$ can be obtained without extrapolation, since the region of titration of the phenolic groups encompasses the pH at which $\bar{Z} = 0$.

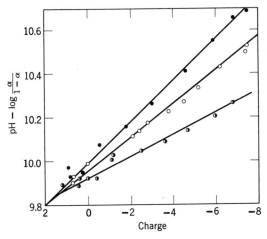

Fig. 30–7. Titration data for the three normal phenolic groups of ribonuclease, plotted according to equation 30-2. Ionic strength is 0.01 for the top curve, 0.03 for the middle curve and 0.15 for the lowest curve. (Data of Tanford et al.[17])

The value of pK_{int} obtained from the intercept is close to 9.95. There is a small but detectable difference in the pK_{int} value obtained at each of the ionic strengths used.

Clearly the results obtained for the carboxyl and phenolic groups of ribonuclease agree as well with the theory as the many crude approximations of the theory can lead us to expect. The extent of agreement between theory and experiment is about the same when the imidazole and amino groups are considered. (The pK_{int} values for these groups are shown in Table 30–1.) When other small proteins are examined the same kind of result is obtained: rough agreement with the theory, but always with a definite though small quantitative deviation from the expected slope or pK_{int} or both.

Some of the pK_{int} values which have been observed for carboxyl groups are shown in Table 30–3. It is noted that a somewhat low pK_{int} is observed

TABLE 30–3. Intrinsic Dissociation Constants of Side Chain
Carboxyl Groups

(a) Appropriate Small Molecules[5]	
$ROOC-(CH_2)_2-COOH$	$pK = 4.52$
$ROOC-(CH_2)_4-COOH$	4.60
$CH_3-CO-(CH_2)_2-COOH$	4.59

(b) From Approximate Treatment of Titration Curves	
Polymethacrylic acid[15]	$pK_{int} = 4.85$
β-Lactoglobulin[19]	4.6
Ribonuclease[10]	4.7
Insulin[20]	4.7
Ovalbumin[21]	4.3
Serum albumin[14]	3.95

TABLE 30–4. Thermodynamic Data for Dissociation of Phenolic Groups

	pK_{int}	ΔH°_{int} kcal/mole	ΔS°_{int} cal/degree mole
(a) Appropriate Small Molecules			
Phenol[5]	9.78	6.1	−24
Tyrosine[22]	9.66[a]	6.0	−24
Tyrosylarginine[5]	9.6	6.0	−24
(b) From Approximate Treatment of Titration Curves			
Polytyrosine[23]	9.5	—	—
Insulin[24]	9.7	7.5	−19
Pepsin[24]	(9.5)	6.0	−23
Ribonuclease[17]	9.95	7.0	−22
Serum albumin[22]	10.35	11.5	−9
Lysozyme[25]	10.8	—	—

[a] Corrected for electrostatic effects.

for ovalbumin and an even lower one for serum albumin. Intrinsic
dissociation constants for phenolic groups are shown in Table 30–4, and
this table also contains some intrinsic heats and entropies of ionization for
phenolic dissociation. Again serum albumin exhibits exceptional deviation
from the expected result.

At least some of these deviations from the approximate theory will be
explained when a more realistic model for titration curve calculation is
examined later. Others, however, may be due to configurational features.
For instance, it is well known that phenolic OH groups are capable of

forming moderately strong hydrogen bonds with appropriate atoms bearing an unshared pair of electrons. The energy required to dissociate such a bond is usually about 6000 cal/mole. If every phenolic group of a protein were hydrogen bonded in this way, ΔH_{int}° for dissociation of the proton would be *increased* by about 6000 cal, for the hydrogen bond would have to be broken before dissociation could occur. The entropy of dissociation would also be affected, for the formation of a hydrogen bond in a protein must be expected to limit the rotational freedom of the side chain groups involved. (For example, rotation about the single bond between the CH_2 group and the benzene ring in $-CH_2-C_6H_4-OH$ would be accompanied by rotation of the hydroxyl hydrogen atom about an axis parallel to this bond; if the hydrogen atom is part of a hydrogen bond, this rotation is clearly prohibited.) Thus a hydrogen-bonded phenolic group has a lower entropy than a free phenolic group, and ΔS_{int}° for dissociation of the phenolic hydrogen ion is accordingly *more positive* than we would ordinarily expect. Laskowski and Scheraga[26] have estimated that the change in ΔS_{int}° can be as large as 20 to 30 cal/degree mole. The figures just given agree, of course, quite well with the anomalies observed for the phenolic groups of serum albumin. ΔH_{int}° in this protein is anomalously high by 5000 cal/mole; ΔS_{int}° is more positive than normal by about 15 cal/degree mole. Hydrogen bonding thus offers one possible explanation for the thermodynamic data observed for phenolic dissociation in this protein.

The calculation of the preceding paragraph has been included because it represents the only simple published attempt to make a numerical estimate of the effect of an unusual configurational feature on H^+ ion dissociation. It is probable that the calculation is actually incorrect, because the ΔH° and ΔS° values computed are those for a hydrogen bond in a vacuum or a non-polar solvent. In a solvent such as water, a broken internal hydrogen bond involving a phenolic group is replaced by new hydrogen bonds to H_2O, so that the expected anomalies would become considerably smaller than those which we have cited.

As will be shown below, serum albumin undergoes a configurational change in the pH region in which the phenolic groups are titrated. The change is in the direction of an expanded configuration in which the degree of dissociation of all acidic groups is increased. If this change (at constant $\bar{r}$) depends appreciably on temperature, part of the observed ΔH° and ΔS° values represent the effect of temperature on configuration, thus providing an alternative explanation for the anomalous values in Table 30-4. Experiments on the effect of temperature on the configurational change have not so far been performed.

30e. Isoionic and isoelectric pH. We have had occasion to refer to the isoionic and isoelectric states of a protein molecule. We shall briefly discuss here how they are related to the titration curve. For convenience we shall confine the discussion to aqueous solutions.

An *isoionic* protein solution is defined as a solution that contains, in addition to dissolved protein, no ions other than those arising from dissociation of the solvent. Such a protein solution can be obtained experimentally.[27] For instance, if an aqueous protein solution is passed down a mixed-bed ion exchange column all cations and anions other than the bulky protein ions are exchanged for H^+ and OH^- ions, and the resulting solution is by definition isoionic.

A protein molecule is said to be *isoelectric* if its average net charge $\bar{Z}$ is zero. It is possible to determine whether a protein molecule is isoelectric by observing its motion in an electric field (electrophoresis, for example). If it is isoelectric it will undergo no net motion in either direction. Since $\bar{Z}$ and hence electrophoretic mobility depend on pH, we often speak of the *isoelectric pH* or *isoelectric point*, this being determined graphically from a plot of mobility versus pH such as that of Fig. 24–8.

An isoionic protein solution (in pure water) will often be close to isoelectric. This follows from the condition of neutrality in an isoionic solution,

$$C_P \bar{Z} + C_{H^+} = C_{OH^-} \qquad (30\text{–}6)$$

where C_P is the molar concentration of protein, together with two other relations, one being the fact that the product $C_{H^+} C_{OH^-}$ is equal to the ionization constant of water, and the other that $\bar{Z}$ is a function of pH (i.e., of C_{H^+}) through equation 30–1. In most typical proteins the isoionic pH lies between pH 3 and pH 11 (e.g., ribonuclease in Fig. 30–5). Thus $|C_{H^+} - C_{OH^-}| \leqslant 10^{-3}M$. If C_P is also $\sim 10^{-3}M$, then, by equation 30–6, $|\bar{Z}| \leqslant 1$. An isoionic protein solution will fail to be close to isoelectric only if the isoelectric pH falls outside the pH region just given (e.g., pepsin, which is isoelectric near pH 1) or if the protein concentration C_P is decreased to low values. It should also be noted that an isoionic protein solution can not in general be *exactly* isoelectric, unless the isoelectric point in water is (at 25° C) *exactly* at pH 7.00.

Equation 30–6 shows that a decrease in the protein concentration of an isoionic solution will increase $|\bar{Z}|$. The pH will adjust itself also so that the equilibrium between the protein and H^+ is maintained. In other words, there is no unique isoionic pH—it depends on concentration. Its value can be calculated, if we so wish, by simultaneous solution of equations 30–1 and 30–6 and the condition that $C_{H^+} C_{OH^-} = K_w$. (The solutions with which we are dealing are often so close to ion-free that we can equate activities and concentrations. Activity coefficients can be introduced in any case if necessary.)

By contrast, the isoelectric pH is independent of protein concentration in so far as equation 30–1 is independent of concentration. In the approximate discussions of the preceding sections we have assumed that this is the case. In practice this assumption is not necessarily true because of electrostatic interaction between protein ions.

It is of interest to consider the effect of the addition of a salt such as NaCl to an isoionic or isoelectric protein solution. For simplicity we consider a protein molecule which is isoelectric say between pH 5 and pH 9, and at sufficiently high concentration so that the isoionic protein in water is essentially also isoelectric. The pH is given by equation 30–1 with $Z = 0$ and $\bar{r}$ equal to the number, $\bar{r}_0$, of protons which must have become dissociated so that Z becomes zero. When salt is added $\bar{r}$ cannot change appreciably because there is no way to add or remove an appreciable number of hydrogen ions. Z, however, can change if the protein combines with either the cation or anion of the added salt. Clearly, from equation 30–1, if $\bar{r}$ is to remain equal to $\bar{r}_0$, e^{2wZ}/a_{H^+} or its logarithm, $0.868wZ + \text{pH}$, must remain essentially constant during the addition of salt. Thus, if Z changes, the pH must change also. If $(\text{pH})_0$ is the original isoionic pH and $\bar{\nu}_i$ is the number of salt ions of charge z_i bound to each protein molecule at a given salt concentration, then

$$\text{pH} - (\text{pH})_0 = -0.868w \sum_i \bar{\nu}_i z_i \qquad (30\text{–}7)$$

We note that the pH rises when anions are bound and falls when cations are bound. Equation 30–7 has been tested very carefully by Scatchard and coworkers.[7] They measured the binding of several univalent anions to serum albumin (some of this work is described in section 29) and also determined the pH in all the solutions they used. Figure 30–8 shows a plot of the measured pH values versus $0.868w\bar{\nu}_{X^-}$, w being calculated by equation 29–32. The line drawn through the data has unit slope, so that equation 30–7 is exactly confirmed.

If, under the conditions of these experiments, cations as well as anions were bound, then the proper abscissa for such a plot would be $\bar{\nu}_{X^-} - \bar{\nu}_{Na^+}$ (all the salts used were sodium salts). Since the various anions are bound with very different affinity, $\bar{\nu}_{X^-}$ has equal values at very different concentrations of NaX and therefore at very different values of $\bar{\nu}_{Na^+}$. The fact that the points for all the anions fall on the same line when plotted against $\bar{\nu}_{X^-}$ alone is proof that $\bar{\nu}_{Na^+}$ must be negligibly small in all these experiments.

The binding of ions must also affect the isoelectric point. For the originally isoelectric protein molecules have an average charge of $\sum_i \bar{\nu}_i z_i$ after ion binding. Accordingly either acid or base must be added to restore the isoelectric condition. If anions are bound, as in serum albumin, acid must be added, and the isoelectric pH is therefore *reduced*. The new isoelectric pH can be found from the condition that $\bar{r}$ must be changed from $\bar{r}_0$ to $\bar{r}_0 - \sum_i \bar{\nu}_i z_i$ to keep $\bar{Z}$ equal to zero. It can be calculated from equation 30–1 with $\bar{Z} = 0$. Experimental confirmation of these conclusions have been obtained by several authors.[28]

One final observation is of interest. By means of equation 30–6 it is possible to determine the average net charge $\bar{Z}$ of an originally isoionic protein preparation, a quantity which will usually be very small, as pointed out above. From the titration curve, on the other hand, we can

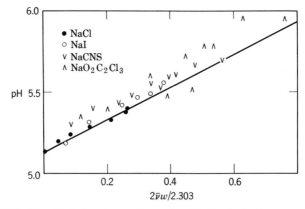

Fig. 30–8. The effect of various salts on the pH of an isoionic serum albumin solution [(pH)$_0$ = 5.15]. Binding experiments indicate that the anions of the salts are bound to the protein molecules; the abscissa of the figure represents experimental values of $0.868w\bar{\nu}$ in each case. Since $z_i = -1$, this is the right-hand side of equation 30–7. (Scatchard, Coleman, and Shen.[7])

obtain the maximum number of H$^+$ ions which can be bound to the isoionic protein molecule. The sum of these two quantities is clearly the maximum positive charge, due to H$^+$ ions, which the protein molecule can attain. This number, however, is equal to the total number of cationic sites (NH$_3$$^+$, etc.) per molecule. This number is therefore determined by the titration curve even though some of the cationic sites (the guanidyl groups) are not titrated in the pH range accessible to the experiment. This was the method used to obtain (by difference) the number of guanidyl groups listed for ribonuclease in Table 30–1.

30f. Evidence from titration curves concerning changes in configuration.[6] It was pointed out in section 28 that configuration changes may accompany the titration of proteins, as a result of the electrostatic free energy built up as the charge increases. Evidence for such changes in configuration is often obtained from titration curves.

For instance, Figs. 30–9 and 30–10 show plots of $pH - \log[\alpha/(1 - \alpha)]$ versus Z for carboxyl, phenolic, and amino groups of serum albumin.[14] Unlike the corresponding plots for ribonuclease, these curves are linear only over a short range of Z. The plots curve towards the horizontal for $Z > 5$ (Fig. 30–9) and also for Z more negative than about 50 (Fig. 30–10). Over the short linear portions w is very close to the value calculated by equation 29–32. Where the curvature is observed w must decrease well below this value. If this decrease in w is to be interpreted as a decrease in

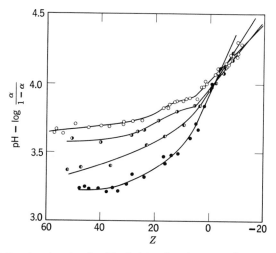

Fig. 30–9. Titration data for the side chain carboxyl groups of serum albumin, plotted according to equation 30–2. Ionic strengths (top curve to bottom curve) are 0.15, 0.08, 0.03, and 0.01, respectively. (Tanford et al.[14])

the factor A of equation 29–28, then, by equation 29–25, W_{el} is also decreased. As was seen in section 26 (Tables 26–1 and 26–2, for example) this most probably implies an expansion of the molecule, and its penetration by salt ions of the solvent. That such an expansion occurs in this protein has already been concluded from hydrodynamic data in section 28; and the titration curve therefore confirms that observation. In both figures, the value of Z at which curvature first sets in is very close to the value of Z at which the viscosity first begins to rise (Fig. 28–3).

If an expansion of the protein molecule has occurred, the appropriate equation for W_{el} is equation 26–42 rather than equation 26–30, and a corresponding equation for w can be obtained to replace equation 29–32. If this equation is used to compute the apparent radius of the expanded molecule, we obtain results which are close in magnitude to those obtained from hydrodynamic data.[29] For instance, at pH 2 and ionic strength 0.15, w gives a radius of about 42 Å, $[\eta]$ a radius of about 43 Å, and the sedimentation constant a radius of about 45 Å. The use of w to estimate a radius in this

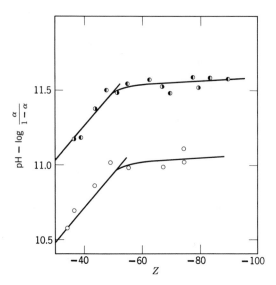

Fig. 30–10. Titration data for the phenolic groups (upper curve) and side chain amino groups (lower curve) of serum albumin, plotted according to equation 30-2. The ionic strength is 0.15. (Tanford et al.[14])

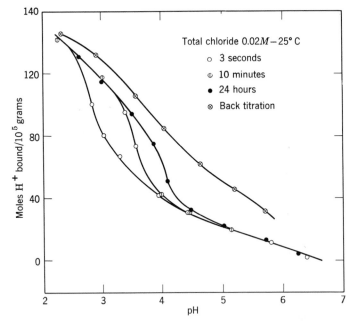

Fig. 30–11. The acid portion of the titration curve of hemoglobin, an extreme example of a time-dependent and irreversible titration curve. The data in the figure are for hemoglobin with carbon monoxide bound to the iron atom. (Steinhardt and Zaiser.[30])

way is, however, generally a dubious procedure, not only because the equations for W_{el} are so approximate but also because in the calculation we have to assume that pK_{int} remains unchanged, an assumption which is not necessarily true when configuration changes are occurring.

An example of an irreversible configuration change, as reflected in a titration curve, has already been noted in ribonuclease (Fig. 5–8). A more spectacular example occurs in hemoglobin[30] and is shown in Fig. 30–11. The reversible part of the titration curve in this case includes relatively few of the groups which would normally titrate on the acid side. A much larger number is released in a time-dependent irreversible step at pH 3.5. That a configuration change occurs at this pH is also revealed by changes in absorption spectrum. The nature of the change has not yet been elucidated.

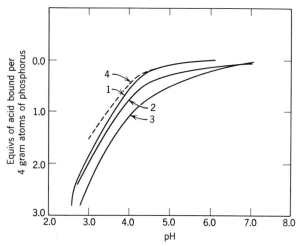

Fig. 30–12. Titration curve of DNA. The portion shown presents the course of titration of the three amino groups (per four nucleotide monomer units). Curve 1 represents titration with acid at 0.4°C, beginning at pH 7. Curve 2 is the back titration at the same temperature. Curve 3 is another titration at 0.4°C with acid but for a sample previously titrated to pH 2.5 at 25°C. Curve 4 shows the probable course of titration in the absence of configurational change. (Cox and Peacocke.[31])

30g. The titration of DNA. The titration curves of nucleic acids, which might ordinarily be expected to be simpler than those of proteins, because nucleic acids contain a smaller variety of titratable groups, are also complicated by irreversible changes. Some typical curves obtained by Peacocke and coworkers[31] for a preparation of deoxyribose nucleic acid (DNA) are shown in Fig. 30–12.

DNA is built up of four different nucleotides (section 2f) in almost exactly equimolar ratios. They contain, respectively, the purine and pyrimidine bases adenine, guanine, cytosine, and thymine. Between them they provide (for each four nucleotides) three titratable amino groups with pK_{int} near 4 and two titratable hydroxyl groups with pK_{int} near 11. Each nucleotide also contributes one phosphate group, which, however, is a *strong* acid group and thus is in its anionic form throughout the accessible pH range. As a result DNA has one negative charge per four nucleotides even at the most acid pH which can be reached. Z becomes -4 per four nucleotides near pH 6 and -6 after the two hydroxyl groups have been dissociated.

As Fig. 30–12 shows, the titration of the amino groups in the acid region does not occur reversibly, and the same is true in the alkaline region for the titration of the hydroxyl groups. This is clear indication that a change in configuration accompanies the titration of these groups. That such a change actually occurs was already noted in section 28. It should be noted (Fig. 30–12) that the extent of irreversible change produced by this process depends on the temperature at which the titration is carried out. This result is in agreement with the finding based on viscosity changes (section 28) that the change in configuration occurs by means of several steps, some of which are reversible. The extent of irreversible change will then become very sensitive to the temperature of the experiment, the length of exposure to low pH, and similar factors.

30h. More exact theory of titration curves. A more exact treatment of titration curves must take into account the fact that all titratable sites, even those which are identical in their *intrinsic* properties, must differ in their positions with respect to other sites, hence in their response to the influence of charge. A parallel requirement is that the electrostatic free energy must be calculated by a method which recognizes the charges as discrete charges. Discussions of titration curves which are based on this more realistic picture have been given by Harris and Rice[32] for polyelectrolytes and by Tanford[33] for compact proteins. We shall briefly describe here the more exact treatment for proteins.

We suppose the protein molecule may be represented by an impenetrable sphere of radius R, of internal dielectric constant $D_i = 2$. We let n be the total number of dissociable sites and suppose that they are all located 1 Å below the surface of the sphere. This assumption probably places these sites in about the same position relative to the solvent as the position of the acidic sites of small organic molecules (see section 26g). With this assumption we are also forced to assign the same intrinsic properties to each dissociable site of a given chemical kind; i.e., we cannot allow any variation in pK_{int} such as that *apparently* observed (Tables 30–3 and 30–4) when the approximate theory is applied. Of course, we must recognize that some of the acidic sites of a protein molecule may be buried in the interior and are not accessible to titration at all (like half the phenolic groups of ribonuclease) and that hydrogen bonding[26] may occur with resulting changes in intrinsic properties.

We consider now *separately* the following forms of the protein molecule: PH_n, the form from which no dissociable protons have been removed; PH_{n-1} the form from which one proton has been removed; etc., down to the form P from which all possible hydrogen ions have been dissociated. We distinguish further among the subspecies of each of these forms. Thus $PH_{n-1}^{(1)}$ may represent a species of PH_{n-1} in which the proton dissociated came from a side chain carboxyl group; $PH_{n-1}^{(2)}$ may represent a species in which it came from an α-carboxyl group; etc. In general we shall speak of a form $PH_\nu^{(i)}$ which is defined as carrying ν dissociable protons, specifically such that ν_1 are attached to, say, α-carboxyl groups, ν_2 to side chain carboxyl groups, and in general ν_j to groups of a particular chemical kind. Obviously $\sum_j \nu_j = \nu$.

It is necessary to recognize that each such species will consist of many possible *configurations*. Let there be n_1 α-carboxyl groups, n_2 side chain carboxyl groups, and in general n_j groups of chemical variety j (cf. Table 30–1). Then there are clearly $n_j!/\nu_j!\,(n_j - \nu_j)!$ ways of placing ν_j protons on the sites available to them, and the resulting total number of configurations for a species $PH_\nu^{(i)}$ is thus

$$\Omega_\nu^{(i)} = \prod_j \frac{n_j!}{\nu_j!\,(n_j - \nu_j)!} \tag{30–8}$$

the product being extended over all possible kinds of dissociable groups. It should be noted here that the forms P and PH_n are uniquely defined, so that the corresponding numbers of configurations, Ω_0 and Ω_n, are equal to unity.

Now let $\Delta F_\nu^{\circ(i)}$ be the standard free energy change, and $k_\nu^{(i)}$ the corresponding *association* constant for the reaction

$$P + \nu H^+ \rightleftharpoons PH_\nu^{(i)}$$

$\Delta F_\nu^{\circ(i)}$ will be a sum of intrinsic free energy terms (one for each proton bound) and an electrostatic free energy term. The former will clearly be independent of configuration, for the numbers ν_1, ν_2, $\cdots$, are the same for all configurations of $PH_\nu^{(i)}$, but W_{el} will depend on configuration. Let $\overline{W_\nu^{(i)}}$ be the average value of W_{el} over all configurations, each weighted by a Boltzmann probability factor $e^{-W_\nu^{(i)}/kT}$, where $W_\nu^{(i)}$ is the value of

W_{el} for a particular configuration. It is then more or less obvious (and rigorously proved by Tanford and Kirkwood[33]) that

$$\Delta F_{\nu}^{\circ(i)} = \sum_j \nu_j (\Delta F_{int}^{\circ})_j + \overline{W_{\nu}^{(i)}} - W_0 \qquad (30\text{–}9)$$

where W_0 is the electrostatic free energy of the form P, and $(\Delta F_{int}^{\circ})_j$ is the intrinsic free energy of *association* for the jth kind of site. $(\Delta F_{int}^{\circ})_j = -2.303 \mathscr{R}T(pK_{int})_j$ where $(K_{int})_j$ is the corresponding intrinsic dissociation constant.

The evaluation of $\overline{W_{\nu}^{(i)}}$, using equation 26–52 for the $W_{\nu}^{(i)}$ of particular configurations, is clearly a formidable task and will not be described here. It suffices to say that the calculation is possible,[33] by expansion of $\overline{W_{\nu}^{(i)}}$ in terms of unweighted averages. In this way the various possible $\Delta F_{\nu}^{\circ(i)}$ are evaluated. They correspond, of course, to the individual equilibrium constants K_A, K_B, K_C, K_D discussed for the dissociation of simple dibasic acids at the beginning of this chapter. It should be noted, however, that species which have the same constant K have been grouped together as different configurations of a single species, i.e., the number of constants to be calculated is far fewer than the number $2^n - 1$ calculated on p. 532.

Where $\Delta F_{\nu}^{\circ(i)} = -\mathscr{R}T \ln k_{\nu}^{(i)}$, we now have for $\bar{\nu}$ the relation

$$\bar{\nu} = \frac{\sum\limits_{\nu,(i)} \nu k_{\nu}^{(i)} a_{H^+}^{\nu}}{1 + \sum\limits_{\nu,(i)} k_{\nu}^{(i)} a_{H^+}^{\nu}} \qquad (30\text{–}10)$$

the summation extending over all values of ν and, for each value of ν, over each species of kind i.

We may also obtain the average degree of ionization of any particular kind of group j, such as the phenolic groups,

$$\bar{\nu}_j = \frac{\sum\limits_{\nu,(i)} \nu_j k_{\nu}^{(i)} a_{H^+}^{\nu}}{1 + \sum\limits_{\nu,(i)} k_{\nu}^{(i)} a_{H^+}^{\nu}} \qquad (30\text{–}11)$$

or the averages of the squares of such quantities, e.g.,

$$\overline{\nu^2} = \frac{\sum\limits_{\nu,(i)} \nu^2 k_{\nu}^{(i)} a_{H^+}^{\nu}}{1 + \sum\limits_{\nu,(i)} k_{\nu}^{(i)} a_{H^+}^{\nu}} \qquad (30\text{–}12)$$

No attempt has yet been made to calculate the titration curve of any protein by the method just outlined, because the precise manner in which protein molecules are folded into a compact configuration has not been established for any known protein, so that the precise locations of the dissociable sites cannot be given. Calculations have been made, however, for simple models which reproduce some of the features of protein configurations. The titration curves calculated for these models were cast into the form employed in the calculations by the approximate methods discussed in section 30d, i.e., into the form of plots of $\log[\alpha/(1-\alpha)]$ versus Z for a particular kind of site. A typical result is given in Fig. 30–13. It represents the titration of four carboxyl groups on a molecule also containing four positively charged groups (e.g., $-NH_3^+$). The radius of

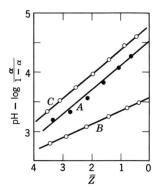

Fig. 30–13. Calculated titration data for four carboxyl groups in the presence of four $-NH_3^+$ groups. The calculations were based on equations 30–11 and 30–9, with W calculated by equation 26–52 and pK_{int} assumed equal to 4.60. All dimensions were kept constant, only the arrangement of carboxyl and $-NH_3^+$ groups being varied, as shown by Models A, B, and C in Fig. 30–14. (Tanford.[33])

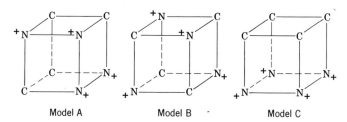

| Model A | Model B | Model C |

Fig. 30–14. Arrangements of carboxyl groups (C) and amino groups (N^+) used to obtain the data of Fig. 30–13.

the protein sphere was taken as 10 Å; the sites were located 1 Å below the surface and arranged as shown in Fig. 30–14. The carboxyl groups were assumed to have a pK_{int} of 4.60.

The results of this and similar calculations may be summarized as follows:

(1) The value of w obtained by the approximate treatment can be either greater or less than that calculated by equation 29–32. In particular, if the number of sites is very small, w tends to be smaller than the calculated value just as was observed for the phenolic groups of ribonuclease (there

are only three of them titrated) in Table 30–2. If the distribution of sites with respect to those of opposite charge is non-uniform so that the probability of reaction at different sites can be expected to differ considerably, w tends to be larger than the calculated value. This is almost certainly one of the factors and probably the only one which produces relatively large values of w for the carboxyl groups of ribonuclease (Table 30–2), for the amino acid sequence of this protein has been largely determined[18] and quite clearly shows carboxyl groups which differ widely in their nearness to positively charged sites. (These differences will persist no matter how the protein is folded.)

(2) The intercept of logarithmic plots (e.g., Figs. 30–6 and 30–7) at $Z = 0$ is not necessarily equal to pK_{int}, as Fig. 30–13 clearly shows, pK_{int} being 4.60 for each of the curves in that figure. Thus the differences in the apparent value of pK_{int} shown in Tables 30–3 and 30–4 may well reflect merely corresponding differences in the distribution of cationic sites with respect to anionic sites. They do not necessarily represent real differences in pK_{int}.

(3) The predicted effect of ionic strength on titration curves is little different from that predicted by the approximate theory. This too may be observed in the data for ribonuclease given in Table 30–2. Thus, between ionic strengths 0.03 and 0.15, we calculate by the approximate theory $\Delta w = 0.034$. For both carboxyl and phenolic groups we observe $\Delta w = 0.032$, though the absolute values of w differ considerably from the calculated values.

30i. Charge fluctuations. Since all the various species $PH_\nu^{(i)}$ are in equilibrium with one another, there will be present at any pH molecules differing from one another in the value of ν and therefore in their net charge Z. As a corollary, since all protein molecules in the solution are identical, we can say that each individual molecule in the solution will, in course of time, take on different values of ν and, hence, of Z. As a result of this the average value of ν^2, either for all molecules in the solution at a given instant or for one molecule over a period of time, differs from the square of the average value of ν, and similarly $\overline{Z^2}$ differs from $(\bar{Z})^2$, etc. These and similar differences can be calculated from the equations of the preceding section. For example the difference between $\overline{\nu^2}$ and $(\bar{\nu})^2$ is given by equations 30–10 and 30–12.

It is easily shown that $\overline{Z^2} - (\bar{Z})^2$ at any pH is equal to $\overline{\nu^2} - (\bar{\nu})^2$. For, if Z_0 represents the charge of a protein molecule with $\nu = 0$, i.e., of the species P, then the charge Z of the species PH_ν is $Z_0 + \nu$. Thus

$$\overline{Z^2} - (\bar{Z})^2 = (Z_0^2 + 2Z_0\bar{\nu} + \overline{\nu^2}) - (Z_0 + \bar{\nu})^2 = \overline{\nu^2} - (\bar{\nu})^2 \quad (30\text{–}13)$$

Equation 30–13 assumes, of course, that no ions other than hydrogen ions determine the charge. However, it is extended with no difficulty to cases in which the binding of other ions is important.

Equations 30–10 and 30–13 provide a simple means for the experimental determination of $\overline{v^2} - (\overline{v})^2$, for differentiation of equation 30–10 leads at once to the result

$$a_{H^+} \frac{\partial \overline{v}}{\partial a_{H^+}} = - \frac{1}{2.303} \frac{\partial \overline{v}}{\partial pH} = \frac{\sum\limits_{v,(i)} v^2 k_v^{(i)} a_{H^+}{}^v}{1 + \sum\limits_{v,(i)} k_v^{(i)} a_{H^+}{}^v} - \left[\frac{\sum\limits_{v,(i)} v k_v^{(i)} a_{H^+}{}^v}{1 + \sum\limits_{v,(i)} k_v^{(i)} a_{H^+}{}^v} \right]^2$$

$$= \overline{v^2} - (\overline{v})^2 \tag{30–14}$$

The difference is thus determined by the *slope* of a titration curve such as that of Fig. 30–4. In this way we obtain, for instance, for ribonuclease at the isoelectric point $(\overline{Z} = 0)$, $\overline{Z^2} - (\overline{Z})^2 = \overline{Z^2} = 3.65$.

If a titration curve has been analyzed in terms of the approximate equation, equation 30–1, $\partial \overline{v}/\partial a_{H^+}$ can be given analytical form. Differentiation of equation 30–1 leads at once to

$$\overline{v^2} - (\overline{v})^2 = \sum_j n_j \frac{k_{int}^{(j)} e^{-2wZ} a_{H^+}}{(1 + k_{int}^{(j)} e^{-2wZ})^2}$$

$$= \sum_j n_j \theta_j (1 - \theta_j) = \sum_j n_j \alpha_j (1 - \alpha_j) \tag{30–15}$$

where n_j is the number of groups of type j; $k_{int}^{(j)}$ is the association constant, $k_{int}^{(j)} = 1/K_{int}^{(j)}$; and θ_j, α_j, and w have the meanings given earlier.

The considerations of this section apply, of course, to any polyampholyte, not merely to proteins.

31. COMPETITIVE AND OTHER COMPLEX EQUILIBRIA

If two substances A and B compete for the same binding site we must distinguish between *three* forms in which the site may exist. The site may be combined with A, or with B, or it may be free. We let θ_A, θ_B, and $1 - \theta_A - \theta_B$, respectively, represent the fractions of all sites in these three forms. If the free concentrations of A and B in solution are C_A and C_B, then, by equation 29–15,

$$\frac{\theta_A}{1 - \theta_A - \theta_B} = k_A C_A$$

$$\frac{\theta_B}{1 - \theta_A - \theta_B} = k_B C_B \tag{31–1}$$

In general, k_A and k_B will not be constants but will depend on the extent of occupation of other sites, as indicated by equation 29–18 or 29–22.

If equations 31–1 are combined, we get

$$\theta_A = \frac{\bar{\nu}_A}{n} = \frac{k_A C_A}{1 + k_A C_A + k_B C_B} \tag{31-2}$$

or

$$\frac{\theta_A}{1 - \theta_A} = \frac{\bar{\nu}_A}{n - \bar{\nu}_A} = \frac{k_A C_A}{1 + k_B C_B} \tag{31-3}$$

In these equations $\bar{\nu}_A$ is the average number of combined molecules (or ions) A and n the total number of binding sites per macromolecule. The characteristic feature of these equations is the appearance of the concentration of the competing substance (in this case B) in the denominator. When $C_B = 0$ the competition disappears, and equation 31–3 reduces to equation 29–14 or 29–15.

Equations 31–2 or 31–3 are often used for approximate calculations (especially in the analysis of rates of enzyme-catalyzed reactions; cf. section 35) as if k_A and k_B were constants. In the examples to be cited below, however, the variation of these "constants" with the extent of binding will have to be taken into account.

31a. The combination of proteins with metal ions.[34] It is a well-known fact that metal ions form complexes in solution with *basic* substances, i.e., with substances which also can combine with hydrogen ions. The reaction is therefore a competitive one. In fact, one of the best ways of measuring the equilibrium in a reaction such as $M^{+z} + NH_3 \rightleftharpoons M(NH_3)^{+z}, M(NH_3)_2^{+z}$ etc., is to do it indirectly by observing the displacement of the corresponding hydrogen ion equilibrium, e.g., $NH_4^+ \rightleftharpoons NH_3 + H^+$. This method has been used with particular success by Bjerrum.[2]

The same principles apply to the combination of metal ions with macromolecules. Ordinarily, the combination occurs at basic sites in competition with hydrogen ions. This may be seen at once from the very pronounced effect of pH on these reactions, as shown in Fig. 31–1. Very few quantitative studies have been made, and all of them have involved proteins. We illustrate the principles involved by discussing the combination of Zn^{++} with serum albumin, as studied by Gurd and Goodman.[37] The study was largely confined to a narrow region of pH in which the only sites for combination with the metal turn out to be the basic imidazole groups.

We proceed now to write the general equations for this type of situation, applicable whenever a single type of binding site is involved. Since both substances competing for these sites are ionic we suppose as before that electrostatic interaction is the only kind of interaction between sites. By equations 29–22 and 29–27 we then have for the association of metal ions, each of charge z_M,

$$k_M = (k_{int})_M e^{-2wz_M \bar{Z}} \tag{31-4}$$

and for the association with hydrogen ions

$$k_{H+} = (k_{int})_{H+} e^{-2w\bar{Z}} = e^{-2w\bar{Z}}/(K_{int})_{H+} \tag{31-5}$$

where $(K_{int})_{H+} = 1/(k_{int})_{H+}$ is the intrinsic hydrogen ion *dissociation* constant for the site under consideration, and w the electrostatic interaction factor defined in section 29c. The hydrogen ion titration curve of the

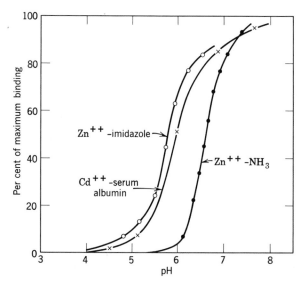

Fig. 31–1. Competition between metal ions and H$^+$ for the same binding sites, showing the characteristic effect of pH on the extent of binding under otherwise identical conditions. The binding sites for the Cd^{++}–serum albumin reaction are believed to be the imidazole groups of the protein molecule. (Data were taken from refs. 2, 35, and 36.)

protein has to have been determined previously, so that w and $(K_{int})_{H+}$ are known quantities. Introducing equations 31–4 and 31–5 into equation 31–3 we get

$$\frac{\theta_M}{1 - \theta_M} = \frac{\bar{\nu}_M}{n - \bar{\nu}_M} = \frac{(k_{int})_M e^{-2wz_M \bar{Z}} C_M}{1 + a_{H+}/(K_{int})_{H+} e^{2w\bar{Z}}} \tag{31-6}$$

where C_M is the free metal ion concentration and a_{H+} the activity of hydrogen ions.

For the combination of Zn^{++} with serum albumin $z_M = +2$. Gurd and Goodman, in their study of this reaction, made up a series of solutions containing various amounts of protein, metal, and acid or base. They determined $\bar{v}_M$ and C_M by dialysis equilibrium and the number, $\bar{r}$, of hydrogen ions dissociated in the usual way by pH measurement. Knowing $\bar{v}_M$ and $\bar{r}$ they could evaluate Z. Hence all parameters of equation 31–6 were known, and $(k_{int})_M$ could be evaluated. The results are given in Table 31–1. The constancy of k_{int} is proof of their assumption that only imidazole groups are involved in this reaction and of the applicability of equation 31–6 to the data.

TABLE 31–1. The Equilibrium between Zn^{++} and Serum Albumin[37]

$C_M \times 10^3$, moles/liter	pH	$\bar{v}_M$	$\log (k_{int})_M$
0.60	5.74	2.0	2.91
0.47	6.09	2.8	2.87
1.32	5.65	2.85	2.88
0.33	6.58	4.0	2.79
4.46	5.48	4.45	2.82
1.06	6.32	4.9	2.71
2.75	5.79	4.9	2.79
4.43	5.58	5.0	2.80
2.63	6.08	6.2	2.73
5.97	5.63	6.5	2.84
4.14	5.98	7.2	2.76
5.74	5.88	8.2	2.84

The values of $(k_{int})_M$ were calculated from a relation equivalent to equation 31–6, with the assumption that the imidazole groups of the protein molecule are the only binding sites.

As further proof of the correctness of their conclusion, the thermodynamic association constant k_1 for the reaction between Zn^{++} and imidazole, Im,

$$Zn^{++} + Im \rightleftharpoons Zn(Im)^{++}$$

was determined under similar conditions of ionic strength and temperature.[35] Our definition of $(k_{int})_M$ for the protein reaction requires that $(k_{int})_M$ should have close to the same magnitude as k_1. The result obtained, at 25°C and ionic strength 0.15, was $\log k_1 = 2.76$, compared with the average value of $\log (k_{int})_M = 2.82$ (from Table 31–1).

An alternative procedure is possible: to calculate the binding data from the difference in hydrogen ion titration curves in the presence and absence of metal, without a direct determination of $\bar{v}_M$ at all. This method is identical in principle with Bjerrum's method[2] for determining the formation constants of metal ammines. Being a method employing *differences* in a measured quantity it is less accurate than a method in which $\bar{\nu}_M$ is measured directly.

We suppose again that the titration curve in the absence of metal is known and that it has been represented by equation 30–1. In the presence of metal all the terms of this equation are unchanged, with one exception: the term which represents the metal binding sites now applies to only $n - \bar{\nu}_M$ of the n sites of this kind which each molecule contains. Where $\bar{r}'$ is the sum of all the *unchanged* terms of equation 30–1, we thus obtain

$$\bar{r} = \bar{r}' + (n - \bar{\nu}_M) \frac{(K_{int})_{H^+} e^{2wZ}/a_{H^+}}{1 + (K_{int})_{H^+} e^{2wZ}/a_{H^+}} \tag{31–7}$$

for the total number of H^+ ions dissociated from all sites which participate in the H^+ equilibrium. (The value of $\bar{r}$ referred to the fully protonated molecule is, of course, $\bar{\nu}_M$ *plus* the value given by equation 31–7 since each acidic group removed from participation in the H^+ equilibrium by metal binding has been deprived of its H^+ ion.)

Equation 31–7 expresses the experimentally observable value of $\bar{r}$, at any given pH and metal concentration C_M, in terms of the unknown $\bar{\nu}_M$ and in terms of intrinsic constants, group numbers, and w, all of which are known from the titration curve in the absence of metal. To use equation 31–7 we first estimate a value of $\bar{\nu}_M$ solely for the purpose of computing Z. With this value of Z we calculate $\bar{r}'$ by equation 30–1 and, hence, evaluate $\bar{\nu}_M$ by equation 31–7. This value of $\bar{\nu}_M$ is then used to get a better approximation to Z, and so forth, until a constant value of $\bar{\nu}_M$ is reached. The value of $(k_{int})_M$ is then obtained by equation 31–6.

A more complicated situation arises in the combination of Zn^{++} with insulin.[20] By analogy with serum albumin we should expect this reaction to involve imidazole groups as binding sites. This proves indeed to be the case, but each Zn^{++} ion appears to react with *two* imidazole groups instead of one, at least when the zinc concentration is low. It is again possible to write general equations for this kind of situation, as follows.

Suppose that there are n binding sites for the metal, each consisting of a *pair* of groups capable of binding H^+. For simplicity, we may take the two sites of each pair to be identical, as they indeed are in the zinc–insulin reaction. There is now a fraction $\theta_M = \bar{\nu}_M/n$ of these sites occupied by metal ions. The same fraction of acidic groups is removed from participation in the acid-base equilibrium. The groups which remain to participate in this equilibrium are divided as follows:

A fraction α,

$$\alpha = \frac{(K_{int})_{H^+} e^{2wZ}/a_{H^+}}{1 + (K_{int})_{H^+} e^{2wZ}/a_{H^+}} \tag{31–8}$$

bears no protons, whereas a fraction $(1 - \alpha)$ does bear protons. Since each metal binding site involves *two* acidic groups, we have the following fractional composition:

A fraction θ_M bears a metal ion.

A fraction $(1 - \theta_M)(1 - \alpha)^2$ bears no metal ion and two H^+ ions.

A fraction $2(1 - \theta_M)\alpha(1 - \alpha)$ bears no metal ion and one H^+ ion.

A fraction $(1 - \theta_M)\alpha^2$ bears no metal ion and no H^+ ions.

We thus get, in place of equation 31–1, again using equation 31–4 for k_M

$$\frac{\theta_M}{(1 - \theta_M)\alpha^2} = (k_{\text{int}})_M e^{-2wz_M Z} C_M \tag{31-9}$$

or

$$\frac{\theta_M}{1 - \theta_M} = \frac{\bar{v}_M}{n - \bar{v}_M} = (k_{\text{int}})_M \alpha^2 e^{-2wz_M Z} C_M \tag{31-10}$$

Again a simultaneous measurement of $\bar{r}$ (to give Z) and of $\bar{v}_M$ will yield a value for $(k_{\text{int}})_M$ if $(K_{\text{int}})_H$ of equation 31–8 and w are known from a previously determined titration curve in the absence of metal.

The equilibrium constant can also be determined, as before, from the titration curve alone. In so doing we must recall that the number of hydrogen ion binding groups is twice the number of metal binding sites and that each bound metal ion removes two groups from the acid-base equilibrium; i.e., if equation 31–7 is used and n is the number of metal binding sites, the last term in the equation must be multiplied by two.

These equations were found to reproduce both direct metal binding data and the titration curve of insulin in the presence of zinc,[20] and $(k_{\text{int}})_M$ for the reaction was evaluated. The value obtained was log $(k_{\text{int}})_M \simeq$ 6.0. It should be noted that this value is about twice that obtained for the combination of Zn^{++} with serum albumin. This difference is entirely in accord with expectation. For whereas k_{int} was formerly of the same order of magnitude as the equilibrium constant for the reaction $A + M \rightleftharpoons AM$, M representing the metal and A a small molecule resembling the reactive site, it should now be comparable instead to the equilibrium constant for the reaction $M + 2A \rightleftharpoons MA_2$. For imidazole and Zn^{++} Edsall and coworkers[35] obtained $Zn^{++} + Im \rightleftharpoons Zn(Im)^{++}$, log $k = 2.8$; $Zn^{++} + 2Im \rightleftharpoons Zn(Im)_2^{++}$, log $k = 5.1$.

Another protein with a complex binding site is conalbumin,[38] which binds iron at a site composed of *three* phenolic groups. In this reaction, as a further complication, there are two sites per molecule which strongly interact so that binding at one site greatly facilitates binding at the second site.

It should be noted, finally, that the equations of the present section are in no way limited to the combination of metals with proteins but apply equally well to other reactions involving competition with hydrogen ions.

31b. Equilibria with pH optima. The preceding considerations can be extended to the combination of any substance A with a binding site involving two different acidic or basic groups, one of which must have a proton attached whereas the other must have its proton dissociated. The latter will ordinarily be the more acidic site. Let its intrinsic dissociation constant be $K_{\text{int}}^{(1)}$, and let that of the less acidic site be $K_{\text{int}}^{(2)}$. Let $\bar{\nu}_A$ be the number of binding sites (total number n) combined with A, so that $n - \bar{\nu}_A$ acidic groups of each kind participate in the equilibrium with hydrogen ion, their degrees of dissociation, α_1 and α_2, being given by equation 31–8.

Of the $n - \bar{\nu}_A$ free binding sites a fraction $\alpha_1\alpha_2$ bears no protons at all. A fraction $(1 - \alpha_1)(1 - \alpha_2)$ bears protons on both acidic groups. A fraction $(1 - \alpha_1)\alpha_2$ bears a proton on the more acidic group and none on the more basic group. This fraction would be expected to be small at any pH unless $K_{\text{int}}^{(1)} \simeq K_{\text{int}}^{(2)}$. Finally, a fraction $\alpha_1(1 - \alpha_2)$ bears a proton on the less acidic group and none on the more acidic group. Only this fraction of free sites is capable of combination with A. This fraction has a maximum at a pH somewhere between $pK_{\text{int}}^{(1)}$ and $pK_{\text{int}}^{(2)}$, the exact variation with pH being also affected by electrostatic interactions.

We now obtain, in place of equation 31–10

$$\frac{\theta_A}{1 - \theta_A} = \frac{\bar{\nu}_A}{n - \bar{\nu}_A} = (k_{\text{int}})_A \alpha_1(1 - \alpha_2)e^{-2wz_A Z}C_A \qquad (31\text{--}11)$$

Figure 31–2 shows the application of these equations to a hypothetical example. It is supposed that the substance A is uncharged, and that there is a single binding site involving groups with pK_{int} 4.3 and 6.7. The binding site is taken to be on a protein molecule with a titration curve resembling that of ovalbumin at ionic strength 0.033. We have placed $(k_{\text{int}})_A C_A = 1$.

There have been no thermodynamic studies of an equilibrium of the kind here described. However, the existence of this kind of equilibrium is sometimes assumed in the analysis of the rates of enzyme-catalyzed reactions. When this is done the approximation is sometimes made that α can be represented by equation 29–17, neglecting electrostatic interaction. With this approximation

$$\alpha_1(1 - \alpha_2) = \frac{K^{(1)}a_{H^+}}{(K^{(1)} + a_{H^+})(K^{(2)} + a_{H^+})} \qquad (31\text{--}12)$$

The equilibrium constants $K^{(1)}$ and $K^{(2)}$ are determined empirically to fit the kinetic data.

The dashed line of Fig. 31–2 shows an attempt to fit the data (calculated with equation 31–8 for α_1 and α_2) by means of equation 31–12. We see that a fair fit can be obtained on the alkaline side. The reason for this is that the assumed titration curve of the protein is very flat in this region of pH, so that $\bar{Z}$ varies little with pH. On the acid side, however, equation 31–12 gives a result differing widely from the result calculated by use of the correct equations. It should be noted that the empirical values of $K^{(1)}$ and $K^{(2)}$ differ from the true intrinsic constants on both sides.

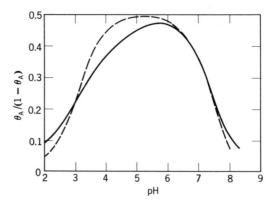

Fig. 31–2. Equilibrium with a pH optimum. The solid line is calculated for combination of an uncharged molecule A with binding sites consisting of acidic groups with pK_{int} 4.3 and 6.7, one of which has to be in its acidic form and the other in its basic form if binding is to occur. Equation 31–11 was used for the calculation, with $(k_{int})_A C_A = 1$, and with electrostatic effects included in the calculation of α_1 and α_2. The dashed line is an attempt to reproduce the same curve by equation 31–12, in which electrostatic effects are neglected. The curve shown is the best obtainable, using $pK^{(1)} = 3.4$ and $pK^{(2)} = 7.1$.

An additional precaution is necessary in applying these considerations to the kinetics of enzyme-catalyzed reactions. If a curve of velocity versus pH like that of Fig. 31–2 is obtained for such a reaction, it can be due to inability of the substrate to combine with any form of the enzyme molecule except that in which two crucial groups are in a particular state of ionization, in which case the foregoing treatment applies. An alternative possibility is that the substrate can combine with all forms of the enzyme, but that the only reactive complex is that in which two crucial groups are in a particular state of ionization. In that event the reaction velocity depends on the dissociation constants of the two groups in the complex, as well as on the constants $K^{(1)}_{int}$ and $K^{(2)}_{int}$. The empirical values of $K^{(1)}$ and $K^{(2)}$, which are obtained by applying equation 31–12 to the experimental data, may then differ considerably from the intrinsic constants appropriate to the enzyme molecule itself. Any serious attempt to interpret the pH-dependence of an enzyme-catalyzed reaction requires a detailed kinetic analysis, such as that discussed in section 35.

31c. Internal hydrogen bonding. Another type of competitive equilibrium occurs if the reactive sites of a molecule are capable of reacting *with one another* to form hydrogen bonds. A detailed discussion of the effect of such a possibility has been given by Laskowski and Scheraga[26] and may be illustrated by a single example. Suppose that all the carboxyl groups of a macromolecule can be arranged in pairs, such that the members of each pair can form hydrogen bonds of the type

$$O \cdots H\text{---}O$$

(structure: $-C$ ... $C-$ with $O \cdots H\text{---}O$ on top and $O\text{---}H \cdots O$ on bottom)

Let the fraction of all pairs existing under a given set of conditions in this form be termed β. The fraction of pairs not hydrogen bonded is then $1 - \beta$. The non-hydrogen-bonded pairs are just single carboxyl groups, and their state of ionization is given by equation 30–2. Where α is the average degree of ionization of a carboxyl group, as given by this equation, we have, as in the formally similar example discussed in the preceding section, the following fractional compositions:

A fraction β in the form —COOH $\cdots$ HOOC--.
A fraction $(1 - \beta)(1 - \alpha)^2$ with separate groups, —COOH, —COOH.
A fraction $2(1 - \beta)(1 - \alpha)\alpha$ with separate groups, —COOH, —COO⁻.
A fraction $(1 - \beta)\alpha^2$ with separate groups, —COO⁻, —COO⁻.

We have assumed that the equilibrium constant for formation of singly hydrogen-bonded structures, such as

(structures shown) or

is sufficiently small so that they need not be considered. (The more complete treatment, allowing for such forms, is given by Laskowski and Scheraga.[26])

The equilibrium constant for the formation of hydrogen bonds can now be written

$$\frac{\beta}{(1 - \beta)(1 - \alpha)^2} = k \qquad (31\text{--}13)$$

With our usual simplification, that only coulombic interaction between charges occurs, k will be a constant independent of $\bar{Z}$.

The quantity experimentally observable from titration curves is the average number $\bar{r}$ of protons dissociated from all the carboxyl groups of the molecule. Let there be n of them, of which, from the previous definition of β, βn will be incorporated in hydrogen-bonded dimers. Of the

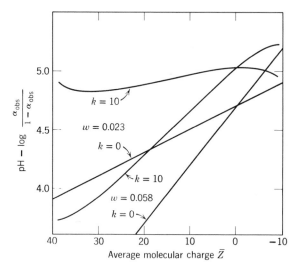

Fig. 31–3. The effect of carboxyl–carboxyl hydrogen bonds on the titration curve of the carboxyl groups of a protein molecule containing fifty carboxyl groups and forty positively charged groups. Titration curves were calculated by equation 31–12 with $pK_{int} = 4.70$, $w = 0.058$ (low ionic strength) or 0.023 (high ionic strength), and the hydrogen bond formation constant equal to zero or 10. The data were then plotted logarithmically as in Figs. 30–6, 30–7, etc.

remainder a fraction α is dissociated, so that $\bar{r} = \alpha n(1 - \beta)$. The experimentally observed degree of dissociation of all carboxyl groups, which we shall call α_{obs}, is thus equal to $\alpha(1 - \beta)$. Introducing equations 31–8 and 31–13 we get

$$\frac{\alpha_{obs}}{1 - \alpha_{obs}} \, a_{H^+} = \frac{K_{int} e^{2w\bar{Z}}(1 + K_{int} e^{2w\bar{Z}}/a_{H^+})}{1 + k + K_{int} e^{2w\bar{Z}}/a_{H^+}} \tag{31-14}$$

Figure 31–3 shows a plot of $\log [\alpha_{obs}/(1 - \alpha_{obs})]$ according to equation 31–14, with reasonable assumed values of k, K_{int}, and w, and of the relation between $\bar{Z}$ and a_{H^+}. It is seen that the effect of hydrogen bonding is in the opposite direction to the normal electrostatic effect; i.e., if one —COO⁻ group of a pair is converted to —COOH, the affinity of the other —COO⁻

groups for H^+ is increased because of the ability to form a carboxyl–carboxyl hydrogen bond. At low ionic strength, where electrostatic effects are large, the over-all result is a diminution of the slope of a plot such as that of Fig. 31–3 and an increase in the apparent value of pK_{int} which we would obtain if the data of Fig. 31–3 were interpreted according to equation 30–2, for by that equation pK_{int} is the value of $pH - \log[\alpha/(1 - \alpha)]$ at $\bar{Z} = 0$. (Thus a possible alternative explanation is provided for some of the differences found in Table 30–3. It was shown in section 30h that they might equally well be due to errors arising from use of approximate equations for the electrostatic effects.) At higher ionic strengths, i.e., smaller w, the effect of hydrogen bonding is more drastic and the curves of $pH - \log[\alpha_{obs}/(1 - \alpha_{obs})]$ versus $\bar{Z}$ take on a character quite different from those of Figs. 30–6 and 30–7.

31d. Concluding remarks. We have discussed in this section a number of complex equilibria which appear to be well understood in terms of an easily visualized mechanism. It should be emphasized, however, that they are isolated examples. There remain numerous reactions for which experimental data exist but for which there is at present no real explanation.

Among them are the reaction between hemoglobin and oxygen, to which reference was made in section 29. Here we know the number of reacting sites and their location, but we do not know the environmental factors which produce the reactivity of the sites and lead to the strong interaction between sites shown in Fig. 29–2.

Another intriguing reaction is that between serum albumin and the two isomers of the optically active dye, α-(N-p-aminobenzoyl) aminophenylacetate. Data obtained for this reaction by Karush[39] are shown in Fig. 31–4. It is seen that the L-isomer of this dye behaves like a typical anion (cf. Fig. 29–4), the curvature of the plot shown being due to the combined effects of electrostatic interaction and the existence of binding sites of different affinity. For the D-isomer, on the other hand, the initial slope is almost zero. This must mean that *cooperative* interaction plays a role in the binding of the first few ions of the D-dye. For, in the absence of interaction a plot such as that of Fig. 31–4 would be a straight line with an intercept at $\bar{\nu} = 0$ equal to kn and with a slope of $-k$, whereas interaction in which binding reduces the affinity of other sites produces a curvature (as seen for the L-dye) opposite in direction to that observed for the D-dye.

This reaction has been interpreted by Karush as follows: Serum albumin initially has a rather rigid structure in which the complex sites required to combine with α-(N-p-aminobenzoyl) aminophenylacetate are better suited for binding of the L-isomer than they are for binding of the D-isomer, as indicated by the initially higher values of $\bar{\nu}/C$ for the L-isomer.

The binding of the first few dye molecules, however, loosens the structure so that the reacting sites become more adaptable. This has little effect on the binding of the L-isomer but greatly facilitates the binding of subsequent molecules of the D-isomer. Obviously, the binding mechanism is complicated, and it is not possible to decide whether binding at one particular site is necessary to cause the loosening of the molecular structure or whether all

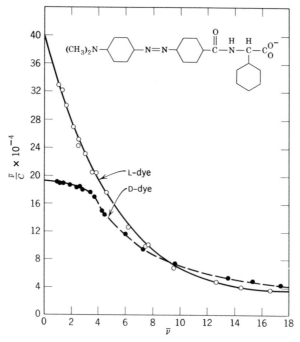

Fig. 31–4. Binding of D- and L-isomers of α-(N-*p*-aminobenzoyl) aminophenylacetate to serum albumin. The data are plotted according to equation 29–14. (Karush.[39])

the available sites are equivalent in this respect. Nor is it possible to exclude the possibility that what we have here vaguely termed "interaction" is in fact a form of competitive equilibrium, involving hydrogen bonds or other internal bonding between the binding sites.

It should be mentioned, finally, that pH-dependent reversible configuration changes, such as those which occur in serum albumin (section 28) and in ferrihemoglobin,[40] are presumably also complex competitive equilibria. The nature of these reactions has not yet been elucidated sufficiently to permit even a tentative formulation in terms of equilibrium constants.

General References

1. G. Scatchard, "The Attractions of Proteins for Small Molecules and Ions," *Ann. N. Y. Acad. Sci.*, **51**, 660 (1949).
2. J. Bjerrum, *Metal Ammine Formation in Aqueous Solution*, P. Haase and Son, Copenhagen, 1941.
3. I. M. Klotz, "Protein Interactions" in H. Neurath and K. Bailey (ed.), *The Proteins*, Academic Press, New York, 1953.
4. C. Tanford, "Hydrogen Ion Titration Curves of Proteins" in T. Shedlovsky (ed.), *Electrochemistry in Biology and Medicine*, John Wiley and Sons, New York, 1955, Ch. 13.
5. J. T. Edsall, "Dipolar Ions and Acid-Base Equilibria," "Some Relations between Acidity and Chemical Structure," "Proteins as Acids and Bases" in E. J. Cohn and J. T. Edsall, *Proteins, Amino Acids and Peptides*, Reinhold Publishing Corp., New York, 1943, Chs. 4, 5, 20.
6. J. Steinhardt and E. M. Zaiser, "Hydrogen Ion Equilibria in Native and Denatured Proteins," *Advances in Protein Chem.*, **10**, 151 (1955).

Specific References

7. G. Scatchard, J. S. Coleman, and A. L. Shen, *J. Am. Chem. Soc.*, **79**, 12 (1957).
8. C. W. Carr in T. Shedlovsky (ed.), *Electrochemistry in Biology and Medicine*, John Wiley and Sons, New York, 1955, Ch. 14.
9. J. Wyman, Jr., *Advances in Protein Chem.*, **4**, 407 (1948).
10. C. Tanford and J. D. Hauenstein, *J. Am. Chem. Soc.*, **78**, 5287 (1956).
11. R. E. Benesch and R. Benesch, *J. Am. Chem. Soc.*, **77**, 5877 (1955); G. Gorin, *ibid*, **78**, 767 (1956).
12. J. G. Kirkwood and F. H. Westheimer, *J. Chem. Phys.*, **6**, 506, 513 (1938)
13. F. Karush and M. Sonenberg, *J. Am. Chem. Soc.*, **71**, 1369 (1949).
14. C. Tanford, S. A. Swanson, and W. S. Shore, *J. Am. Chem. Soc.*, **77**, 6414 (1955).
15. R. Arnold and J. Th. G. Overbeek, *Rec. trav. chim.*, **69**, 192 (1950).
16. A. Katchalsky and I. R. Miller, *J. Polymer Sci.*, **13**, 57 (1954).
17. C. Tanford, J. D. Hauenstein, and D. G. Rands, *J. Am. Chem. Soc.*, **77**, 6409 (1955).
18. C. H. W. Hirs, S. Moore, and W. H. Stein, *J. Biol. Chem.*, **219**, 623 (1956); **221**, 151 (1956).
19. R. K. Cannan, A. H. Palmer, and A. C. Kibrick, *J. Biol. Chem.*, **142**, 803 (1942).
20. C. Tanford and J. Epstein, *J. Am. Chem. Soc.*, **76**, 2163, 2170 (1954).
21. R. K. Cannan, A. C. Kibrick, and A. H. Palmer, *Ann. N. Y. Acad. Sci.*, **41**, 243 (1941).
22. C. Tanford and G. L. Roberts, Jr., *J. Am. Chem. Soc.*, **74**, 2509 (1952).
23. E. Katchalski and M. Sela, *J. Am. Chem. Soc.*, **75**, 5284 (1953).
24. G. L. Roberts, Jr., unpublished data.
25. C. Tanford and M. L. Wagner, *J. Am. Chem. Soc.*, **76**, 3331 (1954).
26. M. Laskowski, Jr., and H. A. Scheraga, *J. Am. Chem. Soc.*, **76**, 6305 (1954).
27. H. M. Dintzis, Ph.D. Thesis, Harvard University, 1952.
28. E.g., L. G. Longsworth and C. F. Jacobsen, *J. Phys. and Colloid Chem.*, **53**, 126 (1949); R. A. Alberty, *ibid.*, **53**, 114 (1949).

29. C. Tanford, J. G. Buzzell, D. G. Rands, and S. A. Swanson, *J. Am. Chem. Soc.*, **77**, 6421 (1955).
30. J. Steinhardt and E. Zaiser, *J. Biol. Chem.*, **190**, 197 (1951).
31. W. A. Lee and A. R. Peacocke, *J. Chem. Soc.*, **1951**, 3361; R. A. Cox and A. R. Peacocke, *ibid.*, **1956**, 2499; **1957**, 4724.
32. F. E. Harris and S. A. Rice, *J. Phys. Chem.*, **58**, 725, 733 (1954).
33. C. Tanford and J. G. Kirkwood, *J. Am. Chem. Soc.*, **79**, 5333 (1957); C. Tanford, *ibid.*, **79**, 5340 (1957).
34. F. R. N. Gurd and P. E. Wilcox, *Advances in Protein Chem.*, **11**, 312 (1956).
35. J. T. Edsall, G. Felsenfeld, D. S. Goodman, and F. R. N. Gurd, *J. Am. Chem. Soc.*, **76**, 3054 (1954).
36. C. Tanford, *J. Am. Chem. Soc.*, **74**, 211 (1952).
37. F. R. N. Gurd and D. S. Goodman, *J. Am. Chem. Soc.*, **74**, 670 (1952).
38. R. C. Warner and I. Weber, *J. Am. Chem. Soc.*, **75**, 5094 (1953).
39. F. Karush, *J. Phys. Chem.* **56**, 70 (1952).
40. J. Steinhardt and E. M. Zaiser, *J. Am. Chem. Soc.*, **75**, 1599 (1953); **76**, 1788 (1954).
41. I. Langmuir, *J. Am. Chem. Soc.*, **38**, 2221 (1916); **40**, 1361 (1918).

9

KINETICS OF
MACROMOLECULAR
REACTIONS

A powerful tool for understanding any chemical reaction is the study of the kinetics of that reaction. This term signifies the measurement of the rate with which the reaction proceeds and the study of the manner in which this rate is influenced by such factors as the concentration of reagents (or other added substances), the temperature at which the reaction is carried out, and the nature of the solvent. The kinetics of a reaction are related to its mechanism, i.e., to the sequence of steps which occur in the transformation of reagents to products, and the principal purpose of rate studies is often to make deductions about the nature of the reaction mechanism.

The general principles of chemical kinetics are well known and require no repetition here. The principles as formulated in textbooks of physical chemistry for reactions between small molecules apply with essentially no change to macromolecular reactions; i.e., there are no new principles which need be formulated. Many macromolecular reactions, in fact, cannot be distinguished as such at all by their kinetic behavior.

An example of such a reaction is the slow dimerization of serum albumin (Alb—SH) in the presence of Hg^{++} ion[8]: $2Alb—SH + Hg^{++} \rightarrow Alb—S—Hg—S—Alb + 2H^{+}$. The kinetic behavior of this reaction falls easily into one of the general schemes for reactions of the type $2A + B \rightarrow$ products. No novel features result from the fact that A in this case represents a very large molecule.

The objective of this chapter will be to examine the kinetic behavior of macromolecular reactions, which, by their very nature, can have no counterpart in the chemistry of small molecules. Included in this group are the reactions by which polymers are formed from small molecules, those in which they are degraded to smaller molecules, and those in which large molecules undergo drastic changes in their configuration in space.

32. KINETICS OF POLYMERIZATION[2,3]

A reaction which obviously can have no counterpart in the chemistry of small molecules is the reaction by which polymers are synthesized from suitable monomers. We shall examine in this section some examples of such reactions and shall see that kinetic studies lead to considerable insight into the polymerization process.

32a. Condensation polymerization. Polycondensation reactions are those in which H_2O, CO_2, NH_3 or some other substance is eliminated during the polymerization. A typical example is polyester formation,

$$x\text{HO}-(\text{CH}_2)_n-\text{COOH} \rightarrow \text{HO}[-(\text{CH}_2)_n-\text{COO}]_x-\text{H} + (x-1)\text{H}_2\text{O} \quad \text{(I)}$$

or the corresponding reaction starting with a glycol and a dicarboxylic acid,

$$\tfrac{1}{2}x\text{HO}-(\text{CH}_2)_n-\text{OH} + \tfrac{1}{2}x\text{HOOC}-(\text{CH}_2)_{n'}-\text{COOH} \rightarrow$$
$$\text{HO}[-(\text{CH}_2)_n-\text{OOC}-(\text{CH}_2)_{n'}-\text{COO}]_{\frac{1}{2}x}-\text{H} + (x-1)\text{H}_2\text{O} \quad \text{(II)}$$

It is reasonable to suppose that these reactions occur by a straightforward molecular mechanism, similar to that involved in ordinary esterification. The steps of the polymerization process would then be of the type

$$\text{HO}-\text{X}-\text{OH} + \text{HOOC}-\text{Y}-\text{COOH} \rightarrow$$
$$\text{HO}-\text{X}-\text{OOC}-\text{Y}-\text{COOH} + \text{H}_2\text{O}$$
$$\text{HO}-\text{X}-\text{OOC}-\text{Y}-\text{COOH} + \text{HO}-\text{X}-\text{OH} \rightarrow$$
$$\text{HO}-\text{X}-\text{OOC}-\text{Y}-\text{COO}-\text{X}-\text{OH} + \text{H}_2\text{O}$$
$$\text{HO}(-\text{X}-\text{OOC}-\text{Y}-\text{COO})_x-\text{H} + \text{HO}(-\text{X}-\text{OOC}-\text{Y}-\text{COO})_y-\text{H} \rightarrow$$
$$\text{HO}(-\text{X}-\text{OOC}-\text{Y}-\text{COO})_{x+y}-\text{H} + \text{H}_2\text{O} \quad \text{etc.} \quad \text{(III)}$$

The distinguishing feature of such a mechanism is that all the molecules in the reaction mixture are equally reactive. Polymer molecules contain two reactive groups, of the same kind as those in the monomer molecule, and there is thus no unique reaction product. The reaction cannot be "completed" in a unique sense until all the material in the reaction mixture is incorporated into a single molecule.

The ordinary esterification reaction between low-molecular-weight acids and alcohols requires an acid catalyst,[1] so that such a catalyst might also be expected to be essential in polyester formation. The rate of any reaction of type III would then be

$$\frac{d[\text{product species}]}{dt} = \nu k[\text{catalyst}][\text{reactant species 1}][\text{reactant species 2}]$$

$$(32\text{--}1)$$

where square brackets represent concentrations. The rate constant k is that for reaction between a long chain molecule with a single —OH group and another such molecule with a single —COOH group; i.e., it is the rate constant for the reaction R—OH + R′—COOH → R—OOC—R′ +H_2O. The factor ν is a multiplicity factor which takes into account the fact that reacting species with more than one —OH or —COOH group react more rapidly in proportion to the number of different encounters between —OH and —COOH groups which can lead to reaction. Thus ν is equal to 2 if the reacting molecules are HO—X—COOH + HO—Y—COOH, for they can react in two ways, at the —OH group of the first molecule and the —COOH group of the second, or vice versa. Similarly, ν equals 2 for a reaction HO—X—OH + HO—Y—COOH or for HOOC—X—COOH + HO—Y—COOH. Finally, ν is equal to 4 if the reacting molecules are HO—X—OH + HOOC—Y—COOH.

If the two reactant species are identical, e.g., HO—X—OOC—Y—COOH + HO—X—OOC—Y—COOH, an additional statistical factor of $\frac{1}{2}$ enters into equation 32–1, since, if we merely substitute [reactant]^2 for the product [reactant 1][reactant 2], every reactive collision would be counted twice. The reason for this is that in a mixture of molecules, A, B, C, etc., at concentrations [A], [B], [C], etc., the relative numbers of collisions between A and A, A and B, A and C, B and B, etc., are in the ratio [A]^2, 2[A][B], 2[A][C], [B]^2, etc.

Under suitable conditions the rate constant k may be assumed independent of chain length. The pertinent conditions are: (a) that the number of methylene groups, n or n', in reactions I and II, be reasonably large, so that the functional groups are essentially isolated from one another; and (b) that the reaction be carried out in concentrated solution or in the liquid reagents in the absence of solvent, so that the process by which the functional groups seek one another out is one which involves motion only of the heads or tails of the reacting molecules. (These same conditions are in any case desirable, for another reason. This is to prevent the formation of low-molecular-weight cyclic inner esters, which would be incapable of further reaction. The formation of these undesired by-products would be favored by increased dilution of the reagents.)

The concentrations of the numerous individual species which arise and disappear during polymerization are, of course, not measurable, nor is there a single species which can be defined as reaction product. The rate of reaction cannot therefore be measured by following the rate of appearance or disappearance of particular molecular species. It is easy, however, to follow the progress of the reaction by means of titration of the free carboxyl groups, the number of which progressively falls as the reaction proceeds. If the monomer is an ω-hydroxy carboxylic acid, then the number of free carboxyl groups is also the number of free hydroxyl groups, as well as the total number of all molecules, for each molecule present has a single terminal carboxyl group and a single terminal hydroxyl group. The same correspondence applies to the polymerization of a mixture of a glycol and a dicarboxylic acid if the two reagents are originally introduced in equimolar amounts, for in that case, although some molecules will have —COOH groups at both ends, an equal number will have —OH groups at both ends.

Let C be the concentration, at time t, of free carboxyl groups and thus also the concentration of hydroxyl groups and of all molecules (other than water or solvent, if present). Let C_0 be the corresponding initial concentration before polymerization has commenced. Then C_0/C is clearly the number-average degree of polymerization, $\bar{x}_n$, since C molecules have C_0 monomer molecules incorporated within them. Introducing the parameter p (cf. p. 140), which represents the fraction of all functional groups which have reacted,

$$p = (C_0 - C)/C \tag{32-2}$$

we have, at any time t,

$$\bar{x}_n = C_0/C = 1/(1 - p) \tag{32-3}$$

If the definition of $\bar{x}_n$ here used is applied to the polymerization of glycols with dicarboxylic acids (reaction II), the residues of either reagent are counted as a single polymer segment. A degree of polymerization x means the presence of $x/2$ glycol residues and $x/2$ acid residues, with an excess of a single residue of one kind if x is odd.

The equation for the rate of change of the concentration C may now be written down at once, as the total rate at which processes of type III occur, for a carboxyl group disappears in each such step regardless of the state of polymerization of the reacting species. Since k is also assumed independent of this factor and since it has been defined as the rate constant for reaction of a single carboxyl group with a single hydroxyl group, we may write

$$-d[\text{—COOH}]/dt = k[\text{catalyst}][\text{—COOH}][\text{—OH}]$$

or

$$-dC/dt = k[\text{catalyst}]C^2 = k'C^2 \tag{32-4}$$

where $k' = k$ [catalyst]. Since the catalyst is not used up during the reaction, k' is a constant, independent of time.

Integrating equation 32–4 we get $1/C = k't + 1/C_0$, and combining this result with equation 32–3 gives $\bar{x}_n$ or p as a function of time,

$$\bar{x}_n = C_0/C = 1 + k'C_0t \qquad (32\text{--}5)$$

$$p = k'C_0t/(1 + k'C_0t) \qquad (32\text{--}6)$$

That these equations are in fact obeyed by reactions of this type is shown, for instance, by Fig. 32–1, which gives data obtained by Flory[9] for

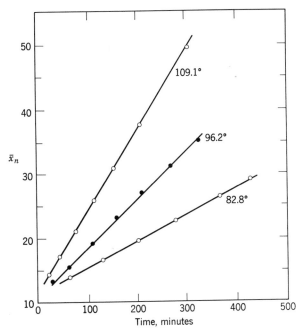

Fig. 32–1. A plot of $\bar{x}_n$ versus time for the polymerization of decamethylene glycol and adipic acid, with p-toluenesulfonic acid as catalyst. The sample was partially polymerized (to $\bar{x}_n \simeq 10$) before measurements were begun. (Flory.[9])

the polymerization of decamethylene glycol and adipic acid, in the presence of a small amount of p-toluenesulfonic acid as catalyst. There are no corresponding data for esterification of long chain simple acids and alcohols, so that the assumption that the rate constant for such a reaction should be the same as the rate constant for polyesterification cannot be tested. The activation energy obtained from the data of Fig. 32–1, $E_a = -\mathscr{R} \, d \ln k/d(1/T)$, is 11,200 cal. This figure is roughly the same as the activation energy of 10,000 cal observed for the esterification of aliphatic acids with methanol, catalyzed by HCl.

Polyesterification may also be carried out without added catalyst, the catalytic role being filled by the carboxyl groups present in the reaction mixture. In that case the catalyst concentration is no longer a constant but becomes instead equal to C. Equation 32–4 becomes

$$-dC/dt = kC^3$$

and, upon integration and combination with equation 32–3,

$$(\bar{x}_n)^2 = C_0^2/C^2 = 1 + 2kC_0^2 t \qquad (32\text{--}7)$$

This equation has also been experimentally confirmed by Flory,[10] for the same polyesterification as above, but without the added catalyst. In this case parallel studies of ordinary esterification were carried out, and the expected similarity in rate constants was observed. The value of k (with concentrations expressed in terms of moles per 1000 grams of reaction mixture and time in minutes) was found to be 163 at 202° C for the poly-merization reaction. At the same temperature the corresponding value of k for the simple esterification reaction between lauryl alcohol and adipic acid was found to be 157 and that between lauryl alcohol and lauric acid was found to be 250.

32b. Kinetic derivation of the distribution function. In Chapter 3 we introduced the concept of a molecular distribution function, an expression which gives the relative number of molecules of any given degree of poly-merization, x, in a mixture of average degree of polymerization, $\bar{x}_n$. An equation for such a distribution function, applicable to condensation polymerization, was derived statistically in section 8. In general, however, this is not the best way to obtain the distribution function, for the distri-bution arises during the polymerization reaction as a result of the relative rate of formation of products of various degrees of polymerization, and the logical way to derive a distribution function is from the kinetics of the reaction. Such a derivation will now be presented for the case of poly-esterification. The result will be found to be identical with that obtained statistically in section 8. (The kinetic derivation was first presented by Dostal and Raff.[11])

Polyesterification is usually carried out in a mixture of a glycol with a dicarboxylic acid (reaction II). The derivation of the distribution function is however less cumbersome for reaction I, involving a single monomer, and it is for this reaction that the present derivation will be carried out. Essentially the same result is obtained if the derivation is carried through for reaction II.

Let M_1 represent a molecule of monomer, M_2 a molecule of dimer, etc., and let $[M_1]$, $[M_2]$, etc., represent the concentrations of these species.

From the definition of C, given earlier,

$$\sum_{x=1}^{\infty} [M_x] = C \qquad (32\text{-}8)$$

We wish to consider the rate of change of concentration of each of the individual species M_x, in the situation where an acid catalyst is present. Beginning with M_1 we note that it may disappear (according to reaction III) by reaction with any other species M_x. The rate is given by equation 32–1, and, since all the molecules under consideration are of the type HO—X—COOH, the statistical factor ν in that equation is equal to 2. For the reaction $M_1 + M_1 \rightarrow M_2$ the statistical factor for formation of M_2 is unity for the reason given earlier. However, two molecules of M_1 disappear in that reaction, so that the rate of disappearance of M_1 by the reaction again requires that $\nu = 2$. The over-all rate of change of $[M_1]$ is therefore

$$d[M_1]/dt = -2k'[M_1] \sum_{x=1}^{\infty} [M_x] = -2k'[M_1]C \qquad (32\text{-}9)$$

where, as before, $k' = k$ [catalyst].

For any species M_x other than M_1 there will be reaction steps in which the species is formed from less polymerized ones, as well as steps by which it disappears. Formation of M_x will occur whenever a species M_j ($j < x$) reacts with M_{x-j}. The rate at which it occurs is given as $k' \sum_{j=1}^{x-1} [M_j][M_{x-j}]$. The coefficient is k' rather than $2k'$ because the summation counts any of the possible reactions, say that between M_i and M_k, twice: once with $j = i$ and once with $j = k$. (Where $j = x/2$ the coefficient is k' for the reason given earlier.) The rate of disappearance of M_x is given by an expression similar to equation 32–9, so that the net result is

$$d[M_x]/dt = k' \sum_{j=1}^{x-1} [M_j][M_{x-j}] - 2k'[M_x]C \qquad (32\text{-}10)$$

Equation 32–9 is solved readily if it is expressed in terms of $\bar{x}_n$ rather than t. By equation 32–5 $C = C_0/\bar{x}_n$ and $dt = d\bar{x}_n/k'C_0$, so that equation 32–9 becomes

$$-d[M_1]/[M_1] = 2\, d\bar{x}_n/\bar{x}_n$$

The solution, with the condition that, at $t = 0$, $\bar{x}_n = 1$ and $[M_1] = C_0$, is

$$[M_1] = C_0/(\bar{x}_n)^2 \qquad (32\text{-}11)$$

Equation 32–10 with $x = 2$ may now be solved for $[M_2]$. Again expressing the equation in terms of $\bar{x}_n$ rather than t we get

$$\frac{d[M_2]}{d\bar{x}_n} = \frac{[M_1]^2}{C_0} - 2\frac{[M_2]}{\bar{x}_n}$$

and, upon substitution of equation 32–11 for $[M_1]$, this leads to

$$\frac{d[M_2]}{d\bar{x}_n} + \frac{2[M_2]}{\bar{x}_n} = \frac{C_0}{(\bar{x}_n)^4} \tag{32-12}$$

Equation 32–12 is a linear differential equation of the first order, and it has the general solution (see Appendix A)

$$[M_2](\bar{x}_n)^2 = \int C_0 \, d\bar{x}_n / (\bar{x}_n)^2 \tag{32-13}$$

Thus

$$[M_2](\bar{x}_n)^2 = K - C_0/\bar{x}_n$$

where K is an integration constant. At $t = 0$, $[M_2] = 0$ and $\bar{x}_n = 1$, so that $K = C_0$ and

$$[M_2] = \frac{C_0}{(\bar{x}_n)^2}\left(1 - \frac{1}{\bar{x}_n}\right) \tag{32-14}$$

Continuing in the same way we find that the differential equation for each $[M_x]$ is linear, differing from equation 32–12 only in the term on the right-hand side. The solution is given by equation 32–13 with a different integral for each value of x. The integrals, after inserting the appropriate values for K, turn out to be C_0 times the binomial expansions of $(1 - 1/\bar{x}_n)^{x-1}$, so that the general solution becomes

$$[M_x] = \frac{C_0}{(\bar{x}_n)^2}\left(1 - \frac{1}{\bar{x}_n}\right)^{x-1} \tag{32-15}$$

In terms of the extent of reaction of functional groups, the factor p given by equation 32–2, equation 32–15 becomes

$$[M_x] = C_0(1 - p)^2 p^{x-1} \tag{32-16}$$

To convert equation 32–15 into a distribution function we again use the fact that $\sum [M_x] = C$, so that the mole fraction X_x of molecules of degree of polymerization x becomes $[M_x]/C$. By equations 32–16 and 32–3 we then get

$$X_x = (1 - p)p^{x-1} \tag{32-17}$$

This result is identical with that obtained statistically in section 8 and given by equation 8–4. As section 8 showed the corresponding weight distribution function is

$$W_x = x(1 - p)^2 p^{x-1} \tag{32-18}$$

That the distribution predicted by these equations actually occurs experimentally has been shown, for a polyamide (nylon) by Taylor.[12] A graphical representation of the distribution is shown in Fig. 32–8.

32c. The progress of polymerization. Equations 32–17 or 32–18 together with equation 32–6 for the time dependence of p give a clear picture of how

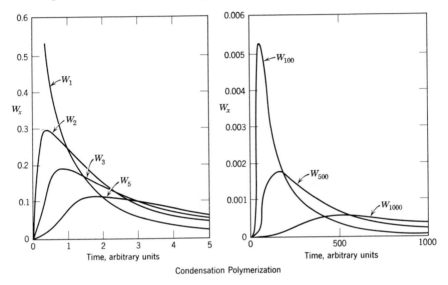

Condensation Polymerization

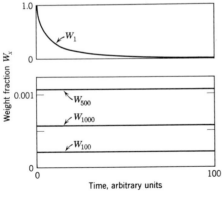

Free Radical Polymerization

Fig. 32–2. The progress of polymerization with time, for condensation polymerization (upper diagrams) and for free radical polymerization (lower diagram). The data for the latter are for the case of thermal initiation.

condensation polymerization proceeds as a function of time. Some calculations based on these relations are shown in Fig. 32–2. It is seen that the polymerization is a progressive one. First the monomer disappears, and dimers and trimers appear instead. They in turn are replaced by higher and higher polymers. Theoretically the reaction will cease only when the entire contents of the reaction mixture become a single molecule. However, as equation 32–4 shows, the reaction rate decreases very rapidly with time (as C decreases), and really high degrees of polymerization ($\bar{x}_n > 10,000$) are consequently very difficult to achieve in practice.

32d. Free radical polymerization. Some of the commonest polymeric materials, such as polystyrene and polyethylene, are obtained by polymerization of monomers containing double bonds. The mechanism of this process is usually a chain reaction involving free radicals, and, as a result, the kinetic characteristics are very different from those of condensation polymerization.

As do all free radical chain reactions, the mechanism requires three kinds of steps: the introduction of free radicals (chain initiation), progressive reaction without change in the number of free radicals (chain propagation), and the eventual destruction of free radicals (chain termination). An additional process, chain transfer, must often occur; it will be discussed in section 32f. In the present discussion we shall confine ourselves to the simplest possible mechanism.

Chain initiation is most often accomplished by means of an *initiator*, such as benzoyl peroxide, which is a substance, added in small amounts, which can steadily produce free radicals during the course of the polymerization. (This substance is sometimes called a catalyst. This is a misnomer, since the initiator is slowly used up during the polymerization.) If the production of free radicals from the initiator is a first-order process, the rate of production of free radicals, in terms of concentration per unit time, will be given by

$$v_i = k_1[A] \tag{32–19}$$

where [A] represents the initiator concentration. An alternative possibility is that free radicals are produced by reaction of the initiator with the monomer, in which case

$$v_i = k_1[A][M] \tag{32–20}$$

Chains may also be initiated without any added initiator, either photochemically, in which case

$$v_i = k_1 I \tag{32–21}$$

where I is the light intensity, or thermally, i.e., by heating the monomer. The formation of free radicals in that event is often of the second order in the monomer concentration i.e.,

$$v_i = k_1[M]^2 \tag{32–22}$$

Chain propagation then occurs by a series of successive steps, all of which may (in analogy with the similar assumption made for condensation polymerization) be assumed governed by the same rate constant, k_2, independent of chain length. The first step is

$$\text{R} \cdot + \ \overset{\diagdown}{\underset{\diagup}{\text{C}}} = \overset{\diagup}{\underset{\diagdown}{\text{C}}} \ \rightarrow \text{R} - \overset{|}{\underset{|}{\text{C}}} - \overset{|}{\underset{|}{\text{C}}} \cdot$$

where $\text{R} \cdot$ represents the free radical formed in the initiation reaction, the dot representing the unpaired electron which makes it a free radical. Representing the monomer by M and the resulting free radical by $\text{RM} \cdot$, we write this reaction as

$$\text{R} \cdot + \text{M} \rightarrow \text{RM} \cdot$$

and subsequent reactions in the chain propagation become

$$\text{RM} \cdot + \text{M} \rightarrow \text{RM}_2 \cdot$$
$$\text{RM}_2 \cdot + \text{M} \rightarrow \text{RM}_3 \cdot$$
$$\cdot$$
$$\cdot \qquad\qquad\qquad\qquad\qquad \text{(IV)}$$
$$\cdot$$
$$\text{RM}_i \cdot + \text{M} \rightarrow \text{RM}_{i+1} \cdot$$
$$\text{etc.}$$

These propagation steps are very much faster than any of the chain initiation reactions, because of the reactivity of the unpaired electron. Since the monomer M is the only molecule which has a double bond and is thus, in principle (cf., however, section 32f), the only molecule capable of rapid reaction with free radicals, the successive propagation steps involve continuous addition of monomer, always to the same growing polymer molecule. The growth of the molecule stops only when chain termination occurs, usually as a result of combination between two free radicals

$$\text{RM}_i \cdot + \text{RM}_j \cdot \rightarrow \text{RM}_{i+j} \text{R} \qquad\qquad\qquad \text{(V)}$$

The product of this reaction, in contrast to the product molecules in condensation polymerization, is incapable of further reaction.

If the mechanism here given actually occurs, polymer molecules are formed just a few at a time. Polymerized material can be separated from the reaction mixture (e.g., by precipitation) even when only a small portion of monomer has reacted. The product molecules so obtained will have as high a degree of polymerization as the product molecules removed at a later time, when more of the monomer has reacted (cf. Fig. 32–2). This behavior is quite different, of course, from that which occurs in condensation polymerization, where all the material in the reaction mixture reacts slowly and simultaneously to form products with ever-increasing chain length.

To obtain the rate equations for the polymerization we make use of the fact that the successive free radicals formed are unstable. Thus, after a brief initial build-up period, they disappear as fast as they are formed, so that a steady state is reached. Any small drift in the concentration of free radicals which may then occur is much smaller (in terms of moles per unit time) than the rates of the reactions by which the radicals are formed or destroyed. The rate of this drift may therefore be set equal to zero. This yields the following equations,

$$d[\text{R·}]/dt = 0 = v_i - k_2[\text{R·}][\text{M}] - k_3[\text{R·}]\sum_{j=0}^{\infty}[\text{RM}_j\text{·}]$$

$$d[\text{RM·}]/dt = 0 = k_2[\text{R·}][\text{M}] - k_2[\text{RM·}][\text{M}] - k_3[\text{RM·}]\sum_{j=0}^{\infty}[\text{RM}_j\text{·}]$$

$$\cdot \qquad\qquad\qquad\qquad\qquad\qquad\qquad\qquad (32\text{--}23)$$

$$\cdot$$

$$\cdot$$

$$d[\text{RM}_i\text{·}]/dt = 0 = k_2[\text{RM}_{i-1}\text{·}][\text{M}] - k_2[\text{RM}_i\text{·}][\text{M}]$$

$$- k_3[\text{RM}_i\text{·}]\sum_{j=0}^{\infty}[\text{RM}_j\text{·}]$$

In these equations k_2 represents the rate constant of reactions of type IV and k_3 that of reactions of type V. Both k_2 and k_3 are assumed independent of chain length. The last term in each of the equations represents chain termination by reaction with all other possible free radicals, including R·, which is represented in the summation by the term with $j = 0$. (In this summation, when $i = j$, the usual factor of $\frac{1}{2}$ occurs in the rate of reactive collisions. In the equations 32–23 it is balanced by the fact that two radicals of the type RM· disappear in each such collision.)

All the equations 32–23 may be added together, leading to the condition that the sum of all radical concentrations reaches a steady state. When this is done, all of the chain-propagating steps cancel, leaving

$$v_i = k_3\left(\sum_{i=0}^{\infty}[\text{RM}_i\text{·}]\right)\left(\sum_{j=0}^{\infty}[\text{RM}_j\text{·}]\right) = k_3\left(\sum_{j=0}^{\infty}[\text{RM}_j\text{·}]\right)^2 \quad (32\text{--}24)$$

Equation 32–24 could, of course, have been obtained directly without equations 32–23. It merely states that, when there is no change in total free radical concentration, the rate of chain initiation must equal the over-all rate of chain termination.

An equation may now be written for the rate of disappearance of monomer. Since thousands of propagation steps occur for each initiation reaction, any loss of monomer which occurs in the latter (as, for instance, in thermal initiation) can be neglected so that

$$-d[\text{M}]/dt = k_2[\text{M}] \sum_{j=0}^{\infty} [\text{RM}_j\cdot]$$

which, by combination with equation 32-24, becomes

$$-d[\text{M}]/dt = (k_2/k_3^{1/2})v_i^{1/2}[\text{M}] \tag{32-25}$$

We may also write an equation for the formation of stable products, which occurs only by the chain-terminating reactions, since the propagating steps lead only to successively longer free radicals. If we let [P] be the concentration of all product molecules at time t, then

$$d[\text{P}]/dt = \tfrac{1}{2}k_3 \left(\sum_{i=0}^{\infty} [\text{RM}_i\cdot] \right) \left(\sum_{j=0}^{\infty} [\text{RM}_j\cdot] \right) \tag{32-26}$$

with the factor $\tfrac{1}{2}$ again correcting for the fact that, if it were omitted, each reaction step would be counted twice. Combining equation 32-26 with equation 32-24 we get

$$d[\text{P}]/dt = v_i/2 \tag{32-27}$$

As was pointed out earlier, free radical polymerization produces fully polymerized product at all stages of the reaction. The average degree of polymerization of the product which is being formed at any instant can then be obtained by combination of equations 32-25 and 32-27 as the ratio of the rate at which monomer disappears to the rate at which product molecules are formed. This ratio will give the average value of $i + j$ (equation V) in the product. If the R groups at the ends of the chain are counted as polymer segments, the degree of polymerization is $i + j + 2$ and

$$\bar{x}_n = 2 + \frac{-d[\text{M}]/dt}{d[\text{P}]/dt} = 2 + \frac{2k_2[\text{M}]}{(k_3 v_i)^{1/2}} \tag{32-28}$$

The exact form taken by the preceding equations depends, of course, on the nature of the initiation process. Consider first the case of thermally initiated chains, with v_i given by equation 32-22. In this case equation 32-25 becomes

$$-d[\text{M}]/dt = (k_1/k_3)^{1/2}k_2[\text{M}]^2 \tag{32-29}$$

and equation 32-28 becomes

$$\bar{x}_n = 2 + 2k_2/(k_1 k_3)^{1/2} \tag{32-30}$$

It is seen that the degree of polymerization of the product is fixed by the rate constants and is independent of time; i.e., $\bar{x}_n$ is exactly the same at the beginning of the reaction and at the end, when most of the monomer is used up.

These conclusions have been experimentally confirmed by Schulz, Dinglinger, and Husemann.[13] Figure 32–3 shows their data on the rate of the thermally initiated polymerization of styrene in toluene solution. The result is seen to be in fair agreement with equation 32–29. The molecular weight of the product was also measured in these experiments. It was found that $\bar{x}_n$ is independent both of time and the initial styrene concentration, but only if the latter is large. The reason for this is that equation 32–22 actually represents the initiation reaction only when [M] is large.

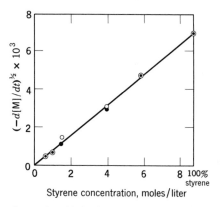

Fig. 32–3. Test of equation 32–29 for the thermally initiated polymerization of styrene. Open circles represent experiments with cyclohexane as solvent; filled circles represent experiments with toluene as solvent. The temperature is 100° C. (Data of Schulz et al.[13])

At low styrene concentrations the initiation becomes predominantly unimolecular (it occurs through collision with solvent molecules) so that $\bar{x}_n$ increases as $[M]^{1/2}$.

More commonly an initiator is used to start polymerization chains. With v_i given by equation 32–19 this gives for equations 32–25 and 32–28

$$-d[M]/dt = (k_1/k_3)^{1/2} k_2 [M][A]^{1/2} \tag{32–31}$$

$$\bar{x}_n = 2 + 2\{k_2/(k_1 k_3)^{1/2}\}[M]/[A]^{1/2} \tag{32–32}$$

In this case it is clearly possible to adjust $\bar{x}_n$ to any desired value by adjusting the relative concentrations of monomer and initiator. With low initiator concentrations, degrees of polymerization of one million or more are, in fact, easily attained. Whether $\bar{x}_n$ changes with time, as monomer and initiator are used up, depends on the particular reaction being studied. The ratio $[M]/[A]^{1/2}$ may remain constant or increase or decrease as the reaction progresses.

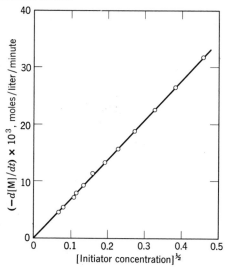

Fig. 32–4. Test of equation 32–31 for the polymerization of methyl methacrylate, with α′-α-azobisisobutyronitrile as initiator. The rates are initial rates, in liquid methyl methacrylate, so that [M] is essentially constant. (Data of Arnett.[14])

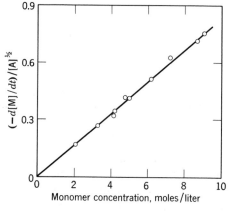

Fig. 32–5. A further test of equation 32–31 for the polymerization of methyl methacrylate, with α′-α-azobisisobutyronitrile as initiator. In this case the monomer was dissolved in benzene at various initial concentrations, so that both [M] and [A] were varied. (Data of Arnett.[14])

Equation 32–31 has been amply confirmed by experimental data, at least for the initial stages of appropriate polymerization reactions. Figure 32–4, for example, shows the linear dependence of the rate of polymerization of methyl methacrylate, with α′-α-azobisisobutyronitrile as initiator, on $[A]^{1/2}$, as predicted by equation 32–31. Figure 32–5 shows the dependence of the rate of the same reaction on the first power of the monomer concentration.

The most stringent test for equation 32–32 is obtained when $[A]^{1/2}$ is eliminated between this equation and equation 32–31 with the result that

$$\frac{1}{\bar{x}_n - 2} = \frac{k_3}{2k_2[M]^2}\left(\frac{-d[M]}{dt}\right) \tag{32–33}$$

Equation 32–33 shows that (for large $\bar{x}_n$) a plot of $1/\bar{x}_n$ versus rate of loss of monomer, in a series of experiments in which [M] is kept constant,

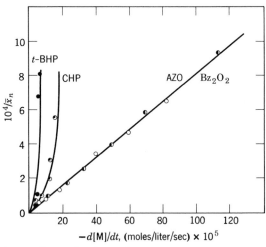

Fig. 32–6. A test of equation 32–33 for the polymerization of methyl methacrylate. The equation is obeyed when α'-α-azobisisobutyronitrile (AZO) or benzoyl peroxide (Bz$_2$O$_2$) is used as initiator but not if cumene hydroperoxide (CHP) or t-butyl hydroperoxide (t-BHP) is used. (Baysal and Tobolsky.[17])

should be linear and independent of the nature of the initiator used in the reaction. It is found that this result is obtained experimentally for only certain cases. Figure 32–6, for example, shows data obtained by Baysal and Tobolsky[17] for the polymerization of methyl methacrylate. Equation 32–33 is seen to be obeyed if α'-α-azobisisobutyronitrile or benzoyl peroxide is used. If a hydroperoxide is used $\bar{x}_n$ falls much more rapidly to low values than equation 32–33 would lead us to expect. A possible explanation for this observation will be given in section 32f.

As a final example of rate studies on free radical polymerization, Figure 32–7 represents an instance of a photochemically initiated polymerization, that of vinyl acetate.[15] As predicted by combination of equations 32–21 and 32–25, the rate depends on the square root of the light intensity.

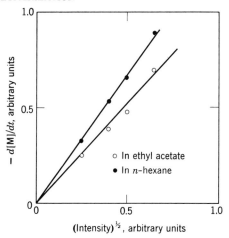

(Intensity)$^{1/2}$, arbitrary units

Fig. 32–7. Photochemical initiation of the polymerization of vinyl acetate. The result is in agreement with equations 32–21 and 32–25. (Burnett and Melville.[15])

32e. Distribution function for free radical polymerization. Molecular weight distribution functions for free radical polymerization can be derived from the preceding kinetic schemes. We first solve equations 32–23 to obtain the steady-state concentrations of the various free radicals, making use of equation 32–24. The result is that each $[RM_i\cdot]$ is related in the same way to the concentration of the free radical $[RM_{i-1}\cdot]$ with one less monomer unit,

$$[RM_i\cdot] = \alpha[RM_{i-1}\cdot] \qquad (32\text{–}34)$$

where α is independent of i and given by the relation

$$\alpha = \frac{k_2[M]}{k_2[M] + (k_3 v_i)^{1/2}} \qquad (32\text{–}35)$$

Equation 32–34 is valid down to $i = 1$; i.e., $[RM\cdot] = \alpha[R\cdot]$. We can obtain the value of $[R\cdot]$ from the first of the equations 32–23, but it is not needed in deriving the distribution function.

It follows from equation 32–34 that each $[RM_i\cdot]$ is related to $[R\cdot]$ by the equation

$$[RM_i\cdot] = \alpha^i[R\cdot] \qquad (32\text{–}36)$$

According to equation 32–28 $k_2[M]$ must be much larger than $(k_3 v_i)^{1/2}$ if the product of polymerization is to have a high value of $\bar{x}_n$. This means that the factor α defined by equation 32–35 must always be close to unity and that $[R\cdot] \simeq [RM\cdot] \simeq [RM_2\cdot]$, etc. However, when i becomes sufficiently large, α^i will become small even if α is close to unity, so that the concentrations of the larger free radicals will still become small in comparison to the shorter ones.

The rate of formation of polymer molecules of chain length x may now be computed. Including the terminal residues as chain segments, the rate of formation of x-mers is the rate of formation of all molecules of the form R—M_{x-2}—R; i.e., it is equal to (cf. equation 32–26)

$$\tfrac{1}{2}k_3 \sum_{i=0}^{x-2} [RM_i \cdot][RM_{x-2-i} \cdot]$$

which, on substituting equation 32–36, becomes

$$\tfrac{1}{2}k_3[R \cdot]^2 \sum_{i=0}^{x-2} \alpha^i \alpha^{x-2-i} = \tfrac{1}{2}k_3(x-1)\alpha^{x-2}[R \cdot]^2$$

Now the composition of the polymerization product which is formed at any instant in the reaction will be determined by the rate at which molecules of different degrees of polymerization are being formed. Thus the mole fraction of x-mer, in the product being formed at a given time, is proportional to $(x-1)\alpha^{x-2}$; i.e.,

$$X_x = B(x-1)\alpha^{x-2}$$

where B is a proportionality constant. Solving for B by the condition that $\sum_{x=2}^{\infty} X_x = 1$, we get, with the aid of the binomial expansions of Appendix A, $B = (1-\alpha)^2$. Thus

$$X_x = (1-\alpha)^2(x-1)\alpha^{x-2} \tag{32–37}$$

The mass present as x-mer is proportional to xX_x, so that the weight fraction of x-mer is

$$W_x = \frac{xX_x}{\sum_{x=2}^{\infty} xX_x} = \tfrac{1}{2}(1-\alpha)^3 x(x-1)\alpha^{x-2} \tag{32–38}$$

and the weight and number averages of x may now be computed as

$$\bar{x}_n = \sum_{x=2}^{\infty} xX_x = 2/(1-\alpha) \tag{32–39}$$

$$\bar{x}_w = \sum_{x=2}^{\infty} xW_x = (2+\alpha)/(1-\alpha) \tag{32–40}$$

and

$$\bar{x}_w/\bar{x}_n = \bar{M}_w/\bar{M}_n = (2+\alpha)/2 \tag{32–41}$$

We have used the binomial expansions of Appendix A in getting equations 32–38 to 32–40. Equation 32–39 could also have been obtained by combination of equations 32–28 and 32–35.

It should be noted that equations 32–39 and 32–40 indicate a unique relation between α and $\bar{x}_n$ or $\bar{x}_w$. Solving for α and substituting in equations 32–37 and 32–38 will therefore also lead to a unique molecular weight

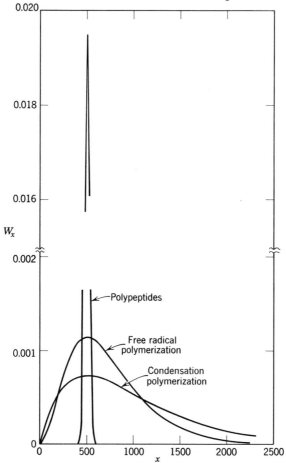

Fig. 32–8. A comparison of the weight distribution functions which are predicted for three types of polymerization kinetics. The three curves represent equations 32–18, 32–38, and 32–55.

distribution in terms of $\bar{x}_n$ or $\bar{x}_w$; i.e., this distribution will be independent of the method of initiation. A typical distribution calculated in this way is shown in Fig. 32–8. The distribution is seen to be considerably sharper than that calculated for condensation polymers.

That the theoretical distribution for free radical polymerization (according to the kinetic scheme here used) is sharper than that for condensation polymerization may also be seen from equation 32–41. It was pointed out earlier that α must be very close to unity if $\bar{x}_n$ is to be large, so that equation 32–41 predicts that the ratio $\bar{x}_w/\bar{x}_n$ or $\bar{M}_w/\bar{M}_n$ will be equal to $\frac{3}{2}$. The corresponding ratio for condensation polymers is $\bar{x}_w/\bar{x}_n = 2$ (cf. p. 147).

Figure 32–9 is an experimental distribution obtained by Baxendale, Bywater, and Evans[16] by fractionation of a sample of polymethyl methacrylate. It is seen to agree quite well with that calculated by equation 32–38. For polymethyl methacrylate $\bar{M}_w/\bar{M}_n$ also generally falls close

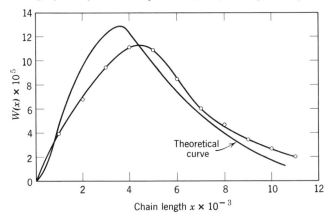

Fig. 32–9. The weight distribution of a sample of polymethyl methacrylate, as determined by fractionation. The theoretical curve is drawn according to equation 32–83. (Baxendale et al.[16])

to 1.5. For polystyrene on the other hand, the resulting distribution is generally not that predicted by equation 32–38. Instead, a random distribution is usually obtained, and $\bar{M}_w/\bar{M}_n$ is close to 2.0 (cf. Tables 16–1 and 17–1). Possible reasons for this discrepancy will appear in the following section.

32f. Chain transfer. It has been shown in section 32d that the rate of disappearance of monomer in free radical polymerization generally agrees very closely with the result predicted by theory. Despite this the observed values of $\bar{x}_n$ are often lower than the theory predicts (cf. Fig. 32–6, for instance) and the molecular weight distribution is often broader.

Another observation not in accord with the mechanism of section 32d was shown on p. 83, where infrared spectra were seen to indicate the occurrence of considerable chain branching in the final reaction product.

We cannot discuss here the various causes which might account for these discrepancies in individual cases. It is likely, however, that the majority of these discrepancies may be due to the occurrence of a *chain transfer* reaction. This is a reaction in which the growth of a polymer chain (reactions IV) ceases, not by chain termination, but by transfer of the odd electron of the growing free radical to another molecule. If a solvent is used, the other molecule may be a solvent molecule (SH); i.e., the transfer reaction may be

$$RM_i \cdot + SH \rightarrow RM_iH + S\cdot \quad \text{(VI)}$$

The intervening molecule could also be a molecule of initiator. If the initiator is benzoyl peroxide, Bz_2O_2, for example, we could have

$$RM_i \cdot + Bz_2O_2 \rightarrow RM_iBz + BzO_2\cdot \quad \text{(VII)}$$

In a similar way, the chain could be transferred to a monomer molecule, or to a previously "completed" polymer molecule

$$RM_i \cdot + R{-}M_k{-}CHR'{-}M_l{-}R \rightarrow$$
$$RM_iH + R{-}M_k{-}\dot{C}R'{-}M_l{-}R \quad \text{(VIII)}$$

It is easy to see that none of these reactions will affect the rate of disappearance of monomer. For they produce no change in the total concentration of free radicals and therefore no change in the rate at which monomer disappears by the chain propagation steps (reactions IV, equation 32–25).

At the same time, however, reactions VI and VII do reduce $\bar{x}_n$ and broaden the molecular weight distribution. That they reduce $\bar{x}_n$ is easily shown. Whereas each *two* chain initiations lead by the mechanism of section 32d to *one* product molecule, the occurrence of a single intervening chain transfer step of type VI or VII will always lead to formation of a second product molecule from the same two chain initiations.

Reaction VIII will not change $\bar{x}_n$ from the calculated value, for the new product molecule is balanced by the destruction of another. It will, however, clearly lead to chain branching, as addition of monomer will start from the odd electron in the middle of the chain.

32g. Ionic polymerization.[1] An alternative way to induce the polymerization of monomers containing double bonds is to create unstable ions rather than free radicals. Boron trifluoride monohydrate, for example, may be used to initiate the polymerization of isobutylene by forming a carbonium ion,

$$BF_3 \cdot OH_2 + CH_2{=}C(CH_3)_2 \rightarrow (CH_3)_3C^+ + BF_3OH^-$$

Monomer molecules may add to this ion without loss of the active ionic center and chain growth, termination, and transfer may occur in much the same way as in free radical polymerization. Similarly, sodium or potassium dissolved in liquid ammonia lead to the formation of NH_2^- which may add to certain monomers to form carbanions,

$$NH_2^- + CH_2\!\!=\!\!CRR' \rightarrow NH_2\!\!-\!\!CH_2\!\!-\!\!C(RR')^-$$

with further chain growth occurring by repeated monomer addition.

32h. Polymerization of N-carboxyanhydrides of amino acids. Both condensation and addition polymerization lead to products with relatively broad distribution of molecular weight. It is possible, however, to carry out polymerization reactions which lead to a product which is highly uniform in molecular weight. A reaction which can be of this type if no complicating factors arise is the synthesis of polypeptides by polymerization of N-carboxyanhydrides of amino acids, the simplest possible mechanism of which is

$$
\begin{array}{c}
\text{O} \\
\text{RCH}\!-\!\text{C} \\
\quad\quad\;\;\diagdown \\
\quad\quad\quad \text{O} + \text{R'H} \rightarrow\;
\begin{array}{c}
\text{RCH}\!-\!\text{CO}\!-\!\text{R'} \\
\mid \\
\text{NH}_2
\end{array}
\; + \text{CO}_2 \\
\quad\quad\;\;\diagup \\
\text{NH}\!-\!\text{C} \\
\quad\quad\diagdown \\
\quad\quad\;\;\text{O}
\end{array}
$$

$$\text{(IX)}$$

$$
\begin{array}{c}
\text{O} \\
\text{RCH}\!-\!\text{C} \\
\quad\quad\;\;\diagdown \\
\quad\quad\quad \text{O} +\;
\begin{array}{c}
\text{RCH}\!-\!\text{CO}\!-\!\text{R'} \\
\mid \\
\text{NH}_2
\end{array}
\;\;\rightarrow \\
\quad\quad\;\;\diagup \\
\text{NH}\!-\!\text{C} \\
\quad\quad\diagdown \\
\quad\quad\;\;\text{O}
\end{array}
$$

$$
\begin{array}{c}
\text{RCH}\!-\!\text{CO}\!-\!\text{NH}\!-\!\text{RCH}\!-\!\text{CO}\!-\!\text{R'} \\
\mid \\
\text{NH}_2
\end{array}
\; + \text{CO}_2
$$

$$\text{etc.,}$$

where R'H is a suitable initiator, such as a primary amine or an alcohol, which can break open the anhydride ring. For simplicity we can assume that all reactions of type IX, regardless of whether the second reactant is R'H or a peptide of any length, have the same rate constant.

This simple mechanism is seen to be one involving successive addition of monomer to the same growing chain. However there is no termination reaction. The reaction ends of its own accord only when all the monomer has been used up.

To derive the distribution function (following Flory[18]) we let M_x represent a molecule $H(-NH-CHR-CO)_{x-1}-R'$, M_1 therefore representing the initiator $R'H$. Let $[M_x]$ represent the corresponding concentrations at any time, and let $[A]$ and $[A_0]$ be the concentrations of anhydride at time t and at the beginning of the polymerization, respectively. Let C_0 be the initial concentration of $R'H$. The mechanism suggested by equations IX conserves molecules of type M_x; i.e., each molecule M_i destroyed is replaced by a molecule M_{i+1}. Therefore C_0 always represents the total number of molecules,

$$\sum_{x=1}^{\infty} [M_x] = C_0 \tag{32-42}$$

The rate of change of $[A]$ is given as

$$-d[A]/dt = k[A]\sum_{x=1}^{\infty}[M_x] = kC_0[A] \tag{32-43}$$

so that

$$[A] = [A_0]e^{-kC_0t} \tag{32-44}$$

From these equations we can get $\bar{x}_n$ as a function of time, for $\bar{x}_n$ is simply one more than the average number of anhydride molecules introduced into each molecule; i.e.,

$$\bar{x}_n = 1 + ([A_0] - [A])/C_0 \tag{32-45}$$

$$= 1 + ([A_0]/C_0)(1 - e^{-kC_0t}) \tag{32-46}$$

The equations for the rate of change of the concentrations of the M_x are simply

$$d[M_1]/dt = -k[A][M_1] \tag{32-47}$$

and, for $x > 1$,

$$d[M_x]/dt = k[A]\{[M_{x-1}] - [M_x]\} \tag{32-48}$$

The species M_x are not free radicals, and we cannot therefore apply the condition used in disussing free radical polymerization, that all $d[M_x]/dt = 0$. Instead, each of the rate equations must be solved to give each $[M_x]$ as a function of time. To do so we first transform the equations so that $v = \bar{x}_n - 1$ is the independent variable instead of t. Differentiating equation 32-45 and combining with equation 32-43 we get

$$dv = d\bar{x}_n = -d[A]/C_0 = k[A]\,dt$$

so that equations 32-47 and 32-48 can be written as

$$d[M_1]/dv + [M_1] = 0 \tag{32-49}$$

$$d[M_x]/dv + [M_x] = [M_{x-1}] \tag{32-50}$$

These are simple linear equations (Appendix A). With the condition that at the beginning of the experiment $\bar{x}_n = 1$, $v = 0$, and $[M_1] = C_0$, equation 32–49 gives

$$[M_1] = C_0 e^{-v} \tag{32-51}$$

and the general solution for $x > 1$ is

$$[M_x]e^v = \int [M_{x-1}]e^v \, dv \tag{32-52}$$

The integration constant which arises when this integration is performed can be obtained by setting $[M_x] = 0$ when $v = 0$.

With $[M_1]$ given by equation 32–51, equation 32–52 gives

$$[M_2] = C_0 v e^{-v}$$

and, with this value of $[M_2]$,

$$[M_3] = C_0(v^2/2)e^{-v}$$

The general solution is clearly

$$X_x = [M_x]/C_0 = e^{-v}v^{x-1}/(x-1)! \tag{32-53}$$

where X_x is the mole fraction of x-mer.

By equation 32–53

$$\sum_{x=1}^{\infty} x X_x = e^{-v}(1 + 2v + 3v^2/2! + 4v^3/3! + \cdots)$$
$$= e^{-v}(1 + v)(1 + v + v^2/2! + v^3/3! + \cdots)$$
$$= 1 + v = \bar{x}_n \tag{32-54}$$

so that the weight distribution is easily obtained as

$$W_x = \frac{x X_x}{\sum\limits_{x=1}^{\infty} x X_x} = \frac{x e^{-v} v^{x-1}}{(1+v)(x-1)!} \tag{32-55}$$

A calculated distribution obtained from this equation is shown in Fig. 32–8, and it is seen to be extremely sharp. Another way of indicating the sharpness of the distribution is to compute the ratio of weight-average to number-average molecular weight. By equation 32–55

$$\bar{x}_w = \sum_{x=1}^{\infty} x W_x = \frac{e^{-v}}{1+v} \sum_{x=1}^{\infty} \frac{x^2 v^{x-1}}{(x-1)!} \tag{32-56}$$

It is seen that the summation of equation 32–56 is equal to $(v^2 + 3v + 1)e^v$ so that

$$\bar{x}_w = (v^2 + 3v + 1)/(1 + v)$$

and, with equation 32–54,

$$\bar{x}_w/\bar{x}_n = (v^2 + 3v + 1)/(v + 1)^2$$

Since $v = \bar{x}_n - 1$ we are interested only in relatively large values so that $\bar{x}_w/\bar{x}_n \simeq 1$.

That such a sharp distribution can indeed be obtained experimentally has been shown in the case of at least one polypeptide, polysarcosine, $(-NCH_3-CH_2-CO)_x-$, prepared by reactions IX from the appropriate anhydride.[19] Fessler and Ogston[20] examined products of this preparation by sedimentation and diffusion. They found that the sharpness of the sedimentation peaks indicated that the preparations were essentially homogeneous and that the diffusion patterns corresponded closely to material with a single diffusion coefficient. The molecular weights, $\bar{M}_w$ obtained for several samples were essentially identical with number-average molecular weights determined by titration.

33. POLYMER DEGRADATION

The dissociation of polymer molecules into smaller units is known as *degradation*. This process, like its opposite, polymerization, may be studied kinetically, and such kinetic studies are particularly useful for naturally occurring macromolecules the polymerization of which cannot be studied in the laboratory. In the case of the polymers discussed in the preceding section we can acquire confidence in our idea concerning their structure and method of synthesis by studying the kinetics of polymerization. For natural macromolecules we cannot hope to obtain from *in vitro* experiments any information on the method of synthesis in the living system, but we can by studying the degradation decide whether our ideas concerning the way monomer units are joined in the polymer molecule are reasonable.

In the simplest situation all bonds of a macromolecular chain are equally susceptible to rupture, and bonds are broken at a rate proportional to the number of intact bonds remaining. This situation would be expected to prevail, for instance, in the degradation of simple condensation polymers. When these polymers are synthesized the formation of bonds appears to occur in a completely random fashion. When they are degraded the rupture of bonds might likewise be expected to be a random process.

The other extreme of degradation can be exemplified by the enzymatic hydrolysis of proteins. In this case each bond between monomer units is a peptide bond, but the side chains extending from any pair of adjacent monomer units are in general different. The ability of an enzyme to attack a peptide bond will in general depend on the nature of these side chains, so that bond rupture will occur preferentially at certain specific spots along the polypeptide chains of protein molecules.

33a. Statistics of random scission.[21] The simplest possible degradation occurs when we begin with a polymer sample formed by random polymerization and subject it to non-specific degradation, so that bonds are broken at random. Suitable polymers might be condensation polymers, polymerization of which leads, as we have seen, to the statistical molecular weight distribution of section 8b, or addition polymers which, owing to chain transfer, have the same kind of distribution. Degradation of such polymers can be carried out by hydrolysis or alcoholysis (polyesters and polyamides) or by heating or ozone treatment (addition polymers).

The progress of degradation is best considered in terms of Fig. 8–1. Consider a mixture of polymer molecules originally formed from N_0 monomer molecules. The monomer units may be thought of as arranged in a giant circle, and the mixture may be described in terms of the fraction of possible bonds which have formed. Let this fraction in the initial mixture be p_0; i.e., there are initially in Fig. 8–1 $N_0 p_0$ bonds and $N_0(1 - p_0)$ unbonded positions, and they will be randomly located. The number of unbonded positions is also the total number of molecules in the mixture. If degradation consists of random scission of bonds, it may be simply described in terms of an increase in the number of unbonded positions, i.e., in terms of a decrease in p. All the results of section 8b and 8d apply; i.e., at any time t,

$$X_x = p^{x-1}(1 - p) \tag{33-1}$$

$$W_x = xp^{x-1}(1 - p)^2 \tag{33-2}$$

$$\bar{M}_n = M_0\bar{x}_n = M_0/(1 - p) \tag{33-3}$$

$$\bar{M}_w = M_0\bar{x}_w = M_0(1 + p)/(1 - p) \tag{33-4}$$

where M_0 is the contribution of each monomer unit to the polymer molecular weight. The initial values of these parameters, $(X_x)_0$, $(\bar{M}_n)_0$, etc., can be obtained by placing $p = p_0$ in equations 33–1 to 33–4.

By following analytically the appearance of new end groups or the disappearance of the reagent added to bring about bond rupture, it is often possible to measure directly the number of bonds which have been broken. If we let this number be qN_0 (i.e., the number of bonds broken is again expressed as a fraction of the total number of bonds which could be formed by all the monomer units in the system), then

$$q = p_0 - p \qquad (33\text{–}5)$$

This relation may be combined with equations 33–3 and 33–4 to express $\bar{x}_n$ or $\bar{x}_w$ as a function of q,

$$\frac{1}{\bar{x}_n} = 1 - p = (1 - p_0) + (p_0 - p) = \frac{1}{(\bar{x}_n)_0} + q \qquad (33\text{–}6)$$

$$\frac{1}{\bar{x}_w + 1} = \frac{1 - p}{2} = \frac{(1 - p_0)}{2} + \frac{(p_0 - p)}{2} = \frac{1}{(\bar{x}_w)_0 + 1} + \frac{q}{2} \qquad (33\text{–}7)$$

Equation 33–6 expresses the obvious fact that the number of new molecules formed at any time is exactly equal to the number of bonds broken and applies thus to the degradation of any simple polymer chain. Equation 33–7, however, applies only to random degradation of an initially random polymer sample. For the initial stages of degradation, when $\bar{x}_w$ is large, equation 33–7 predicts that $1/\bar{x}_w$ will, like $1/\bar{x}_n$, be a linear function of q.

With random scission the rate at which bonds are broken is proportional to the total number of intact bonds; i.e., since $N_0 p$ is the number of bonds at time t,

$$-d(N_0 p)/dt = k N_0 p \qquad (33\text{–}8)$$

The rate constant k in this equation includes the concentration of any reagent which may be used to produce degradation. The reagent is assumed added in excess, so that its concentration remains essentially constant.

Equation 33–8 is, of course the equation which imposes the condition of randomness on the progress of the reaction with time. For non-random scission the $N_0 p$ bonds of the system would have to be divided into several classes, each of which is ruptured at a different rate.

The solution of equation 33–8, with the condition that $p = p_0$ at zero time is simply that

$$p = p_0 e^{-kt} \qquad (33\text{–}9)$$

Combining equation 33–9 with equations 33–5 to 33–7 gives the number of bonds broken and the average degree of polymerization as a function of time.

Degradation is often followed experimentally by determination of the molecular weight as a function of time, using light scattering or viscosity as a measure of molecular weight. These techniques can be applied only as long as the molecular weight remains large; i.e., in terms of the number of bonds broken, these methods involve a study of the initial stages of the

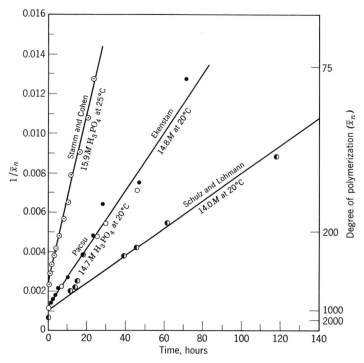

Fig. 33–1. Test of equation 33–11 for the initial stages of the hydrolysis of cellulose in phosphoric acid. (Mark and Tobolsky.[22])

reaction only. Under these conditions p must remain close to unity throughout the reaction, so that e^{-kt} must be close to unity; i.e., we can set $e^{-kt} = 1 - kt$. Thus $p = p_0(1 - kt)$ and q, $1/\bar{x}_n$ and $1/\bar{x}_w$ all become linear functions of time,

$$q = p_0 kt \simeq kt \tag{33–10}$$

$$\frac{1}{\bar{x}_n} = \frac{1}{(\bar{x}_n)_0} + kt \tag{33–11}$$

$$\frac{1}{\bar{x}_w} = \frac{1}{(\bar{x}_w)_0} + \frac{kt}{2} \tag{33–12}$$

A natural polymer which might be expected to resemble a synthetic condensation polymer is cellulose. Figure 33–1 shows degradation data obtained for the initial stages of the hydrolysis of this substance by phosphoric acid, and they are indeed seen to obey equation 33–11, with a rate constant which increases with increasing concentration of phosphoric acid.

It should be noted that Fig. 33–1 does not prove that all bonds between the glucose residues of cellulose are necessarily equally susceptible to hydrolysis. At the end of each experiment $\bar{x}_n$ is still 75, and seventy-four of every seventy-five bonds remain unbroken. It is possible that some of them are more resistant to rupture, a fact which would be indicated by a decrease in the rate constant for scission as lower values of $\bar{x}_n$ are attained. There is, in fact, evidence that this occurs in the hydrolysis of cellulose by sulfuric acid,[23] the results suggesting that the terminal linkage at the reducing end of the molecule is more labile than other bonds.

Schulz and Husemann[24] have observed results at variance with Fig. 33–1, presumably reflecting a difference in the structure of the cellulose which they were using. They found that they were unable to fit the initial stages of the reaction by using the equations given in this section, with a single rate constant for bond scission. Moreover, they found that the product of partial hydrolysis, upon fractionation, did not have the molecular weight distribution predicted by equation 33–1. An abnormally high content of material with $x \simeq 500$ was found upon fractionation, a result which was interpreted as meaning that bonds of relatively high susceptibility to hydrolysis occur at regular intervals of roughly 500 segments along the original cellulose chains.

33b. Kinetic treatment of random scission.[25] A purely kinetic treatment of random scission may be based on the fact that a molecule M_x, of degree of polymerization x, can be formed in two different ways from any larger molecule M_y ($y > x$), by rupture of a bond x segments from either end of the y-mer. The rate of formation of x-mers from any y-mer is therefore twice the rate of rupture of single bonds at concentration of the y-mer; i.e.,

$$d[M_x]/dt = 2k \sum_{y=x+1}^{\infty} [M_y]$$

The rate of disappearance of x-mer, on the other hand, to form shorter chains, can occur by rupture of any of its $x - 1$ bonds, so that

$$-d[M_x]/dt = (x - 1)k[M_x]$$

The over-all rate of change of concentration of x-mer is therefore

$$d[M_x]/dt = -k(x - 1)[M_x] + 2k \sum_{y=x+1}^{\infty} [M_y] \qquad (33\text{–}13)$$

The solutions to this set of linear equations depends on the initial conditions i.e., on the initial molecular weight distribution. If the initial distribution is a random one, we get the same equations as those obtained statistically in the previous section. We shall present here the solution for an initially uniform polymer; i.e., we shall suppose that at $t = 0$ all

$[M_x]_0 = 0$ unless $x = n$. $[M_n]_0$ is then the total concentration of polymer molecules at the beginning of the experiment. No molecules with $x > n$ appear at any time.

Equation 33–13 for $x = n$ then becomes

$$d[M_n]/dt = -k(n - 1)[M_n]$$

and

$$[M_n] = [M_n]_0 e^{-k(n-1)t} \qquad (33\text{–}14)$$

Equation 33–13 is now solved for $[M_{n-1}]$. There is only one term in the summation, that with $y = n$. Substituting equation 33–14 and placing $[M_{n-1}]_0 = 0$, we get

$$[M_{n-1}] = [M_n]_0 e^{-k(n-2)t}(1 - e^{-kt})(2) \qquad (33\text{–}15)$$

Continuing in this fashion we get

$$[M_{n-2}] = [M_n]_0 e^{-k(n-3)t}(1 - e^{-kt})(3 - e^{-kt})$$
$$[M_{n-3}] = [M_n]_0 e^{-k(n-4)t}(1 - e^{-kt})(4 - 2e^{-kt})$$

and, in general (except where $x = n$),

$$[M_x] = [M_n]_0 e^{-k(x-1)t}(1 - e^{-kt})\{2 + (n - x - 1)(1 - e^{-kt})\} \qquad (33\text{–}16)$$

If we use N_0 as before as the total number of monomer units (per unit volume) and if the $[M_x]$ are expressed in terms of the number of molecules per unit volume, then

$$N_0 = n[M_n]_0 = \sum_{x=1}^{n} x[M_x] \qquad (33\text{–}17)$$

If, as before, pN_0 represents the number of bonds in the system (per unit volume), then equations 33–8 and 33–9 still apply; i.e., $p = p_0 e^{-kt}$. Each bond broken means an increase by one in the number of molecules, so that the total number of molecules at time t is

$$\sum_{x=1}^{n} [M_x] = [M_n]_0 + N_0(p_0 - p)$$
$$= N_0[1/n + p_0(1 - e^{-kt})]$$

Thus

$$\bar{x}_n = N_0 \Big/ \sum_{x=1}^{n} [M_x] = [1/n + p_0(1 - e^{-kt})]^{-1} \qquad (33\text{–}18)$$

For all degradations in which $\bar{x}_n$ remains reasonably large at the end of the experiment we can, as before, place $e^{-kt} = 1 - kt$. Furthermore $p_0 \simeq 1$ so that

$$\frac{1}{\bar{x}_n} = \frac{1}{n} + kt \qquad (33\text{–}19)$$

i.e., a linear relation between $1/\bar{x}_n$ and t continues to be the expected result.

(Thus the validity of the discussion of the preceding section, concerning the degradation of cellulose, does not depend on the nature of the initial distribution.)

To evaluate the behavior of $\bar{x}_w$ as a function of time is a little more cumbersome. By definition (equation 8–18)

$$\bar{x}_w = \sum_{x=1}^{n} x^2[M_x] / \sum_{x=1}^{n} x[M_x]$$

$$= \sum_{x=1}^{n} x^2[M_x]/N_0 \qquad (33\text{–}20)$$

By equation 33–16, placing

$$1 - e^{-kt} = q/p_0 = \alpha \qquad (33\text{–}21)$$

and

$$e^{-kt} = 1 - \alpha = \beta \qquad (33\text{–}22)$$

gives

$$\sum_{x=1}^{n} x^2[M_x] = [M_n]_0 \left\{ [2 + (n-1)\alpha]\alpha \cdot \sum_{x=1}^{n} x^2\beta^{x-1} - \alpha^2 \sum_{x=1}^{n} x^3\beta^{x-1} \right\} \qquad (33\text{–}23)$$

To evaluate the finite sums which occur in this equation we consider the geometrical series

$$I_0 \equiv \sum_{x=1}^{n} \beta^x \equiv (\beta^{n+1} - \beta)/(\beta - 1)$$

By differentiation

$$dI_0/d\beta = \sum_{x=1}^{n} x\beta^{x-1} = (\beta^{n+1} - \beta)/(\beta - 1)^2 + [(n+1)\beta^n - 1]/(\beta - 1)$$

and we can now evaluate as functions of β

$$I_1 \equiv \beta \, dI_0/d\beta \equiv \sum_{x=0}^{n} x\beta^x$$

$$dI_1/d\beta = \sum_{x=1}^{n} x^2\beta^{x-1}$$

$$I_2 \equiv \beta \, dI_1/d\beta \equiv \sum_{x=1}^{n} x^2\beta^x$$

$$dI_2/d\beta = \sum_{x=1}^{n} x^3\beta^{x-1}$$

The expressions for $dI_1/d\beta$ and $dI_2/d\beta$ are the ones needed in equation 33–23. With them we get $\sum_{x=1}^{n} x^2[M_x]$ as a function of α and β and, hence, obtain by equation 33–20,

$$\bar{x}_w = \frac{n\alpha^2 + 2(1-\alpha)[(1-\alpha)^n + n\alpha - 1]}{n\alpha^2} \qquad (33\text{–}24)$$

This expression cannot be simplified, even for the initial stages of degradation. For although equation 33–21 shows that α is essentially equal to q, so that initially $\alpha \ll 1$ and $(1 - \alpha) \simeq 1$, $n\alpha$ can become large and $(1 - \alpha)^n$ quite small since n may be very large. Thus $1/\bar{x}_w$ is never a linear function of time for random degradation of a uniform starting material.

33c. The enzymatic degradation of deoxyribose nucleic acid. An interesting application of these considerations is involved in the degradation of deoxyribose nucleic acid (DNA). In a study of the enzymatic degradation

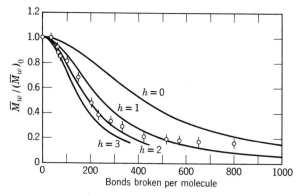

Fig. 33–2. Weight-average molecular weight as a function of the number of bonds broken in the enzymatic degradation of DNA. (Thomas.[26])

of this molecule (using a specific enzyme deoxyribonuclease) Thomas[26] measured the number of bonds broken at any time by observing the rate at which base had to be added to keep the pH constant. (Each bond broken leads to the formation of an ionic phosphate group and the release of a proton.) At the same time he measured $\bar{M}_w/(\bar{M}_w)_0$ by means of light scattering. The number of bonds broken was found to be a linear function of time, indicating (equation 33–10) that random scission was occurring. Also the DNA preparation used was thought to be close to uniform in molecular weight. However, the molecular weight measurements, shown in Fig. 33–2, clearly do not at all obey equation 33–24. (Nor do they obey equation 33–12 for an initially random molecular weight distribution.) Figure 33–2 shows, in fact, that over 100 bonds can be broken per molecule without appreciable change in $\bar{M}_w$.

Thomas explained these results in terms of the Watson-Crick structure (p. 68) for DNA. If this structure is correct, each DNA molecule consists of two parallel chains, each segment of one chain being joined by hydrogen bonding to the opposite segment of the other chain. The structure is represented schematically in Fig. 33–3. In this structure scission of many bonds of the two polynucleotide chains can occur without change in molecular weight, for the hydrogen bonds can continue to hold fragments of the two chains together.

The mathematical treatment of this situation assumes that the equations of the preceding sections apply to the scission of bonds in the individual polynucleotide chains. Thus q represents, as before, the probability that

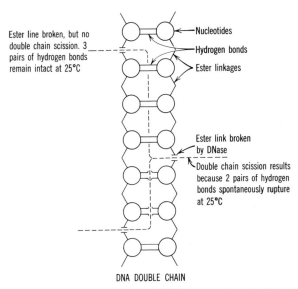

Ester line broken, but no double chain scission. 3 pairs of hydrogen bonds remain intact at 25°C

Nucleotides

Hydrogen bonds

Ester linkages

Ester link broken by DNase

Double chain scission results because 2 pairs of hydrogen bonds spontaneously rupture at 25°C

DNA DOUBLE CHAIN

Fig. 33–3. Kinetics of degradation of DNA, assuming the Watson-Crick double chain structure shown in Fig. 4–13. Each circle represents a monomer unit. A break in the double chain is assumed to occur whenever two or fewer pairs of hydogen bonds provide the only impediment. No break occurs when three or more pairs of hydrogen bonds must be broken. (Thomas.[26])

any bond of either chain is broken, and, as has been said, it is a linear function of time in the initial stages of the degradation. The molecular weight equation, equation 33–24, depends, however, on the probability s of a complete break in the DNA double chain; i.e., the term α in this equation, which for a single chain is essentially equal to q, now becomes equal to s.

To obtain $\bar{x}_w$ as a function of time (or of q) it is only necessary to relate s to q. This relation will depend on the ability of the hydrogen bonds to hold the two chains together. If the hydrogen bonds are very strong, a molecular scission will occur at a given position of the chain only if *both* individual chains are broken at this position. Under these conditions the probability of a molecular break is clearly equal to the square of the probability of chain scission; i.e.,

$$s = q^2 \qquad (33\text{-}25)$$

On the other hand, hydrogen bonding between a single pair of opposite nucleotides may not suffice to prevent a break. In this case a molecular break occurs at a given position if there is rupture of one chain at that position and rupture of the second chain either at the same position or at one of the immediately adjacent positions. Thus the probability of a molecular break is increased by a factor of three, and

$$s = 3q^2 \qquad (33\text{-}26)$$

Strictly speaking we should allow for the possibility that there occurs a break at more than one of the three possible positions of the second chain. Thus s becomes equal to $q(3q + 3q^2 + q^3)$, the second and third terms in the parentheses representing respectively the probabilities of two and three simultaneous breaks in the second chain. When q is small, however, this relation reduces to equation 33-26.

Continuing in the same way we may suppose that hydrogen bonding between $h + 1$ successive pairs of opposite nucleotides are required to prevent a break occurring, so that a break will occur if, in the second chain, there is scission at the position directly opposite the point of rupture of the first chain or at any of the h next neighboring positions in either direction. If h is small, this clearly leads to

$$s = (2h + 1)q^2 \qquad (33\text{-}27)$$

Any of these values of s may now be substituted for α in equation 33-24 to give a theoretical equation for the fall in molecular weight. As Fig. 33-2 shows, equation 33-27 with $h = 2$ is able to account satisfactorily for the initial stages of the degradation. The kinetic data are thus compatible with the double-stranded helical structure for DNA, provided it is reasonable to assume that five successive hydrogen bond cross-links will maintain the structure intact even if a primary internucleotide bond is broken.

It is important to note, however, that the interpretation here given is by no means the only one which will account for the experimental results. At least two other mechanisms may be postulated which will account for the progress of the reaction *regardless of the structure which DNA molecules may possess.*

For instance, it may be supposed that the enzyme used has a strong preference for attacking the terminal bonds of a DNA chain (which could for this purpose be either a single or a double chain). Then q will still be a linear function of time, for the number of polymer molecules will remain constant and each will always provide two terminal bonds for the enzyme to attack. The product of each scission, however, will be a monomer molecule and a polymer molecule shorter by only one residue than the parent molecule. In the initial stages of such a process there will be virtually no change in weight-average molecular weight. If, for instance, each original DNA molecule had 10,000 nucleotide units, then, after 100 scissions per molecule, there would remain, for each initial molecule, one product molecule with $x \simeq 9900$ plus 100 monomer molecules ($x = 1$). The weight-average degree of polymerization (equation 8–14) would be

$$\bar{x}_w = [(9900)^2 + 100(1)^2]/[(9900) + (100)] = 9800$$

i.e., this mechanism would lead to just the type of molecular weight decrease shown in Fig. 33–2. If we allow an occasional random scission to occur in the middle of the molecules, an adjustable parameter (relative rate of terminal to random scission) would be provided by which an exact fit of the data could undoubtedly be provided.

Another possibility is that the enzyme which catalyzes DNA degradation in fact performs two functions, one of them being to "denature" the DNA molecules, i.e., to alter their secondary structure (whatever it may be) and to produce perhaps a randomly coiled molecule. The second function would be to break the internucleotide bonds. It would be probable that the rate of scission in "denatured" molecules would be much faster than in the original molecules. (This is true in the case of proteins; "denatured" proteins are hydrolyzed more readily than protein molecules of compact configuration.) Thus, once a DNA molecule would be "denatured" it would be rapidly degraded all the way to monomer, whereas all the molecules which retain their original structure would remain intact. The fall in $\bar{M}_w$ would then follow essentially the course followed in free radical degradation. The result, shown in Fig. 33–4, is again similar to that observed experimentally.

These alternative mechanisms of the reaction are not proposed here to throw doubt upon the double-stranded helical structure for DNA but simply to illustrate a weakness inherent to all attempts to provide proof of a structure by kinetic means. Kinetic experiments can rarely provide such proof. They can only show that the course of a reaction is *compatible* with a particular structure. The course of reaction is usually compatible with other structures also, and other means have to be found to exclude such structures.

It is of interest in this connection that Dekker and Schachman have proposed a modified double-stranded helical structure for DNA, in which the helical structure is interrupted at intervals. The basis for this structure is not primarily kinetic. This structure is not compatible with Thomas' results if the enzyme used by him attacks DNA randomly. The two alternative mechanisms here suggested will, however, allow this interrupted helical structure, as well as any other.

33d. Degradation of addition polymers. Detailed theoretical kinetic schemes for the degradation of addition polymers by a free radical mechanism have been set up by Simha and coworkers.[27] A number of experimental studies for a variety of polymers have been made,[28] We shall not discuss them here except to note that, if degradation occurs by an exact reversal of the simple polymerization mechanism described in section 32d, a simple result is obtained which is quite different from that of random degradation.

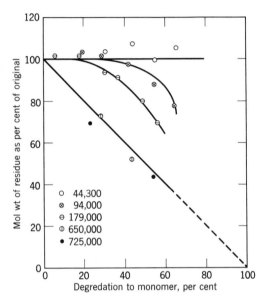

Fig. 33–4. Molecular weight of the residual polymer in the thermal degradation of polymethyl methacrylate. The initial molecular weight for each experiment is indicated. (Melville and Grassie.[29])

For, in the polymerization process of section 32d, polymer molecules are formed very rapidly, a few at a time by successive addition of monomer to chains terminating in an unpaired electron. The polymerized material obtained at the beginning of the reaction can easily have essentially the same molecular weight distribution as that which is formed as most of the monomer becomes used up. In the reverse process, then, polymer molecules would form long free radical chains by a single chain scission. Each molecule affected would be degraded completely to monomer (by the reverse of reaction IV on p. 597), and this would happen very rapidly compared to the formation of new radical chains. The reaction mixture at any time during degradation would then consist of a mixture of monomer and of undergraded original polymer.

That this situation may actually occur has been shown by studies of Melville and Grassie[29] on the depolymerization by heat of polymethyl methacrylate samples of relatively low molecular weight. Some of the data are shown in Fig. 33–4, and it is seen that, for the sample of lowest molecular weight, as much as 70% of the polymer material can be degraded to monomer without change in the molecular weight of the residue.

The situation described by Fig. 33–4 is, however, rare. At the elevated temperatures used for degradation chain transfer reactions tend to become rapid so that the chain propagation reaction does not proceed to complete conversion to monomer. Thus new polymer molecules of lower molecular weight are formed and the molecular weight of the residue decreases steadily as depolymerization proceeds, with the most rapid decrease of average molecular weight occurring at the beginning of the experiment.

33e. Degradation of proteins. In the polymers discussed so far, all the bonds which join one segment to the next have been essentially identical, and random scission has been the logical starting point of the discussion. To a certain extent the same is true of proteins. All the bonds in the chain are peptide links, and under some conditions they are undoubtedly equally susceptible to attack. However, protein molecules have a variety of side chains, and there will ordinarily be few bonds in the molecule which join segments having the same pair of side chains. The chief interest in protein degradation has therefore been to take advantage of this situation and to find reagents which will specifically or preferentially split peptide bonds joining residues with particular side chains.

Hydrolytic enzymes have been most useful in this respect. The enzyme *trypsin*, for instance, splits peptide bonds, which may be generally represented as $-NH-CHR_1-CO-NH-CHR_2-CO-$, essentially only if R_1 represents an arginine or lysine side chain. For example, in oxidized ribonuclease (a protein which contains 124 amino acid residues) only twelve of the 123 peptide bonds fall into this category. If this protein is hydrolyzed by trypsin[31] the production of new terminal amino groups $-CO-NH- \rightarrow -COO^- + {}^+H_3N-$ proceeds uniformly until these twelve bonds have been broken, after which time essentially no further reaction occurs. When the products of the hydrolysis are examined they consist of thirteen peptides in good yield and no other fragments.

Another enzyme, *chymotrypsin*, preferentially attacks peptide bonds where R_1 contains an aromatic ring. Some additional peptide bonds seem to be split more slowly, for the reaction between chymotrypsin and ribonuclease breaks about ten bonds rapidly and about six more at a slower rate.[31]

In the case of proteins even hydrolysis by strong acids is somewhat specific. Peptide bonds which have an alcoholic side chain in the position of R_2 are split more rapidly than other peptide bonds, though all bonds are broken on prolonged hydrolysis. A possible mechanism for this preference is that the initial reaction is

$$-NH-CHR_1-CO-NH-\underset{\underset{\displaystyle HO-CH_2}{|}}{CH}-CO- + H^+ \rightarrow$$

$$-NH-CHR_1-CO\ \underset{\underset{\displaystyle O-CH_2}{|}}{\overset{\overset{\displaystyle NH_3^+}{|}}{CH}}-CO-$$

leading to an ester link of greater susceptibility to hydrolysis than the initial peptide bond.

The specificity of protein hydrolysis is of vital importance in the determination of the sequence of amino acids in the polypeptide chains of proteins. For a review of some of the pioneering work on this subject the reader should consult an article by Sanger.[30]

34. KINETICS OF PROTEIN DENATURATION

It was pointed out in section 7 that many proteins (the globular proteins) exist in the solid state as compact, rigid, specifically coiled molecules. Many of these proteins continue to have such a structure in aqueous solution and presumably have the same structure in the living systems from which they are taken. When these proteins are heated, however, or when they are dissolved in other than aqueous media, or, in aqueous solution, when they are subjected to a pH far from the isoelectric point, their molecular configuration changes. Several examples of such changes have been cited earlier, for instance on pp. 515–519.

A configuration change of this kind is called *denaturation*. The term applies to similar reactions in nucleic acids, viruses, etc., and to proteins which may originally be rodlike rather than compactly folded, as well as to the globular proteins. The bulk of experimental work has been done on the globular proteins, however, and this section will be confined to their denaturation.

As was explained in section 7 we have as yet little definite knowledge of the *native* state of globular proteins. Specifically, the precise nature of the chemical forces which maintain the molecules in their compactly folded condition is unknown. At the same time we also have little knowledge of the denatured states. In many instances the denatured state resembles a configuration which is partly randomly coiled (e.g., p. 518), but the extent of the molecular domain is ordinarily less than we would expect for completely random coiling so that some intramolecular bonding must persist in the denatured molecules.

Kinetic studies of these reactions (if they are slow enough) are undertaken with the hope that they may throw some light upon the intramolecular forces. These studies, however, follow a very different course from, for instance, studies of polymerization and degradation. In the latter both initial and final states are well characterized, and we are usually able to formulate in advance a likely kinetic scheme. If the experiments run counter to prediction, it usually leads to a modification of the predicted scheme, rather than to its abandonment. In protein denaturation, on the other hand, neither the initial nor final state is clearly characterized, nor is it possible in general to formulate a reaction mechanism in advance. Thus studies of protein denaturation tend to be broad, exploring the effect of numerous variables on the reaction rate, and, to date, none of them has led to a convincing detailed mechanism.

34a. General characteristics. A few generalizations may be made which apply to all protein denaturation reactions.

(1) Different denaturing agents lead to different configurational changes. Any individual protein is thus subject to several different denaturation reactions. For example, it was shown in section 28 that serum albumin becomes denatured in acid solution; i.e., it takes on an expanded configuration. Serum albumin may also be denatured by heat,[32] leading to a product which becomes insoluble under certain standard conditions where the native molecules are soluble. (Whether the initial product of this reaction is in an expanded configuration has not been investigated.) The rate of the thermal reaction is found to decrease under conditions where the extent of acid expansion increases; i.e., denaturation by acid clearly does not lead to the same product as denaturation by heat.

(2) The same reagent may affect different globular proteins in quite different ways. As has already been mentioned acid produces an expansion of the serum albumin molecule. The reaction is very rapid and reversible. Acids will also cause an expansion of hemoglobin, the reaction being slow and reversible.[33] Ovalbumin is also denatured by acid, but a lower pH is required, and the reaction cannot be reversed.[34] Ribonuclease, finally is not affected by acid at all.

(3) A given denaturing agent may lead to two or more successive reactions. The denaturation of ovalbumin by urea, for example, which will

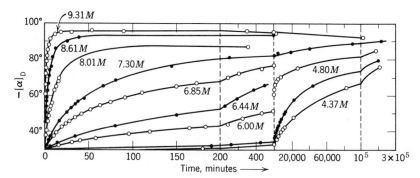

Fig. 34–1. The rate of change of the specific optical rotation of ovalbumin in aqueous urea solutions at 30° C. The urea concentration which corresponds to each curve is given. The time scale changes at 200, 500, and 10^5 minutes. (Simpson and Kauzmann.[35])

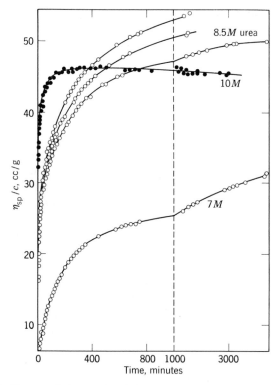

Fig. 34–2. The rate of change of η_{sp}/c for ovalbumin in aqueous urea solutions at 30° C. Three independent experiments were performed at each concentration of urea: at $8.5M$ urea the three experiments did not agree, so that three separate curves are given. The ovalbumin concentration was 1 gram per 100 cc of solution in each experiment. The time scale changes at 1000 minutes. (Frensdorff et al.[35])

be discussed in detail below, consists of a slow unfolding, followed, even more slowly, by aggregation. This characteristic requires that denaturation reactions be studied simultaneously by several different experimental methods, so that the rates of change of molecular weight, configuration, etc., can be observed separately.

In the following sections a detailed account will be given of a single denaturation reaction, that of ovalbumin by urea. It will serve as illustration of the type of data which are obtained and of the type of mechanism often proposed.

34b. The denaturation of ovalbumin by urea. The reaction between urea and ovalbumin has been studied in considerable detail by Kauzmann and coworkers.[35] Two properties were measured, optical rotation, representing internal order (for instance helix content; cf. section 6), and viscosity, representing effective hydrodynamic volume. Typical experimental data at 30° C in phosphate buffer (pH 7 to 8) at different urea concentrations are shown in Figs. 34–1 and 34–2.

It is at once apparent that the reaction proceeds in at least two separate stages. At $8.5M$ urea, for instance, there is a change in optical rotation, from about -30 to about $-90°$, which is complete in about 50 minutes. During the same period there is a sharp increase in reduced viscosity from an initial value near 4.0 to about 30. The increase in viscosity continues, however, and is not complete even after 3000 minutes, although no change in optical rotation occurs in this time. Kauzmann and coworkers interpret these two stages of reaction as follows:

The first stage of the reaction clearly involves destruction of elements of internal order, which, by section 6, can be interpreted as an unfolding of helical structures. Such a change would be expected to be part of a general expansion of the molecule and, hence, would be expected to be accompanied by an increase in intrinsic viscosity. The measurements of Fig. 34–2 are of η_{sp}/c at a 1 % protein concentration and thus are somewhat greater than intrinsic viscosities (cf. section 23), but, at least semi-quantitatively, they may be taken to reflect changes in intrinsic viscosity. The first stage is thus an intramolecular reaction, involving structural changes within individual molecules. In confirmation of this, the progress of the first stage, in terms of the optical rotation *per mole* of protein, was found to be independent of protein concentration; i.e., the reaction is a first-order reaction.

The second stage of the reaction involves an increase in viscosity without change in optical rotation. It could be interpreted as a further expansion, but, as Fig. 34–3 shows, the rate of this change depends very strongly on protein concentration. This means that several protein molecules must be involved in the reaction, so that the reaction must be an aggregation reaction. Further evidence for this comes from the fact that, at the highest protein concentrations used, the solutions are eventually transformed into a gel.

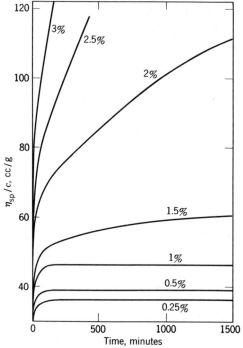

Fig. 34–3. The effect of protein concentration on η_{sp}/c for the denaturation of ovalbumin in $10M$ urea at 30° C. Comparison with Fig. 34–2 shows that all the data here presented were obtained after the first stage of the reaction is essentially complete. The first stage is independent of protein concentration, and, at $10M$ urea, is virtually complete after a few minutes. (Frensdorff et al.[35])

It should be noted that the conclusion that the first stage of the reaction is purely intramolecular whereas the second stage is purely an aggregation could have been made unequivocal if a third method had been used to follow the reaction, the property studied being one (such as light scattering) which could give a clear-cut indication of molecular weight changes.

34c. The first stage of the reaction. The first stage of the reaction has a number of startling kinetic features. The effect of urea concentration, as shown by Fig. 34–1, is tremendous. The effect of temperature, shown by

Fig. 34–4, is quite different from that of any known reaction involving small molecules, with a sharp minimum in the reaction rate at about 20° C. There is a pronounced effect of pH and of added inorganic salts. Finally, although the dependence on protein concentration indicates a first-order reaction, the time dependence during a single run does not correspond to first order. These phenomena will now be examined in some detail and will be seen to be capable of explanation in terms of a reasonable mechanism.

To discuss first the time dependence let [N] be the concentration of native protein at time t and [D] be the concentration of the denatured

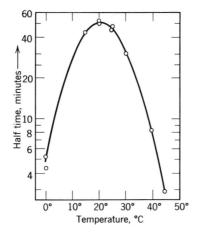

Fig. 34–4. The effect of temperature on the "half time" of the first stage of the urea denaturation of ovalbumin. The half time is equal to $0.69/\bar{k}$, where $\bar{k}$ is the rate constant defined in the text. (Simpson and Kauzmann.[35])

protein, assuming them to be species with the same molecular weight. Let [P] be the total concentration of protein so that, throughout the reaction,

$$[P] = [N] + [D] \tag{34–1}$$

If $[\alpha]_N$ and $[\alpha]_D$ represent the specific optical rotations of the two forms, the observed rotation at time t will be

$$[\alpha] = ([\alpha]_N[N] + [\alpha]_D[D])/[P] \tag{34–2}$$

At the end of the first stage of the reaction $[P] = [D]$ and $[N] = 0$ and $[\alpha] = [\alpha]_D$. Thus, where $[\alpha]_\infty$ is the value of $[\alpha]$ at the end of the first stage of reaction

$$[\alpha] - [\alpha]_\infty = [\alpha] - [\alpha]_D = ([\alpha]_N - [\alpha]_D)[N]/[P] \tag{34–3}$$

If the reaction is a first-order reaction, then $-d[N]/dt = d[D]/dt = k[N]$, and $\log[N]$ is a linear function of time, By equation 34–3, therefore, $\log([\alpha] - [\alpha]_\infty)$ should be a linear function of time. Figure 34–5, however,

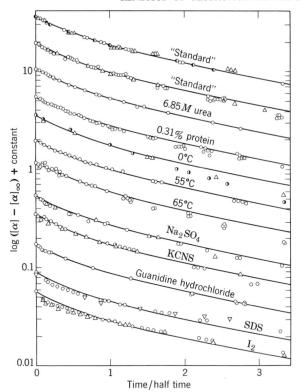

Fig. 34–5. Test of the first-order rate law for the first stage of the denaturation of ovalbumin. The first-order rate law is not obeyed, despite the fact that initial rates are independent of protein concentration. The observed deviation from linearity (i.e., from the first-order rate law) is, however, the same under a variety of conditions. [Protein concentration in all the data was 2 to 3 grams/ 100 cc except for the curve labeled "0.31 %," in which the concentration was 0.31 gram/100 cc. This curve and the "standard" curves represent 7.3M urea at 30° C; the 6.85M urea curve is at 30° C; the 0° C curve is from experiments at 6.08 and 5.58M urea; the 55° C curve is at 4.44M urea; the 65° C curve at 1.52M urea; the curve labeled "guanidine hydrochloride" used 2.47M guanidine hydrochloride in place of urea; the other four curves contained Na_2SO_4, KCNS, sodium dodecyl sulfate (SDS), and I_2 in addition to urea and to the phosphate buffer present in all the experiments.] (Simpson and Kaufmann.[35])

shows that this is not what is observed. Since the initial slopes of Fig. 34–1 are independent of concentration, the non-linearity of Fig. 34–5 cannot be explained on the basis of supposing the reaction to be of some order other than the first. The same situation arises often in the study of radioactive decay, a process which is necessarily of first order, and it is well known that there are two possible explanations. The first is that the starting material is inhomogeneous, so that we are actually measuring the rate of a number of simultaneous first-order reactions, each with a different rate constant. The material with larger rate constants is used up faster, so that the undenatured protein remaining contains an increasing proportion of material with lower rate constants, and this is reflected in a decrease in the average value of k with time. The second possible explanation is that the reaction may proceed in a series of first-order steps, e.g., $N \rightarrow D' \rightarrow D'' \rightarrow D$, each with a rate constant of the same order of magnitude. If the rate equations for such a scheme are set up, a plot of the type of Fig. 34–5 results. The second of these explanations is more likely to be correct since ovalbumin ordinarily behaves as a homogeneous material. In terms of mechanism it might mean that only a portion of the protein molecule unfolds at a time.

If either of the preceding explanations for the non-linearity of Fig. 34–5 is specifically taken into account in the reaction scheme, the rate of the first stage of the reaction in each experiment must be described by more than one rate constant. The several rate constants which would have to be used would, however, all have the same order of magnitude, and, moreover, their relative values would be essentially the same for each experiment; i.e., the variation of each rate constant with urea concentration, temperature, etc., would be essentially the same. Kauzmann and coworkers accordingly decided to substitute a single average rate constant $\bar{k}$, obtaining its value from the half time of the reaction; i.e., they replaced the actual data of Fig. 34–5 by straight lines tangent to the observed curve when the reaction has gone to 50% completion.

Considering next the effect of urea concentration, Fig. 34–1 shows this effect to be very large, at least at the temperature of 30° C. At this temperature a doubling of the urea concentration leads to a rate increase by a factor of roughly 30,000, corresponding to a reaction order of about 15 with respect to urea concentration. Data at several temperatures are shown in Fig. 34–6. Over the limited range of concentration used $\log \bar{k}$ is close to a linear function of $\log C$, where C is the urea concentration, so that $\bar{k}$ may be replaced by kC^m where k and m are independent of C but do depend on temperature. Thus we may write, for the effective reaction rate,

$$-d[N]/dt = \bar{k}[N] = k[N]C^m \qquad (34\text{-}4)$$

Figure 34–6 shows that m decreases with increasing temperature. At 0° C, $m = 15$; at 30°, $m = 12.5$; at 50°, $m = 8$; and at 65° $m = 3$.

A simultaneous attack by as many as fifteen urea molecules on a single protein molecule is out of the question, so that the simplest possible explanation of equation 34–4 is in terms of a mechanism which involves successive equilibria preceding the actual reaction step. Kauzmann suggested

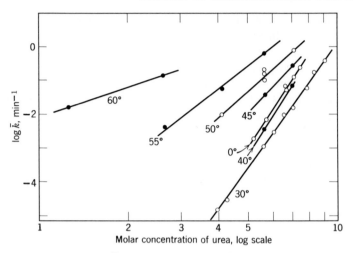

Fig. 34-6. A plot of $\log \bar{k}$ versus the logarithm of the urea concentration, at different temperatures. (Simpson and Kauzmann.[35])

that a native protein molecule N_0 devoid of urea molecules can combine with up to n urea molecules and that the successive complexes NU, NU_2, etc., can undergo denaturation with increased facility. The reaction scheme is thus

$$N_0 \xrightarrow{k_0} D$$

$$N_0 + U \underset{\longleftarrow}{\overset{K_1}{\rightleftharpoons}} NU \xrightarrow{k_1} D$$

$$N_0 + 2U \underset{\longleftarrow}{\overset{K_2}{\rightleftharpoons}} NU_2 \xrightarrow{k_2} D$$

$$\vdots$$

$$N_0 + nU \underset{\longleftarrow}{\overset{K_n}{\rightleftharpoons}} NU_n \xrightarrow{k_n} D$$

where each K_i is an equilibrium constant of the type

$$K_i = [NU_i]/[N_0]C^i \tag{34-5}$$

and the k_i are first-order rate constants increasing in value as i increases. The total concentration [N] of native ovalbumin can now be written as

$$[N] = [N_0] + \sum_{i=1}^{n} [NU_i] = [N_0]\left(1 + \sum_{i=1}^{n} K_i C^i\right) \tag{34-6}$$

while the over-all rate of denaturation is

$$-d[N]/dt = k_0[N_0] + \sum_{i=1}^{n} k_i[NU_i] = [N_0]\left(k_0 + \sum_{i=1}^{n} k_i K_i C^i\right) \tag{34-7}$$

[N_0] can be eliminated between equations 34–6 and 34–7 to yield

$$\frac{-d[N]}{dt} = \frac{k_0 + \sum_{i=1}^{n} k_i K_i C^i}{1 + \sum_{i=1}^{n} K_i C^i} [N] \qquad (34\text{–}8)$$

Equation 34–8 must be identical, over the limited range of urea concentrations at which rate measurements were made, with the experimental rate equation, i.e., equation 34–4. Thus

$$\bar{k} = k C_m = \frac{k_0 + \sum_{i=1}^{n} k_i K_i C^i}{1 + \sum_{i=1}^{n} K_i C^i} \qquad (34\text{–}9)$$

Equations 34–8 or 34–9 contain enough adjustable parameters so that any experimental result can be accounted for. If such an equation is to be a test for the mechanism suggested, the majority of the constants must be assigned calculated values, leaving as few adjustable constants as possible. The following reasonable assumptions achieve this.

The combination of ovalbumin with urea can be assumed to occur at n identical and independent sites, so that each K_i can be expressed in terms of a single equilibrium constant K. Each K_i as defined by equation 34–5 is a product of i of the equilibrium constants used in section 29a, such that $K_i = k_1 k_2 \cdots k_i$. (These k's should not be confused with the rate constants used in the present section.) By equations 29–7 and 29–9 we thus have

$$K_i = n! \, K^i / (n - i)! \, i! \qquad (34\text{–}10)$$

It may further be guessed that each successive urea molecule increases the rate constant k by the same factor, so that $k_1 = ak_0$, $k_2 = ak_1$, etc.; i.e., in general,

$$k_i = a^i k_0 \qquad (34\text{–}11)$$

This leaves four arbitrary parameters, n, K, a, and k_0 by which the effect of urea concentration is to be described. One of them, k_0, can be obtained from quite unrelated data if it is supposed that urea denaturation leads to the same product as heat denaturation. This assumption was made by Kauzmann and coworkers. Rate constants for heat denaturation, obtained at higher temperatures in the absence of urea, were extrapolated (see equation 34–16) to the experimental temperatures and used to represent k_0.

Substitution of equations 34–10 and 34–11 into equation 34–9 and subsequent combination with equation 29–11 gives

$$\bar{k} = kC^m = \frac{k_0 \left[1 + \sum_{i=1}^{n} n! \, (aKC)^i/(n-i)! \, i! \right]}{\left[1 + \sum_{i=1}^{n} n! \, (KC)^i/(n-i)! \, i! \right]}$$

$$= k_0 \left(\frac{1 + aKC}{1 + KC} \right)^n \tag{34-12}$$

from which the experimental parameter m can be obtained,

$$m = \frac{d \ln \bar{k}}{d \ln C} = C \frac{d \ln \bar{k}}{dC} = \frac{naKC}{1 + aKC} - \frac{nKC}{1 + KC} \tag{34-13}$$

It will be seen later that KC is quite small, so that m is largely determined by the first term on the right-hand side of equation 34–13. The observed decrease in m with increasing temperature (Fig. 34–6) will therefore require that the product aK must also decrease with increasing temperature. The value of m is, of course, not independent of urea concentration. The fact that the data of Fig. 34–6 fall on straight lines, with constant m, is due to the narrow range of concentration encompassed by the experiments at any one temperature.

It can be seen from equation 29–13 that the last term in equation 34–13 is the average number of molecules of urea bound to the undenatured protein, which we may designate as v. The term $naKC/(1 + aKC)$ is of the same form, with the substitution of aK for K. It can easily be shown to represent the average number of urea molecules in the activated complex of the denaturation. We may call it v'. The apparent order of reaction is thus merely the difference between the average number of urea molecules in the activated complex and the native protein:

$$m = v' - v \tag{34-14}$$

where

$$v' = \frac{naKC}{1 + aKC} \qquad v = \frac{nKC}{1 + KC} \tag{34-15}$$

Since v will be found to be small, $m \simeq v'$.

It was noted above that aK must decrease with increasing temperature to account for the experimental variation in m with temperature. Since aK has now been interpreted as an intrinsic binding constant for the addition of urea to the activated complex for denaturation, a fall in its value with increasing temperature is entirely reasonable, since increasing temperature quite generally leads to increased dissociation of addition complexes.

The effect of temperature on the reaction rate, as shown by Fig. 34–4, is an unusual one. At low temperature $\bar{k}$ decreases sharply as temperature increases. The apparent activation energy, $E_a = \mathscr{R}T^2(\partial \ln \bar{k}/\partial T)_c$, is about $-28,000$ cal/mole at $0°$ C. Near $20°$ C $\bar{k}$ becomes independent of temperature so that E_a is zero. At the higher temperatures, on the other hand, $\bar{k}$ increases with temperatures, and E_a approaches a value of $+50,000$ cal/mole.

To understand this effect, we must examine the temperature dependence of the parameters n, k_0, K, and a. The first of them is independent of temperature. The rate constant k_0, however, depends on temperature in the usual way and may be represented as

$$k_0 = A\,e^{-E_0/\mathscr{R}T} \tag{34–16}$$

where E_0 is the Arrhenius activation energy. As was mentioned above A and E_0 are taken from experimental data on the thermal denaturation of ovalbumin in the absence of urea, $\log A = 79.1$, $E_0 = 130,000$ cal/mole.

All other k_i must also have the form (constant) $e^{-E/\mathscr{R}T}$, so that, by equation 34–11, a must be of this form, too. Thus we may write

$$a = B\,e^{\delta E/\mathscr{R}T} \tag{34–17}$$

The exponential is written with a positive exponent since all k_i are greater than k_0, so that the activation energy for the ith reaction is presumably less than E_0.

Finally K, the intrinsic association constant between ovalbumin and urea, may be expressed in terms of the standard intrinsic heat and entropy of association; i.e., $\ln K = \Delta S°/\mathscr{R} - \Delta H°/\mathscr{R}T$, or

$$K = L\,e^{-\Delta H°/\mathscr{R}T} \tag{34–18}$$

where

$$L = e^{\Delta S°/\mathscr{R}}$$

All these relations may be used to evaluate the apparent activation energy, $E_a = \mathscr{R}T^2(\partial \ln \bar{k}/\partial T)_c$, by differentiation of equation 34–12. Making use of the fact that

$$n\left[\frac{\partial \ln(1 + KC)}{\partial T}\right]_c = \frac{nC}{1 + KC}\frac{dK}{dT} = \frac{nKC}{1 + KC}\frac{d\ln K}{dT} = v\,\frac{d\ln K}{dT}$$

and a similar relation for the derivative of $\ln(1 + aKC)$ we get

$$E_a = E_0 + v'(\Delta H° - \delta E) - v\,\Delta H° \tag{34–19}$$

It was noted earlier that v is always small so that the effect of equation 34–19 will come principally from the first two terms on the right-hand side. Furthermore it was shown that aK must decrease with increasing temperature, so that, by equations 34–17 and 34–18, $\Delta H° - \delta E$ must be negative. If $\Delta H° - \delta E$ is about $-10{,}500$, the experimental values of E_a are well accounted for. At $0°$ C $v' \simeq m = 15$, and E_a becomes $-28{,}000$ cal. At $50°$ C, on the other hand, $v' \simeq m = 8$, and E_a becomes $+46{,}000$ cal/mole.

The best values for the adjustable parameters, obtained from a consideration of all the available data, were $n = 16$, $\log BL = -0.7$, and $\Delta H° - \delta E = -10{,}000$ cal/mole. The contribution of KC to any of the equations was found to be too small to permit the determination of best values for L or $\Delta H°$.

34d. Interpretation in terms of structure. The primary purpose of investigating denaturation kinetics is to gain an insight into the structural features of the native and denatured molecules. The question which arises is thus: what kind of a rigid structural unit might a protein molecule possess which could combine reversibly with sixteen urea molecules and which would have an increasing tendency to become unraveled as its urea content increases? Unfortunately, this question is not easy to answer, for no such structures are known in the chemistry of small molecules. Kauzmann has speculated that an answer might be given in terms of a helical structure, such as the Pauling α-helix (Fig. 4–7). He has supposed that the native molecule contains several short lengths of such helical structures, as illustrated by Fig. 34–7a, and that the breakdown of the native structure must begin with the unfolding of one coil of such a helix, as shown in Fig. 34–7b. Each such coil is fixed in space with respect to its neighbor on either side by four hydrogen bonds; i.e., altogether eight such bonds hold it in place. Urea might be able to break these bonds by the reaction

$$
\begin{array}{c}
-C- \\
\parallel \\
O \\
\cdot \\
\cdot \\
\cdot \\
H \\
| \\
-N-
\end{array}
\quad + \ 2O{=}C(NH_2)_2 \rightarrow \quad
\begin{array}{c}
| \\
C{=}O \cdots H_2N{-}CO{-}NH_2 \\
| \\
\\
+ \\
\\
| \\
N{-}H \cdots O{=}C(NH_2)_2 \\
|
\end{array}
$$

Thus up to sixteen urea molecules could be accommodated, each progressively weakening the forces which keep the helix coil in position.

Whether this speculative explanation is correct cannot, on the basis of present knowledge, be decided. It has already been pointed out (section 7) that —C=O ⋯ HN— hydrogen bonds should not differ greatly in strength from —C=O ⋯ H$_2$O and H$_2$O ⋯ HN— hydrogen bonds. If the reaction in the absence of urea involves an activation step in which sixteen water molecules perform the function ascribed to urea in Fig. 34–7,

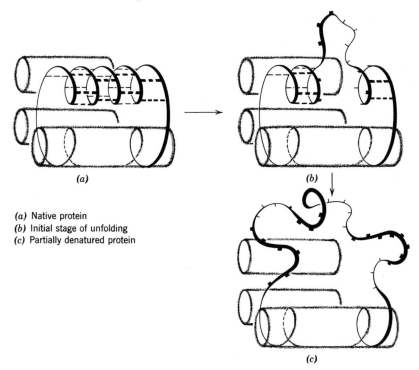

(*a*)   (*b*)

(*a*) Native protein
(*b*) Initial stage of unfolding
(*c*) Partially denatured protein

(*c*)

Fig. 34–7. Speculative mode for the denaturation of ovalbumin. (Kauzmann.[4])

it is difficult to see why the activation energy (E_0) should be as large as 130,000 cal. It is equally difficult to see how each urea molecule can contribute as much as 10,000 cal to the lowering of the activation energy.

Some auxiliary experiments described by Simpson and Kauzmann[35] bear on this question but do not greatly clarify the situation. The rate of the unfolding reaction was found to be independent of pH between pH 7 and 9, so that the reaction appears not to depend on the electrostatic charge of the molecules. It also appears not to depend on the dielectric constant of the reaction medium, at least in the range of dielectric constant from 78 to 126. Thus a primarily electrostatic mechanism appears to be ruled out.

Despite this, added salts have a pronounced effect on the reaction rate. Thus $0.1M$ $CaCl_2$ increases the rate of reaction by a factor of 3.5, $0.1M$ KI increases it by a factor of 2.5, $0.1M$ sodium acetate decreases it by a factor of 1.2, and $0.1M$ Na_2SO_4 decreases it by a factor of 2.5. No simple explanation for these effects is possible.

Many organic substances have relatively little effect, but some, e.g. p-nitrophenol, greatly increase the rate of reaction. Sucrose decreases the rate. Reagents which affect the reactivity of sulfhydryl groups have little or no influence.

34e. The second stage of the reaction. The second stage of the denaturation of ovalbumin is characterized by an increase in reduced viscosity and eventual gelling of the solution. The most prominent feature is the very large effect of protein concentration, shown in Fig. 34–3. This effect suggests that the reaction is an aggregation. If the ovalbumin molecules, at the end of the first stage of denaturation, are randomly coiled (even if only in part), then such association would result (cf. section 23) in the sizable viscosity increases actually observed.

In contrast to the first stage of the reaction such an association is easily explained in terms of a chemical reaction which occurs in small molecules as well as large ones. This is the exchange reaction between organic sulfide ions and disulfide bonds,

$$R_1S^- + R_2S\!-\!SR_3 \rightarrow R_2S^- + R_1S\!-\!SR_3$$

Each ovalbumin molecule contains two —S—S— bonds and five —SH groups which, at the alkaline pH of the denaturation studies, will be partly dissociated to —S⁻. The disulfide exchange reaction between an —S⁻ group of one molecule and an —S—S— bond of another will lead to formation of a dimer with four —S—S— bonds,

which can then react further to form higher polymers. That this reaction is not observed in alkaline ovalbumin solutions in the absence of urea is not surprising since the disulfide bonds of the molecule which has not undergone the first unfolding stage of the reaction are presumably rigidly fixed in the interior of the molecule and inaccessible to the —S⁻ groups of other molecules.

That the disulfide exchange reaction here postulated is indeed that responsible for the second stage of the aggregation of ovalbumin is shown by the fact that no second stage is observed in the presence of p-chloromercuribenzoate, a reagent which combines with —S$^-$ groups and blocks any further action on their part. As Fig. 34–8 shows, viscosity changes in the presence of this reagent exactly parallel optical rotation changes.

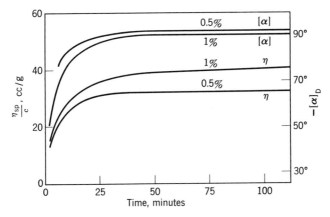

Fig. 34–8. Comparison between the rates of change of optical rotation and of η_{sp}/c in the presence of p-chloromercuribenzoate (PCMB), at $7.5M$ urea concentration. Two experiments, at different protein concentrations, are shown. The second stage of the denaturation, as seen in Figs. 34–1 and 34–2, is no longer observed. (Frensdorff et al.[35])

(The value of the reduced viscosity at the end of the reaction is higher at 1% concentration because of the concentration dependence of this quantity.)

35. KINETICS OF ENZYME ACTION

The complex chemistry of life processes is made possible by the existence of highly specific catalysts which abound in living matter. These substances are known as *enzymes*. Without exception they are protein molecules.

These enzymes permit chemical reactions to occur in living cells which either could not occur at all in the absence of the enzymes' catalytic action or would occur only slowly or only under drastic conditions. That the catalytic action lies in the enzyme molecules and does not require participation by the entire living cell is easily demonstrated, for, if an enzyme is extracted, purified, and crystallized, it is found to be capable of catalyzing in the laboratory the same reactions which it catalyzes in the living cell.

Among especially well-known enzymes are the proteolytic enzymes (chymotrypsin, trypsin, papain, etc.) which catalyze the degradation of proteins into smaller peptides at ordinary temperatures in aqueous solutions, often near pH 7. (These enzymes have already been mentioned in section 33.) In the absence of enzyme catalysis this degradation requires high temperature and the presence of strong acid or base. Another well-known group is the group of oxidizing enzymes, which permit the oxidation of sugars (for example) to occur by a series of well-regulated steps, with the gradual release of energy.

A general characteristic of enzymes is their specificity. That of proteolytic enzymes has already been noted in section 33. Another example is the enzyme *fumarase*, which will be discussed below in some detail. It catalyzes the addition of water to fumaric acid, forming *l*-malic acid, and the reverse process, the dehydration of *l*-malic acid. As far as is known it cannot catalyze the transfer of water to or from any other molecule.

Superficially, enzyme reactions are quite simple. For example, if the reaction is a simple rearrangement, of the form S → P, then the mechanism often involves just two steps, the combination of enzyme with the starting material S, usually called the *substrate*, followed by the conversion of the substrate-enzyme complex into the product P plus regenerated free enzyme. The resulting rate equation (see section 35a) is identical with that which applies to countless catalytic processes which occur in the chemistry of small molecules, such as acid catalysis of organic reactions. If the reaction catalyzed is of the type S + A → products, the mechanism and rate equation become slightly more complicated, and the same is true, for instance, if other substances are present which can combine with the enzyme and thus prevent some of the enzyme molecules from being available for catalysis. In neither case, however, does any process occur which does not have its parallel in the chemistry of small molecules.

To dismiss enzyme kinetics as a straightforward extension of ordinary catalysis is, however, unwarranted, for the fact remains that enzymes are invariably proteins (i.e., that they are always large molecules) and that their action is often so highly specific. It would seem possible that these two characteristics are connected and that a deeper probe into the reaction kinetics might reveal features of the mechanism which explain the necessity for the large molecular size as well as the phenomenon of high specificity. The bulk of this section will be devoted to a single example of enzyme action, in which a study of the effect of pH has led to at least a partial solution of the problem just posed.

No attempt will be made here at all to survey, to summarize, or even to discuss selected features of the vast literature of enzyme kinetics in general. For general reviews of this subject the reader is referred to appropriate treatises.[6]

35a. Steady-state rate equations for enzyme action. The bulk of all kinetic measurements are confined to the *steady-state* reaction velocity. Such a velocity may be defined for any reaction in which the progress from reactants to products involves intermediate products only as transient forms which are never present in high concentration. The steady-state velocity is that which persists at all times during the reaction except for the brief period at the end of the reaction when these forms decay. The present section is devoted entirely to the steady-state velocity of enzyme reactions. (The study of the formation and disappearance of transient forms will be considered in section 36.)

We consider first the simplest possible situation, involving just a single substrate molecule, S; a single product, P; and a single intermediate complex between enzyme, E, and substrate, which we designate as ES. Since any chemical reaction must be capable of proceeding in either direction and since an enzyme must catalyze both reactions, we may describe this simple situation in terms of the mechanism

$$S + E \underset{k_2}{\overset{k_1}{\rightleftharpoons}} ES$$

$$ES \underset{k_4}{\overset{k_3}{\rightleftharpoons}} E + P$$

The complex between enzyme and substrate could be written ES, EP, or EX, depending on whether the S molecule in the complex has a structure resembling its original structure or resembling the product P or some other structure. It is ordinarily not possible to distinguish between these possibilities, and the symbol ES is employed without implied commitment concerning the actual structure of the complex.

The intermediate ES in this reaction scheme is a transient intermediate because the total amount of enzyme in any reaction mixture is small, so that no appreciable fraction of S molecules can be in the form ES at any one time. The steady-state rate equation is thus obtained by imposing the condition $d[ES]/dt$ (and, hence, $d[E]/dt$) may be set equal to zero in comparison with the rates of the individual steps of the reaction; i.e., we write

$$-d[E]/dt = d[ES]/dt = k_1[E][S] - (k_2 + k_3)[ES] + k_4[E][P] = 0$$
$$(35\text{--}1)$$

Equation 35-1 may be solved to relate the concentration of E to that of ES; i.e.,

$$[E] = (k_2 + k_3)[ES]/(k_1[S] + k_4[P]) \qquad (35\text{-}2)$$

Furthermore, use may be made of the fact that the total concentration of enzyme does not change, since the enzyme is a true catalyst which is not used up in the reaction. Thus, if $[E]_0$ represents the concentration of enzyme added originally to the reaction mixture, then

$$[E]_0 = [E] + [ES] \qquad (35\text{-}3)$$

The rate v of the reaction in the forward direction may now be written down at once: $v = d[P]/dt = k_3[ES] - k_4[E][P]$, which, with the aid of equations 35-2 and 35-3, becomes

$$v = \frac{k_1 k_3[S] - k_2 k_4[P]}{k_1[S] + k_4[P] + k_2 + k_3} [E]_0 \qquad (35\text{-}4)$$

Equation 35-4 may be considerably simplified for a reaction in which the state of equilibrium is far to the right, so that conversion of P to S does not occur under the conditions used to study the reaction $S \rightarrow P$. If the kinetic reason for this is that k_4 is negligibly small, equation 35-4 reduces to the well-known equation of Michaelis and Menten,[36]

$$v = \frac{k_3[S][E]_0}{K_m + [S]} \qquad (35\text{-}5)$$

where

$$K_m = (k_2 + k_3)/k_1$$

It is important to note that the same equation applies to the *initial rate* of any reaction obeying equation 35-4, since $k_4[P]$ is necessarily zero when $[P] = 0$, regardless of the magnitude of k_4. The parameter K_m of equation 35-5 is often called the Michaelis constant, and it is sometimes mistakenly taken to be the equilibrium constant for the dissociation of ES, ES $\rightleftharpoons$ E + S. This is, of course, correct only if $k_3 \ll k_2$ so that $K_m = k_2/k_1$.

At a constant enzyme concentration, the velocity given by equation 35-5 increases with substrate concentration, at first linearly, but then less rapidly (cf. Fig. 35-1). Finally, as $[S] \gg K_m$, the velocity becomes independent of substrate concentration, attaining a constant value of $k_3[E]_0$. This is often represented by the symbol v_{max}; i.e., the Michaelis-Menten equation is often written in the form

$$v = v_{max}[S]/(K_m + [S]) \qquad (35\text{-}6)$$

This may also be written as

$$\frac{1}{v} = \frac{1}{v_{\max}} + \frac{K_m}{v_{\max}}\frac{1}{[S]} \tag{35-7}$$

If the reaction obeys the Michaelis-Menten equation, a plot of $1/v$ versus $1/[S]$ is linear, and the parameters $v_{\max}$ and K_m may be obtained from the intercept and slope of the plot.

Steady-state rate equations which are applicable to more complicated reactions or more elaborate mechanisms may be derived equally simply. In the reaction $S \rightarrow P$ we could, for instance, allow for the presence of two

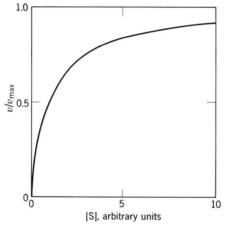

Fig. 35-1. The effect of substrate concentration on the velocity of an enzyme-catalyzed reaction obeying the Michaelis-Menten equation.

distinct intermediates, ES and EP. More complicated reactions can be considered, of the type $S + X \rightarrow P + Q$, with intermediates such as ES, ESX, EPQ, and EP. Allowance can be made for participation by extraneous substances, such as activators, inhibitors, and coenzymes. Many of these possibilities have been considered in the literature, and a summary of typical equations which have been obtained has been given by Alberty.[37]

We shall consider in the present section only one of the more complex situations, that of a reaction of the type $S + X \rightleftharpoons P$, under conditions where the reaction proceeds readily in both directions, the equilibrium constant $[P]_{eq}/[S]_{eq}$ being not far from unity. (This example has been chosen because it applies to the hydration of fumarate, a reaction which forms the main part of the subsequent discussion.) A reasonable mechanism for such a reaction would involve two intermediates and the following set of reactions:

$$E + S \underset{k_2}{\overset{k_1}{\rightleftharpoons}} ES$$

$$X + ES \underset{k_4}{\overset{k_3}{\rightleftharpoons}} EP$$

$$EP \underset{k_6}{\overset{k_5}{\rightleftharpoons}} E + P$$

The general rate equation which corresponds to this mechanism is most unwieldy. The expressions for the *initial* rate of reaction, in either direction, are reasonably simple, however. If we begin with the substrate S, then, for the initial rate, reaction 6 may be ignored. Proceeding exactly as before we now get two independent equations from the condition that $d[E]/dt$, $d[ES]/dt$, and $d[EP]/dt$ are negligibly small when the steady state has been reached; i.e.,

$$d[ES]/dt = 0 = k_1[E][S] - (k_2 + k_3[X])[ES] + k_4[EP]$$
$$d[EP]/dt = 0 = k_3[X][ES] - (k_4 + k_5)[EP]$$

In place of equation 35–3 we write

$$[E]_0 = [E] + [ES] + [EP]$$

and, combining these equations, we get, for the initial rate of the reaction in the forward direction, $v_F = d[P]/dt$,

$$v_F = \frac{k_5[E]_0}{1 + K_A/[S] + K_B/[X] + K_C/[S][X]} \tag{35–8}$$

where $K_A = k_5/k_1$
$K_B = (k_4 + k_5)/k_3$
$K_C = k_2(k_4 + k_5)/k_1 k_3$

In many cases the reagent X is present at constant concentration throughout the course of the reaction. This happens, for instance, if X represents O_2 and the O_2 concentration is controlled by a constant oxygen pressure above the solution or if X represents a solvent molecule, e.g. water in a hydrolysis reaction. (In the latter case [X] is generally expressed in activity units and placed equal to unity.) In all such situations [X] becomes a constant, and equation 35–8 reduces to the Michaelis-Menten equation, equation 35–6, with

$$(v_{max})_F = k_5[E]_0[X]/(K_B + [X])$$
$$(K_m)_F = (K_C + K_A[X])/(K_B + [X]) \tag{35–9}$$

the subscript F denoting that these relations apply to the initial rate in the forward direction.

For the initial rate of the reverse reaction we neglect reaction 1 rather than reaction 6 and obtain for the reaction rate, $v_R = d[S]/dt$,

$$v_R = \frac{k_D[E]_0}{1 + K_E/[P] + K_F[X]/[P] + K_G[X]} \qquad (35\text{-}10)$$

where $k_D = k_2 k_4/(k_2 + k_4)$
$K_E = k_2(k_4 + k_5)/k_6(k_2 + k_4)$
$K_F = k_3 k_5/k_6(k_2 + k_4)$
$K_G = k_3/(k_2 + k_4)$

Like equation 35-9, this equation reduces to equation 35-6 when [X] is constant, with [P] replacing [S] and with

$$(v_{\max})_R = k_D[E]_0/(1 + K_G[X])$$
$$(K_m)_R = (K_E + K_F[X])/(1 + K_G[X]) \qquad (35\text{-}11)$$

35b. The relation between rate constants and equilibrium constants. Both of the reaction schemes which were discussed above may proceed in both forward and reverse directions. The final state of the system will be determined by the thermodynamics of the system. Thus, in the reaction, $S \rightleftharpoons P$, the final equilibrium composition will be such that

$$[P]_{eq}/[S]_{eq} = K_{eq}$$

is the equilibrium constant for the reaction.

The condition for equilibrium may also be stated kinetically as that composition at which no further reaction occurs. By equation 35-4, therefore, $k_1 k_3[S]_{eq} - k_2 k_4[P]_{eq} = 0$, so that

$$[P]_{eq}/[S]_{eq} \equiv K_{eq} = k_1 k_3/k_2 k_4$$

It is seen that, by placing $k_4 = 0$ in the derivation of the Michaelis-Menten equation, we have assumed that K_{eq} is infinite.

For the reaction $S + X \rightleftharpoons P$ the equilibrium constant is

$$K_{eq} = [P]_{eq}/[S]_{eq}[X]_{eq} \qquad (35\text{-}12)$$

and the kinetic condition, $d[P]/dt = 0$, $d[S]/dt = 0$, $d[E]/dt = 0$, etc., gives

$$k_1[E]_{eq}[S]_{eq} = k_2[ES]_{eq}$$
$$k_3[X]_{eq}[ES]_{eq} = k_4[EP]_{eq}$$
$$k_5[EP]_{eq} = k_6[E]_{eq}[P]_{eq}$$

Eliminating the enzyme concentrations from these equations, we get

$$K_{eq} = \frac{k_1 k_3 k_5}{k_2 k_4 k_6} \tag{35-13}$$

This relation must hold true regardless of additional assumptions which may be made.

The condition for equilibrium in this reaction could also have been obtained by specifying that the forward rate be equal to the reverse rate. It is important to note, however, that equations 35–8 and 35–10 for v_F and v_R could not have been used for the forward and reverse rates since these equations apply only to the *initial* rates in the two directions.

35c. The action of fumarase. The enzyme fumarase, a crystalline protein of molecular weight 220,000, catalyzes the reversible reaction,

$$\begin{array}{ccc}
{}^{-}\text{OOC}-\text{CH} & & {}^{-}\text{OOC}-\text{CHOH} \\
\parallel & + \text{H}_2\text{O} \rightleftharpoons & \mid \\
\text{HC}-\text{COO}^{-} & & \text{H}_2\text{C}-\text{COO}^{-} \\
\text{\small Fumarate} & & \text{\small {\it l}-Malate}
\end{array}$$

Detailed kinetic studies of its action have been made by Alberty and coworkers.[38] It is not unlikely that the reaction obeys the mechanism assigned in section 35a to a reaction of the type $S + X \rightleftharpoons P$. If so, the initial rates in both the forward and reverse directions should obey equation 35–6, since X, which is water in the present case, is present in very large excess. That it does so is shown by the sample data of Fig. 35–2, which are plotted according to equation 35–7. These data were obtained in solutions containing phosphate buffer at pH 7.4. Alberty and coworkers have made similar studies in phosphate and in other buffers over a wide range of pH. In each case the data agree equally well with equations 35–6 or 35–7, although the values of $v_{\max}$ and K_m depend both on pH and on the nature of the buffer; i.e., the rate constants defined by the reaction scheme on p. 644 are constants only under a particular set of conditions.

The pH dependence of the rate constants is particularly striking, as illustrated, for instance, by Fig. 35–3, which shows the effect of pH on $(v_{\max})_F$ and $(v_{\max})_R$. An analysis of this effect will be made in the following section.

It was pointed out in section 35b that the rate constants for a reversible reaction must be related to the equilibrium constant for that reaction. In the present case, using equations 35–9 and 35–11 and placing [X] = activity of $\text{H}_2\text{O} = 1$, we note that

$$\frac{(v_{\max})_F (K_m)_R}{(v_{\max})_R (K_m)_F} = \frac{k_5(K_E + K_F)}{k_D(K_A + K_C)} = \frac{k_1 k_3 k_5}{k_2 k_4 k_6} \tag{35-14}$$

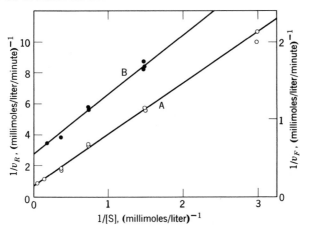

Fig. 35–2. Initial rates for the fumarase-catalyzed reaction, fumarate + $H_2O \rightarrow l$-malate (curve B), and for the reverse reaction (curve A). The data are plotted according to equation 35-7. (Alberty and coworkers.[38])

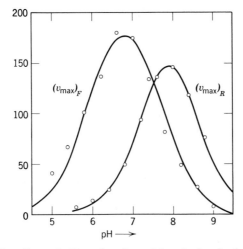

Fig. 35–3. The effect of pH on $(v_{max})_F$ and $(v_{max})_R$ for the fumarase reaction. The curves are calculated curves. (Alberty and coworkers.[38])

which, by equation 35-13, is the desired equilibrium constant; i.e.,

$$\frac{(v_{max})_F (K_m)_R}{(v_{max})_R (K_m)_F} = \frac{[l\text{-malate}]_{eq}}{[\text{fumarate}]_{eq}} = K_{eq} \qquad (35\text{-}15)$$

the equilibrium concentrations required being those determined in dilute aqueous solution ($a_{H_2O} = 1$).

This relation provides a useful test for the consistency of the kinetic studies, for the equilibrium constant K_{eq} can be determined quite independently from the kinetic experiments, by spectrophotometric examination of the end product of reaction under a variety of conditions. (Kinetic experiments, it will be recalled, were confined to the initial stages of the reaction.) The results of such a determination (Bock and Alberty[38]) are that K_{eq} is independent of pH between pH 6 and 9, its value lying between 4.0 and 5.0, depending on ionic strength. The same result is obtained when equation 35–15 is used to compute the equilibrium constant from the kinetic data. Despite the very large variation which v_F and v_R undergo in this pH range, the ratio $(v_{max})_F (K_m)_R / (v_{max})_R (K_m)_F$ is constant and equal to about 4.8, in agreement with the directly determined value of K. (The same kind of consistency is obtained for data at pH values below pH 6 if account is taken of the fact that both fumarate and l-malate become partially converted below that pH to the univalent ions hydrogen fumarate and l-hydrogen malate.)

35d. The effect of pH on the fumarase reaction. Fumarase is a protein molecule and, like all protein molecules, will undergo changes in its state of ionization with pH (section 30). Such changes in turn produce changes in the net charge of the molecule, and they in turn will undoubtedly affect the rates of any reaction of the enzyme with ions. However, this effect of electrostatic charge should be relatively small, and it is highly improbable that it could account for the striking pH effects illustrated by Figs. 35–3 and 35–4. Such sharp changes in kinetic parameters with pH almost certainly means a direct influence of the state of ionization of particular acidic groups on the rate of reaction.

The type of effect illustrated by Figs. 35–3 and 35–4 is observed not only in fumarase but in a variety of other enzymes. It has been generally agreed, since 1911 (Michaelis and coworkers[39]), that the simplest explanation requires the direct involvement of two dissociable groups, one of which must be in its basic form and the other in its acidic form if the enzyme is to be active. It was pointed out in section 31 that such a situation will lead to an activity curve of the type of Fig. 35–3, with a sharp pH maximum if the two groups involved differ little in pK, and with a broad plateau (Fig. 31–2) if their pK values are very different.

This explanation has been adopted by Alberty and coworkers (for a review see Alberty[40]). However, they did not make direct application of the treatment of section 31 to the activity curves, for to do so would imply that the dissociable groups are characterized by unique dissociation constants throughout the reaction. Such an assumption is clearly unwarranted. If we adopt the explanation that the state of ionization of the dissociable groups has a profound effect on the course of the reaction, we cannot avoid the corollary that the progress of the reaction must influence the state of ionization. In terms of the reaction mechanism of p. 644, therefore, we must assign different dissociation constants to the free enzyme E and to the two complexed forms ES and EP.

Let $^+$HEH$^+$ represent the form of the *free* enzyme E in which both dissociable groups have protons attached. Let EH$^+$ be the corresponding molecule with one proton attached, making no distinction between the two possible forms of this molecule. Finally, let E$_-$ represent the form with

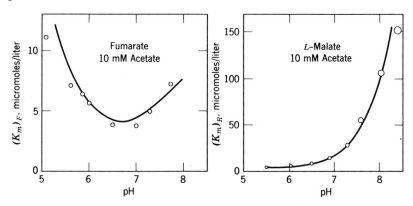

Fig. 35–4. The effect of pH on $(K_m)_F$ and $(K_m)_R$ for the fumarase reaction. The curves are theoretical curves. (Alberty and coworkers.[38])

both protons removed. If [E] continues to be used to designate the total concentration of free enzyme, the following equations must hold

$$[E] = [^+HEH^+] + [EH^+] + [E_-]$$
$$[EH^+]a_{H^+}/[^+HEH^+] = K_{1E} \qquad (35–16)$$
$$[E_-]a_{H^+}/[EH^+] = K_{2E}$$

where a_{H^+} is the hydrogen ion activity and K_{1E} and K_{2E} are the acid dissociation constants of $^+$HEH$^+$ and EH$^+$, respectively.

The fact that no distinction is made between the two possible forms of EH$^+$, i.e., the fact that the proton may be on either of the two important dissociable groups, has no effect on the equations which are obtained, for the relative concentrations of the two forms is necessarily independent of pH. Unless the two forms are in fact identical, it will, however, affect the interpretation to be placed on the constants K_{1E} and K_{2E}.

Another assumption which is made in connection with equation 35–16 is that K_{1E} and K_{2E} are independent of pH. As was shown in section 31 (Fig. 31–2) this assumption is a poor one if the pH range under consideration is one in which the charge on the protein molecule is changing rapidly. The pH range over which the fumarase reaction was studied is one in which the charge changes only slowly so that the assumption is reasonable here. (Even so there are small deviations between calculated and observed data towards the ends of the bell-shaped curves, such as those of Fig. 35–3, and they probably arise as a result of the neglect of the electrostatic interactions with the charged groups not immediately involved in the reaction.)

The dissociation of the two important acidic groups of ES and EP can be taken into account in the same way. Where $^+HESH^+$, ESH^+, ES_-, $^+HEPH^+$, EPH^+, and EP_- are the forms involved and K_{1ES}, K_{2ES}, K_{1EP}, and K_{2EP} the corresponding constants, we get, with the same assumptions as before

$$[ES] = [^+HESH^+] + [ESH^+] + [ES_-]$$

$$[ESH^+]a_{H^+}/[^+HESH^+] = K_{1ES} \qquad (35\text{--}17)$$

$$[ES_-]a_{H^+}/[ESH^+] = K_{2ES}$$

and a corresponding set of equations for EP.

The crucial hypothesis of the mechanism is that only ESH^+ and EPH^+ can be interconverted. The forms of ES and EP in which both dissociable groups bear protons, or in which both have lost their protons, are inactive. Thus reactions 3 and 4 of the mechanism on p. 644 can occur only for those molecules of ES and EP which are in the forms ESH^+ and EPH^+. By equation 35–17 we can evaluate the fractions,

$$f_{ES} = \frac{[ESH^+]}{[ES]} = \frac{1}{1 + a_{H^+}/K_{1ES} + K_{2ES}/a_{H^+}} \qquad (35\text{--}18)$$

$$f_{EP} = \frac{[EPH^+]}{[EP]} = \frac{1}{1 + a_{H^+}/K_{1EP} + K_{2EP}/a_{H^+}} \qquad (35\text{--}19)$$

where [ES] and [EP] represent total concentrations of all forms of ES and EP. Then, if k_3^0 and k_4^0 are the rate constants for interconversion of ES and EP which would obtain if *all* ES and EP molecules were in their active form,

$$k_3 = k_3^0 f_{ES} \qquad (35\text{--}20)$$

$$k_4 = k_4^0 f_{EP} \qquad (35\text{--}21)$$

It remains to consider the pH dependence of the other rate constants, k_1, k_2, k_5, and k_6. It is most unlikely that these constants can be pH independent. Considering k_1, for example, it is to be expected that fumarate ions will not combine equally readily with $^+HEH^+$, EH^+, and E_-. Since k_1 is the rate constant for the combination of fumarate with all forms of the free enzyme, its value will therefore change as the pH affects the relative proportions of the three individual forms.

This situation will greatly complicate the rate equations to be derived, except under one favorable set of circumstances. If, when the steady state for the forward reaction is reached, the system $E + S \rightleftharpoons ES$ is in *equilibrium*, then k_1/k_2 represents an equilibrium constant, and its pH dependence is determined entirely by the other equilibrium constants (K_{1E}, etc.) of the system. The complete equilibrium can be set up in terms of a single pair of pH-independent rate constants, k_1^0 and k_2^0, as follows

$$
\begin{array}{ccc}
S + {}^+HEH^+ & & {}^+HESH^+ \\
K_{1E} \big\updownarrow & & \big\updownarrow K_{1ES} \\
S + EH^+ & \underset{k_2^0}{\overset{k_1^0}{\rightleftharpoons}} & ESH^+ \\
K_{2E} \big\updownarrow & & \big\updownarrow K_{2ES} \\
S + E_- & & ES_-
\end{array}
$$

where k_1^0 and k_2^0 could, of course, have been equally well assigned to the reactions $S + {}^+HEH^+ \rightleftharpoons {}^+HESH^+$ or $S + E_- \rightleftharpoons ES_-$ without in any way affecting the result to be obtained. An exactly parallel situation exists for the reverse reaction. If, when the steady state is reached, the system $E + P \rightleftharpoons EP$ is in equilibrium, then the effect of pH on k_5 and k_6 is completely described in terms of a pair of constants, k_5^0 and k_6^0, which may be assigned to the reaction $EPH^+ \rightleftharpoons EH^+ + P$.

Whether this simplifying situation can actually be applied to the system depends on the relative values of k_2 and k_3 and of k_4 and k_5. If $k_2 \gg k_3$, the system $S + E \rightleftharpoons ES$ will be in equilibrium for the forward reaction; if $k_5 \gg k_4$, the system $E + P \rightleftharpoons EP$ will be in equilibrium for the reverse reaction. We shall accordingly assume that this is the case, without, however, replacing k_1^0/k_2^0 and k_6^0/k_5^0 by single equilibrium constants. By keeping the separate rate constants k_1^0, etc., in the kinetic scheme we shall be able to test the validity of the assumptions made by finding whether the values which these constants must have to fit the experimental data are such that $k_2^0 \gg k_3^0$ and $k_5^0 \gg k_4^0$. If so, the assumption made will be justified.

It may be noted that the analysis of the problem as presented here differs formally from the analysis given in the original papers of Alberty and coworkers. We have chosen to make use of information not obtained directly from the experimental data, e.g., we have considered it intrinsically unlikely that S can combine rapidly with EH^+ unless it can also combine rapidly (but with a different rate constant) with ${}^+HEH^+$ and E_-. To obtain a simple relation between v and pH, however, we want, if possible, to use a single pH-independent rate constant for reactions 1, 2, 5, and 6. To achieve this, we are forced to assume that equilibrium among S, E, and H^+ is maintained throughout the reaction. Alberty and coworkers have not made any assumption of this kind but have limited themselves to seeking the simplest possible steady-state formulation which will account for the experimental data *without other considerations*. They conclude that the experimental data may be described by use of pH-independent rate constants for reactions 1, 2, 5, and 6, but they do not feel that an explanation is warranted.

In principle, the problem which arises here could be solved unequivocally by use of the rapid flow technique discussed in section 36a. This technique would permit direct determination of k_1, k_2, k_5, and k_6 and their dependence on pH. Unfortunately the rates of these reactions are too fast for application of the flow method.

With the assumptions made we can write for the rate constants for the over-all rates of formation of ES from E and S, etc.,

$$\begin{aligned}
k_1 &= k_1{}^0 f_E \\
k_2 &= k_2{}^0 f_{ES} \\
k_5 &= k_5{}^0 f_{EP} \\
k_6 &= k_6{}^0 f_E
\end{aligned} \tag{35–22}$$

where, analogous to equations 35–18 and 35–19,

$$f_E = \frac{[EH^+]}{[E]} = \frac{1}{1 + a_{H^+}/K_{1E} + K_{2E}/a_{H^+}} \tag{35–23}$$

where [E] is the concentration of all forms of the free enzyme.

Combining equation 35–9, with $[X] = 1$, with equations 35–20 to 35–22, we find that f_{ES} and f_{EP} cancel in the ratio $(v_{max})_F/(K_m)_F$; i.e.,

$$\frac{(v_{max})_F}{(K_m)_F} = \frac{k_1 k_3 k_5 [E]_0}{k_2 k_4 + k_2 k_5 + k_3 k_5} = \frac{k_1{}^0 k_3{}^0 k_5{}^0 [E]_0 f_E}{k_2{}^0 k_4{}^0 + k_2{}^0 k_5{}^0 + k_3{}^0 k_5{}^0} \tag{35–24}$$

and, similarly, by equation 35–11,

$$\frac{(v_{max})_R}{(K_m)_R} = \frac{k_2{}^0 k_4{}^0 k_6{}^0 [E]_0 f_E}{k_2{}^0 k_4{}^0 + k_2{}^0 k_5{}^0 + k_3{}^0 k_5{}^0} \tag{35–25}$$

Thus the ratio v_{max}/K_m, both in the forward and reverse direction, depends only on f_E, i.e., by equation 35–23, only on K_{1E} and K_{2E}. The pH dependence of this ratio should therefore be the same in both directions (a result which could also have been obtained from equation 35–15), and from it K_{1E} and K_{2E} can be determined. The required experimental plots are shown in Fig. 35–5. The values of the constants (cf. Table 35–1) are $pK_{1E} = 6.2$, $pK_{2E} = 6.8$.

The remaining constants of the system can now be determined from the pH dependence of v_{max} and K_m (Figs. 35–3 and 35–4), as described in detail by Alberty and Peirce.[41] It is found that unique values cannot be

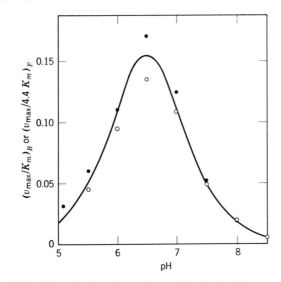

Fig. 35–5. The effect of pH on the ratio v_{max}/K_m for the fumarase reaction. Solid circles are for the forward reaction and open circles for the reverse reaction. According to equations 35–24 and 35–25, the pH dependence should be the same for both reactions, in essential agreement with the experimental result. (Alberty.[40])

TABLE 35–1. Typical Values for the Rate and Equilibrium Constants of the Fumarase Reaction at $25°$ C[a]

pK_{1E}	6.2	k_2^0	$>27 \times 10^3$ sec^{-1}
pK_{2E}	6.8	k_3^0	2.4×10^3 sec^{-1}
pK_{1ES}	<5.3	k_4^0	1.8×10^3 sec^{-1}
pK_{2ES}	7.3	k_5^0	$>46 \times 10^3$ sec^{-1}
pK_{1EP}	6.6	k_1^0	$>11 \times 10^9$ (M/L)$^{-1}$ sec^{-1}
pK_{2EP}	>8.5	k_6^0	$>5 \times 10^9$ (M/L)$^{-1}$ sec^{-1}

[a] These values apply to the reaction in buffers prepared from tris-(hydroxymethyl)-amino methane and acetate, at ionic strength 0.01. Values in other buffer systems, or at other ionic strengths, show minor differences.

obtained for all the constants. At the summary in Table 35–1 shows, only maximum or minimum values can be evaluated for some of them.

An important conclusion which can be drawn from the values of pK_{1E} and pK_{2E} is that the two critical dissociable groups of the active site of the enzyme are likely to be identical and that they are probably imidazole groups contained on histidine side chains of the enzyme. For the dissociation of two identical groups may be represented by four identical dissociation constants (neglecting, as we have done, all electrostatic effects) for the reactions $^+$HEH$^+ \rightleftharpoons {}^+$HE + H$^+$, $^+$HEH$^+ \rightleftharpoons$ EH$^+$ + H$^+$, $^+$HE $\rightleftharpoons$ E$_-$ + H$^+$, EH$^+ \rightleftharpoons$ E$_-$ + H$^+$. In the kinetic scheme used EH$^+$ and $^+$HE have been treated as a single species, so that $K_{1E} = ([^+$HE] + [EH$^+$])$a_{H^+}/[^+$HEH$^+$] should have twice the value of the single constant common to the four separate dissociations. Similarly, K_{2E} should have half the value of this common constant. Thus $K_{1E} = 4K_{2E}$ or $pK_{2E} - pK_{1E} = 0.60$, exactly as found. That the sites are probably imidazole groups follows from the fact that the isoelectric point of fumarase, under the conditions of the experiment, lies somewhere between pH 6 and 7. The net charge of the protein is therefore low throughout the pH region of interest, so that the pK values correspond closely to the *intrinsic* pK values for the groups involved (cf. section 30). Only imidazole groups are typically found to have intrinsic pK values near 6.5.

Table 35–1 shows that the two dissociable groups are no longer identical in the complexes ES and EP; i.e., $pK_{2ES} - pK_{1ES}$ and $pK_{2EP} - pK_{1EP}$ are considerably greater than 0.6. No attempt has yet been made to give a quantitative interpretation to these differences and to the difference between the dissociation constants of the complexes and pK_{1E} or pK_{2E}.

Two important conclusions may be drawn from the values of the rate constants which appear in Table 35–1. One of them is that even the minimum values of $k_2{}^0$ and $k_5{}^0$ are much larger than the values of $k_3{}^0$ and $k_4{}^0$. The assumption made earlier, that the pH dependence of the formation of ES from E and S, and of EP from E and P, can be evaluated from equilibrium constants alone, is therefore proved to be valid.

Also noteworthy are the very large values of k_1 and k_6. They indicate that a large fraction of the collisions between E and S (or E and P) must lead to formation of ES (or EP).

35e. Detailed mechanism for the reaction. The results presented above allow formulation of a tentative detailed mechanism for the reaction, which is shown in Fig. 35–6. The active site in the free enzyme must contain two dissociable groups (assumed to be imidazole groups), and, in the pH range in which the enzyme is active, they will spend part of their time in the form ImH$^+$ and part in the form Im. The active site must contain other as yet unknown structural features which permit the binding of fumarate or malate ion. The high values of k_1 and k_6 suggest that the entire spatial configuration is rigidly fixed, so that combination with fumarate and malate can occur without the necessity for rearrangement at the site. In the free enzyme water molecules presumably occupy the space later filled by fumarate or malate.

In the form ES the two dissociable sites are no longer equivalent, a fact which may be explained by the incorporation of a single water molecule (that required for the hydration reaction) within the complex. The very low value found for pK_{1ES} (Table 35–1) may be due to the fact that this water molecule can fit into the available space only if its H atoms are directed towards the imidazole groups. Such an orientation would tend to repel an additional proton at this site. (Such an orientation would also greatly facilitate the subsequent steps of the reaction.)

An attractive hypothesis to account for the fact that conversion from ES to EP can occur only if one of the imidazole groups is in the form ImH$^+$ while the other is in its basic form, Im, is to suppose that these

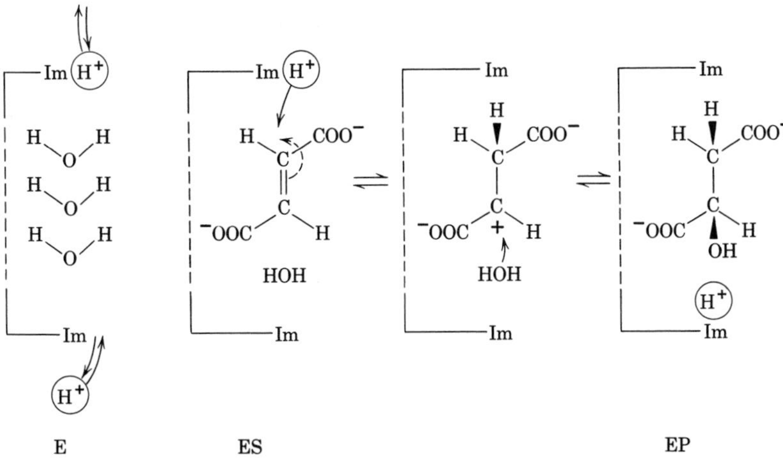

Fig. 35–6. Schematic formulation of a detailed mechanism for the fumarase reaction. The line joining the two imidazole groups represents the adjacent portions of the enzyme molecule.

groups participate directly in the reaction mechanism, most simply by transfer of the proton from ImH$^+$ to Im. To accomplish this requires a further intermediate which can be regarded as the activated complex for the reaction. A carbonium ion structure is the most reasonable form for this intermediate. (Experiments involving deuterium incorporation[42] suggest that this intermediate may be a stable intermediate rather than the activated complex. The over-all kinetics would not be affected if this is the case.)

The detailed mechanism here suggested provides a possible answer to the questions raised at the beginning of this section. Why are protein molecules alone able to act as enzymes? Why are they so specific in their choice of substrates? The mechanism suggests that the enzymatic site must be highly rigid in its spatial configuration. A large rigid structure is therefore needed to act as a framework on which the various parts of the active site can be mounted. Of all the macromolecules which have been encountered in this book proteins alone would seem able to fill this role. The rigidly specified dimensions of the active site also account for the specificity. A spatial configuration precisely tailored for the fumarate–malate reaction will obviously not be able to accommodate other molecules which may resemble fumarate or l-malate chemically but will differ in their space requirements.

It is interesting in this connection that the fumarase reaction is stereospecific. Only l-malate is formed as product of the forward reaction, and only l-malate is capable of undergoing the reverse reaction. The difference between the isomers l-malate and d-malate is illustrated by the structural formulas

$$
\begin{array}{cc}
\mathrm{C} & \mathrm{C} \\
| & | \\
\mathrm{C} & \mathrm{C} \\
{}^{-}\mathrm{OOC}\diagup\ \ \diagdown\mathrm{H} & {}^{-}\mathrm{OOC}\diagup\ \ \diagdown\mathrm{OH} \\
\mathrm{OH} & \mathrm{H}
\end{array}
$$

and that only one of them can fit into the reaction scheme is a logical consequence of the specific structure of the enzymatic site. It will be shown in section 36b that the reaction at the opposite end of the molecule (i.e., the proton transfer from $\mathrm{ImH^+}$ to fumarate) is also stereospecific.

36. SPECIAL TECHNIQUES

All the kinetic studies reported in the preceding sections have consisted experimentally of the preparation of a suitable homogeneous mixture of reagents and the determination, over a period of time, of the changing composition of the mixture. Theoretical analysis of the results has involved the assumption that a steady state is reached early in the reaction, before any measurements have been made.

This procedure is undoubtedly the most widely used method in the study of reaction kinetics. However, it has its drawbacks, chiefly because there are often several plausible reaction mechanisms which could lead to the same behavior in the steady state. Examples of the kind of uncertainty which can arise are provided by two of the reactions studied in the earlier sections of this chapter. In the degradation of DNA (section 33c) three quite different mechanisms were found to be capable of explaining the progress of the over-all reaction. In the study of the action of fumarase (section 35) a range of rate constants for formation of the enzyme substrate complexes could be accommodated by the observed data.

For this reason special procedures are often employed which provide a more intimate view of the progress of a reaction. Two such procedures of general applicability will be briefly described in this section: (*a*) the flow technique for measurement of the rate of very rapid reactions and (*b*) the use of isotopes in rate studies.

36a. Kinetics of fast reactions.[7] Certain chemical reactions are so rapid that ordinary methods of mixing of solutions and of following the progress of reaction fail. For such reactions a flow technique was developed in 1923 by Hartridge and Roughton,[43] which, with modifications, continues to be the principal method by which very fast reactions are studied. The same method can be used for observing the rapid approach to the steady state in reactions which, once the steady state is attained, proceed sufficiently slowly to be studied in the ordinary way. (It is of interest to note that the

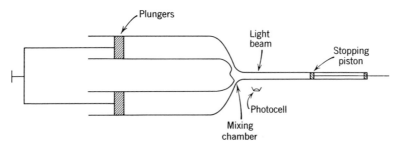

Fig. 36–1. Schematic diagram of the apparatus used to measure the rates of very rapid reactions.

method was originally developed for the study of a macromolecular reaction, the combination of hemoglobin with oxygen.)

The principle of the method is illustrated by Fig. 36–1. The simplest form of the apparatus consists of two syringes, which lead into a mixing chamber designed so as to produce a maximum of intermixing between the entering solutions. A narrow observation tube leads from the mixing chamber past a point at which observations can be made. The commonest technique is to use light absorption as the method of observation. A light beam traverses the observation tube at the chosen position, and its intensity is measured by a photocell mounted behind the tube. Electronic devices may be employed to record the rate at which the syringe plungers are depressed and to record the optical density at the observation point either as a function of time or, by use of a scanning procedure, as a function of wavelength.

When the syringe plungers are depressed the solutions in the two syringes are placed in motion, at first slowly, because of the inertia of the system, but eventually more rapidly. The two solutions are mixed and forced down the observation tube. If this tube has a narrow diameter, the rate of travel down it can be quite rapid, say 1000 cm/sec. The freshly mixed solution can therefore reach the observation point, which may be placed one or a few centimeters from the mixing chamber, in about 0.001 sec. As long as flow continues, the solution at the point of observation will be continuously replaced by fresh solution, which, because of the progressive acceleration of the syringe plungers, will have taken a progressively shorter time to reach the observation point. Once the maximum flow velocity has been reached the time of travel between the mixing chamber and the observation point remains constant, and the solution under observation retains identical composition. It may be examined for a relatively "long" period (many milliseconds), the only limitation being the length of the syringes and the need to exercise economy in the use of reagents.

For most kinetic studies flow is stopped when the maximum flow velocity is reached. This can be done by halting the depression of the syringe plungers. More rapid stopping of the flow can be provided by a stopping piston at the end of the observation tube (cf. Fig. 36–1), which automatically halts the flow when the tube is full. Following the stopping of flow, light absorption or other measurements at the observation point can be made over as long a time period as is desired.

Spectrophotometry is not the only method which has been used for determining the state of reaction at the observation point. Alternatives are: heat evolution, measured with the aid of thermocouples; pH or oxidation-reduction potential, measured with the aid of suitable electrodes; polarography; etc. The apparatus may also be modified to allow for the introduction of several different reagents in succession.

A recent review by Roughton and Chance[7] describes in some detail the various types of apparatus which have been used. A discussion of the theory has been given by Chance.[44] Fast reactions in general have been the subject of a recent Faraday Society discussion.[45]

36b. Kinetics of the enzyme action of peroxidase.[46,47,48] The most intensive use of the rapid reaction technique has been by Chance in his investigation of the kinetics of action of catalases and peroxidases. These enzymes are heme proteins, and they catalyze the reduction of peroxides. Being heme proteins, they absorb visible light (cf. section 5f). Their absorption spectra are sensitive to changes in the oxidation state of the heme iron atom and to the nature of ions or molecules which may be attached to the iron atom. It is thus to be expected and confirmed by experiment that spectral changes will accompany the enzymatic process during which, as described in section 35, the enzyme becomes transformed into one or more enzyme-substrate complexes.

We shall briefly summarize here the work on one of these enzymes, *horseradish peroxidase*. It catalyzes irreversible reactions of the type

$$HOOR + AH_2 \rightarrow A + H_2O + ROH$$

where HOOR may represent H_2O_2 or an alkyl peroxide, and AH_2 can be one of a number of reducing agents such as an amine, a phenol, or an enediol. The description of the reaction here given will not only serve as an illustration of the fast reaction method but will also provide a second example of an enzyme-catalyzed reaction, one which will be seen to differ in many respects from the example discussed in section 35.

Chance's kinetic studies provide direct evidence for the kind of mechanism for enzyme action which was postulated in section 35. The spectrum changes immediately upon initiation of the reaction, indicative of the formation of an enzyme-substrate complex, and returns to its original state at the end of the reaction, indicative of the ultimate release of unchanged enzyme molecules. In the peroxidase reaction (as in the fumarase reaction) there must in fact be two separate enzyme-substrate complexes, for the change in optical density follows a different course at different wavelengths so as to require at least two absorbing species different from the free enzyme E. We shall call them ES_I and ES_{II}, postponing for the moment a positive identification of the molecular changes involved.

The experiments to be described in Figs. 36–2 and 36–3, using H_2O_2 with ascorbic acid as reducing agent, were performed under conditions of substrate excess, so that no free enzyme, E, is present at the steady state. Throughout the steady state, as Fig. 36–3 shows, all the enzyme is present as ES_{II}.

Figure 36–2 shows the approach of the system to the steady state, as measured by the spectral changes at the *isosbestic* points of the system. These points occur at the wavelengths at which two of the enzyme species involved in the reaction have identical molar absorbancies, and they are found experimentally, by trial and error, as wavelengths at which no change in absorption occurs during a portion of the reaction. Such isosbestic points occur in the present system at 395 mμ (ES_I and ES_{II} indistinguishable), at 410 mμ (E and ES_{II} indistinguishable) and at 427 mμ (E and ES_I indistinguishable). Changes in absorption at each of these wavelengths clearly represent the appearance or disappearance of a single species, interconversion of the other two being accompanied by no spectral change. The spectral records shown for illustration in Fig. 36–2 were obtained in dilute solutions in which the rate of reaction is so slow that no appreciable reaction occurs during the period of accelerated flow. All records are stopped flow records, the initial state being free enzyme, observed a few milliseconds after mixing with substrate.

The trace at 395 mμ shows the disappearance of free enzyme and no further change during the period during which ES_I is converted to ES_{II}.

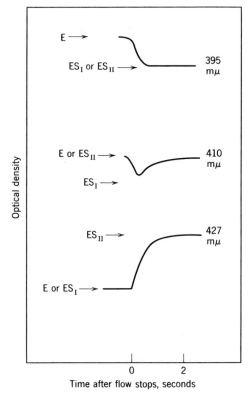

Fig. 36–2 Initial stages of the decomposition of H_2O_2 catalyzed by horseradish peroxidase. Each wavelength is that of an isosbestic point, i.e., a point where two of the forms, E, ES_I, and ES_{II}, have identical absorption coefficients. The wavelength 410 mμ is the isosbestic point for E and ES_{II}, and the trace at that wavelength clearly shows the transient appearance of ES_I. (Chance.[46])

This reaction is observed in the absence of the donor molecule AH_2 as well as in its presence. Furthermore, the rate of formation is directly proportional to the peroxide concentration. Thus ES_I is clearly an enzyme-peroxide complex, and its formation a second-order process. Since the active site of the peroxidase molecule is presumably the iron atom of its heme group and since a water molecule is believed to be attached to the iron atom in the free enzyme, the first step of the reaction can thus be represented as

$$E(Fe{-}OH_2) + H_2O_2 \underset{k_2}{\overset{k_1}{\rightleftharpoons}} E(Fe{-}O_2H_2) + H_2O$$

This reaction is written as a reversible reaction since reactions of this type are in principle always reversible. In actual fact, however, the reverse reaction can never be observed. The subsequent reaction of ES_I to form ES_{II} is always faster than reaction 2 (i.e., that for which k_2 is the rate constant).

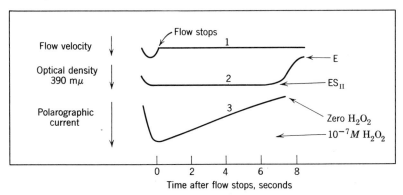

Fig. 36–3. Kinetic data for the same reaction as in Fig. 36–2, obtained at much higher concentration. Curve 1 shows the liquid flow velocity in the apparatus; curve 2 shows the absorption at 390 mμ (ES_I and ES_{II} indistinguishable); curve 3 shows the concentration of H_2O_2, measured polarographically. The initial stage of the reaction is too fast to be detected here, and what the data show is the disappearance of H_2O_2 while ES_{II} maintains a constant, steady-state concentration. At the end of the reaction ES_{II} disappears, and free peroxidase is regenerated. (Chance.[46])

The 427 mμ trace of Fig. 36–2 represents the formation of ES_{II}, since E and ES_I are indistinguishable at this wavelength. This reaction proceeds slowly in the absence of AH_2, with an impurity or a second molecule of H_2O_2 presumably acting as reducing agent. In the presence of donor molecules this reaction is much more rapid, and its rate is proportional to the concentration of AH_2. It is therefore a second-order process involving AH_2, and it is thought to represent the process

$$E(Fe{-}O_2H_2) + AH_2 \overset{k_3}{\longrightarrow} E(Fe{-}OH\cdot)AH\cdot + H_2O$$

the two dots in the product representing unpaired electrons. The reaction is written as a one-electron reduction since a second molecule of AH_2 is required to complete the reduction process. (For a full discussion see Chance and Ferguson.[48])

Reaction 3 is very fast, and this is reflected in the 410 mμ trace of Fig. 36–2. At this wavelength E and ES_{II} have the same absorbance, so that the trace represents the extent to which ES_I is present. As Fig. 36–2 shows, the first intermediate is present for a very brief period only.

Figure 36–3 shows kinetic data for the same reaction at considerably higher concentrations. Under these conditions the reactions described above have already occurred during the flow to the point of observation, and all the enzyme is in the form ES_{II} when flow is stopped. The major portion of the figure represents the steady-state stage of the reaction, and during it the spectrophotometric trace, which is a measure of the relative concentrations of E and ES_{II}, remains unchanged. There is a change only at the end of the reaction, when all the substrate has been used up. The change there clearly shows the reappearance of the free enzyme.

Figure 36–3 shows not only a spectrophotometric trace but also a record of the polarographic current measured at a potential of $+0.8$ volt. This current is a measure of the concentration of peroxide, which is seen to disappear at a constant rate during the steady-state period. This rate is a measure of the rate of the final step of the reaction mechanism. For during the steady state $d[E]/dt = 0$, so that $-d[E]/dt$ by the net effect of reactions 1 and 2, which is necessarily equal to $-d[H_2O_2]/dt$, must be exactly equal to the rate at which free enzyme is regenerated. Variation in the concentration of AH_2 (in this case ascorbic acid) shows that the steady-state rate of disappearance of H_2O_2 at a constant enzyme concentration is proportional to the concentration of AH_2. The final stage of the reaction therefore presumably represents the process

$$E(Fe-OH\cdot)AH\cdot + AH_2 \xrightarrow{k_4} E(Fe-OH_2) + AH_2 + A$$

The individual rate constants of the reaction may be calculated from data such as those of Figs. 36–2 and 36–3. The value of k_4, for example, can be calculated from the rate of disappearance of H_2O_2 during the steady state, the concentration of ES_{II} being (under the experimental conditions shown) equal to the total enzyme concentration, and AH_2 being present at known concentration. The rate constant of this reaction can also be obtained independently from the spectrophotometric trace at the end of the reaction in Fig. 36–3, since this trace is a direct measure of the rate of disappearance of ES_{II} after all the H_2O_2 has been used up. The data of Fig. 36–3 give $k_4 = 8600$ and 9100 (moles/liter)$^{-1}$ sec^{-1}, respectively, by the two methods of calculation.

The rate constant k_1 is determined directly from the initial phases of the 395 mμ trace of Fig. 36–2. The entire trace at this wavelength involves also the reverse of reaction 1, i.e., the rate constant k_2. It appears, however, that k_2 is too small for this reaction to occur to a sufficient extent to influence the observed curve; i.e., the entire curve can be fitted with a single value of k_1 and with $k_2 = 0$.

The rate constant k_3 cannot be determined from the rate of reaction 3, as given by the 427 mμ trace of Fig. 36–2, since over most of the time covered by that trace $[ES_I] \simeq 0$ and the rate of reaction 3 is determined by the rate of reaction 1. This rate constant can, however, be obtained from the 410 mμ trace, which (with k_2 set equal to zero) represents the function

$$k_1[E][H_2O_2] - k_3[ES_1][AH_2].$$

For horseradish peroxidase and H_2O_2 Chance obtained $k_1 = 9 \times 10^6$ (moles/liter)$^{-1}$ sec^{-1}. The rate of formation of the initial enzyme-substrate complex is thus seen to be much slower in this reaction than the corresponding rate for the fumarase reaction (cf. Table 35–1). The fact that the rate of reaction 3 is generally faster than the rate of reaction 2 is also a point of distinction between this reaction and the fumarase reaction. In the latter the enzyme-substrate complex dissociates and reforms many times before transition to the second complex occurs.

The values of k_3 and k_4 depend, of course, on which reducing agent is being employed. Generally k_3 is larger (otherwise the complex ES_{II} would not be present at high concentration in the steady state), typical values being 5×10^4 and 2×10^3 (moles/liter)$^{-1}$ sec^{-1}, respectively, for k_3 and k_4.

Perhaps the most striking difference between the action of peroxidase and that of fumarase is that all the rate constants for peroxidase are independent of pH over a wide range of pH, probably over the entire range of stability of the protein. This is presumably a reflection of the fact that the active site of the enzyme is the iron atom of its heme group and that none of the dissociable groups of the enzyme are directly involved. It also means that none of the reaction steps involve participation of H^+ or OH^- ions. Another difference is that peroxidase is less specific than fumarase. Not only organic molecules but also inorganic ions such as NO_2^- and I^- can fill the role of AH_2 in the reaction scheme. Moreover, a whole series of alkyl peroxides can be utilized.

The features of the peroxidase reaction which have been discussed provide (again in contrast to fumarase) no easy clue as to why a macromolecule is needed to carry out the catalysis. This problem is further complicated by the fact that the enzyme catalase,[47] which, like peroxidase, can catalyze reactions of the type $HOOR + AH_2 \rightarrow ROH + A + H_2O$ and which is also a heme protein of the same valence and bond type as peroxidase (cf. p. 92), prefers different reducing agents from peroxidase. Moreover, its mechanism involves a single step two-electron transfer instead of two successive one-electron transfers.[48] It is probable that a better understanding of the structure of iron–porphyrin compounds in general will be a necessary prerequisite for a full explanation of the action of these enzymes.

36c. Kinetics of isotope exchange. Another useful aid to the kineticist is the availability of isotopes of some of the elements, which may be introduced into the reacting molecules at specified positions. Molecules so labeled can be utilized in two ways: (1) If the light or heavy atom in question has no opportunity for being exchanged for a normal atom at some stage of the reaction, it can be used as a tracer. Its position in the product molecule can be identified, and we can then be certain of the origin of that particular atom of the product molecule. (2) If, on the other hand, the opportunity for exchange exists, we can gain useful information from the kinetics of exchange.

We shall discuss here only the second kind of experiment and confine ourselves to the exchange between hydrogen and deuterium atoms in water solution, i.e., to the exchange which may occur when a molecule containing H atoms is placed in D_2O or when a molecule containing D atoms is placed in H_2O. The general principle which governs such exchange is that atoms which belong to acid or basic groups (i.e., H or D atoms attached to oxygen or nitrogen) normally exchange very rapidly whereas H or D atoms attached to carbon do not exchange except at elevated temperatures. The acidic or basic groups need not have measurable dissociation constants in the ordinary pH range for rapid exhange to occur. Thus in aliphatic alcohols the equilibrium $-OH \rightleftharpoons -O^- + H^+$ lies virtually completely to the left under all ordinary experimental conditions, and in amides the equilibrium $-CO-NH_3^+ \rightleftharpoons -CO-NH_2 + H^+$ lies virtually completely to the right and the equilibrium $-CO-NH_2 \rightleftharpoons -CO-NH + H^+$ virtually completely to the left. However, these equilibria are dynamic; reaction in both directions takes place all the time so that exchange is fast.

An example of the application of these principles is the method used by Alberty and coworkers[49] to prove that the proton addition to fumarate (i.e., the addition of the proton at the top of the formulas in Fig. 35–6) is stereospecific. To do this they carried out the reaction, starting with normal fumarate, in D_2O rather than H_2O. The reaction was allowed to proceed to equilibrium. During the relatively long period required to attain equilibrium most of the substrate molecules must pass through several cycles of the reaction fumarate $\rightleftharpoons$ l-malate.

In D_2O as solvent, all the acidic and basic groups of the protein enzyme will acquire D atoms in place of H atoms. The first conversion of fumarate to l-malate will thus lead to transfer of a D^+ ion in place of the H^+ ion shown in Fig. 35–6. The hydroxyl group added at the other end of the molecule comes from the solvent and will, of course, be an OD group, so that the first product is $^-OOC—CDH—CHOD—COO^-$. The addition and removal of the hydroxyl group is stereospecific (otherwise the product molecule would be a mixture of l- and d-malate). If the same is true for the hydrogen ion, then, the first dehydration of the product will necessarily lead to removal of the same ion as that originally added. The D^+ ion will be the one removed and fumarate containing no D isomer will be reformed. Each cycle of the reaction will have the same result, and, at equilibrium, the final products will be $^-OOC—CH{=}CH—COO^-$ and $^-OOC—CDH—CHOD—COO^-$. To determine the deuterium content of the products all adhering D_2O must be removed by washing with H_2O. During this process the OD group of l-malate is converted to an OH group, leaving $^-OOC—CDH—CHOH—COO^-$.

On the other hand, the addition of the hydrogen ion may not be stereospecific. This would mean that the carboxyl group at the top of the molecule could lie on either side of the C—C bond during the addition. In that case the first dehydration would remove an H^+ ion from some of the molecules of $^-OOC—CDH—CHOD—COO^-$ and a D^+ ion from others, forming a mixture of $^-OOC—CD{=}CH—COO^-$ and $^-OOC—CH{=}CH—COO^-$. Since the solvent is pure D_2O, the H^+ ions which can be introduced into the solvent in this kind of reaction form a negligible fraction of the total H and D content, so that the second and all subsequent hydration reactions will still introduce D_2O into all but a negligible fraction of the molecules. The product of the second hydration will then be a mixture of $^-OOC—CD_2—CHOD—COO^-$, $^-OOC—CHD—CDOD—COO^-$, and $^-OOC—CHD—CHOD—COO^-$. The last two forms, on dehydration, may again lose H^+ ions rather than D^+ ions, and it is clear that continued repetition of the cycle will lead to virtually complete elimination of H^+. After isolation of the equilibrium mixture and washing with H_2O, the expected products are $^-OOC—CD{=}CD—COO^-$ and $^-OOC—CD_2—CDOH—COO^-$.

The experimental result was that the fumarate isolated from the equilibrium mixture contained less than 10^{-4} D atoms per molecule, while the l-malate contained 0.97 D atoms per molecule. The result proves unequivocally that the reaction is stereospecific.

36d. Deuterium–hydrogen exchange by globular proteins. Another interesting application of deuterium exchange has been developed by Linderstrøm Lang.[50] Protein molecules are dissolved in D_2O and kept in contact with it for a long period, during which all H atoms attached to nitrogen or oxygen should be replaced by D atoms. The protein is then frozen and dried by exhaustive evacuation. During this process all D_2O is removed.

The protein is then redissolved in H_2O, and the rate of appearance of D atoms in the H_2O is measured.

When this experiment is carried out with polypeptides or proteins which are in a more or less randomly coiled configuration (e.g., ribonuclease which has been denatured by guanidine), all the theoretically expected D atoms are exchanged instantaneously. In other words, the number of D atoms released immediately is equal to the number of protons ordinarily attached to oxygen or nitrogen atoms.[51]

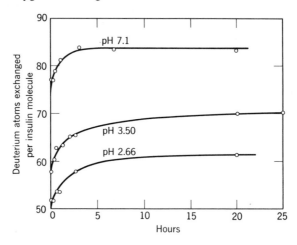

Fig. 36–4. Deuterium–hydrogen exchange for insulin. The number of exchangeable D^+ ions is theoretically 83 at pH 7.1, and about 90 at the two lower pH's. (Hvidt and Linderstrøm-Lang.[52])

However, a different result is obtained for proteins which are in their native compact configurations and also for polypeptides which exist as helical rods. Only a portion of the exchangable D atoms in these molecules is released instantaneously. The remainder is released slowly, over a period of 24 hours or more.

Figure 36–4, for example, shows exchange data for insulin.[52] The isoelectric form of this molecule should have eighty-five D atoms per molecule, of which forty-eight belong to the —CO—NH— bonds of the two polypeptide chains, six to the $-NH_3^+$ groups at the ends of the chains, and the remaining thirty-one to side chain groups. If the protein is acid to its isoelectric point, the expected number of exchangeable D atoms is increased (to a maximum of ninety-one at pH 2) because of the greater binding of H^+ or D^+ ions at lower pH. Alkaline to the isoelectric point, for the same reason, the theoretical number of atoms is decreased, e.g., to eighty-three at pH 7.1. The number of D atoms initially released, as shown by the figure, is much smaller, and the release of the remainder is seen to proceed very slowly.

The protons which fail to exchange quickly are presumably prevented by the compact structure of the molecule from direct contact with solvent. It is believed that most such protons belong to —CO—NH— peptide bonds and that their inaccessibility to the solvent is due at least in part to helical structure of the polypeptide backbone. Optical rotation data for insulin (section 6) indicate that about 50% of the polypeptide content of this molecule is in helical form. Thus about twenty-four peptide hydrogen atoms might become inaccessible to the solvent, leaving about sixty positions at which instantaneous exchange can occur. This agrees at least roughly with the observed result.

The pH effect illustrated by Fig. 36–4, showing a faster rate of exchange as the pH increases, is observed in several proteins. No simple explanation for it exists. The effect cannot be correlated with structural changes in the proteins, which implies that it must be due to direct participation of OH^- in the exchange reaction. So far, however, no reasonable mechanism has been suggested which would explain why OH^- should be better able to produce D–H exchange than H_2O molecules.

General References

1. P. J. Flory, *Principles of Polymer Chemistry*, Cornell University Press, Ithaca, 1953 Chs. III, IV.
2. H. Mark and A. V. Tobolsky, *Physical Chemistry of High Polymeric Systems*, 2nd ed., Interscience Publishers, New York, 1950, Chs. XI, XII, XIII.
3. G. M. Burnett, *Mechanism of Polymer Reactions*, Interscience Publishers, New York, 1954.
4. W. Kauzmann, "Denaturation of Proteins and Enzymes" in W. D. McElroy and B. Glass, *The Mechanism of Enzyme Action*, Johns Hopkins Press, Baltimore, 1954.
5. F. W. Putnam, "Protein Denaturation" in H. Neurath and K. Bailey (ed.), *The Proteins*, vol. IB, Academic Press, New York, 1953.
6. P. D. Boyer, H. Lardy and K. Myrbäck (ed.), *The Enzymes*, 2nd. ed., Academic Press, New York, 1959.
7. F. J. W. Roughton and B. Chance, "Rapid Reactions" in A. Weissberger (ed.), *Technique of Organic Chemistry*, vol. VIII, Interscience Publishers, New York, 1953. Ch. X.

Specific References

8. H. Edelhoch, E. Katchalski, R. H. Maybury, W. L. Hughes, Jr., and J. T. Edsall, *J. Am. Chem. Soc.*, **75**, 5058 (1953).
9. P. J. Flory, *J. Am. Chem. Soc.*, **62**, 2261 (1940).
10. P. J. Flory, *J. Am. Chem. Soc.*, **61**, 3334 (1939).
11. H. Dostal and R. Raff, *Z. physik. Chem.*, **B32**, 117 (1936); *Monatsh*, **68**, 188 (1936).
12. G. B. Taylor, *J. Am. Chem, Soc.*, **69**, 638 (1947).
13. G. V. Schulz, A. Dinglinger, and E. Husemann, *Z. physik. Chem.*, **B43**, 385 (1939).

14. L. M. Arnett, *J. Am. Chem. Soc.*, **74**, 2027 (1952).
15. G. M. Burnett and H. W. Melville, *Proc. Roy. Soc.*, **A189**, 456, 494 (1947).
16. J. H. Baxendale, S. Bywater, and M. G. Evans, *Trans. Faraday Soc.*, **42**, 675 (1946).
17. B. Baysal and A. V. Tobolsky, *J. Polymer Sci.*, **8**, 529 (1952).
18. P. J. Flory, *J. Am. Chem. Soc.*, **62**, 1561 (1940).
19. S. G. Waley and J. Watson, *Proc. Roy. Soc.*, **A199**, 499 (1949).
20. J. H. Fessler and A. G. Ogston, *Trans. Faraday Soc.*, **47**, 667 (1951).
21. E. W. Montroll and R. Simha, *J. Chem. Phys.*, **8**, 721 (1940).
22. Ref. 2, p. 469.
23. K. Freudenberg, W. Kuhn, W. Dürr, F. Bolz, and G. Steinbrunn, *Ber.*, **63**, 1510 (1930).
24. G. V. Schulz and E. Husemann, *Z. Naturforsch.*, **1**, 268 (1946).
25. R. Simha, *J. Applied Phys.*, **12**, 569 (1941).
26. C. A. Thomas, Jr., *J. Am. Chem. Soc.*, **78**, 1861 (1956).
27. R. Simha, L. A. Wall, and P. J. Blatz, *J. Polymer Sci.*, **5**, 615 (1950); R. Simha and L. A. Wall, *ibid.*, **6**, 39 (1951).
28. Ref. 3, p. 331 et seq.
29. H. W. Melville and N. Grassie, *Bull. soc., chim. Belge.*, **57**, 142 (1948); *Proc. Roy. Soc.*, **A199**, 1, 14 (1949).
30. F. Sanger, *Advances in Protein Chem.*, **7**, 1 (1952).
31. C. H. W. Hirs, W. H. Stein and S. Moore, *J. Biol. Chem.*, **221**, 151 (1956).
32. M. Levy and R. C. Warner, *J. Phys. Chem.*, **58**, 106 (1954).
33. J. Steinhardt and E. M. Zaiser, *Advances in Protein Chem.*, **10**, 186 et seq. (1955).
34. R. J. Gibbs, M. Bier, and F. F. Nord, *Arch. Biochem. Biophys.*, **35**, 216 (1952).
35. R. B. Simpson and W. Kauzmann, *J. Am. Chem. Soc.*, **75**, 5139 (1953); H. K. Frensdorff, M. T. Watson, and W. Kauzmann, *ibid.*, **75**, 5157 (1953).
36. L. Michaelis and M. L. Menten, *Biochem. Z.*, **49**, 333 (1913).
37. R. A. Alberty, *Advances in Enzymology*, **17**, 1 (1956).
38. R. M. Bock and R. A. Alberty, *J. Am. Chem. Soc.*, **75**, 1921 (1953); R. A. Alberty, V. Massey, C. Frieden, and A. R. Fuhlbrigge, *ibid.*, **76**, 2485 (1954); C. Frieden and R. A. Alberty, *J. Biol. Chem.*, **212**, 859 (1955); C. Frieden, R. G. Wolfe, Jr., and R. A. Alberty, *J. Am. Chem. Soc.*, **79**, 1523 (1957).
39. L. Michaelis and H. Davidsohn, *Biochem. Z.*, **35**, 386 (1911); L. Michaelis and H. Pechstein, *ibid.*, **59**, 77 (1914).
40. R. A. Alberty, *J. Cellular Comp. Physiol.*, **47**, 245 (1956).
41. R. A. Alberty and W. H. Peirce, *J. Am. Chem. Soc.*, **79**, 1526 (1957).
42. R. A. Alberty, W. G. Miller, and H. F. Fisher, *J. Am. Chem. Soc.*, **79**, 3973 (1957).
43. H. Hartridge and F. J. W. Roughton, *Proc. Roy. Soc.*, **A104**, 376 (1923).
44. B. Chance, *J. Franklin Inst.*, **229**, 455, 613 (1940).
45. *Discussions Faraday Soc.*, No. 17, 1954.
46. B. Chance, *Arch. Biochem.*, **21**, 416 (1949); **22**, 224 (1949); **24**, 389 (1949).
47. B. Chance, *Advances in Enzymology*, **12**, 153 (1951).
48. B. Chance and R. R. Ferguson in W. D. McElroy and B. Glass (ed.), *The Mechanism of Enzyme Action*, Johns Hopkins Press, Baltimore, (1954).
49. H. F. Fisher, C. Frieden, J. S. M. McKee, and R. A. Alberty, *J. Am. Chem. Soc.*, **77**, 4436 (1955).
50. K. Linderstrøm-Lang, *Chem. Soc.*, (*London*), *Spec. Publ.*, No. 2, 1 (1955).
51. A. Hvidt, G. Johansen, K. Linderstrøm-Lang, and F. Vaslow, *Compt. rend. trav. lab. Carlsberg, Sér. chim.*, **29**, 129 (1954); I. M. Krause and K. Linderstrøm-Lang, *ibid.*, **29**, 367 (1955); A. Hvidt and K. Linderstrøm-Lang, *ibid.*, **29**, 385 (1955).
52. A. Hvidt and K. Linderstrøm-Lang, *Biochim. et Biophys. Acta*, **18**, 306 (1955).

APPENDICES

A

MATHEMATICAL FORMULAE

1. INTEGRALS

$$\int_0^\infty e^{-ax^2} dx = \frac{1}{2}\left(\frac{\pi}{a}\right)^{1/2}$$

$$\int_0^\infty x^2 e^{-ax^2} dx = \frac{1}{4a}\left(\frac{\pi}{a}\right)^{1/2}$$

$$\int_0^\infty x^4 e^{-ax^2} dx = \frac{3}{8a^2}\left(\frac{\pi}{a}\right)^{1/2}$$

$$\int xe^{ax} dx = \frac{e^{ax}(ax-1)}{a^2}$$

It will be noted that $\int_1^{-1} xe^{ax} dx = (e^a - e^{-a})/a^2 - (e^a + e^{-a})/a$, so that equation 6-6 on p. 108 becomes $(e^a + e^{-a})/(e^a - e^{-a}) - 1/a = \coth a - 1/a$. This function is known as the Langevin function, $\mathscr{L}(a)$. For small a, expressing the exponentials as infinite series, $\mathscr{L}(a) = a/3 - a^3/45 + \cdots$.

2. BINOMIAL EXPANSION

For $x < 1$, $(1-x)^{-n}$ can be expanded as an infinite series,

$$(1-x)^{-n} = 1 + nx + \frac{n(n+1)}{2!}x^2 + \cdots + \frac{(n+i-1)!}{(n-1)!\,i!}x^i + \cdots$$

This relation has been used to obtain the following results, applicable to derivations in sections 8, 9, and 32:

$$1/(1 - x) = 1 + x + x^2 + x^3 + \cdots$$

$$1/(1 - x)^2 = 1 + 2x + 3x^2 + 4x^3 + \cdots = \sum_{i=2}^{\infty} (i - 1)x^{i-2}$$

$$1/(1 - x)^3 = 1 + 3x + 6x^2 + 10x^3 + \cdots = \tfrac{1}{2} \sum_{i=2}^{\infty} (i^2 - i)x^{i-2}$$

$$1/(1 - x)^4 = 1 + 4x + 10x^2 + 20x^3 + \cdots = \tfrac{1}{6} \sum_{i=2}^{\infty} (i^3 - i)x^{i-2}$$

$$(1 + x)/(1 - x) = 1 + 2x + 2x^2 + 2x^3 + \cdots$$

$$(1 + x)/(1 - x)^3 = 1 + 4x + 9x^2 + 16x^3 + \cdots = \sum_{i=1}^{\infty} i^2 x^{i-1}$$

$$(2 + x)/(1 - x)^4 = 2 + 9x + 24x^2 + 50x^3 + \cdots = \tfrac{1}{2} \sum_{i=2}^{\infty} (i^3 - i^2)x^{i-2}$$

3. FIRST-ORDER LINEAR DIFFERENTIAL EQUATIONS

The solution of any equation of the type

$$dy/dx + P(x)y = Q(x)$$

where $P(x)$ and $Q(x)$ are arbitrary functions of x, is

$$y = Ce^{-\int P(x)\,dx} + e^{-\int P(x)\,dx} \int e^{\int P(x)\,dx}\, Q(x)\, dx$$

where C is the integration constant. This equation is used several times in section 32.

B

VECTOR NOTATION

Elementary physical quantities are either scalars or vectors. The former, represented by *italic* letters, are quantities which can be specified by a single number. The latter, written as *bold face* letters, require both a number (the *magnitude* of the vector) and a direction for complete specification.

If **A** is a vector, we represent its magnitude by the symbol A. Its direction may be represented graphically, as in Fig. B–1, or analytically by the relation

$$\mathbf{A} = A_x\mathbf{i} + A_y\mathbf{j} + A_z\mathbf{k}$$

where **i**, **j**, and **k** are vectors of unit length in the positive directions of the x, y, and z axes of a rectangular coordinate system, and A_x, A_y, and A_z are the *components* of **A** along these directions.

The following relations define various operations which may be performed on vectors, and Fig. B–1 provides graphical representations of the same operations. In all these relations **A** and **B** are vectors with components A_x, A_y, and A_z, and B_x, B_y, and B_z, respectively. The symbol n represents a scalar quantity.

$$-\mathbf{A} = (-A_x)\mathbf{i} - (A_y)\mathbf{j} - (A_z)\mathbf{k}$$

$$n\mathbf{A} = nA_x\mathbf{i} + nA_y\mathbf{j} + nA_z\mathbf{k}$$

$$\mathbf{A} + \mathbf{B} = (A_x + B_x)\mathbf{i} + (A_y + B_y)\mathbf{j} + (A_z + B_z)\mathbf{k}$$

$$n(\mathbf{A} + \mathbf{B}) = n\mathbf{A} + n\mathbf{B}$$

$$\mathbf{A} - \mathbf{B} = \mathbf{A} + (-\mathbf{B})$$

Two kinds of product have been defined. One of them is the scalar product

$$\mathbf{A} \cdot \mathbf{B} = AB \cos \theta$$

673

where θ is the angle between the positive directions of $\mathbf{A}$ and $\mathbf{B}$. A special case of the scalar product is $\mathbf{A} \cdot \mathbf{A}$. Since the angle θ is in this case $0°$, we have

$$\mathbf{A} \cdot \mathbf{A} = A^2 = A_x^2 + A_y^2 + A_z^2$$

Another special case is that of vectors $\mathbf{A}$ and $\mathbf{B}$ which are perpendicular to one another. For these vectors $\mathbf{A} \cdot \mathbf{B} = 0$.

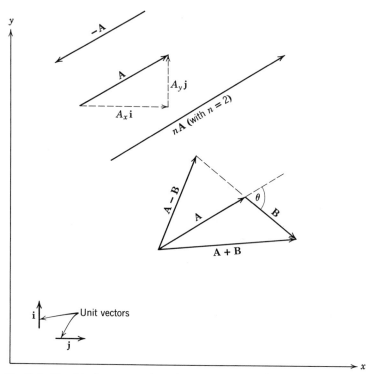

Fig. B–1. Graphical representation of a vector and of some vector operations. In this two-dimensional figure both $\mathbf{A}$ and $\mathbf{B}$ lie in the xy plane, so that A_z and B_z are zero.

The following relations pertaining to the scalar product are readily verified graphically or analytically:

$$\mathbf{A} \cdot \mathbf{B} = \mathbf{B} \cdot \mathbf{A}$$

$$\mathbf{A} \cdot (\mathbf{B} + \mathbf{C}) = \mathbf{A} \cdot \mathbf{B} + \mathbf{A} \cdot \mathbf{C}$$

On the other hand $\mathbf{A}(\mathbf{B} \cdot \mathbf{C})$, the product of the vector $\mathbf{A}$ with the scalar quantity $\mathbf{B} \cdot \mathbf{C}$ is not the same as $(\mathbf{A} \cdot \mathbf{B})\mathbf{C}$. It is to be noted that we can without ambiguity omit the parentheses in this expression, i.e., $\mathbf{AB} \cdot \mathbf{C} \equiv \mathbf{A}(\mathbf{B} \cdot \mathbf{C})$.

The other product of vector multiplication is the *vector product*,

$$\mathbf{A} \times \mathbf{B} = (AB \sin \theta)\mathbf{p}$$

which is itself a vector, $\mathbf{p}$ representing a unit vector which is perpendicular to the plane of $\mathbf{A}$ and $\mathbf{B}$. By taking this plane as the horizontal plane, $\mathbf{p}$ is directed downward if the rotation from $\mathbf{A}$ to $\mathbf{B}$ through the angle θ is clockwise. It is directed upward if the rotation is counterclockwise. It follows from this definition that $\mathbf{A} \times \mathbf{A} = 0$ and that $\mathbf{A} \times \mathbf{B} = -\mathbf{B} \times \mathbf{A}$. (Vector products are not used in any of the detailed derivations in this book, so that a complete set of relations analogous to those for the scalar product need not be given.)

A special symbol which needs to be defined is the vector differential operator

$$\nabla = \mathbf{i}\frac{\partial}{\partial x} + \mathbf{j}\frac{\partial}{\partial y} + \mathbf{k}\frac{\partial}{\partial z}$$

which may be treated mathematically as if it were itself a vector. Thus $\nabla \cdot \mathbf{A}$ and $\nabla \cdot \nabla$ are scalar quantities. Since $\mathbf{i} \cdot \mathbf{j} = \mathbf{j} \cdot \mathbf{k} = \mathbf{k} \cdot \mathbf{i} = 0$ and $\mathbf{i} \cdot \mathbf{i} = \mathbf{j} \cdot \mathbf{j} = \mathbf{k} \cdot \mathbf{k} = 1$,

$$\nabla \cdot \mathbf{A} = \partial A_x/\partial x + \partial A_y/\partial y + \partial A_z/\partial z$$

$$\nabla \cdot \nabla = \nabla^2 = \frac{\partial^2}{\partial x^2} + \frac{\partial^2}{\partial y^2} + \frac{\partial^2}{\partial z^2}$$

These relations may be used to explain some of the terms used in equation 19–10, on p. 324. Where u_x, u_y, and u_z are the components of the vector $\mathbf{u}$

$$\nabla\nabla \cdot \mathbf{u} = \mathbf{i}\left(\frac{\partial^2 u_x}{\partial x^2} + \frac{\partial^2 u_y}{\partial x\,\partial y} + \frac{\partial^2 u_z}{\partial x\,\partial z}\right) + \begin{array}{l}\text{symmetrical terms}\\ \text{in } \mathbf{j} \text{ and } \mathbf{k}\end{array}$$

$$\nabla \cdot \nabla\mathbf{u} = \mathbf{i}\left(\frac{\partial^2 u_x}{\partial x^2} + \frac{\partial^2 u_y}{\partial y^2} + \frac{\partial^2 u_z}{\partial z^2}\right) + \begin{array}{l}\text{symmetrical terms}\\ \text{in } \mathbf{j} \text{ and } \mathbf{k}\end{array}$$

C

OPTICAL METHODS
FOR DETERMINING
CONCENTRATION
GRADIENTS

In the measurement of diffusion, sedimentation, and electrophoresis, the experimental methods lead to solutions in which the solute concentration c depends on the location along a given coordinate x. It is necessary to determine c or dc/dx as a function of x in order to obtain the desired

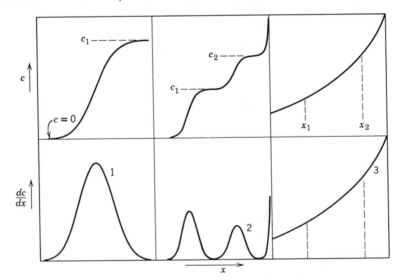

Fig. C–1. Typical plots of c versus x and corresponding plots of dc/dx versus x.

physical quantities. The concentration c does not necessarily represent that of a single homogeneous solute; i.e., several different solute species may be present.

If the refractive index increment $d\tilde{n}/dc$ is the same for all solute species which are present and if $d\tilde{n}/dc$ is independent of c within the range of c present in the solution, then

$$\frac{d\tilde{n}}{dx} = \frac{d\tilde{n}}{dc}\frac{dc}{dx}$$

$$\tilde{n} = \tilde{n}_0 + \frac{d\tilde{n}}{dc}c$$

where $\tilde{n}_0$ is the refractive index of the solvent. In this situation, therefore, $\tilde{n}$ or $d\tilde{n}/dx$ provides a direct measure of c or dc/dx.

The technique used most frequently is to determine $d\tilde{n}/dx$ by measuring the bending of light rays which are passed through the solution almost perpendicular to the x-direction, as shown in Fig. C–2. Applying Snell's law to a light ray entering at any position x_1, we have

$$\tilde{n}_1 \sin \alpha_1 = \tilde{n}_e \sin \alpha_0$$

where $\tilde{n}_e$ is the refractive index of the external medium and $\tilde{n}_1$ the refractive index of the solution at x_1. The angles α_0 and α_1 are defined in the figure.

At any point within the solution the deflection of the ray is determined by the angle γ, which is the complement of the angle α. (See Fig. C–2.) With $\tilde{n}$ and γ both continuous functions of x, Snell's law may be written as $\tilde{n} \sin \gamma = $ constant, or, since α is the complement of γ,

$$\tilde{n} \cos \alpha = \text{constant}$$

Differentiating with respect to x (along the path of the ray) gives

$$\tilde{n} \sin \alpha \, (d\alpha/dx) = \cos \alpha (d\tilde{n}/dx)$$

Since $\sin \alpha / \cos \alpha = \tan \alpha = dx/dy$, this may be written as

$$\frac{d\alpha}{dy} = \frac{1}{\tilde{n}}\frac{d\tilde{n}}{dx}$$

If x_2 and x_1 are sufficiently close to one another, both $\tilde{n}$ and $d\tilde{n}/dx$ may be considered as constants, so that

$$\alpha_2 - \alpha_1 = \frac{1}{\tilde{n}}\frac{d\tilde{n}}{dx}a$$

where $\tilde{n}$ is the same as $\tilde{n}_1$ and $a = y_2 - y_1$ is the thickness of the medium.

Applying Snell's law to the outgoing ray, we have (with $\tilde{n}_2 = \tilde{n}_1$)

$$\tilde{n}_1 \sin \alpha_2 = \tilde{n}_e \sin \alpha_f$$

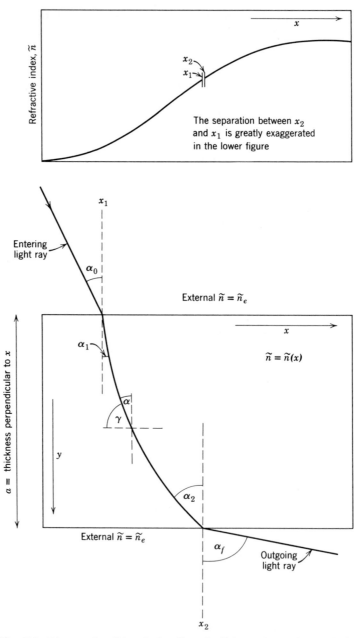

Fig. C-2. Diagram describing the bending of a light ray by a refractive index gradient. The lower figure shows the xy plane (containing the light ray) of the medium. The refractive index depends on x only, as shown in the upper figure.

Finally, the condition that x_2 and x_1 are very close together implies that both the entering and outgoing rays are actually very close to the vertical (i.e., the x direction), so that α_0, α_1, α_2, and α_f are all small angles. We may thus write $\sin \alpha = \alpha$ and obtain

$$\alpha_f - \alpha_0 = \frac{a}{\tilde{n}_e} \frac{d\tilde{n}}{dx}$$

Where the external medium is air, $\tilde{n}_e = 1$. It is easily shown that $\tilde{n}_e$ may also be set equal to 1 if the solution of interest is separated from air by a layer of glass on each side.

The preceding equation shows that the bending of the light ray is a direct measure of $d\tilde{n}/dx$ at the value of x at which the light ray enters the solution. Any of a number of optical techniques may be used to magnify the value of $\alpha_f - \alpha_0$ and to produce a two-dimensional record, one dimension of which represents the value of x, the other the magnified value of $\alpha_f - \alpha_0$.

The method in commonest use is the schlieren method, developed by Philpot and Svenson. Alternative techniques are a method of Lamm which depends on the displacement of the image of the lines on a ruled glass scale and an interference method originally suggested by Gouy. These methods are described in Gosting's review of diffusion measurements (ref. 7, p. 451), and this paper should be consulted for reference to the history and theory of each technique.

With the assumptions stated earlier each of these methods provides a record of $K(dc/dx)$ as a function of x, where the constant K depends both on the value of $d\tilde{n}/dc$ for the particular solute or solute mixture which is being examined and on the magnification factor by which the deflection $\alpha_f - \alpha_0$ is increased.

It is to be noted that the methods here cited can determine not only $K(dc/dx)$ but also integrals of this quantity. Thus the area under curve 1 of Fig. C–1 is equal to $\int (dc/dx)\,dx$ between $x = 0$ and the bottom of the cell, and this is clearly equal to c_1. The area of the corresponding plot of $d\tilde{n}/dx$ versus x is just K times this integral, i.e., Kc_1. Similarly, from the area corresponding to curve 2, we can evaluate $K(c_2 - c_1)$, and, in general (curve 3), we can determine $K(c_2 - c_1)$ between any points x_1 and x_2.

The constant K which occurs in these relations can in principle be determined by experiment, but this is usually not necessary. Ratios of concentrations can always be evaluated without knowledge of the constant K; e.g., if $K\,dc/dx$ and $K(c_2 - c_1)$ have been measured, $(dc/dx)/(c_2 - c_1)$ is known regardless of the value of K. As the text has shown, the desired physical properties can usually be evaluated from a knowledge of such ratios.

It should be noted in conclusion that two methods are available for the direct determination of the curve of c versus x. One of them is the Rayleigh interference method (described by Gosting). This method yields a curve of $\bar{n}$ versus x, and thus depends again on the assumption of a unique value of $d\bar{n}/dc$ for all solutes. The other method is the light absorption method (described by Schachman, ref. 10 p. 451) in which we measure absorption at a suitable wavelength as a function of x. This method will yield a curve of concentration versus x if the absorbance, a, for any solute is independent of c and if (in a mixture) all solutes have the same absorbance.

Of course, any of the methods listed may be applied to solute mixtures in which $d\bar{n}/dc$ or a is different for each component. They will then lead to curves of *apparent* concentration or concentration gradient, the contribution of each component being weighted in proportion to its value of $d\bar{n}/dc$ or a. Where $d\bar{n}/dc$ or a has been independently evaluated for each component, the apparent concentrations may sometimes be convertible to true concentrations.

AUTHOR INDEX

Reference to an author is often made in the text by means of a reference number alone. This is indicated in the Index by having the reference number in parentheses after the page number. Boldface page numbers refer to the bibliographical listings.

681

684

SUBJECT INDEX

Information is listed primarily under two headings: by subject matter (e.g., light scattering) and by specific compound name (e.g., hemoglobin). There is no listing under generic compound names (e.g., proteins), but important subject headings have subheadings for such names (e.g., light scattering, of proteins).